Ning Zhong
Jiming Liu

Intelligent Technologies for Information Analysis

With 161 Figures and 48 Tables

 Springer

Prof. Ning Zhong
Department of Information Engineering
Maebashi Institute of Technology
Maebashi, Japan
e-mail: zhong@maebashi-it.ac.jp

Prof. Jiming Liu
Department of Computer Science
Hong Kong Baptist University
Kowloon Tong, Hong Kong
e-mail: jiming@comp.hkbu.edu.hk

ACM Computing Classification (1998):
H.3.3, H.2.8, I.2, G.3

ISBN 978-3-642-07378-6

Springer is a part of Springer Science+Business Media
springeronline.com
© Springer-Verlag Berlin Heidelberg 2010
Printed in Germany

Cover design: KünkelLopka, Heidelberg

Printed on acid-free paper 33/3142/YL - 5 4 3 2 1 0

Ning Zhong
Jiming Liu

Intelligent Technologies
for Information Analysis

Preface

Intelligent Information Technology (iIT) encompasses the theories and applications of artificial intelligence, statistical pattern recognition, learning theory, data warehousing, data mining and knowledge discovery, Grid computing, and autonomous agents and multi-agent systems in the context of today's as well as future IT, such as Electronic Commerce (EC), Business Intelligence (BI), Social Intelligence (SI), Web Intelligence (WI), Knowledge Grid (KG), and Knowledge Community (KC), among others.

The multi-author monograph presents the current state of the research and development in intelligent technologies for information analysis, in particular, advances in agents, data mining, and learning theory, from both theoretical and application aspects. It investigates the future of information technology (IT) from a new intelligent IT (iIT) perspective, and highlights major iIT-related topics by structuring an introductory chapter and 22 survey/research chapters into 5 parts: (1) emerging data mining technology, (2) data mining for Web intelligence, (3) emerging agent technology, (4) emerging soft computing technology, and (5) statistical learning theory. Each chapter includes the original work of the author(s) as well as a comprehensive survey related to the chapter's topic.

This book will become a valuable source of reference for R&D professionals active in advanced intelligent information technologies. Students as well as IT professionals and ambitious practitioners concerned with advanced intelligent information technologies will appreciate the book as a useful text enhanced by numerous illustrations and examples.

To meet the strong demands for participation and the growing interests in iIT, in particular Web and agent intelligence, the Web Intelligence Consortium (WIC) was formed in spring 2002. The WIC (http://wi-consortium.org/) is an international non-profit organization dedicated to promoting world-wide scientific research and industrial development in the era of Web and agent intelligence. The scope of its activities covers encouraging deep collaborations among the WIC Research Centers around the world, supporting organizational and individual members, steering advanced research and technology showcases at the international IEEE/WIC/ACM conferences and workshops, and producing official WIC publications.

This book is recommended by the WIC as the first book on intelligent technologies for information analysis, and as a book related to the *Web Intelligence* monograph published by Springer-Verlag in spring 2003. It is a collaborative effort involving many leading researchers and practitioners who have contributed chapters in their areas of expertise. We wish to express our gratitude to all the authors and reviewers for their contributions.

We would like to thank Yiyu Yao for his continuing collaborations in the development of Web Intelligence (WI). We are very grateful to people who joined or supported the WIC-related research activities, and, in particular, the WIC Advisory Board members, Edward A. Feigenbaum, Setsuo Ohsuga, Benjamin Wah, Philip Yu, and Lotfi A. Zadeh, all the WIC Technical Committee members, as well as the IEEE TCCI Chair, Xindong Wu. We thank them for their strong support.

Last, but not least, we thank Ralf Gerstner of Springer-Verlag and Jia Hu for their help in coordinating the publication of this monograph and editorial assistance.

Maebashi, Japan   *Ning Zhong*
Hong Kong   *Jiming Liu*
Spring 2004

Table of Contents

1. **The Alchemy of Intelligent IT (iIT): A Blueprint for Future Information Technology**
   Ning Zhong and Jiming Liu 1

   1.1 What Is iIT? 1
   1.2 Why iIT? 1
   1.3 iIT as the New Generation of Information Technology 2
   1.4 iIT for e-Business Intelligence 5
       1.4.1 Virtual Industry Park: An Example of Enterprise Portals 5
       1.4.2 Web Mining and Farming 6
       1.4.3 Semantic Social Networks for Intelligent Enterprise
             Portals 7
       1.4.4 Data Mining Grids for Web-Based Targeted Marketing 8
       1.4.5 Wisdom Web-Based Computing 9
   1.5 Dimensions of iIT Research 10
   1.6 An Overview of This Book 11
   References 14

Part I. Emerging Data Mining Technology

2. **Grid-Based Data Mining and Knowledge Discovery**
   Mario Cannataro, Antonio Congiusta, Carlo Mastroianni, Andrea
   Pugliese, Domenico Talia, and Paolo Trunfio 19

   2.1 Introduction 19
   2.2 Knowledge Discovery on Grids 21
   2.3 The *Knowledge Grid* Architecture 26
       2.3.1 *Knowledge Grid* Services 26
   2.4 An XML-Based Metadata Model for the *Knowledge Grid* 28
       2.4.1 Data Mining Software 29
       2.4.2 Data Sources 30
       2.4.3 Abstract and Concrete Resources 32
       2.4.4 Execution Plans 32
       2.4.5 Data Mining Models 33
       2.4.6 Metadata Management 33

2.5   Design of a PDKD Computation 34
    2.5.1   Task Composition 36
    2.5.2   Task Consistency Checking 38
    2.5.3   Execution Plan Generation 39
2.6   Execution of a PDKD Computation 42
2.7   Conclusions....................................... 43
References 44

3. The MiningMart Approach to Knowledge Discovery in Databases
Katharina Morik and Martin Scholz 47

3.1   Introduction: Acquiring Knowledge from Existing Databases . 47
3.2   The MiningMart Approach 51
    3.2.1   The Meta-Model of Metadata M4 52
    3.2.2   Editing the Conceptual Data Model 54
    3.2.3   Editing the Relational Model 55
    3.2.4   The Case and Its Compiler 56
3.3   The Case Base 59
3.4   Conclusions....................................... 61
References 64

4. Ensemble Methods and Rule Generation
Yongdai Kim, Jinseog Kim, and Jongwoo Jeon................... 67

4.1   Introduction 67
4.2   Ensemble Algorithms: A Review....................... 69
    4.2.1   Bagging 69
    4.2.2   Boosting 70
    4.2.3   Convex Hull Ensemble Machine: CHEM 75
    4.2.4   Comments on Roles of Base Learners 80
4.3   Rule Generation...................................... 82
4.4   Illustration 84
4.5   Conclusions... 86
References 86

5. Evaluation Scheme for Exception Rule/Group Discovery
Einoshin Suzuki 89

5.1   Introduction 89
5.2   Exception Rule/Group Discovery 90
    5.2.1   Notice 90
    5.2.2   Use of Domain Knowledge....................... 91
    5.2.3   Hypothesis-Driven Discovery...................... 95
5.3   Evaluation Scheme and Classification of Existing Methods ... 102
    5.3.1   Generality....................................... 102

        5.3.2 Break of Monotonicity in Discovered Patterns 103
        5.3.3 Evaluation of Reliability............................ 103
        5.3.4 Search Range 104
        5.3.5 Interpretation of the Evaluation Measure 104
        5.3.6 Use of Domain Knowledge......................... 105
        5.3.7 Successes in Real Applications 105
    5.4 Conclusions... 106
    References 106

6. Data Mining for Targeted Marketing
   Ning Zhong, Yiyu Yao, Chunnian Liu, Jiajin Huang, and Chuangxin
   Ou 109

    6.1 Introduction 109
    6.2 The Process of Targeted Marketing 110
    6.3 Problems in Targeted Marketing 110
    6.4 Target Selection Algorithms 112
        6.4.1 Segmentation Models 112
        6.4.2 Response Models 112
    6.5 Mining Market Value Functions 118
        6.5.1 Utility Functions 119
        6.5.2 Attribute Weighting 121
    6.6 Evaluation of the Learning Algorithms 123
    6.7 Experimental Results 124
    6.8 New Directions...................................... 125
        6.8.1 Web-Based Targeted Marketing 125
        6.8.2 Multi-Aspect Analysis in Multiple Data Sources 127
        6.8.3 Building a Data Mining Grid 127
    6.9 Conclusions... 128
    References 129

Part II. Data Mining for Web Intelligence

7. Mining for Information Discovery on the Web: Overview
   and Illustrative Research
   Hwanjo Yu, AnHai Doan, and Jiawei Han 135

    7.1 Introduction 135
    7.2 Finding Information on the Web 136
        7.2.1 Exploring and Navigating the Web 136
        7.2.2 Querying with Information Processing Systems 138
    7.3 Web Page Classification from Positive Examples............ 141
        7.3.1 Related Work 142
        7.3.2 SVM Margin-Maximization Property 143
        7.3.3 The Mapping-Convergence (M-C) Algorithm 144

      7.3.4   Experimental Results 148
  7.4   Object Matching Across Disparate Data Sources 153
      7.4.1   Problem Definition 156
      7.4.2   The PROM Approach 156
      7.4.3   Empirical Evaluation 158
      7.4.4   Related Work 161
      7.4.5   Summary 162
  7.5   Conclusions 163
  References 163

8. Mining Web Logs for Actionable Knowledge
Qiang Yang, Charles X. Ling, and Jianfeng Gao 169

  8.1   Introduction 169
  8.2   Web Log Mining for Prefetching 170
      8.2.1   Data Cleaning on Web Log Data 170
      8.2.2   Mining Web Logs for Path Profiles 171
      8.2.3   Web Object Prediction 171
      8.2.4   Learning to Prefetch Web Documents 173
  8.3   Web Page Clustering for Intelligent User Interfaces 175
  8.4   Web Query Log Mining 178
      8.4.1   Web Query Logs 178
      8.4.2   Mining Generalized Query Patterns 181
      8.4.3   A Bottom-Up Generalization Algorithm 182
      8.4.4   Improvement 1: A Hierarchy over Keywords 183
      8.4.5   Improvement 2: Flexible Generalizations 184
      8.4.6   Improvement 3: Morphology Conversion 185
      8.4.7   Improvement 4: Synonym Conversion 185
      8.4.8   Implementations 186
      8.4.9   Simulation Experiments 187
      8.4.10 Analyses of the Results 187
      8.4.11 Relation to Previous Work 188
  8.5   Conclusions 189
  References 190

**9. Discovery of Web Robot Sessions Based on Their
Navigational Patterns**
Pang-Ning Tan and Vipin Kumar 193

  9.1   Introduction 193
  9.2   Web Robot Detection: Overview 197
      9.2.1   Limitations of Current Robot Detection Techniques ... 197
      9.2.2   Motivation for Proposed Robot Detection Technique .. 200
  9.3   Methodology 202
      9.3.1   Data Source and Preprocessing 202
      9.3.2   Feature Vector Construction 204

9.3.3   Session Labeling 207
9.3.4   Classification 209
9.3.5   Identifying Mislabeled Sessions 210
9.4   Experimental Evaluation 211
9.4.1   Experimental Data Set 211
9.4.2   Correlation Analysis 212
9.4.3   Classifier Performance 215
9.4.4   Finding Mislabeled Data 217
9.5   Conclusions 219
References 221

10. Web Ontology Learning and Engineering: An Integrated Approach
Roberto Navigli, Paola Velardi, and Michele Missikoff 223

10.1   Introduction 223
10.2   The OntoLearn System 227
10.2.1   Identification of Relevant Domain Terminology 228
10.2.2   Semantic Interpretation of Terms 230
10.2.3   Creating a Specific Domain Ontology 234
10.2.4   Creating a Domain Ontology from WordNet 237
10.3   Evaluation of the OntoLearn System 237
10.4   Conclusions 239
References 241

11. Browsing Semi-Structured Texts on the Web Using Formal Concept Analysis
Richard Cole, Florence Amardeilh, and Peter Eklund 243

11.1   Introduction: Information Extraction and the Web 243
11.1.1   Definitions – Information Extraction 244
11.1.2   Web Documents and Text Diversity 244
11.1.3   Architecture and Components 245
11.1.4   Related Work 246
11.1.5   The Interaction Paradigm and Learning Context 247
11.2   Formal Concept Analysis and RDBMSs 248
11.3   Web-Robot for Extracting Structured Data from
       Unstructured Sources 252
11.4   RFCA – the Web-Based FCA Interface 255
11.5   Reusing CEM for Nesting and Zooming 259
11.6   Conclusions 262
References 263

12. Graph Discovery and Visualization from Textual Data
Vincent Dubois and Mohamed Quafafou 265

12.1 Introduction 265
12.2 Graph Structure Discovery 266
    12.2.1 Graph Structure Semantic 266
    12.2.2 Network Evaluation 267
    12.2.3 Methods and Algorithms 268
12.3 Visualization Constrains Discovery 269
    12.3.1 Graph Structure Visualization 269
    12.3.2 Incremental and Dynamic Properties 271
12.4 Experimental Results 272
    12.4.1 Results on Corpus 272
    12.4.2 Convergence Problem 274
    12.4.3 Effect of Initial Position on Result 275
    12.4.4 Effect of New Data Insertion 275
    12.4.5 Influence of Data Order 276
12.5 Application: Web Content Mining 277
    12.5.1 Introduction 277
    12.5.2 General Architecture 281
    12.5.3 Related Work 282
    12.5.4 Dynamic Search and Crawling 283
    12.5.5 Integration and Implementation Issues 284
    12.5.6 Experimental Results 285
12.6 Conclusions 287
References 287

Part III. Emerging Agent Technology

**13. Agent Networks: Topological and Clustering
Characterization**
Xiaolong Jin and Jiming Liu 291

13.1 Introduction 291
    13.1.1 Small World Phenomena 291
    13.1.2 Agent Networks 291
    13.1.3 Satisfiability Problems 292
    13.1.4 Problem Statements 293
    13.1.5 Organization of This Chapter 293
13.2 Topologies of Agent Networks 293
    13.2.1 Representation I 294
    13.2.2 Representation II 297
13.3 Discussions 299
    13.3.1 Complexities in Different Representations 299

13.3.2 Balanced Complexities in Intra- and Inter-Agent
        Computations.................................... 300
13.3.3 A Guiding Principle 300
13.4 Average Value of the Clustering Coefficient 301
13.5 Lower Bounds of the Clustering Coefficient 303
13.6 Conclusions... 306
References 308

14. Finding the Best Agents for Cooperation
Francesco Buccafurri, Luigi Palopoli, Domenico Rosaci, and Giuseppe
M.L. Sarnè... 311

14.1 Introduction 311
14.2 Related Work 312
14.3 The Knowledge Bases 314
    14.3.1 An Ontology for Describing the Domain of Interest ... 314
    14.3.2 The Local Knowledge Base 315
14.4 Extraction of the Semantic Properties.................... 315
    14.4.1 Local Properties: Similarity....................... 316
    14.4.2 Global Properties: Interest and Attractiveness........ 317
    14.4.3 Choice Lists 321
    14.4.4 Reactive Properties.............................. 322
14.5 Local Knowledge Base Integration 324
14.6 The SPY System Architecture 325
14.7 Experiments 329
14.8 Conclusions.. 330
References 330

**15. Constructing Hybrid Intelligent Systems for Data
Mining from Agent Perspectives**
Zili Zhang and Chengqi Zhang.............................. 333

15.1 Introduction 333
15.2 Data Mining Needs Hybrid Solutions..................... 335
15.3 Agent Perspectives Are Suitable for Modeling Hybrid
     Intelligent Systems 337
15.4 Agent-Oriented Methodologies 338
    15.4.1 The Role Model................................. 339
    15.4.2 The Interaction Model 342
    15.4.3 Organizational Rules............................. 343
    15.4.4 The Agent Model 343
    15.4.5 The Skill Model 344
    15.4.6 The Knowledge Model 344
    15.4.7 Organizational Structures and Patterns 345
15.5 Agent-Based Hybrid Systems for Data Mining.............. 346
    15.5.1 Analysis and Design 346

      15.5.2 Implementation 352
      15.5.3 Case Study 355
   15.6 Evaluation... 357
   15.7 Conclusions.. 357
   References 358

16. Making Agents Acceptable to People

Jeffrey M. Bradshaw, Patrick Beautement, Maggie R. Breedy, Larry
Bunch, Sergey V. Drakunov, Paul J. Feltovich, Robert R. Hoffman,
Renia Jeffers, Matthew Johnson, Shriniwas Kulkarni, James Lott,
Anil K. Raj, Niranjan Suri, and Andrzej Uszok 361

   16.1 Introduction 361
   16.2 Addressing Agent Acceptability Through the Use of Policy .. 367
      16.2.1 What Is Policy? 368
      16.2.2 Distinguishing Policy from Related Concepts......... 369
      16.2.3 Types of Policy 370
      16.2.4 Autonomy and Policy 371
      16.2.5 Benefits of Policy Management 375
      16.2.6 Applications of Policy Using KAoS and Nomads...... 377
   16.3 Technical Aspects of Agent Acceptability 379
      16.3.1 Examples of Policy Types Relating to Technical
             Aspects of Agent Acceptability..................... 380
   16.4 Social Aspects of Agent Acceptability 386
      16.4.1 Examples of Policy Types Relating to Social Aspects
             of Agent Acceptability 387
      16.4.2 Cognitive and Robotic Prostheses 394
   16.5 Conclusions.. 398
   References 398

Part IV. Emerging Soft Computing Technology

17. Constraint-Based Neural Network Learning for Time Series Predictions

Benjamin W. Wah and Minglun Qian 409

   17.1 Introduction 409
   17.2 Previous Work in Time Series Modeling 410
      17.2.1 Linearity 411
      17.2.2 Piecewise Chaos................................. 412
      17.2.3 Random Noise 412
      17.2.4 Artificial Neural Networks 415
   17.3 Predictions of Noise-Free Stationary and Piecewise Chaotic
      Time Series 417
      17.3.1 Recurrent FIR Neural Networks 417

17.3.2 Constrained Formulations for ANN Learning......... 418
17.3.3 Violation-Guided Backpropagation Algorithm 420
17.3.4 Experimental Results 422
17.4 Predictions of Noisy Time Series with High Frequency
Random Noise 423
17.4.1 Review on Financial Time Series Predictions......... 424
17.4.2 Constraint in the Lag Period....................... 424
17.4.3 Experimental Results 425
17.5 Conclusions... 428
References 429

18. Approximate Reasoning in Distributed Environments
Andrzej Skowron 433

18.1 Introduction 433
18.2 Information Granules 439
18.2.1 Rough Sets and Approximation Spaces 439
18.2.2 Syntax and Semantics of Information Granules 441
18.2.3 Granule Inclusion and Closeness.................... 447
18.2.4 Rough–Fuzzy Granules............................. 451
18.2.5 Classifiers as Information Granules 451
18.3 Rough-Neural Computing: Weights Defined by
Approximation Spaces 452
18.4 Rough-Neural Computing: Rough Mereological Approach.... 457
18.4.1 Distributed Systems of Agents 458
18.4.2 Approximate Synthesis of Complex Objects.......... 461
18.5 Extracting AR-Schemes from Data and Background
Knowledge 464
18.5.1 Granule Decomposition 465
18.6 Conclusions... 469
References 472

**19. Soft Computing Pattern Recognition, Data Mining and
Web Intelligence**
Sankar K. Pal, Sushmita Mitra, and Pabitra Mitra 475

19.1 Introduction 475
19.2 Soft Computing Pattern Recognition 478
19.2.1 Relevance of Fuzzy Set Theory in Pattern Recognition 478
19.2.2 Relevance of Neural Network Approaches............ 479
19.2.3 Genetic Algorithms for Pattern Recognition 481
19.2.4 Relevance of Rough Sets 481
19.2.5 Integration and Hybrid Systems 482
19.3 Knowledge Discovery and Data Mining.................... 485
19.3.1 Data Mining...................................... 485
19.4 Soft Computing for Data Mining 486

19.4.1 Fuzzy Sets 486
19.4.2 Neural Networks 489
19.4.3 Neuro-Fuzzy Computing 490
19.4.4 Genetic Algorithms 491
19.4.5 Rough Sets 492
19.5 Web Mining 492
19.5.1 Web Mining Components and the Methodologies 493
19.6 Soft Computing for Web Mining 494
19.6.1 Fuzzy Logic for Web Mining 494
19.6.2 Neural Networks and Learning Systems for Web
        Mining 497
19.6.3 Genetic Algorithms for Web Mining 502
19.6.4 Rough Sets for Web Mining 503
19.7 Conclusions 505
References 506

20. Dominance-Based Rough Set Approach to Knowledge Discovery (I): General Perspective
Salvatore Greco, Benedetto Matarazzo, and Roman Slowinski..... 513

20.1 Introduction: Three Types of Prior Knowledge to Be Included
     in Knowledge Discovery 513
20.2 The Influence of Preference Order in Data on Granular
     Computing 516
20.3 Dominance-Based Rough Set Approach (DRSA) 518
20.3.1 Granular Computing with Dominance Cones 518
20.3.2 Induction of Decision Rules 523
20.3.3 Illustrative Example 525
20.4 DRSA for Multicriteria Choice and Ranking 529
20.4.1 Pairwise Comparison Table (PCT) as a Preferential
        Information and a Learning Sample 530
20.4.2 Rough Approximation of Outranking and Non-outranking
        Relations Specified in PCT 532
20.4.3 Induction of Decision Rules from Rough Approxima-
        tions of Outranking and Non-outranking Relations.... 534
20.4.4 Use of Decision Rules for Decision Support 536
20.4.5 Illustrative Example 537
20.4.6 Summary 539
20.5 DRSA with Missing Values of Attributes and Criteria 540
20.5.1 Generalized Indiscernibility Relation 540
20.5.2 Illustrative Example 542
20.5.3 Generalized Dominance Relation 543
20.5.4 Illustrative Example 546
20.6 Conclusions 547
References 548

21. Dominance-Based Rough Set Approach to Knowledge Discovery (II): Extensions and Applications
Salvatore Greco, Benedetto Matarazzo, and Roman Slowinski..... 553

21.1 Introduction 553
21.2 Fuzzy Set Extensions of DRSA 553
    21.2.1 Fuzzy DRSA for Multicriteria Classification 554
    21.2.2 Fuzzy DRSA for Multicriteria Choice and Ranking ... 560
    21.2.3 Gradual Rules and Fuzzy Rough Approximations
    Without Fuzzy Logical Connectives 564
21.3 DRSA for Decision Under Risk 575
    21.3.1 DRSA Based on Stochastic Dominance 576
    21.3.2 Illustrative Example 577
21.4 DRSA for Hierarchical Decision Making 580
    21.4.1 Rough Set Approach for Attribute Subset Values and
    Interval Order 582
    21.4.2 Propagation of Inconsistencies and Application of
    Decision Rules 587
    21.4.3 Illustrative Example 589
21.5 Comparison of DRSA with Other Paradigms 596
    21.5.1 Axiomatic Foundations of Multicriteria Classification
    Problems and Associated Preference Models 596
    21.5.2 Conjoint Measurement for Multicriteria Classification
    Problems with Inconsistencies 605
    21.5.3 Summary 606
21.6 Conclusions 606
References 607

Part V. Statistical Learning Theory

22. Bayesian Ying Yang Learning (I): A Unified Perspective for Statistical Modeling
Lei Xu 615

22.1 Introduction: Basic Issues of Statistical Learning 615
22.2 Dependence Among Samples from One-Object World 616
22.3 Dependence Among Samples from a Multi-Object World 620
    22.3.1 Dependence Among Samples from a Multi-Object
    World 620
    22.3.2 Mining Dependence Structure Across Invisible
    Multi-Object 621
22.4 A Systemic View on Various Dependence Structures 623
22.5 Bayesian Ying Yang System 627
22.6 BYY Harmony Learning 631

22.6.1 Kullback Divergence, Harmony Measure, and
Z-Regularization 631

22.6.2 BYY Harmony Learning 634

22.6.3 A Further Extension: From $\ln(r)$ to Convex Function . 637

22.7 Ying-Yang Alternative Procedure for Parameter Learning ... 640

22.8 Learning Implementation: From Optimization Search to
Accumulation Consensus 643

22.9 Main Results and Bibliographic Remarks 647

22.9.1 Main Results on Typical Learning Problems 647

22.9.2 Bibliographic Remarks on BYY System with KL
Learning 650

22.9.3 Bibliographic Remarks on Computing Techniques 654

22.10 Conclusions 654

References 655

**23. Bayesian Ying Yang Learning (II): A New Mechanism
for Model Selection and Regularization**
Lei Xu.. 661

23.1 Introduction: A Key Challenge and Existing Solutions 661

23.2 Existing Solutions 663

23.2.1 Efforts in the First Stream 663

23.2.2 Efforts in the Second Stream...................... 664

23.3 Bayesian Ying Yang Harmony Learning 667

23.3.1 Bayesian Ying Yang Harmony Learning 668

23.3.2 Structural Inner Representations 672

23.4 Regularization Versus Model Selection 675

23.4.1 ML, HL, and Z-Regularization 675

23.4.2 KL-λ-HL Spectrum 678

23.5 An Information Transfer Perspective 681

23.6 BYY Harmony Learning Versus Related Approaches 684

23.6.1 Relation and Difference to the Bits-Back Based MDL
and Bayesian Approaches........................... 684

23.6.2 Relations to Information Geometry, Helmholtz
Machine and Variational Approximation 686

23.6.3 A Projection Geometry Perspective 688

23.7 Bibliographic Remarks 693

23.7.1 On BYY Harmony Learning (I): Model Selection
Criteria vs. Automatic Model Selection.............. 693

23.7.2 On BYY Harmony Learning (II): Model Selection
Criteria 694

23.7.3 On BYY Harmony Learning (III): Automatic Model
Selection 696

23.7.4 On Regularization Methods........................ 698

23.8 Conclusions.. 700

References 700

Author Index 707

Subject Index 709

List of Contributors

Florence Amardeilh
School of Information Technology
and Electrical Engineering
University of Queensland
St. Lucia, Queensland 4072
Australia
amardeilh@uq.edu.au

Patrick Beautement
Intelligent Distributed Systems E109
Distributed Technology Group
QinetiQ Ltd.
Malvern Technology Centre
St Andrews Road
Great Malvern WR14 3PS
UK
pbeautement@qinetiq.com

Jeffrey M. Bradshaw
Institute for Human and
Machine Cognition
40 S. Alcaniz St.
Pensacola, FL 32502
USA
jbradshaw@ihmc.us

Maggie R. Breedy
Institute for Human and
Machine Cognition
40 S. Alcaniz St.
Pensacola, FL 32502
USA
mbreedy@ihmc.us

Francesco Buccafurri
DIMET Department
Università Mediterranea of
Reggio Calabria
Via Graziella Loc. Feo di Vito
89060 Reggio Calabria
Italy
bucca@unirc.it

Larry Bunch
Institute for Human and
Machine Cognition
40 S. Alcaniz St.
Pensacola, FL 32502
USA
lbunch@ihmc.us

Mario Cannataro
Informatics and Biomedical Eng.
University Magna
Græcia of Catanzaro
Via T. Campanella, 115
88100 Catanzaro
Italy
cannataro@unicz.it

Antonio Congiusta
DEIS
University of Calabria
Via P. Bucci 41C, Cubo 41/C
87036 Rende (CS)
Italy
congiusta@icar.cnr.it

Richard Cole
School of Information Technology
and Electrical Engineering
University of Queensland
St. Lucia, Queensland 4072
Australia
rcole@itee.uq.edu.au

AnHai Doan
Department of Computer Science
Thomas M. Siebel Center for
Computer Science
201 N. Goodwin
Urbana, IL 61801-2302
USA
mflesner@cs.uiuc.edu

Sergey V. Drakunov
Department of Electrical Eng.
and Computer Science
Tulane, University
New Orleans, LA 70118-5674
USA
drakunov@tulane.edu

Vincent Dubois
IAAI - Institute Advanced
Applications of the Internet
Marseille
France
vincent.dubois@iaai.fr

Peter Eklund
School of Information Technology
and Computer Science
University of Wollongong
Northfields Avenue, NSW 2522
Australia
peklund@uow.edu.au

Paul J. Feltovich
Institute for Human and
Machine Cognition
40 S. Alcaniz St.
Pensacola, FL 32502
USA
pfeltovich@ihmc.us

Jianfeng Gao
Microsoft Research Asia
5F, Beijing Sigma Center
No. 49, Zhichun Road
Haidian District, Beijing 100080
China
jfgao@microsoft.com

Salvatore Greco
Faculty of Economics
University of Catania
Corso Italia 55
95129 Catania
Italy
salgreco@unict.it

Jiawei Han
Department of Computer Science
Thomas M. Siebel Center for
Computer Science
201 N. Goodwin
Urbana, IL 61801-2302
USA
hhall@cs.uiuc.edu

Robert R. Hoffman
Institute for Human and
Machine Cognition
40 S. Alcaniz St.
Pensacola, FL 32502
USA
rhoffman@ihmc.us

Jiajin Huang
College of Computer Science
Beijing University of Technology
No. 100, Lepingyuan, Chaoyang
Beijing 100022
China
hjj@emails.bjpu.edu.cn

Renia Jeffers
Institute for Human and
Machine Cognition
40 S. Alcaniz St.
Pensacola, FL 32502
USA
rjeffers@ihmc.us

Jongwoo Jeon
Department of Statistics
Seoul National University
Seoul, 151-741
South Korea
jwjeon@plaza.snu.ac.kr

Xiaolong Jin
Department of Computer Science
Hong Kong Baptist University
Kowloon Tong
Hong Kong
jxl@comp.hkbu.edu.hk

Matthew Johnson
Institute for Human and
Machine Cognition
40 S. Alcaniz St.
Pensacola, FL 32502
USA
mjohnson@ihmc.us

Jinseog Kim
Department of Statistics
Seoul National University
Seoul, 151-741
South Korea
jskim@stats.snu.ac.kr

Yongdai Kim
Department of Statistics
Seoul National University
Seoul, 151-741
South Korea
ydkim@stats.snu.ac.kr

Shriniwas Kulkarni
Institute for Human and
Machine Cognition
40 S. Alcaniz St.
Pensacola, FL 32502
USA
skulkarni@ihmc.us

Vipin Kumar
University of Minnesota
4-192 EE/CSci Building
200 Union Street SE
Minneapolis, MN 55455
USA
kumar@cs.umn.edu

Charles X. Ling
Department of Computer Science
University of Western Ontario
London, Ontario N6A 5B7
Canada
cling@csd.uwo.ca

Chunnian Liu
College of Computer Science
Beijing University of Technology
No. 100, Lepingyuan, Chaoyang
Beijing 100022
China
ai@bjut.edu.cn

Jiming Liu
Department of Computer Science
Hong Kong Baptist University
Kowloon Tong
Hong Kong
jiming@comp.hkbu.edu.hk

James Lott
Institute for Human and
Machine Cognition
40 S. Alcaniz St.
Pensacola, FL 32502
USA
jlott@ihmc.us

Carlo Mastroianni
ICAR-CNR
Italian National Research Council
Via P., Bucci, 87036 Rende (CS)
Italy
mastroianni@icar.cnr.it

Benedetto Matarazzo
Faculty of Economics
University of Catania
Corso Italia 55
95129 Catania
Italy
matarazz@unict.it

Michele Missikoff
Istituto di Analisi dei Sistemi
ed Informatica Antonio Ruberti
CNR Viale Manzoni
30 00185, Roma
Italy
missikoff@iasi.rm.cnr.it

Pabitra Mitra
Machine Intelligence Unit
Indian Statistical Institute
Kolkata 700108
India
pabitra_r@isical.ac.in

Sushmita Mitra
Machine Intelligence Unit
Indian Statistical Institute
Kolkata 700108
India
sushmita@isical.ac.in

Katharina Morik
University Dortmund
Computer Science VIII
Baroperstr.301, D-44227, Dortmund
Germany
morik@ls8.cs.uni-dortmund.de

Roberto Navigli
Dipartimento di Informatica
Università di Roma
"La Sapienza"
Via Salaria 113
00198 Roma
Italy
navigli@di.uniroma1.it

Chuangxin Ou
College of Computer Science
Beijing University of Technology
No. 100, Lepingyuan, Chaoyang
Beijing 100022
China
ocx@emails.bjpu.edu.cn

Sankar K. Pal
Distinguished Scientist
Indian Statistical Institute
Calcutta 700108
India
sankar@isical.ac.in

Luigi Palopoli
DIMET Department
Università Mediterranea of
Reggio Calabria
Via Graziella Loc. Feo di Vito
89060 Reggio Calabria
Italy
palopoli@ing.unirc.it

Andrea Pugliese
DEIS
University of Calabria
Via P. Bucci 41C, Cubo 41/C
87036 Rende (CS)
Italy
apugliese@si.deis.unical.it

Minglun Qian
Dept. of Electrical and
Computer Engineering
University of Illinois
Urbana-Champaign
1308 West Main Street
Urbana, IL 61801
USA
m-qian@uiuc.edu

Mohamed Quafafou
IAAI - Institute Advanced
Applications of the Internet
Marseille
France
mohamed.quafafou@iaai.fr

Anil K. Raj
Institute for Human and
Machine Cognition
40 S. Alcaniz St.
Pensacola, FL 32502
USA
araj@ihmc.us

Domenico Rosaci
DIMET Department
Università Mediterranea of
Reggio Calabria
Via Graziella Loc. Feo di Vito
89060 Reggio Calabria
Italy
domenico.rosaci@unirc.it

Giuseppe M.L. Sarnè
DIMET Department
Università Mediterranea of
Reggio Calabria
Via Graziella Loc. Feo di Vito
89060 Reggio Calabria
Italy
sarne@ing.unirc.it

Martin Scholz
University Dortmund
Computer Science VIII
Baroperstr.301, D-44227, Dortmund
Germany
scholz@ls8.cs.uni-dortmund.de

Andrzej Skowron
Institute of Mathematics
Faculty of Mathematics
Computer Science and Mechanics
Warsaw University
Banacha 2, 02-097 Warsaw
Poland
skowron@mimuw.edu.pl

Roman Slowinski
Institute of Computing Science
Poznan University of Technology
Street Piotrowo 3A
60-965 Poznan
Poland
slowinsk@sol.put.poznan.pl

Niranjan Suri
Institute for Human and
Machine Cognition
40 S. Alcaniz St.
Pensacola, FL 32502
USA
nsuri@ihmc.us

Einoshin Suzuki
Electrical and Computer Eng.
Yokohama National University
79-5, Tokiwadai, Hodogaya
Yokohama, 240-8501
Japan
suzuki@ynu.ac.jp

Domenico Talia
DEIS
University of Calabria
Via P. Bucci 41C, Cubo 41/C
87036 Rende (CS)
Italy
talia@deis.unical.it

Pang-Ning Tan
Dept. of Computer Science
and Engineering
Michigan State University
3115 Engineering Building
East Lansing, MI 48824
USA
ptan@cse.msu.edu

Paolo Trunfio
DEIS
University of Calabria
Via P. Bucci 41C, Cubo 41/C
87036 Rende (CS)
Italy
trunfio@deis.unical.it

Andrzej Uszok
Institute for Human and
Machine Cognition
40 S. Alcaniz St.
Pensacola, FL 32502
USA
auszok@ihmc.us

Paola Velardi
Dipartimento di Informatica
Università di Roma La
Sapienza, Via Salaria 113
00198 Roma
Italy
velardi@di.uniroma1.it

Benjamin W. Wah
Dept. of Electrical and
Computer Engineering
University of Illinois
Urbana-Champaign
1308 West Main Street
Urbana, IL 61801
USA
wah@uiuc.edu

Lei Xu
Dept. of Computer Science
and Engineering
Chinese University of Hong Kong
Hong Kong
lxu@cse.cuhk.edu.hk

Qiang Yang
Department of Computer Science
Hong Kong University of Science
and Technology
Clearwater Bay, Kowloon
Hong Kong
qyang@cs.ust.hk

Yiyu Yao
Department of Computer Science
University of Regina
Regina, Saskatchewan, S4S 0A2
Canada
yyao@cs.uregina.ca

Hwanjo Yu
Department of Computer Science
Thomas M. Siebel Center for
Computer Science
201 N. Goodwin
Urbana, IL 61801-2302
USA
hwanjoyu@uiuc.edu

Chengqi Zhang
Faculty of Information Technology
University of Technology, Sydney
PO Box 123, Broadway NSW 2007
Australia
chengqi@it.uts.edu.au

Zili Zhang
School of Information Technology
Deakin University
Geelong, VIC 3217
Australia
zzhang@deakin.edu.au
AND
Faculty of Computer and
Information Science
Southwest China Normal University
Chongqing 400715
China
zhangzl@swnu.edu.cn

Ning Zhong
Dept. of Systems & Information Eng.
Maebashi Institute of Technology
460-1 Kamisadori-Cho
Maebashi-City 371-0816
Japan
zhong@maebashi-it.ac.jp

1. The Alchemy of Intelligent IT (iIT): A Blueprint for Future Information Technology

Ning Zhong[1] and Jiming Liu[2]

[1] Maebashi Institute of Technology, Japan
[2] Hong Kong Baptist University, Hong Kong

Abstract

This chapter investigates intelligent technologies for information analysis from a new intelligent IT (iIT) perspective in the context of today's as well as future IT. An overview of the book devoted to advanced intelligent information technologies shows that this coherently written multi-author monograph presents the current state of research and development in iIT, as well as both theoretical and application aspects of iIT.

1.1 What Is iIT?

The *intelligent* information technology (iIT) is a newly emerged field that encompasses the theories and applications of artificial intelligence, statistical pattern recognition, learning theory, data warehousing, data mining and knowledge discovery, Grid computing, and autonomous agents and multi-agent systems in the context of today's as well as future IT, such as Electronic Commerce (EC), Business Intelligence (BI), Social Intelligence (SI), Web Intelligence (WI), Knowledge Grid (KG), and Knowledge Community (KC), among others.

1.2 Why iIT?

There are several reasons why we need to investigate the future of information technology (IT) from a new *intelligent* IT (iIT) perspective.

First, we now live in an information age, and are in a fast transition to emerging information media and technologies. Information has become a very important commodity. In particular, with the popularization of Internet and WWW, every second hundreds of thousands of new records of information are generated. This information needs to be summarized and synthesized in order to support effective problem-solving and decision-making.

Massive data sets have driven research, applications, and tool development in business, science, government, and academia. The continued growth in data collection in all of these areas ensures that the fundamental problem that the iIT addresses, namely how one understands and uses one's data, will continue to be of critical importance across a large swath of organizations. We are experiencing a strong demand for more powerful, intelligent computing paradigms for large-scale data handling, exploration, and management.

Second, we have seen a sea change in both the content and the methodology of research in AI [1.32]. It is now more common to build on existing theories than to propose brand new ones, to base claims on rigorous theorems or hard experimental evidence rather than on intuition, and to show relevance to *real world* applications rather than toy examples.

The iIT is a typical interdisciplinary field. In order to solve real world problems, enabling techniques developed not only from the AI community, but also from other related communities, such as computer science, statistics, and cognitive science, are required. Furthermore, the iIT also facilitates human and social intelligence development and understanding.

Third, information processing should be data-centric. Data is the source of human knowledge. By analyzing and using data, we can

- turn raw data into money; that is, data is used to make strategic and tactical business decisions, to win new customers, and to retain existing customers, as well as to reduce the cost (and waste) of doing business.
- turn raw data into new science and technology; that is, data is used to invent new scientific laws, to protect and develop the environment and space of human living, and to create new science and technology.

The iIT will facilitate the development of new cultures, new business models, new science and technology, and even the understanding of the universe that we live in.

1.3 iIT as the New Generation of Information Technology

Although various IT techniques were developed long ago, advanced techniques for intelligent information processing are not yet mature. The iIT can be regarded as the new generation of information technology. The development of iIT should be based on the analyses of the following three issues:

1. What new requirements are being put forward from an increasing number of real world application areas, such as e-business (include e-commerce), e-service, e-science, e-learning, e-government, and e-community?
2. What new architecture and features should data repositories and data mining have to fit the requirements of intelligent information processing?
3. What are the new platform and methods to fit the requirements of intelligent information processing in large-scale, multiple distributed data sources?

The answers to these questions help to develop new iIT methodologies and goals. The key issue is how to deal with the scalability and complexity of the real world, efficiently and effectively. We consider there to be at least three major methodologies to deal with such real world issues.

The first is *hybridization*. Here "hybridization" means the way of combining many advantages of existing methods, and avoiding their disadvantages or weaknesses when the existing methods are used separately. There are ongoing efforts to integrate logic (including non-classical logic), artificial neural networks, probabilistic and statistical reasoning, fuzzy sets, rough sets, genetic algorithms, and other disciplines in the soft computing paradigm [1.5, 6.45].

The second methodology is the *multi-phase process*. For instance, the data mining process is usually a *multi-phase* process involving numerous steps like data preparation, preprocessing, search for hypothesis generation, pattern formation, knowledge evaluation, representation, refinement, and management. Furthermore, the process may iterate previous steps if the mined result is not satisfactory [1.7, 6.47].

Although the process-centric view has recently been widely accepted by researchers in the data mining community as an important methodology for knowledge discovery from real-life data [6.45], few data mining systems provide the capabilities that a more complete process should possess. Furthermore, a key shortcoming of the existing data mining process is that it relies on human beings to plan and control the exact steps for solving the problem and on the careful distribution of the task to individual data mining systems. In order to increase both *autonomy* and *versatility* of a data mining system, we are confronted with the following issues:

- How to plan, organize, control, and manage the data mining process dynamically for different mining tasks;
- How to get the system to use what it knows by imparting it the knowledge to decide what tools are appropriate for what problems and when they should be employed.

Solving such issues needs the development of *meta levels* for the data mining process by modeling such a process and using agent technologies on a new platform. We argue that modeling of the data mining process constitutes an important and new research area of data mining including formal specification of the process, its planning, scheduling, controlling, management, evolution, and reuse for dynamically organizing data mining processes and for multi-aspect analysis across multiple data sources.

The third methodology is *distributed and parallel processing*, more specifically, *distributed and parallel data mining* [1.13, 6.48]. Distributed and parallel data mining faces unique challenges and, to cope with them, needs architectural support on a new infrastructure and platform [1.18, 6.48]. The requirements of architectural support for distributed and parallel data mining can be summarized as follows:

- *Mining on data of huge size*: Gigabytes or even terabytes of data have been accumulated in large organizations. Mining on such a large scale of data has the following implications on data mining architecture:
  - We need large computational power (high-performance servers) for mining tasks, and visualization tools for data analysis and model analysis.
  - Mining operations should be run close to databases, because it is not practical to move vast data between the sites of individual analysts. This requirement can be supported either by mobile mining components traveling to the multiple distributed database sites and executing there, or by setting up high-performance servers close to databases.
  - A user should be allowed to browse and sample data during planning and editing his/her mining tasks.
- *Mining on diverse and distributed data sources*: Because various types of data are accumulated on many sites in a large organization, a user may need to access multiple data sources. Thus, the data mining system must support distributed mining and combining partial results into a meaningful total.
- *Data mining process planning*: There are several stages in a data mining process (the three major stages are pre-processing, model-building and model analysis and refinement). For each stage, there is a large number of available data mining techniques and algorithms. Some of them may soon be out-of-date, while new ones come continuously. So, a good combination of data mining techniques and easy integration with new techniques is very desirable, and this demands careful planning of data mining tasks. Note that different data mining techniques are needed from different kinds of data resources, so the planning involves browsing and sampling data.
- *Interactions among data mining roles*: Because a data mining process is iterative through the cycle of data-selection, pre-processing, model building, and model analysis and refinement, a high degree of interactions among analysts, knowledge engineers, and the end-users is needed.
- *Flexibility*: A wide range of configuration options is needed to fulfill the different needs of large organizations, so that applications can be scaled from a few client workstations to high-performance server machines.
- *Open-endedness* for future extension.
- *Conceptual and architectural simplicity* is important in designing such a complex system to ensure/enhance its correctness, flexibility, and openness.

The ideal infrastructure, platform, and technology to support such a distributed and parallel data mining architecture are Grid computing and agent technologies [6.5, 6.6]. We need to study data mining, intelligent agents, and Grid computing, as the key techniques of iIT, in a uniform way.

1.4 iIT for e-Business Intelligence

There is a great potential for iIT to make useful contributions to the development of intelligent *enterprise portals* for e-business intelligence. An enterprise portal enables a company or an organization to create a *virtual enterprise* where key production steps are outsourced to partners. Many organizations are implementing a corporate portal first and are then growing this solution into more of an intelligent B2B portal. By using a portal to tie in back-end enterprise systems, a company can manage the complex interactions of the virtual enterprise partners through all phases of the value and supply chain.

1.4.1 Virtual Industry Park: An Example of Enterprise Portals

As an example for developing enterprise portals by using iIT technology, we discuss here how to construct an intelligent Virtual Industry Park (VIP), which our group has been developing [6.53]. The VIP portal is a website in which all of the contents related to the mid-sized/small-scale companies in Maebashi city can be accessed.

The construction process can be divided into three phases. We first constructed a basic system including the fundamental functions such as the interface for dynamically registering/updating enterprise information, the database for storing the enterprise information, the automatic generation/modification of enterprise homepages, and the domain-specific, keyword-based search engine. When designing the basic system, we also started by analyzing customer performance: what each customer bought over time, total volumes, trends, and so on.

Although the basic system can work as a whole, we now need to know not only past performance on the business front, but also how the customer or prospect enters our VIP portal, in order to target products and manage promotions and marketing campaigns. To the already demanding requirement to capture transaction data for further analysis, we now also need to apply Web usage mining techniques to capture the clicks of the mouse that define where the visitor has been on our website. What pages has he or she visited? What is the semantic association between the pages he or she visited? Is the visitor familiar with the Web structure? Or is he or she a new user or a random one? Is the visitor a Web robot or "other" user? In search for the holy grail of "stickiness," we know that a prime factor is *personalization* for:

- making a dynamic recommendation to a Web user based on the user profile and usage behavior,
- automatic modification of a website's contents and organization,
- combining Web usage data with marketing data to give information about how visitors used a website to marketers.

Hence, we need to extend the basic VIP system by adding more advanced functions such as Web mining, the ontologies-based search engine, as well as automatic email filtering and management.

Finally, a portal for e-business intelligence can be implemented by adding e-business related application functions such as customer relationship management (CRM), targeted marketing, electronic data interchange (EDI), and security solution.

1.4.2 Web Mining and Farming

The enterprise portal-based e-business activity that involves the end user is undergoing a significant revolution [6.33]. The ability to track users' browsing behavior down to individual mouse clicks has brought the vendor and end customer closer than ever before. It is now possible for a vendor to personalize his product message for individual customers at a massive scale. This is called *targeted marketing* (or *direct marketing*) [6.32]. Web mining and Web usage analysis play an important role in e-business for customer relationship management (CRM) and targeted marketing.

Web mining is the use of data mining techniques to automatically discover and extract information from large Web data repositories such as Web documents and services [1.15, 1.26, 6.33, 1.37, 6.48]. Web mining research is at the crossroads of research from several research communities, such as database, information retrieval, and artificial intelligence, especially the subareas of machine learning and natural language processing. Web mining can be divided into four classes of data available on the Web:

- Web content: the data that constitutes the Web pages and conveys information to the users, i.e., html, graphical, video, and audio files of a Web page.
- Web structure: the data that formulates the hyperlink structure of a website and the Web, i.e., various HTML tags used to link one page to another and one website to another website.
- Web usage: the data that reflects the usages of Web resources, i.e., entries in Web browser's history and Internet temporary files, proxy server, and Web server logs.
- Web user profile: the data that provides demographic information about users of the website, i.e., users' registration data and customers' profile information.

Furthermore, Web content, structure, and usage information, in many cases, are co-present in the same data file. For instance, the file names appearing in the log files and Web structure data contain useful content information. One may safely assume that a file named "WebLogMining.html" must contain information about Web log mining. Similarly, the categories of Web mining cannot be considered exclusive or isolated from each other. Web content mining sometimes must utilize Web structure data in order to classify a Web page. In the same way, Web usage mining sometimes has to make use of Web content data and Web structure information.

A challenge is to explore the connection between Web mining and the related agent paradigm such as Web farming that is the systematic refining of information resources on the Web for business intelligence [1.11]. Web farming extends Web mining into an evolving breed of information analysis in a whole process of Web-based information management including seeding, breeding, gathering, harvesting, refining, and so on.

1.4.3 Semantic Social Networks for Intelligent Enterprise Portals

Developing intelligent enterprise portals needs the study of both centralized and distributed information structures on the Web. Information/knowledge on the Web can be either globally distributed throughout the Web over the multi-layer infrastructure of Web protocols, or located locally, centralized on an intelligent portal providing Web services (i.e. the intelligent service provider) that is integrated with its own cluster of specialized intelligent applications. However, each approach has a serious flaw. As pointed out by Alesso and Smith [1.1], the intelligent portal approach limits uniformity and access, while the global semantic Web approach faces combinational complexity limitations.

A way to solve the above issue is to develop and use the *Problem Solver Markup Language* (PSML) for collecting globally distributed contents and knowledge from Web-supported, semantic social networks and incorporating them with locally operational knowledge/data bases in an enterprise or community for local centralized, adaptable intelligent Web services [1.49, 6.52].

The core of PSML is distributed inference engines that can perform automatic reasoning on the Web by incorporating contents and meta-knowledge autonomically collected and transformed from the semantic Web with locally operational knowledge/data bases. A feasible way to implement such PSML is to use an existing Prolog-like logic language with dynamic contents and meta-knowledge collection and transformation agents.

In our experiments, we use KAUS for representation of local information sources and for inference and reasoning. KAUS is a knowledge-based system developed in our group which involves knowledge bases on the basis of Multi-Layer Logic and databases based on the relational data model [1.28, 1.29, 1.35]. KAUS enables representation of knowledge and data in the first-order logic with data structure in multi-level and can be easily used for inference and reasoning as well as transforming and managing both knowledge and data.

By using this information transformation approach, the dynamic, global information sources on the Web can be combined with the local information sources in an enterprise portal for decision-making and e-business intelligence.

1.4.4 Data Mining Grids for Web-Based Targeted Marketing

Targeted marketing (also called direct marketing) is a new marketing discipline for e-business intelligence.

First, e-business intelligence needs the function of Web-based direct marketing, which is integrated with other functions of Web Intelligence (WI), such as Web mining and farming, the ontologies-based search engine, personalized recommendation, and automatic email filtering and management [6.51, 6.52]. The e-business activity that involves the end user is undergoing a significant revolution [6.33]. The ability to track users' browsing behavior down to individual mouse clicks has brought vendors and end customers closer than ever before. It is now possible for a vendor to personalize its product message for individual customers at a massive scale on the Web. Web mining and Web usage analysis play an important role in e-business for Web-based customer relationship management (CRM) and Web-based direct marketing.

Second, customer data, in general, can be obtained from multiple customer touchpoints [6.13]. In response, multiple data sources that are obtained from multiple customer touchpoints, including the Web, wireless, call center, and brick-and-mortar store data, need to be integrated into a single data warehouse that provides a multi-faceted view of customers including their preferences, interests, and expectations, for multi-aspect analysis. Hence, a multi-strategy and multi-agent data mining framework is required [6.47, 6.48].

The main reason for developing a multi-agent data mining system is that we cannot expect to develop a single data mining algorithm that can be used to solve all direct marketing problems due to the complexity of real-world applications. Hence, various data mining agents need to be cooperatively used in the multi-step data mining process for performing multi-aspect analysis as well as multi-level conceptual abstraction and learning.

Another reason for developing a multi-agent data mining system is that when performing multi-aspect analysis for complex direct marketing problems, a data mining task needs to be decomposed into sub-tasks. Thus these sub-tasks can be solved by using one or more data mining agents that are distributed over different computers. Thus, the decomposition problem leads us to the problem of distributed cooperative system design.

Third, a new infrastructure and platform is required as middleware in order to enable Web-based direct marketing for multi-aspect analysis from multiple data sources. One way is to create a Grid-based, organized society of data mining agents, called a *Data Mining Grid* on the grid computing platform (e.g. the Globus toolkit) [6.5, 6.6, 6.9, 6.10, 6.19, 1.53]. This means:

– Developing various data mining agents for various services oriented multi-aspect data analysis and targeted marketing tasks;
– Organizing the data mining agents into a multi-layer Grid such as data-grid, mining-grid, and knowledge-grid, under the OGSA (Open Grid Services Architecture) that firmly aligns with service-oriented architecture and

Web services, understands a user's questions, transforms them to data mining issues, discovers resources and information about the issues, and gets a composite answer or solution;
- Using a conceptual model with three level workflows, namely data-flow, mining-flow, and knowledge-flow, with respect to the data-grid, the mining-grid, and the knowledge-grid, respectively, for managing the grid of data mining agents for multi-aspect analysis in distributed, multiple data sources and for organizing the dynamic, status-based business processes.

That is, the data mining Grid is made of many smaller components that are called *data mining agents*. Each agent by itself can only do some simple things. Yet, when we join these agents in a *Grid*, we can implement more complex targeted marketing and business intelligence tasks.

The e-business intelligence example shows that we need to study and use iIT systematically to deal with the scalability and complexity of the real world.

1.4.5 Wisdom Web-Based Computing

In our previous paper, we argued that developing the *Wisdom Web* will become a tangible goal for Web Intelligence (WI) research [1.25, 6.51]. The paradigm of Wisdom Web-based computing is aimed at providing not only a medium for seamless information exchange and knowledge sharing but also a type of man-made resource for sustainable knowledge creation, and scientific and social evolution. The Wisdom Web will reply on *grid-like service agencies* that self-organize, learn, and evolve their courses of actions, in order to perform service tasks, as well as their identities and interrelationships in communities. They will also cooperate and compete among themselves in order to optimize their own as well as others' resources and utilities.

One of the key features in wisdom agents is *ubiquity*. Ubiquitous agents are autonomous entities that automatically explore and exploit WI-based Web services, and are capable of self-improving their performance in dynamically changing and unpredictable task environments. The core of those techniques is the notion of synthetic or emergent autonomy based on behavioral self-organization. There has been some earlier work performed in the domains of e-business and e-commerce. Related issues and applications in which various tasks and objectives were achieved can be found in [1.21].

In this area of research, a reasonable challenge to begin with would be to develop and demonstrate a *Ubiquitous Agent Community* (UAC), i.e., an intelligent infrastructure that enables agents to look ahead, plan, and deliver what a user wants [1.24]. It works like a personal agency. For instance, it can help a user to effectively manage tedious daily routine activities, such as processing emails, placing orders, making meeting arrangements, downloading news, etc.

Furthermore, in order to effectively develop the new generation iIT systems, we need to define *benchmark* applications, i.e., a new *Turing Test*, that will capture and demonstrate the Wisdom Web capabilities.

Take the Wisdom Web based computing benchmark as an example. We can use a service task of compiling and generating a market report on an existing product or *a potential market report on a new product*. In order to get such service jobs done, an information agent on the Wisdom Web will mine and integrate available Web information, which will in turn be passed on to a market analysis agent. Market analysis will involve the quantitative simulations of customer behavior in a marketplace, instantaneously handled by other serviced agencies, involving a large number of Grid agents [6.5]. Since the number of variables concerned may be in the order of hundreds or thousands, it can easily cost a single system years to generate one predication. Developing advanced enterprise portals will be a good demonstration of the Wisdom Web for the new Turing Test.

1.5 Dimensions of iIT Research

In order to study iIT systematically, we can engage iIT research from the following dimensions:

- iIT Foundation (e.g., autonomic, Grid, and ubiquitous computational architectures; methods and concepts; ontology models; languages; Web services; Wisdom Web).
- iIT Functionality Development (e.g., data mining; adaptive interface agents; profiling; personalization; Web information retrieval and filtering; Web mining and farming; visualization).
- iIT Problem Identification and Representative Applications (e.g., wireless communication; e-business; e-science; e-learning; e-government; e-services, e-community; customer relationship management and targeted marketing; digital library; information markets; intelligent portals).
- iIT Real-World Experience, Evaluation, and Management (e.g., scalability; interoperability; security; availability; manageability; complexity; Web-supported cooperative work; resource planning; workflow; business process engineering; social network intelligence).

Today, researchers have already studied the Web as a new medium for business intelligence and social intelligence, as well as an analogy of how our brain functions. New advances in such areas as data mining, intelligent agents, and Grid computing will soon provide even richer metaphors and experiments to expand our understanding of new social, universe, and biological realities.

1.6 An Overview of This Book

The aim of this volume, as the first monograph devoted to intelligent technologies for information analysis, is to present the current state of research and development in iIT, as well as both theoretical and application aspects of iIT. It highlights major iIT related topics by structuring one introductory chapter and 22 research/survey chapters into 5 parts: emerging data mining technology, data mining for Web intelligence, emerging agent technology, emerging soft computing technology, and statistical learning theory. Each chapter includes the original work of author(s) as well as a comprehensive survey related to the chapter's topic.

Part 1 presents emerging data mining technology.

In Chap. 2, Mario Cannataro and colleagues describe the *Knowledge Grid* architecture and discuss some related systems and models recently proposed for knowledge discovery on Grids. They present also how to design and implement distributed data mining applications by using the *Knowledge Grid* tools starting from searching Grid resources, composing software and data elements, and executing the resulting application on a Grid.

In Chap. 3, Katharina Morik and Martin Scholz present their data mining methodology and system called MiningMart. The MiningMart methodology/system focuses on setting up and reusing best-practice cases of preprocessing data stored in very large databases. A meta-data model named M4 is used to declaratively define and document both, all steps of such a preprocessing chain and all the data involved.

In Chap. 4, Yongdai Kim and colleagues investigate possibilities of using ensemble methods for generating useful rules, which help understanding the data set as well as the decision. They give an extensive review of three ensemble algorithms - bagging, boosting, and CHEM, and propose a novel algorithm of rule generation with CHEM.

In Chap. 5, Einoshin Suzuki classifies various exception rule/group discovery methods into those that employ domain knowledge and those that don't, and gives a brief survey. He then proposes an evaluation scheme for evaluating various exception rule/group discovery methods in a uniform way.

In Chap. 6, Ning Zhong and colleagues provide an overview on the recent development in data mining applications for targeted marketing that is a new marketing discipline for e-business intelligence.

Part 2 presents data mining for Web intelligence.

In Chap. 7, Hwanjo Yu and colleagues discuss research on finding targeted information on the Web, in particular on two key issues: mining to impose structures over Web data to aid in Web navigation, and mining to build information processing systems. They also describe two recent Web mining projects that illustrate the use of mining techniques to address the above two key issues, and point out novel research opportunities.

In Chap. 8, Qiang Yang and colleagues first provide an overview of the current research on Web log mining, and then present their own approaches to actionable Web log mining to mine *actionable knowledge* that can be immediately applied to the operation of the website.

In Chap. 9, Pang-Ning Tan and Vipin Kumar analyze the navigational patterns for various types of Web robots, and propose a novel approach for discovering Web robot sessions by using the navigational patterns in the click-stream data to determine if they are due to a robot. Their experimental results show that the proposed approach is able to discover many camouflaging and previously unknown Web robots, as well as that highly accurate robot classification models can be induced from the access features of the Web sessions.

In Chap. 10, Roberto Navigli and colleagues present an approach and a tool called OntoLearn for extracting domain concepts and for detecting semantic relationships among them from websites and, more generally, from documents shared among the members of virtual organizations to support the construction of a domain ontology.

In Chap. 11, Richard Cole and colleagues give first an overview on Information Extraction (IE) techniques and their application on the Web, and then present a novel approach for browsing semi-structured Web-texts by using Formal Concept Analysis (FCA).

In Chap. 12, Vincent Dubois and Mohamed Quafafou present a novel approach for producing dynamic and incremental graph structure by mixing text extraction and visualization.

Part 3 presents emerging agent technology.

In Chapter 13, Xiaolong Jin and Jiming Liu investigate agent networks by studying the topological and clustering characteristics. Specifically, the authors examine two multi-agent-based representations: agents representing clauses and agents representing variables, for satisfiability problems (SATs).

In Chap. 14, Buccafurri and colleagues investigate a basic issue to face in multi-agent systems: among a (possibly large) number of agents, how dose one detect agents which are promising candidates for cooperation? The authors propose a new framework for dealing with this problem by defining a formal model for representing agents and a number of semantic properties exploited for detecting fruitful cooperation. On the basis of this framework, the authors design a multi-agent system that allows the user to find the best agents for cooperation in the network of agents.

In Chap. 15, Zili Zhang and Chengqi Zhang propose an agent-based framework, and apply this framework to construct hybrid intelligent systems for data mining. By combining these cutting-edge technologies together, it is expected that more and more difficult real-world problems can be solved.

In Chap. 16, Jeffrey Bradshaw and colleagues investigate the technical and social challenges in the problem of making agents acceptable to people.

They give examples and an explanation of how a policy-based approach might be used to address some of those challenges.

Part 4 presents emerging soft computing technology.

In Chap. 17, Benjamin W. Wah and Minglun Qian provide a survey of previous work in predicting noise-free piecewise chaotic time series and noisy time series with high frequency random noise, and propose a novel approach that incorporates constraints on predicting low-pass data in the lag period. The new constraints enable active training in the lag period that greatly improves the prediction accuracy in that period. Their experimental results show significant improvements in prediction accuracy on standard benchmarks and stock-price time series.

In Chap. 18, Andrzej Skowron investigates a general scheme of approximate reasoning by agents for rough neuro-computation based on knowledge granulation ideas using rough mereological tools.

In Chap. 19, Sankar K. Pal and colleagues provide a new perspective of soft computing paradigm for Pattern Recognition, Data Mining, Web Intelligence, and their integration.

In Chaps. 20 and 21, Salvatore Greco and colleagues are concerned with the issue of prior knowledge about preference semantics in patterns to be discovered. In particular, the authors propose a new approach called Dominance-based Rough Set Approach (DRSA) to deal with preference-ordered data. That is, given a set of objects described by at least one condition attribute with preference-ordered scale and partitioned into preference-ordered classes, the new rough set approach is able to approximate this partition by means of dominance relations.

Part 5 presents statistical learning theory as a foundation of knowledge discovery and data mining.

In Chaps. 22 and 23, Lei Xu investigates the methodologies for mining dependence structures from statistical learning and independence analysis perspectives, respectively. The author proposes the Bayesian Ying-Yang (BYY) system, as a unified framework, for various dependence structures, and BYY harmony learning as a promising tool for making learning on a finite size of samples with model selection ability.

In summary, this monograph presents the current state of research and development in both theoretical and application aspects of intelligent information analysis. It will become a valuable source of reference for R&D professionals active in iIT. Students as well as IT professionals and ambitious practitioners concerned with iIT will appreciate the book as a useful text enhanced by numerous illustrations and examples.

References

1.1 H.P. Alesso, C.F. Smith: *The Intelligent Wireless Web* (Addison-Wesley, 2002)

1.2 F. Berman: From TeraGrid to Knowledge Grid, *CACM*, 44, 27-28 (2001)

1.3 T. Berners-Lee, J. Hendler, O. Lassila: The Semantic Web, *Scientific American*, 284, 34-43 (2001)

1.4 David Shepard Associates: *The New Direct Marketing* (McGraw-Hill, 1999)

1.5 S.K. Pal, S. Mitra, P. Mitra: Rough Fuzzy MLP: Modular Evolution, Rule Generation, and Evaluation, *IEEE Tran. Knowledge and Data Engineering*, Vol. 15, No. 1, 14-25 (2003)

1.6 M. Cannataro and D. Talia: The Knowledge Grid, *CACM*, Vol. 46, 89-93 (2003)

1.7 U.M. Fayyad, G. Piatetsky-Shapiro, P. Smyth: From Data Mining to Knowledge Discovery: an Overview, *Advances in Knowledge Discovery and Data Mining* (MIT Press, 1996) pp. 1-36

1.8 U.M. Fayyad, G. Piatetsky-Shapiro, P. Smyth, R. Uthurusamy (eds.): *Advances in Knowledge Discovery and Data Mining* (AAAI/MIT Press, 1996)

1.9 I. Foster and C. Kesselman: *The Grid: Blueprint for a New Computing Infrastructure* (Morgan Kaufmann, 1999)

1.10 I. Foster and C. Kesselman: *The Grid 2: Blueprint for a New Computing Infrastructure* (Morgan Kaufmann, 2004)

1.11 R.D. Hackathorn: *Web Farming for the Data Warehouse* (Morgan Kaufmann, 2000)

1.12 S.Y. Hwang, E.P. Lim, J.H. Wang, J. Srivastava: *Proc. PAKDD 2002 Workshop on Mining Data across Multiple Customer Touchpoints for CRM* (2002)

1.13 H. Kargupta and P. Chan (eds.): *Advances in Distributed and Parallel Knowledge Discovery* (AAAI/MIT Press, 1996)

1.14 W. Klosgen and J.M. Zytkow: *Handbook of Data Mining and Knowledge Discovery* (Oxford University Press, 2002)

1.15 R. Kosala and H. Blockeel: Web Mining Research: A Survey, *ACM SIGKDD Explorations Newsletter*, 2, 1-15 (2000)

1.16 R. Kumar, P. Raghavan, S. Rajagopalan, A. Tomkins: The Web and Social Networks. *IEEE Computer*, Special Issue on Web Intelligence, 35(11) 32-36 (November 2002)

1.17 V. Kumar, S. Tsumoto, N. Zhong, P.S. Yu, X. Wu (eds.): *Proceedings of 2002 IEEE International Conference on Data Mining* (IEEE Computer Society Press, California, USA, 2002)

1.18 C. Liu and N. Zhong: A Multi-Agent Based Architecture for Distributed KDD Process, Z.W. Ras and S. Ohsuga (eds.) *Foundations of Intelligent Systems*, LNAI 1932 (Springer, 2000) pp. 591-600

1.19 J. Liu, N. Zhong (eds.): *Intelligent Agent Technology: Systems, Methodologies, and Tools* (World Scientific, 1999)

1.20 J. Liu: *Autonomous Agents and Multiagent Systems* (World Scientific, 2001)

1.21 J. Liu and Y. Ye (eds.): *Advances in E-commerce Agents: Brokerage, Negotiation, Security, and Mobility* (Springer, 2001)

1.22 J. Liu, N. Zhong, Y.Y. Tang, P.S.P. Wang (eds.): *Agent Engineering* (World Scientific, 2001)

1.23 J. Liu: Web Intelligence (WI): What Makes Wisdom Web? *Proc. Eighteenth International Joint Conference on Artificial Intelligence (IJCAI-03)* (2003) pp. 1596-1601

1.24 J. Liu and C.Y. Yao: Rational Competition and Cooperation in Ubiquitous Agent Communities. *Knowledge Based Systems* (Elsevier, 2004).

1.25 J. Liu, N. Zhong, Y.Y. Yao, Z.W. Ras: The Wisdom Web: New Challenges for Web Intelligence (WI), *Journal of Intelligent Information Systems*, 20(1) 5-9 (Kluwer, 2003)

1.26 Z. Lu, Y.Y. Yao, N. Zhong: Web Log Mining. In N. Zhong, J. Liu, Y.Y. Yao (eds.) *Web Intelligence*, 172-194 (Springer, 2003)

1.27 J. Nabrzyski, J.M. Schopf, J. Weglarz, *Grid Resource Management* (Kluwer, 2004)

1.28 S. Ohsuga and H. Yamauchi: Multi-Layer Logic - A Predicate Logic Including Data Structure as Knowledge Representation Language. *New Generation Computing*, 3(4) 403-439 (Springer, 1985)

1.29 S. Ohsuga: Framework of Knowledge Based Systems - Multiple Meta-Level Architecture for Representing Problems and Problem Solving Processes. *Knowledge Based Systems*, 3(4) 204-214 (Elsevier, 1990)

1.30 Z. Pawlak: *Rough Sets, Theoretical Aspects of Reasoning about Data*, (Kluwer, 1991)

1.31 P. Raghavan: Social Networks: From the Web to the Enterprise, *IEEE Internet Computing*, 6(1), 91-94 (2002)

1.32 S.J. Russell and P. Norvig: *Artificial Intelligence - A Modern Approach* Prentice Hall, Inc. (1995).

1.33 A.R. Simon, S.L. Shaffer: *Data Warehousing and Business Intelligence for e-Commerce* (Morgan Kaufmann, 2001)

1.34 J. Srivastava, R. Cooley, M. Deshpande, P. Tan: Web Usage Mining: Discovery and Applications of Usage Patterns from Web Data. *SIGKDD Explorations, Newsletter of SIGKDD*, 1, 12-23 (2000)

1.35 H. Yamauchi and S. Ohsuga: Loose coupling of KAUS with existing RDBMSs. *Data & Knowledge Engineering*, 5(4) 227-251 (North-Holland, 1990)

1.36 Y.Y. Yao, N. Zhong, J. Liu, S. Ohsuga: Web Intelligence (WI): Research Challenges and Trends in the New Information Age. In: N. Zhong, Y. Y. Yao, J. Liu, S. Ohsuga (eds.) *Web Intelligence: Research and Development*, LNAI 2198, (Springer, 2001) pp. 1-17

1.37 Y.Y. Yao, N. Zhong, J. Huang, C. Ou, C. Liu: Using Market Value Functions for Targeted Marketing Data Mining, *International Journal of Pattern Recognition and Artificial Intelligence*, 16(8) 1117-1131 (World Scientific, 2002)

1.38 Y.Y. Yao and N. Zhong: Granular Computing Using Information Tables, Lin, T.Y., Yao, Y.Y., and Zadeh, L.A. (eds.) *Data Mining, Rough Sets and Granular Computing* (Physica-Verlag, 2002) pp. 102-124

1.39 N. Zhong and L. Zhou (eds.): *Methodologies for Knowledge Discovery and Data Mining*, LNAI 1574 (Springer, 1999)

1.40 N. Zhong, A. Skowron, S. Ohsuga (eds.): *New Directions in Rough Sets, Data Mining, and Granular-Soft Computing*, LNAI 1711 (Springer, 1999).

1.41 N. Zhong, J.Z. Dong, C. Liu, S. Ohsuga: A Hybrid Model for Rule Discovery in Data, *Knowledge Based Systems*, Vol. 14, No. 7, 397-412 (Elsevier, 2001)

1.42 N. Zhong and S. Ohsuga: Automatic Knowledge Discovery in Larger Scale Knowledge-Data Bases, in C. Leondes (ed.) *the Handbook of Expert Systems*, Vol. 4, 1015-1070 (Academic Press, 2001)

1.43 N. Zhong, C. Liu, S. Ohsuga: Dynamically Organizing KDD Processes, *International Journal of Pattern Recognition and Artificial Intelligence*, Vol. 15, No. 3, 451-473 (World Scientific, 2001)

1.44 N. Zhong: Knowledge Discovery and Data Mining, *The Encyclopedia of Microcomputers*, 27(Supplement 6) 235-285 (Marcel Dekker, 2001)

1.45 N. Zhong, Y.Y. Yao, J. Liu, S. Ohsuga (eds.): *Web Intelligence: Research and Development*, LNAI 2198 (Springer, 2001)

1.46 N. Zhong, C. Liu, S. Ohsuga: Dynamically Organizing KDD Process in a Multi-Agent Based KDD System, in J. Liu, N. Zhong, Y.Y. Tang, P. Wang (eds.) *Agent Engineering* (World Scientific, 2001) pp. 93-122

1.47 N. Zhong, J. Liu, S. Ohsuga, J. Bradshaw (eds.): *Intelligent Agent Technology: Research and Development* (World Scientific, 2001)

1.48 N. Zhong, J. Liu, Y.Y. Yao (eds.): Special issue on Web Intelligence (WI), *IEEE Computer*, Vol. 35, No. 11 (2002).

1.49 N. Zhong: Representation and Construction of Ontologies for Web Intelligence, *International Journal of Foundations of Computer Science*, 13(4) 555-570 (World Scientific, 2002)

1.50 N. Zhong, J. Liu, Y.Y. Yao: In Search of the Wisdom Web, *IEEE Computer*, Vol. 35, No. 11, 27-31 (2002)

1.51 N. Zhong, J. Liu, Y.Y. Yao (eds.): *Web Intelligence* (Springer, 2003)

1.52 N. Zhong: Toward Web Intelligence, in E. Menasalvas Ruiz, J. Segovia, P.S. Szczepaniak (eds.): *Advances in Web Intelligence*, LNAI 2663 (Springer, 2003) pp. 1-14

1.53 N. Zhong, J.L. Wu, C. Liu: Building a Data Mining Grid for Multiple Human Brain Data Analysis, *Proc. International Workshop on Knowledge Grid and Grid Intelligence*, Halifax, Canada (2003) pp. 29-35

Part I

Emerging Data Mining Technology

2. Grid-Based Data Mining and Knowledge Discovery

Mario Cannataro[1], Antonio Congiusta[2], Carlo Mastroianni[3],
Andrea Pugliese[2], Domenico Talia[2], and Paolo Trunfio[2]

[1] Informatics and Biomedical Engineering
   University Magna Græcia of Catanzaro, Italy
[2] DEIS, Università della Calabria, Italy
[3] ICAR-CNR, Italian National Research Council, Italy

Abstract

The increasing use of computers in all the areas of human activities is resulting in huge collections of digital data. Databases are common everywhere and are used as repositories of every kind of data. Knowledge discovery techniques and tools are used today to analyze those very large data sets to identify interesting patterns and trends in them. When data is maintained over geographically distributed sites the computational power of distributed and parallel systems can be exploited for knowledge discovery in databases. In this scenario the Grid can provide an effective computational support for distributed knowledge discovery on large data sets. To this purpose we designed a system called *Knowledge Grid*. This chapter describes the *Knowledge Grid* architecture and discusses some related systems and models recently proposed for knowledge discovery on Grids. The chapter presents also how to design and implement distributed data mining applications by using the *Knowledge Grid* tools starting from searching Grid resources, composing software and data elements, and executing the resulting application on a Grid.

2.1 Introduction

Enlarging our knowledge about the world and the secrets of life is one of the strongest motivations of the human activities. The use of computers is changing our way to make discoveries and is improving both speed and quality of the discovery processes. Advances in electronic data gathering, storage, and distribution technologies have far outpaced computational advances in techniques for analyzing and understanding data. This has created the need for models, tools, and techniques for automated data mining and *Knowledge Discovery in Databases (KDD)*. These terms indicate the automated analysis of large volumes of data stored in computers, looking for the relationships and knowledge that are implicit in large volumes of data and are "interesting" for users.

To manage the very large amount of data available today, computer scientists are working on efficient systems, algorithms, and applications that can handle and analyze very large data repositories. Intensive data-consuming applications are running on massive amounts of data with the task of extracting valuable knowledge. Data mining is one of the key technologies in this scenario. However, these intensive data-consuming applications suffer from performance problems and single database sources.

Distributed architectures supported by high performance networks and knowledge-based middleware offer parallel and distributed databases a great opportunity to support cost-effective everyday applications. Moreover, using distributed computing systems and tools allows users to share large data sources, the mining process building, and the extracted knowledge. Large communities of users can pool their resources from different sites of a single organization or from a large number of institutions and perform all the steps of the KDD process from remote sites according to a cooperative approach.

Grid computing is an innovative distributed computational model focusing on large-scale resource sharing, innovative applications, and high-performance orientation. Grids can be used today as effective infrastructures for distributed high-performance computing and data processing [2.1]. Grid application areas are shifting from scientific computing towards industry and business applications. To meet those needs *data Grids* are designed to store, move, and manage large data sets located in remote sites. Data Grids represent an enhancement of computational Grids, driven by the need to handle large data sets without constant, repeated authentication, aiming to support the implementation of distributed data-intensive applications. Significant examples are the EU DataGrid [2.2], the Particle Physics Data Grid [2.3], the Japanese Grid DataFarm [2.4] and the Globus Data Grid [2.5].

As an advancement of the data Grid concept, it is imperative to develop knowledge-based Grids that may offer tools and environments to support the process of analysis, inference, and discovery over data available in many scientific and business areas. These environments will support scientists and engineers in the implementation and use of Grid-based Problem-Solving Environments (PSEs) for modeling, simulation, and analysis of scientific experiments. The same can occur in industry and commerce, where analysts need to be able to mine the large volumes of information that can be distributed over different plants to support corporate decision making.

Knowledge Grids offer high-level tools and techniques for the distributed mining and extraction of knowledge from data repositories available on the Grid. The development of such an infrastructure is the main goal of our research, focused on the design and implementation of an environment for geographically distributed high-performance knowledge discovery applications called *Knowledge Grid*. The *Knowledge Grid* can be used to perform data mining on very large data sets available over Grids, to make scientific discoveries, improve industrial processes and organization models, and uncover business valuable information.

The outline of the chapter is as follows. Section 2.2 discusses related work. Section 2.3 briefly describes the *Knowledge Grid* architecture and the main features of its components. Section 2.4 describes the metadata model adopted to describe the resources used for developing the data mining applications.

Sections 2.5–2.6 discuss how the tools of the *Knowledge Grid* support a user in designing, building, and executing a distributed data mining application. Section 2.7 concludes the chapter.

2.2 Knowledge Discovery on Grids

Berman [2.6], Johnston [2.7], and some of us [2.8, 2.9] claimed that the creation of knowledge Grids on top of computational Grids is the enabling condition for developing high-performance knowledge discovery processes and meeting the challenges posed by the increasing demand of power and abstractness coming from complex Problem-Solving Environments. The design of knowledge Grids can benefit from the layered Grid architecture, with lower levels providing middleware support for higher level application-specific services.

Whereas some high-performance parallel and distributed knowledge discovery (*PDKD*) systems recently appeared [2.10] (see also [2.8]), there are few projects attempting to implement and/or support knowledge discovery processes over computational Grids. A main issue here is the integration of two main demands: synthesizing useful and usable knowledge from data, and performing sophisticated large-scale computations leveraging the Grid infrastructure. Such integration must pass through a clear representation of the knowledge base used to translate moderately abstract domain-specific queries into computations and data analysis operations able to answer such queries by operating on the underlying systems [2.6].

In the remainder of this section we review the most significant systems oriented at supporting knowledge discovery processes over distributed/Grid infrastructures. The systems discussed here provide different approaches to supporting knowledge discovery on Grids. We discuss them starting from general frameworks, such as the TeraGrid infrastructure, then outlining data-intensive oriented systems, such as DataCutter and InfoGrid, and, finally, describing KDD systems similar to the *Knowledge Grid*, such as Discovery Net, and some significant data mining testbed experiences.

The *TeraGrid* project is building a powerful Grid infrastructure, called *Distributed TeraScale Facility* (*DTF*), connecting four main sites in the USA (the San Diego Supercomputer Center, the National Center for Supercomputing Applications, Caltech, and Argonne National Laboratory). Recently, the NSF funded the integration into the DTF of the *TeraScale Computing System* (*TCS-1*) at the Pittsburgh Supercomputer Center; the resulting Grid environment will provide, besides tera-scale data storage, 21 TFLOPS of computational capacity [2.11]. Furthermore, the TeraGrid network connections, whose bandwidth is in the order of tenths of Gbps, have been designed in such a way that all resources appear as a single physical site. The connections have also been optimized to support peak requirements rather than an average load, as is natural in Grid environments. The TeraGrid adopts

Grid software technologies and, from this point of view, appears as a "virtual system" in which each resource describes its own capabilities and behavior through *Service Specifications*. The basic software components are called *Grid Services*, and are organized into three distinct layers. The *Basic* layer comprises authentication, resource allocation, data access, and resource information services; the *Core* layer comprises services such as advanced data management, single job scheduling, and monitoring; the *Advanced* layer comprises superschedulers, resource discovery services, repositories, etc. Finally, *TeraGrid Application Services* are built using Grid Services. The definition of such services is still under discussion, but they should comprise, for example, the support of on-demand/interactive applications, the support of GridFTP interface to data services, etc.

The most challenging application on the TeraGrid will be the synthesis of knowledge from very large scientific data sets. The development of knowledge synthesis tools and services will enable the TeraGrid to operate as a Knowledge Grid. A first application is the establishment of the Biomedical Informatics Research Network to allow brain researchers at geographically distributed advanced imaging centers to share data acquired from different subjects and using different techniques. Such applications make a full use of a distributed data Grid with hundreds of terabytes of data online, enabling the TeraGrid to be used as a knowledge Grid in the biomedical domain. The use of the *Knowledge Grid* services can be potentially effective in these applications [2.12].

InfoGrid is a service-based data integration middleware engine designed to operate on Grids. Its main objective is to provide information access and querying services to knowledge discovery applications [2.13]. The information integration approach of InfoGrid is not based on the classical idea of providing a "universal" query system: instead of abstracting everything for users, it gives a personalized view of the resources for each particular application domain. The assumption here is that users have enough knowledge and expertise to handle the absence of "transparency". In InfoGrid the main entity is the *Wrapper*; wrappers are distributed on a Grid and each node publishes a directory of the wrappers it owns. A wrapper can wrap information sources and programs, or can be built by composing other wrappers (*Composite Wrapper*). Each wrapper provides: (*i*) a set of query construction interfaces that can be used to query the underlying information sources in their native language, and (*ii*) a set of administration interfaces that can be used to configure its properties (access metadata, linkage metadata, configuration files). In summary, InfoGrid puts the emphasis on delivering metadata describing resources and providing an extensible framework for composing queries.

DataCutter is another middleware infrastructure that aims to provide specific services for the support of multi-dimensional range querying, data aggregation, and user-defined filtering over large scientific data sets in shared distributed environments [2.14]. DataCutter has been developed in the context of the *Chaos* project at the University of Maryland; it uses and extends features of the *Active Data Repository* (*ADR*), that is a set of tools for the optimization of storage, retrieval, and processing of very large multidimensional data sets. In ADR, data processing is performed at the site where data is stored, whereas in Grid environments this is naturally unfeasible, due to inherent data distribution and resource sharing at servers, which may lead to inefficiencies.

To overcome this, in the DataCutter framework an application is decomposed into a set of processes, called *filters*, that are able to perform a rich set of queries and data transformation operations. Filters can execute anywhere but are intended to run on a machine close (in terms of connectivity) to the storage server. DataCutter supports efficient indexing. In order to avoid the construction of a huge single index that would be very costly to use and keep updated, the system adopts a multi-level hierarchical indexing scheme, specifically targeted at the multi-dimensional data model adopted.

Different from the two environments discussed above, the *Datacentric Grid* is a system directed at knowledge discovery on Grids designed for mainly dealing with immovable data [2.15]. The system consists of four kinds of entities. The nodes at which computations happen are called *Data/Compute Servers* (*DCS*). Besides a compute engine and a data repository, each DCS comprises a *metadata tree*, that is a structure for maintaining relationships among raw data sets and models extracted from them. Furthermore, extracted models become new data sets, potentially useful at subsequent steps and/or for other applications.

The *Grid Support Nodes* (*GSNs*) maintain information about the whole Grid. Each GSN contains a directory of DCSs with static and dynamic information about them (e.g., properties and usage), and an execution plan cache containing recent plans along with their achieved performance. Since a computation in the Datacentric Grid is always executed on a single node, execution plans are simple. However, they can start at different places in the model hierarchy because, when they reach a node, they may find already computed models. The *User Support Nodes* (*USNs*) carry out execution planning and maintain results. USNs are basically proxies for user interface nodes (called *User Access Points* or *UAPs*). This is because user requests (i.e., task descriptions) and their results can be small in size, so in principle UAPs could be simple devices not always online, and USNs could interact with the Datacentric Grid when users are not connected.

An agent-based data mining framework, called *ADaM* (*Algorithm Development and Mining*), has been developed at the University of Alabama [2.16]. Initially, this framework was adopted for processing large data sets for geophysical phenomena. More recently, it has been ported to the NASA's *Information Power Grid* (*IPG*) environment, for the mining of satellite data [2.17]. In this system, the user specifies *what* is to be mined (data set names and locations) and *how* and *where* to perform the mining (sequence of operations, required parameters, and IPG processors to be used). Initially, "thin" agents are associated with the sequence of mining operations; such agents acquire and combine the needed mining operations from repositories that can be public or private, i.e., provided by mining users or private companies. Data is acquired "on-the-fly", through SRB/MCAT and GridFTP, in order to minimize storage requirements at the mining site. ADaM comprises a moderately rich set of interoperable operation modules, comprising *data readers* and *writers* for a variety of formats, *preprocessing modules*, for example for data subsetting, and *analysis modules* providing data mining algorithms.

The InfoGrid system mentioned before has been designed as an application-specific layer for constructing and publishing knowledge discovery services. In particular, it is intended to be used in the *Discovery Net* (*D-NET*) system. D-NET is a project of the Engineering and Physical Sciences Research Council at Imperial College [2.18, 2.19] whose main goal is to design, develop, and implement an infrastructure to effectively support scientific knowledge discovery processes from high-throughput informatics. In this context, a series of testbeds and demonstrations are being carried out for using the technology in the areas of life sciences, environmental modeling, and geo-hazard prediction.

The building blocks in Discovery Net are the so-called *Knowledge Discovery Services* (*KDSs*), comprising *Computation Services* and *Data Services*. The former typically comprise algorithms, e.g., data preparation and data mining, while the latter define relational tables (as queries) and other data sources. Both kinds of services are described by means of *adapters* that provide information about input and output data types, allowed parameters, location, platform constraints, and available *factories*, i.e. objects allowing for the creation of service instances. KDSs are used to compose moderately complex data-pipelined processes. The composition may be carried out by means of a GUI which provides access to a library of services. The XML-based language used to describe processes is called *Discovery Process Markup Language* (*DPML*). Each composed process can be deployed and published as a new process. Typically, process descriptions are not bound to specific servers since the actual resources are later resolved by lookup servers (see below).

Discovery Net is based on an open architecture using common protocols and infrastructures such as the Globus Toolkit. Servers are distinguished into (*i*) *Knowledge Servers*, allowing storage and retrieval of knowledge (meant as raw data and knowledge models) and processes; (*ii*) *Resource Discovery Servers*, providing a knowledge base of service definitions and performing resource resolution; and (*iii*) *Discovery Meta-Information Servers*, used to store information about the *Knowledge Schema*, i.e., the sets of features of known databases, their types, and how they can be composed with each other.

Finally, we outline here some interesting data mining testbeds developed at the National Center for Data Mining (NCDM) at the University of Illinois at Chicago (UIC) [2.20]:

- *The Terra Wide Data Mining Testbed (TWDM).* TWDM is an infrastructure for the remote analysis, distributed mining, and real time exploration of scientific, engineering, business, and other complex data. It consists of five geographically distributed nodes linked by optical networks through *StarLight* (an advanced optical infrastructure) in Chicago. These sites include StarLight, the Laboratory for Advanced Computing at UIC, SARA in Amsterdam, and Dalhousie University in Halifax. In 2003 new sites will be connected, including Imperial College in London. A central idea in TWDM is to keep generated predictive models up-to-date with respect to newly available data in order to achieve better predictions (as this is an important aspect in many "critical" domains, such as infectious disease tracking). TWDM is based on *DataSpace*, another NCDM project for supporting real-time streaming data. In DataSpace the *Data Tranformation Markup Language (DTML)* is used to describe how to update "profiles", aggregate data which are inputs of predictive models, on the basis of new "events", i.e., new bits of information.
- *The Terabyte Challenge Testbed.* The Terabyte Challenge Testbed is an open, distributed testbed for DataSpace tools, services, and protocols. It involves a number of organizations, including the University of Illinois at Chicago, University of Pennsylvania, University of California at Davis, Imperial College. The testbed consists of ten sites distributed over three continents connected by high performance links. Each site provides a number of local clusters of workstations which are connected to form wide area *meta-clusters* maintained by the *National Scalable Cluster Project.* So far, meta-clusters have been used by applications in high energy physics, computational chemistry, nonlinear simulation, bioinformatics, medical imaging, network traffic analysis, and digital libraries of video data. Currently, the Terabyte Challenge Testbed consists of approximately 100 nodes and two terabytes of disk storage.
- *The Global Discovery Network (GDN).* The GDN is a collaboration between the Laboratory for Advanced Computing of the National Center for Data Mining and the Discovery Net project (see above). It will link the Discovery Net to the Terra Wide Data Mining Testbed to create a combined global testbed with a critical mass of data.

In summary, many of the recent knowledge discovery-oriented systems have been designed for specific domains, and have later been extended to support more general applications. Some of such systems are essentially advanced interfaces for integrating, accessing, and elaborating large data sets. Furthermore, they provide specific functionalities for the support of typical knowledge discovery processes. The *Knowledge Grid* we designed is one of the first attempts to build a domain-independent knowledge discovery environment on the Grid. Moreover, our system provides services specifically designed for the integration of parallel and sequential data mining algorithms, and the management of base data sets and extracted knowledge models.

2.3 The *Knowledge Grid* Architecture

The *Knowledge Grid* architecture [2.8] is defined on top of Grid toolkits and services, i.e., it uses basic Grid services to build specific knowledge extraction services. Following the *Integrated Grid Architecture* approach [2.21], these services can be developed in different ways using the available Grid toolkits and services. The current implementation is based on the Globus toolkit [2.22]. As in Globus, the *Knowledge Grid* offers global services based on the cooperation and combination of local services. We designed the *Knowledge Grid* architecture so that more specialized data mining tools are compatible with lower-level Grid mechanisms and also with the *Data Grid* services. This approach benefits from "standard" Grid services that are more and more utilized and offers an open parallel and distributed knowledge discovery architecture that can be configured on top of Grid middleware in a simple way.

2.3.1 *Knowledge Grid* Services

The *Knowledge Grid* services are organized in two hierarchic levels: the Core K-grid layer and the High level K-grid layer depicted in Fig. 2.1. The figure shows layers (as implemented on top of Globus services), the *Knowledge Grid* data, and metadata repositories. In the following the term *K-grid node* will denote a Globus node implementing the *Knowledge Grid* services.

The core K-grid layer offers the basic services for the definition, composition, and execution of a distributed knowledge discovery computation over the Grid. Its main goal is the management of all metadata describing features of data sources, third party data mining tools, data management, and data visualization tools and algorithms. Moreover, this layer coordinates the application execution by attempting to fulfill the application requirements and the available Grid resources. The core K-grid layer comprises two main services:

– The *Knowledge Directory Service* (*KDS*) extends the basic Globus MDS service and is responsible for maintaining metadata describing data and tools used in the *Knowledge Grid*. They comprise repositories of data to be mined (data sources), tools, and algorithms used to extract, analyze, and manipulate data, distributed knowledge discovery execution plans, and knowledge obtained as result of the mining process, i.e., learned models and discovered patterns. The metadata information is represented by *eXtensible Markup Language* (*XML*) documents stored in a *Knowledge Metadata Repository* (*KMR*).

– The *Resource Allocation and Execution Management Service (RAEMS)* is used to find the best mapping between an execution plan and available resources, with the goal of satisfying the application requirements (computing power, storage, memory, database, network bandwidth, and latency) and Grid constraints. The mapping is obtained by co-allocating resources. After the execution plan activation, this layer manages and coordinates the application execution and the storing of knowledge results in the *Knowledge Base Repository (KBR)*. Resource requests of each single data mining process are expressed using the *Resource Specification Language (RSL)* [2.23].

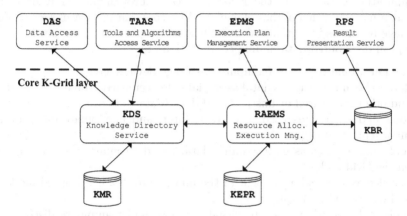

Fig. 2.1. The *Knowledge Grid* architecture

The high-level K-grid layer includes services used to compose, validate, and execute a parallel and distributed knowledge discovery computation. Moreover, the layer offers services to store and analyze the discovered knowledge. Main services here are:

– The *Data Access Service (DAS)* is responsible for the search, selection (data search services), extraction, transformation, and delivery (data extraction services) of data to be mined.
– The *Tools and Algorithms Access Service (TAAS)* is responsible for searching, selecting, and downloading data mining tools and algorithms.
– The *Execution Plan Management Service (EPMS)*. An execution plan is represented by a graph describing interactions and data flows between data sources, extraction tools, DM tools, and visualization tools. The Execution Plan Management Service is a semi-automatic tool that takes data and programs selected by the user, and generates a set of different possible execution plans that meet user, data, and algorithms requirements and

constraints. Execution plans are stored in the *Knowledge Execution Plan Repository (KEPR)*.

- The *Results Presentation Service (RPS)* specifies how to generate, present, and visualize the knowledge models extracted (e.g., association rules, clustering models, classifications). The results metadata are stored in the KMR to be managed by the KDS.

2.4 An XML-Based Metadata Model for the *Knowledge Grid*

The large heterogeneity of the resources involved in a distributed data mining computation is tackled in the *Knowledge Grid* through the definition of a suitable metadata model that is exploited by core and high level services to manage resources in a standard and efficient way. The resources involved in a typical distributed data mining computation are:

- computational resources (computers, storage devices, etc.);
- data to be mined, such as databases, plain files, semi-structured documents, and other structured or unstructured data (data sources);
- tools and algorithms used to extract, filter, and manipulate data (data management tools);
- tools and algorithms used to mine data, i.e., data mining tools available on the Grid nodes;
- knowledge obtained as a result of the mining process, i.e., learned models and discovered patterns;
- tools and algorithms used to visualize, store, and manipulate discovered models.

Heterogeneity arises mainly from the large variety of resources within each category. For instance, software can run only on some particular host machines whereas data can be extracted from different data management systems such as relational databases, semi-structured databases, plain files, etc. The management of such heterogeneous resources requires an intense use of metadata, whose purpose is to provide information about the features of resources and their effective use. Since metadata represents a key element for effective resource discovery and utilization, Grids need to use mechanisms and models that define rich metadata schemas able to represent the variety of resources involved.

The *Knowledge Grid* uses the Globus MDS, and therefore the LDAP protocol, to publish, discover, and manage information about the generic resources of the underlying Grid (e.g., CPU performance, memory size, etc.). However, the complexity of the information associated with specific *Knowledge Grid* resources (data sources, mining algorithms, models), led us to design a more effective model to represent and manage the corresponding metadata.

The basic objectives that guided us through the definition of the *Knowledge Grid* metadata model are the following:

- Metadata should document in a simple and human-readable fashion the features of a data mining application and of the resources involved.
- Metadata should allow an effective search of resources.
- Metadata should provide an efficient way to access resources.

To satisfy such requirements, we chose to express metadata in XML, as it provides a set of functionalities and capabilities that are making it a common model for data description and exchange.

XML metadata documents are defined according to a set of XML schemas properly defined to describe and categorize the different classes of resources. In the remainder of this section, we will give some hints on the definition of metadata related to software, data sources, data mining tools, and discovered knowledge. Furthermore, we will show how metadata can be used to distinguish abstract from concrete resources, and we will introduce execution plans, also defined with an XML formalism, used by the *Knowledge Grid* to manage complex data mining applications. More information on the topical issue of metadata management in the *Knowledge Grid* can be found in [2.24].

2.4.1 Data Mining Software

Categorization of data mining software is based on the following classification parameters [2.25]:

- the kind of data sources the software works on;
- the kind of knowledge that is to be discovered by the software;
- the type of techniques that the software uses in the data mining process; and
- the driving method, i.e., whether the mining process is autonomous, driven by data or queries, or driven by the user (interactive).

As an example, Fig. 2.2 reports the XML metadata related to the data mining software *AutoClass*. The XML document is composed of two parts. The first part is the software Description, the second one is the software Usage. The Description section specifies, for each classification parameter, one or more values that characterize the software. The Usage section contains all the information that can be used by a client to access and use the software. This section is composed of a set of subsections, among which are Syntax, Hostname, ManualPath, and DocumentationURL. The Syntax subsection describes the format of the command that the client should use to invoke the software. This subsection is defined as a tree, where each node is an Arg element and the root is the name of the software itself. The root children specify the arguments that should follow the software name in the software invocation, and these arguments can in turn have children, i.e., sub-arguments, and so on. Each `Arg` element has the following attributes: the `description` attribute, which is a textual description of the argument and the `type` attribute, which specifies whether the argument is `optional`, `required`, or `alternative`. In the last case, all the sibling arguments should have the same value for this attribute, meaning that only one of the siblings should be used in the software invocation. Finally, the `value` attribute (optional) specifies the fixed value of the argument. If the `value` attribute is omitted, the value is to be provided by the client. In the example shown in Fig. 2.2, in the `AutoClass` execution command the executable name should be followed by the `-search` argument, to ask for a classification, or by the `-reports` argument, to obtain the model file. If the `-search` argument is chosen, it should be followed by four sub-arguments, all `required`. Therefore, `AutoClass` can be invoked with the command `/usr/autoclass/autoclass -search aFile.db2 aFile.hd2 aFile.model aFile.s-params`.

```
<DataMiningSoftware name="AutoClass">
  <Description>
    <KindOfData>flat file</KindOfData>
    <KindOfKnowledge>clusters</KindOfKnowledge>
    <KindOfTecnique>statistics</KindOfTecnique>
    <DrivingMethod>autonomous knowledge miner</DrivingMethod>
  </Description>
  <Usage>
    ...
    <Syntax>
      <Arg description="executable" type="required" value="/usr/autoclass/autoclass">
        <Arg description="make a classification" type="alternative" value="-search">
          <Arg description="a .db2 file" type="required"/>
          <Arg description="a .hd2 file" type="required"/>
          <Arg description="a .model file" type="required"/>
          <Arg description="a .s-params file" type="required"/>
        </Arg>
        <Arg description="create a report" type="alternative" value="-reports">
          <Arg description="a .results-bin file" type="required"/>
          ...
        </Arg>
        ...
      </Arg>
    </Syntax>
    <Hostname>icarus.cs.icar.cnr.it</Hostname>
    <ManualPath>/usr/autoclass/read-me.text</ManualPath>
    <DocumentationURL>http://ic-www.arc.nasa.gov/ic/projects/...</DocumentationURL>
    ...
  </Usage>
</DataMiningSoftware>
```

Fig. 2.2. An extract from an XML metadata sample for the AutoClass software

2.4.2 Data Sources

Data sources are the input on which data mining algorithms work to extract new knowledge. They can be provided by relational databases, plain files, and other structured and semi-structured documents. In spite of the wide variety of the possible data source types, we aim to define a common structure of data source metadata in order to standardize access and search operations on such resources.

The common structure of source metadata is composed of two parts: (*i*) an `Access` section including information for retrieving the data source (location, size, etc.), and (*ii*) a `Structure` section providing information about the data source logical and/or physical structure.

As an example, Fig. 2.3 shows an XML metadata document for a flat file that can be used as an input by the `AutoClass` software. The `Structure` section includes two subsections, `Format` and `Attributes`. The `Format` subsection contains information about the physical structure of the flat file, e.g., the strings that are used to separate the records and the attributes within a record. The `Attributes` subsection contains information about the logical structure, i.e., it lists the table attributes and provides the relative specifications (such as the name of the `Attribute`, its type, etc.).

The high-level metadata model is the same for all kinds of data sources, but some details can be added or omitted depending on the specific kind of data to be represented. For instance, in a relational database no information is needed to describe the structure of data, since it is directly managed by the database management system.

```
<FlatFile>
  <Access>
    <Location>/usr/share/imports-85c.db2</Location>
    <Size>26756</Size>
    ...
  </Access>
  <Structure>
    <Format>
      <AttributeSeparatorString>,</AttributeSeparatorString>
      <RecordSeparatorString>#</RecordSeparatorString>
      <UnknownTokenString>?</UnknownTokenString>
      ...
    </Format>
    <Attributes>
      <Attribute name="symboling" type="discrete">
        <SubType>nominal</SubType>
        <Parameter>range 7</Parameter>
      </Attribute>
      <Attribute name="normalized-loses" type="real">
        <SubType>scalar</SubType>
        <Parameter>zero_point 0.0</Parameter>
        <Parameter>rel_error 0.01</Parameter>
      </Attribute>
      ...
    </Attributes>
  </Structure>
</FlatFile>
```

Fig. 2.3. An extract from an XML metadata sample for a flat file

2.4.3 Abstract and Concrete Resources

The resources discussed so far (software, data sources, models) can be either *concrete* or *abstract*. A concrete resource is a resource, published on a KMR of a *Knowledge Grid* node, which is completely specified by its metadata. In abstract resource metadata, some features are expressed as constraints and not as well known values: an abstract resource is not used to identify a specific resource, but to define a set of requirements on the resources that should be discovered by the KDS service.

For instance, whereas the metadata described in Fig. 2.2 describes the concrete software `Autoclass` available on a given node, the metadata document shown in Fig. 2.4 describes an abstract data mining software able to perform a clustering computation on flat files.

An abstract resource can be instantiated into an existing concrete resource whose metadata matches the specified constraints.

```
<DataMiningSoftware name="genericSoftware">
  <Description>
    <KindOfData>flat file</KindOfData>
    <KindOfKnowledge>clusters</KindOfKnowledge>
  </Description>
</DataMiningSoftware>
```

Fig. 2.4. An XML metadata sample for a generic clustering software

2.4.4 Execution Plans

A distributed data mining computation is a process composed of several steps which are executed sequentially or in parallel. In the *Knowledge Grid* framework, the management of complex data mining processes is carried out by defining an execution plan. An execution plan is a graph that describes interactions and data flows between data sources, data mining tools, visualization tools, and output models.

Execution plans are also described through an XML formalism. Such descriptions contain information about relationships among atomic tasks (e.g., computations and data transfers), along with references to the XML documents that describe the involved resources.

Execution plans can be either *abstract* or *instantiated*. An abstract execution plan contains at least one abstract resource, whereas an instantiated execution plan contains only concrete resources. Such a distinction is made to take into account the dynamic nature of a Grid environment, in which resources fail and become available, data gets deleted, software gets updated, etc. In general, a user builds an abstract execution plan, and the RAEMS attempts to transform it into an instantiated execution plan, by mapping abstract resources into concrete resources. Such an action is performed by a scheduler that allows the generation of a suitable execution plan. From an abstract execution plan, different instantiated execution plans could be generated, depending on the resources that are available on the *Knowledge Grid* at different times. This topic will be further discussed in Sect. 2.5, where a sample execution plan will be shown.

2.4.5 Data Mining Models

The knowledge discovered through the data mining process is represented by "data mining models". Whereas so far no common models have been defined for the definition of the data mining resources discussed before, a standard model called *Predictive Model Markup Language* (*PMML*) has been defined to describe data mining results. PMML is an XML language which provides a vendor-independent method for defining data mining models [2.26]. The PMML provides a *Document Type Definition* (*DTD*) to describe different kinds of models such as classification rules and association rules. We use it to define data mining models in the *Knowledge Grid*.

2.4.6 Metadata Management

As stated before, the metadata management process is a key aspect in the development of data mining applications over the *Knowledge Grid*. A typical life cycle of metadata consists of the following steps:

1. Resource metadata are published on the KMRs of the corresponding nodes.
2. The user specifies the features of the resources needed to design a data mining application.
3. The DAS and TAAS services search the KMRs of the *Knowledge Grid* nodes for the requested resources, using the core level KDS service (which directly interacts with local and remote KMRs) to manage resource metadata.
4. Metadata describing the resources of interest are delivered by such services to the requesting user. Figure 2.5 shows the flows of resource-related metadata in the *Knowledge Grid* architecture.
5. Metadata related to software, data, and operations are combined into an execution plan to design a complete data mining application. The execution plan metadata are managed by the EPMS service; they are accessed through the core level RAEMS service and stored in the KEPR (see Fig. 2.6).
6. After application execution, results and related model metadata are stored in the KBR and processed by the RPS service. Moreover, metadata related to new and/or modified resources are published in the corresponding KMRs for future use (see Fig. 2.7).

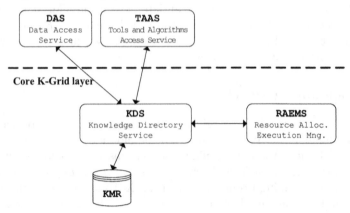

Fig. 2.5. Resource metadata flows and services involved

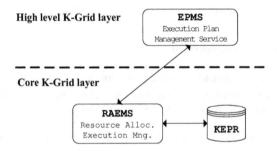

Fig. 2.6. Execution plan metadata flows and services involved

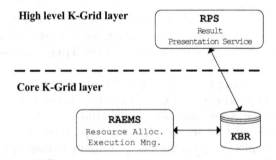

Fig. 2.7. Knowledge metadata flows and services involved

2.5 Design of a PDKD Computation

The design of a PDKD computation on the *Knowledge Grid* is performed as shown in Fig. 2.8. The process starts by searching and selecting the re-

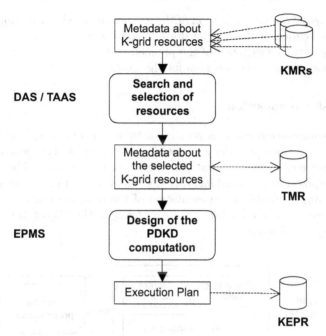

Fig. 2.8. The PDKD computation design process

sources to be used in a PDKD computation. This step is accomplished by means of the DAS and TAAS tools that analyze the XML metadata documents stored in the KMRs of the participant K-grid nodes, which represent the resources available on those nodes. Such an analysis attempts to find information about useful resources (e.g., software implementing a specific data mining algorithm, data sources about a specific argument, etc.), and is carried out on the basis of the search parameters and selection filters chosen by the user. Metadata about the resources selected for the computation (i.e., those satisfying the searching and filtering criteria) are then stored in the *Task Metadata Repository* (*TMR*), a local storage space that contains information about resources (computational nodes, data sources, and software) selected to perform a computation. The TMR is organized as a set of directories: each one is named with the fully qualified hostname of a Grid node, and contains metadata documents about its resources.

The design of a PDKD computation is performed by means of the EPMS. To allow a user to build the computation in a simple way, we developed a toolset named *VEGA* (*Visual Environment for Grid Applications*). Its architecture is depicted in Fig. 2.9. VEGA integrates functionalities of the EPMS and other K-grid services; in particular, it provides the following EPMS operations:

- *task composition*, i.e., definition of the entities involved in the computation and specification of the relationships among them;
- *checking* of the consistency of the planned task;
- *generation* of the execution plan for the task.

2.5.1 Task Composition

The task composition phase is performed by means of a graphical interface (see Fig. 2.10), which provides a user with a set of graphical objects representing the resources (data sets, data mining tools, Grid nodes). These objects can be composed using visual facilities that allow a user to insert links among them, forming a graphical representation of the computation. In particular, such a phase is realized by the *Resource Manager*, the *Object Manager*, and the *Workspace Manager*.

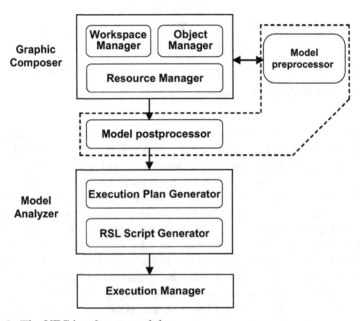

Fig. 2.9. The VEGA software modules

The Resource Manager permits a user to browse the TMR in order to search and choose the resources to be used in the computation. Selected hosts are displayed into the *Hosts* panel, and the user can explore resources of each one by clicking on its label. Those resources are displayed by categories in the *Resources* panel.

The Object Manager deals with the graphical objects during the visual composition. Each graphical object is associated with information about the related resources; this information is used for the creation of the internal model and for the execution plan generation. The Object Manager handles three kinds of objects: data, software, and hosts. It allows the user to drag the objects presented in the hosts and resources panels into a workspace. After this, those objects can be linked to indicate the interactions among them. Links can represent different actions, such as data transfer, program execution, and input and output relationships. The Object Manager performs the labeling of the links and the attribution of other properties characterizing them. The *data transfer* link is used to move resources among different locations of the Grid. The *execute* link is used to run an application on a Grid host; the *input* and *output* links are used to respectively indicate input and output of a program. For each link type it is possible to set related parameters (e.g., protocol and destination path of the data transfer, job manager of the execution, etc.).

A complex computation is composed of several jobs. The design environment is organized into *workspaces*. Jobs present in a given workspace can be executed concurrently, whereas workspaces are executed sequentially. To this end, an ordering between the workspaces is defined. In addition, the Workspace Manager handles an internal model of the graphical representation shown to the user. We describe here an example that shows the task composition process.

A user logged on the K-grid node g1.isi.cs.cnr.it intends to perform a data mining application composed of two data mining steps, clustering and classification, on the data set Unidb stored on the same node. The data set must be clustered using three different algorithms running in parallel on three copies of the data set. Clustering results must be analyzed by a classification algorithm that will be executed in parallel on three different nodes, generating three classification models of the same data set. Finally, the three different models will be shown to the user, who will select the more accurate one. The user has located the K-grid nodes k1.deis.unical.it, k2.deis.unical.it, and k3.deis.unical.it offering, respectively, the clustering algorithms K-Means [2.27], Intelligent Miner [2.28], and AutoClass [2.29], and the node g2.isi.cs.cnr.it, offering the C5.0 classifier [2.30].

Figures 2.10–2.13 show the sequence of the four workspaces composed by the user to design such a computation:

- *Workspace 1* (Fig. 2.10). The data set Unidb (which is located on the node g1) and the classifier C5.0 (which is located on g2) are copied to the nodes k1, k2, and k3.
- *Workspace 2* (Fig. 2.11). The data set Unidb is analyzed on k1 by K-Means, producing as output K-Means.out; on k2, the data set Unidb is analyzed by Intelligent Miner, producing Iminer.out; on k3, the data set Unidb is analyzed by AutoClass, producing Autoclass.out.
- *Workspace 3* (Fig. 2.12). On k1, K-Means.out is analyzed by C5.0, producing K-Means_c5.out; on k2, IMiner.out is analyzed by C5.0, produc-

38   M. Cannataro et al.

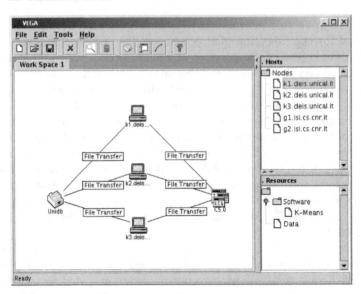

Fig. 2.10. VEGA: Example workspace 1

ing `Iminer_c5.out`; on `k3`, `Autoclass.out` is analyzed by C5.0, producing
`Autoclass_c5.out`.
– *Workspace 4* (Fig. 2.13). `K-Means_c5.out`, `IMiner_c5.out`, and
`Autoclass_c5.out` are moved from `k1`, `k2`, and `k3` to `g1`.

Since the set of workspaces represents a unique logical computation, the
Workspace Manager must deal with the case in which a task in a given
workspace needs to operate on resources generated by tasks in previous
workspaces. Such resources are not physically generated when the user starts
composing a subsequent workspace of the same computation because all the
workspaces are processed for the execution only at the end of the design session. The Workspace Manager recognizes such a situation during the composition of a workspace, generates the needed *virtual* resources, and makes them
available, through the Resource Manager, to the subsequent workspaces. For
instance, in *workspace 1* (Fig. 2.10), the data set `Unidb` is copied to the node
`k1.deis.unical.it`; therefore, a new metadata document is created for `Unidb`
and stored in the directory `k1.deis.unical.it` of the TMR. The document
specifying the new location of `Unidb` is marked as temporary until the data
transfer is performed. However, in *workspace 2* (Fig. 2.11), the data set `Unidb`
is displayed as already available under the resources of `k1.deis.unical.it`.

2.5.2 Task Consistency Checking

The goal of this phase is to obtain a correct and consistent model of the computation. The validation process is performed by means of two components:
the model preprocessor the model postprocessor.

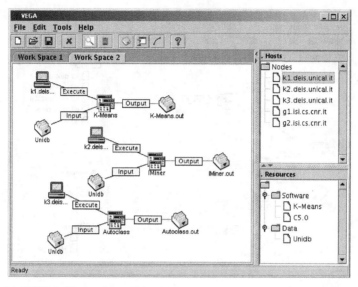

Fig. 2.11. VEGA: Example workspace 2

The preprocessing of the computation model takes place during the graphical composition. The model preprocessor verifies the composition consistency, allowing the user, with a context-sensitive control, to create links only if they represent actions that can really be executed. For instance, given a data set, it allows the user to link it to a software through an input/output link, not to a host through an execution link.

The checking is completed by the model postprocessor, which is responsible for catching error occurrences that cannot be recognized during the preprocessing phase. For instance, it indicates accordingly if the graphical composition in a workspace does not contain at least one host.

2.5.3 Execution Plan Generation

In this phase, the computation model is translated into an execution plan represented by an XML document. This task is performed by the *Execution Plan Generator*.

Basically, the Execution Plan Generator is a parser that analyzes the computation model produced during the graphical composition, and is able to generate its equivalent XML representation, taking into account the properties of the involved resources and the parameters of the links. The XML execution plan describes a data mining computation at a high level, containing physical information neither about resources (which are identified by metadata references) nor about the status and current availability of such resources. In fact, specific information about the resources involved will be

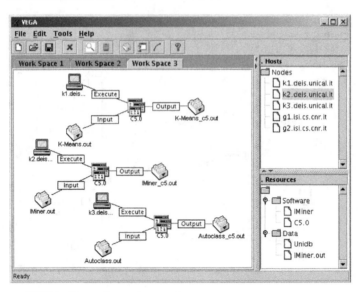

Fig. 2.12. VEGA: Example workspace 3

included in the RSL generation phase, when the computation model is translated into this language. Figure 2.14 shows an extract of the execution plan for the example described above. The execution plan gives a list of tasks and task links, which are specified using the XML tags Task and TaskLink, respectively. The label attribute of a Task element identifies each basic task in the execution plan, and is used in linking various basic tasks to form the overall task flow.

Each Task element contains a task-specific sub-element, which indicates the parameters of the particular task represented. For instance, the task identified by the ws1_dt2 label contains a DataTransfer element, indicating that it is a data transfer task. The DataTransfer element specifies Protocol, Source, and Destination of the transfer. The href attributes of such elements specify the location of metadata about protocol, source, and destination objects. In this example, metadata about the source of the data transfer in the ws1_dt2 task are provided by the Unidb.xml file stored in the directory named g1.isi.cs.cnr.it of the TMR, whereas metadata about the destination are provided by the Unidb.xml file stored in the directory named k2.deis.unical.it of the same TMR. The first of such XML documents provides metadata about the Unidb data set when stored on g1.isi.cs.cnr.it, whereas the second one provides metadata about Unidb when, after the data transfer, it is stored on k2.deis.unical.it. The TaskLink elements represent the relationships among the tasks of the execution plan. For instance, the TaskLink shown indicates that the ws2_c2 task follows ws1_dt2, as specified by its from and to attributes.

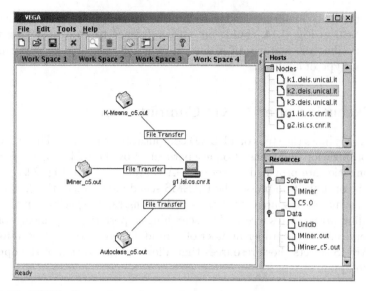

Fig. 2.13. VEGA: Example workspace 4

```
<ExecutionPlan>
...
<Task ep:label="ws1_dt2">
 <DataTransfer>
  <Protocol ep:href="g1../GridFTP.xml"
            ep:title="GridFTP on g1.isi.cs.cnr.it"/>
  <Source ep:href="g1../Unidb.xml"
          ep:title="Unidb on g1.isi.cs.cnr.it"/>
  <Destination ep:href="k2../Unidb.xml"
               ep:title="Unidb on k2.deis.unical.it"/>
 </DataTransfer>
</Task>
...
<Task ep:label="ws2_c2">
 <Execution>
  <Program ep:href="k2../IMiner.xml"
           ep:title="IMiner on k2.deis.unical.it"/>
  <Input ep:href="k2../Unidb.xml"
         ep:title="Unidb on k2.deis.unical.it"/>
  ...
  <Output ep:href="k2../IMiner.out.xml"
          ep:title="IMiner.out on k2.deis.unical.it"/>
 </Execution>
</Task>
...
<TaskLink ep:from="ws1_dt2" ep:to="ws2_c2"/>
...
</ExecutionPlan>
```

Fig. 2.14. The extract of an execution plan

As stated in Sect. 2.4, execution plans are abstract if some of the resources are not completely determined. They are specified by means of one or more constraints. In this case, the Execution Plan Generator produces an abstract execution plan, i.e., an XML document similar to the example shown in Fig. 2.14, except that at least one resource is an abstract resource. For instance, the XML `Program` element within the description of the computation task `ws2_c2`, in Fig. 2.14, could contain a reference to an abstract resource which specifies the features of a data mining software able to perform the requested type of analysis.

2.6 Execution of a PDKD Computation

Figure 2.15 shows the steps of a PDKD computation execution. The execution plan optimization and translation are performed by the RAEMS, whose basic functionalities are provided by the VEGA components (see Fig. 2.9).

In the optimization phase, the RAEMS scheduler searches the *Knowledge Grid* for concrete resources that satisfy the constraints specified in the corresponding abstract resources. The searching is performed by means of the KDS service. Once a proper number of candidate resources have been found, the RAEMS selects those resources that allow the generation of the optimal execution plan.

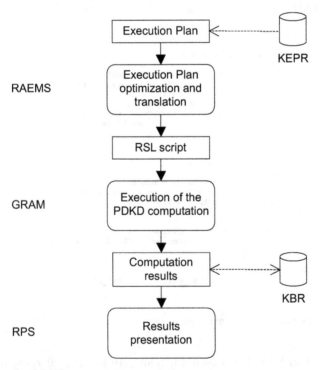

Fig. 2.15. The PDKD computation execution process

Currently, VEGA integrates an *RSL Generator* module, which produces an RSL script that can be directly submitted to the *Globus Resource Alloca-*

tion Manager (GRAM) of a Grid node running Globus. The RSL (Resource Specification Language) is a structured language with which resource requirements and parameters can be outlined by a user [2.23]. In contrast with the XML execution plan, the RSL script entirely describes an instance of the designed computation, i.e., it specifies all the physical information needed for the execution (e.g., name and location of resources, software parameters, etc.). Figure 2.16 shows an extract of a sample RSL script.

```
+
  ...
(&(resourceManagerContact=g1.isi.cs.cnr.it)
  (subjobStartType=strict-barrier)
  (label=ws1_dt2)
  (executable=$(GLOBUS_LOCATION)/bin/globus-url-copy)
  (arguments=-vb -notpt gsiftp://g1.isi.cs.cnr.it/.../Unidb
             gsiftp://k2.deis.unical.it/.../Unidb
  )
)
  ...
(&(resourceManagerContact=k2.deis.unical.it)
  (subjobStartType=strict-barrier)
  (label=ws2_c2)
  (executable=.../IMiner)
  ...
  )
)
  ...
```

Fig. 2.16. The extract of a sample RSL script

The execution of the computation is performed by means of the VEGA *Execution Manager* module. The Execution Manager allows the system to authenticate a user to the Grid, using the Globus GSI (Grid Security Infrastructure) services, and to submit the RSL script to the Globus GRAM for its execution. The Execution Manager is also responsible for the monitoring of the jobs that compose the overall data mining computation during their life cycle. Finally, the Execution Manager collects results of the PDKD computation and presents them to the user.

2.7 Conclusions

Grid computing is enlarging its scope and is going to be more and more complete and complex both in the number of developed tools and in the variety of supported applications. Deployed Grids are growing up very quickly and support high-performance applications in science, industry, and business. According to this trend, Grid services are shifting from generic computation-oriented services to high-level information management and knowledge discovery services. It is vitally important to design and develop knowledge-based systems for supporting sophisticated Grid applications. The *Knowledge Grid* system we discussed here is a significant system that, according to this approach, aims to support the implementation of knowledge discovery processes on Grids. It integrates and completes the data Grid services by supporting distributed data analysis and knowledge discovery and management services that will enlarge the application scenario and the community of Grid computing users [2.6].

In this chapter we also presented the *Knowledge Grid* features and tools by showing how a user through a step-by-step process can compose and execute a knowledge discovery application on a Grid in a simple way. Besides completing the *Knowledge Grid* implementation we are developing real distributed data mining applications in different domains that exploit the system features and give us feedback on the user needs in terms of capabilities and performance. These experiments will help us to improve and extend the system functionalities.

Acknowledgements

This work has been partially funded by the project MIUR Fondo Speciale SP3: Grid Computing: Tecnologie abilitanti e applicazioni per eScience.

References

2.1 I. Foster, C. Kesselman, S. Tuecke: The Anatomy of the Grid: Enabling Scalable Virtual Organizations. Int. J. of Supercomputing Applications, 15(3) (2001)

2.2 W. Hoschek, J.J. Martinez, A. Samar, H. Stockinger, K. Stockinger: Data Management in an International Data Grid Project. *Proc. IEEE/ACM Int. Workshop on Grid Computing, Grid 2000* (LNCS Vol. 1971, Springer Verlag) pp. 77-90

2.3 P. Avery, I. Foster: GriPhyN Project Description. Available at http://www.griphyn.org/info/index.html

2.4 Y. Morita et al.: Grid Data Farm for Atlas Simulation Data Challenges. *Proc. of Int. Conf. on Computing of High Energy and Nuclear Physics, 2001* pp. 699-701

2.5 A. Chervenak, I. Foster, C. Kesselman, C. Salisbury, S. Tuecke: The Data Grid: Towards an Architecture for the Distributed Management and Analysis of Large Scientific Data sets. Journal of Network and Computer Applications, 23, 187-200 (2001)

2.6 F. Berman: From TeraGrid to Knowledge Grid. Communications of the ACM, 44(11), 27-28 (2001)

2.7 W.E. Johnston: Computational and Data Grids in Large-Scale Science and Engineering. Future Generation Computer Systems, 18 (8), 1085-1100 (2002)

2.8 M. Cannataro, D. Talia, P. Trunfio: Knowledge Grid: High Performance Knowledge Discovery Services on the Grid. *Proc. GRID 2001* (Springer-Verlag, 2001) pp. 38-50

2.9 M. Cannataro, A. Congiusta, D. Talia, P. Trunfio: A Data Mining Toolset for Distributed High-performance Platforms. *Proc. Conf. Data Mining 2002* (Wessex Inst. Press, Bologna, Italy, 2002)

2.10 H. Kargupta, P. Chan (eds.): *Advances in Distributed and Parallel Knowledge Discovery* (AAAI/MIT Press, 2000)

2.11 C. Catlett: The TeraGrid: a Primer. Available at
      http://www.teragrid.org/
2.12 F. Berman: Private communication (November 2001)
2.13 N. Giannadakis, A. Rowe, M. Ghanem, Y. Guo: InfoGrid: Providing Informa-
      tion Integration for Knowledge Discovery. To Appear in the Journal of Infor-
      mation Science
2.14 The DataCutter project:
      http://www.cs.umd.edu/projects/hpsl/chaos/ResearchAreas/dc/
2.15 D. Skillicorn, D. Talia: Mining Large Data Sets on Grids: Issues and Prospects.
      Computing and Informatics, 21, 347-362 (2002)
2.16 The ADaM system. http://datamining.itsc.uah.edu/adam/.
2.17 T. Hinke, J. Novonty: Data Mining on NASA's Information Power Grid. *Proc.
      Ninth IEEE Int. Symposium on High Performance Distributed Computing, 2000*
2.18 Discovery Net. http://ex.doc.ic.ac.uk/new/
2.19 V. Curcin, M. Ghanem, Y. Guo, M. Kohler, A. Rowe, J. Syed, P. Wendel:
      Discovery Net: Towards a Grid of Knowledge Discovery. *Proc. Eighth ACM
      SIGKDD Int. Conf. on Knowledge Discovery and Data Mining* (Edmonton,
      Canada, 2002)
2.20 Testbeds, National Center for Data Mining, Laboratory for Advanced Com-
      puting, University of Illinois at Chicago.
      http://www.ncdm.uic.edu/testbeds.htm
2.21 I. Foster: Building the Grid: An Integrated Services and Toolkit Architecture
      for Next Generation Networked Applications. Technical Report, available at
      http://www.gridforum.org/building_the_grid.htm (2000)
2.22 I. Foster, C. Kesselman: Globus: a metacomputing infrastructure toolkit. Int.
      J. of Supercomputing Applications, 11, pp. 115-128 (1997)
2.23 The Globus Project. The Globus Resource Specification Language RSL v1.0.
      Available at http://www.globus.org/gram/rsl_spec1.html
2.24 C. Mastroianni, D. Talia, P. Trunfio: Managing Heterogeneous Resources in
      Data Mining Applications on Grids Using XML-based Metadata. To appear on
      Heterogeneous Computing Workshop (HCW 2003, Nice, France, April 2003)
2.25 M.S. Chen, J. Han, P.S. Yu: Data Mining: An Overview from a Database Per-
      spective. IEEE Trans. Knowledge and Data Engineering, 8(6), 866-883 (1996)
2.26 R.L. Grossman, M.F. Hornick, G. Meyer: Data Mining Standard Initiatives.
      Communications of the ACM, 45(8) (August 2002)
2.27 J. MacQueen: Some Methods for Classification and Analysis of Multivari-
      ate Observations. *Proc. 5th Symp. on Mathematical Statistics and Probability,
      1967*, pp. 281-297
2.28 IBM Intelligent Miner. http://www.software.ibm.com/data/iminer/
2.29 P. Cheeseman, J. Stutz: Bayesian Classification (AutoClass): Theory and Re-
      sults. In: U.M. Fayyad, G.P. Shapiro, P. Smyth, R. Uthurusamy (eds.), *Ad-
      vances in Knowledge Discovery and Data Mining* (AAAI Press/MIT Press,
      1996) pp. 61-83
2.30 J.R. Quinlan: See5/C5.0, version 1.16.
      http://www.rulequest.com/see5-info.html (2002)

3. The MiningMart Approach to Knowledge Discovery in Databases

Katharina Morik and Martin Scholz

Department of Computer Science, University of Dortmund, Germany

Abstract

Although preprocessing is one of the key issues in data analysis, it is still common practice to address this task by manually entering SQL statements and using a variety of stand-alone tools. The results are not properly documented and hardly re-usable. The MiningMart system presented in this chapter focuses on setting up and re-using best practice cases of preprocessing data stored in very large databases. A metadata model named M4 is used to declaratively define and document both, all steps of such a preprocessing chain and all the data involved. For data and applied operators there is an abstract level, understandable by human users, and an executable level, used by the metadata compiler to run cases for given data sets. An integrated environment allows for rapid development of preprocessing chains. Adaptation to different environments is supported simply by specifying all involved database entities in the target DBMS. This allows reuse of best practice cases published on the Internet.

3.1 Introduction: Acquiring Knowledge from Existing Databases

The use of very large databases has evolved in the last years from supporting transactions to, additionally, reporting business trends. The interest in analyzing data has increased. One important topic is customer relationship management with the particular tasks of customer segmentation, customer profitability, customer retention, and customer acquisition (e.g., by direct mailing). Other tasks are the prediction of sales in order to minimize stocks, and the prediction of electricity consumption or telecommunication services at particular times of the day in order to minimize the use of external services or to optimize network routing, respectively. The health sector demands several analysis tasks for resource management, quality control, and decision making. Existing databases which were designed for transactions, such as billing and booking, are now considered a mine of information, and digging knowledge from the already gathered data is considered a tool for building up an organizational memory. Managers of an institution want to be informed about states and trends in their business. Hence, they demand concise reports from the database department.

On-line Analytical Processing (OLAP) offers interactive data analysis by aggregating data and counting the frequencies. This already answers questions like the following:

- What are the attributes of my most frequent customers?
- Which are the frequently sold products?
- How many returns did I receive after my last direct mailing action?
- What is the average duration of stay in my hospital?

Reports that support managers in decision making need more detailed information. Questions are more specific, for instance:

- Which customers are most likely to sell their insurance contract back to the insurance company before it ends?
- How many sales of a certain item do I have to expect in order not to offer empty shelves to customers and at the same time minimize my stock?
- Which group of customers best answers to direct mailing advertising a particular product?
- Who are the most cost-intensive patients in my hospital?

Knowledge Discovery in Databases (KDD) can be considered a high-level query language for relational databases that aims at generating sensible reports such that a company may enhance its performance. The high-level question is answered by a *data mining* step. Several data mining algorithms exist. However, their application is still a cumbersome process. Several reasons explain why KDD has not yet become a standard procedure. We list here the three obstacles that – in our view – are the most important ones and then discuss them.

- Most tools for data mining need to handle the data internally and cannot access the database directly. Sampling the data and converting it into the desired format increases the effort for data analysis.
- Preprocessing of the given data is decisive for the success of the data mining step. Aggregation, discretization, data cleaning, the treatment of null values, and the selection of relevant attributes are steps that still have to be programmed (usually in SQL) without any high-level support.
- The selection of the appropriate algorithm for the data mining step as well as for preprocessing is not yet well understood, but remains the result of a trial and error process.

The conversion of given data into the formats of diverse data mining tools is eased by toolboxes which use a common representation language for all the tools. Then, the given data need to be transformed only once and can be input into diverse tools. A first approach to such a toolbox was the development of a Common Knowledge Representation Language (CKRL), from which translators to several learning algorithms were implemented in the European *Machine Learning Toolbox* project [3.3, 3.11]. Today, the *WEKA* collection of learning algorithms implemented in JAVA with a common input format offers the opportunity to apply several distinct algorithms on a data set [3.15]. However, these toolboxes do not scale up to real-world databases naturally[1]. In contrast, database management systems offer basic statistical or OLAP procedures on the given data, but do not yet provide users with more sophisticated data mining algorithms. Building upon the database facilities and integrating data mining algorithms into the database environment will be the synergy of both developments. We expect the first obstacle for KDD applications to be overcome very soon.

The second obstacle is the most important one. If we inspect real-world applications of knowledge discovery, we realize that up to 80 percent of the efforts are spent on the clever preprocessing of the data. Preprocessing has long been underestimated, both in its relevance and in its complexity. If the data conversion problem is solved, the preprocessing is not at all done. Feature generation and selection[2] (in databases this means constructing additional columns and selecting the relevant attributes for further learning) is a major challenge for KDD [3.9]. Machine learning is not restricted to the data mining step, but is also applicable in preprocessing. This view offers a variety of learning tasks that are not as well investigated as are learning classifiers. For instance, an important task is to acquire events and their duration (i.e., a time interval) on the basis of time series (i.e., measurements at time points). Another example is the replacement of null values in the database by the results of a learning algorithm. Given attributes A_i without null values, we may train our algorithm to predict the values of attribute A_j on those records which do have a value for A_j. The learning result can then be applied in order to replace null values in A_j. Records without null values are a prerequisite for the application of some algorithms. These algorithms become applicable as the data mining step because of the learning in the preprocessing. With respect to preprocessing, we are just beginning to explore our opportunities. It is a field of greatest potential.

The third obstacle, the selection of the appropriate algorithm for a data mining task, has long been on the research agenda of machine learning. The main problem is that nobody has yet been able to identify reliable rules predicting when one algorithm should be superior to others. Beginning with the *Mlt-Consultant* [3.13], there was the idea of having a knowledge-based system support the selection of a machine learning method for an application. The *Mlt-Consultant* succeeded in differentiating the nine learning methods of the Machine Learning Toolbox with respect to specific syntactic properties of the input and output languages of the methods. However, there was little success in describing and differentiating the methods on an application level that went beyond the well known classification of machine learning systems into classification learning, rule learning, and clustering. Also, the European *Stat-*

---

[1] Specialized in multi-relational learning algorithms, the *ILP toolbox* from Stefan Wrobel (to be published on the network ILPnet2) allows one to try several logic learning programs on a database.

[2] Specialized in feature generation and selection, the toolbox YALE offers the opportunity to try and test diverse feature sets for learning with the support vector machine [3.6]. However, the YALE environment does not access a database.

log-Project [3.10], which systematically applied classification learning systems to various domains, did not succeed in establishing criteria for the selection of the best classification learning system. It was concluded that some systems have generally acceptable performance. In order to select the best system for a certain purpose, they must each be applied to the task and the best selected through a test method such as cross-validation. Theusinger and Lindner [3.14] are in the process of re-applying this idea of searching for statistical dataset characteristics necessary for successful application of tools. An even more demanding approach was started by Engels [3.4]. This approach not only attempts to support the selection of data mining tools, but to build a knowledge-based process planning support for the entire knowledge discovery process. To date this work has not led to a usable system [3.5]. The European project *MetaL* now aims at learning how to combine learning algorithms and datasets [3.2]. Although successful in many respects, there is not enough knowledge available in order to propose the correct combination of preprocessing operations for a given dataset and task. The *IDEA* system now tries the bottom-up exploration of the space of preprocessing chains [3.1]. Ideally, the system would evaluate all possible transformations in parallel, and propose the most successful sequence of preprocessing steps to the user. For short sequences and few algorithms, this approach is feasible. Problems like the collection of all data concerning one customer (or patient) from several tables, or the generation of most suitable features, enlarge the preprocessing sequences considerably. Moreover, considering learning algorithms as preprocessing steps enlarges the set of algorithms per step. For long sequences and many algorithms, this principled approach of *IDEA* becomes computationally infeasible.

If the pairing of data and algorithms is all that difficult, can we support an application developer at all? The difficulty of the principled approaches to algorithm selection is that they all start from scratch. They apply rules that pair data and algorithm characteristics, or plan a sequence of steps, or try and evaluate possible sequences for each application anew. However, there are similar applications where somebody has already done the cumbersome exploration. Why not use these efforts to ease the new application development? Normally, it is much easier to solve a task if we are informed about the solution of a similar task. This is the basic assumption of case-based reasoning and it is the basis of the MiningMart approach. A successful case of a full KDD process is described at the meta-level. This description at the meta-level can be used as a blueprint for other, similar cases. In this way, the MiningMart project[3] eases preprocessing and algorithm selection in order to make KDD an actual high-level query language accessing real world databases.

---

[3] The MiningMart project is supported by the European Union under the contract IST-1999-11993.

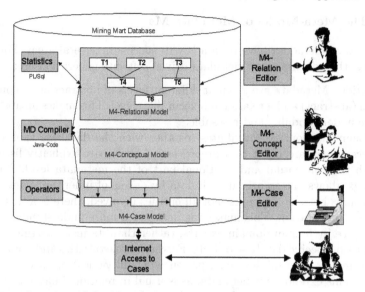

Fig. 3.1. Overview of the MiningMart system

3.2 The MiningMart Approach

Now that we have stated our goal of easing the KDD process, we may ask: What is MiningMart's path to reaching the goal? A first step is to implement operators that perform data transformations such as discretization, handling of null values, aggregation of attributes into a new one, or collecting of sequences from time-stamped data. The operators directly access the database and are capable of handling large masses of data.

Given database-oriented operators for preprocessing, the second step is to develop and collect successful cases of knowledge discovery. Since most of the time is used to find chains of operator applications that lead to good answers to complex questions, it is cumbersome to develop such chains over and over again for very similar discovery tasks and data. Currently, in practice even the same task on data of the same format is implemented anew every time new data is to be analyzed. Therefore, the reuse of successful cases speeds up the process considerably. The particular approach of the MiningMart project is to allow the reuse of cases by means of metadata, also called *ontologies*. Metadata describe the data as well as the operator chains. A compiler generates the SQL code according to the metadata.

Several KDD applications have been considered when developing the operators, the method, and the meta-model. In the remaining part of this chapter, we shall first present the metadata together with their editors and the compiler. We then describe the case base. We conclude the chapter by summarizing the MiningMart approach and relating it to other approaches.

3.2.1 The Meta-Model of Metadata M4

Ontologies or metadata have been a key to success in several areas. For our purposes, the advantages of metadata driven software generation are:

Abstraction: Metadata are given at different levels of abstraction, a conceptual (abstract) and a relational (executable) level. This makes an abstract case understandable and re-usable.

Data documentation: All attributes together with the database tables and views, which are input to a preprocessing chain, are explicitly listed at both the conceptual and relational part of the metadata level. An ontology allows one to organize all data by means of inheritance and relationships between concepts. For all entities involved, there is a text field for documentation. This makes the data much more understandable, for instance by human domain experts, rather than its just referring to the names of specific database objects. Furthermore, statistics and important features for data mining (e.g., presence of null values) are accessible as well. This extends the metadata, as is usual in relational databases, and gives a good impression of the data sets at hand.

Case documentation: The chain of preprocessing operators is documented, as well. First of all, the declarative definition of an executable case in the M4 model can already be considered to be documentation. Furthermore, apart from the opportunity to use "speaking names" for steps and data objects, there are text fields to document all steps of a case together with their parameter settings. This helps to quickly figure out the relevance of all steps and makes cases reproducible. In contrast, the current state of documentation is most often the memory of the particular scientist who developed the case.

Ease of case adaptation: In order to run a given sequence of operators on a new database, only the relational metadata and their mapping to the conceptual metadata has to be written. A sales prediction case can for instance be applied to different kinds of shops, or a standard sequence of steps for preparing time series for a specific learner might even be applied as a template in very different mining contexts. The same effect eases the maintenance of cases when the database schema changes over time. The user just needs to update the corresponding links from the conceptual to the relational level. This is especially easy, having all abstract M4 entities documented.

The MiningMart project[4] has developed a model for metadata together with its compiler, and has implemented human-computer interfaces that allow database managers and case designers to fill in their application-specific metadata. The system will support preprocessing and can be used stand-alone or in combination with a toolbox for the data mining step.

---

[4] http://mmart.cs.uni-dortmund.de

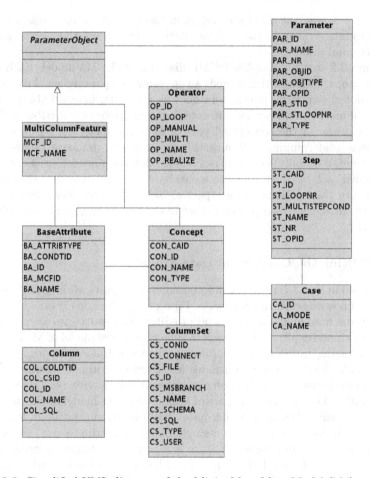

Fig. 3.2. Simplified UML diagram of the MiningMart Meta Model (M4)

This section gives an overview of how a case is represented at the meta-level, how it is practically applied to a database, and which steps need to be performed when developing a new case or adapting a given one.

The form in which metadata are to be written is specified in the meta-model of metadata, M4. It is structured along two dimensions, topic and abstraction. The *topic* is either the data or the case. The data is to be analyzed. The case is a sequence of (preprocessing) steps. The *abstraction* is either conceptual or relational. Wherever the conceptual level is expected to be the same for various applications, the relational level actually refers to the particular database at hand. The conceptual data model describes concepts like Customer and Product, and relationships between them like Buys. The relational data model describes the business data that is analyzed. Most often it already exists in the database system in the form of the database schema. The metadata written in the form as specified by M4 are themselves stored in a relational database.

Figure 3.2 shows a simplified UML diagram of the M4 model. Each case contains steps, each of which embeds an operator and parameters. Apart from values, not shown here, parameters may be concepts, base attributes, or a multi-column feature, i.e. a feature containing multiple base attributes. This part is a subset of the conceptual part of M4. The relational part contains columnsets and columns. Columnsets refer either to database tables or to virtual (metadata only) or database views. Each columnset consists of a set of columns, each of which refers to a database attribute. On the other hand, columns are the relational counterpart of base attributes. For columns and base attributes, there is a predefined set of data types, which is also omitted in Fig. 3.2.

3.2.2 Editing the Conceptual Data Model

As depicted in Fig. 3.1, there are different kinds of experts working at different ends of a knowledge discovery process. First of all a domain expert will define a conceptual data model using a concept editor. The entities involved in data mining are made explicit by this expert. The conceptual model of M4 is about *concepts* having *features*, and *relationships* between these concepts.

Examples for concepts are Customer and Product. Although at the current stage of development concepts refer to either database views or tables, they should rather be considered as part of a more abstract model of the domain. Concepts consist of features, either base attributes or multi-column features. A base attribute corresponds to a single database attribute, e.g., the name of a customer. A multi-column feature is a feature containing a fixed set of base attributes. This kind of feature should be used when information is split over multiple base attributes. An example is to define a single multi-column feature for the amount and the currency of a bank transfer, which are both represented by base attributes.

Relationships are connections between concepts. There could be a relationship named Buys between the concepts Customer and Product, for example. At the database level, one-to-many relationships are represented by foreign key references and many-to-many relationships make use of cross tables. However, these details are hidden from the user at the abstract conceptual level.

To organize concepts and relationships, the M4 model offers the opportunity to use inheritance. Modelling the domain in this fashion, the concept Customer could have *subconcepts* like Private Customer and Business Customer. Subconcepts inherit all features of their superconcept. The relationship Buys could for instance have a subrelationship Purchases on credit.

Figure 3.3 shows a screenshot of the concept editor while it is used to list and edit base attributes. The right part of the lower window states, that

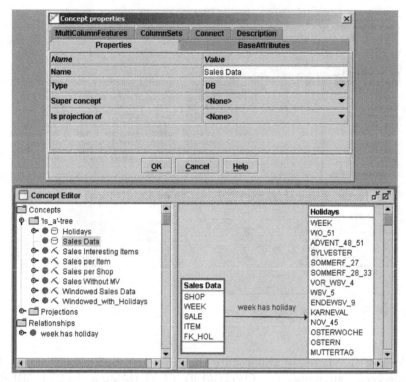

Fig. 3.3. The Concept Editor

the selected concept Sales Data is connected to another concept Holidays by a relationship week has holiday.

3.2.3 Editing the Relational Model

Given a conceptual data model, a database administrator maps the entities involved to corresponding database objects. The relational data model of M4 is capable of representing all the relevant properties of a relational database. The most simple mapping from the conceptual to the relational level is given if concepts directly correspond to database tables or views. This can always be achieved manually by inspecting the database and creating a view for each concept. However, more sophisticated ways of graphically selecting features in the database and aggregating them to concepts increase the acceptance of the system by end users and ease the adaptation of cases to other environments. In the MiningMart project, the relational editor is intended to support this kind of activity. In general, it should be possible to map all reasonable representations of entities to reasonable conceptual definitions. A simple mapping of the concept Customer, containing the features Customer ID, Name, and

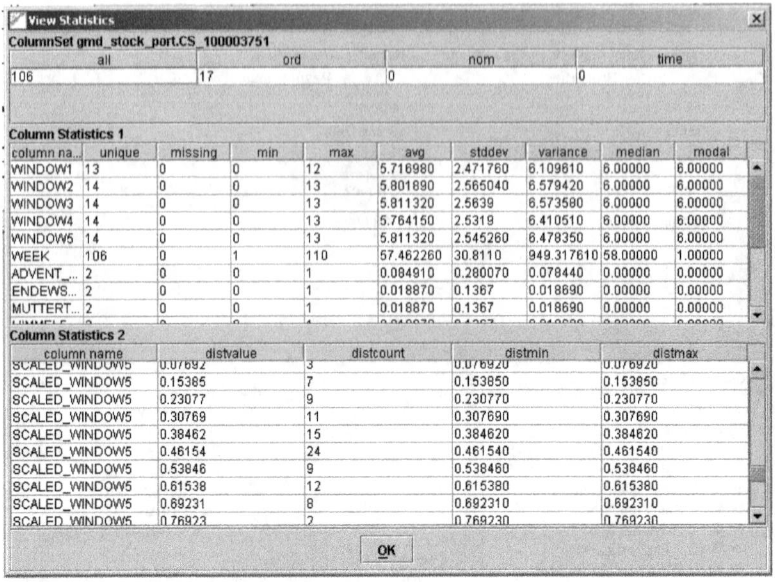

Fig. 3.4. Statistics of a database view

Address of the database would state that the table CUSTOMER holds all the necessary attributes, i.e., CUSTOM_ID, CUST_NAME and CUST_ADDR. Having the information about name and address distributed over different tables (e.g., sharing the key attribute CUSTOM_ID) is an example for more complex mappings. In this case, the relational editor should be able to use a join operation.

Apart from connecting conceptual entities to database entities, the relational editor offers a data viewer and is capable of displaying statistics of connected views or tables. Figure 3.4 shows an example of the statistics displayed. For each view or table the number of tuples and the numbers of nominal, ordinal, and time attributes are counted. For numerical attributes the number of different and missing values is displayed, the minimum, maximum, average, median, and modal value are calculated together with the standard deviation and variance. For ordinal and time attributes, the most reasonable subset of this information is given. Finally, we have information on the distribution of the values for all attributes.

3.2.4 The Case and Its Compiler

All the information about the conceptual descriptions and about the database objects involved are represented within the M4 model and stored within relational tables. M4 *cases* denote a collection of steps, basically performed sequentially, each of which changes or augments one or more concepts. Each

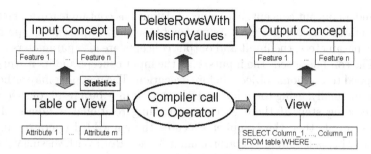

Fig. 3.5. An illustration of the coupling of abstract conceptual and executable levels

step is related to exactly one M4 *operator*, and holds all its input arguments. The M4 compiler reads the specifications of steps and executes the operator, passing it all the necessary inputs. This process requires the compiler to translate the conceptual entities, like input concepts of a step, to the corresponding relational entities, like database table name, the name of a view, or the SQL definition of a virtual view, which are only defined as relational metadata in the M4 model.

Two kinds of operators are distinct:, manual and machine learning operators. Manual operators just read the M4 metadata of their input and add an SQL-definition to the metadata for their output, establishing a virtual table. Currently, the MiningMart system offers 20 manual operators for selecting rows, selecting columns, handling time data, and generating new columns for the purposes of, e.g., handling null values, discretization, moving windows over time series, and gathering information concerning an individual (e.g., customer, patient, shop).

External machine learning operators on the other hand are invoked by using a wrapper approach. Currently, the MiningMart system offers learning of decision trees, k-means, and the support vector machine as learning preprocessing operators[5]. The necessary business data are read from the relational database tables, converted to the required format, and passed to the algorithm. After execution, the result is read by the wrapper, parsed, and either stored as an SQL-function or materialized as additional business data.

In any case, the M4 metadata will have to be updated by the compiler. A complex machine learning tool to replace missing values is an example for operators altering the database. In contrast, for operators like a join it is sufficient to *virtually* add the resulting view together with its corresponding SQLstatement to the metadata.

Figure 3.5 illustrates how the abstract and the executable or relational level interact. First of all, just the upper sequence is given, an input concept, a

---

[5] Of course, the algorithms may also be used in the classical way, as data mining step operators.

step, and an output concept. The concept definitions contain features, and the step contains an operator with its parameter settings. Apart from operator-specific parameters, the input and output concepts are also parameters of the step. The compiler needs the inputs, e.g., the input concept and its features to be mapped to relational objects before execution. The mapping may either be defined manually, using the relational editor, or it may be a result of executing the preceding step. If there is a corresponding relational database object for each input, then the compiler executes the embedded operator. In the example, this is a simple operator named "DeleteRowsWithMissingValues". The corresponding executable part of this operator generates a view definition in the database and in the relational metadata of M4. The latter is connected to the conceptual level, so that afterward there is a mapping from the output concept to a view definition. The generated views may be used as inputs to subsequent steps, or they may be used by other tools for the data mining step.

Following the overall idea of declarative knowledge representation of the project, known preconditions and assertions of operators are formalized in the M4 schema. Conditions are checked at runtime, before an operator is applied. Assertions help to decrease the number of necessary database accesses, because necessary properties of the data can be derived from formalized knowledge, saving expensive database scans. A step replacing missing values might be skipped, for instance, if the preceding operator is known not to produce any missing values. If a user applies linear scaling to an attribute, then all values are known to lie in a specific interval. If the succeeding operator requires all values to be positive, then this pre-condition can be derived from the formalized knowledge about the linear scaling operator, rather than by recalculating this property by another database scan.

The task of a case designer, ideally a data mining expert, is to find sequences of steps resulting in a representation well suited for the given data mining task. This work is supported by a special tool, the *case editor*. Figure 3.6 shows a screenshot of a rather small example case edited by this tool. Typically, a preprocessing chain consists of many different steps, usually organized as a directed acyclic graph, rather than as a linear sequence as in the example case shown in Fig. 3.6. To support the case designer, a list of available operators and their overall categories, e.g., feature construction, clustering, or sampling, is part of the conceptual M4 case model. The idea is to offer a fixed set of powerful preprocessing operators, in order to offer a comfortable way of setting up cases on the one hand, and ensuring re-usability of cases on the other. By modeling real world cases in the scope of the project, further useful operators will be identified, implemented, and added to the repository.

For each step the case designer chooses an applicable operator from the collection, sets all of its parameters, assigns the input concepts, input attributes and/or input relations, and specifies the output. To ease the process

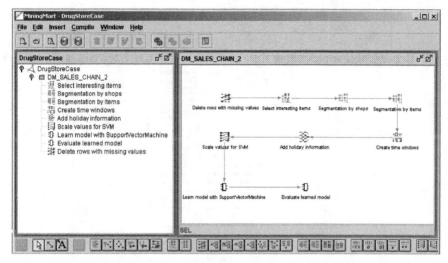

Fig. 3.6. A small example case in the case editor

of editing cases, applicability constraints on the basis of metadata are provided as formalized knowledge and are automatically checked by the human-
computer interface. This way only valid sequences of steps can be produced
by a case designer. Furthermore, the case editor supports the user by automatically defining output concepts of steps according to the metadata constraints, and by offering property windows tailored to the demands of chosen
operators.

A sequence of many steps, namely a *case* in M4 terminology, transforms
the original database into another representation. Each step and its ordering
is formalized within M4, so the system is automatically keeping track of the
performed activities. This enables the user to interactively edit and replay a
case or parts of it.

As soon as an efficient chain of preprocessing has been found, it can easily
be exported and added to an Internet repository of best practice MiningMart
cases. Only the conceptual metadata is submitted, so even if a case handles
sensitive information, as given for most medical or business applications, it
is still possible to distribute the valuable metadata for reuse, while hiding all
the sensitive data and even the local database schema.

3.3 The Case Base

One of the project's objectives is to set up a case-base of successful cases on
the Internet. The shared knowledge allows all Internet users to benefit from
a new case. Submitting a new case of best-practices is a safe advertisement

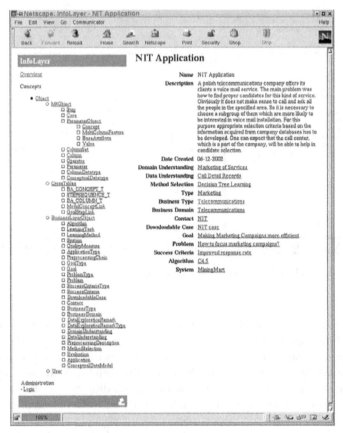

Fig. 3.7. The Internet interface to the case base visualizes all cases, their steps, embedded operators, and parameters in HTML format. Entities related in the M4 schema are connected by hyperlinks. Additionally, a business level is part of the interface. It describe the available cases in terms like the addressed business goals of the data analysis. After deciding for a case with the help of conceptual M4 and business layer descriptions, the user can simply download the one addressing the most similar problem. The case adaption facilities of The MiningMart system helps to quickly adjust the case to the user's environment

for KDD specialists or service providers, since the relational data model is kept private.

To support users in finding the most relevant cases, their inherent structure will be exploited. An Internet interface will be accessible, visualizing the conceptual metadata. It will be possible to navigate through the case-base and to investigate single steps, i.e., which operators were used on which kinds of concepts. The Internet interface is supposed to read the data directly from the M4 tables in the database, eliminating additional efforts and redundancies. Figure 3.7 shows a screenshot of a case's business level description.

In addition to the data explicitly represented in M4, a business level has been added. This level aims at relating the case to business goals and at giving several kinds of additional descriptions, such as which success criteria were important for the case. For instance, the sales prediction answers the question "How many sales of a particular item do I have to expect?" where the business goal is that it must not happen that the item is sold out, but that the stock should be minimized. A particular application need is that the forecast can only be used if it predicts the sales four weeks ahead because of delivery times. The more informal descriptions should especially help decision makers to find a case tailored for their specific domains and problems. The additional information is stored in an XML-representation directly connected to the M4 entities. On the Internet these connections are represented by hyperlinks. Figure 3.8 shows the ontology of the business level.

It is possible to start the search for a case at any category of the business level or conceptual level. In this sense, the cases are indexed by all the categories that are a part of the conceptual M4 model and the business model. If a user considers a case useful, then its conceptual data can be downloaded from the server. The downloadable case is itself a category in the XML framework. The locally installed MiningMart system offers an import facility, installing the metadata into the user's M4 tables. If problems arise or further help is necessary, the business level holds a category for the case designer or the company providing the service.

The project has developed four cases:

- analysis of insurance data for direct mailing [3.7, 3.8],
- call center analyisis for marketing,
- analysis of data about calls and contracts for fraud detection in telecommunication, and
- analysis of sales data for sales prediction [3.12].

3.4 Conclusions

The relevance of supporting not only single steps of data analysis but sequences of steps has long been underestimated. While a large variety of excellent tools exist which offer algorithms for a data mining step, only very few approaches exist to tackle the task of making clever choices during preprocessing and combining these choices into an effective and efficient sequence. The *Clementine* system offers processing chains to users. However, its focus lies on the data mining step, not on the preprocessing chain. The common data format in tool boxes such as, e.g., *SPSS* or *WEKA*, provides users with the prerequisites to formulate their own sequences [3.15]. However, the user is programming the sequence and has to do this anew for very similar tasks.

Zhong et al. have proposed an agent system, *GLS*, which supports the overall KDD process, i.e., preprocessing, knowledge elicitation, and refinement of the result [3.16, 3.17]. In some aspects, this system is similar to

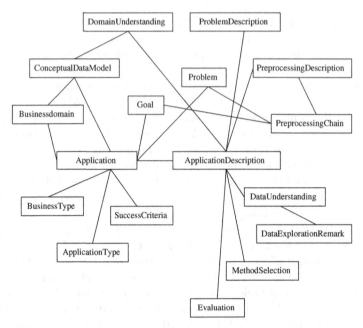

Fig. 3.8. The ontology of the business layer, used to describe M4 cases in business terms

MiningMart. Its agents are our operators, its controller corresponds to our compiler, and both systems describe data and operators at the meta-level. Whereas in MiningMart the operator description entails applicability conditions and pointers to the resulting table, in *GLS* the pre- and post-conditions for the application of an agent are stated. The hierarchy of agents in *GLS* corresponds to the inheritance hierarchy of operators, as exploited in MiningMart. In addition, MiningMart offers an even more abstract level for the description of a case in business terms. The planning approach of *GLS* is also similar to the use of applicability constraints, as in MiningMart. In contrast to *IDEA*, no comparison of quality is performed for alternative chains of operators. Hence, both MiningMart and *GLS* produce valid sequences of steps, and none of them performs experiments – as does *IDEA* – in order to decide between several algorithms or agents. In spite of the similarities, the two systems do, of course, also differ. First, the set of algorithms (operators or agents) is different. Feature generation and selection – a focus of MiningMart – is less developed within *GLS*. Data mining algorithms are less complete in the MiningMart. However, both systems allow for easily enhancing the set of operators. Second, the relationship with the database is different. The interaction between *GLS* and the database is not the primary focus of research [3.16, 3.17]. In contrast, the MiningMart resides to a large degree within the database, compiles metadata into SQL code, and many of its operators are directly integrated into the database. This allows work on very large databases. Third, the use of human expertise is different. In *GLS*, some user interaction is required to optimize the automatically generated valid sequences. However, the notion of a complete case at the meta-level is not part of the meta-model. This means that the diverse trials to establish an optimal sequence of agent activities are not documented. Hence, experience of failed selections, groupings, or parameter settings cannot prevent users from again doing so. Experience of successful cases is not stored at the meta-level. There is no mechanism to apply a successful chain to similar but different databases. In contrast, MiningMart compiles a successful case together with a meta-model of new data into a new running KDD case. We believe that the reuse of best practice cases and the case documentation is extremely important.

The recent *IDEA* system is also similar to the MiningMart approach [3.1]. Chains of operators are composed according to a ranking of algorithms in order to detect the best choice of an algorithm given data characteristics. Metadata describing the data as well as the algorithms are used in order to check the validity of operator sequences or to incorporate an additional step which allows the application of the best operator. The difference lies first in MiningMart's orientation towards very large databases. *IDEA* uses the *WEKA* data format and, hence, is restricted to smaller files. The data transformations and aggregations incorporated as manual operators in the MiningMart system are not taken into account in *IDEA*, because they are not needed in the single table, small sample representation of *WEKA* tools. The second distinction is the approach to determining the best sequence of preprocessing. Although the MiningMart system exploits applicability conditions of operators in order to check the validity of sequences, it does not aim at planning the best sequence or at performing a ranking of possible algorithms at each step of an operator chain, as *IDEA* can do. Instead, MiningMart exploits the findings of expert case designers. Real-world application of knowledge discovery comprises hundreds of steps in a KDD run (including manual operators) and ranking every algorithm at each of the steps would exhaust computing capacity. We feel that the adaptation of excellently solved KDD problems best combines human expertise and computational power.

We can now summarize the characteristics of the MiningMart:

Very large databases: It is a database-oriented approach which easily interacts with all SQL-databases and scales up to real-world databases without any problems. Several operators have been re-implemented in order to make them ready for very large data sets.

Sophisticated operators for preprocessing: Preprocessing can make good use of learning operators as does the data mining step. For instance, a learning result can be used to replace missing values by the learned (predicted) values. Feature generation and selection in the course of preprocessing enhances the quality of data that are the input to the data mining step.

Metadata driven code generation: The MiningMart approach relies on meta-driven software generation. Metadata about operators and data are used by the compiler in order to generate a running KDD application.

Case documentation: Metadata about a case document the overall KDD process with all operator selections and their parameter settings. In addition, a business layer offers the case description in less technical terms so that end-users of the KDD process are kept informed.

Case adaptation: The notion of a complete case in the meta-model allows the application of given expert solution to a new database. The user only needs to provide the system with a new data model and the compiler generates the new case. For fine-tuning the new application, the human-computer interface offers easy access to the meta-model with all operators.

References

3.1 A. Bernstein, S. Hill, F. Provost: An Intelligent Assistant for the Knowledge Discovery Process. Technical Report IS02-02, New York University, Leonard Stern School of Business (2002)

3.2 P. Brazdil: Data Transformation and Model Selection by Experimentation and Meta-Learning. In: C.G. Carrier, M. Hilario (eds.), *Workshop Notes-Upgrading Learning to the Meta-Level: Model Selection and Data Transformation* (Technical University Chemnitz, April 1998), number CSR-98-02 in Technical Report, pp. 11-17

3.3 K. Causse, M. Csernel, K. Morik, C. Rouveirol: MLT Deliverable 2.2: Specification of the Common Knowledge Representation Language of the MLToolbox. GMD (German Natl. Research Center for Computer Science, P.O.Box 1240, W-5205 St. Augustin 1, Germany, September 1990)

3.4 R. Engels: Planning Tasks for Knowledge Discovery in Databases; Performing Task-Oriented User-Guidance. In: *Proc. of th 2nd Int. Conf. on Knowledge Discovery in Databases, August 1996*

3.5 R. Engels, G. Lindner, R. Studer: A Guided Tour through the Data Mining Jungle. In: *Proceedings of the 3rd International Conference on Knowledge Discovery in Databases (KDD-97)* pp. 14-17 (August, 1997)

3.6 S. Fischer, R. Klinkenberg, I. Mierswa, O. Ritthoff: Yale: Yet Another Learning Environment-Tutorial. Technical Report CI-136/02, Collaborative Research Center 531, University of Dortmund, Dortmund, Germany, 2002. ISSN 1433-3325

3.7 J.U. Kietz, R. Züecker, A. Fiammengo, G. Beccari: Data Sets, Metadata and Preprocessing Operators at Swiss Life and CSELT. Deliverable D6.2, IST Project MiningMart, IST-11993 (2000)

3.8 J.U. Kietz, R. Züecker, A. Vaduva: Mining Mart: Combining Case-Based Reasoning and Multi-Strategy Learning into a Framework to reuse KDD-Application. In: R.S. Michalski, P. Brazdil (eds.), *Proceedings of the fifth International Workshop on Multistrategy Learning (MSL2000)* (Guimares, Portugal, May 2000)

3.9 H. Liu, H. Motoda: *Feature Selection for Knowledge Discovery and Data Mining* (Kluwer Academic Publishers, 1998)

3.10 D. Michie, D.J. Spiegelhalter, C.C. Taylor: *Machine Learning, Neural and Statistical Classification* (Ellis Horwood, New York u.a., 1994)

3.11 K. Morik, K. Causse, R. Boswell: A Common Knowledge Representation Integrating Learning Tools. In: *Proc. of the 1st International Workshop on Multistrategy Learning* (Harpers Ferry, 1991)

3.12 S. Rüeping: Zeitreihenprognose für Warenwirtschaftssysteme unter Berücksichtigung asymmetrischer Kostenfunktionen. Master's thesis, Universität Dortmund (1999)

3.13 D. Sleeman, R. Oehlman, R. Davidge: Specification of Consultant-0 and a Comparision of Several Learning Algorithms. Deliverable D5.1, Esprit Project, pp. 2154 (1989)

3.14 C. Theusinger, G. Lindner: Benutzerunterstützung eines KDD-Prozesses anhand von Datencharakteristiken. In: F. Wysotzki, P. Geibel, K. Schädler (eds.), *Beiträge zum Treffen der GI-Fachgruppe 1.1.3 Machinelles Lernen (FGML-98)* (Technical University Berlin, 1998) volume 98/11 of *Technical Report*

3.15 I. Witten, E. Frank: *Data Mining-Practical Machine Learning Tools and Techniques with JAVA Implementations* (Morgan Kaufmann, 2000)

3.16 N. Zhong, C. Liu, S. Ohsuga: A Way of Increasing both Autonomy and Versatility of a KDD System. In: Z.W. Ras, A. Skowron (eds.), *Foundations of Intelligent Systems* (Springer, 1997) pp. 94-105

3.17 N. Zhong, C. Liu, S. Ohsuga: Dynamically Organizing KDD Processes. International Journal of Pattern Recognition and Artificial Intelligence, 15(3), 451-473 (2001)

4. Ensemble Methods and Rule Generation

Yongdai Kim, Jinseog Kim, and Jongwoo Jeon

Seoul National University, Korea

Abstract
Ensemble methods have received much attention recently for their significant improvements in classification accuracy. However, ensemble algorithms do not provide any information about how the final decision is made. That is, ensemble methods improve classification accuracy at the expense of interpretability. In this chapter, we investigate possibilities of using ensemble methods for generating useful rules, which help understanding the data set as well as the decision. Extensive review of three ensemble algorithms - bagging, boosting, and CHEM is presented and the algorithm of rule generation with CHEM is proposed. The proposed rule generation algorithm is illustrated with a real data set.

4.1 Introduction

Ensemble methods are learning algorithms (prediction algorithms or classification methods) which construct many classifiers called base learners and combine them to make a final decision. Ensemble methods have shown great success in machine learning and statistics for their significant improvements in classification accuracy. Examples of ensemble algorithms are bagging [4.2], boosting [4.16], arcing [4.4], random forests [4.5]), and CHEM [4.17].

One disadvantage of ensemble methods is that improvements in classification accuracy are often obtained at the expense of interpretability. In many situations, not only accurate prediction but also interpretation of data is an important ingredient. Processes for interpreting a given data set include feature selection, visualization, rule generation, etc. In particular, rule generation is the most efficient way of understanding the data. For a given data $-n$ input-output pairs $\{(y_1, \mathbf{x}_1), \ldots, (y_n, \mathbf{x}_n)\}$ where $\mathbf{x}_i \in R^p$ and $y_i \in \{1, \ldots, J\}$, a rule is a combination of an any conjunction of conditions on the input value $\mathbf{x} = (x_1, \ldots, x_p)$ given by $\cap_{i=1}^{p} \{x_i \in A_i\}$ where the A_i's are subsets of the domains of x_i's together with their corresponding decision support and confidence. Here, the support is the number of observations satisfying the rule and the confidence is the proportion of observations correctly specified by the rule. For example, for the credit scoring model, a rule can be *if credit history is bad, then the credit status is classified as "bad" with support 8.9% and confidence 59.6%*. The conjunction of conditions of the input value is *"credit history is bad"* and the decision is that the credit status is *"bad"*. The support 8.9% means that 8.9% of the customers in the data set has bad credit history and the confidence 59.6% implies that among the customers whose credit history is bad, 59.6% of customers really have bad credit status (i.e., correctly classified).

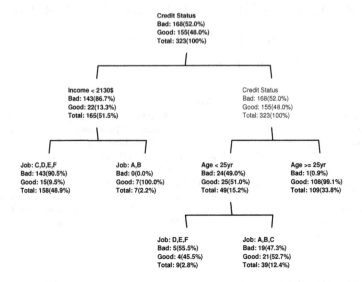

Fig. 4.1. An examplified tree for credit scoring

Table 4.1. Rules generated from the tree in Fig. 4.1

Conditions on input	Decision	Support	Confidence
Income $\geq$ 2130\$	"good"	33.8%	99.1%
Age $\geq$ 25 yr			
Income < 2130\$	"bad"	48.9%	90.5%
Job is either C,D,E, or F			

Decision trees such as CART [4.6] and C4.5 [4.18] are the most powerful and popular tools for rule generation in classification problems. Since each node represents a rule, the final model of a decision tree can be thought to be a collection of rules. For example, Fig. 4.1 has a decision tree with 5 terminal nodes for the credit scoring model. For given support 10% and confidence 90%, the rules generated from the tree in Fig. 4.1 are listed in Table 4.1.

This nice property of decision trees - rule generation, however, soon disappears when many decision trees are combined to make an ensemble model, for the final ensemble model is not of a tree structure any more. That is, ensemble methods improve predictability, that at a considerable sacrifice of interpretability. In this chapter, we investigate possibilities of using ensemble methods not only for improving classification accuracy but also for improving interpretability by generating various useful rules. The main idea of using ensemble methods for improving interpretability is to extract rules in the decision trees constructed as base learners. To implement this simple idea, however, at least two questions should be answered. The first question is which ensemble algorithm is most appropriate for rule generation. Understanding roles of base learners are crucial for answering this question and extensive reviews of three ensemble algorithms - bagging, boosting and CHEM are presented in Sect. 4.2 for this purpose. The second question is how to extract useful and dissimilar rules from the set of rules generated from the base learners. It is typical that similar base learners appear several times during the update of the ensemble model. Hence, in the set of all rules generated from the base learners, many rules appear repeatedly. To resolve the problem, a measurement for the degree of similarity of two rules is introduced, and an algorithm to select useful and dissimilar rules from a given set of rules is developed. Details are given in Sect. 4.3.

This chapter is organized as follows. Review of various ensemble algorithms is presented in Sect. 4.2. In Sect. 4.3, the process of rule generation with CHEM is described in detail. In Sect. 4.4, illustrations of the algorithm proposed in Sect. 4.3 are presented. Discussions follow in Sect. 4.5.

4.2 Ensemble Algorithms: A Review

In this section, we review the ensemble algorithms and investigate the roles of base learners. This review will give a guidance on which ensemble algorithm is most appropriate for a purpose of rule generation. The three ensemble algorithms - bagging, boosting and CHEM are considered. From what follows, we assume that the training samples

$$\mathcal{L} = \{(y_1, \mathbf{x}_1), \ldots, (y_n, \mathbf{x}_n)\}$$

are independent realizations of a random vector $(Y, \mathbf{X})$ whose probability measure is P. We use, unless otherwise stated, decision trees as base learners.

4.2.1 Bagging

Bagging [4.2], *"Bootstrap Aggregating"*, is a simple but very efficient method of improving unstable estimators by use of the bootstrap [4.10]. The idea of bagging is to make many training samples by bootstrap and construct many learners on the bootstrapped samples. And the final model is constructed by use of the majority vote. The algorithm of bagging is described on the next page.

Algorithm 1 Bagging

1: Make M many bootstrap samples: $\{\mathcal{L}^{(m)}\}$ from $\mathcal{L}$, $m = 1, \ldots, M$
2: Form M many predictors: $\{f(\mathbf{x}, \mathcal{L}^{(m)})\}, m = 1, \ldots, M$
3: Combine the M many predictors to construct the final one by use of voting.

Even if the algorithm of bagging is simple, its improvement of classification accuracy is noticeable. Table 4.2 (quoted from [4.2]) shows considerable evidence demonstrating the effectiveness of bagging. Detailed description of the data sets is in [4.2]. However, it is not clear why such huge improvements can be achieved by bagging. Breiman [4.2] explains it by arguing that bagging reduces the variance and hence bagging is most effective for unstable learners such as decision trees and neural networks. Consider the regression problem. Let $H_A(\mathbf{x})$ be the aggregated estimator defined by

$$H_A(\mathbf{x}) = e_{\mathcal{L}}(f(\mathbf{x}, \mathcal{L})).$$

Then, Jensen's inequality implies that

$$e_{\mathcal{L}} e_{Y,\mathbf{X}}(Y - f(\mathbf{X}, \mathcal{L}))^2 - e_{Y,\mathbf{X}}(Y - H_A(\mathbf{X}))^2$$
$$\geq e_{\mathbf{X}} \left\{ e_{\mathcal{L}} f^2(\mathbf{X}, \mathcal{L}) - [e_{\mathcal{L}} f(\mathbf{X}, \mathcal{L})]^2 \right\}$$
$$= e_{\mathbf{X}} \left\{ \mathrm{Var}_{\mathcal{L}}(f(\mathbf{X}, \mathcal{L})) \right\}.$$

Breiman [4.2] argued that $H_B(\mathbf{x}) = \sum_{i=1}^{M} f(\mathbf{x}, \mathcal{L}^{(m)})/M$ is thought of being an estimator of $H_A(\mathbf{x})$ and so the population mean squared error of the bagged estimator is smaller than that of the original estimator. Also, he argued that the improvement due to bagging is large when $\mathrm{Var}_{\mathcal{L}}(f(\mathbf{x}, \mathcal{L}))$ is large. Though his arguments provide some theoretical insight into bagging, the underlying reasons as to when and why $\mathrm{Var}_{\mathcal{L}}(f(\mathbf{x}, \mathcal{L}))$ is large are still unclear. Recently, there are several papers which explain the success of bagging. Friedman and Hall [4.13] expand a smooth estimator $f(\mathbf{x}, \mathcal{L})$ as

$$f(\mathbf{x}, \mathcal{L}) = \frac{1}{n} \sum_{i=1}^{n} \phi(\mathbf{x}; \mathbf{x}_i) + \text{high order terms},$$

and show that bagging reduces variability of the high order terms. Similar explanation for U-statistics is given by Buja and Stuetzle [4.9]. Bühlman and Yu [4.7] consider non-smooth estimators such as decision trees and linear models with variable selection, and they prove that bagging reduces the variance around the decision boundaries. For classification problems, no concrete explanation of the success of bagging is available. Partial answers can be found in Friedman [4.11].

The above explanation of bagging gives a clue on how to construct base learners. Since bagging reduces only the variance, the base learner should have large variance and small bias. Large variance and small bias is achieved via deliberate overfitting. For decision trees, full grown trees without pruning are most effective for bagging.

4.2.2 Boosting

Boosting is known to be the most effective ensemble algorithm. It improves classification accuracy significantly and is resistant to overfit. While bagging

Table 4.2. Classification Tree. $\bar{e}_S$ and $\bar{e}_B$ is the averaged test set misclassification errors of the single tree and bagged tree

Data Set	Samples	Variables	Classes	$\bar{e}_S$	$\bar{e}_B$	Decrease
waveform	300	21	3	29.0	19.4	33%
heart	1395	16	2	10.0	5.3	47%
breast cancer	699	9	2	6.0	4.2	30%
ionosphere	351	34	2	11.2	8.6	23%
diabetes	1036	8	2	23.4	18.8	20%
glass	214	9	6	32.0	24.9	22%
soybean	307	35	19	14.5	10.6	27%

mainly reduces the variance, boosting reduces bias as well as variance. Since it has been proposed [4.16], various variants of boosting have been suggested. Examples are real boosting [4.24], logit boosting [4.14], and gradient boosting [4.12].

Boosting refers to a general and provably effective method of producing a very accurate prediction rule by combining rough and moderately inaccurate classifiers in a certain way. The idea of boosting was initiated by Valiant [4.25] and a first practical and simple boosting algorithm was developed by Schapire [4.21]. Schapire showed that a weak learner could always improve its performance by training two additional classifiers on filtered versions of the input data stream. A weak learner is an algorithm for producing a two-class classifier with performance guaranteed (with high probability) to be significantly better than a coin flip. Schapire [4.21] proved that for a given weak learner f we can construct a learner f^* based on f such that f^* improves performance over f. Freund [4.15] proposed a progressed version of Schapire's boosting, and [4.16] finally developed more adaptive and realistic boosting algorithm AdaBoost.

We first consider a two-class classification problem (i.e., $J = 2$). Without loss of generality, we assume that $y \in \{-1, 1\}$. The AdaBoost algorithm proposed by Freund and Schapire [4.16] is described on the next page.

Many empirical studies on AdaBoost including Bauer and Kohavi [4.1], Ditterich [4.27] and Quinlan [4.19] have been done to show that AdaBoost is superior in reducing classification accuracy than bagging. Bauer and Kohavi [4.1] argued that better classification accuracy of AdaBoost comes from the fact that it reduces bias as well as variance while bagging reduces mainly variance. However, Ditterich [4.27], Quinlan [4.19], and Breiman [4.5] presented empirical results which show that when noise in output exists, performance of Adaboost can be seriously degraded. Rätsch, Onoda, and Müller [4.28] proposed a version of AdaBoost which is less sensitive to noise in data.

Along with classification accuracy, Bauer and Kohavi [4.1] and Quinlan [4.19] noted that AdaBoost is resistant to overfit. Resistance to overfit means that no matter how many base learners are added into the ensemble

---

Algorithm 2 AdaBoost

1: Start with weights $w_i = 1/n$, $i = 1, \ldots, n$.
2: Repeat for $m = 1, \ldots, M$;
    (a) Fit the classifier $f_m(\mathbf{x}) \in \{-1, 1\}$ using weights w_i on the data.
    (b) Compute err_m by

$$\text{err}_m = \frac{\sum_{i=1}^{n} w_i I(y_i \neq f_m(\mathbf{x}_i))}{\sum_{i=1}^{n} w_i}.$$

    (c) Set $c_m = \log((1 - \text{err}_m)/\text{err}_m)$.
    (d) Update w_i by

$$w_i = w_i \exp\left(c_m I(y_i \neq f_m(\mathbf{x}_i))\right).$$

3: Output the classifier sign $\left(\sum_{m=1}^{M} c_m f_m(\mathbf{x})\right)$.

---

model, the generalization error is not degraded. In particular, even after the training error reaches 0, the test error keeps decreasing as more base learners are added.

For resistance to overfit as well as excellent classification accuracy, Schapire et al. [4.23] gave an analysis of AdaBoost in terms of the margin of the training examples. The margin of an observation $(y, \mathbf{x})$ is defined to be

$$\frac{y \sum_m c_m f_m(\mathbf{x})}{\sum_m c_m}.$$

Schapire et al [4.23] proved that with high probability the test error is at most

$$\mathrm{P}_{\mathcal{L}}\left[\text{margin}(Y, \mathbf{X}) \leq \theta\right] + O\left(\sqrt{\frac{d}{n\theta^2}}\right), \quad \text{for all } \theta > 0,$$

where d is the VC-dimension of the set of base learners and $\mathrm{P}_{\mathcal{L}}$ is the empirical measure on the training sample $\mathcal{L}$. In addition, they proved that AdaBoost is particularly aggressive at reducing the margin since it concentrates on the examples with the smallest margins. Also, continual decreasing of the test error after the training error reaches 0 can be explained by the margin that keeps increasing. [4.3] argued that this explanation is incomplete since the upper bound is too loose to have any practical meaning. Instead, [4.5] conjectured that the success of AdaBoost might be due to the fact that the procedure of constructing base learners is ergodic.

Interpretations of AdaBoost via a gradient descent method have been proposed by Schapire and Singer [4.24], Friedman et al. [4.14] and Friedman [4.12]. Friedman et al. [4.14] show that AdaBoost is equivalent to forward stagewise additive modelling using the loss function $L(y, f(\mathbf{x})) = \exp(-yf(\mathbf{x}))$. Suppose $H_{m-1}(\mathbf{x})$ is the current model and we seek an updated model $H_m(\mathbf{x})$ given by $H_{m-1}(\mathbf{x}) + \beta f(\mathbf{x})$ where $f(\mathbf{x}) \in \{-1, 1\}$. Using the exponential loss, one must solve

$$(\beta_m, f_m) = \arg\min_{\beta, f} \sum_{i=1}^{n} \exp\left(-y_i(H_{m-1}(\mathbf{x}) + \beta f(\mathbf{x}))\right),$$

and update $H_m(\mathbf{x})$ by $H_{m-1}(\mathbf{x}) + \beta_m f_m(\mathbf{x})$. The above equation can be expressed as

$$(\beta_m, f_m) = \arg\min_{\beta, f} \sum_{i=1}^{n} w_i^{(m)} \exp(-\beta y_i f(\mathbf{x}_i)), \tag{4.1}$$

with $w_i^{(m)} = \exp(-y_i H_{m-1}(\mathbf{x}_i))$. The solution of Eq. (4.1) can be obtained in two steps. First, for any value of $\beta > 0$, the solution to Eq. (4.1) for f_m is

$$f_m(\mathbf{x}) = \arg\min_f = \sum_{i=1}^{m} w_i^{(m)} I(y_i \neq f(\mathbf{x}_i)),$$

which is the classifier that minimizes the weighted error rate in predicting y. This can be easily seen by expressing the criterion in Eq. (4.1) as

$$e^{-\beta} \sum_{y_i = f(\mathbf{x}_i)} w_i^{(m)} + e^{\beta} \sum_{y_i \neq f(\mathbf{x}_i)} w_i^{(m)},$$

which in turn can be written as

$$\left(e^{\beta} - e^{-\beta}\right) \sum_{i=1}^{n} w_i^{(m)} I(y_i \neq f(\mathbf{x}_i)) + e^{-\beta} \sum_{i=1}^{n} w_i^{(m)}. \tag{4.2}$$

To minimize Eq. (4.2), it suffices to minimize the weighted misclassification error with weights $\{w_i^{(m)}\}$. This can be achieved by fitting a classifier f_m. Once f_m is constructed, it can easily shown that β_m given by

$$\beta_m = \frac{1}{2} \log \frac{1 - \text{err}_m}{\text{err}_m}$$

minimizes Eq. (4.1) where err_m is the minimized weighted error rate

$$\text{err}_m = \frac{\sum_{i=1}^{n} w_i^{(m)} I(y_i \neq f_m(\mathbf{x}_i))}{\sum_{i=1}^{n} w_i^{(m)}}.$$

The updated model H_m becomes $H_m(\mathbf{x}) = H_{m-1}(\mathbf{x}) + \beta_m f_m(\mathbf{x})$ which causes the weights for the next iteration to be

$$w_i^{(m+1)} = w_i^{(m)} e^{\beta_m y_i f_m(\mathbf{x}_i)}. \tag{4.3}$$

Using the fact that $y_i f_m(\mathbf{x}_i) = 2I(y_i \neq f_m(\mathbf{x}_i)) - 1$, Eq. (4.3) becomes

$$w_i^{(m+1)} = w_i^{(m)} e^{c_m I(y_i \neq f_m(\mathbf{x}_i))} e^{-\beta_m}, \tag{4.4}$$

where $c_m = 2\beta_m$. The factor $e^{-\beta_m}$ can be ignored since it multiplies all weights by the same value. Finally, the updated model $H_m(\mathbf{x})$ and $\sum_{k=1}^{m} c_k f_k(\mathbf{x})$ have the same sign and so the decisions are the same.

The interpretation of the forward stagewise modelling for AdaBoost yields various versions of boosting algorithms. If we let

$$f(\mathbf{x}) = \frac{1}{2} \log \frac{p(\mathbf{x})}{1 - p(\mathbf{x})}$$

instead of $f(\mathbf{x}) \in \{-1, 1\}$, where $p(\mathbf{x}) = \Pr(Y = 1|\mathbf{X} = \mathbf{x})$, the resulting algorithm is RealBoost by Schapire and Singer [4.24]. With the same $f(\mathbf{x})$ as RealBoost, using the negative binomial log-likelihood as a loss function

$$L(y, f(\mathbf{x})) = -\log\left(1 + e^{-2yf(\mathbf{x})}\right)$$

and the Newton-Raphson method leads us to LogitBoost [4.14]. The same $f(\mathbf{x})$ and loss function with the gradient descent method results in Gradient-Boost [4.12]. The algorithms of these various boosting methods are given in Algorithms 3, 4, and 5.

Algorithm 3 RealBoost

1: Start with weights $w_i = 1/n$, $i = 1, \ldots, n$.
2: Repeat for $m = 1, \ldots, M$;
    (a) Fit the classifier to obtain a class probability estimate $p_m(\mathbf{x}) = \hat{P}_w(y = 1|\mathbf{x})$ using weights w_i on the training data.
    (b) Set $f_m(\mathbf{x}) = \frac{1}{2} \log p_m(\mathbf{x})/(1 - p_m(\mathbf{x}))$.
    (c) Update $w_i = w_i \exp(-y_i f_m(\mathbf{x}_i))$, $i = 1, \ldots, n$, and renormalize so that $\sum_i w_i = 1$.
3: Output the classifier sign $(\sum_{m=1}^{M} f_m(\mathbf{x}))$.

Algorithm 4 LogitBoost

1: Let $y = (y + 1)/2$.
2: Start with weights $w_i = 1/n$, $i = 1, \ldots, n$, $H(\mathbf{x}) = 0$ and probability estimates $p(x_i) = 1/2$.
3: Repeat for $m = 1, \ldots, M$;
    (a) Compute the working response and weights

$$z_i = \frac{y_i - p(\mathbf{x}_i)}{p(\mathbf{x}_i)(1 - p(\mathbf{x}_i))},$$

$$w_i = p(\mathbf{x}_i)(1 - p(\mathbf{x}_i)).$$

    (b) Fit the function $f_m(\mathbf{x})$ by a weighted least-squares regression of z_i to x_i using weights w_i.
    (c) Update $H(\mathbf{x}) = H(\mathbf{x}) + f_m(\mathbf{x})/2$ and $p(\mathbf{x}) = e^{H(\mathbf{x})}/\left(e^{H(\mathbf{x})} + e^{-H(\mathbf{x})}\right)$.
4: Output the classifier sign $(H(\mathbf{x}))$.

Algorithm 5 GradientBoost

1: Let $\bar{y} = \sum_{i=1}^{n} y_i/n$ and $H(\mathbf{x}) = \frac{1}{2}\frac{1+\bar{y}}{1-\bar{y}}$.
2: Repeat for $m = 1, \ldots, M$;
   (a) Let $z_i = \frac{2y_i}{1+\exp(2y_i H(\mathbf{x}_i))}$.
   (b) Fit the function $f_m(\mathbf{x})$ by a weighted least-squares regression of z_i to x_i
       by using the regression tree.
   (c) The prediction value γ_l of the l-th terminal node of $f_m(\mathbf{x})$ by

$$\gamma_l = \frac{\sum_{\mathbf{x}_i \in R_l} z_i}{\sum_{\mathbf{x}_i \in R_l} |z_i|(2 - |z_i|)}$$

   where R_l is the set of observations in the l-th terminal node of $f_m(\mathbf{x})$.
   (d) Update $H(\mathbf{x}) = H(\mathbf{x}) + f_m(\mathbf{x})$.
3: Output the classifier sign $(H(\mathbf{x}))$.

There are several methods of extending boosting to the multi-class problem. Some of them are given in Algorithms 6, 7, and 8. Schapire and Singer [4.24] provide several generalizations of AdaBoost for the multi-class case and refer to other proposals [4.16, 4.22]. Among various methods, AdaBoost.MH algorithm is dominated in the empirical studies. The main ideas of AdaBoost.MH is to reduce the multi-class problem to a larger binary problem. That is, AdaBoost.MH creates a set of binary problems, for each example $\mathbf{x}$ and each possible label y, of the form: "For example $\mathbf{x}$, is the correct label y or is it one of the other labels?" Friedman et al. [4.14] and Friedman [4.12] developed a different approach of dealing with multi-class problems by use of the symmetric multiple logistic transformation. Let $p_j(\mathbf{x}) = \Pr(Y = j|\mathbf{X} = \mathbf{x})$. The symmetric multiple logistic transformation is defined by

$$f_j(\mathbf{x}) = \log p_j(\mathbf{x}) - \frac{1}{J}\sum_{k=1}^{J} \log p_k(\mathbf{x}),$$

which is equivalent to

$$p_j(\mathbf{x}) = \frac{e^{f_j(\mathbf{x})}}{\sum_{k=1}^{J} e^{f_k(\mathbf{x})}}$$

with $\sum_{k=1}^{J} f_k(\mathbf{x}) = 0$. They devised the natural generalizations of boosting algorithms for two-class to the multi-class case with the symmetric multiple logistic regression and negative multinomial log-likelihood as a loss function. Detailed derivation of the algorithm is in Friedman et al. [4.14] for LogitBoost and Friedman [4.12] for GradientBoost.

4.2.3 Convex Hull Ensemble Machine: CHEM

Kim [4.17] developed a new ensemble method called "*Convex Hull Ensemble Machine* (CHEM)". CHEM successively updates the ensemble model so

---

Algorithm 6 AdaBoost.MH

1: Expand the original n observations into $n \times J$ pairs
$\{(y_{i1}, (\mathbf{x}_i, 1)), \ldots, (y_{iJ}, (\mathbf{x}_i, J))\}, \quad i = 1, \ldots, n$. Here, y_{ij} is 1 if $y_i = j$ and -1 otherwise.
2: Apply AdaBoost to the augmented data set, producing a function $H :$
$R^p \times (1, \ldots, J) \to R; H(\mathbf{x}, j) = \sum_{m=1}^{M} f_m(\mathbf{x}, j)$.
3: Output the classifier arg $\max_j H(\mathbf{x}, j)$.

Algorithm 7 LogitBoost for multi-class

1: Starts with weights $w_{ij} = 1/n$, $i = 1, \ldots, n$, $j = 1, \ldots, J$, $H_j(\mathbf{x}) = 0$ and $p_j(\mathbf{x}) = 1/J$, $j = 1, \ldots, J$.
2: Repeat for $m = 1, \ldots, M$;
   (a) Repeat for $j = 1, \ldots, J$;
      (i) Compute working responses and weights in the jth class

$$z_{ij} = \frac{y_{ij}^* - p_j(\mathbf{x}_i)}{p_j(\mathbf{x}_i)(1 - p_j(\mathbf{x}_i))},$$

$$w_{ij} = p_j(\mathbf{x}_i)(1 - p_j(\mathbf{x}_i))$$

     where $y_{ij}^* = I(y_i = j)$.
      (ii) Fit the function $f_{mj}(\mathbf{x})$ by a weighted least-squares regression of z_{ij} to $\mathbf{x}_i$ with weights w_{ij}.
   (b) Set

$$f_{mj}(\mathbf{x}) = \frac{J - 1}{J}(f_{mj}(\mathbf{x}) - \frac{1}{J}\sum_{k=1}^{J} f_{mk}(\mathbf{x}))$$

   and

$$H_j(\mathbf{x}) = H_j(\mathbf{x}) + f_{mj}(\mathbf{x}).$$

   (c) Update $p_j(\mathbf{x})$ by

$$p_j(\mathbf{x}) = \frac{e^{H_j(\mathbf{x})}}{\sum_{k=1}^{J} e^{H_k(\mathbf{x})}}$$

3: Output the classifier arg $\max_j H_j(\mathbf{x})$.

---

that it converges to the optimal model inside the convex hull of the set of base learners. Empirical studies revealed that CHEM has superior classification performance than boosting; CHEM does not overfit even when boosting overfits seriously, and CHEM is much more robust to output noise [4.17].

CHEM begins by investigating instability observed in many learning algorithms, such as decision trees and neural networks. Figure 4.2 shows the instability of decision trees. Two decision trees are constructed from two bootstrap samples of the same original training data. The structures of the two trees are completely different even though the two bootstrap samples are thought to be similar. One way of explaining this kind of severe instability is that there are many different base learners which explain the data similarly. Figure 4.3 describes this situation. The shaded area ($\mathcal{F}$) represents the set of base learners and f^* is the true model. There are at least three models

---

Algorithm 8 GradientBoost for multi-class

1: Let $H_j(\mathbf{x}) = 0$, $j = 1, \ldots, J$.
2: Repeat for $m = 1, \ldots, M$;
   (a) Let

$$p_j(\mathbf{x}) = \frac{e^{H_j(\mathbf{x})}}{\sum_{k=1}^{J} e^{H_k(\mathbf{x})}}$$

   for $j = 1, \ldots, J$.
   (b) Repeat for $j = 1, \ldots, J$;
      (i) Let $z_{ij} = y_{ij}^* - p_j(\mathbf{x}_i)$ where $y_{ij}^* = 1$ if $y_i = j$ and 0 otherwise.
      (ii) Fit the function $f_{mj}(\mathbf{x})$ by a weighted least-squares regression of z_{ij} to
          x_i by using the regression tree.
      (iii) Compute the prediction value γ_l of the l-th terminal node of $f_{mj}(\mathbf{x})$
   by

$$\gamma_l = \frac{J-1}{J} \frac{\sum_{\mathbf{x}_i \in R_l} z_{ij}}{\sum_{\mathbf{x}_i \in R_l} |z_{ij}|(1-|z_{ij}|)}$$

     where R_l is the set of observations in the l-th terminal node of $f_{mj}(\mathbf{x})$.
     (iv) Update $H_j(\mathbf{x}) = H_j(\mathbf{x}) + f_{mj}(\mathbf{x})$.
3: Output the classifier arg $\max_j H_j(\mathbf{x})$.

---

f_1, f_2, and f_3 located at similar distance from the true model. Suppose the data is given as $f^* + \epsilon$. In this case, a small change in ϵ results in a large change of the optimal model (the model closest to $f^* + \epsilon$). Two important facts can be deduced from Fig. 4.3. First, the set of base learners is not convex. Second, the true model is located inside a convex hull of the set of base learners, which means that the true model can be represented by the convex combination of base learners. CHEM searches for such convex combinations via forward-stage manner.

Remark. Decision trees are suitable as base learners for CHEM. This is because (i) the set of decision trees is not convex and (ii) the convex combination of decision trees covers $L_2(P)$ space where $L_2(P)$ is the set of all functions such that $\int f^2 dP < \infty$. That is, as long as the true model f belongs to $L_2(P)$, it can be represented by the convex combination of (possibly infinitely many) decision trees.

We first consider CHEM in $L_2(P)$ space. CHEM constructs a sequence of convex combinations of base learners (ensemble models) as follows. Suppose H_m is the current ensemble model given by

$$H_m(\mathbf{x}) = \frac{\sum_{i=1}^{m} w_i f_i(\mathbf{x})}{\sum_{i=1}^{m} w_i}$$

where $f_1(\mathbf{x}), \ldots, f_m(\mathbf{x})$ are base learners and $w_1, \ldots, w_m$ are nonnegative constants. Let $\mathcal{F}$ be the class of base learners. First, CHEM finds the model f_{m+1} in $\mathcal{F}$ where

$$f_{m+1} = \operatorname{argmin}_{f \in \mathcal{F}_m} \|f - f^*\|$$

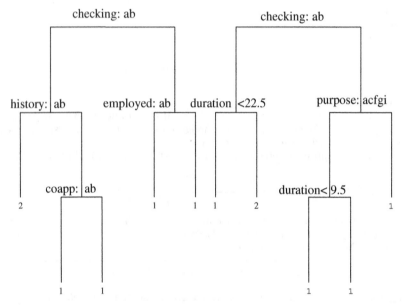

Fig. 4.2. Two trees from two bootstrap samples of the German credit data with node size 5

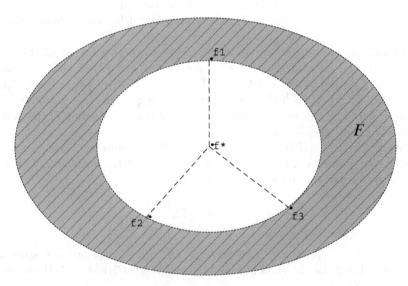

Fig. 4.3. Geometry for instability

and $\mathcal{F}_m = \{f \in \mathcal{F} : f - f^* \perp f - H_m\}$. Here, $\|f\|^2 = <f, f>$ and $<f, g>$ is the inner product of two functions f and g in $L_2(P)$. That is, f_{m+1} is the closest model from f^* with $f - f^* \perp f - H_m$. After constructing f_{m+1}, CHEM updates the ensemble mode by

$$H_{m+1} = \frac{u_m H_m + w_{m+1} f_{m+1}}{u_m + w_{m+1}}$$

where u_m and w_{m+1} are chosen so that $\|H_{m+1} - f^*\|$ is minimized. The resulting H_{m+1} is shown to be a form of

$$H_{m+1}(\mathbf{x}) = \frac{\sum_{i=1}^{m+1} w_i f_i(\mathbf{x})}{\sum_{i=1}^{m+1} w_i}$$

where $w_i = 1/\|f_{m+1} - f^*\|$. It can be shown that $\|H_m - f^*\| \geq \|H_{m+1} - f^*\|$. That is, the updated ensemble model is closer to the true model, which can explain why CHEM never overfits. Also, under regularity conditions, it can be shown that $\|H_m - f^*\| \to 0$ as $m \to \infty$.

To translate the CHEM algorithm for the two-class classification problem, we assume the logistic model:

$$\Pr(Y = 1 | \mathbf{X} = \mathbf{x}) = \frac{\exp(f(\mathbf{x}))}{1 + \exp(f(\mathbf{x}))},$$

and embed the classification problem into the function estimation problem (i.e., estimation of f). In this set-up, two devices should be developed. One is how to measure the distance between a given base learner and the true model (i.e., $\|f - f^*\|$ for a given $f \in \mathcal{F}$) and the other is how to construct f_{m+1} for a given H_m. The distance from a given base learner f to the true model can be approximated by the square root of the deviance of f, defined by

$$d(f) = \sum_{i=1}^{n} l(y_i, f(\mathbf{x}_i))/n$$

for some loss function l. For the loss function, the negative log-likelihood of the binomial distribution is used, which is given by

$$l(y, f(\mathbf{x})) = -\log\left(1 + e^{-yf}\right)$$

for $y \in \{-1, 1\}$. To construct a base learner f_{m+1} orthogonal to H_m, the model ϕ based on the residuals of H_m is constructed. Then, $\phi - f^*$ and $H_m - f^*$ are expected to be nearly orthogonal since the residuals and H_m are orthogonal. Then, we set $f_{m+1} = \eta\phi(\mathbf{x})$ where

$$\eta = \operatorname{argmin}_\delta d(\delta\phi).$$

That is, we first find the appropriate direction ϕ and construct the optimal model on that direction. For residuals, the Pearson residual, defined by

$$r_i = \frac{y_i - P_m(\mathbf{x}_i)}{\sqrt{P_m(\mathbf{x}_i)(1 - P_m(\mathbf{x}_i))}}$$

where $P_m(\mathbf{x}) = \exp(H_m(\mathbf{x}))/(1 + \exp(H_m(\mathbf{x})))$ and $H_m(\mathbf{x})$ is the mth ensemble model, is used. For some technical reasons, however, we do not use the residuals $\{r_i\}$ as a response variable. Instead, we use $|r_i|$ as a weight of the ith observation, and construct ϕ on a weighted bootstrap sample of the original sample $\{(\mathbf{x}_1, y_1), \ldots, (\mathbf{x}_n, y_n)\}$ with weights $\{|r_i|\}$. In summary, the algorithm of CHEM for the two-class problem is given below.

Algorithm 9 CHEM algorithm for two-class classification problem

---

1: Initialization : let the weights $v_i = 1/n$ for $i = 1, \ldots, n$.
2: Repeat $m = 1, \ldots, M$;
   (a) Make a bootstrap sample $\mathcal{L}^B$ from $\mathcal{L}$ with weights $\{v_i\}$.
   (b) Estimate $p(\mathbf{x}) = \hat{P}(Y = 1 | \mathbf{X} = \mathbf{x})$ using $\mathcal{L}^B$ with a given class of base learners.
   (c) Let $\phi(\mathbf{x}) = \log(p(\mathbf{x})/(1 - p(\mathbf{x})))$.
   (d) Calculate the correction factor η by

$$\eta = \mathrm{argmin}_\delta d(\delta\phi).$$

   (e) Let $f_m(\mathbf{x}) = \eta\phi(\mathbf{x})$.
   (f) Update the ensemble model $H_m(\mathbf{x}) = \sum_{i=1}^{m} w_i f_i(\mathbf{x}) / \sum_{i=1}^{m} w_i$ where $w_m = 1/d(f_m)$.
   (g) Update the weights $\{v_i\}$ by

$$v_i = \left| \frac{y_i - P_m(\mathbf{x}_i)}{\sqrt{P_m(\mathbf{x}_i)(1 - P_m(\mathbf{x}_i))}} \right|,$$

   where $P_m(\mathbf{x}) = \exp(H_m(\mathbf{x}))/(1 + \exp(H_m(\mathbf{x})))$.
3: For a new input $\mathbf{x}$, assign it to class 1 if $H_M(\mathbf{x}) > 0$ and to class -1 otherwise.

For multi-class problems (i.e., $Y \in \{1, \ldots, J\}, J > 2$), we assume the symmetric logistic model:

$$\Pr(Y = k | \mathbf{X} = \mathbf{x}) = \frac{\exp(f_k(\mathbf{x}))}{\sum_{j=1}^{J} \exp(f_j(\mathbf{x}))}.$$

The extension of CHEM to multi-class problems (i.e., estimation of $\underline{f} = (f_1, \ldots, f_J)$ with fitting separate base learners for each class) is as follows. We use the negative log-likelihood of the multinomial distribution as a loss function.

4.2.4 Comments on Roles of Base Learners

In the previous subsections, three important ensemble algorithms have been reviewed. In this section, the roles of base learners in the three ensemble

---

Algorithm 10 CHEM algorithm for multi-class

1: Initialization :

    (a) Set weights $\{v_{ij}\}$ by $v_{ij} = 1/n$ for $i = 1, \ldots, n$ and $j = 1, \ldots, J$.

    (b) Set $y_{ij}^* = I(y_i = j)$ for $i = 1, \ldots, n$ and $j = 1, \ldots, J$.

2: Repeat for $m = 1, \ldots, M$;

    (a) Repeat for $j = 1, \ldots, J$;

        (i) Make a bootstrap sample $\mathcal{L}_j^B$ from $\{(y_{1j}^*, \mathbf{x}_1), \ldots, (y_{nj}^*, \mathbf{x}_n)\}$ with weights $\{v_{1j}, \ldots, v_{nj}\}$.

        (ii) Estimate $p_j(\mathbf{x}) = \hat{P}(Y_j^* = 1 | \mathbf{X} = \mathbf{x})$ using $\mathcal{L}_j^B$ with a given class of base learners.

        (iii) Set $\phi_j(\mathbf{x}) = \log(p_j(\mathbf{x})/(1 - p_j(\mathbf{x})))$.

    (b) Set $\phi_j(\mathbf{x}) = \phi_j(\mathbf{x}) - \sum_{k=1}^{J} \phi_k(\mathbf{x})/J$ for $j = 1, \ldots, J$.

    (c) Calculate the correction factor η by

$$\eta = \mathrm{argmin}_\delta d(\delta \underline{\phi})$$

    where $\underline{\phi} = (\phi_1, \ldots, \phi_J)$.

    (d) Let $\underline{f}_m(\mathbf{x}) = \eta \underline{\phi}(\mathbf{x})$.

    (e) Update the ensemble model $\underline{H}_m(\mathbf{x}) = \sum_{i=1}^{m} w_i \underline{f}_i(\mathbf{x}) / \sum_{i=1}^{m} w_i$

    where $w_m = 1/d(\underline{f}_m)$.

    (f) For $j = 1, \ldots, J$, update the weights $\{v_{ij}\}$ by

$$v_{ij} = \left| \frac{y_{ij}^* - P_{mj}(\mathbf{x}_i)}{\sqrt{P_{mj}(\mathbf{x}_i)(1 - P_{mj}(\mathbf{x}_i))}} \right|$$

    where $P_{mj}(\mathbf{x}) = \exp(H_{mj}(\mathbf{x}))/(\sum_{k=1}^{J} \exp(H_{mk}(\mathbf{x})))$.

3: Assign a new data with input variable $\mathbf{x}$ to class $\mathrm{argmax}_j H_{Mj}(\mathbf{x})$.

---

algorithms are discussed. This discussion helps us to decide which ensemble algorithm is most appropriate for rule generation.

In bagging, base learners should have large variance and small bias. For this purpose, deliberately overfitted base learners are used. That is, in bagging, performance of each base learner is not important, but the closeness of the average of base learners to the true model is a key of success. Hence, rules in each base learner do not have any meaning at all.

In boosting, the complexity of base learners determines the level of dominant interaction of the final ensemble model [4.14]. Let the true model (or the decision boundary) f^* be decomposed as

$$f^*(\mathbf{x}) = \sum_j f_j(x_j) + \sum_{j,k} f_{j,k}(x_j, x_k) + \sum_{j,k,l} f_{j,k,l}(x_j, x_k, x_l) + \cdots.$$

Then, the interaction level of boosting is limited by the tree size T, the number of terminal nodes. Namely, no interaction effects of level greater than $T-1$ are possible. With stumps as base learners, the decision boundary includes only the main effect; no interaction is permitted. With $T = 3$, two-variable interactions are allowed, and so on. Hence, building larger trees may increase the error rate of the final ensemble model because the resulting approximation involves high-order interactions among the features. Boosting is an additive model, and only the final ensemble model approximates the decision boundary, and the complexity of base learners controls the level of dominant interactions.

Base learners in CHEM have very clear roles. Each base learner explains the data from a different angle. In this view, we can say that CHEM combines strong learners at each angle to produce a stronger one. Hence, for rule generation, CHEM is the most appropriate ensemble algorithm, and hence used in the sequel.

4.3 Rule Generation

In this section, the process of rule generation from CHEM is described. We only consider a two-class problem, but multi-class problems can be treated similarly. For a given rule R, define $\mathtt{cond}(R), \mathtt{deci}(R), \mathtt{supp}(R)$, and $\mathtt{conf}(R)$ to be the conjunction of conditions of the input values, decision, support, and confidence respectively. In order to be meaningful for a given rule R, its support and confidence should be larger than given thresholds. Let $s_{\min}$ be the minimum allowable support. That is, a rule whose support is less than $s_{\min}$ is discarded. Also, let $c_{\min}$ be the minimum allowable confidence. If $\mathtt{conf}(R) < c_{\min}$, then the rule R is discarded. In summary, we are to generate rules whose supports are large and confidences are close to 1. The minimum allowable support $s_{\min}$ and the minimum allowable confidence $c_{\min}$ are supposed to be given from outside.

For the proposed algorithm, a group of interest (decision) should be chosen beforehand. Once the group of interest is chosen, rules which are meaningful (large support and high confidence) to explain the given group are generated. The first step of rule generation is to construct decision trees as base learners by use of CHEM. The typical procedure of constructing a decision tree consists of three parts : growing, pruning, and selection. This procedure, however, is not fitted for our purpose of rule generation. First, it needs too much computation. For selecting the best tree among the sequence of pruned trees, cross-validation (CV) is mostly used, which increases computation substantially. While the amount of computation for CV is not a problem for constructing a single best tree, it is really a matter of great concern for the ensemble model, since many trees should be constructed. The second problem of using the standard procedure for constructing a decision tree is that many terminal nodes may have observations less than $s_{\min}$. Because of these two reasons, we use a different method of constructing decision trees as base learners in CHEM. The growing part is the same as the standard one except that the tree stops growing once the conditions predefined have been met. The stopping conditions include the number of observations (support) and the probability of the group of interest (confidence) at the node. If either the support is less than or equal to $s_{\min}$ or the confidence is greater than $c_{\min}$, the growing stops.

Once many decision trees are constructed by CHEM, the next step is to extract useful and dissimilar rules. Let $\mathcal{R}$ be the set of candidate rules with the decision of interest, which is made by the following algorithm. Suppose there are M trees $T_1, \ldots, T_M$. For a given tree T, we let $|T|$ be the number of nodes in T and we abuse the notation slightly to let T be the set of all nodes in T. We assume that the nodes in T are assigned the numbers $1, \ldots, |T|$ from the root node to the terminal nodes.

Algorithm 11 Making the set of candidate rules

1: Initialization: $\mathcal{R} = \emptyset$.
2: Repeat for $m = 1, \ldots, M$;
   (a) For $i = 1, \ldots, |T_m|$
      If the node i of T_m has the support more than $s_{\min}$ and the confidence
      higher than $c_{\min}$, the rule corresponding to the node i is included to $\mathcal{R}$.
   (b) Continue for i.
3: Continue for m.

After the set $\mathcal{R}$ of the candidate rules are constructed, the next step is to eliminate repeated rules from $\mathcal{R}$. For this purpose, a measure of similarity of two given rules is necessary to compare the two rules in order to check whether the two given rules are similar or not. Let R_1 and R_2 be two rules. The similarity measure $M(R_1, R_2)$ is defined as follows. Suppose $\text{cond}(R_1) = \cap_{i=1}^{p}\{x_i \in A_{1i}\}$ and $\text{cond}(R_2) = \cap_{i=1}^{p}\{x_i \in A_{2i}\}$. Then

$$M(R_1, R_2) = \max_{i=1,\ldots,p} \frac{n(x_i \in A_{1i} \cap A_{2i})}{n(x_i \in A_{1i} \cup A_{2i})},$$

where $n(x_i \in A)$ is the number of observations in the data set such that $x_i \in A$. Now, the algorithm for eliminating similar rules for the set of candidate rules is as follows. Let m be the maximum allowable similarity. Suppose $\mathcal{R}$ has N rules.

Algorithm 12 Eliminating similar rules

1: Initialization: $\mathcal{O} = \emptyset$.
2: Sort the rules in $\mathcal{R}$ by their confidences.
3: Let the rule with the largest confidence be R_1, the rule with the second largest
   confidence be R_2, and so on up to R_N.
4: Repeat for $i = 1, \ldots, N$;
   (a) If $\mathcal{R}^* = \emptyset$, add R_i into $\mathcal{O}$.
   (b) Else add R_i into $\mathcal{O}$, if $M(R_i, R_k) < m$ for all $k = 1, \ldots, i - 1$.
5: Output the final set of rules $\mathcal{O}$.

In summary, the proposed rule generation algorithm has four input arguments – the group of interest, the minimum allowable support $s_{\min}$, the minimum allowable confidence $c_{\min}$, and the maximum allowable similarity m. Once these input arguments are given, the algorithm first makes the set of candidate rules and then eliminates repeated rules from it to yield the pruned set $\mathcal{O}$ of useful (large support and high confidence) and dissimilar (not repeated) rules as a final result.

4.4 Illustration

In this section, we illustrate the proposed rule generation algorithm with the German credit data, which is available at the UCI machine learning depository. The data has 21 variables, one for response and 20 for input features. The response variable has two values, "good" and "bad", for credit status.

We compare the rules from the proposed algorithm with those from the best single decision tree. We set the support 5% and the confidence 50% for the "bad" customer and 90% for the "good" customer. The different confidences are given because the proportions of the bad and good customers in the data set are unbalanced (30% for "bad" and 70% for "good"). Table 4.3 summarizes the rules generated from the best single tree and Table 4.4 lists all the rules generated from the proposed algorithm. First, it is obvious that the proposed algorithm generates much more diverse rules than the single decision tree. Also, all rules generated from the proposed algorithm contain more or less different information. Second, the decision tree can omit some simple but useful rules. For example, the rule "credit history is bad" is not detected by the decision tree while it is detected by the proposed algorithm. Third, the rules from the proposed algorithm include the rules from the decision tree. Fourth, the rules from the proposed algorithm are not mutually exclusive. That is, an observation can belong to more than one rule. Furthermore, the decisions of the rules to which a given observation belongs can be different. For example, a customer is classified as a "bad" customer for a small credit amount but at the same time he is classified as a "good" customer for a large checking amount. In this case, the final decision should be made carefully by compromising all information. If the single decision tree is used, only one rule would be available and therefore a hasty decision could be made. However, since the proposed algorithm may generate the different rules, we can avoid such a decision. In summary, the proposed algorithm provides a way of explaining a given data set from various angles while the single decision tree can portray the data from only one direction.

Table 4.3. Rules generated from the single decision tree

Credit Status	Rule	Support	Confidence
"bad"	Checking amount ≤ 200 Duration > 22.5 mo Saving amount < 500 Purpose is neither radio/TV nor business	16.8%	66.7%
"good"	Checking amount > 200 Purpose is either car, appliance, or business	5.7%	100.0%

Table 4.4. Rules generated from CHEM

Credit Status	Rule	Support	Confidence
"bad"	Checking amount ≤ 200 Duration > 22.5 mo Saving amount < 500 Purpose is neither radio/TV nor business	16.8%	66.7%
"bad"	Credit history is good Age < 37.5 yr Duration > 26.5 Saving amount < 500 Employment duration < 7 or unemployed	9.2%	59.8%
"bad"	Credit history is bad	8.9%	59.6%
"bad"	Age < 49.5 Credit amount < 5043.5 Duration > 15.5 Purpose is either car, furniture, or others	6.3%	58.7%
"good"	Checking amount > 200 Purpose is either car, appliance, or business	5.7%	100.0%
"good"	Credit amount > 2319 Other installment plan is blank Credit history is good 30.5 yr < Age < 38.5	5.2%	100.0%
"good"	Credit amount < 4049.5 People liable to maintenance < 1.5 Employment duration < 7 Present residence duration > 1.5 Checking amount > 200 Age > 39.5	7.7%	98.7%
"good"	Single or married males Purpose is either car, appliance, or business	6.0%	98.3%

4.5 Conclusions

In this chapter, we proposed an algorithm of rule generation based on the ensemble algorithm CHEM. We have seen that the proposed algorithm generates various useful nonexclusive rules which could be missed in the best single decision tree, which makes it possible to interpret the data from various angles.

Boosting can be used for rule generation. Cohen and Singer [4.8] developed an algorithm SLIPPER for this purpose. From each base learner, SLIPPER makes only one rule, and improvement of classification accuracy is not significant over the single decision tree. In contrast, the proposed rule generating algorithm generates various rules from each base learner and does not hamper the classification accuracy of the ensemble algorithm at all.

The rough set theory has been received much attention (see [4.26] and references therein). The proposed algorithm can be fine-tuned and extended further by employing rough set theories. In particular, a better method of constructing base learners than decision trees for rule generation can be developed. For this, a device to compare the quality of rules generated from two different algorithms should be developed.

Acknowledgement

This reserach is supported in part by US Air Force Research Grant F62562-02-P-0547 and in part by KOSEF through Statistical Research Center for Complex Systems at Seoul National University, Korea.

References

4.1 E. Bauer, R. Kohavi: An Empirical Comparison of Voting Classification Algorithms: Bagging, Boosting and Variants. Machine Learning, 36, 105-139 (1999)
4.2 L. Breiman: Bagging Predictors. Machine Learning, 24, 123-140 (1996)
4.3 L. Breiman: Arcing the Edge. Technical Report (1997)
4.4 L. Breiman: Arcing Classifiers. Annals of Statistics, 26, pp. 801-846 (1998)
4.5 L Breiman: Random Forests. Machine Learning, 45, pp. 5-32 (2001)
4.6 L. Breima, J. Friedman, J. Olshen, C. Stone: *Classification and Regression Trees* (Wadsworth, Belmont, CA 1984)
4.7 P. Bühlman, B. Yu: Boosting with the L2 Loss: Regression and Classification. Techincal Report (2001)
4.8 W. Cohen, Y. Singer: A Simple, Fast and Effective Rule Learner. *Proceedings of the Sixtennth National Conference on Artificial Intelligience, 1999*
4.9 A. Buja, W. Stuetzle: The Effect of Bagging on Variance, Bias and Mean Squared Error. Techincal Report (2000)
4.10 B. Efron: Bootstrap Methods: Another Look at the Lackknife. Annals of Statistics, 7, pp. 1-26 (1979)

4.11 J. Friedman: On Bias, Variance, 0-1 Loss and the Curse of Dimensionality. Journal of Data Mining and Knowledge Discovery, 1, pp. 55-77 (1997)

4.12 J. Friedman: Geedy Function Approximation: A Gradient Boosting Machine. Annals of Statistics (2001)

4.13 J. Friedman, P. Hall: On Bagging and Nonlinear Estimation. Techincal Report (1999)

4.14 J. Friedman, J.H. Hastie, R. Tibschirani: Additive Logistic Regression: A Statistical View of Boosting. Annals of Statistics, 38, pp. 337-374 (2000)

4.15 Y. Freund: Boosting a Weak Learning Algorithm by Majority. Information and Computation, 121, pp. 256-285 (1995)

4.16 Y. Freund, R. Schapire: A Decision-Theoretic Generalization of On-Line Learning and an Application to Boosting. Journal of Computer and System Science, 55, pp. 119-139 (1997)

4.17 Y. Kim: Convex Hulle Ensemble Machine. *Proceedings of 2002 IEEE International Conference on Data Mining* pp. 243-249 (2002)

4.18 J. R. Quinlan: *C4.5: Programs for Machine Learning* (Morgan Kaufmann, 1994)

4.19 J.R. Quinlan: Bagging, Boosting and C4.5. *Proceedings of the Thirteen National Conference an Artificial Intelligience* pp. 725-730 (1996)

4.20 G. Rätsch, T. Onoda, K.R. Müller: Soft Margins for AdaBoost. Machine Learning, 42, pp. 287-320 (2001)

4.21 R. Schapire: The Strength of Weak Learnability. Machine Learning, 5, 197-227 (1990)

4.22 R. Schapire: Using Output Codes to Boost Multiclass Learning Problems. *Proceedings of the Fourteenth International Conference on Machine Learning* pp. 313-321 (1997)

4.23 R. Schapire, Y. Freund, P. Bartlett, W. Lee: Boosting the Margin: A New Explanation for the Effectiveness of Voting Methods. Annals of Statistics, 26, pp. 1651-1686(1998)

4.24 R. Schapire, Y. Singer: Improved Boosting Algorithm Using Confidence-Rated Predictions. Machine Learning, 37, pp. 297-336 (1999)

4.25 L.G. Valiant: A Theory of the Learnable. Communication of the ACM, 27, pp. 1134-1142 (1984)

4.26 N. Zhong: Rough Stes in Knowledge Discovery and Data Mining. Journal of Japan Society for Fuzzy Theory and Systems, 13, pp. 581-591 (2001)

4.27 T.G. Dietterich: An experimental comparison of three methods for constructing ensembles of decision trees: Bagging, Boosting and Randomization. Machine Learning, 40, pp. 139-157 (2000)

4.28 G. Rätsch, T. Onoda, K.R. Müller: Soft Margins for AdaBoost. Machine Learning, 42, pp. 287-320 (2001)

5. Evaluation Scheme for Exception Rule/Group Discovery

Einoshin Suzuki

Department of Electrical and Computer Engineering,
Yokohama National University, Japan

Abstract

In this chapter, we propose an evaluation scheme for exception rule/group discovery. Exception rule/group discovery, which aims at discovering a set of rules/groups different from most of the rest, has been gaining increasing attention as a promising approach for discovering interesting patterns. We classify various exception rule/group discovery methods into those that employ domain knowledge and those that don't, and give a brief survey. We then propose an evaluation scheme that consists of seven evaluation criteria and apply the scheme to the methods.

5.1 Introduction

The goal of Knowledge Discovery in Databases (KDD) is to identify valid, novel, potentially useful, and ultimately understandable patterns in data [5.5]. Since these evaluation criteria are typically hard to measure, "interestingness", which represents their possibility, has been employed as an important criterion in KDD. Since interestingness contains ambiguity, however, there remains a long way to establish an effective evaluation method of KDD methods.

Consider expressions, each of which employs "attribute = value"; then, in KDD, a rule[1] typically represents a probabilistic constraint among expressions, and a group represents a set of examples which can be specified with an expression. Here, we consider an attribute c on which we focus our attention, such as a class in learning from examples, and an event $c = v$ in which v represents a value. Note that the premise of a rule of which the conclusion is $c = v$ can be regarded as a group most of which examples satisfy $c = v$. As we can see from this example, a rule and a group are deeply related to each other in KDD.

As Mitchell points out, optimization of decision is more important than prediction in data mining [5.16]. Throughout this chapter we intuitively define an exception as something which differs from most of the rest. An exception example that has been neglected as noise in a typical prediction problem due to its small influence on performance is often important in a decision problem. For instance, suppose the behavior of wealthy customers which represent 0.1% of the population is significantly different from the behavior of

---

[1] In this chapter we deal with probabilistic rules.

ordinary customers which represent 40% of the population. Neglecting the wealthy customers would typically yield good performance such as accuracy in a prediction problem, but might yield miserable performance such as profit in a decision problem.

Exception rule/group discovery, which aims at discovering a set of rules/ groups different from most of the rest, has been gaining increasing attention in KDD. As described earlier, one of the goals in this research domain is discovery of interesting rules/groups. However, since interestingness is ambiguous, effective evaluation methods for methodologies are far from being established. There are numerous methods in exception rule/group discovery. The reason for this situation comes not only from the fact that this research domain is important but also from the fact that its evaluation method remains unfixed. This situation is unfortunate from the viewpoint of the high demand for exception rule/group discovery from the application side, and the promising contribution of exception rule/group discovery to the goal of KDD.

Currently articles, each of which proposes an exception rule/group discovery method, typically evaluate the methods by a decrease in the number of discovered rules or success in a case study. A decrease in the number of discovered rules represents an important quantitative criterion which typically results in reduction of inspection costs for the output of data mining. However, this evaluation criterion has a serious deficiency that it neglects interestingness of discovered rules. Successes in case studies, on the other hand, can be an encouraging factor for one who wishes to apply the method to his/her problem, but it is desirable that the reasons for the successes be available. If the reasons are unavailable, it is desirable to know how other methods performed with the problem. Moreover, the evaluations of a success can be classified into best case analysis, which considers only the best results, and average case analysis, which considers all the results [5.30]. Although the former analysis fits the objectives of discovery more than the latter, it does not necessarily guarantee that the proposed method is effective in various domains. The latter analysis has the opposite strong points and weak points.

In order to evaluate various exception rule/group discovery methods in a uniform way, we propose an evaluation scheme in this chapter. The proposal of the evaluation scheme also contributes to deeply understand the existing methods. Sect. 5.2 briefly surveys existing methods. We propose the evaluation scheme in Sect. 5.3 and evaluate the methods, and Sect. 17.5 concludes.

5.2 Exception Rule/Group Discovery

5.2.1 Notice

As we have described in the previous section, we give an intuitive but general definition of our problem since there are many methods in exception rule/group discovery. We define an exception as something different from most of the rest. The objective of an exception rule/group discovery task is to obtain a set of exception rules/groups $p_1, p_2, \cdots, p_K$ given data D and additional information α. Here, α typically represents domain-specific criteria or knowledge, such as expected profit, and K represents a natural number. We also consider methods in which the user is involved in the discovery process.

From the next section, we briefly overview exception rule/group discovery methods. As we have described in the previous section, we omit classification methods, even if the classifier is represented by a set of rules[2], since we are interested in KDD methods for discovering interesting patterns. Although methods for Jumping Emerging Patterns [5.10] are devoted to discovery of exceptions, we skip them since their objectives are maximization of accuracy. Methods for maximizing accuracy of a specific class such as those in cost-sensitive learning [5.3], information retrieval, and information filtering are neglected as well. We exclude outlier detection methods such as [5.9] and conventional rule discovery methods such as [5.1, 5.21], since the obtained patterns are different from exception rules/groups.

We explain methods which employ domain knowledge in Sect. 5.2.2, and methods which do not employ domain knowledge in Sect. 5.2.3. It should be noted that the methods explained in the following sections are non-exhaustive.

5.2.2 Use of Domain Knowledge

Subjective Measure of Interestingness by Silberschatz and Tuzhilin. As a general discovery method which employs domain knowledge, Silberschatz and Tuzhilin [5.19, 5.20] proposed evaluation functions for subjective interestingness of a discovered pattern. In this work, they mainly consider "unexpectedness" and try to evaluate it assuming that the user describes pieces of knowledge as a set of beliefs. Since their methods are useful in discovering rules which are unexpected from the beliefs, the methods can be regarded as exception rule discovery methods.

Beliefs are classified in "soft beliefs", each of which degree is subject to change according to discovered patterns, and "hard beliefs", each of which always holds. The degree of a soft belief represents the extent to which the user believes the belief. Conditional probabilities in the Bayesian approach and certainty factors in Dempster-Shafer theory can be considered as natural candidates for such degrees. Consider the case of being supplied a new fact E when another fact ξ supports a soft belief α; then the degree $P(\alpha|E, \xi)$ associated with α is updated based on Bayes' theorem in the Bayesian approach:

$$P(\alpha|E, \xi) = \frac{P(E|\alpha, \xi)P(\alpha|\xi)}{P(E|\alpha, \xi)P(\alpha|\xi) + P(E|\overline{\alpha}, \xi)P(\overline{\alpha}|\xi)}.$$

---

[2] A detailed survey is available in [5.6].

The interestingness of a discovered pattern (e.g. rule), p, is represented by its degree of influence to the set of beliefs. If p contradicts a set of hard beliefs, p is regarded as interesting. If p contradicts a set B of soft beliefs, the interestingness $I(p, B)$ of p is given as follows [5.19]:

$$I(p, B) \equiv \sum_{\alpha \in B} \frac{|P(\alpha|p, \xi) - P(\alpha|\xi)|}{P(\alpha|\xi)}.$$

Silberschatz and Tuzhilin [5.20] consider a normalized weight, w_i, for each belief, α_i. Subjective measure of interestingness, though no rule discovery algorithm is considered, represents pioneering work for discovering exception rules based on domain knowledge.

Discovery of Unexpected Rules by Padmanabhan and Tuzhilin. Later, Tuzhilin, together with Padmanabhan, proposed a method for discovering unexpected rules [5.17]. This method can be regarded as discovering a kind of exception rules, since it defines unexpectedness as logical contradiction of a rule to a set of beliefs. Let XA represent $X \wedge A$; then, given a belief $X \rightarrow Y$, the method first discovers all rules $XA \rightarrow B$, each of which satisfies the conditions for association rules (support $\Pr(XAB)$ and confidence $\Pr(B|XA)$ are greater than their respective thresholds [5.1]), and B contradicts to Y. Next, the method obtains more general and more unexpected rules $X'A \rightarrow B$ by generalizing X to X'.

For instance, given a belief "professional $\rightarrow$ weekend" (i.e. a professional tends to go shopping during weekends rather than weekdays), the method might discover "December $\wedge$ professional $\rightarrow$ weekday" (i.e. a professional tends to go shopping during weekdays rather than weekends in December), and then "December $\rightarrow$ weekday" (i.e. one tends to go shopping during weekdays rather than weekends in December). This method has been applied to various data, including a marketing data set of 87,437 examples and 36 attributes. In the application to the marketing data set, approximately 600 rules were discovered based on 15 beliefs, and some of the discovered rules were judged truly interesting by domain experts.

Fuzzy Matching-based Interestingness by Liu et al. In the previous methods, domain knowledge is used only in a relatively simple manner. Liu et al. have proposed a method which ranks rules according to their interestingness based on fuzzy matching as a post-processing of rule discovery [5.12]. In this method, only a fixed attribute is allowed in the conclusions of rules, and the attribute is called a class attribute. In this case, a rule is called a classification rule, and predicts the value of the class attribute when its premise holds.

Since domain knowledge is given by a set of rules, each of which is associated with fuzzy patterns in this framework, the similarity and dissimilarity of a discovered rule to domain knowledge can be obtained by fuzzy matching. Intuitively, the similarity between a pair of rules is defined as a combination of similarities between the corresponding pair of the premises and the corresponding pair of the conclusions. Similarly, a pair of rules are judged dissimilar either when (1) the premises are similar but the conclusions are different; (2) the conclusions are identical, the attributes in the premises are the same, but their values are different; or (3) the conclusions are identical but the attributes in the premises are different. For the calculation method, please refer to [5.12]. Since there are several kinds of similarities and dissimilarities, there are several kinds of rankings of rules. In this method, a rule which is dissimilar to domain knowledge can be regarded as an exception rule by its definition. This method has been applied to data sets in the UCI Machine Learning Repository [5.2], and about ten subjects confirmed its effectiveness. The data sets were at most 9,366 examples and 35 attributes.

As Liu et al. admit, describing domain knowledge as a set of rules with fuzzy patterns is a difficult task for a subject. Data mining, however, represents a complex process which iterates pre-processing, pattern discovery, and postprocessing by trial and error guided by user interaction. Liu et al. observed a process in which users gradually describe their domain knowledge explicitly in a series of experiments with the "credit" data set (125 examples, 10 attributes, and 2 classes). This process can be considered as mandatory despite its difficulty.

General Impressions by Liu et al. In order to remedy this difficulty, Liu et al. have proposed a language in which a user can express domain knowledge as "impressions", and an algorithm which evaluates the interestingness of a discovered rule based on a set of impressions [5.11]. Similar to their method in the previous section, this method serves as a post-processing of rule discovery. Contrary to their previous method which employs relatively concrete domain knowledge such as "monthly salary $\geq$ \$5,000 $\rightarrow$ loan = approved", this method employs relatively abstract domain knowledge, such as "high monthly salary often implies loan approval". A user can employ impressions including the following formats, where C_j, a, and C_{sub} represent a class, an attribute, and a set of classes, respectively.

1. $a <\rightarrow C_j$: if the value of a is small then C_j is likely to occur.
2. $a >\rightarrow C_j$: if the value of a is large then C_j is likely to occur.
3. $a <<\rightarrow C_j$: if the value of a is within a certain range then C_j is likely to occur.
4. $a| \rightarrow C_{sub}$: there is a relation between a and C_{sub}.
5. $a[S] \rightarrow C_j$: if the value of a is an element of a set S then C_j is likely to occur.

It is possible to specify an impression which involves multiple attributes. For instance, an impression "savings $>$ $\wedge$ age $<<$ $\rightarrow$ loan = approved" represents that the loan is likely to be approved if the value of the saving attribute is large and the value of the age attribute is within a certain range. Furthermore, a user can specify an impression in which only a part of its conditions hold.

Discovered rules are ranked according to the result of their matching to a set of impressions. As in the method in the previous section, several kinds of rankings exist and a rule which violates a set of impressions can be regarded as an exception rule. This method has been applied to several data sets in the UCI Machine Learning Repository. Liu et al. [5.11] show an example of applying this method to the "credit" data set.

Gamberger and Lavrač's Subgroup Discovery. Subgroup discovery dates at least from W. Klösgen's EXPLORA [5.8]. Recently, Gamberger and Lavrač have proposed a subgroup discovery method which considers a specific class under the guidance of a domain expert [5.14, 5.15]. According to their definition, a subgroup discovery problem is: "given a population of individuals and a property of those individuals we are interested in, find population subgroups that are statistically 'most interesting', e.g., are as large as possible and have the most unusual statistical (distributional) characteristics with respect to the property of interest". This definition shows that this method represents an exception group discovery method.

We represent the target class as "Pos" in the confusion matrix

	Pos	Neg
P	TP	FP
N	FN	TN

where Pos and Neg represent actual classes, and P and N represent predicted classes. Consider the examples which are predicted as belonging to the target class; then, in the table, TP and FP represent the number of examples which actually belong to the target class and which do not belong to the target class, respectively. This method employs $q_g = TP/(FP + g)$ as an evaluation criterion of a subgroup, where g is called a generalization parameter and its value is specified by a user. This method performs beam search using q_g as its evaluation criterion. The user can modify the value of g as well as select a set of attributes in order to discover an interesting subgroup. When the value of g is large, the method tends to discover general subgroups, each of which consists of many examples and includes examples of other classes. The opposite situation happens when the value of g is small. This method can describe discovered results using visualization techniques as well as statistical methods, and can discover a set of subgroups that show diversity.

In discovering a set of rules/groups with a target attribute, it is common to consider benefits associated with TP and costs associated with FP. Therefore, $q_c = TP - c*FP$ seems intuitive and appropriate as an evaluation criterion, where a parameter c is introduced in order to evaluate the benefits and the costs in a uniform way. According to them, however, q_g is superior to q_c since the latter, employed by beam search, tends to discover a subgroup of which TP is small.

This method has been applied to an early detection problem of coronary heart disease (CHD). The data set consists of 238 patients, and the target class represents those with CHD. Gamberger and Lavra [5.14] show results discovered with the help of experts, together with several lessons learned from the endeavor.

5.2.3 Hypothesis-Driven Discovery

Suzuki and Shimura's MEPRO. We have defined our exception rule[3] as a rule which contradicts a strong rule, and proposed several hypothesis-driven methods, each of which obtains a set of pairs of an exception rule and a strong rule [5.22, 5.23, 5.24, 5.25, 5.29, 5.30, 5.31]. Here, a strong rule represents a rule which holds for many examples with high probability. The input to our approach is a table-formatted data set with n examples, each of which is described by m propositional attributes. Let an atom be an event representing a value or a range assignment to an attribute; then, a discovered pattern is represented by a rule pair $r(x, x', Y, Z)$:

$$r(x, x', Y, Z) \equiv (Y \rightarrow x, Y \wedge Z \rightarrow x'), \tag{5.1}$$

where each of Y and Z represents a conjunction of atoms, and each of x and x' represents an atom. Here x and x' have the same attribute but different values. We call $Y \rightarrow x$ a strong rule, $Y \wedge Z \rightarrow x'$ an exception rule, and $Z \rightarrow x'$ a reference rule.

Smyth and Goodman [5.21], in their rule discovery system ITRULE, have proposed the quantity $J(x; y)$ of information compressed by a rule $y \rightarrow x$ as a measure, J, of interestingness [5.21]:

$$J(x; y) = \Pr(y) \, j(x; y) \tag{5.2}$$

$$\text{where } j(x; y) = \Pr(x|y) \log_2 \frac{\Pr(x|y)}{\Pr(x)} + \Pr(\overline{x}|y) \log_2 \frac{\Pr(\overline{x}|y)}{\Pr(\overline{x})}. \tag{5.3}$$

We initially defined our measure of interestingness of a rule pair as a product of J-measure of a strong rule and J-measure of an exception rule [5.22]. We have then derived an upper bound of our measure and proposed an efficient algorithm which performs a branch-and-bound search based on it. Our method MEPRO has been applied to the "mushroom" data set (8,124 examples, 22 attributes, two classes) and the "Congressional Voting Records" data set (435 examples, 16 attributes) in the UCI Repository and discovered exception rules, each of which is interesting at least from the statistical point of view. The upper bound turned out to be useful since it improved speed more than five times for the "Congressional Voting Records" data set. We have introduced several probabilistic constraints since, when $\Pr(x'|Z)$ is large, our exception rule exhibits low unexpectedness [5.23]. We, together with Kodratoff, have considered unexpectedness from a different perspective and proposed a novel probabilistic criterion which mainly considers the number of counter-examples [5.25].

Suzuki's Simultaneous Reliability Evaluation. We then proposed a method in which we specify thresholds θ_1^S, θ_1^F, θ_2^S, θ_2^F, and θ_2^I for probabilistic criteria of a rule pair. Since a rule pair discovered from 10,000 examples exhibits different reliability from another rule pair discovered from 100

---

[3] Our exception rule represents a subclass of an exception rule defined in Sect. 9.1.

examples, it is inappropriate to use a ratio $\widehat{\Pr}(\cdot)$ in a data set as a probabilistic criterion. Therefore, we considered a true probability $\Pr(\cdot)$ for each probabilistic criterion, and obtained a set of rule pairs, each of which satisfies discovery conditions with the significance level δ [5.24, 5.31]. In the following, $\text{MIN}(x, y)$ and $\text{MAX}(x, y)$ represent the smaller and the larger of x and y, respectively.

$$
\begin{aligned}
&Pr[Pr(Y) \geq \theta_1^S, Pr(x|Y) \geq MAX(\theta_1^F, \widehat{Pr}(x)), Pr(YZ) \geq \theta_2^S, \\
&Pr(x'|YZ) \geq MAX(\theta_2^F, \widehat{Pr}(x')), \\
&Pr(x'|Z) \leq MIN(\theta_2^I, \widehat{Pr}(x'))] \geq 1 - \delta.
\end{aligned}
\tag{5.4}
$$

Calculating (5.4) is difficult due to two reasons. First, obtaining a value of a true probability requires assumptions. Second, calculating (5.4) for a rule pair numerically is time-consuming, since (5.4) contains five true probabilities. This method overcomes these difficulties by obtaining analytical solutions based on simultaneous reliability estimation of true probabilities. Let the number of examples in the data set be n, and $(nPr(xYZ),\ nPr(x'YZ),\ nPr(\overline{x}\overline{x'}YZ),\ nPr(xY\overline{Z}),\ nPr(\overline{x}Y\overline{Z}),\ nPr(\overline{x'}YZ),\ \text{and}\ nPr(x'\overline{Y}Z))$ follow a multi-dimensional normal distribution; then (5.4) is equivalent to (5.5) – (5.9):

$$
G(Y,\ \delta,\ k)\widehat{\Pr}(Y) \geq \theta_1^S,
\tag{5.5}
$$

$$
F(Y,\ x,\ \delta,\ k)\widehat{\Pr}(x|Y) \geq \theta_1^F,
\tag{5.6}
$$

$$
G(YZ,\ \delta,\ k)\widehat{\Pr}(YZ) \geq \theta_2^S,
\tag{5.7}
$$

$$
F(YZ,\ x',\ \delta,\ k)\widehat{\Pr}(x'|YZ) \geq \theta_2^F, and
\tag{5.8}
$$

$$
F'(Z,\ x',\ \delta,\ k)\widehat{\Pr}(x'|Z) \leq \theta_2^I,
\tag{5.9}
$$

$$
\text{where}\quad G(a,\ \delta,\ k) \equiv 1 - \beta(\delta, k)\sqrt{\frac{1 - \widehat{\Pr}(a)}{n\widehat{\Pr}(a)}},
\tag{5.10}
$$

$$
F(a,\ b,\ \delta,\ k) \equiv 1 - \beta(\delta, k)\varphi(a, b),
$$
$$
F'(a,\ b,\ \delta,\ k) \equiv 1 + \beta(\delta, k)\varphi(a, b), and
\tag{5.11}
$$

$$
\varphi(a,\ b) \equiv \sqrt{\frac{\widehat{\Pr}(a) - \widehat{\Pr}(a, b)}{\widehat{\Pr}(a, b)\{(n + \beta(\delta, k)^2)\widehat{\Pr}(a) - \beta(\delta, k)^2\}}},
\tag{5.12}
$$

where $\beta(\delta, k)$ represents a positive value which is related to the confidence region and is obtained by numerical integration. k represents the number of true probabilities each of which is satisfied by at least an example in the data set minus 1. This author has also proposed an efficient discovery algorithm based on pruning.

The effectiveness of the method has been confirmed by experiments with the "census" data set (48,842 examples, 14 attributes, two classes) and the "mushroom" data set in the UCI Repository. The pruning method turned out to be highly efficient, and improved speed 70 times for the "mushroom" data set.

Application to Meningitis Data by Suzuki and Tsumoto. Here, we show effectiveness of the method in the previous section in terms of discovery of interesting rules. A data mining contest represents a systematic attempt to evaluate discovery methods of participants with a common data set or a common problem. A data mining contest can be considered important in evaluating various data mining methods.

We have participated in various data mining contests with several methods such as the one presented in the previous section. We briefly explain our endeavor in a contest with the meningitis data set [5.27]. The data set consists of 140 patients, each of whom is described by 38 attributes, and has been made public as a benchmark problem to the data mining community. Our method has discovered 169 rule pairs from a preprocessed version of this data set. These rule pairs were inspected by Tsumoto et al. (Dr. Tsumoto is a domain expert), and each rule pair was assigned a five rank score for the following evaluation criteria [5.32]:

- validness: the degree that the discovered pattern fits domain knowledge
- novelty: the degree that the discovered pattern does not exist in domain knowledge
- usefulness: the degree that the discovered pattern is useful in the domain
- unexpectedness: the degree that the discovered pattern partially contradicts domain knowledge

For the scores, five and one represent the best score and the worst score, respectively. We show the results classified by the attributes in the conclusions in Table 5.1.

Table 5.1. Average performance of the proposed method with respect to attributes in the conclusion. The column "#" represents the number of discovered rule pairs

attribute	#	validness	novelty	unexpectedness	usefulness
(all)	169	2.9	2.0	2.0	2.7
CULTURE	2	1.0	1.0	1.0	1.0
C_COURSE	1	1.0	1.0	1.0	1.0
RISK	1	1.0	1.0	1.0	1.0
CT_FIND	36	3.3	3.0	3.0	3.2
EEG_FOCUS	11	3.0	2.9	2.9	3.3
Course (G)	8	1.8	2.0	2.0	1.8
FOCAL	18	3.1	2.2	2.7	3.0
LOC_DAT	11	2.5	1.8	1.8	2.5
Diag2	72	3.0	1.1	1.1	2.6
CULT_FIND	4	3.3	4.0	4.0	3.5
KERNIG	4	2.0	3.0	3.0	2.0
SEX	1	2.0	3.0	3.0	2.0

From Table 5.1, we see that the average scores of the discovered rule pairs are high for several attributes in the conclusions. As Tsumoto et al. admit, this is due to the fact that the structure of a rule pair is useful for discovery of interesting patterns. According to him, our method discovered the most interesting results in the data mining contest [5.32].

Our method has been also applied to the 1994 bacterial test data set (20,919 examples, 135 attributes, 2 classes) [5.28]. We have found that discovery of interesting patterns from the data set requires further preprocessing that considers distribution of attribute values and cause and effect relationships. However, this application shows that our method is adequate in terms of efficiency in exception rule mining from a relatively large-scale data set.

Suzuki's Threshold Scheduling. The method in Suzuki's simultaneous reliability evaluation requires appropriate specification of values for five thresholds, θ_1^S, θ_2^S, θ_1^F, θ_2^F, and θ_2^I. A strict specification for a data set with a small number of exception rules can result in no discovery of exception rules. On the other hand, a loose specification for a data set with a large number of exception rules can result in discovery of many exception rules and the computation process is typically time-consuming. These problems come from the fact that there are five evaluation criteria, $\widehat{\Pr}(Y)$, $\widehat{\Pr}(YZ)$, $\widehat{\Pr}(x|Y)$, $\widehat{\Pr}(x'|YZ)$, and $\widehat{\Pr}(x'|Z)$, of a discovered pattern.

For this problem, we have first invented a data structure which can manage discovered patterns with multiple criteria [5.26]. The data structure is based on a height-balanced tree. In our method, a height-balanced tree is assigned to each evaluation criterion, and a node of a tree represents a pointer to a discovered pattern. Then we have proposed an algorithm which updates thresholds and discovers at most η rule pairs [5.26]. In this method, each time an $(\eta+1)$th rule pair is discovered, the worst rule pair in terms of the current criterion is deleted and the value of the criterion is updated according to the deleted rule pair. Each time this process occurs, the current criterion is replaced by another criterion. With experiments using four data sets, we have confirmed that specification of values for five thresholds can be replaced by an easier procedure of specifying the largest number of discovered rule pairs.

Meta Pattern by Suzuki and Żytkow. In Suzuki and Shimora's ME-PRO and Suzuki's simultaneous reliability evaluation, the discovered pattern is restricted to Eq. (5.1). Suzuki and Żytkow classified exception/deviation structures for discovery of interesting patterns based on a meta pattern and proposed an efficient algorithm which discovers all structures [5.29].

In the approach, an exception/deviation structure is defined as a rule triple $t(y, x, \alpha, \beta, \gamma, \delta)$, which represents the meta pattern, using literals x, y, z. A strong rule, an exception rule, and a reference rule are defined as $y \rightarrow x$, $\alpha \nrightarrow \beta$, and $\gamma \rightarrow \delta$, respectively.

$$t(y, x, \alpha, \beta, \gamma, \delta) = (y \rightarrow x, \ \alpha \nrightarrow \beta, \ \gamma \rightarrow \delta), \tag{5.13}$$

where each of α, β, γ, and δ represents a meta variable which is instantiated by variables x, y, and z, resulting in definitions of various exception/deviation

type1   type2   type3   type4   type5   type6   type7   type8   type9   type10   type11

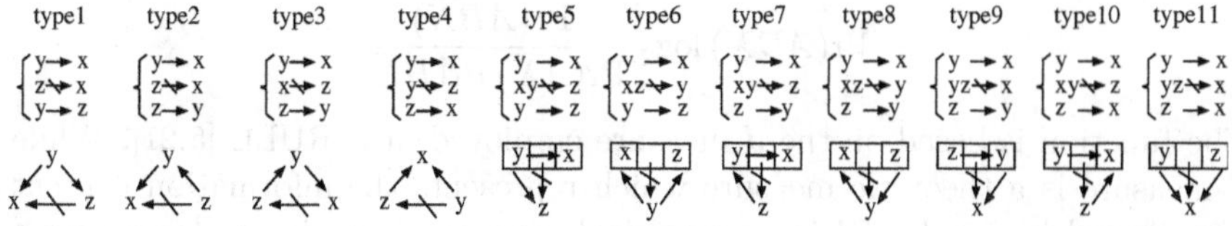

$$\begin{cases} y \to x \\ z \to x \\ y \to z \end{cases} \quad \begin{cases} y \to x \\ z \to x \\ z \to y \end{cases} \quad \begin{cases} y \to x \\ x \to z \\ z \to y \end{cases} \quad \begin{cases} y \to x \\ y \to z \\ z \to x \end{cases} \quad \begin{cases} y \to x \\ xy \to z \\ y \to z \end{cases} \quad \begin{cases} y \to x \\ xz \to y \\ z \to y \end{cases} \quad \begin{cases} y \to x \\ xy \to z \\ z \to y \end{cases} \quad \begin{cases} y \to x \\ xz \to y \\ z \to y \end{cases} \quad \begin{cases} y \to x \\ yz \to x \\ z \to y \end{cases} \quad \begin{cases} y \to x \\ xy \to z \\ z \to x \end{cases} \quad \begin{cases} y \to x \\ yz \to x \\ z \to x \end{cases}$$

Fig. 5.1. Classification of the rule triple. A rectangle on the top center for a rule triple represents a conjunction of literals in the top right and left. For instance, the three rectangles in type 11 represent, from the left to the right, "y", "$y \wedge z$", and "z"

structures. Here $y \to x$ represents a rule and shows that $\widehat{\mathrm{Pr}}(y)$ and $\widehat{\mathrm{Pr}}(x|y)$ are greater than their respective thresholds. On the other hand, $\alpha \not\to \beta$ represents a negative rule and shows that $\widehat{\mathrm{Pr}}(\alpha)$ is greater than its threshold and $\widehat{\mathrm{Pr}}(\beta|\alpha)$ is smaller than its threshold.

Under appropriate assumptions, our exception/deviation structures can be classified into the eleven structures shown in Fig. 5.1. Our algorithm efficiently searches rule triples with pruning. Experiments using 15 UCI data sets show that the pruning method is effective and exception/deviation structures which seem interesting (types 2, 5, 8, 9, and 11) are rarely discovered.

Relative Interestingness Measure by Hussain et al. Hussain et al. proposed an information-theoretic measure based on Suzuki and Shimora's MEPRO [5.7]. This method discovers a set of the following simplified rule pairs where each of A, B, and X represents a conjunction of "attribute = value" and $\overline{X}$ represents the negation of X.

$$A \to X \quad \text{common sense rule (strong pattern)} \tag{5.14}$$
$$\text{(hight support, high confidence)}$$

$$AB \to \overline{X} \quad \text{exception rule (weak pattern)} \tag{5.15}$$
$$\text{(low support, high confidence)}$$

$$B \to \overline{X} \quad \text{reference rule} \tag{5.16}$$
$$\text{(low support and/or low confidence)}$$

The measure RI represents relative interestingness of an exception rule $AB \to X$ to rules $A \to X$ and $B \to X$:

$$RI = RI_c^{AB} + RI_s^{AB} , \tag{5.17}$$

$$\text{where } RI_c^{AB} = \Pr(X|AB) \log_2 \frac{\Pr(X|AB)}{\Pr(X|A)\Pr(X|B)}$$

$$+ \Pr(\overline{X}|AB) \log_2 \frac{\Pr(\overline{X}|AB)}{\Pr(\overline{X}|A)\Pr(\overline{X}|B)} \quad and$$

$$RI_s^{AB} = \Pr(ABX) \log_2 \frac{\Pr(ABX)}{\Pr(AX)\Pr(BX)}$$

$$+ \Pr(AB\overline{X}) \log_2 \frac{\Pr(AB\overline{X})}{\Pr(A\overline{X}) \Pr(B\overline{X})} \ .$$

The function is based on the J-measure employed in ITRULE [5.21]. While J-measure is a theoretic measure which represents the information content compressed by a rule, RI is an empirical measure. Hussain et al. propose a search algorithm which first obtains all common sense rules, and then promising exception rules [5.7].

Their paper shows experimental results using the "credit" data set and the "mushroom" data set in the UCI Machine Learning Repository [5.2]. According to the results, the method discovers almost the same rule pairs as Suzuki and Shimora's MEPRO, but their rankings are different.

Direction Setting Rules by Liu et al. Liu et al. have proposed a method for discovering a set of interesting rules based on statistical tests [5.13]. They modified a condition, $\widehat{\Pr}(x|y) \geq \theta_c$, of the conditions for association rule $y \rightarrow x$ discovery to existence of correlation of the premise and the conclusion based on a χ^2 test. A rule which satisfies the other condition, $\widehat{\Pr}(xy) \geq \theta_s$, is classified into the following three.

1. Positive correlation (direction 1): if x and y are correlated, and $\widehat{\Pr}(xy) / (\widehat{\Pr}(x)\widehat{\Pr}(y)) > 1$.
2. Negative correlation (direction -1): if x and y are correlated, and $\widehat{\Pr}(xy) / (\widehat{\Pr}(x)\widehat{\Pr}(y)) < 1$.
3. No correlation (direction 0): if x and y are not correlated.

This method deals with the problem of discovering interesting rules $yz \rightarrow x$ of which yz and x are positively correlated. If the directions of $y \rightarrow x$, $z \rightarrow x$, and $yz \rightarrow x$ are either of (1) 0, 0, 1, (2) -1, -1, 1, (3) -1, 0, 1, and (4) -1, 1, 1, then $yz \rightarrow x$ is considered to be possibly interesting. Liu et al. define a direction setting rule as a rule which satisfies one of the above four conditions, and propose an algorithm for discovering a set of direction setting rules. A direction setting rule can be regarded as a kind of exception rule. Experiments using 30 data sets show that the method can reduce the number of discovered rules considerably.

Yugami et al.'s DIG. Yugami et al. proposed to discover a set of exception rules each of which shows a high confidence value though simpler rules do not show high confidence values [5.33]. For instance, this method DIG (Discover Interesting rules with Grouping attribute values) discovers the following rule.

$$\text{cap\_color} \in \{\text{brown, red}\} \wedge \text{stalk\_root} = \text{bulbous} \rightarrow \text{edible} :$$
$$\text{confidence value } 100\%$$
$$\text{where } \text{cap\_color} \in \{\text{brown, red}\} \rightarrow \text{edible} : \text{confidence value } 50\%$$
$$\text{stalk\_root} = \text{bulbous} \rightarrow \text{edible} : \text{confidence value } 51\%$$

As Yugami admits, his exception rule is highly related to Suzuki's simultaneous reliability evaluation, but should be interpreted differently since each atom in the premise has little influence in predicting the conclusion.

DIG employs the relative degree of the confidence of the rule compared to the case in which no interaction exists among conditions in the premise as a measure of interestingness. As we can see from the above example, the number of candidates of a rule is large since a premise of a rule is represented by a conjunction of "attribute ∈ a set of values". Due to an efficient algorithm, DIG obtains all rules each of which satisfies user-specified conditions on the number of conditions in the premise, support, confidence, and the degree of interestingness.

DIG has been applied to the "mushroom", "satimage" (6,435 examples, 36 attributes), and "letter recognition" (20,000 examples, 16 attributes) data sets in the UCI Machine Learning Repository [5.2]. The results show that DIG can discover rules, each of which is at least interesting from the statistical viewpoint, in practical time.

Peculiarity Rules by Zhong et al. Zhong et al. have proposed a method for discovering a set of peculiarity rules, each of which holds for a small number of examples [5.34]. Such an example is called peculiar data. Since a different tendency typically holds for peculiar data, a peculiarity rule can be considered as a kind of an exception rule.

This method first evaluates the degree of peculiarity of a value x_i of an attribute by a peculiarity factor $PF(x_i)$.

$$PF(x_i) = \sum_{j=1}^{n} \sqrt{N(x_i, x_j)} \ , \tag{5.18}$$

where $N(x_i, x_j)$ represents a conceptual distance between an attribute value x_i and an attribute value x_j, and can be obtained by domain knowledge. If no domain knowledge is available, a conceptual distance between two values of a continuous attribute is given by $N(x_i, x_j) = \|x_i - x_j\|$. This method then collects examples, each of whose peculiarity factor is large as peculiar data. Finally, this method discovers a set of peculiarity rules based on relevance between peculiar data. Zhong et al. have proposed a novel discovery method from multiple databases, but we omit it since it is outside the scope of this chapter.

This method has been applied to Japan census data, amino acid sequence and thermodynamic measurement data set (35 examples, 238 attributes), and chronic hepatitis data. The results show that this method has discovered several patterns from the amino acid sequence and thermodynamic measurement data set that are interesting for a domain expert, and several examples that are useful in further analysis by a domain expert.

5.3 Evaluation Scheme and Classification of Existing Methods

The evaluation scheme, which is described below, is invented from our experience on exception rule discovery. We admit that it is influenced by our bias to some extent. In the evaluation, a "low" or a "middle" does not imply ineffectiveness of the corresponding method. This scheme just serves as a basis for selecting an appropriate method for a problem.

We also evaluate several methods presented in the previous section based on this scheme. Silberschatz and Tuzhilin's subjective measure of interestingness (abbreviation: subjective measure), Padmanabhan and Tuzhilin's discovery of unexpectedness rules (abbreviation: unexpectedness rule), Liu et al.'s interestingness based on fuzzy matching (abbreviation: fuzzy), Liu et al.'s general impression (abbreviation: general impression), Gamberger and Lavrač's subgroup discovery (abbreviation: subgroup), Suzuki and Shimura's MEPRO (abbreviation: MEPRO), Suzuki's simultaneous reliability evaluation (abbreviation: simultaneous reliability evaluation), Hussain et al.'s relative interestingness measure (abbreviation: relative interestingness measure), Liu et al.'s direction setting rule (abbreviation: direction setting rule), Yugami et al.'s DIG (abbreviation: DIG), and Zhong et al.'s peculiarity rule discovery (abbreviation: peculiarity rule).

5.3.1 Generality

When we try to apply a data mining method to a problem, it is a crucial matter if the method is not applicable or the application requires considerable effort. The former case occurs when several conditions that are considered necessary for the method do not hold in the problem; for instance, a method which discovers knowledge based on expected profit cannot be applied to a problem which has no such notions. The latter case typically occurs for data mining methods which employ domain knowledge, since such methods typically require efforts to encode domain knowledge from scratch.

Generality represents the possibility that a data mining method can be applied to a problem with a small amount of effort, and has been chosen as an evaluation criterion. We have settled the categories of generality as low (cannot be applied to many domains), middle (employs domain knowledge), and high (no domain knowledge required). Note that a method which employs domain knowledge belongs to the category "high" if it can be executed without domain knowledge. It should be noted that a method which belongs to the category "high" typically discovers knowledge based on probabilities related to the data set.

According to our definition, no methods belong to "low". Subjective measure, unexpected rule, fuzzy, and general impression belong to "middle". Subgroup, MEPRO, simultaneous reliability evaluation, relative interestingness measure, direction setting rule, DIG, and peculiarity rule belong to "high".

5.3.2 Break of Monotonicity in Discovered Patterns

If an exception rule is represented by $yz \to x$, the value of $\Pr(x|yz)$ typically differs from those of $\Pr(x|y)$ and $\Pr(x|z)$ considerably. For instance, when the value of $\Pr(x|yz)$ is nearly 1, the values of $\Pr(x|y)$ and $\Pr(x|z)$ are often nearly 0. In such a case, the exception rule $yz \to x$ is said to break the monotonicity of the rules $y \to x$ and $z \to x$ [4].

Since an exception rule/group typically exhibits different statistics from a normal rule/group, we have chosen this break of monotonicity in order to measure such degrees. We have settled the categories of this measure as low (unknown), middle (one direction), and high (multiple directions). Here unknown does not imply that there is no break of monotonicity but represents the fact that there might be no break of monotonicity. One direction means that at least either the value of $\Pr(x|y)$ or the value of $\Pr(x|z)$ largely differs from the value of $\Pr(x|yz)$. By multiple directions, we mean that, for an exception rule $y_1 y_2 \cdots y_K \to x$, at least two values of $\Pr(x|y_i)$ largely differ from the value of $\Pr(x|y_1 y_2 \cdots y_K)$. It should be noted that a method which belongs to the category "low" typically discovers knowledge based on a measure or a process different from break of monotonicity. On the other hand, a method which belongs to the category "middle" or "high" typically discovers knowledge based on break of monotonicity.

For instance, the rule "antibiotics ∧ staphylococci → death" can be considered as being discovered by a method in any category since neither antibiotics nor staphylococci are related to death [5]. On the other hand the rule "antibiotics ∧ Ebola fever → death" can be discovered by a method in "middle" but not "high", since Ebola fever is highly related to death. This rule can be thus considered as a touchstone for exception rule/group discovery methods. A method which belongs to the category "high" can be considered effective since a rule discovery method typically outputs a glut of patterns.

According to our definition, subjective measure, unexpectedness rule, relative interestingness measure, and peculiarity rule belong to "low". Fuzzy, general impression, subgroup, MEPRO, and direction setting rule belong to "middle". MEPRO belongs to the category "high" if it is employed with its probabilistic constraints. Simultaneous reliability evaluation and DIG belong to "high".

5.3.3 Evaluation of Reliability

As described in Sect. 5.2.3, a pattern discovered from 100 examples and a pattern discovered from 10,000 examples differ in their reliability. The latter can be considered as relatively reliable, but the former might be an unreliable pattern which has occurred by chance. This kind of problem can be typically resolved by a statistical approach.

---

[4] Thanks to A. Tuzhilin.

[5] This rule was related to the discovery of hospital infection.

Since an exception rule/group holds for a small number of examples, reliability evaluation is considered important. A pattern has multiple evaluation criteria such as generality, $\Pr(y)$, and accuracy, $\Pr(x|y)$, for a rule $y \to x$. We have settled the categories of this measure as low (none), middle (done for a criterion), and high (done for multiple criteria). For rule discovery, a method which evaluates the reliability of generality based on Chernoff bound has been proposed [5.18]. Reliability evaluation for multiple criteria is difficult since it involves estimation of a confidence region for simultaneous occurrence of probabilistic variables.

According to our definition, subjective measure, unexpectedness rule, fuzzy, general impression, subgroup, MEPRO, relative interestingness measure, direction setting rule, DIG, and peculiarity rule belong to "low". No method belongs to "middle" and simultaneous reliability evaluation belongs to "high".

5.3.4 Search Range

Since an exception rule/group holds for a small number of examples, their candidates are typically large in number, causing combinatorial explosion with its large search range. However, it is empirically known that an exception rule/group with good statistics is rare, and nothing is discovered with a small search range.

We consider this fact important and have chosen search range as an evaluation criterion. We have settled the categories of this measure as low (partial), middle (all), and high (all, but adjusted by the size of the discovery problem). It should be noted that a method which belongs to the category "low" typically employs heuristic search. Although such a method can be applied to a large-scale problem, it might discover nothing and the core process might be so inefficient that it cannot employ an exhaustive search process. A method which belongs to the category "middle" or "high" typically employs a speed-up technique, such as pruning or branch-and-bound, so that it can employ exhaustive search even for a large-scale problem. It is straightforward for such a method to replace exhaustive search with heuristic search.

According to our definition, subgroup belongs to "low". Unexpected rule, MEPRO, simultaneous reliability evaluation, relative interestingness measure, and DIG belong to "middle". Note that simultaneous reliability evaluation belongs to "high" if it is employed with the threshold scheduling. Direction setting rule and peculiarity rule belong to "high". Note that subjective measure, fuzzy, and general impression are out of the scope of this criterion since each of them represents a post-processing of rule discovery.

5.3.5 Interpretation of the Evaluation Measure

If the evaluation measure had a clear interpretation, it would be easy to explain discovered exception rules/groups with it. An evaluation measure can be classified into theoretical and empirical, and the latter is often difficult to interpret. Since the meaning of discovered patterns is an important issue, we have chosen interpretation of the evaluation measure as an evaluation criterion. We have settled the categories of this measure as middle (empirical) and high (interpretable).

According to our definition, unexpected rule, fuzzy, subgroup, MEPRO, relative interestingness measure, and peculiarity rule belong to "middle". Subjective measure, general impression, simultaneous reliability evaluation, direction setting rule, and DIG belong to "high".

5.3.6 Use of Domain Knowledge

In discovering interesting exception rules/groups using domain knowledge, it is desirable that the use necessitates less effort. However, we must be prudent since the use might hinder us from discovering interesting exception rules/groups. In order to handle these aspects, we have chosen the use of domain knowledge as an evaluation criterion.

We include intervention of a domain expert to a discovery system in this criterion. The importance of such intervention is widely recognized in the KDD community, and this fact applies to discovery of interesting exception rules/groups. We assume that such intervention can be realized by use of domain knowledge, such as pruning in a search process, and exclude other kinds of intervention from our evaluation.

We have settled the categories of this measure as low (usage is possible), middle (usage is direct), and high (usage is sophisticated). Note that even a method presented in the previous section as hypothesis-driven can employ domain knowledge such as in pruning.

According to our definition, MEPRO, simultaneous reliability evaluation, relative interestingness measure, direction setting rule, and DIG belong to "low". Unexpected rule and subgroup belong to "middle". Subjective measure, fuzzy, general impression, and peculiarity rule belong to "high".

5.3.7 Successes in Real Applications

As we described in Sect. 9.1, successes in case studies can be an encouraging factor for one who wishes to apply the method to his/her problem, but it is desirable that the reason for the successes be also available. If the reason is unavailable, it is desirable to know how other methods performed with the problem. It should be noted that simply observing decrease in the number of discovered patterns, even in a real application, cannot be regarded as a success. An application is judged a success if a set of interesting patterns is discovered for a domain expert. We have established the categories of this measure as low (none), middle (one domain), and high (multiple domains).

According to our definition, subjective measure, general impression, ME-PRO, relative interestingness measure, direction setting rule, and DIG belong to "low". Unexpected rule, subgroup, fuzzy, and simultaneous reliability evaluation belong to "middle". Peculiarity rule belongs to "high".

5.4 Conclusions

We have proposed seven evaluation criteria for exception rule/group discovery methods, and applied them to eleven methods. As we described previously, these criteria are not meant to rank these methods but can serve as a basis for selecting an appropriate method for a problem.

Data mining has originated from discovery of rules/groups related to a large number of examples, but increasing attention has been paid to discovery of exception rules/groups. Since interestingness is regarded as an important target of data mining, this trend can be considered as natural. It is our wish that the evaluation scheme proposed in this chapter serves positively for this trend.

Acknowledgement

This work was partially supported by the grant-in-aid for scientific research on priority area "Active Mining" from the Japanese Ministry of Education, Culture, Sports, Science, and Technology.

References

5.1 R. Agrawal et al.: Fast Discovery of Association Rules, Advances in Knowledge Discovery and Data Mining (AAAI/MIT Press, Menlo Park, Calif. 1996) pp. 307-328
5.2 C. Blake, C.J. Merz, E. Keogh: UCI Repository of Machine Learning Databases. http://www.ics.uci.edu/~mlearn/MLRepository.html, Univ. of Calif. Irvine, Dept. Information and CS (current May 5, 1999)
5.3 O. Boz: Cost-Sensitive Learning Bibliography. http://home.ptd.net/~olcay/cost-sensitive.html (current Nov. 7, 2002)
5.4 U.M. Fayyad, G.P. Shapiro, P. Smyth: From Data Mining to Knowledge Discovery: An Overview, Advances in Knowledge Discovery and Data Mining (AAAI/MIT Press, Menlo Park, Calif. 1996) pp. 1-34
5.5 W.J. Frawley, G.P Shapiro, C.J. Matheus: Knowledge Discovery in Databases: An Overview, Knowledge Discovery in Databases (AAAI/MIT Press, Menlo Park, Calif. 1991) pp. 1-27
5.6 J. Fürnkranz: Separate-and-Conquer Rule Learning. Artificial Intelligence Review, 13, pp. 3-54 (1999)
5.7 F. Hussain, H. Liu, E. Suzuki, H. Lu: *Exception Rule Mining with a Relative Interestingness Measure, Knowledge Discovery and Data Mining* (LNAI 1805 PAKDD, Springer, 2000) pp. 86-97

5.8 W. Klösgen: Explora: *A Multipattern and Multistrategy Discovery Approach, Advances in Knowledge Discovery and Data Mining* (AAAI/MIT Press, Menlo Park, Calif. 1996) pp. 249-271

5.9 E.M. Knorr, R.T. Ng: Algorithms for Mining Distance-Based Outliers in Large Datasets. *Proc. 24th Ann. Int'l Conf. Very Large Data Bases, VLDB 1998* pp. 392-403

5.10 J. Li, G. Dong, K. Ramamohanarao: Making Use of the Most Expressive Jumping Emerging Patterns for Classification. Knowledge and Information Systems, 3, pp. 131-145 (2001)

5.11 B. Liu, W. Hsu, S. Chen: Using General Impressions to Analyze Discovered Classification Rules. *Proc. Third Int'l Conf. on Knowledge Discovery and Data Mining, KDD 1997* pp. 31-36

5.12 B. Liu, W. Hsu, L.F. Mun, H.Y. Lee: Finding Interesting Patterns Using User Expectations, IEEE Trans. Knowledge and Data Eng., 11, pp. 817-832 (1999)

5.13 B. Liu, W. Hsu, Y. Ma: Pruning and Summarizing the Discovered Associations. *Proc. Fifth ACM SIGKDD Int'l Conf. on Knowledge Discovery and Data Mining, KDD 1999* pp. 125-134

5.14 D. Gamberger, N. Lavrač: Descriptive Induction through Subgroup Discovery: A Case Study in a Medical Domain. *Proc. Nineteenth Int'l Conf. on Machine Learning, ICML 2002* pp. 163-170

5.15 D. Gamberger, N. Lavrač: Generating Actionable Knowledge by Expert-guided Subgroup Discovery. *Principles of Data Mining and Knowledge Discovery* (LNAI 2431 PKDD, Springer, 2002) pp. 163-174

5.16 T. M. Mitchell: Machine Learning and Data Mining. CACM, 42, pp. 31-36 (1999)

5.17 B. Padmanabhan, A. Tuzhilin: A Belief-driven Method for Discovering Unexpected Patterns. *Proc. Fourth Int'l Conf. on Knowledge Discovery and Data Mining, KDD 1998* pp. 94-100

5.18 A. Siebes: Homogeneous Discoveries Contain no Surprises: Inferring Risk-profiles from Large Databases. *Proc. AAAI-94 Workshop on Knowledge Discovery in Databases, 1994* pp. 97-107

5.19 A. Silberschatz, A. Tuzhilin: On Subjective Measure of Interestingness in Knowledge Discovery. *Proc. First Int'l Conf. on Knowledge Discovery and Data Mining, KDD 1995* pp. 275-281

5.20 A. Silberschatz, A. Tuzhilin: What Makes Patterns Interesting in Knowledge Discovery Systems. IEEE Trans. Knowledge and Data Eng., 8, pp. 970-974 (1996)

5.21 P. Smyth, R.M. Goodman: An Information Theoretic Approach to Rule Induction from Databases. IEEE Trans. Knowledge and Data Eng., 4, pp. 301-316 (1992)

5.22 E. Suzuki, M. Shimura: Exceptional Knowledge Discovery in Databases Based on Information Theory. *Proc. Second Int'l Conf. Knowledge Discovery and Data Mining, KDD 1996* pp. 275-278

5.23 E. Suzuki: Discovering Unexpected Exceptions: A Stochastic Approach. *Proc. Fourth Int'l Workshop on Rough Sets, Fuzzy Sets, and Machine Discovery, RSFD 1996* pp. 225-232

5.24 E. Suzuki: Autonomous Discovery of Reliable Exception Rules. *Proc. Third Int'l Conf. Knowledge Discovery and Data Mining, KDD 1997* pp. 259-262

5.25 E. Suzuki, Y. Kodratoff: *Discovery of Surprising Exception Rules Based on Intensity of Implication, Principles of Data Mining and Knowledge Discovery* (LNAI 1510, PKDD, Springer, 1998) pp. 10-18

5.26 E. Suzuki: *Scheduled Discovery of Exception Rules, Discovery Science* (LNAI 1721, DS, Springer, Berlin 1999) pp. 184-195

5.27 E. Suzuki, S. Tsumoto: *Evaluating Hypothesis-driven Exception-rule Discovery with Medical Data Sets, Knowledge Discovery and Data Mining* (LNAI 1805, PAKDD, Springer 2000) pp. 208-211

5.28 E. Suzuki: Mining Bacterial Test Data with Scheduled Discovery of Exception Rules. *Proc. Int'l Workshop of KDD Challenge on Real-world Data, KDD Challenge 2000* pp. 3440

5.29 E. Suzuki, J.M. Żytkow: *Unified Algorithm for Undirected Discovery of Exception Rules, Principles of Data Mining and Knowledge Discovery* (LNAI 1910, PKDD, Springer 2000) pp. 169-180

5.30 E. Suzuki: *In Pursuit of Interesting Patterns with Undirected Discovery of Exception Rules, Progresses in Discovery Science* (LNCS 2281, State-of-the-Art Surveys, Springer 2002) pp. 504-517

5.31 E. Suzuki: Undirected Discovery of Interesting Exception Rules. International Journal of Pattern Recognition and Artificial Intelligence, 16, pp. 1065-1086 (2002)

5.32 S. Tsumoto et al.: Comparison of Data Mining Methods using Common Medical Datasets. ISM Symp.: Data Mining and Knowledge Discovery in Data Science (1999) pp. 63-72

5.33 N. Yugami, Y. Ohta, S. Okamoto: Fast Discovery of Interesting Rules, Knowledge Discovery and Data Mining (LNAI 1805, PAKDD, Springer 2000) pp. 17-28

5.34 N. Zhong, Y.Y. Yao, S. Ohsuga: Peculiarity Oriented Multi-database Mining, Principles of Data Mining and Knowledge Discovery (LNAI 1704, PKDD, Springer 1999) pp. 136-146

6. Data Mining for Targeted Marketing

Ning Zhong[1], Yiyu Yao[2], Chunnian Liu[3], Jiajin Huang[3], and Chuangxin Ou[3]

[1] Maebashi Institute of Technology, Japan
[2] University of Regina, Canada
[3] Beijing University of Technology, China

Abstract

Targeted marketing is a new business model of interactive one-to-one communication between marketer and customer. There is great potential for data mining to make useful contributions to the marketing discipline for business intelligence. This chapter provides an overview of the recent development in data mining applications for targeted marketing.

6.1 Introduction

In marketing, there are two main different approaches to communication: *mass marketing* and *targeted marketing* (or *direct marketing*) [6.18, 6.32, 6.53]. Mass marketing uses mass media such as print, radio, and television for the public, without discrimination. Targeted marketing involves the identification of customers having potential market value by studying the customers' characteristics and needs (the past and the future), and selects certain customers for promotion. Targeted marketing aims at obtaining and maintaining direct relationships between suppliers and buyers based on one or more product/market combinations. Targeted marketing becomes more and more popular because of the increased competition and cost.

Furthermore, the scope of targeted marketing can be expanded, from considering only how products are distributed, to include the enhancing of the relationships between an organization and its customers [6.15] because of the strategic importance of long-term relationships with customers. In other words, once customers are acquired, customer retention becomes the target. Retention through customer satisfaction and loyalty can be greatly improved by acquiring and exploiting knowledge about these customers and their needs. Such targeted marketing is called "targeted relationship marketing" or "customer relationship management" (CRM for short) [6.34].

Targeted marketing is an important area of applications for data mining, data warehousing, statistical pattern recognition, and intelligent agents [6.21]. Although standard data mining methods may be applied for the purpose of targeted marketing, many specific algorithms need to be developed and applied for the direct marketer to make decisions effectively.

The chapter investigates the recent development in data mining applications for targeted marketing. Section 2 gives a sample process of targeted marketing. Section 3 discusses the main problems of applying traditional data mining techniques for targeted marketing. Section 4 introduces major target selection algorithms and proposes our market value functions model. Section 5 describes how to mine the market value functions. Section 6 provides the evaluation methods for targeted marketing data mining. Section 7 shows experimental results. Section 8 discusses new directions in data mining for targeted marketing. Finally, we conclude the chapter in Sect. 9.

6.2 The Process of Targeted Marketing

The process of data mining for targeted marketing is a more specific version of a general data mining process that is usually a *multi-phase* process involving numerous steps, such as data preparation, preprocessing, hypothesis generation, pattern formation, knowledge evaluation, representation, refinement, and management. The process of data mining for targeted marketing should have at least the following steps (see Fig. 6.1).

Data preparation and preprocessing: Collecting potential customers' related data from multiple data sources. Such customer data include all sales, promotion, and customer service activities that have occurred as a result of the customer's relationship with a company, as well as personal related information such as name, address, age, sex, hobby, income, occupation, employment status, and marital status. The data are transformed into the format a data mining system needs.

Finding patterns: Splitting the potential customer data into training dataset and testing dataset, applying learning algorithms to a training dataset for finding patterns, and evaluating the patterns on the testing dataset. If the patterns are not satisfactory, the process may iterate to a previous step.

Promoting clients: Using the patterns to predict and promoting the likely buyers.

6.3 Problems in Targeted Marketing

Although traditional algorithms of data mining may be applied to targeted marketing and provide a good background for our research, no satisfactory solutions have yet been found to solve real-world problems such as the following:

– The traditional algorithms cannot be used for the very low response rate situation in marketing databases. In other words, the distribution of such data is unbalanced, and the coverage of rules mined from such data is usually very low.
  The problem of unbalanced data can be formalized as follows. A data set D contains n objects, m attributes, and k classes. Focusing on the class $C_i(1 \leq i \leq k)$ that is of interest, let the class C_i be a positive class $C+$ and

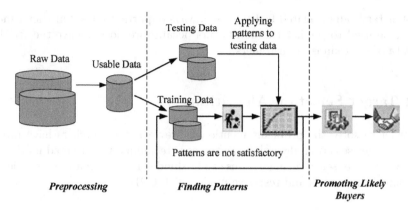

Fig. 6.1. The process of targeted marketing

others are considered as a negative class $C-$. Thus, $|C+| \ll |C-|$ and many objects in $C+$ and $C-$ are indiscernible. Most existing algorithms cannot produce rules with high coverage and accuracy from such data sets. Hence, they do not work well on such datasets and the results cannot be used to predict the likely buyers.

- The predictive accuracy cannot be used as a suitable evaluation criterion for targeted marketing. One reason is that classification errors must be dealt with differently. Another reason is that the predictive accuracy is too weak for the purpose of customer targeting [6.18]. It cannot provide the flexibility in choosing a certain percent of likely buyers for promotion.
- Most of the methods only consider response. In practice, a direct marketer is more interested in maximal profit rather than maximal response. Sometimes, the maximal response is not equal to the maximal profit. We may get more profit from the low probability responders.
- Most of the methods do not consider the different actions of responders. It is not realistic that each of customers will respond in the same way. In practice, the actions of response are different for the same promotion material. Some responders may buy cheap goods, but a few of the responders may choose luxury goods. Thus the direct marketers get different profits from the two kinds of responders.
- Features may be difficult to choose from the datasets if the features are numerous. Existing learning algorithms may not work well in targeted marketing datasets since they have a large number of variables (features) and enormous amounts of data. This is because of the extremely unbalanced class distribution. For instance, we cannot get the preferable features if we choose the method of information gain to choose the features. We found that the values are very near to each other when calculating the information gain for each of the features. Hence, it is difficult to find the key features. The RFM (Recency of the last purchase, Frequency of purchase,

Monetary value) related features are always important ones, but they cannot be used to predict the non-buyers because we do not have the RFM data of such kinds of customers.

6.4 Target Selection Algorithms

In order to solve the problems mentioned previously, researchers have proposed target selection algorithms as the core technique for targeted marketing. Such target selection algorithms can be divided into two major categories: segmentation models and response models [6.4, 6.21].

6.4.1 Segmentation Models

Segmentation models are designed to assign people or geographic areas to groups or clusters on the basis of the similarities in characteristics or attributes that describe them rather than on the basis of some specific action such as response to a mailing [6.4]. They are as homogeneous as possible within segments and as heterogeneous as possible between segments in response behaviors [6.37]. The group with the highest probability to respond is selected to promote.

The major segmentation techniques are cluster analysis such as AID, CHAID, and CART [6.3, 6.4]. The result of the three methods is a decision tree, and each node in the tree is a group in which every individual is homogeneous. In practice, the segmentation is often performed by calculating the RFM-score and dividing the total list into different segments.

The segmentation cannot work well on a dataset with extremely unbalanced class distribution because the rate of positive instances is always low in marketing databases. It is difficult to construct a suitable tree and the models may only discover simply unusable rules or patterns. In such segmentation, the models cannot rank the individuals, and all clients in the same segment are treated alike.

6.4.2 Response Models

Response models use some models to calculate the response probability of each individual and then choose the highest probability responders to promote. Predictive models can be regarded as a class of response models to predict some outcome, e.g., response to mailing, returns, bad debts, sales volume, and so on [6.4]. In the following discussion, we will introduce several response models, including the logit/probit model, the genetic model, the neural network model, the rough set model, the association model, the peculiarity oriented model, and the market value functions model.

The Logit/Probit Model. The logit and probit model can handle discrete response data [6.4]. It can be used to calculate the target score of each individual in marketing databases. The model assumes that every individual i has a certain tendency in responding to a mailing received at time t. More specifically, the tendency r_{it}^* can be computed as:

$$r_{it}^* = \beta_0 + \beta_1 X_{1it} + \beta_2 X_{2it} + \ldots + \beta_n X_{nit} + \varepsilon_{it}. \tag{6.1}$$

where $\mathbf{X}$ are the predictor variables related to response, β are the "weights" or coefficients, and ε denotes the cumulative distribution function of a logistic distribution (i.e., the logit model) or a normal distribution (i.e., the probit model).

If the tendency r_t^* is larger than 0, we assume the individual will respond; otherwise, the individual will not respond. The individuals whose probabilities of the predicted responses are sufficiently high are selected for promotion.

However, the formulation of the model has some drawbacks. Firstly, it assumes that the customers will spend the same amount of money on the same promotion material according to the response probability. Secondly, although it can predict the maximal response, the model cannot predict the maximal profit. In order to solve this problem, Bult and Wansbeek added a cutoff point to the logit model so that only those individuals will be selected for which profit is maximized [6.14]. The formula is defined as follows:

$$\Pi_i = rR_i - c \tag{6.2}$$

where Π_i is the revenue for the firm generated by individual i, r is the revenue from positive reply, c is the mail cost, and R_i is a binary variable indicating response by individual i.

The Genetic Model. The genetic model in targeted marketing is used to build models that maximize the expected response and profit from solicitations. The genetic model we refer here is the GMAX model [6.25], which is a hybrid AI-statistic method. The GMAX model uses genetic modeling as the optimization technique for targeted marketing. Each model in the genetic modeling has an associated fitness value. A model with higher fitness value solves the problem better than a model with lower fitness value, and survives and reproduces at a higher rate.

The advantage of a genetic model is that such a model is assumption-free, robust, nonparametric, and performs well on both large and small samples. It can be used to learn complex relationships. However, the genetic model has some problems that need to be solved. The first is to find a fitness function so that the genetic model performs well with the dataset. The second is to set the genetic model parameters: population size and reproduction, crossover and mutation probabilities. Sometimes, even with correct parameter settings, genetic models cannot guarantee the optimal solution.

The Neural Network Model. One advantage of neural networks is that they can adapt to nonlinearity in the data to capture complex relations. This is an important motivation for applying neural networks for target selection. The neural network sets a threshold, and only ones that are over the given threshold are selected as targets to promote.

The complexity of the neural network depends on the degree of nonlinearity of the problems to be solved. Generally, the feed-forward neural networks are sufficient for target selection problems. Important parameters that determine the complexity of feed-forward neural networks are the number of hidden layers and the number of neurons in each layer. In target selecting problems, a feed-forward network with one hidden layer can provide models with sufficient accuracy [6.22].

If we use a feed-forward neural network with back propagation, we may encounter the problem of local minima. Because neural networks are gradient-descent algorithm, chances are that you will end up with a local minimum instead of a global minimum. It is difficult to initialize a neural network given a certain problem. One way to solve this problem is to use a genetic algorithm to determine the initial weight in the neural network [6.4].

Neural networks, which can model non-linear/complex relationships, are a useful technique for pattern discovery, and are non-parametric and robust for noise. The neural network model can give us a target score for each client. We can rank the clients by the scores and send promotion material to the high probability clients. However, the neural network model has also some drawbacks: firstly, it generates a complex formula that is too difficult to interpret; secondly, the neural network is difficult to configure.

The Rough Set Model. Rough set theory constitutes a sound basis for data mining. It offers useful tools to discover patterns hidden in data in many aspects [6.20]. It can be used in different phases of the knowledge discovery process like attribute selection, attribute extraction, data reduction, decision rule generation, and pattern extraction (templates, association rules).

The ProbRough system, which combines basic principles of rough-set-based rule induction systems with the flexibility of statistical techniques for classification, is a data mining system of rough classifier generation [6.35]. It can be used to analyze the potential customers and purchase prediction. The algorithm of rough classifier generation consists of two phases. The first phase is the global segmentation of the attribute space. The algorithm tries to minimize the average global cost of the decision-making. The second phase is the reduction of the number of decision rules. In this phase, the algorithm minimizes the number of rules of the classifier.

The ProbRough algorithm behaves well on training data that involve redundant attributes. It can deal with two key problems: prior probabilities and unequal misclassification costs. The resultant rough classifiers are not sensitive to outliers in the data and can accept databases with noisy and inconsistent information. Users of the ProbRough system are not required to adhere to assumptions about the data or model that is produced.

The GDT-RS system proposed by Zhong *et al.* is a rough-set-based hybrid approach that can be used to produce predictive rules for targeted marketing [6.45]. The system is based on a hybridization of *Generalization Distribution Table (GDT)* and the *Rough Set* methodology. The GDT-RS system can generate, from noisy and incomplete training data, a set of rules with the minimal (semi-minimal) description length, having large strength, and covering all instances. It can deal with unbalanced data by cooperatively using several rough-set-based hybrid approaches such as RSH for attribute subset selection, RSBR for discretization of real-valued attributes, and so on, in a multi-step process [6.49, 6.50]. Recently, the GDT-RS system has been extended to mine a new type of rule called *ordering rules* [6.28].

The Association Model. The association model is used to mine association rules from a large transaction dataset for targeted marketing [6.1, 6.2]. If a strong association between two sets of items A and B is observed, i.e., a customer buying A tends to buy B, it may be beneficial to send advertisements of B, or recommend item B, to customers buying A. In fact, using association rules to promote related products is a common practice of targeted marketing.

Two measures, called the support and the confidence, are used to mine association rules:

$$support(A \Rightarrow B) =$$
$$\frac{the\ number\ of\ tuples\ containing\ both\ A\ and\ B}{total\ number\ of\ tuples} \tag{6.3}$$
$$confidence(A \Rightarrow B) =$$
$$\frac{the\ number\ of\ tuples\ containing\ both\ A\ and\ B}{the\ number\ of\ tuples\ containing\ A}. \tag{6.4}$$

By specifying threshold values of support and confidence, one can obtain all association rules whose support and confidence are above the thresholds. An association rule has the requirement of large support.

The Peculiarity-Oriented Model. Unlike the ordinary association mining stated above, the peculiarity-oriented model focuses on a relatively small number of peculiar data and searches their relevance for discovering peculiarity rules [6.48, 6.54]. Roughly speaking, data are *peculiar* if they represent a relatively small number of objects, and if furthermore those objects are very different from other objects in the dataset.

Zhong *et al.* have developed the *Peculiarity Factor*, $PF(x_{ij})$, as a measure for evaluating such peculiarity:

$$PF(x_{ij}) = \sum_{k=1}^{n} N(x_{ij}, x_{kj})^{\alpha} \tag{6.5}$$

where N denotes the conceptual distance, α is a parameter which can be adjusted by a user, and $\alpha = 0.5$ is used as default. Eq. (6.5) evaluates whether x_{ij} has a low frequency and is very different from other values x_{kj}. There are two merits of the measure: (1) it can handle both continuous and symbolic attributes based on a unified semantic interpretation, and (2) background knowledge represented by binary neighborhoods can be used to evaluate the peculiarity. Furthermore, a peculiarity rule is discovered by searching the relevance among peculiar data. A peculiarity rule has a low support value.

Patterns (rules) with low support have been considered by many researchers. Two examples are the studies of exception rules [6.30] and emerging patterns [6.7]. They share a common feature with peculiarity rules in the sense that all describe a relatively small number of objects. They differ in the way in which peculiar data are collected, interpreted, and used, as well as in the interpretation of corresponding rules. An exception rule is an amendment to, or a clarification of, a more general rule. The peculiar data covered by an exception rule is obtained from the subset of data covered by the general rule. On the other hand, a peculiarity rule stands on its own and has a commonsense interpretation, like ordinary association rules [6.2, 6.30]. Dong and Li proposed a framework for discovering emerging patterns [6.7]. Their method is essentially a study of the change of supports in different datasets. A large change suggests an interesting emerging pattern. Since emerging patterns with large supports are perhaps well-known facts, they concentrated on emerging patterns with small supports. In particular, they suggested that patterns with low support, such as 1%-20%, can give useful new insights about data. Unfortunately, such patterns are difficult to discover by traditional association rule mining methods. Their approach provides another use of peculiar data. In general, it may be desirable to have a unified framework within which many different uses of peculiar data can be studied for targeted marketing.

The Market Value Functions Model. Let us consider now another type of targeted marketing problem. Suppose there is a health club that needs to expand its operation by attracting more members. Assume that each existing member is described by a finite set of attributes. It is natural to examine existing members in order to identify their common features. Information about the health club may be sent to non-members who share the same features of members or similar to those of members. Other examples include promotion of special types of phone services and marketing of different classes of credit cards. In this case, we explore the relationships (similarities) between people (objects) based on their attribute values. The underlying assumption is that *similar types of people tend to make similar decisions and to choose similar services*. Techniques for mining association rules may not be directly applicable to this type of targeted marketing. One may produce too many or too few rules. The selection of a good set of rules may not be an easy task.

Furthermore, the use of the derived rules may produce too many or too few potential new members.

This section describes the market value functions model for targeted marketing by focusing on the issues of knowledge representation and computation of market values [6.42, 6.44]. More specifically, we assume that each object is represented by its values on a finite set of attributes. We further assume that market values of objects can be computed using a linear market value function. Thus, we may consider the proposed model to be a *linear* model, which is related to, but different from, the linear model for information retrieval.

Let U be a finite universe of objects. Elements of U may be customers or products we are interested in market-oriented decision making. The universe U is divided into three pair-wise disjoint classes, i.e., $U = P \cup N \cup D$. The sets P, N, and D are called *positive, negative,* and *don't know* instances, respectively. Take the earlier health club example; P is the set of current members, N is the set of people who had previously refused to join the club, and D is the set of the rest. The set N may be empty. A targeted marketing problem may be defined as finding elements from D, and possibly from N, that are similar to elements in P, and possibly dissimilar to elements in N. In other words, we want to identify elements from D and N that are more likely to become new members of P. We are interested in finding a market value function so that elements of D can be ranked accordingly.

Information about objects in a finite universe is given by an information table [6.20, 6.43]. The rows of the table correspond to objects of the universe, the columns correspond to attributes, and each cell is the value of an object with respect to an attribute. Formally, an information table is a quadruple:

$$S = (U, At, \{V_a \mid a \in At\}, \{I_a \mid a \in At\})$$

where U is a finite nonemptyset of objects, At is a finite nonempty set of attributes, V_a is a nonempty set of values for $a \in At$, $I_a : U \rightarrow V_a$ is an information function for $a \in At$. Each information function I_a is a total function that maps an object of U to exactly one value in V_a. An information table represents all available information and knowledge. Objects are only perceived, observed, or measured by using a finite number of properties [6.20].

A straightforward solution for the targeted marketing problem is the mining of characteristic rules for both P and N, or discriminant rules for differentiating elements of P and N. The mined rules can then be used to classify elements of D. There are extensive studies on such techniques. Some of the difficulties with this straightforward solution have been mentioned in Sect. 6.3. We therefore focus our attention on an alternative approach.

A market value function is a real-valued function from the universe to the set of real numbers, $r : U \longrightarrow \Re$. In the context of information retrieval, the values of r represent the potential usefulness or relevance of documents with respect to a query. Documents are ranked according to the values of r. For the targeted marketing problem, a market value function ranks objects according to their potential market values. For the health club example, a market value function ranks people according to their likelihood of becoming members of the health club. The likelihood may be estimated based on its similarity to a typical member of P.

We study the simplest form of market value functions, i.e., the linear discriminant functions. Let $u_a : V_a \longrightarrow \Re$ be a utility function defined on V_a for an attribute $a \in At$. The utility $u_a(\cdot)$ may be positive, negative, or zero. For $v \in V_a$, if $u_a(v) > 0$ and $I_a(x) = v$, i.e., $u_a(I_a(x)) > 0$, then attribute a has a positive contribution to the overall market value of x. If $u_a(I_a(x)) < 0$, then a has a negative contribution. If $u_a(I_a(x)) = 0$, then a has no contribution. The pool of contributions from all attributes is computed by a linear market value function of the following form:

$$r(x) = \sum_{a \in At} w_a u_a(I_a(x)) \tag{6.6}$$

where w_a is the weight of attribute a. Similarly, the weight w_a may be positive, negative, or zero. Attributes with larger weights (absolute value) are more important, and attributes with weights close to zero are not important. The overall market value of x is a weighted combination of utilities of all attributes. By using a linear market value function, we have implicitly assumed that contributions made by individual attributes are independent. Such an assumption is commonly known as utility independence assumption. Implications of utility independence assumption can be found in the literature of multi-criteria decision making [6.8].

The market value model proposes a linear model to solve the target selection problem by drawing and extending results from information retrieval [6.42, 6.44]. It is assumed that each object is represented by values of a finite set of attributes. A market value function is a linear combination of utility functions on attribute values, which depend on two parts: *utility function* and *attribute weighting*.

The market value function has some advantages: firstly, it can rank individuals according to their market value instead of classification; secondly, the market value function is interpretable; thirdly, the system using the market value function can perform without expertise. In the next section, we discuss how to mine *market value functions* in detail.

6.5 Mining Market Value Functions

The potential usefulness and effectiveness of the proposed market value function model depend, to a large extent, on the estimation of the individual utility functions and the attribute weights, i.e., the coefficients of the linear market value function. This section investigates various methods for estimating and mining market value functions. The estimation of utility functions draws from probabilistic models of information retrieval [6.26, 6.27, 6.39]. The estimation of attribute weights is based on information-theoretic measures of attribute importance [6.24, 6.36, 6.38, 6.40, 6.41].

6.5.1 Utility Functions

Utility functions can be defined based on either positive instances or both positive and negative instances.

Estimation from Positive Instances. There are several situations that require estimation of the market value function from only positive examples. It may happen that the negative examples are not available. Although negative examples may be available, they should be used very cautiously. For instance, people who are not in a health club, or previously refused to join, perhaps will join the club. Thus, one should not rule out the possibility that people similar to them may join the club. In other words, we use positive examples for including potential new members, and we do not use negative examples to exclude potential new members. It may also happen that one can easily find a regularity (structure) explaining why an element belongs to P, and cannot find a structure explaining why an element does not belong to P, as there may be a great diversity of reasons for the latter.

Consider an attribute $a \in At$ taking its value from V_a. For $v \in V_a$, let $m(a = v \mid P) = m(v \mid P)$ be the subset of P defined by:

$$m(v \mid P) = \{x \in U \mid x \in P, I_a(x) = v\}. \tag{6.7}$$

It consists of elements from P whose value on a is v. Let $|m(v \mid P)|$ denote the cardinality of the set $m(v \mid P)$. For two values $v, v' \in V_a$, if $|m(v \mid P)| > |m(v' \mid P)|$, then more elements from P have v as their value than have v' as their value. Intuitively, one may say that an object having v as its value is more likely to belong to P than another element having v' as its value. Based merely on attribute a, for two elements, $x, y \in U$, with $I_a(x) = v$, $I_a(y) = v'$, and $|m(v \mid P)| > |m(v' \mid P)|$, we may say that the market value of x is more than that of y. This suggests that the value of the utility function $u_a : V_a \longrightarrow \Re$ at $v \in V_a$ should be proportional to the size of the set $m(v \mid P)$. We therefore choose the following utility function:

$$u_a^1(v) = |m(v \mid P)|. \tag{6.8}$$

The values of $u_a(\cdot)$ are between 0 and $|P|$, which is based on a simple counting of elements having the value v in P. The set of elements P may be considered to be a sub-population of U. One may also use a probabilistic version of the utility function:

$$u_a^2(v) = Pr(a = v \mid P) = Pr(v \mid P) = \frac{|m(v \mid P)|}{|P|} = \frac{u_a^1(v)}{|P|}. \tag{6.9}$$

Since $|P|$ is a constant independent of any attribute, u_a^1 and u_a^2 will produce the same result in the linear model.

In general, one would expect an attribute to contribute more towards the market value of an element if its value is concentrated in the sub-population P. This can be done by comparing the conditional probability $Pr(v \mid P)$ and the unconditional probability:

$$Pr(a = v) = Pr(v) = \frac{|m(v)|}{|U|} \tag{6.10}$$

where

$$m(v) = \{x \in U \mid I_a(x) = v\}. \tag{6.11}$$

For simplicity, we assume that $m(v) \neq \emptyset$; otherwise, we can delete v from V_a. The corresponding utility can be defined by:

$$u_a^3(v) = \frac{Pr(v \mid P)}{Pr(v)} = \frac{|m(v \mid P)| \, |U|}{|m(v)| \, |P|}. \tag{6.12}$$

If $u_a^3(v) > 1$, the value v is concentrated more on the sub-population P; if $u_a^3(v) < 1$, the value v is not concentrated on P. One would expect a positive contribution for the former case and a negative contribution for the latter case. To achieve this, we use the logarithm transformation of u_a^3 as follows:

$$u_a^4(v) = \log u_a^3(v) = \log \frac{|m(v \mid P)| \, |U|}{|m(v)| \, |P|}. \tag{6.13}$$

It follows that $u_a^4(v) > 0$ if and only if $u_a^3(v) > 1$, $u_a^4(v) < 0$ if and only if $u_a^3(v) < 1$, and $u_a^4(v) = 0$ if and only if $u_a^3(v) = 1$. In a practical situation, it may happen that $m(v \mid P) = \emptyset$. The utility function u_a^4 is not defined. In this case, we may use the point-5 formula, as was done in information retrieval [6.26]:

$$u_a^4(v) = \log \frac{(|m(v \mid P)| + 0.5)(|U| + 1.0)}{(|m(v)| + 1.0)(|P| + 0.5)}. \tag{6.14}$$

This implicitly assumed that there is a notional sample of size one divided equally into P and N.

The quantity $|U|/|P|$ is a constant independent of any attribute, and will not effect the ranking. It can therefore be removed from the utility function, and the value $|m(v \mid P)|/|m(v)|$ can be used.

Estimation from Positive and Negative Instances. With both positive and negative instances, we have two sub-populations P and N. The estimation methods presented earlier can be modified to take into consideration of the distribution of attribute values in both P and N.

For an attribute value $v \in V_a$, it contributes more, or positively, to the market value of an object if v appears more in the sub-population P than in the sub-population N; otherwise, it contributes less, or negatively, to the market value of the object. Similar to utility functions u_a^3 and u_a^4, we define two new utility functions:

$$u_a^5(v) = \frac{Pr(v \mid P)}{Pr(v \mid N)} = \frac{|m(v \mid P)| \, |N|}{|m(v \mid N)| \, |P|} \qquad (6.15)$$

$$u_a^6(v) = \log u_a^5(v) = \log \frac{|m(v \mid P)| \, |N|}{|m(v \mid N)| \, |P|} \qquad (6.16)$$

where

$$m(v \mid N) = \{x \in U \mid x \in N, I_a(x) = v\}. \qquad (6.17)$$

The point-5 formula of u_a^6 is given by:

$$u_a^6(v) = \log \frac{(|m(v \mid P)| + 0.5)(|N| + 0.5)}{(|m(v \mid N)| + 0.5)(|P| + 0.5)}. \qquad (6.18)$$

Since P and N are disjoint subsets of U, the new utility functions are not a simple replacement of U by N in u_a^3 and u_a^4. The ratio $|N|/|P|$ is a constant independent of any attribute; it can be removed from the utility functions.

6.5.2 Attribute Weighting

For the computation of attribute weights, we adopt information-theoretic measures [6.40, 6.41]. For an attribute a, its Shannon entropy $H_P(a)$ in the population P is defined by:

$$H_P(a) = H_P(Pr(\cdot \mid P)) = - \sum_{v \in V_a} Pr(v \mid P) \log Pr(v \mid P) \qquad (6.19)$$

where $Pr(\cdot \mid P)$ denotes the probability distribution of attribute values in P. We define $0 \log 0$ to be 0 by extending function $x \log x$ to the origin by continuity. The entropy is a nonnegative function, i.e., $H_P(a) \geq 0$. It may be interpreted as a measure of the information content of, or the uncertainty about, a random variable a taking values from V_a. The entropy reaches the maximum value $\log |V_a|$ for the *uniform* distribution, i.e., $Pr(v) = 1/|V_a|$ for all $v \in V_a$. The minimum entropy value 0 is obtained when the distribution focuses on a particular value v_0, i.e., $Pr(v_0 \mid P) = 1$ and $Pr(v \mid P) = 0$ for all $v \in V_a$ and $v \neq v_0$. One may also interpret the entropy value as representing the degree of structure or diversity of a probability distribution [6.24, 6.36].

A lower entropy value indicates a higher degree of structuredness. If an attribute has a lower entropy value, we can say that the distribution of its values is uneven in the population P. Thus, the attribute may be more informative in predicting if an object belongs to P. On the other hand, an attribute with a larger entropy is less informative, as the values of the attribute a are distributed more evenly in P. A measure for weighting attributes can be designed so that it is inversely proportional to the entropy value. An attribute weighting formula, adopted from information retrieval [6.40], is given below:

$$w_a^1 = 1 - \frac{H_p(a)}{\log |V_a|}. \qquad (6.20)$$

Clearly, $0 \le w_a^1 \le 1$. We have also assumed that $|V_a| > 1$. Otherwise, every object would have the same value on the attribute a, and there is no point in using this attribute.

The entropy of attribute a in the entire population U is given by:

$$H(a) = H(Pr(\cdot)) = -\sum_{v \in V_a} Pr(v) \log Pr(v). \tag{6.21}$$

It reflects the structuredness of the distribution of a's values in U. For a more informative attribute, we would expect that it shows less structuredness in U than in the subpopulation P. We may use another weighting formula involving both $H_P(a)$ and $H(a)$:

$$w_a^2 = \frac{H(a) - H_P(a)}{\log |V_a|}. \tag{6.22}$$

The weight, $-1 \le w_a^2 \le 1$, gives the change of entropy values as we move from the entire population to a sub-population. A positive value suggests that attribute a shows more structuredness in P than in U, and a negative value suggests the reverse. It can also be seen that w_a^1 is a special case of w_a^2, where $H(a)$ takes the maximum value of $\log |V_a|$.

The well-known Kullback-Leibler *divergence* measure offers another attribute weighting formula [6.16]:

$$w_a^3 = D(Pr(\cdot \mid P)||Pr(\cdot)) = \sum_{v \in V_a} Pr(v \mid P) \log \frac{Pr(v \mid P)}{Pr(v)}. \tag{6.23}$$

It measures the degree of deviation of the probability distribution $Pr(\cdot \mid P)$ from the distribution $Pr(\cdot)$. From an information-theoretic point of view, the divergence can be interpreted as the difference between the information contained in distribution $Pr(\cdot \mid P)$ and that contained in $Pr(\cdot)$ about $Pr(\cdot \mid P)$. The measure is nonnegative, i.e., $w_a^3 > 0$. This quantity becomes minimum 0 if $Pr(v \mid P) = Pr(v)$ for all $v \in V_a$. The maximum value is realized when $Pr(v_0 \mid P) = 1$ for a particular v_0 for which $Pr(v_0)$ is the smallest [6.36].

We can also use both positive and negative instances for attribute weighting. An attribute is informative if sub-populations P and N are different from each other from the view point of the attribute. In this case, we have three sub-populations, P, N, and $P \cup N$. Let $H_P(a)$, $H_N(a)$, and $H_{P \cup N}(a)$ denote the entropy values of attribute a in the three sub-populations, respectively. If distributions of attribute values in P and N are similar, then both of them should be similar to the distribution in $P \cup N$. We would expect a small difference of entropy values of a in P, N, and $P \cup N$. On the other hand, if distributions of attribute values in P and N are different, we would expect a large difference. For this purpose, we adopt the following weighting formula:

$$w_a^4 = H_{P \cup N}(a) - \left[\frac{|P|}{|P \cup N|} H_P(a) + \frac{|N|}{|P \cup N|} H_N(a) \right]$$
$$= H_{P \cup N}(a) - [\lambda_P H_P(a) + \lambda_N H_N(a)] \tag{6.24}$$

where $\lambda_P + \lambda_N = 1$. The second term in the formula is the average of entropy values in two sub-populations P and N. For any attribute value v, we have:

$$Pr(v \mid P \cup N) = \lambda_P Pr(v \mid P) + \lambda_N Pr(v \mid N). \tag{6.25}$$

Since $-x \log x$ is a concave function, the Jensen inequality immediately implies that the lower bound of w_a^4 is 0, i.e., $w_a^4 \geq 0$. It reaches the minimum value 0 when the two distributions in P and N are *identical*. It can also be shown that w_a^4 reaches the maximum value, $-[\lambda_P \log \lambda_P + \lambda_N \log \lambda_N]$, if the distributions are *totally different*, namely $Pr(v \mid P) \neq 0$ whenever $Pr(v \mid N) = 0$ and $Pr(v \mid P) = 0$ whenever $Pr(v \mid N) \neq 0$.

In terms of Kullback Leibler divergence, w_a^4 can be expressed by [6.38]:

$$w_a^4 = \lambda_P D(Pr(\cdot \mid P)\|Pr(\cdot)) + \lambda_N D(Pr(\cdot \mid N)\|Pr(\cdot)). \tag{6.26}$$

The weighting formula w_a^3 uses the first term of w_a^4. Thus, w_a^4 is an expected divergence considering two sub-populations. It may be considered as a more generalized version of w_a^3.

The entropy function is determined only by probability values in a probability distribution. It is independent of how these probability values are assigned to different attributes. Different probability distributions may produce the same entropy value. For instance, the following distributions produce the same entropy value, although they are totally different distributions:

$$Pr(v_1 \mid P) = 0.5, \quad Pr(v_2 \mid P) = 0.5,$$
$$Pr(v_3 \mid P) = 0.0, \quad Pr(v_4 \mid P) = 0.0;$$

$$Pr(v_1 \mid N) = 0.0, \quad Pr(v_2 \mid N) = 0.0,$$
$$Pr(v_3 \mid N) = 0.5, \quad Pr(v_4 \mid N) = 0.5.$$

The difference between $H_P(a)$ and $H_N(a)$ cannot tell us if P and N are similar based on attribute a. This mainly stems from the fact that there is no inherent relationships between probability distributions $Pr(\cdot \mid P)$ and $Pr(\cdot \mid N)$. On the other hand, the proposed measures w_a^3 and w_a^4 do not suffer from this problem. In those formulas, we use probability distributions from related populations.

6.6 Evaluation of the Learning Algorithms

Since the predictive accuracy cannot be used as a suitable evaluation criterion for the targeted marketing process as mentioned in previous section, a new evaluation criterion needs to be developed. The main reason is that classification errors (false negative, false positive) must be dealt with differently. So far, many evaluation methods instead of predictive accuracy have been employed in targeted marketing. Among them, *decile analysis* and *lift measure* are two well-known ones.

Decile analysis is a tabular display of model performance [6.25]. If the model represents the regulations well, we will see more responders in the top decile than in the bottom decile, and the value of the Cum Response lift will distribute in descending order in the table.

From the table of the decile we can evaluate a learning algorithm. In the KDD-97-Cup Competition, two measurements were used. One was the number of responders in the top decile; the second was the number of responders in the top four deciles. They are not suitable ways of measuring learning algorithms. Ling and Li propose a solution to the problem [6.18]. They calculate a weighted sum of the items in the lift table as the evaluation. Assume the 10 deciles in the lift table are $S_1, S_2, \ldots, S_{10}$; then the lift index is defined as:

$$S_{lift} = (1.0 \times S_1 + 0.9 \times S_2 + \ldots + 0.1 \times S_{10}) / \sum_{i=1}^{10} S_i \qquad (6.27)$$

where S_i denotes the number of positive examples in the ith decile.

The S_{lift} value shows the degree of how well the model solves the problem concerned. A higher S_{lift} value indicates that more responders distribute in the top deciles than those in the bottom deciles, and it is preferable to those models with low values.

6.7 Experimental Results

A data set on potential members of a club has been used for our approach. There are 58,102 potential customers in this data set, and 3,856 of them are existing members of the club. Furthermore, each customer is described by 96 attributes, such as sex, age, hobby, and income.

We randomly select 18,963 instances (894 positive instances) as the training set and use two testing sets: the first one consists of 24,134 instances and the second one uses all the data. The training set is used to learn the market-value function, and then the learned function is used to rank the testing examples.

The lift index defined in Eq. (6.27) is used as the evaluation criterion [6.18], since it has a better intuitive meaning for the targeted marketing problem. The lift index shows the distribution of the positive examples in the ranked list for the intuitive evaluation. After all testing examples are ranked using the proposed algorithm, we divide the ranked list into 10 equal deciles, and see how the original responders distribute in the 10 deciles. If regularities are found, we will see more responders in the top deciles than in the bottom deciles.

Table 6.1 shows the results (S_{lift} values) of three linear combinations of utility functions and the comparison with the Naive Bayesian method. The distribution of the positive examples in the two ranked testing sets are illustrated in Figs. 6.2 and 6.3, respectively. In the figures, the x-axis is the number of the deciles, and the y-axis is the number of the positive instances in each decile [1].

By comparing with the result of the Naive Bayesian method, as shown in Table 6.1, we can see that the linear models are acceptable. Figures 6.2 and 6.3 show the same results. That is, the results of the linear models are better than those of the Naive Bayesian method according to the distribution of the positive examples if we select the top 20% of customers in the testing set 1. From the same figures, we can also observe that the distribution of the positive examples in the ranked testing sets is reasonable because the most positive examples are in the top deciles.

The results of the linear models can also be more easily explained than those of the Naive Bayesian method. In the linear models, we can determine which of the original attributes is more predictive according to the attribute weighting value based on the information retrieval theory. So, we can select more important attributes to build our linear model in order to reduce the dimension of the attributes. It will benefit our further decision.

Table 6.1. S_{lift} values of three linear combinations of utility functions and the comparison with the Naive Bayesian method

Data Set No.	w_a^1 and u_a^1	w_a^2 and u_a^2	w_a^3 and u_a^3	Naive Bayesian
Testing set 1	67.3%	67.3%	67.1%	67.6%
Testing set 2	66.0%	66.0%	66.1%	66.9%

6.8 New Directions

This section discusses new directions in data mining for targeted marketing.

6.8.1 Web-Based Targeted Marketing

The topic of targeted marketing can be studied in the context of an emerging new research discipline known as Web Intelligence (WI) [6.51, 6.52, 6.53, 6.55]. Web Intelligence is a new direction for scientific research and development that explores the fundamental roles as well as practical impacts of Artificial Intelligence (AI), such as knowledge representation, planning, knowledge discovery and data mining, intelligent agents, and social network

---

[1] In the figures, the count of the positive examples in the fifth decile is more than in the fourth decile. In order to solve the problem, we can use the formula according to the Odds ratio formula in information retrieval.

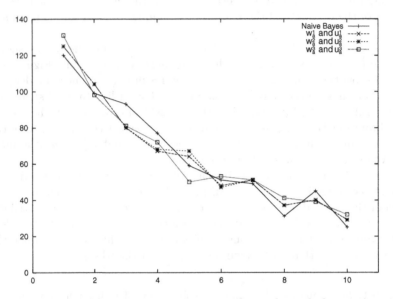

Fig. 6.2. The distribution of the positive examples in the ranked testing-set-1

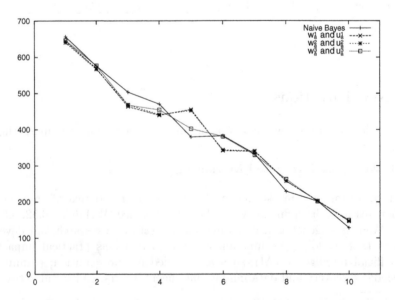

Fig. 6.3. The distribution of the positive examples in the ranked testing-set-2

intelligence, as well as advanced Information Technology (IT), such as wireless networks, ubiquitous devices, social networks, and data/knowledge grids, on the next generation of Web-empowered products, systems, services, and activities. Web-based targeted marketing can be viewed as an important sub-area of Web Intelligence.

Web-based targeted marketing is a key capability for developing intelligent enterprise portals [6.55]. E-business intelligence needs the function of Web-based targeted marketing, which is integrated with other functions of Web intelligence such as Web mining, the ontologies-based search engine, personalized recommendation, as well as automatic email filtering and management [6.51, 6.52]. The e-business activity that involves the end user is undergoing a significant revolution [6.33]. The ability to track users' browsing behavior down to individual mouse clicks has brought the vendor and end-customer closer than ever before. It is now possible for a vendor to personalize his product message for individual customers at a massive scale on the Web. This is called Web-based targeted marketing. Web mining and Web usage analysis play an important role in e-business for Web-based customer relationship management (CRM) and Web-based targeted marketing.

6.8.2 Multi-Aspect Analysis in Multiple Data Sources

Generally speaking, customer data can be obtained from multiple customer touchpoints [6.13]. In response, multiple data sources that are obtained from multiple customer touchpoints, including the Web, wireless, call center, and brick-and-mortar store data, need to be integrated into a single data warehouse that provides a multi-faceted view of customers, including their preferences, interests, and expectations, for multi-aspect analysis. Hence, a multi-strategy and multi-agent data mining framework is required [6.47, 6.48].

The main reason for developing a multi-agent data mining system is that we cannot expect to develop a single data mining algorithm that can be used to solve all targeted marketing problems due to the complexity of real-world applications. Hence, various data mining agents need to be cooperatively used in the multi-step data mining process for performing multi-aspect analysis as well as multi-level conceptual abstraction and learning.

Another reason for developing a multi-agent data mining system is that when performing multi-aspect analysis for complex targeted marketing problems, a data mining task needs to be decomposed into sub-tasks. These sub-tasks can be solved by using one or more data mining agents that are distributed over different computers. Thus, the decomposition problem leads us to the problem of distributed cooperative system design.

6.8.3 Building a Data Mining Grid

In order to implement Web-based targeted marketing by using multiple data sources for multi-aspect analysis, a new infrastructure and platform as middleware is required. Our methodology is to create a Grid-based, organized society of data mining agents, called a *Data Mining Grid* on the grid computing platform (e.g., the Globus toolkit) [6.5, 6.6, 6.9, 6.10, 6.19]. This means:

- Developing various data mining agents for various services oriented multi-aspect data analysis and targeted marketing tasks;
- Organizing the data mining agents into a multi-layer Grid such as data-grid, mining-grid, and knowledge-grid, under the OGSA (Open Grid Services Architecture) that firmly aligns with service-oriented architecture and Web services, understands a user's questions, transforms them to data mining issues, discovers resources and information about the issues, and gets a composite answer or solution;
- Using a conceptual model with three level workflows, namely data-flow, mining-flow, and knowledge-flow, with respect to the data-grid, the mining-grid, and the knowledge-grid, respectively, for managing the grid of data mining agents for multi-aspect analysis in distributed, multiple data sources and for organizing the dynamic, status-based business processes.

That is, the data mining Grid is made of many smaller components that are called *data mining agents*. Each agent by itself can only do some simple things. Yet, when we join these agents in a *Grid*, we can implement more complex targeted marketing and business intelligence tasks.

6.9 Conclusions

Targeted marketing is a new business model in which data mining and marketing databases are used for personalization and business intelligence. Although various individual data mining algorithms may be applied in this field, the specific issues of targeted marketing need new integrated methodologies, platforms, and systems to be developed for mining in enormous amounts of multiple data sources and for multi-aspect analysis. Furthermore, targeted marketing must be expanded to Web-based targeted marketing and to enhancing strategic customer relationship management as key functions for e-business portals.

Acknowledgments

This work was supported by the grant-in-aid for scientific research on priority area "Active Mining" from the Japanese Ministry of Education, Culture, Sports, Science, and Technology, as well as the Natural Science Foundation of China (60173014) and the Beijing Municipal Natural Science Foundation (4022003).

References

6.1 R. Agrawal, T. Imielinski, A. Swami: Mining Association Rules Between Sets of Items in Large Aatabases. *Proc. ACM SIGMOD International Conference on the Management of Data* (ACM Press, 1993) pp. 207-216

6.2 R. Agrawal, H. Mannila, R. Srikant, H. Toivonen, A.I. Verkamo: Fast Discovery of Association Rules, *Advances in Knowledge Discovery and Data Mining* (MIT Press, 1996) pp. 307-328

6.3 L. Breiman, J.H. Friedman, R.A. Olshen, C.J. Stone: *Classification and Regression Trees* (Wadsworth, 1984)

6.4 David Shepard Associates: *The New Direct Marketing* (McGraw-Hill, 1999)

6.5 F. Berman: From TeraGrid to Knowledge Grid, *CACM*, 44, 27-28 (2001)

6.6 M. Cannataro and D. Talia: The Knowledge Grid, *CACM*, 46, 89-93 (2003)

6.7 G. Dong and J. Li: Efficient Mining of Emerging Patterns: Discovering Trends and Differences. *Proc. 5th ACM SIGKDD International Conference on Knowledge Discovery and Data Mining (KDD-99)* (ACM Press, 1999) pp. 43-52

6.8 P.C. Fishburn: Seven Independence Concepts and Continuous Multiattribute Utility Functions. *Journal of Mathematical Psychology*, 11, 294-327 (1974)

6.9 I. Foster and C. Kesselman: *The Grid: Blueprint for a New Computing Infrastructure* (Morgan Kaufmann, 1999)

6.10 I. Foster and C. Kesselman: *The Grid 2: Blueprint for a New Computing Infrastructure* (Morgan Kaufmann, 2004)

6.11 J. Han, Y. Cai, N. Cercone: Data-Driven Discovery of Quantitative Rules in Relational Databases. *IEEE Transaction on Knowledge and Data Engineering*, 5, 29-40 (1993)

6.12 J. Han, M. Kamber: *Data Mining: Concepts and Techniques* (Morgan Kaufmann, 2001)

6.13 S.Y. Hwang, E.P. Lim, J.H. Wang, J. Srivastava: *Proc. PAKDD 2002 Workshop on Mining Data across Multiple Customer Touchpoints for CRM* (2002)

6.14 J.J. Jonker, P.H. Franses, N. Piersma: Evaluating Direct Marketing Campaigns; Recent Findings and Future Research Topics. Erasmus Research Institute of Management (ERIM), Erasmus University Rotterdam in Its Series Discussion Paper with Number 166 (2002)

6.15 W. Klosgen, J.M. Zytkow: *Handbook of Data Mining and Knowledge Discovery* (Oxford University Press, 2002)

6.16 S. Kullback and R.A. Leibler: On Information and Sufficiency. *Annals of Mathematical Statistics*, 22, 79-86 (1951)

6.17 D.B. Leake: *Case-Based Reasoning* (AAAI Press, 1996)

6.18 C.X. Ling, C. Li: Data Mining for Direct Marketing: Problems and Solutions. *Proc. 4th International Conference on Knowlege Discovery and Data Mining (KDD'98)* (AAAI Press, 1998) pp. 73-79

6.19 J. Nabrzyski, J.M. Schopf, J. Weglarz: *Grid Resource Management* (Kluwer, 2004)

6.20 Z. Pawlak: *Rough Sets, Theoretical Aspects of Reasoning about Data* (Kluwer, 1991)

6.21 P. Van Der Putten: Data Mining in Direct Marketing Databases, W. Baets (ed.) *Complexity and Management: A Collection of Essays* (World Scientific, 1999)

6.22 R. Potharst, U. Kaymak, W. Pijls: Neural Networks for Target Selection in Direct Marketing, In K.A. Smith, J.N.D. Gupta (eds.) *Networks in Business: Techniques and Applications* (Idea Group Publishing, 2001)

6.23 J.R. Quinlan: *C4.5: Programs for Machine Learning* (Morgan Kaufmann, 1993)

6.24 C.R. Rao: Diversity and Dissimilarity Coefficients: a Unified Approach. *Theoretical Population Biology*, 21, 24-43 (1982)

6.25 B. Ratner: Finding the Best Variables for Direct Marketing Models. *Journal of Targeting Measurement and Analysis for Marketing*, 9, 270-296 (2001)

6.26 S.E. Robertson: On Relevance Weight Estimation and Query Expansion. *Journal of Documentation*, 42, 182-188 (1986)

6.27 S.E. Robertson, K. Sparck Jones: Relevance Weighting of Search Terms. *Journal of the American Society for Information Science*, 27, 129-146 (1976)

6.28 Y. Sai, Y.Y. Yao, N. Zhong: Data Analysis and Mining in Ordered Information Tables. *Proc. 2001 IEEE International Conference on Data Mining (ICDM'01)* (IEEE Computer Society Press, 2001) pp. 497-504

6.29 G. Salton and M.H. McGill: *Introduction to Modern Information Retrieval* (McGraw-Hill, 1983)

6.30 E. Suzuki: Autonomous Discovery of Reliable Exception Rules, *Proc Third International Conference on Knowledge Discovery and Data Mining (KDD-97)* (AAAI Press, 1997) pp. 259-262

6.31 K.S. Jones, P. Willett: *Readings in Information Retrieval* (Morgan Kaufmann, 1997)

6.32 A.R. Simon, S.L. Shaffer: *Data Warehousing and Business Intelligence for e-Commerce* (Morgan Kaufmann, 2001)

6.33 J. Srivastava, R. Cooley, M. Deshpande, P. Tan: Web Usage Mining: Discovery and Applications of Usage Patterns from Web Data. *SIGKDD Explorations, Newsletter of SIGKDD*, 1, 12-23 (2000)

6.34 R. Stone: *Successful Direct Marketing Methods*, 6th ed. (NTC Business Books, 1996)

6.35 D. Van den Poel, Z. Piasta: Purchase Prediction in Database Marketing with the ProbRough System, In L. Polkowski, A. Skowron (eds.) *Rough Sets and Current Trends in Computing*, LNAI 1424, 593-600 (Springer, 1998)

6.36 S. Watanabe: Pattern Recognition as a Quest for Minimum Entropy. *Pattern Recognition*, 13, 381-387 (Elsevier, 1981)

6.37 M. Wedel, W.A. Kamakura: Market Segmentation: Conceptual and Methodological Foundations (Kluwer, 1999)

6.38 S.K.M. Wong, Y.Y. Yao: A Probability Distribution Model for Information Retrieval. *Information Processing and Management*, 25, 39-53 (1989)

6.39 S.K.M. Wong, Y.Y. Yao: A Generalized Binary Probabilistic Independence Model. *Journal of the American Society for Information Science*, 41, 324-329 (1990)

6.40 S.K.M. Wong, Y.Y. Yao: An Information-Theoretic Measure of Term Specificity. *Journal of the American Society for Information Science*, 43, 54-61 (1992)

6.41 Y.Y. Yao, S.K.M. Wong, C.J. Butz: On Information-Theoretic Measures of Attribute Importance. In N. Zhong, L. Zhou (eds.) *Methodologies for Knowledge Discovery and Data Mining*, LNAI 1574 (Springer, 1999) pp. 479-488

6.42 Y.Y. Yao, N. Zhong: Mining Market Value Functions for Targeted Marketing. *Proc. 25th IEEE International Computer Software and Applications Conference (COMPSAC'01)* (IEEE Computer Society Press, 2001) pp. 517-522

6.43 Y.Y. Yao, N. Zhong: Granular Computing Using Information Tables. T.Y. Lin, Y.Y. Yao, L.A. Zadeh (eds.) *Data Mining, Rough Sets and Granular Computing* (Physica-Verlag, 2002) pp. 102-124

6.44 Y.Y. Yao, N. Zhong, J. Huang, C. Ou, C. Liu: Using Market Value Functions for Targeted Marketing Data Mining. *International Journal of Pattern Recognition and Artificial Intelligence*, 16 (8) 1117-1131 (World Scientific, 2002)

6.45 N. Zhong, J.Z. Dong, C. Liu, S. Ohsuga: A Hybrid Model for Rule Discovery in Data. *Knowledge Based Systems*, 14 (7) 397-412 (Elsevier, 2001)

6.46 N. Zhong, S. Ohsuga: Automatic Knowledge Discovery in Larger Scale Knowledge-Data Bases. In C. Leondes (ed.) *the Handbook of Expert Systems*, 4, 1015-1070 (Academic Press, 2001)

6.47 N. Zhong, C. Liu, S. Ohsuga: Dynamically Organizing KDD Processes, *International Journal of Pattern Recognition and Artificial Intelligence*, 15 (3) 451-473 (World Scientific, 2001)

6.48 N. Zhong, Y.Y. Yao, M. Ohshima, S. Ohsuga: Interestingness, Peculiarity, and Multi-Database Mining. *Proc. 2001 IEEE International Conference on Data Mining (ICDM'01)* (IEEE Computer Society Press, 2001) pp. 566-573

6.49 N. Zhong, J.Z. Dong, S. Ohsuga: Using Rough Sets with Heuristics to Feature Selection. *Journal of Intelligent Information Systems*, 16 (3) 199-214 (Kluwer, 2001)

6.50 N. Zhong, A. Skowron: A Rough Sets Based Knowledge Discovery Process. *International Journal of Applied Mathematics and Computer Science*, 11 (3) 101-117 (Technical University Press, 2001)

6.51 N. Zhong, J. Liu, Y.Y. Yao (eds.): In Search of the Wisdom Web, *IEEE Computer*, 35 (11) 27-31 (2002)

6.52 N. Zhong, J. Liu, Y.Y. Yao (eds.): *Web Intelligence* (Springer, 2003)

6.53 N. Zhong: Toward Web Intelligence. In E.M. Ruiz, J. Segovia, P.S. Szczepaniak (eds.) *Advances in Web Intelligence*, LNAI 2663 (Springer, 2003) pp. 1-14

6.54 N. Zhong, Y.Y. Yao, M. Ohshima: Peculiarity Oriented Multi-Database Mining, *IEEE Transaction on Knowlegde and Data Engineering*, 15 (4) 952-960 (2003)

6.55 N. Zhong: Developing Intelligent Portals by Using WI Technologies, *Proc. the Second International Conference on Active Media Technology (AMT'04)* (World Scientific, 2004)

Part II

Data Mining for Web Intelligence

7. Mining for Information Discovery on the Web: Overview and Illustrative Research

Hwanjo Yu, AnHai Doan, and Jiawei Han

Department of Computer Science
Thomas M. Siebel Center for Computer Science, USA

Abstract

The Web has become a fertile ground for numerous research activities in mining. In this chapter, we discuss research on finding targeted information on the Web. First, we briefly survey the research area. We focus in particular on two key issues: (a) mining to impose structures over Web data, by building taxonomies and portals for example, to aid in Web navigation, and (b) mining to build information processing systems, such as search engines, question answering systems, and data integration systems. Next, we describe two recent Web mining projects that illustrate the use of mining techniques to address the above two key issues. We conclude by briefly discussing novel research opportunities in the area of mining for information discovery on the Web.

7.1 Introduction

Web mining seeks to discover and exploit knowledge on the Web in order to improve user task performance. The past few years have seen an explosion of such mining activities, as conducted in the database, AI, Web, IR, and other research communities [7.10, 7.37, 7.53, 7.68]. The bulk of the research falls roughly into three groups. Works in the first group focus on *finding targeted information on the Web*, either by browsing and navigating, or by querying an information processing system such as a search engine, a question answering system, or a data integration system. The works develop mining techniques that facilitate such activities. Works in the second group aim at *creating and mining structured data*, for example, by transforming HTML data into structured form, so that more expressive queries can be made and conventional data mining can be conducted on top of the structured data. Finally, works in the third group target *building communities and leveraging user activities*. Examples of such works include mining Web usage to improve network performance, collaborative filtering to better predict user preferences, and mass collaboration to build Web artifacts.

Due to the huge body of research on Web mining, in this chapter we focus only on works in the first group: information discovery on the Web. Several excellent survey efforts [7.10, 7.37, 7.53, 7.68] and recent books on data mining and Web mining [7.11, 7.44] provide detailed discussions of the whole body of Web mining research, including those on creating structured data and leveraging user activities.

We begin by giving a broad overview of mining techniques for information discovery on the Web. We focus in particular on two key areas: (a) mining to *impose structures over the data*, for example to build taxonomies and portals, thus helping users explore and find the desired information, and (b) mining to *build information processing systems*, such as search engines, data integration systems, and digital libraries, which in turn help users find information efficiently.

Next, we describe in detail our two recent Web mining projects in these areas. The first project deals with mining to impose structures. It addresses the important topic of classifying Web pages given only positive training examples. Such classification settings arise commonly on the Web, but have not received much attention. The second project deals with mining to build Web information processing systems. It addresses the fundamental problem of matching objects across disparate data sources: deciding if two given objects (e.g., two tuples) refer to the same real-world entity. This problem has recently received much attention in both the database and AI communities.

Our goal with the chapter is to recall attention to the topic of information discovery on the Web, and to illustrate with our research novel opportunities for Web mining in this exciting area. The rest of the chapter is organized as follows. In the next section, we briefly overview the topic of information discovery. The overview is not intended to be comprehensive, but rather to provide a reference frame from a bird's-eye perspective. In Sect. 7.3, we describe the PEBL project on classifying Web pages using only positive examples. In Sect. 7.4, we describe the project on matching objects across disparate data sources. We conclude with a forward look on mining for information discovery in Sect. 7.5.

7.2 Finding Information on the Web

Finding the desired information on the Web is a difficult task that has received much attention. Within this task, we focus on two problems. In the first problem, we employ mining techniques to help users *navigate and explore the Web* by building taxonomies, directories, and portals. In the second one, we use mining to help build information processing systems, which allow users to *make targeted queries on the Web*. Examples of such systems include search engines, comparison shopping sites, data integration systems, data warehouses, and the emerging peer data management systems. We now survey mining activities in each problem.

7.2.1 Exploring and Navigating the Web

To help users explore a topic or zoom in on entities of interest, much research has dealt with building directories and portals on the Web, over Web documents as well as Web sources.

Building Directories of Web Documents: The bulk of research on exploration and navigation builds directories of Web documents. At the beginning, such directories were constructed manually (e.g., *yahoo.com*), but it soon became clear that manual construction would not scale to Web proportions. Hence, several alternatives were explored. One approach is to enlist an army of volunteers to build taxonomies (e.g., Open Directories at *dmoz.org*). Another popular approach is to employ learning techniques to automatically construct taxonomies [7.13, 7.14, 7.26, 7.52, 7.70, 7.73, 7.91]. The key idea is to manually construct a taxonomy, assign to each taxonomic node a set of Web documents (called the training set for that node), use the training sets to build a classifier or set of classifiers, and finally use the classifier(s) to assign new Web documents to appropriate nodes in the taxonomy.

There are several key challenges to this approach. The first challenge is to find relevant new Web documents. Suppose we are to build a taxonomy of Web pages on machine learning. Obviously we do not want to crawl and classify the *entire* Web. A much more efficient way is to crawl intelligently to retrieve only Web pages that are likely to be about machine learning. Such *focused crawling* (also known as *topic distillation*) has been the subject of much recent research [7.12, 7.71, 7.91]. The crawler can be guided by supervised learning techniques, in the form of a Web document classifier [7.12, 7.91], or by reinforcement learning techniques [7.71]. A well-known example of a taxonomy constructed from such focused crawling was the machine learning portal at *cora.justresearch.com*.

The second challenge in building taxonomies is to *accurately* classify Web documents. Existing works have addressed this challenge in several ways. Since the labels (i.e., the taxonomic nodes) form a hierarchy, several works have exploited this structure in the label space, typically by performing hierarchical classification with many classifiers, one at each taxonomic node [7.13, 7.52]. Furthermore, Web documents are typically related via hyperlinks, and such relations can intuitively help in classification: if all "neighbor" documents of a document A are about machine learning, then A is also likely to be about machine learning. This idea was exploited in several recent works [7.14, 7.33, 7.80, 7.92]. Finally, some works have developed new, improved algorithms for classifying text documents [7.52].

Since acquiring training data (i.e., labeled Web documents in our case) tends to be rather labor intensive, another key challenge is to minimize the amount of training data required, and yet maximize its impact. Several works addressed this problem by developing techniques to learn from both a small set of labeled documents and a large set of unlabeled documents [7.81, 7.82, 7.100]. In Sect. 7.3, we describe the PEBL project which develops such a technique.

Building Directories of Data Sources: Besides HTML pages, the Web also contains a huge number of data sources, such as *amazon.com* and *realestate.com*. Thus, it is also important to build directories of these sources, to help users navigate to and query the relevant sources.

Most current works in this area have focused on *source modeling*, that is, learning source characteristics, in order to assign sources accurately into a taxonomy [7.9, 7.47]. In [7.47], the authors focus on learning a keyword distribution of text sources, by repeatedly "probing" sources with queries and counting the word distributions of the returned results. These works have focused on *text* sources. Several recent works have begun to consider learning the characteristics of so-called "Deep-Web" sources, that is, sources that export structured data via a query interface [7.17, 7.45].

7.2.2 Querying with Information Processing Systems

In the previous section, we have discussed mining techniques that help users explore and navigate the Web. We now discuss mining techniques that help build and improve information processing systems that allow users to pose targeted queries. Examples of such systems include search engines, data integration systems, and question answering systems.

The basic objective underlying these systems is to provide a *uniform* query interface to Web data. They can be roughly classified into *query-routing* and *query-answering systems*. A query-routing system takes a user query and returns the set of data sources deemed most relevant to the query, often in decreasing order of relevance. The user then explores the sources to obtain desired information. Examples include search engines and comparison shopping systems. (Many query-routing systems also provide a taxonomy of the sources, to further aid user navigation.) A query-answering system takes a user query, interacts with the sources, and combines data from the source to obtain the exact answers to the query. Examples include data integration systems [7.3, 7.19, 7.40, 7.43, 7.48, 7.56, 7.61], data warehousing systems [7.41], and question answering systems [7.58, 7.64].

The described systems form a natural spectrum. At one end of the spectrum are search engines, which have limited query interfaces that allow for ease of querying, but have also very limited data processing capabilities: they can only *route* queries, not *access* and *combine* Web data to obtain the desired answers. The key advantage is that they are relatively easy to build and maintain. At the other end of the spectrum are systems that answer natural language questions. Clearly, such systems have very easy-to-use query interfaces and powerful data processing capabilities, but are extremely hard to build. Situated between search engines and question answering systems are systems that handle structured data, such as data integration and data warehousing data, which have relatively difficult-to-learn query interfaces and strong data processing capabilities. They are more difficult to build and maintain than search engines, but not as difficult as question answering systems.

In what follows, we describe some major challenges in building and maintaining these information processing systems. These challenges have benefited or can benefit from mining techniques.

Resource Discovery. Search engines must crawl the Web (i.e., "discover" pages) to build a central index for Web pages. Several works have developed techniques to crawl the Web efficiently, and to detect and model changes to pages so that the crawler can focus on pages that change frequently [7.20]. For domain-specific search engines, focused crawling to retrieve Web pages in the domain has been studied [7.12, 7.71, 7.91], as we mentioned in Sect. 7.2.1. However, focused crawling is not yet easy to deploy and adapt across domains. Thus, much more work is still needed in this direction.

Resource Modeling. Once resources have been discovered, information processing systems must *model* the sources for later query processing. Search engines do not have any notable difficulty in this aspect, as they consider a Web page as simply a bag of words. In contrast, modeling data sources for more complex systems (such as data integration ones) is very difficult, and has received much attention [7.2, 7.9, 7.22, 7.27, 7.36, 7.47, 7.55]. We have mentioned modeling text sources in Sect. 7.2.1 [7.9, 7.47]. Modeling a structured data source typically means recovering the *source schema* via examining the HTML pages exported by the source. This process is referred to as *wrapper construction*, and has been researched actively.

Construction of the Query Interface. Search engines and question answering systems have simple query interfaces. In contrast, the query interface for data integration systems can be fairly complex, often in the form of a relational or object-oriented schema (referred to as the *mediated schema*) or an ontology. Constructing such a mediated schema from a set of given source schemas, or from data and schema in a domain, is a very difficult problem that has not received much attention, and that can benefit much from mining techniques. Some recent works (e.g., [7.17, 7.45]) provide techniques that can be considered for the above problem.

Relating the Query Interface and the Source Models. To answer a user query, an information processing system must be able to relate the query to the source models. This problem tends not to arise in search engines, which regard both queries and Web documents as bags of words. However, it becomes a key problem in data integration and warehousing systems. In such system, the relationships are often captured with a set of *semantic mappings* between the query interface and the source schemas [7.31, 7.87, 7.96]. Manually constructing the mappings is extremely labor intensive and error prone. Hence, numerous works have leveraged a broad variety of techniques, including mining ones, to automatically create semantic mappings (e.g., [7.5, 7.6, 7.15, 7.35, 7.62, 7.67, 7.74, 7.75, 7.76, 7.77, 7.79, 7.83, 7.84, 7.86]; see also [7.4, 7.87] for surveys).

Object Matching and Fusion. When passing data across sources or collating data from different sources to answer a user query, an information system often must detect and "fuse" duplicate objects to make the answers more compact and comprehensible to the user. A well-known example of such duplicate elimination is the removal of duplicate Web pages in the ranked answers produced by *Google*. Object matching has been studied extensively in a variety of communities (e.g., databases, data mining, AI, Web) [7.1, 7.7, 7.21, 7.39, 7.42, 7.46, 7.57, 7.72, 7.88, 7.90, 7.95, 7.98]. Earlier solutions employ manually specified rules to match objects [7.46]. Many subsequent solutions attacked the problem using a range of mining techniques [7.7, 7.24, 7.72, 7.90, 7.95]. The commonality underlying these solutions is that they match objects by comparing the shared attributes. A recent work [7.32] extends these previous solutions by adding another layer that utilizes the correlations among the disjoint attributes to maximize matching accuracy. We describe this work in detail in Sect. 7.4.

Evaluating and Ranking the Result. It is important that information systems evaluate and rank the results in a way that displays the most important results first, to allow the user to quickly locate them. Many works have addressed this important topic. The well-known works include the hub-and-authority method [7.51] and Page Rank [7.85].

Displaying the Result. Information systems may also want to *cluster* or *re-organize* the answers in certain ways, to ease the cognitive load on the user and help the user zoom in faster to the desired result. Several recent works have focused on clustering search engine results [7.18, 7.101, 7.102].

Maintaining the System Over Time. The Web is in constant evolution; thus, maintaining information processing systems by keeping them in tune with the highly dynamic environment is an extremely challenging task. We have noted that search engines must constantly recrawl the Web to maintain up-to-date indexes. Maintaining a data integration system is very costly because of the added complexity. One must constantly monitor and update source schemas. When a source changes, one must reestablish the semantic mappings and possibly recompute source characteristics. Thus, efficient maintenance of Web information systems is a crucial task, which can benefit much from mining techniques. This topic has only recently received some attention, with several works on using learning techniques to detect and repair wrappers [7.54, 7.59].

Finally, we note several other topics in Web information systems that also benefit from mining techniques. Several works have dealt with building special-purpose search engines and data integration systems. Information filtering and publish/subscribe research have also benefited from mining techniques. Well-known example of works in this area are the use of learning techniques to filter spam mails and to classify mails into folders.

Query optimization in a Web information system context can certainly benefit from knowing source statistics, which again can be gleaned using a variety of mining techniques. We note also the close similarity among Web information systems as described above and softbots, agents, and personal assistants.

In this section we have provided an overview of mining techniques for information discovery on the Web. In the next two sections we will describe two recent projects that we have conducted in this area: the PEBL project on classifying Web pages using only positive training examples, and a project on matching objects across disparate data sources. The two projects serve to illustrate the issues discussed in the overview, and to highlight novel research issues in information discovery on the Web that can benefit from mining techniques.

7.3 Web Page Classification from Positive Examples

As discussed in Sect. 7.2.1, classifying Web pages of an interesting class is often the first step in building information discovery infrastructures on the Web.

However, constructing a classifier for an interesting class requires laborious preprocessing, such as collecting positive and negative training examples. For instance, in order to construct a "homepage" classifier, one needs to collect a sample of homepages (positive examples) and a sample of non-homepages (negative examples). In particular, *collecting negative training examples* is especially delicate and arduous because (1) negative training examples must uniformly represent the universal set excluding the positive class (e.g., the sample of non-homepages should represent the Internet uniformly, excluding the homepages) and it thus involves laborious manual classifications, and (2) manually collected negative training examples could be biased because of unintentional human's prejudice, which could be detrimental to classification accuracy.

In this chapter, we presents a framework, called *Positive Example Based Learning (PEBL)*, for Web page classification which eliminates the need for manually collecting negative training examples in preprocessing and constructs a classifier from positive and unlabeled examples. (Unlabeled examples are automatically collected by a random sampler.) Figure 7.1 illustrates the difference between a typical learning framework and the PEBL framework for Web page classification. The PEBL framework applies an algorithm, called *Mapping-Convergence (M-C)*, to achieve high classification accuracy (from positive and unlabeled data), as high as that of a traditional SVM (from positive and negative data).

We will first discuss related work in Sect. 7.3.1, and review the margin-maximization property of SVM in Sect. 7.3.2, which is a key to the M-C algorithm. We will present the M-C algorithm with theoretical justification in Sect. 7.3.3 and provide experimental evidences in Sect. 7.3.4.

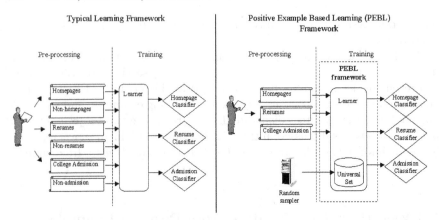

Fig. 7.1. A typical learning framework versus the PEBL framework. Once a sample of the universal set is collected in PEBL, the same sample is reused as unlabeled data for every class

7.3.1 Related Work

Traditional classification approaches using both fully-labeled positive and negative examples, or semi-supervised learning schemes using unlabeled data with labeled data, are not suitable for our problem because: (1) the portions of positive and negative data in feature space are seriously unbalanced without being known (i.e., $Pr(P) << Pr(\overline{P})$), and (2) the absence of negative samples in the labeled data set makes unfair the initial parameters of the model and thus leads to unfair guesses for the unlabeled data.

Learning from positive and unlabeled data, often referred to as *single-class classification* have been attempted by many different approaches. In 1998, F. Denis defined the PAC learning model for positive and unlabeled examples, and showed that k-DNF (Disjunctive Normal Form) is learnable from positive and unlabeled examples [7.29]. After that, some experimental attempts [7.28, 7.60] have been pursued using k-DNF or C4.5. However, these rule-based learning methods are often not applicable to Web page classification because (1) they are not very tolerant with high dimensionality and sparse instance space, which are essential issues for Web page classification; (2) their algorithms require knowledge of the proportion of positive instances within the universal set, which is not available in many problem settings; and (3) they perform poorer than traditional learning schemes given sufficient labeled data.

Recently, a probabilistic method built upon the EM algorithm has been proposed for the text domain [7.65]. The method has several fundamental assumptions: the generative model assumption, the attribute independence assumption, which results in linear separation, and the availability of prior probabilities. PEBL does not require the prior probability of each class, and it can draw nonlinear boundaries using advanced SVM kernels.

For document classification, Manevitz [7.69] compared various single-class classification methods including the neural network method, One-class SVM, nearest neighbor, naive Bayes, and Rocchio, and concluded that One-class SVM and the neural network method were superior to all the other methods, and the two are comparable.

OSVM (One-class SVM), based on the strong mathematical foundation of SVM, distinguishes one class of data from the rest of the feature space, given only a positive data set [7.69, 7.94]. OSVM draws a nonlinear boundary of the positive data set in the feature space using two parameters – ν (to control the noise in the training data) and γ (to control the "smoothness" of the boundary). However, OSVM requires a much larger amount of positive training data to induce an accurate class boundary because its support vectors (SVs) of the boundary come only from the positive data set and thus the small number of positive SVs can rarely cover the major directions of the boundary, especially in high dimensional spaces. Due to the SVs coming from only positive data, OSVM tends to overfit and underfit easily. Tax and Duin proposed a sophisticated method which uses artifically generated unlabeled data to optimize the OSVM's parameters that "balance" between overfitting and underfitting [7.94]. However, their optimization method is infeasibly inefficient in high dimensional spaces, and even with the best parameter setting, its performance still lags far behind bothe the original SVM with negative data and the M-C algorithm without labeled negative data, due to the shortage of SVs which makes the boundary description "incomplete" [7.99].

7.3.2 SVM Margin-Maximization Property

SVM has been widely used in many domains of classification problems [7.34, 7.49, 7.97]. It provides several salient properties, such as margin-maximization and nonlinear transformation of the input space to the feature space using kernel methods [7.8, 7.25]. Here, we briefly review the margin-maximization property of SVM, which is a key to the M-C algorithm.

Consider its simplest form, a linear SVM. A linear SVM is a hyperplane that separates a set of positive data from a set of negative data with *maximum margin* in the feature space. The *margin* indicates the distance from the hyperplane (class boundary) to the nearest positive and negative data in the feature space. (The nearest data points are called *Support Vectors*.) Each feature corresponds to one dimension in the feature space. The distance from the hyperplane to a data point is determined by the strength of each feature of the data. For instance, consider a resume page classifier. If a page has many strong features related to the concept of "resume" (e.g., words "resume" or "objective" in headings), the page would belong to the positive (resume) class in the feature space, and the location of the data point should be far from the class boundary on the positive side. Likewise, another page not having any resume-related features, but having many non-resume related features, should be located far from the class boundary on the negative side. Basically,

SVM computes the class boundary that maximizes the *margin* in the feature space.

7.3.3 The Mapping-Convergence (M-C) Algorithm

In this section, we present the Mapping-Convergence (M-C) algorithm. For convenience of presentation, we use the following notations.

- x is a data instance such that $x \in \mathcal{U}$.
- $\mathcal{P}$ is a subspace for the positive class within $\mathcal{U}$, from which the positive data set P is sampled.
- U (unlabeled data set) is a uniform sample of the universal set.
- $\mathcal{U}$ is the feature space for the universal set such that $\mathcal{U} \subseteq \Re^m$ where m is the number of dimensions.

For example, the universal set is the entire Web, U is a sample of the Web, P is a collection of Web pages of interest, and $x \in \Re^m$ is an instance of a Web page.

We first introduce the notion of "negative strength" to the M-C algorithm.

Let $h(x)$ be the boundary function of the positive class in $\mathcal{U}$ which outputs the distance from the boundary to the instance x in $\mathcal{U}$ such that

$$h(x) > 0 \quad if\ x\ is\ a\ positive\ instance,$$
$$h(x) < 0 \quad if\ x\ is\ a\ negative\ instance,\ \text{and}$$
$$|h(x)| > |h(x')| \quad if\ x\ is\ located\ farther\ than\ x'$$
$$from\ the\ boundary\ in\ \mathcal{U}.$$

Definition 7.3.1 (Strength of negative instances). *For two negative instances x and x' such that $h(x) < 0$ and $h(x') < 0$, if $|h(x)| > |h(x')|$, then x is **stronger** than x'.*

Example 7.3.1. Consider a resume page classification function $h(x)$ from the Web ($\mathcal{U}$). Suppose there are two negative data objects x and x' (non-resume pages) in $\mathcal{U}$ such that $h(x) < 0$ and $h(x') < 0$: x is "how to write a resume" page, and x' is "how to write an article" page. In $\mathcal{U}$, x' is considered more distant from the boundary of the resume class because x has more features relevant to the resume class (e.g., the word "resume" in text), though it is not a true resume page.

The M-C algorithm is composed of two stages: the *mapping stage* and the *convergence stage*. In the mapping stage, the algorithm uses a weak classifier Ψ_1 (e.g., Rocchio or OSVM), which draws an initial approximation of "strong negatives" – the negative data located far from the boundary of the positive class in $\mathcal{U}$ (steps 1 and 2 in Algorithm 13). Based on the initial approximation, the convergence stage runs in iteration using a second base classifier Ψ_2

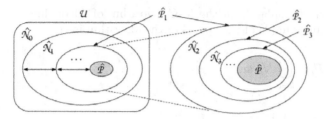

Fig. 7.2. Example of the spaces of the M-C algorithm in U

(e.g., SVM), which maximizes margin to make a progressively better approximation of negative data (steps 3 through 5 in Algorithm. 13). Thus, the class boundary eventually converges to the boundary around the positive data set in the feature space, which also maximizes the margin with respect to the converging negative data.

Algorithm 13 M-C algorithm

1: Input: - positive data set P, unlabeled data set U
2: Output: - a boundary function h_i
   (a) Ψ_1: an algorithm identifying "strong negatives" from U, e.g., Rocchio or OSVM
   (b) Ψ_2: a supervised learning algorithm that maximizes the margin, e.g., SVM
3: Use Ψ_1 to construct a classifier h_0 from P and U which classifies only "strong negatives" as negative and the others as positive
4: Classify U by h_0
   (a) $\hat{N}_0$:= examples from U classified as negative by h_0
   (b) $\hat{P}_0$:= examples from U classified as positive by h_0
5: Set $N := \emptyset$ and $i := 0$
6: Do loop
   (a) $N := N \cup \hat{N}_i$
   (b) Use Ψ_2 to construct h_{i+1} from P and N
   (c) Classify $\hat{P}_i$ by h_{i+1}

   $- \hat{N}_{i+1}$:= examples from $\hat{P}_i$ classified as negative by h_{i+1}
   $- \hat{P}_{i+1}$:= examples from $\hat{P}_i$ classified as positive by h_{i+1}

   (d) $i := i + 1$
   (e) Repeat until $\hat{N}_i = \emptyset$
7: return h_i

The M-C process is illustrated in Algorithm 13. Assume that $\hat{P}$ is a subspace tightly subsuming P within U where the class of the boundary function for $\hat{P}$ is from the algorithm Ψ_2 (e.g., SVM). In Algorithm 13, let $\hat{N}_0$ be the negative space and $\hat{P}_0$ be the positive space within U divided by h_0 (a boundary drawn by Ψ_1), and let $\hat{N}_i$ be the negative space and $\hat{P}_i$ be the positive space within $\hat{P}_{i-1}$ divided by h_i (a boundary drawn by Ψ_2). Then, we can induce the following formulae from the M-C algorithm of Algorithm 13. (Fig. 7.2 illustrates an example of the spaces of the algorithm in $\mathcal{U}$.)

$$\mathcal{U} = \hat{\mathcal{P}}_i + \bigcup_{k=0}^{i} \hat{\mathcal{N}}_k \text{ and} \tag{7.1}$$

$$\hat{\mathcal{P}}_i = \hat{\mathcal{P}} + \bigcup_{k=i+1}^{I} \hat{\mathcal{N}}_k \,, \tag{7.2}$$

where I is the number of iterations in the M-C algorithm.

Theorem 7.3.1 (Boundary Convergence).

Suppose U is uniformly distributed in $\mathcal{U}$. If algorithm Ψ_1 does not gener-ate false negatives, and algorithm Ψ_2 maximizes margin, then (1) the class boundary of the M-C algorithm converges into the boundary that maximally separates P and U outside $\hat{\mathcal{P}}$, and (2) I (the number of iterations) is loga-rithmic to the margin between $\hat{\mathcal{N}}_0$ and $\hat{\mathcal{P}}$.

Proof. $\hat{\mathcal{N}}_0 \cap \hat{\mathcal{P}} = \emptyset$ because a classifier h_0 constructed by the algorithm Ψ_1 does not generate false negatives. A classifier h_1 constructed by the algorithm Ψ_2, trained from the separated space $\hat{\mathcal{N}}_0$ and $\hat{\mathcal{P}}$, divides the rest of the space $(\mathcal{U} - (\hat{\mathcal{N}}_0 + \hat{\mathcal{P}}))$, which is equal to $\cup_{k=1}^{I} \hat{\mathcal{N}}_k$) into two classes with a boundary that maximizes the margin between $\hat{\mathcal{N}}_0$ and $\hat{\mathcal{P}}$. The first part becomes $\hat{\mathcal{N}}_1$ and the other becomes $\cup_{k=2}^{I} \hat{\mathcal{N}}_k$. Repeatedly, a classifier h_{i+1} constructed by the same algorithm Ψ_2, trained from the separated space $\cup_{k=0}^{i} \hat{\mathcal{N}}_k$ and $\hat{\mathcal{P}}$, divides the rest of the space $\cup_{k=i+1}^{I} \hat{\mathcal{N}}_k$ into $\hat{\mathcal{N}}_{i+1}$ and $\cup_{k=i+2}^{I} \hat{\mathcal{N}}_k$ with equal margins. Thus, $\hat{\mathcal{N}}_{i+1}$ always has the margin of half of $\hat{\mathcal{N}}_i$ (for $i \geq 1$). Therefore, I will be logarithmic to the margin between $\hat{\mathcal{N}}_0$ and $\hat{\mathcal{P}}$.

The iteration stops when $\hat{\mathcal{N}}_i = \emptyset$, where there exists no sample of U outside $\hat{\mathcal{P}}$. Therefore, the final boundary will be located between P and U outside $\hat{\mathcal{P}}$ while maximizing the margin between them.

Theorem 7.3.1 proves that, under certain conditions, the final boundary will be located between P and U outside $\hat{\mathcal{P}}$. However, in theorem 7.3.1, we made a somewhat strong assumption, i.e., U is uniformly distributed, to guarantee the boundary convergence. In a more realistic situation where there is some distance δ between classes, *if the margin between h_{i+1} and $\hat{\mathcal{N}}_i$ becomes smaller than δ at some iteration, the convergence stops because $\hat{\mathcal{N}}_{i+1}$ becomes empty.* The margin between h_{i+1} and $\hat{\mathcal{N}}_i$ reduces by half at each iteration as the boundary h_{i+1} approaches $\hat{\mathcal{P}}$, and thus the boundary is not likely to stop converging when it is far from $\hat{\mathcal{P}}$, unless U is severely sparse. Thus, we have the following claim:

Claim. The boundary of M-C is located between P and U outside $\mathcal{P}$ if U and P are not severely sparse and there exist visible gaps between $\mathcal{P}$ and U.

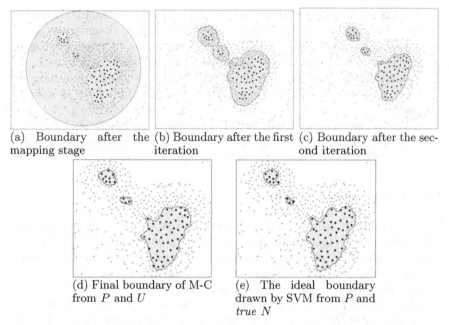

(a) Boundary after the mapping stage

(b) Boundary after the first iteration

(c) Boundary after the second iteration

(d) Final boundary of M-C from P and U

(e) The ideal boundary drawn by SVM from P and *true* N

Fig. 7.3. Intermediate results of M-C. P is big dots, and U is all dots (small and big).

Note that the M-C algorithm is quite general, as long as the component algorithms Ψ_1 and Ψ_2 satisfy the following requirements:

1. **Ψ_1 must not generate false negatives.**
   We can use any reasonable classifier, such as Rocchio or OSVM, and adjust the threshold so that it makes near 100% recall by sacrificing precision. OSVM is used for Ψ_1 for this research, and sets the bias b very low to achieve a high recall. In practice, a small fraction of false negatives can be handled by the soft constraint of Ψ_2 (e.g., SVM). The precision of Ψ_1 does not affect the accuracy of the final boundary as long as it approximates a certain amount of negative data because the final boundary will be determined by Ψ_2.
   Figure 7.3 shows an example of the boundary after each iteration of M-C. The mapping stage only identifies very strong negatives by covering a wide area around the positives (Figure 7.3(a)). Although the precision quality of mapping is poor, the boundary at each iteration converges (Figures 7.3(b) and (c)), and the final boundary is very close to the ideal boundary drawn by SVM on P and the *true* N (Figures 7.3(d) and 7.3(e)). The experiments in Sect. 7.3.4 also show that the final boundary becomes very accurate although the initial boundary of the mapping stage is very rough by the "loose" setting of the threshold of Ψ_1.

2. **Ψ_2 must maximize margin.**

SVM is used for Ψ_2 for this research. With a strong mathematical foundation, SVM automatically finds the margin-maximized boundary without a validation process and without many parameters to tune. The small numbers of theoretically motivated parameters also work well with an intuitive setting. In practice, the soft constraint of SVM is necessary to cope with noise or outliers, though P is unlikely to have a lot of noise in practice since it is usually carefully labeled by users. In the experiments, a low setting (i.e., $\nu = 0.01$ or 0.1) of ν (the parameter to control the rate of noise in the training data) performed well for this reason. (We used ν-SVM for the semantically meaningful parameter [7.16].)

7.3.4 Experimental Results

In this section, we provide empirical evidence that the PEBL framework using positive and unlabeled data performs as well as the traditional SVM using manually labeled (positive and unbiased negative) data. We present experimental results with two different universal sets: the Internet (*Experiment 1*) and university computer science departments (*Experiment 2*). The Internet (*Experiment 1*) is the largest possible universal set on the Web, and CS department sites (*Experiment 2*) constitute a conventionally small universal set. We design these two experiments with the two totally different sizes of universal sets so that we may verify the applicability of the method on various domains of universal sets.

Data Sets and Experimental Methodology. *Experiment 1* (The Internet). The first universal set in the experiments is the Internet. To collect random samples of Internet pages, we used DMOZ[1], which is a free open directory of the Web containing millions of Web pages. Random sampling of a search engine database such as DMOZ is sufficient (we assume) to construct an unbiased sample of the Internet. We randomly selected 2,388 pages from DMOZ to collect unbiased unlabeled data. We also manually collected 368 personal homepages, 192 college admission pages, and 188 resume pages to classify three corresponding classes. (Each class is classified independently.) We used about half of the pages of each class for training and the other half for testing. For testing negative data (for evaluating the classifier), we manually collected 449 non-homepages, 450 non-admission pages, and 533 non-resume pages. (We collected negative data just for evaluating the classifier we construct. The PEBL does not require collecting negative data to construct classifiers.) For instance, for the personal homepage class, we used 183 positive and 2,388 unlabeled data for training, and used 185 positive and 449 negative data for testing.

Experiment 2 (University computer science department). The WebKB data set [7.26] contains 8,282 Web pages gathered from university computer

---

[1] Open Directory Project http://dmoz.org.

science departments. The collection includes the entire computer science department Web sites from various universities. The pages are divided into seven categories: student, project, faculty, course, staff, department, and others. In the experiments, we classify independently the three most common categories: *student, project,* and *faculty,* which contain 1,641, 504, and 1,124 pages respectively. We randomly selected 1052 and 589 student pages, 339 and 165 project pages, and 741 and 383 faculty pages for training and testing, respectively. For testing negative data, we also randomly selected 662 non-student pages, 753 non-project pages, and 729 non-faculty pages. We randomly picked up 4,093 pages from all categories to make a sample universal set, and the same sample is used for all the three classes as unlabeled data. For instance, for faculty page classification, we used 741 positive and 4,093 unlabeled data for training, and used 383 positive and 729 negative data for testing.

We extracted features from different parts of a page – URL, title, headings, link, anchor-text, normal text, and meta tags. Each feature is a predicate indicating whether each term or special character appears in each part, e.g., $\sim$ in URL, or the word 'homepage' in title. In Web page classification, normal text is often a small portion of a Web page, and structural information usually embodies crucial features for Web pages; thus, the standard feature representation for text classification such as TFIDF is not often used in the Web domain because it is tricky to incorporate such representations for structural features. For instance, occurrence of $\sim$ in a URL is more important information than how many times it occurs. For the same reason, we did not perform stopwording and stemming because the common words in text classification may not be common in Web pages. For instance, a common stopword, such as I or my, can be a good indicator of a student homepage. However, this feature extraction method may not be the best way for SVM for Web page classification. Using more sophisticated techniques for preprocessing the features could improve the performance further.

For SVM implementation, we used LIBSVM[2]. We used Gaussian kernels because of its high accuracy. For single-class classification problems, Gaussian kernels perform the best because of their flexible boundary shapes that fit complicated positive concepts [7.93]. We used theoretically motivated parameters for SVM (e.g., $\nu = 0.1$ or 0.01, $\gamma = \frac{1}{m}$), without explicit performing validation since they already perform well.

We report the result with *precision-recall breakeven point* (P-R), a standard measure for binary classification. Accuracy is not a good performance metric because very high accuracy can be achieved by always predicting the negative class. Precision and recall are defined as:

$$Precision = \frac{\#\ of\ correct\ positive\ predictions}{\#\ of\ positive\ predictions}\ \ and$$

---

[2] http://www.csie.ntu.edu.tw/~cjlin/libsvm/

$$Recall = \frac{\#\ of\ correct\ positive\ predictions}{\#\ of\ positive\ data}\ .$$

The *precision-recall breakeven point* (P-R) is defined as *the precision and recall value at which the two are equal*. We adjusted the decision threshold b of the SVM at the end of each experiment to find P-R.

Table 7.1. Precision-recall breakeven points (P-R) showing performance of PEBL, TSVM (Traditional SVM trained from manually labeled data), and OSVM (One-class SVM) in the two universal sets (U). The number of iterations to the convergence in PEBL is shown in parentheses

U	Class	TSVM	PEBL	OSVM
	homepage	88.11	85.95 (7)	43.24
The Internet	admission	93.0	95.0 (8)	51.0
	resume	98.73	98.73 (4)	26.58
	student	94.91	94.74 (14)	61.12
CS Department	project	84.85	83.03 (12)	18.18
	faculty	93.47	92.69 (14)	40.47

Results. We compare three different methods: TSVM, PEBL, and OSVM. (See Table 7.1 for the full names.) We show the performance comparison on the six classes of the two universal sets: the Internet and CS department sites. We first constructed an SVM from positive (P) and unlabeled data (U) using PEBL. On the other hand, we manually classified the unlabeled data (U) to extract unbiased negatives from them, and then built a TSVM (Traditional SVM) trained from P and those unbiased negatives. We also constructed OSVM from P. We tested the same testing documents using those three methods.

Table 7.1 shows the P-Rs (precision-recall breakeven points) of each method, and also the number of iterations to converge in the case of PEBL. In most cases, PEBL without negative training data performs almost as well as TSVM with manually labeled training data. For example, when we collect 1,052 student pages and manually classify 4,093 unlabeled data to extract non-student pages to train TSVM, it gives 94.91% P-R (precision-recall breakeven point). When we use PEBL without doing the manual classification, it gives 94.74% P-R. However, OSVM without the manual classification performs very poorly (61.12% P-R).

Figures 7.4 and 7.5 show the details of performance convergence at each iteration in the experiment of the universal set, the Internet and CS department sites respectively. For instance, consider the first graph (personal homepage class) in Fig. 7.4. The P-R of M-C is 0.65 at the first iteration, and 0.7 at the second iteration. At the seventh iteration, the P-R of M-C is very close to that of TSVM. The performance of M-C is converging rapidly into that of TSVM in all the experiments.

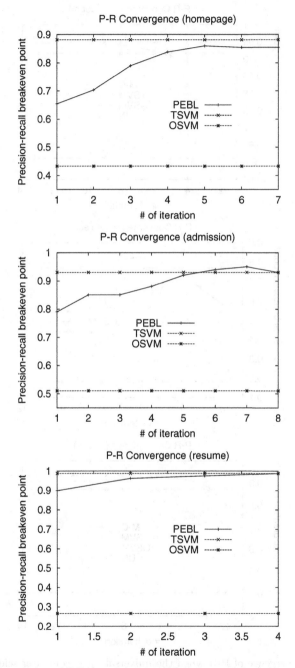

Fig. 7.4. Convergence of P-R (precision-recall breakeven point) when the universal set is the Internet

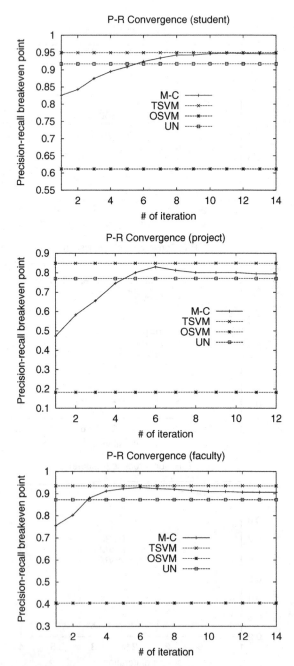

Fig. 7.5. Convergence of P-R when the universal set is computer science department sites. UN indicates the SVM constructed from positive and sample of universal set as a substitute for negative data

The P-R convergence graphs in Fig. 7.5 show one more line (P-R of UN), which is the P-R when using the sample of the universal set (U) as a substitute for negative training data. This is reasonable since U can be thought of as an approximation of negative data. However, UN obviously degrades the performance because a small number of false positive training data significantly affects the set of support vectors which is critical to classification accuracy.

For some classes, the performance starts decreasing at an intermediate point of the iterations in M-C. For instance, the convergence graph of the project class in Fig. 7.5 shows peak performance at the sixth iteration, and the performance starts decreasing from that point. This happens when the positive data set is under-sampled so that it does not cover major directions of positive area in the feature space. In this case, the final boundary of M-C, which fits around the positive data set tightly, tends to end up overfitting the true positive concept space. Finding the best number of iterations in M-C requires a validation process which is used to optimize parameters in conventional classifications. However, M-C is assumed to have no negative examples available, which makes it impossible to use conventional validation methods to find the best number of iterations. Determining the stopping criteria for M-C without negative examples is interesting further work for cases where the positive data is seriously under-sampled.

7.4 Object Matching Across Disparate Data Sources

The second project that we discuss in the chapter is *object matching*: the problem of deciding if two given objects (e.g., two relational tuples) refer to the same real-world entity. Object matching is often used to consolidate information about entities and to remove duplicates when merging multiple information sources. As such, it plays an important role in many information processing contexts, including information integration, data warehousing, information extraction, and text join in databases (e.g., [7.1, 7.7, 7.21, 7.42, 7.46, 7.57, 7.72, 7.90, 7.95, 7.98]).

Numerous solutions to object matching have been proposed, in both the AI and database communities (see the related work section). Virtually all of these solutions make the assumption that the objects under consideration *share the same set of attributes*. They then match objects by comparing the similarity of the shared attributes.

In this section, we consider the more general matching problem where the objects can also have non-overlapping (i.e., *disjoint*) attributes, such as matching tuples that come from two relational tables with schemas (age,name) and (name,salary). We observe that this problem frequently arises in information integration, when querying a data source, or when merging tuples coming from different sources. In information integration, the sources are typically developed in an independent fashion, and therefore are likely

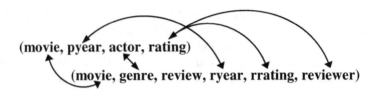

Fig. 7.6. The schemas of two tables in the movie domain. There are many correlations among table attributes (signified with arrows) that can be exploited for object matching

to have overlapping, but different, schemas. When dealing with such sources, prior work has not exploited the disjoint schema portions for the purpose of object matching.

We describe the (Profiler-Based Object Matching) solution that can exploit the disjoint attributes to maximize matching accuracy. The key observation underlying our approach is that the disjoint attributes are often correlated, and that such correlation can be leveraged to perform a "sanity check" for object matching. For example, consider the two tuples (9, "Mike Smith") and ("Mike Smith",200K). Assuming that they match, we would have a "Mike Smith" who is 9 year old and has a salary of 200K. This appears unlikely, based on our knowledge, specifically on the *"profile"* that we have about what constitutes a typical person. This profile tells us that such a relationship between age and salary is unlikely to exist. Thus, the above two tuples are unlikely to match.

As another example that illustrates the approach, consider two relational tables that contain information about movies and their reviews, respectively. Figure 7.6 shows the schemas of the two tables. The meanings of most schema attributes are clear from their names, except perhaps pyear and ryear, which specify the years when the movie was produced and reviewed, respectively, and rrating, which specifies the rating as given by the reviewer.

Given two tuples from the tables, begin by matching the shared attribute movie (i.e., movie name), using any of the existing object matching techniques. If the similarity between the names is low, discard the tuple pair as being without match. Otherwise, apply a set of modules called *profilers* to the tuple pair to perform "sanity check". A profiler contains knowledge about a *specific concept*, such as movie, actor, or review. When given a tuple pair that contains information about the concept, the profiler can examine the pair to decide if it violates any constraint on the concept.

In our movie example, a tuple pair contains information about several concepts in the movie domain, and therefore can be examined by many profilers. For example, a *review profiler* may know that the year in which a review was published must not precede the production year of the reviewed movie. Thus, this profiler can check if the values of the disjoint attributes pyear and ryear satisfy that constraint. This profiler may also know that certain reviewers (e.g., Roger Ebert) have never reviewed any movie with an average rating below 4 (out of 10). Thus, it may also check reviewer and rating for this correlation. An *actor profiler*, on the other hand, may know that a certain actor has never performed in action movies, and would check attributes actor and genre. As yet another example, a *movie profiler* may know that the average rating of a movie tends to be positively correlated. Thus, it may check attributes rating and rrating. Suppose their values are 9 and 2, respectively, then the profiler may conclude that the two tuples probably do not match. knows how to combine the conclusions of the many profilers in order to arrive at a final matching decision for the tuple pair.

A compelling property of profilers is that they contain knowledge about *domain concepts* (e.g., movies, reviews, persons, etc.). Hence, they can be constructed once, and then applied to many object matching tasks, as long as the tasks involve the concepts. They can be constructed by domain experts and users, and can also be learned from the data in the domain (e.g., from all movie tuples in the Internet Movie Database at *imdb.com*). Alternatively, they can be constructed in the context of a specific matching task from the training data for that task. Afterwards, they can also be transferred to other related matching tasks in the domain.

The approach to object matching therefore possesses several desirable characteristics. First, unlike previous approaches, it can exploit disjoint attributes to maximize matching accuracy. Second, it enables the construction and transfer of matching knowledge (in the form of profilers) across matching tasks. Finally, it provides an extensible framework into which to plug newly developed profilers, to further improve matching accuracy. Such frameworks have proven useful for solving other problems, such as schema matching [7.30, 7.31, 7.66] and information extraction [7.26, 7.38], but to our knowledge have not been considered for object matching.

The key challenges are to define, construct, and combine profilers. In the rest of the section, we describe the first steps toward solving these challenges. Specifically, we make the following contributions:

- We introduce the general object matching problem where objects can also have disjoint attributes.
- We describe the solution that exploits the disjoint attributes to maximize matching accuracy. The solution can reuse knowledge from previous matching tasks, and provides an extensible framework into which new matching knowledge can be easily incorporated.
- We present preliminary experimental results on two real-world data sets that show the promise of the approach. The results also show that extending existing matching techniques in a straightforward manner to exploit disjoint attributes may actually decrease rather than increase matching accuracy.

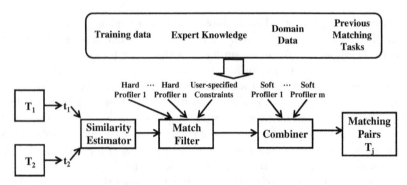

Fig. 7.7. The working of the PROM system

7.4.1 Problem Definition

We now describe the specific object matching problem that we consider in this section. Let T_1 and T_2 be two relational tables. We say two tuples from the tables *match* if they refer to the same real-world entity. A table attribute is called a *shared attribute* if and only if it appears in both tables, and any two tuples that match must agree on the value of that attribute. Other attributes are called *disjoint attributes*. We assume that tables T_1 and T_2 have a non-empty set of shared attributes.

For example, the two tables in Fig. 7.6 share the attribute movie. A matching pair of tuples from the two tables must share the same movie name. In contrast, attributes rating and rrating are not shared, because the same movie may have different ratings.

Given the two tables T_1 and T_2, the matching problem that we consider is to find all matching tuples between them. This is a very general problem setting which arises in many contexts, including data integration [7.95], data warehousing [7.1], and text join in databases [7.42]. In the rest of the section, we shall use the terms "object" and "tuple" interchangeably when there is no ambiguity.

The performance of matching algorithms have typically been evaluated with *matching accuracy* and *runtime efficiency* [7.1, 7.46]. As a first step, we shall in this section focus on improving matching accuracy. Improving runtime is the subject of future research. In the experiment section, we describe our accuracy measures in detail. In the next section, we describe the approach to solving the above object matching problem.

7.4.2 The PROM Approach

Figure 7.7 illustrates the working of the PROM system. Given two tuples, T_1 and T_2, from the input tables, the Similarity Estimator computes a similarity value for the tuple pair and decides if T_1 and T_2 can potentially match. Note that the similarity value is computed *based solely on the shared attributes*. If this module decides that the similarity value is low, it discards the tuple pair; otherwise it passes the pair to the Match Filter.

The Match Filter uses a set of hard profilers to check if the tuple pair could possibly match. A hard profiler contains hard constraints on the concept that it profiles. If *any* hard profiler says no, then the pair is discarded from further consideration. Notice that the Match Filter can also take user-specified hard constraints (treating them as belonging to yet another hard profiler).

Any tuple pair surviving the Match Filter is passed to the Meta Profiler. This module employs a set of soft profilers. Each soft profiler issues a confidence score that indicates how well the tuple pair fits the profile maintained by the profiler. The Combiner merges the confidence scores to obtain a single overall score, then decides based on this score if the tuple pair fits the profile and the Tuples are thus likely to match. If the decision is "yes", the pair is stored in the result table T_J, otherwise it is discarded.

We now describe the module in more detail. We note that the Similarity Estimator can employ any of existing object matching techniques (see the related work section), and hence is not discussed further. In the experiment section, we describe specific instantiations of the architecture for the real-world data sets.

Profilers: As mentioned earlier, a profiler contains a "profile" of a concept, that is, the knowledge about what constitutes a typical instance of the concept. Most importantly, given a tuple pair that includes information about the concept, the profiler can issue a confidence score on how well the pair fits the concept "profile" (in a sense, on how well the data of the two tuples fit together).

A hard profiler specifies "hard" constraints about a concept, that is, constraints that *any* instance of that concept *must* satisfy. An example of such constraints is that the review year must not precede the year the movie is produced. As another example, an actor hard profiler covers actors and may specify that a certain actor has never played in a movie with the average rating less than 4.

Hard profilers can be constructed manually, or automatically by examining the data in the domain, given that the data is *complete*. For example, "hard" constraints about an actor and his/her movie rating can be automatically derived by examining *all* movies that involve the actor.

Note that the user can also specify "hard" constraints about the matching process, and these constraints can be thought of as making up a *temporary* hard profiler (see Fig. 7.7). While other hard profilers cover general concepts, and can thus be transferred across matching tasks, some user-supplied "hard" constraints may be task-specific, and thus not transferable.

A soft profiler also covers a certain concept, but specifies "soft" constraints that any instance of that concept is *likely* to satisfy. A movie soft profiler may specify that the IMDB rating and the Ebert rating of a movie are strongly

Table 7.2. Experimental results on the Citeseer data set

		Baseline	Extended Traditional			PROM			
			Man	AR	extDT	DT	Man+DT	Man+AR	Man+DT+AR
CiteSeer	R	0.99	0.97	0.96	0.91	0.95	0.67	0.96	0.97
	P	0.67	0.83	0.71	0.58	0.78	0.87	0.82	0.86
	F	0.80	0.89	0.81	0.71	0.85	0.76	0.88	0.91

correlated, in that they would differ by less than 3. Most movies, but not all, would satisfy this constraint.

Like hard profilers, soft profilers can also be constructed in several ways. They can be elicited manually from domain experts and users (and then evaluated on some training data to obtain confidence scores for the elicited rules). They can also be constructed from domain data. For example, we can learn a Bayesian network from movie instances in the IMDB database. This Bayesian network would form a soft profiler for movies. Soft profilers can also be constructed directly from training data for a particular matching task. Given a set of matching and non-matching pairs, virtually any learning technique can be applied to construct a classifier that in essence represents a soft profiler.

Combining Profilers: Since the hard profilers issue "yes/no" predictions whereas the soft profilers issue confidence scores, we separate the combination of the two types of profilers, as represented by the Match Filter and the Combiner. We also believe that separating the combination of profilers this way improves matching accuracy over methods that combine all profilers in a single stage; we are in the process of verifying this with current research.

The Match Filter uses an AND combination to merge hard profilers' predictions. That is, if any hard profiler says no, then the overall prediction is no, and the tuple pair is discarded. The Combiner merges soft profilers' predictions by computing the weighted sum of the confidence scores. The weights are currently set manually, based on some experiments on holdout data. In the future, we shall explore methods to set weights automatically, in a fashion similar to that of [7.31].

7.4.3 Empirical Evaluation

We now present preliminary results that demonstrate the utility of exploiting disjoint attributes and the potential of the approach.

Data: We evaluated on two data sets, Citeseer and Movies. The data set Citeseer was obtained from http://citeseer.nj.nec.com/mostcited.html, which lists highly cited authors together with their homepages. An actual line from this page is "J. Gray homepage-url1 ... homepage-url5", where the five homepage URLs were suggested by a search engine. The homepages belong to James Gray at Walker Informatics, Jeffrey Gray at University of Alabama, and so on. Only one homepage actually belongs to the correct Jim Gray (at Microsoft Research). Thus, the object matching problem here is to match author names such as "J. Gray" with their correct homepage URLs.

We downloaded the top 200 authors, together with the suggested home-pages. Since in this first step we only consider matching relational tuples, we manually converted each homepage into a tuple by extracting from the homepage information such as author's name, name and rank of current university, author's position, and graduation year. We removed authors who have no associated homepage, and performed some simple text processing. The final data set consists of 150 author names and 254 homepage tuples, for an average of 1.7 homepage tuples per author.

The data set Movies consists of two tables, with formats (movie-name1, production-year,avg-rating) and (movie-name2,review-year,ebert-rating,review). They are obtained from the Internet Movie Database (imdb.com) and Roger Ebert's Review Page (suntimes.com/ebert/ebertser.html), respectively. The tables consist of about 10,000 tuples and 3,000 tuples, respectively.

Algorithms & Methodologies: We begin by describing algorithms applied to the Citeseer data set. First, we applied Baseline, an algorithm that matches tuples based only on the shared attributes: in our case author's name with homepage owner's name. Baseline converts the values of the shared attributes into a set of tokens, then compares the sets of tokens obtained .

Next, we applied three *extended traditional* algorithms, which extend existing object matching techniques that exploit only shared attributes to also exploit disjoint attributes. Extended-Manual manually specifies the matching rules (e.g., "if similarity(name1,name2) ≥ 0.8 but position=student then the two tuples do not match"). Thus, in a sense this method extends the manual method described in [7.46], which would exploit only shared attributes such as "name1" and "name2". Extended-AR is similar to Extended-Manual, but uses the association rule classification method of [7.63] to guide the process of generating rules. The rules of Extended-AR are then manually picked among the generated rules. In contrast to the above two (semi-)manual methods, Extended-DT is completely automatic. It extends the decision tree method in [7.95] by adding to the training data set all disjoint attributes, and a new attribute that specifies for each tuple pair its similarity value, as computed based on the shared attributes.

For the Similarity Estimator, we used the Baseline algorithm described above. We used no hard profilers. We used three soft profilers: one that consists of several "soft" manually specified rules, one that uses decision tree techniques, and one that uses association rule techniques.

We applied similar algorithms to the Movies data set. We then evaluated matching accuracy using three measures: *recall* (number of correct matching pairs in the join table divided by the total number of correct matching pairs), *precision* (number of correct matching pairs in the join table divided by the total number of pairs in the join table), and *F-value* (defined as $2 * recall * precision/(recall + precision)$). These measures have been used widely in the object matching literature. They also suit our objectives of developing matching methods that maximize precision and recall.

On each data set, we performed 4-fold cross validation, and reported the *average* recall, precision, and F-value. We took care to create folds that are representative of the overall data set (see [7.7] for an example of such fold creation).

Results: Since the results on both data sets are similar, we report only those of Citeseer. Table 7.2 shows the evaluation results for this data set. Each column in the table lists recall, precision, and F-value (in that order) for a specific object matching algorithm.

The results for Baseline (first column) show that it achieves high recall (99%) but low precision (67%), thereby demonstrating that matching based on the shared attributes only (names, in this case) can be quite inaccurate. Extended-Manual (second column) decreases recall slightly (by 2%) but increases precision substantially (by 16%), thus demonstrating that exploiting disjoint attributes (any attribute other than names in this case) can significantly boost matching accuracy. Extended-AR (third column) shows similar, albeit slightly worse, performance to Extended-Manual.

The automatic method Extended-DT (fourth column) shows some surprising results: its precision is substantially lower than that of Baseline (58% vs. 67%). This is unusual because one would expect that Extended-DT improve matching precision by exploiting disjoint attributes. A close inspection reveals that many rules that Extended-DT constructed do not refer to the similarity values of the input tuples at all. In other words, these rules match tuples based *solely on exploiting the correlation among the disjoint attributes*, ignoring the shared attributes. (The two previous methods do not have any such rules because those rules are manually constructed or verified.) It is thus clear that such rules will not be very accurate on the testing data. This surprising result suggests that extending prior matching techniques in a straightforward manner to handle disjoint attributes may actually decrease rather than increase matching accuracy.

Besides examining a method's performance with respect to the baseline and extended algorithms, we also want to know if adding more profilers would be better for accuracy than having fewer profilers. Thus, we ran four variations of the methods (see the last four columns of Table 7.2). The DT variation uses only one soft profiler, which is the decision tree method. Man+DT uses the soft manual profiler and the soft decision tree profiler. Man+AR is similar to the above variation, but replaces the decision tree with the association rule classifier. Finally, Man+DT+AR is the complete algorithm.

The results show that the DT variation beats the Extended-DT. This suggests that extending prior matching techniques to exploit disjoint attributes is promising and potentially better than a straightforward extension of traditional techniques. The results also show that the complete system (last column) achieves the highest F-value (0.91) over any previous method, due to high precision and recall. (In particular, this algorithm found the correct Jim Gray homepage, which the Baseline algorithm could not.) The results suggest that the complete system obtains the best performance and that adding more profilers may improve matching accuracy, because more matching knowledge can be utilized.

In summary, the preliminary results on the two data sets suggest that:

- exploiting disjoint attributes can substantially improve matching accuracy, but
- exploiting them by straightforwardly extending existing techniques may actually decrease rather than increase matching accuracy, and
- the approach to exploit disjoint attributes with domain knowledge can improve accuracy over baseline and extended traditional methods.

Discussion: We are currently experimenting with several new methods to train profilers in these domains (e.g., Naive Bayes as well as methods that do not require training data). We also plan to transfer the profilers constructed in these matching tasks (e.g., the decision tree soft profiler) to other related matching tasks to examine the effect of transferring such knowledge. We are also particularly interested in training profilers from domain data, independently of any matching task (e.g., training movie and actor profilers from *imdb.com*), and then applying these profilers to matching tasks in the domain.

7.4.4 Related Work

Our work builds upon numerous matching solutions that have been developed in the AI, database, and data mining communities (e.g., [7.1, 7.7, 7.21, 7.39, 7.42, 7.46, 7.57, 7.72, 7.88, 7.90, 7.95, 7.98]). Earlier solutions employ manually specified rules to match objects [7.46]. Many subsequent solutions learn matching rules from a set of training data created from the input tables [7.7, 7.90, 7.95]. Several solutions focus on efficient techniques to match strings [7.42, 7.78]. Others also address techniques to scale up to a very large number of objects [7.24, 7.72]. The commonality underlying these solutions is that they match objects by comparing the shared attributes. Our solution extends these previous solutions by adding another layer that utilizes the correlations among the disjoint attributes to maximize matching accuracy. Our use of attribute correlation bears some resemblance to [7.50], in which the authors exploit statistical correlation among schema attributes to find semantic mappings between the attributes of two relational tables.

The topics of knowledge reuse and incorporating prior knowledge have been studied actively in the AI community. More closely related to our approach, several AI works have considered the issue of reusing classifiers that are learned in other domains (e.g., [7.23]). Our work differs from these in several aspects. First, we also reuse knowledge types other than classifiers (e.g., the manual profilers). Second, when reusing classifiers, we do not attempt to reuse *arbitrary* classifiers from other domains. Instead, we advocate building task-independent classifiers and reusing only those. This is possible in our context due to the frequent recurrence of common concepts in matching tasks within a domain. For example, any matching task in the movie domain is likely to involve the concepts of movie, review, actor, and so on.

Recently, knowledge reuse has received increasing attention in the database community, and several works on schema matching [7.6, 7.30, 7.31, 7.66] and data integration (e.g., [7.89]) have investigated the issue. Our work can be seen as a step in this direction. To our knowledge, this is the first work that attempts to reuse knowledge in the context of object matching.

7.4.5 Summary

Object matching plays an important role in a wide variety of information management applications. Previous solutions to this problem have typically assumed a uniform setting where objects share the same attributes. In this section, we have considered a more general setting where objects can have different but overlapping sets of attributes. Such a setting commonly arises in practice, where data sources are independently developed and are thus unlikely to share the same schemas.

We have proposed the solution that builds upon previous work, but exploits the disjoint attributes to substantially improve matching accuracy. To do this, the solution employs multiple profilers, each of which contains information about a certain concept in the matching task. The profilers can be specified by domain experts, trained from training data that is obtained from the input objects, transferred from related matching tasks, or constructed from domain data. Most importantly, the profilers contain task-independent information and thus can be reused once constructed. This makes the approach conserve labor and maximize accuracy on any particular matching task. Preliminary experiments on two real-world data sets show the promise of the approach.

Our approach also suggests a broader knowledge-reuse methodology: within any particular task, isolate knowledge that is task-dependent (e.g., similarity knowledge) from that which is task-independent (e.g., profile knowledge). The latter, once learned, can be reused across tasks. This methodology is clearly not always applicable, but can be effective in appropriate settings, as we have demonstrated. Our future research, besides developing the solution – as discussed in the experiment section – will aim to further explore this idea.

7.5 Conclusions

Information discovery on the Web remains a central theme of Web research into the foreseeable future. Numerous mining techniques have been developed to address this problem. In this chapter we have surveyed these techniques. We have also described two of our recent projects that address different aspects of the problem. The PEBL project develops methods to classify Web pages using only positive training examples, while the project focuses on matching objects across disparate data sources. The two projects specifically illustrate the use of mining techniques for information discovery.

Our overview and the described projects point to several emerging opportunities for research on Web mining. Clearly, there are additional issues to consider in Web page classification, as discussed in Sect. 7.3.4, and there are many opportunities for mining techniques to address building Web information processing systems, as discussed in Sect. 7.2.2. The mining efforts that help build next-generation information processing systems remain both challenging and crucial. These new systems will have the ability to handle both unstructured and structured data, to return the results in the way desired by the user, and to incur only a low cost of ownership.

Acknowledgments

We thank Ying Lu and Yoonkyong Lee for obtaining the data sets and conducting the experiments in the PROM project, and for valuable comments on parts of this chapter.

References

7.1 R. Ananthakrishna, S. Chaudhuri, V. Ganti: Eliminating fuzzy duplicates in data warehouses. In: *Proc. of 28th Int. Conf. on Very Large Databases (2002)*
7.2 N. Ashish, C. Knoblock: Wrapper Generation for Semi-structured Information Sources. In: *Proc. ACM SIGMOD Workshop on Management of Semi-structured Data (1997)*
7.3 R. Avnur, J. Hellerstein: Continuous query optimization. In: *SIGMOD '00 (2000)*
7.4 C. Batini, M. Lenzerini, SB. Navathe: A comparative analysis of methodologies for database schema integration. ACM Computing Survey, 18(4), 323-364 (1986)
7.5 J. Berlin, A. Motro: Autoplex: Automated discovery of content for virtual databases. In: *Proc. of the Conf. on Cooperative Information Systems (CoopIS) (2001)*
7.6 J. Berlin, A. Motro: Database schema matching using machine learning with feature selection. In: *Proc. of the Conf. on Advanced Information Systems Engineering (CAiSE) (2002)*

7.7 M. Bilenko, R. Mooney: Learning to combine trained distance metrics for duplicate detection in databases. Technical Report Technical Report AI 02-296, Artificial Intelligence Laboratory, University of Texas at Austin, Austin, TX (February 2002)

7.8 C.J.C. Burges: A tutorial on support vector machines for pattern recognition. Data Mining and Knowledge Discovery, 2, 121-167 (1998)

7.9 J. Callan, M. Connell, A. Du: Automatic discovery of language models for text databases. In: *Proc. of the ACM SIGMOD Conf. (SIGMOD)* (1999)

7.10 S. Chakrabarti: Data mining for hypertext: A tutorial survey. In: *SIGKDD Explorations: Newsletter of the Special Interest Group (SIG) on Knowledge Discovery and Data Mining, ACM, 1 (2000)*

7.11 S. Chakrabarti: *Mining the Web: Discovering Knowledge from Hypertext Data* (Morgan Kaufmann Publishers (2002))

7.12 S. Chakrabarti, M. Berg, B. Dom: Focused crawling: a new approach to topic-specific Web resource discovery. Computer Networks, (Amsterdam, Netherlands, 1999) 31(11-16), 1623-1640 (1999)

7.13 S. Chakrabarti, B. Dom, R. Agrawal, P. Raghavan: Scalable feature selection, classification and signature generation for organizing large text databases into hierarchical topic taxonomies. Journal of Very Large Data Bases, 7(3), 163-178 (1998)

7.14 S. Chakrabarti, B. Dom, P. Indyk: Enhanced Hypertext Categorization Using Hyperlinks. In: *Proc. of the ACM SIGMOD Conf. (1998)*

7.15 H. Chalupsky: Ontomorph: A Translation system for symbolic knowledge. Principles of Knowledge Representation and Reasoning (2000)

7.16 C.C. Chang, C.J. Lin: Training nu-support vector classifiers: theory and algorithms. Neural Computation, 13, 2119-2147 (2001)

7.17 K. Chang, B. He, C. Li, Z. Zhang: Structured databases on the Web: Observations and implications. Technical Report UIUCDCS-R-2003-2321, Department of Computer Science, UIUC (February 2003)

7.18 H. Chen, S. Dumais: Bringing order to the Web: automatically categorizing search results. In: *Proc. of CHI-00, Human Factors in Computing Systems, Den Haag, NL, 2000* (Forthcoming)

7.19 J. Chen, D. DeWitt, F. Tian, Y. Wang: Niagaracq: A scalable continuous query system for internet databases. In: *SIGMOD '00 (2000)*

7.20 J. Cho, A. Ntoulas: Effective change detection using sampling (2002)

7.21 W. Cohen: Integration of heterogeneous databases without common domains using queries based on textual similarity. In: *Procceedings of SIGMOD-98 (1998)*

7.22 W. Cohen, M. Hurst, L. Jensen: A flexible learning system for wrapping tables and lists in html documents. In: *Proc. of the Int. World-Wide Web Conf. (WWW) (2002)*

7.23 W. Cohen, D. Kudenko: Transferring and retraining learned information filters. In: *Proc. of the AAAI Conf. (AAAI-97) (1997)*

7.24 W. Cohen, J. Richman: Learning to match and cluster entity names. In: *Proc. of 8th ACM SIGKDD Int. Conf. on Knowledge Discovery and Data Mining (2002)*

7.25 C. Cortes, V. Vapnik: Support vector networks. Machine Learning, 30 (3), 273-297 (1995)

7.26 M. Craven, D. DiPasquo, D. Freitag, A. McCallum, T. Mitchell, K. Nigam, S. Slattery: Learning to construct knowledge bases from the World Wide Web. Artificial Intelligence, 118(1-2), 69-113 (2000)

7.27 V. Crescenzi, G. Mecca, P. Merialdo: Roadrunner: Towards automatic data extraction from large Web sites. VLDB Journal, pp. 109-118 (2001)

7.28 F. DeComite, F. Denis, R. Gilleron: Positive and unlabeled examples help learning. In: *Proc. 11th Int. Conf. Algorithmic Learning Theory (ALT'99)* (Tokyo, Japan, 1999) pp. 219-230

7.29 F. Denis: PAC learning from positive statistical queries. In: *Proc. 10th Int. Conf. Algorithmic Learning Theory (ALT'99)* (Otzenhausen, Germany, 1998) pp. 112-126

7.30 H. Do, E. Rahm: Coma: A system for flexible combination of schema matching approaches. In: *Proc. of the 28th Conf. on Very Large Databases (VLDB) (2002)*

7.31 A. Doan, P. Domingos, A. Halevy: Reconciling Schemas of Disparate Data Sources: A Machine Learning Approach. In: *Proc. of the ACM SIGMOD Conf. (2001)*

7.32 A. Doan, Y. Lu, Y. Lee, J. Han: Object matching for data integration: A profile-based approach. In: *Proc. of the IJCAI-03 Workshop on Information Integration on the Web (2003)*

7.33 A. Doan, J. Madhavan, P. Domingos, A. Halevy: Learning to map ontologies on the Semantic Web. In: *Proc. of the World-Wide Web Conf. (WWW-02) (2002)*

7.34 S. Dumais, H. Chen: Hierarchical classification of Web content. In: *Proc. 23rd ACM Int. Conf. on Research and Development in Information Retrieval (SIGIR'00)* (Athens, Greece) pp. 256-263 (2000)

7.35 D. Embley, D. Jackman, L. Xu: Multifaceted exploitation of metadata for attribute match discovery in information integration. In: *Proc. of the WIIW-01 (2001)*

7.36 D. Embley, Y. Jiang, Y. Ng: Record-boundary discovery in Web documents. In: *Proc. of the ACM SIGMOD Conf. (1999)*

7.37 D. Florescu, A. Levy, A. Mendelzon: Database techniques for the World-Wide Web: A survey. SIGMOD Record, 27(3), 59-74 (1998)

7.38 D. Freitag: Multistrategy learning for information extraction. In: *Proc. 15th Int. Conf. on Machine Learning (ICML-98) (1998)*

7.39 H. Galhardas, D. Florescu, D. Shasha, E. Simon: An extensible framework for data cleaning. In: *Proc. of 16th Int. Conf. on Data Engineering (2000)*

7.40 H. Garcia-Molina, Y. Papakonstantinou, D. Quass, A. Rajaraman, Y. Sagiv, J. Ullman, J. Widom: The TSIMMIS project: Integration of heterogeneous information sources. Journal of Intelligent Inf. Systems, 8 (2) (1997)

7.41 C. Giles, K. Bollacker, S. Lawrence: CiteSeer: An automatic citation indexing system. In: *Digital Libraries 98 - The 3rd ACM Conf. on Digital Libraries (1998)*

7.42 L. Gravano, P. Ipeirotis, N. Koudas, D: Srivastava. Text join for data cleansing and integration in an rdbms. In: *Proc. of 19th Int. Conf. on Data Engineering (2003)*

7.43 L.M. Haas, D. Kossmann, E.L. Wimmers, J. Yang: Optimizing queries across diverse data sources. In: *Proc. of VLDB '97 (1997)*

7.44 J. Han, K. Chang: Data mining for Web intelligence. *IEEE Computer, 2002*

7.45 B. He, K. Chang: Statistical schema matching across Web query interfaces. In: *Proc. of the ACM SIGMOD Conf. (SIGMOD) (2003)*

7.46 M. Hernandez, S. Stolfo: The merge/purge problem for large databases. In: *SIGMOD Conf., 1995* pp. 127-138

7.47 P. Ipeirotis, L. Gravano, M. Sahami: Probe, count, and classify: Categorizing hidden Web databases. In: *Proc. of the ACM SIGMOD Conf. (SIGMOD) (2001)*

7.48 Z. Ives, D. Florescu, M. Friedman, A. Levy, D. Weld: An adaptive query execution system for data integration. In: *Proc. of SIGMOD (1999)*

7.49 T. Joachims: Text categorization with support vector machines. In: *Proc. 10th European Conf. on Machine Learning (ECML'98)* (Chemnitz, Germany, 1998) pp. 137-142
7.50 J. Kang, J. Naughton: On schema matching with opaque column names and data values. In: *Proc. of the ACM SIGMOD Int. Conf. on Management of Data (SIGMOD-03) (2003)*
7.51 J. Kleinberg: Authoritative sources in a hyperlinked environment. In: *Proc. 9th ACM-SIAM Symposium on Discrete Algorithms (1998)*
7.52 D. Koller, M. Sahami: Hierarchically classifying documents using very few words. In: *Proc. 14th Int. Conf. on Machine Learning* (Morgan Kaufmann, 1997) pp. 170-178
7.53 R. Kosala, H. Blockeel: Web mining research: A survey. *SIGKDD: SIGKDD Explorations: Newsletter of the Special Interest Group (SIG) on Knowledge Discovery and Data Mining, 2 (2000)*
7.54 N. Kushmerick: Wrapper verification. World Wide Web Journal, 3 (2), 79-94 (2000)
7.55 N. Kushmerick, D. Weld, R. Doorenbos: Wrapper Induction for Information Extraction. In: *Proc. of the Int. Joint Conf. on AI (IJCAI) (1997)*
7.56 E. Lambrecht, S. Kambhampati, S. Gnanaprakasam: Optimizing recursive information gathering plans. In: *Proc. of the Int. Joint Conf. on AI (IJCAI) (1999)*
7.57 S. Lawrence, K. Bollacker, C.L. Giles: Autonomous citation matching. In: *Proc. of the 3rd Int. Conf. on Autonomous Agents (1999)*
7.58 W. Lehnert: A conceptual theory of question answering. In: B. Grosz, K. Jones, B. Webber (eds.), *Natural Language Processing* (Kaufmann, 1986)
7.59 K. Lerman, S. Minton, C. Knoblock: Wrapper maintenance: A machine learning approach. Journal of Artificial Intelligence Research (2003)
7.60 F. Letouzey, F. Denis, R. Gilleron: Learning from positive and unlabeled examples. In: *Proc. 11th Int. Conf. Algorithmic Learning Theory (ALT'00), Sydney, Australia, 2000* pp. 11-30
7.61 A.Y. Levy, A. Rajaraman, J. Ordille: Querying heterogeneous information sources using source descriptions. In: *Proc. of VLDB (1996)*
7.62 W. Li, C. Clifton: SEMINT: A tool for identifying attribute correspondence in heterogeneous databases using neural networks. Data and Knowledge Engineering, 33, 49-84 (2000)
7.63 W. Li, J. Han, J. Pei. CMAR: Accurate and efficient classification based on multiple class-association rules. In: *Proc. of the Int. Conf. on Data Mining (ICDM-01) (2001)*
7.64 M. Light, G. Mann, E. Riloff, E. Breck: Analyses for elucidating current question answering technology. Journal for Natural Language Engineering (2001)
7.65 B. Liu, W. S. Lee, P. S. Yu, X. Li: Partially supervised classification of text documents. In: *Proc. 19th Int. Conf. Machine Learning (ICML'02), Sydney, Australia, 2002* pp. 387-394
7.66 J. Madhavan, P. Bernstein, K. Chen, A. Halevy, P. Shenoy: Matching schemas by learning from a schema corpus. In: *Proc. of the IJCAI-03 Workshop on Information Integration on the Web (2003)*
7.67 J. Madhavan, P.A. Bernstein, E. Rahm: Generic schema matching with cupid. In: *Proc. of the Int. Con. on Very Large Databases (VLDB) (2001)*
7.68 S. Madria, S. Bhowmick, W. Ng, E. Lim: Research issues in Web data mining. In: *Data Warehousing and Knowledge Discovery*, pp. 303-312 (1999)
7.69 L. M. Manevitz, M. Yousef: One-class SVMs for document classification. Journal of Machine Learning Research, 2, 139-154 (2001)

7.70 A. McCallum, K. Nigam, J. Rennie, K. Seymore: A machine learning approach to building domain-specific search engines. In: *Proc. of the Int. Joint Conf. on AI (IJCAI) (1999)*

7.71 A. McCallum, K. Nigam, J. Rennie, K. Seymore: Automating the construction of internet portals with machinelearning. Information Retrieval, 3(2), 127-163 (2000)

7.72 A. McCallum, K. Nigam, L. Ungar: Efficient clustering of high-dimensional data sets with application to reference matching. In: *Proc. 6th ACM SIGKDD Int. Conf. on Knowledge Discovery and Data Mining (2000)*

7.73 A. McCallum, R. Rosenfeld, T. Mitchell, A.Y. Ng: Improving text classification by shrinkage in a hierarchy of classes (Madison, WI, 1998) pp. 359-367,

7.74 D. McGuinness, R. Fikes, J. Rice, S. Wilder: The Chimaera Ontology Environment. In: *Proc. of the 17th National Conf. on Artificial Intelligence (2000)*

7.75 S. Melnik, H.M. Garcia, E. Rahm: Similarity Flooding: A Versatile Graph Matching Algorithm. In: *Proc. of the Int. Conf. on Data Engineering (ICDE) (2002)*

7.76 T. Milo, S. Zohar: Using schema matching to simplify heterogeneous data translation. In: *Proc. of VLDB (1998)*

7.77 P. Mitra, G. Wiederhold, J. Jannink: Semi-automatic Integration of Knowledge Sources. In: *Proc. of Fusion'99 (1999)*

7.78 A. Monge, C. Elkan: The field matching problem: Algorithms and applications. In: *Proc. 2nd Int. Conf. Knowledge Discovery and Data Mining (1996)*

7.79 F. Neumann, CT. Ho, X. Tian, L. Haas, N. Meggido: Attribute classification using feature analysis. In *Proc. of the Int. Conf. on Data Engineering (ICDE) (2002)*

7.80 J. Neville, D. Jensen: Iterative classification in relational data (2000)

7.81 K. Nigam: Using unlabeled data to improve text classification. Ph.D. thesis, Carnegie-Mellon University, School of Computer Science (2001)

7.82 K. Nigam, A. McCallum, S. Thrun, T. Mitchell: Learning to classify text from labeled and unlabeled documents. In: *Proc. of the Nat. Conf. on AI (AAAI) (1998)*

7.83 N.F. Noy, M.A. Musen: PROMPT: Algorithm and Tool for Automated Ontology Merging and Alignment. In: *Proc. of the National Conf. on Artificial Intelligence (AAAI) (2000)*

7.84 N.F. Noy, M.A. Musen: PromptDiff: A fixed-point algorithm for comparing ontology versions. In: *Proc. of the Nat. Conf. on Artificial Intelligence (AAAI) (2002)*

7.85 L. Page, S. Brin, R. Motwani, T. Winograd: The pagerank citation ranking: Bringing order to the Web. Technical report, Stanford Digital Library Technologies Project (1998)

7.86 L. Palopoli, D. Sacca, D. Ursino: Semi-automatic, semantic discovery of properties from database schemes. In: *Proc. of the Int. Database Engineering and Applications Symposium (IDEAS-98), 1998* pp. 244-253

7.87 E. Rahm, P.A. Bernstein: On matching schemas automatically. VLDB Journal, 10(4) (2001)

7.88 V. Raman, J. Hellerstein: Potter's wheel: An interactive data cleaning system. VLDB Journal, pp. 381-390 (2001)

7.89 A. Rosenthal, S. Renner, L. Seligman, F. Manola: Data integration needs an industrial revolution. In: *Proc. of the Workshop on Foundations of Data Integration (2001)*

7.90 S. Sarawagi, A. Bhamidipaty: Interactive deduplication using active learning. In: *Proc. of 8th ACM SIGKDD Int. Conf. on Knowledge Discovery and Data Mining (2002)*

7.91 S. Sizov, M. Theobald, S. Siersdorfer, G. Weikum, J. Graupmann, M. Biwer, P. Zimmer: The Bingo! system for information portal generation and expert Web search. In: *Proc. of the Conf. on Innovative Database Research (CIDR-03) (2003)*

7.92 S. Slattery, T. Mitchell: Discovering test set regularities in relational domains. In: *Proc. of the 17th Int. Conf. on Machine Learning (ICML) (2000)*

7.93 D.M.J. Tax, R.P.W. Duin: Support vector domain description. Pattern Recognition Letters, 20, 1991-1999 (1999)

7.94 D.M.J. Tax, R.P.W. Duin: Uniform object generation for optimizing one-class classifiers. Journal of Machine Learning Research, 2, 155-173 (2001)

7.95 S. Tejada, C. Knoblock, S. Minton: Learning domain-independent string transformation weights for high accuracy object identification. In: *Proc. of the 8th SIGKDD Int. Conf. (KDD-2002) (2002)*

7.96 L.L. Yan, R.J. Miller, L.M. Haas, R. Fagin: Data Driven Understanding and Refinement of Schema Mappings. In: *Proc. of the ACM SIGMOD (2001)*

7.97 Y. Yang, X. Liu: A re-examination of text categorization methods. In: *Proc. 22th ACM Int. Conf. on Research and Development in Information Retrieval (SIGIR'99), Berkeley, CA, 1999* pp. 42-49

7.98 W. Yih, D. Roth: Probabilistic reasoning for entity and relation recognition. In: *Proc. of COLING'02 (2002)*

7.99 H. Yu: SVMC: Single-class classification with support vector machines. In: *Proc. Int. Joint Conf. on Articial Intelligence (IJCAI-03), Acapulco, Maxico (2003)*

7.100 H. Yu, J. Han, K. Chang: PEBL: Positive Example Based Learning for Web page classification using svm. In: *Proc. of the Conf. on Knowledge Discovery and Data Mining, KDD (2002)*

7.101 O. Zamir, O. Etzioni: Web document clustering: A feasibility demonstration. In: *Proc. of the 21st Annual Int. ACM SIGIR Conf. on Research and Development in Information Retrival* (August 1998)

7.102 O. Zamir, O. Etzioni, O. Madani, R.M. Karp: Fast and intuitive clustering of Web documents. In: *Proc. 3rd Int. Conf. Knowledge Discovery and Data Mining* pp. 287-290 (1997)

8. Mining Web Logs for Actionable Knowledge

Qiang Yang[1], Charles X. Ling[2], and Jianfeng Gao[3]

[1] Hong Kong University of Science and Technology, Hong Kong
[2] University of Western Ontario, Canada
[3] Microsoft Research Asia, China

Abstract

Everyday, popular Websites attract millions of visitors. These visitors leave behind vast amounts of Websites traversal information in the form of Web server and query logs. By analyzing these logs, it is possible to discover various kinds of knowledge, which can be applied to improve the performance of Web services. A particularly useful kind of knowledge is knowledge that can be immediately applied to the operation of the Websites; we call this type of knowledge *actionable knowledge*. In this chapter, we present three examples of actionable Web log mining. The first method is to mine a Web log for Markov models that can be used for improving caching and prefetching of Web objects. A second method is to use the mined knowledge for building better, adaptive user interfaces. The new user interface can adjust as the user behavior changes with time. Finally, we present an example of applying Web query log knowledge to improving Web search for a search engine application.

8.1 Introduction

Today's popular Web servers are accessed by millions of Web users each day. These visitors leave behind their visiting behavior in the form of Web logs. By analyzing the Web logs recorded at these servers as well as proxy servers, it is possible to learn the behavior of the Web users themselves. It is further possible to use the learned knowledge to serve the users better.

A particularly useful kind of knowledge is knowledge that can be immediately applied to the operation of the Websites; we call this type of knowledge actionable knowledge. Many previous approaches are aimed only at mining Web log knowledge for human consumption, where the knowledge is presented for humans to decide how to make use of it. In contrast, we advocate an approach to mining actionable knowledge from Web logs. The knowledge is for computers to consume. In this manner, the quality of the mined knowledge can be immediately put to test and the Web services can be said to be truly self-adaptive. The advantage of mining actionable knowledge is similar to that of the Semantic Web [8.5].

In this chapter, we first provide an overview of the current research on Web log mining. We provide a detailed description of our own work in this area. We first describe, in Sect. 8.2, algorithms that acquire user preferences and behavior from the Web log data and show how to design Web page prefetching systems for retrieving new Web pages. We empirically demonstrate that the Web prefetching system can be made self-adaptive over time and can out-perform some state-of-the-art Web caching systems. Then we, in Sect. 8.3, describe our work on dynamically reconfiguring the user interface for individual users to cater to their personal interests. These reconfigured Web pages are presented to the user in the form of index pages. The computer generates these pages by consulting a data mining algorithm for potential clusters of Web objects, and determines the best balancing point according to a cost function. Finally, in Sect. 8.4, we present an approach to mining Web query logs from a Web search engine to relate the semantics of queries to their targets. The knowledge about these query logs can then be used for improving the performance of the search engine.

8.2 Web Log Mining for Prefetching

8.2.1 Data Cleaning on Web Log Data

To learn from Web logs, the first task is to perform data cleaning by breaking apart a long sequence of visits by the users into user sessions. Each user session consists of only the pages visited by a user in a row. Since we are only dealing with Web server logs, the best we can do is to take an intelligent guess as to which pages in a long sequence belong to a single user session. As we will see, a strong indicator is the time interval between two successive visits. When the time exceeds the average time it takes a person to read a Web page, there is a strong indication that the user has left the browser to do other things.

In the Web logs, a user can be identified by an individual IP address, although using the IP address to identify the users is not reliable. Several users may share the same IP address, and the same user may be assigned a different IP address at different times. In addition, the same user may make different sequences of visits at different times. Thus, data cleaning means separating the visiting sequence of pages into visiting sessions. Most work in Web log mining employ a predefined time interval to find the visiting sessions. For example, one can use a two-hour time limit as the separating time interval between two consecutive visiting sessions, because people do not usually spend two hours on a single Web page.

Sometimes it is possible to obtain more information about user sessions than by using a fixed time interval. By learning the grouping of Web pages and Websites, one can find a more meaningful session separator. The work by [8.20] provides a method to use clustering analysis to find the group of related Web servers visited by users from a Web proxy server. If a user jumps from one group of related server to another, then it is highly likely that the user ends one session and starts another. Using this information to pick the Web pages, more accurate user session knowledge can be obtained.

8.2.2 Mining Web Logs for Path Profiles

Once a Web log is organized into separate visiting sessions, it is possible to learn user profiles from the data. For example, Schechter et al. [8.30] developed a system to learn users' path profiles from the Web log data. A user visiting a sequence of Web pages often leaves a trail of the pages' URLs in a Web log. A page's successor is the page requested immediately after that page in a URL sequence. A point profile contains, for any given page, the set of that page's successors in all URL sequences, and the frequency with which that successor occurred. A path profile considers frequent subsequences from the frequently occurring paths.

Both the point and the path profiles help us to predict the next pages that are most likely to occur by consulting the set of path profiles whose prefixes match the observed sequence. For example, Pitkow and Pirolli [8.25] described in their work how to find the longest repeating subsequences from a Web log, and then how to use these sequences to make prediction on users' next likely requests.

Similar to the path-profile-based prediction, Albrecht et al. [8.2] designed a system to learn an nth-order Markov model based not only on the immediately preceding Web page, but also on pages that precede the last pages, as well as the time between successive Web page retrieval activities. Together, these factors should give more information than the path profiles. The objective is still the same: to predict the next page which is most likely to be requested by a user. Markov models are natural candidates for Web traversal patterns, where the majority of the work now has used Web pages as states and hyperlinks as potential transitions between the states.

One particular fact about the Web is that the user sessions follow a Zipf distribution. By this distribution, most visitors view only few pages in each session, while few visitors visit many pages in a sequence. However, experiments show that the longer paths also provide more accurate predictions. To exploit these facts, Su et al. [8.36] developed an algorithm to combine path profiles with different lengths. For any given observed sequence of visits, their cascading model first selects the best prediction by looking for path profiles with long lengths. When such paths do not exist in the model, the system retreats to shorter paths and makes predictions based on these paths. Experiments show that this algorithm can provide a good balance between prediction accuracy and coverage. In this chapter, we develop this n-gram based prediction system into a prefetching system for prefetching new Web pages, which are then stored in the cache. When the prediction system is accurate, the prefetching system is expected to save on network latency.

8.2.3 Web Object Prediction

Given a Web-server browsing log L, it is possible to train a path-based model for predicting future URLs based on a sequence of current URL accesses. This

Table 8.1. A learned example of n-gram model

2-Gram	Prediction
A, B	{ < C, 100% > }
B, C	{ < A, 100% > }
C, A	{ < B, 50% >, < F, 50% > }

can be done on a per-user basis or on a per-server basis. The former requires that the user-session be recognized and broken down nicely through a filtering system, while the latter takes the simplistic view that access on a server is a single long thread. We now describe how to build this model using a single sequence L using the latter view.

We can build an n-gram prediction model based on the object-occurrence frequency. Each substring of length n is an n-gram. These sub-strings serve as the indices of a hash table T that contains the model. During its operation, the algorithm scans through all sub-strings exactly once, recording occurrence frequencies of documents' requests of the next m clicks after the sub-string in all sessions. The maximum occurred request (conditional probability greater than θ, where θ is a threshold set at 0.6 in our experiments) is used as the prediction of next m steps for the sub-string. In this case, we say that the n-gram prediction has a window size m.

We observe that many of the objects are accessed only once or twice, and a few objects are accessed a great many times. Using this well-known fact (also known as the Zipf distribution), we can filter out a large portion of the raw log file and obtain a compressed prediction model. In our filtering step, we removed all URLs that are accessed 10 times or less among all user requests.

As an example, consider a sequence of URLs in the server log: {A, B, C, A, B, C, A, F.} Then, a two-gram model will learn on the two-grams shown in Table 8.1, along with the predicted URLs and their conditional probabilities.

Applying n-gram prediction models has a long tradition in network systems research. Su et al. [8.36] compared n-gram prediction models under different n, and presented a cascading algorithm for making the prediction. Schechter et al. [8.30] provided a detailed statistical analysis of Web log data, pointing out the distribution of access patterns on Web pages.

We utilize the prediction system in [8.36] based on the n-grams in order to calculate the weights of different objects. Normally, there exist simultaneously a number of sessions on a Web server. Based on their access sequences, our prediction model can predict future requests for each particular session. Different sessions will give different predictions to future objects. Since our prediction of an object comes with a probability of its arrival, we can combine these predictions to calculate the future occurrence frequency of an object. Let O_i denote a Web object on the server, S_j be a session on a Web server, and $P_{i,j}$ be the probability predicted by a session S_j for object O_i. If $P_{i,j}=0$, it indicates that object O_i is not predicted by session S_j. Let W_i be the future frequency of requests to object O_i. A weight can be computed according to the following equation:

$$W_i = \sum_j P_{i,j}.$$

To illustrate this equation, consider the three sessions in Fig. 8.1. Each session yields a set of predictions of Web objects. Since sessions are assumed to be independent to each other, we use the equation to compute their weights W_i. For example, object O_1 is predicted by two sessions with a probability of 0.70 and 0.60, respectively. From the weight equation, $W_1 = 1.3$. This means that, probabilistically, object O_1 will be accessed 1.3 times in the near future.

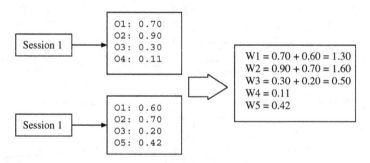

Fig. 8.1. Prediction to frequency weight calculation

8.2.4 Learning to Prefetch Web Documents

When a user requests a sequence of pages that match the left-hand-side of a rule, the right-hand-side can be predicted to occur with the same associated confidence as the probability of occurrence of O_i. The Web prefetching system can then add all such predicted probabilities from all rules in the prediction model as a measure of the potential future frequency estimate for O_i. These scores are compared and the top ranked objects are selected to be fetched into the cache if they are not already there. This prefetching method runs side by side with an existing caching system.

We proposed an integrated caching and prefetching model to further reduce the network latency perceived by users [8.37]. The motivation lies in two aspects. Firstly, from Fig. 8.2, we can see both the hit rate and byte hit rate are growing in a log-like fashion as a function of the cache size. Our results are consistent with those of other researchers [8.4, 8.9]. This suggests that hit rate or byte hit rate does not increase as much as the cache size does, especially when the cache size is large. This fact naturally leads to our thought of separating part (e.g., 10% of its size) of the cache memory for prefetching. By this means, we can trade the minor hit rate loss in caching with the greater reduction of network latency in prefetching. Secondly, almost all prefetching methods require a prediction model. Since we have already embodied an n-gram model into predictive caching, this model can also serve prefetching. Therefore, a uniform prediction model is at the heart of our integrated approach.

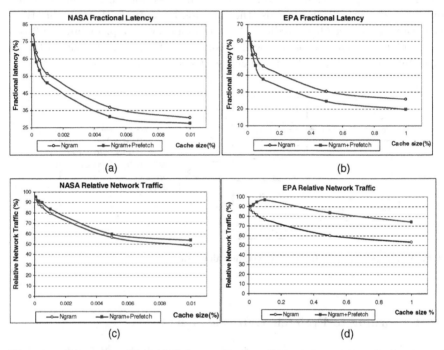

(a)   (b)   (c)   (d)

Fig. 8.2. Comparison of network latency and traffic

In our approach, the original cache memory is partitioned into two parts: cache-buffer and prefetching buffer. A prefetching agent keeps pre-loading the prefetching buffer with documents predicted to have the highest weight W. The prefetching stops when the prefetching buffer is full. The original caching system behaves as before on the reduced cache buffer, except that it also checks for hits in the prefetching buffer. If a hit occurs in the prefetching buffer, the requested object will be moved into the cache buffer according to the original replacement algorithm. Of course, one potential drawback of prefetching is that the network load may be increased. Therefore, there is a need to balance the decrease in network latency and the increase in network traffic. We next describe two experiments that show that our integrated predictive caching and prefetching model does not suffer much from the draw-back.

In our prefetching experiments, we again used the EPA and NASA Web logs to study the prefetching impact on caching. For fair comparison, the cache memory in a cache-alone system equals the total size of the cache buffer and the prefetching buffer in the integrated system. We assume that the prefetching buffer has a size of 20% of the cache memory. Two metrics are used to gauge the network latency and increased network traffic:

Fractional Latency: the ratio between the observed latency with a caching system and the observed latency without a caching system.

Fractional Network Traffic: the ratio between the number of bytes that are transmitted from Web servers to the proxy and the total number of bytes requested.

As can be seen from Fig. 8.2, prefetching does reduce network latency in all cache sizes. On EPA data, when cache size is 1% (approximately 3.1MB), fractional latency has been reduced from 25.6% to 19.7%. On NASA data, when cache size is 0.001% (240KB), fractional latency has been reduced from 56.4% to 50.9%. However, we pay a price for the network traffic, whereby the prefetching algorithm incurs an increase in network load. For example, in the NASA data set, the fractional network traffic increases 6% when cache size is 0.01%. It is therefore important to strike for a balance between the improvement in hit rates and the network traffic. From our result, prefetching strategy performs better in a larger cache size while relatively less additional network traffic is incurred.

8.3 Web Page Clustering for Intelligent User Interfaces

Web logs can be used not only for prefetching, but also to build server-side customization and transformation – operations that convert a Websites, on the server side, into one that is more convenient for users to visit and meet their objectives.

Several researchers have studied the problem of creating Web interfaces that can adapt to user behavior based on Web logs. Examples include path prediction algorithms that guess where the user wants to go next in a browsing session, so that the server can either pre-send documents or shortcut browsing paths. For instance, WebWatcher [8.3] learns to predict what links users will follow on a particular page as a function of their specified interests. A link that WebWatcher believes a particular user is likely to follow will be highlighted graphically and duplicated at the top of the page when it is presented.

One particularly useful application is to compose index pages for individual groups of users automatically. This application corresponds to a kind of personalization on the Web. One of the first systems to do this is the Page-Gather system [8.24], which provides users with shortcuts, which takes as input a Web server access log, where the log records the pages visited by a user at the site. Based on the visiting statistics, the system provides links on each page to visitors' eventual goals, skipping the in-between pages. An index page is a page consisting of links to a set of pages that cover a particular topic (e.g., electric guitars).

Given this terminology, Perkowitz and Etzioni [8.24] define the index-page synthesis problem: given a Website and a visitor access log, create new index pages containing collections of links to related but currently unlinked pages. An access log is a document containing one entry for each page requested of the Web server. Each request lists at least the origin (IP address) of the request, the URL requested, and the time of the request. Related but unlinked pages are pages that share a common topic but are not currently linked at the site; two pages are considered linked if there exists a link from one to the other or if there exists a page that links to both of them. In their PageGather algorithm, Perkowitz and Etzioni [8.24] presented a clustering-based method for generating the contents of the new Web page from server access logs.

Given an access log, a useful task is to find collections of pages that tend to co-occur in visits. Clustering is a natural technique to consider for this task. In clustering, documents are represented in an n-dimensional space formed by term vectors. A cluster is a collection of documents close to each other and relatively distant from other clusters. Standard clustering algorithms partition the documents into a set of mutually exclusive clusters.

The PageGather algorithm uses cluster mining to find collections of related pages at a Website. In essence, PageGather takes a Web server access log as input and maps it into a form ready for clustering; it then applies cluster mining to the data and produces candidate index-page contents as output. The algorithm has six basic steps:

PageGather Algorithm (Server Access Logs)
1. Process the access log into visits.
2. Compute the co-occurrence frequencies between pages and create a similarity matrix.
3. Create the graph corresponding to the matrix, and find maximal cliques (or connected components) in the graph.
4. Rank the clusters found, and choose which to output.
5. For each cluster, create a Web page consisting of links to the documents in the cluster.
6. Present clusters to the Webmaster for evaluation.

Let N be the number of Web pages at the site. This algorithm is thus $O(N^2)$ time, quadratic in the original number of Web pages. We note that due to this high complexity, the algorithm is not suitable for processing large data sets that are typical of today's Web access patterns. For example, everyday, MSN collects millions of visits. Data sets of this scale must be processed with very efficient disk-based algorithms. Thus, one of our intentions is to explore more efficient clustering algorithms for synthesizing index pages.

Another drawback of the PageGather algorithm is that it relies on human Web masters to determine the appropriateness of the generated index pages in a final check. This will likely create a bottleneck for the workflow, especially for sites that have many Web pages to be indexed. A particularly important problem is the question of how many index pages to create, and how many links to include in each index page. We answer this question in their chapter.

Having obtained the clusters, we now turn our attention to the second contribution of the chapter, building Web interfaces that provide useful short cuts for people. We do this by providing index pages, which are table-of-content pages that we can put at the root of a Website. The idea of index pages was first proposed by [8.24], but many open issues still remained. For example, to make the knowledge actionable, it would be nice to eliminate the need for human users to subjectively select the links to compose the index pages. Here, we would like to address the important issue of how to extend their manually evaluated index pages by a novel technique to find an optimal construction of index pages automatically.

In our own work, we improve the PageGather system by automating the index-page creation process completely [8.35]. We do this by introducing a cost function, which we then use to judge the quality of the index pages created.

We first define the cost models of Web browsing. The cost arising from Web browsing can be summarized as the transition costs between Web pages. Various cost models can be used to describe the relation between costs and the number of URLs traversed; in this chapter, we adopt a cost model such that the cost of a browsing session is directly proportional to the number of URLs on the browsing path. Shortcutting by offering index pages should not be considered to be cost free, however. Index pages themselves tax the users' attention by requiring that users flip through extra pages in the process of finding their destinations. Therefore, when the number of index pages increases, the transition costs should decrease while, at the same time, the page cost associated with the need to flip through the index pages increases.

Let $OverallCost$ be the overall cost of browsing the Web pages and the index pages. Let $PageCost$ be the cost of flipping index pages and $TransitionCost$ be the cost of switching from one Web page to another. Then, our cost model is as follows:

$$OverallCost = PageCost + TransitionCost. \tag{8.1}$$

We now describe how to compose index pages in our framework. In their algorithm, Perkowitz and Etzioni first compute clusters from the Web logs and then put all clusters into index pages, so that each cluster will correspond to one index page. In our experience, we have found that each cluster will often contain a large number of index pages. When hundreds of hyperlinks are included in an index page, it is very difficult for a user to find the information he/she is looking for. In addition, we feel that there should be a limited number of index pages; if the user is required to read a huge number of index pages, then the purpose of including the index pages in the first place might be defeated.

Therefore, in index page construction, we will include two parameters. Let the L be the number of hyperlinks we would like to include in each index page, and let M be the number of index pages we wish to build. Algorithm ConstructIndexPages takes the parameters M and L and the clusters constructed by our RDBC algorithm, and produces M index pages as output:

Algorithm ConstructIndexPages(Clusters, M, L)
   1. For j = 1 to M
   2. {
   3. Sort clusters by the frequency count of the top L Web pages;
   4. Extract the top L Web pages from the first cluster and insert their hyperlinks into an index page;
   5. If (No cluster is left or size of each cluster less than L), Stop;
   6. }

More importantly, based on these cost functions, we can find a minimal value M for the *OverallCost* and its corresponding number of index pages to build. This optimal index-page construction process represents another major contribution of our work. What we do is analyze the overall cost as a function of the number of index pages M to construct, based on a fixed value of L. We can then find empirically the best value for M so as to minimize the overall cost of user browsing effort.

Figure 8.3 shows the overall cost function calculated from the combination of the Web page switching cost and the index page cost with the NASA data. As seen in the figure, the overall cost has a minimum value at around 3.5 index pages. This is an indication that when considering all factors, it is the best to include around 3 to 4 index pages, where each page contains L hyperlinks. The optimal number in this example is computed automatically from different statistics, rather then decided subjectively by a human user. In real applications, the Websites can be said to truly reconfigure itself based on the Web log files accumulated over a certain period of time.

8.4 Web Query Log Mining

8.4.1 Web Query Logs

We now describe an application where we apply data mining to an Internet search engine. In particular, we apply data mining to discover useful and implicit knowledge from user logs of the Microsoft Encarta search engine (http://encarta.msn.com) to improve its performance. The user logs keep traces of users, behavior when they use the Encarta search engine. The purpose of the Web log mining is to improve Encarta search engine's performance

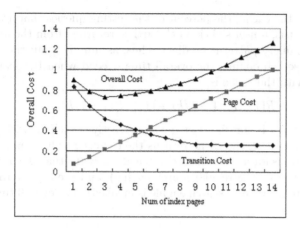

Fig. 8.3. Computing the optimal overall cost

(which is defined precisely later in the chapter) by utilizing the mined knowledge as cache. Indeed, data mining is a promising approach since popular search engines like Encarta get hundreds of thousands of hits each single day, and therefore it would be infeasible for traditional methods or human beings to analyze such logs.

Encarta accepts users' submissions as keyword queries, and returns a list of articles for users to choose from. There are over 40,000 articles in Encarta. Users can also browse articles organized under a hierarchical structure of several levels. The log keeps track of users queries and their clicks on the returned articles, as well as their browsing activities. The log file is quite large: a one day log is over several hundreds megabytes in size.

Our log mining system is an integration of previous techniques (mostly on smaller problems though), but when they are carefully integrated and tuned, they are effective in solving this large-scale real-world problem. The goal is to find useful patterns, that is, rules associating keywords of the queries with the target articles. If many users keying in similar queries clicked on the same article, then generalized query patterns summarizing the queries will provide a direct answer (the article) to similar queries in the future. When a set of such patterns is used as a cache of the Encarta search engine, its overall performance can be dramatically improved.

Let us assume that a cache search is k times faster than a regular search by the search engine. To obtain overall improvement with the cache, we must consider how likely a future user query is to be hit by the cache, and how accurately the cache predicts the article (so that the user does not have to revoke the regular search engine for the correct answer). The first part is measured by the recall (R) (or the hit ratio) of the cache; it measures the percent of the testing queries that are matched with the query patterns in the cache. The second part is the precision (P), which measures, among queries hit by the cache, the percent of the testing queries that are correctly predicted (that is, a user's click on the article returned from the query is the same as what query patterns predict). Thus, assuming a unit time is needed by a regular search engine, the overall time needed when the search engine is equipped with the cache is

$$P \times R/k + R(1 - P)(1 + 1/k) + (1 - R)(1 + 1/k) \qquad (8.2)$$

The first term is for the user query correctly matching with a pattern (with probability PR, the cache returns the correct result within time $1/k$); the second term is for an incorrectly matched query (so that the user requests another search from the regular search engine); and the third term is for no match (so that the regular search engine is invoked). The overall time is equal to

$$1 + 1/k - P \times R. \qquad (8.3)$$

Clearly, the larger the k, and the larger the product of P and R, the less the overall time of the search engine equipped with a cache. Thus, our first goal of mining user logs of Encarta is to improve the product of the precision and recall of the cache.

A major advantage of our caching technique over the previous ones is that our cache consists of generalized query patterns, so that each pattern covers many different user queries. They can even match with new queries that were not asked previously. Most previous caching approaches keep the frequent user queries, or invert lists directly (see [8.34] and the references), limiting the recall of cache with a fixed size dramatically. Some previous work on semantic caching (such as [8.8]) can compose answers of new queries by applying Boolean operations to the previous answers. However, they do not allow generalization of the previous queries as we do. The second goal of the research is to discover comprehensible descriptions on the topics in which users are mostly interested. The Web editors can study those topics, and add more articles on popular topics. Our generalized query patterns are production rule-based, and are easy to comprehend. Therefore, we need to build a user-friendly interface, providing different and clear views of the query patterns discovered.

Past work on Web log mining has been done. However, most has focused on mining to change the Web structure for easier browsing [8.11, 8.33], predicting browsing behaviors for prefetching [8.7], or predicting user preference for active advertising [8.24, 8.26]. Some work has been done on mining to improve search engine's performance. The most notable example is the Google search engine [8.15], in which statistical information is mined from the linkage structure among Web pages for weighting and ranking of the Web pages. However, when users click on Web pages returned, they access the Web pages directly. Therefore, Google does not collect users' feedback on the search results (except when users click on the cached pages on the Google server). Thus, Google's improvement on the search engine is largely syntactic and static, based on linkage relations among Web pages. Our method is semantic and dynamic, since it mines the users, actual clicks (feedback) on the search results, and it dynamically reflects the actual usage of the search engine. See Sect. 8.4 for more reviews of other work and its relation to ours.

8.4.2 Mining Generalized Query Patterns

In this section, we describe in detail how data mining is used to improve the Encarta search engine's performance. The Encarta Website (http://encarta. 'msn.com) consists of a search engine, with 41,942 well-versed articles (written by experts of various domains) as answers to the user queries, and a well-organized hierarchy of articles for browsing (though one cannot use the search engine and the hierarchy interchangeably). The users's activities are recorded in IIS (Internet Information Services) log files, which keep a lot of information about each user's access to the Web server. The log files are large: one day's log files are about 700 megabytes in size.

A log preprocessing program is written to extract user queries and their corresponding articles clicked. The search engine is a keyword-based search engine, and therefore almost all queries are simply lists of keywords. The result of the log preprocessing program is as follows:

$$keyword_1, keyword_2, \ldots, article_1, article_2, \ldots.$$

Each line indicates that a user query contains $keyword_1, keyword_2$, and so on, and the corresponding articles that the user clicks are $article_1, article_2$, etc.

We find that most queries have a very small number of keywords: 52.5% of queries contain only one keyword, 32.5% contain two, 10% contain three, and 5% contain four or more. Also, most users clicked on only one article for their queries: only 0.0619% of user session clicked on two or more articles for a query. We simply convert each of those queries into several queries, each of which contains the same keywords and one article that the user clicked. This introduces an inherit error rate, since the same query is linked to different articles as answers.

One inherent difficulty in our work is that user queries are mostly short, consisting of one or two keywords. On the other hand, user clicks of the articles can vary due to various user intentions. Our data mining algorithm should be able to deal with noise in the data, and be able to find rules associating queries with the most relevant articles.

From two month's Encarta's user logs, we obtained 4.8 million queries, each of which is a list of keywords, and one article clicked. Our data mining starts from this query list, and produces generalized query patterns which summarize similar queries answered by the same article. Again, those query patterns will act as the cache of the Encarta search engine.

8.4.3 A Bottom-Up Generalization Algorithm

A query pattern represents a set of user queries with the same or similar intention, and is thus associated with an article as the answer. It is difficult to capture similar intention in natural language, and we therefore use both syntactic constraints (such as stemming of keywords, discussed later) as well as semantic constraints (such as generalized concepts and synonyms, also discussed later) in our definition of queries with similar intentions.

The simplest form of patterns consists of some keywords, and possibly a "don't care" keyword, in the format of:

$$articleID \leftarrow keyword_1, \ldots, keyword_i, [any]; \text{ accuracy; coverage}$$

where "any" represents any keywords, and is placed in [...] to distinguish it from user input keywords. If a pattern contains a generalized term placed in [...], it is called a generalized pattern; otherwise (if it contains user input keywords only), it is called a simple pattern.

Accuracy and coverage are two measures for evaluating query patterns. (Note that the precision and recall mentioned earlier are measures for testing queries on the whole set of patterns, while accuracy and coverage are for each individual query pattern.) Coverage is the number of original user queries that are correctly covered (or predicted) by this pattern, and accuracy is the percentage of queries that are correctly covered. If error is the number of queries incorrectly covered by a pattern (i.e., when users clicked a different article), then accuracy = coverage/(coverage + error). If the accuracy and coverage of all query patterns are known, and the distributions of the testing set on those patterns are known, then the overall precision and recall of the pattern set (cache) on the test set can be easily derived.

A bottom-up generalization algorithm is designed to find generalized patterns with [any], and the overall process is described below. First, it groups the queries clicked by the same article together, and sorts them by the article frequency, so that queries of the most clicked articles are mined first. From the queries of the same article, two are chosen according to a greedy heuristic (see later) if they have some keywords in common. Then, a tentative pattern is created with those common keywords, and the [any] keyword inserted. Its accuracy and coverage can then be calculated. If the accuracy or the coverage of this pattern is below some thresholds, it is discarded; otherwise, the pattern replaces all queries correctly covered by it, and the process repeats until no new patterns can be formed. Then, the generalized patterns (with the generalized concept [any]) and the simple patterns (the remaining user queries) are returned, and they are all associated with that article. This process is applied to all articles. In the end, all patterns of all articles are sorted according to their coverage, and the top t patterns with highest coverage from all articles are output as the cache. The cache size t was chosen as 5,000 in this chapter, since a query distribution analysis shows that when t is much larger than 5,000, the overall benefit of larger caching is reduced.

Note that the relation between query patterns and articles is many-to-many. On one hand, the same query pattern can associate with more than one article (since a general keyword can be the keyword of several articles). Therefore, several articles can be returned if the user query matches with several query patterns with different articles. On the other hand, the same article can be associated with many query patterns (different keywords can associate with the same article).

To choose two queries for generalization, a greedy heuristic is used. Each pair of queries is tried, and the pair with the maximum coverage – while the accuracy is above some preset threshold – is chosen at each step for generalization.

Experiments (not shown here) indicate that this algorithm does not produce satisfactory results: only a few useful patterns are produced. The first reason is that the generalization step of introducing [any] into patterns is often too bold: such patterns cover too many queries incorrectly, and thus have a low accuracy. Second, many keywords have slightly different spellings (sometimes with a minor spelling error) and are thus not regarded as the same, preventing possible generalization. Third, many keywords are synonyms; but since they are spelled differently, they cannot be treated as the same for possible generalization. Fourth, one-keyword queries are not paired for generalization, since it would inevitably produce an overly generalized pattern article ID [any]. However, most (52.5%) of the queries in the user logs are one word queries. We want generalization to happen more gradually. This would also alleviate the first problem mentioned above.

In the following four subsections we provide solutions to the four problems mentioned above. Most improvements have been published previously (see Sect. 8.4.11), but when they are carefully integrated and tuned, they are shown to be effective to solve this large-scale real-world problem.

8.4.4 Improvement 1: A Hierarchy over Keywords

The first problem mentioned earlier is over-generalization. That is, any two different keywords from two queries will be replaced by [any]. To provide a more graded generalization, a hierarchical structure with the is-a relation would be useful. With this hierarchy, the generalization of two keywords would be the lowest concept that is an ancestor of both keywords in the hierarchy. It is a tedious job to construct such a hierarchy over tens of thousands of keywords in the user queries. We used WordNet [8.23] to generate such a hierarchy automatically. A problem encountered is that most keywords have different senses or meanings, which in turn have different parents in the hierarchy. We adopt the first, or the most frequently used, meaning of each keyword in WordNet. Previous research [8.19, 8.21] found that the most frequently used meanings are accurate enough compared to more sophisticated methods.

For example, if we use "$<$" to represent the *is-a* relation, then WordNet would generalize "alphabet" as

alphabet $<$ character set $<$ list $<$ ... $<$ communication $<$
social relation $<$ relation $<$ abstraction.

Similarly, symbol could be generalized as

symbol $<$ signal $<$ communication... $<$ abstraction.

Then, communication would be the least general concept of alphabet and symbol in the hierarchy.

A problem that follows form gradual generalization using WordNet is that it introduces too many possibilities of generalization from two user queries. A keyword in one query may generalize with any other keyword in the other query using WordNet, introducing an explosive number of combinations. A heuristic is further introduced: two queries can be generalized only when they have the same number of keywords, and only one keyword in the two queries is different. That is, the two queries must be in the format

$keyword_1, \ldots, keyword_i, keyword_j$ and
$keyword_1, \ldots, keyword_i, keyword_k,$

where $keyword_j \neq keyword_k$. A potential pattern is then produced:

$keyword_1, \ldots, keyword_i, [concept_1],$

where $concept_1$ is the lowest concept which is an ancestor of $keyword_j$ and $keyword_k$ in the hierarchy. With the introduction of the concept hierarchy, more interesting patterns can be discovered. For example, from

    Greek, alphabet
    Greek, symbol

a generalized query pattern:

$articleID \leftarrow Greek, [communication]$

is produced. "[communication]" would also cover other keywords such as "document", "letter", and "myth", which might be contained in future user queries.

Obviously, when the generalized query patterns are cached for the search engine Encarta, WordNet API calls must be available for testing whether or not a keyword in a new user query is covered by generalized concepts. This can be optimized to have little impact on the speed of the cache.

8.4.5 Improvement 2: Flexible Generalizations

Due to the introduction of the hierarchy, pairs of one-keyword queries may now be generalized. As to which two queries are paired for generalization, the same greedy heuristic is used: the pair that produces a pattern with the maximum coverage while the accuracy is above a threshold is chosen. All queries covered by the new pattern are removed, and the process repeats until no new patterns can be generated.

As discussed above, the goal of producing generalized query patterns is to have the product of the recall and precision of patterns as large as possible. Therefore, a further generalization is applied here. After a least general keyword in WordNet is obtained from a pair of two keywords, it is further generalized by "climbing" up in the hierarchy of WordNet until just before the accuracy of the pattern falls under the fixed threshold. This would produce patterns with a maximum recall while maintaining the adequate accuracy.

This improvement produces many useful generalized queries, such as the following

Democracy, of, [American state] from

* Democracy, of, California
* Democracy, of, Texas
* Democracy, of, Pennsylvania;

[Weapons system] from

* Gun
* Pistol
* Bullet
* fire arm

8.4.6 Improvement 3: Morphology Conversion

The second reason preventing useful generalization of the two queries is that minor differences in the spelling of same keyword are regarded as different keywords. A simple morphology analysis using WordNet is designed to convert words into their original forms (i.e., stemming). We apply the conversion in our program before generalization has taken place. Currently, we are working on spelling correction on the keywords in user queries.

With this improvement, we see user queries such as

* populations, Glasgow
* population, Edinburgh

which could not previously be generalized (because both keywords in the two queries are different) can now be generalized, and produce a generalized query pattern population, [UK region].

8.4.7 Improvement 4: Synonym Conversion

A similar problem is that many keywords are synonyms: they have very similar meanings but with different spellings. Since any generalization of two queries requires that all keywords except one must be the same, such synonyms should be recognized to allow more possible generalizations.

WordNet is again used for the synonym conversion. A dictionary of keywords appearing in user queries is dynamically created when the user logs are scanned. Before a new keyword is inserted into the dictionary, its synonyms are checked to see if any one of them is already in the dictionary. If any is, it replaces this keyword. This would reduce the size of the dictionary by about 27%. Even though the reduction is not huge, we found that the reduction happens on keywords that are very frequently used in the queries.

With this improvement, more useful generalized queries can be produced, which would not be previously possible, for example,

"DNA, [investigation]" from

```
* DNA, testing
* gene, analysis
* genetic, inquiry
```

Table 8.2. Implementation of the algorithm

QuerySet = IIS log files	/* Original user log of Encarta */
QuerySet = Filtering(QuerySet)	/* Producing list of user keywords and click */
QuerySet = Stemming(QuerySet)	/* Improvement 3 (morphology conversion) */
QuerySet = Synonym(QuerySet)	/* Improvement 4 (synonym conversion) */
QuerySet = sort(QuerySet)	/* Cluster queries by article ID */
For each article ID do	
Repeat	
Maxpattern = " ";	
For each pair of queries with the same length and only one keyword different Temppattern = generalize(pair)	
	/* use WordNet to generalize the keyword */
While	/* Improvement 2 */
accuracy(climbup(Temppattern))> threshold do	

8.4.8 Implementations

The pseudo-code of our log mining algorithm for Encarta with all improvements is presented in Table 8.2. The program is written in Visual C++.

Several further possible generalization methods are currently under investigation, for example, queries with different numbers of keywords, queries with two different keywords in the pair, and so on.

8.4.9 Simulation Experiments

We have constructed an offline mining system as described in the previous section. From the two month user log files of total size about 22 GB, it takes a few hours to mine all the patterns. Most of the time was spent on error calculation of each generalized query pattern, since it requires scanning all of the rest of the queries to see if it covers any "negative examples". The efficiency of the code would be improved since no attempt is made to optimize it. In the end, 5,000 query patterns are chosen whose accuracy is above 0.75 (a threshold we chose for the experiments).

We run several realistic simulations to see how our generalized query patterns help to improve the Encarta search engine, compared to a baseline cache which consists of the same number of the most frequent (un-generalized) user queries. We partition the two-month log files into two batches: a training batch, which is about 80% of all logs from the beginning, and a testing batch, which is the last 20% of the logs. We then run our mining algorithm on the training batch, and test the patterns produced on the testing batch (last 20% of the queries unused in the training), simulating the actual deployment of the cache in the search engine after data mining has been conducted.

Again, recall and precision of the cache are calculated to measure the overall speed of the search engine with the cache of the generalized query patterns. The overall saving would be larger if the product of recall and precision is larger, and if the k (the speed difference between regular search and cache) is larger.

8.4.10 Analyses of the Results

Table 8.3 presents our experiment results. As we can see, the recall (or hit ratio) of the cache does improve dramatically when more improvements in the mining algorithms are incorporated, and is much higher than the baseline cache, which consists of the same number of most frequent original user queries. On the other hand, the precision is virtually the same for different versions. This is actually within our expectation, since patterns are chosen according to their coverage when their accuracy is above a fixed threshold (0.75). Overall, the best version produces the highest product of precision and recall. As we can see, the major improvement in recall comes from the generalization with hierarchy from WordNet (improvements I and II), and from synonym conversion (improvement III).

If k is equal to 100, which is quite common as the speedup of using the cache versus the regular search engine, then the search engine with the cache of generalized query patterns would have an overall user search time of only 30% of the search engine without the patterns. That is, the search engine with the cache of the best generalized patterns is 3.3 times faster overall then the one without using the cache, and 3.1 times faster than the baseline cache of the most frequent user queries.

Table 8.3. Results of our log mining algorithm for the Encarta search engine.

	Recall	Precision	Search Time (W/Cache,100%)
Baseline Cache	7.3%	82.3%	95%
I: with Hierarchy	41.7%	81.7%	95 %
II: I + morphology analysis	52.8%	90.6 %	67%
III: II + synonyms	80.4 %	88.3 %	30 %

8.4.11 Relation to Previous Work

Many techniques used in this chapter have been used before(though mostly on smaller problems). We demonstrate that when those techniques are carefully integrated and tuned, they can solve the large-scale real-world query-log mining problem.

Pattern learning discussed in this chapter is similar to rule learning from labeled examples, a well studied area in machine learning. Many methods, such as C4.5 rule, the AQ family, and RIPPER, have been developed for pattern learning. C4.5 rule [8.27] is a state-of-the-art method for learning production rules. It first builds a decision tree, and then extracts rules from the decision tree. Rules are then pruned by deleting conditions that do not affect the predictive accuracy. However, C4.5 rule is limited to produce rules using only the original features (that is, only "dropping variables" are used in generalization), while our method also introduces generalized concepts in the hierarchy.

The AQ family [8.22] generates production rules directly from data. It uses the "generate and remove" strategy, which is adopted in our algorithm. However, AQ is a top-down induction algorithm, starting from a most general rule and making it more specific by adding more conditions. AQ does not utilize concept hierarchy as we do in our generalization process. It seems difficult to incorporate concept hierarchy in the top-down learning strategy, as a large number of concepts in the hierarchy must be tried for making a rule more specific. The difference between our mining algorithm and RIPPER [8.10] are also similar.

Concept hierarchy has been used in various machine learning and data mining systems. One usage is as background knowledge, as in [8.17]. Concept hierarchies have also been used in various algorithms for characteristic rule

mining [8.16, 8.31], multiple-level association mining [8.16], and classification [8.18]. What makes our work different is that our concept hierarchy is
much larger and more general; it is generated automatically from WordNet
over tens of thousands of keywords.

The improvements over the simple generalization algorithm are similar
to approaches to the term mismatch problem in information retrieval. These
approaches, called dimensionality reduction in [8.13], aim to increase the
chance that a query and a document refer to the same concept using different terms. This can be achieved by reducing the number of possible ways
in which a concept is expressed; in other words, by reducing the "vocabulary" used to represent the concepts. A number of techniques have been
proposed. The most important ones are manual thesauri [8.1], stemming and
conflation [8.14], clustering or automatic thesauri [8.28, 8.32], and Latent Semantic Indexing [8.12]. These techniques propose different strategies of term
replacement. The strategies can be characterized by (1) semantic considerations (manual thesauri), (2) morphological rules (stemming and conflation),
and (3) term similarity or co-occurrence (clustering or Latent Semantic Indexing). In our work, we used strategies similar to (1) and (2). The manual
thesaurus we used is WordNet. We also dealt with synonymous, but unlike
strategy (3), which clusters similar terms based on co-occurrence, we used
clusters of synonymous provided by WordNet directly. Our training data are
user logs, which do not contain full text information of processed documents,
which are necessary for co-occurrence estimation.

8.5 Conclusions

The purpose of this chapter is to advocate the discovery of actionable knowledge from Web logs. Actionable knowledge is particularly attractive for Web
applications because it can be consumed by machines rather than human
developers. Furthermore, the effectiveness of the knowledge can be immediately put to test, making the merits of the type of knowledge and methods
for discovering the knowledge under more objective scrutiny than before. In
this chapter, we presented two examples of actionable Web log mining. The
first method is to mine a Web log for Markov models that can be used for
improving caching and prefetching of Web objects. A second method is to use
the mined knowledge for building better, adaptive user interfaces. A third application is to use the mined knowledge from a query Web log to improve the
search performance of an Internet Search Engine. In our future work, we will
further explore other types of actionable knowledge in Web applications, including the extraction of content knowledge and knowledge integration from
multiple Websites.

References

8.1 J. Aitchison, A. Gilchrist: *Thesaurus Construction: A practical manual* (ASLIB, London, 2nd edition, 1987)

8.2 D. Albrecht, I. Zukerman, A. Nicholson: Pre-sending Documents on the WWW: A Comparative Study. In: *Proceedings of the 1999 International Conference on Artificial Intelligence, IJCAI'99, Sweden* pp. 1274-1279

8.3 R. Armstrong, D. Freitag, T. Joachims, T. Mitchell: Webwatcher: A learning apprentice for the World Wide Web. In: *Working Notes of the AAAI Spring Symposium: Information Gathering from Heterogeneous, Distributed Environments* (Stanford University, 1995. AAAI Press) pp. 6-12

8.4 M. Arlitt, R. Friedrich L. Cherkasova, J. Dilley, T. Jin: Evaluating content management techniques for web proxy caches. In: *HP Technical report, Palo Alto, Apr. 1999*

8.5 T.B. Lee, J. Hendler, O. Lassila: The Semantic Web. Scientific American, 284(5), 34-43 (2001)

8.6 S. Brin, L. Page: The Anatomy of a Large-Scale Hypertextual Web Search Engine. *Proceedings of the 7th WWW, Brisbane, 1998*

8.7 J. Boyan, D. Freitag, T. Joachims: A Machine Learning Architecture for Optimizing Web Search Engines. *Proc. of AAAI Workshop on Internet-Based Information Systems, Portland, Oregon, 1996*

8.8 B. Chidlovskii, C. Roncancio, M.L. Schneider: Semantic cache mechanism for heterogeneous web querying. *Proc. of the 8th Int. World Wide Web Conf. 1999* (Computer Networks) 31(11-16), 1347-1360 (1999)

8.9 P. Cao, S. Irani: Cost-aware WWW proxy caching algorithms. In: *Proceedings of the USENIX Symposium on Internet Technologies and Systems, December 1997* pp. 193-206

8.10 W.W. Cohen: Fast Effective Rule Induction, In: *Proc. of International Conference on Machine Learning, 1995*

8.11 M. Craven, D. DiPasquo, D. Freitag, A. McCallum, T. Mitchell, K. Nigam, S. Slattery: Learning to Construct Knowledge Bases from the World Wide Web. Artificial Intelligence, 118(1-2), 69-113 (2000)

8.12 S. Deerwester, S.T. Dumais, T. Landauer, Harshman: Indexing by Latent Semantic Analysis. Journal of the American Society for Information Science, 41(6), 391-407 (1990)

8.13 C. Fabio: Exploiting the Similarity of Non-Matching Terms at Retrieval Time. Information Retrieval, 2, 25-45 (2000)

8.14 W.B. Frakes: Stemming Algorithm, In: W.B. Frakes, R.B. Yates (eds.), *Information Retrieval: data structures and algorithms* (Prentice Hall, Englewood Cliffs, New Jersey, USA, 1992)

8.15 http://www.google.com.

8.16 J. Han: Mining Knowledge at Multiple Concept Levels. *Proc. of 4th Int. Conf. on Information and Knowledge Management, Maryland, 1995*

8.17 J. Han, Y. Cai, N. Cercone: Data-Driven Discovery of Quantitative Rules in Relational Databases. IEEE Tran. On Knowledge and Data Engineering, 5(1), 29-40 (1993)

8.18 M. Kamber, L. Winstone, W. Gong, S. Cheng, J. Han: Generalization and Decision Tree Induction: Efficient Classification in Data Mining. *Proc. of 1997 Int. Workshop on Research Issues on Data Engineering, Birmingham, England, 1997*

8.19 D. Lin: Using Syntactic Dependency as Local Context to Resolve Word Sense Ambiguity. *Proceedings of ACL/EACL-97* pp. 64-71 (1997)

8.20 W. Lou, G. Liu, H. Lu, Q. Yang: Cut-and-Pick Transactions for Proxy Log Mining. In: *Proceedings of the 2002 Conference on Extending Database Technology, March 24-28 2002, Prague*

8.21 H.T. Ng, H.B. Lee: Integrating Multiple Knowledge Sources to Disambiguate Word Sense: An Exemplar-Based Approach. *Proceedings of 34th Annual Meeting of the Association for Computational Linguistics* pp. 44-47 (1996)

8.22 R.S. Michalski: A Theory and Methodology of Inductive Learning. Artificial Intelligence, 20(2), 111-161 (1983)

8.23 G. Miller: WordNet: an online lexicon database. International Journal of Lexicography, 3(4), 235-312 (1990)

8.24 M. Perkowitz, O. Etzioni: Adaptive Web Sites: Concept and Case Study Artificial Intelligence, 118(1-2), pp 245-275 (2000)

8.25 J. Pitkow, P. Pirolli: Mining Longest Repeating Subsequences to Predict World Wide Web Surfing. In: *Second USENIX Symposium on Internet Technologies and Systems, Boulder, C0, 1999*

8.26 J. Pei, J. Han, B.M. asl, H. Zhu: Mining Access Patterns efficiently from Web Logs. *Proc. 2000 Pacific-Asia Conf. on Knowledge Discovery and Data Mining, Kyoto, Japan, 2000*

8.27 J.R. Quinlan: *C4.5: Programs for Machine Learning* (San Mateo, Morgan Kaufmann, 1993)

8.28 E. Rasmussen: Clustering Algorithm. In: W. B. Frakes, R. Baeza-Yates (eds.), *Information Retrieval: data structures and algorithms, Chapter 16* (Prentice Hall, Englewood Cliffs, New Jersey, USA, 1992)

8.29 J. Srivastava, R. Cooley, M. Deshpande, P. Tan: Web Usage Mining: Discovery and Applications of Usage Patterns from Web Data. ACM SIGKDD Explorations, (1)2 (2000)

8.30 S. Schechter, M. Krishnan, M. D. Smith: Using Path Profiles to Predict HTTP Requests. In: *Proc. 7th International World Wide Web Conference, Brisbane, Qld., Australia, April 1998* pp. 457-467

8.31 R. Srikant, R. Agrawal: Mining Generalized Association Rules. *Proc. 1995 Int. Conf. Very Large Data Bases, Zurich, Switzerland, 1995*

8.32 P. Srinivasan: Thesaurus Construction. In: W.B. Frakes, R. Baeza-Yates (Eds.), Information Retrieval: data structures and algorithms, Prentice Hall, Englewood Cliffs, New Jersey, USA, 1992

8.33 N. Sundaresan, J. Yi: Mining the Web for Relations. *The Ninth International WWW Conference, Amsterdam, The Netherlands, 2000*

8.34 P.C. Saraiva, E.S. Moura, N. Ziviani, W. Meira, R. Fonseca, B.R. Neto: Rank-Preserving Two-Level Caching for Scalable Search Engines. *ACM SIGIR 24th Int. Conference on Information Retrieval* pp. 51-58 (2001)

8.35 Z. Su, Q. Yang, H. Zhang, X. Xu, Y. Hu, S.P. Ma: Correlation-based Web-Document Clustering for Web Interface Design. International Journal Knowledge and Information Systems (Springer-Verlag London Ltd) 4, 141-167 (2002)

8.36 Z. Su, Q. Yang, H. Zhang: A Prediction System for Multimedia Pre-fetching in Internet. In: *Proc. 2000 Intl ACM Conf. on Multimedia, Los Angeles, California, 2000*

8.37 Q. Yang, H. Zhang: Integrating Web Prefetching and Caching Using Prediction Models. World Wide Web Journal, Kluwer Academic Publishers, 4(4)

9. Discovery of Web Robot Sessions Based on Their Navigational Patterns

Pang-Ning Tan[1] and Vipin Kumar[2]

[1] Michigan State University, USA
[2] University of Minnesota, USA

Abstract

Web robots are software programs that automatically traverse the hyperlink structure of the World Wide Web in order to locate and retrieve information. There are many reasons why it is important to identify visits by Web robots and distinguish them from other users. First of all, e-commerce retailers are particularly concerned about the unauthorized deployment of robots for gathering business intelligence at their Websites. In addition, Web robots tend to consume considerable network bandwidth at the expense of other users. Sessions due to Web robots also make it more difficult to perform clickstream analysis effectively on the Web data. Conventional techniques for detecting Web robots are often based on identifying the IP address and user agent of the Web clients. While these techniques are applicable to many well-known robots, they may not be sufficient to detect camouflaging and previously unknown robots. In this paper, we propose an alternative approach that uses the navigational patterns in the clickstream data to determine if it is due to a robot. Experimental results on our Computer Science department Web server logs show that highly accurate classification models can be built using this approach. We also show that these models are able to discover many camouflaging and previously unidentified robots.

9.1 Introduction

Web robots are software programs or agents that automatically traverse the hyperlink structure of the World Wide Web in order to locate and retrieve information. The emergence of the World Wide Web as an information dissemination medium, along with the availability of many Web robot authoring tools, has resulted in the rapid proliferation of Web robots unleashed onto the Internet today. These robots are sent out to scour the Web for various purposes. For instance, they can be used to collect statistics about the structure of the World Wide Web [9.8]. As another example, Internet search engines such as Google and Altavista rely on the documents retrieved by Web robots to build their index databases. Web administrators employ Web robots to perform site maintenance tasks such as mirroring and checking for broken hyperlinks. Web robots are also used by business organizations to collect email addresses and online resumes, monitor product prices, despatch corporate news, etc.

There are many situations in which it is desirable to identify visits by Web robots and distinguish them from other users. First, e-commerce retailers are particularly concerned about the unauthorized deployment of Web robots for gathering business intelligence at their Websites. In such a situation, the e-commerce site may want to block HTTP requests coming from the unauthorized robots [9.7]. For example, eBay filed a lawsuit against an auction aggregator site last year for using unauthorized shopbots to retrieve auction information from their Website[1].

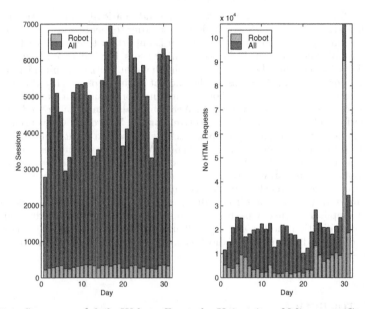

Fig. 9.1. Summary of daily Web traffic at the University of Minnesota Computer Science department Web server. The above figure shows the fraction of total sessions and HTML requests due to Web robots. The anomaly on day 30 for the right-hand figure is due to HTML requests from a site mapping robot called linbot

Secondly, many e-commerce Websites perform Web traffic analysis in order to infer the demographic and browsing behavior of their customers. Unfortunately, such analysis can be severely distorted by the presence of traffic due to Web robots. For example, Fig. 9.1 shows the total number of sessions and HTML pages requested at the University of Minnesota Computer Science department Website between January 1, 2001 and January 31, 2001.

---

[1] An auction aggregator collects information from various on-line auction sites and lists the integrated results at their own Website. As a result, prospective buyers can buy products from auction sellers without visiting the site where the auction was originally posted. This is of great concern to many auction site operators because buyers and sellers may eventually stop visiting their Website and use the services of aggregator sites instead.

Table 9.1. List of IP addresses and user agents for several known Web clients.

Client's Type	IP address	User agent
Browser(Netscape)	160.94.178.152	Mozilla/4.7 [en] (X11;I; Linux 2.2.14-5.0 i686)
Browser(IE)	160.94.178.205	Mozilla/4.0 (compatible; MSIE 5.01; Windows NT)
Browser(Opera)	160.94.103.248	Opera/5.01 (Windows NT & Opera 5.0; U)
Search Engine	199.172.149.184	ArchitextSpider
Email Harvester	4.41.77.204	EmailSiphon
Link Checker	130.237.234.90	LinkChecker/1.0
Search Engine (looksmart.com)	207.138.42.10	Mozilla/4.5 [en] (Win95; I)

On average, about 5% of the total sessions are due to visits by Web robots. However, Web robot sessions may account for as many as 85% of the total number of HTML pages requested. If these robot sessions are not identified and eliminated, an analyst may end up making wrong inferences about the site visitors.

Thirdly, the deployment of Web robots usually comes at the expense of other users because they often consume considerable network bandwidth and server resources. Poorly designed robots may tie up these resources and overload the Web server. In this situation, it will be desirable to detect the disruptive robots and reduce their priority of service immediately.

Fourthly, Web robot accesses could be indicative of fraudulent behavior. For example, there are many click-through payment programs established on the Web, in which an advertiser (i.e., the target site) would reward the referring Website for every visitor who reaches the target site by clicking on the referrer's advertisement banner. Unscrupulous referrer site owners can easily abuse such a payment scheme by inflating the click-through rate using Web robots. Hence, detection of Web robot sessions is absolutely necessary to protect the target site owner from such malpractice.

Even though Web robot detection is a widely recognized problem, there are very few published papers in this area. A standard way to detect robots is by examining the HTTP request messages sent to a Web server [9.9, 9.20]. For example, accesses by many well-known robots can be detected by comparing the IP address and user agent fields of their HTTP request messages against those of known robots (cf. Table 9.1). However, since Web robots can be easily constructed and deployed, it is impossible to keep a comprehensive list of IP addresses and user agents for all robots. This problem is exacerbated by robots that attempt to disguise their identities by using user agents that are very similar to conventional browsers (e.g., the last entry of Table 9.1). As a result, this approach may fail to detect the presence of such robots.

In this paper, we offer an alternative solution by building classification models that distinguish robot sessions from non-robot sessions. Our main assumption is that the navigational patterns of Web robots are inherently different from those of human users.

Building classification models from Web usage data is a challenging problem due to the following reasons. First, the clickstream data is sequential and temporal in nature. Therefore, the models must take into account both the static and dynamic properties of each session. Furthermore, since the length of a Web session varies from one session to another, it is not sufficient to generate a single classification model for all the sessions. Instead, the models must be built incrementally, after each request made by the Web client. Secondly, it is not trivial to construct a reliable training set for the classification task. This requires a dependable preprocessing step to convert the raw clickstream data into individual sessions. In addition, the class label for each session must be determined. Manual assignment of the class labels can be a laborious task, while an automated procedure may introduce labeling errors into the training data. Thirdly, it would be desirable to detect the Web robots swiftly, after looking at just a few requests. This is a formidable goal because the smaller the number of requests of a session, the less information it contains to correctly predict whether it is due to a Web robot. Hence, building classification models for Web robots involves a tradeoff between early detection and model accuracy.

The main contributions of this paper are summarized below:

1. We analyze the navigational patterns for various types of Web robots and show that these patterns are quite different from those for human users.
2. We propose a robust session identification technique to preprocess the Web server logs. Our technique can identify sessions with multiple IP addresses and user agents.
3. We present a procedure for labeling the training and test data sets, and a technique for identifying the mislabeled samples. We show that this technique was able to discover many camouflaging and previously unknown Web robots.
4. We show that highly accurate robot classification models can be induced from the access features of the Web sessions.

The rest of the paper is organized as follows. In Sect. 9.2, we discuss the limitations of some of the existing techniques used for detecting Web robots. Section 9.3 describes the preprocessing steps needed to convert the raw clickstream data into server sessions. A discussion about how to derive the session features and class labels is also presented. This is followed by our experimental results in Sect. 9.4, while Sect. 9.5 concludes with suggestions for future work.

9.2 Web Robot Detection: Overview

9.2.1 Limitations of Current Robot Detection Techniques

In this section, we present some of the common techniques used to detect Web robots, and describe the limitations of each technique:

1. **Robots.txt Access Check**
   The Robot Exclusion Standard [9.11, 9.13] was proposed to allow Web administrators to specify which part of their Website is off-limits to visiting robots. According to this Standard, whenever a robot visits a Website, say at www.xyz.com, it should first examine the file http://www.xyz.com/robots.txt. This file contains a list of access restrictions for the Website as specified by the Web administrator. For example, the following entry in the robots.txt file forbids all robots from accessing the file http://www.xyz.com/A.html.

   > User-agent: *
   > Disallow: /A.html

   This suggests that Web robots should be easily detected from sessions that access the robots.txt file. Indeed, this is a reasonably good heuristic because many Websites do not provide a direct hyperlink to this file from any other HTML page. As a result, most users are unaware of the existence of this file. However, one should not rely on this criterion alone because compliance with the Robot Exclusion standard is voluntary, and many robots simply do not follow the proposed standard.

2. **User Agent Check**
   It is commonly agreed that poor implementation of the Web robots can lead to serious network and server overload problems. Thus, a guideline is needed to ensure that both the Web robot and the Web server can cooperate with each other in a way that is beneficial to both parties. Under the proposed ethical guidelines for robot designers [9.6, 9.12, 9.14], a cooperative robot must declare its identity to a Web server via its user agent field. For instance, the user agent field of Web robots should contain the name of the robot, unlike the user agent field of Web browsers, which often contains the name Mozilla (cf. Table 9.1). Figure 9.2 shows an example where an Internet Explorer browser, identified by its user agent field, Mozilla/4.0 (compatible; MSIE 5.01), was used to request the HTML page http://www.xyz.com/A.html.
   In practice, not all robot designers adhere to these guidelines. Some robots (and browsers) would use multiple user agent fields within the same session. For example, an offline browser called Teleport Pro has an empty user agent field when accessing the robots.txt file, but uses Teleport Pro when retrieving other documents. Even conventional Web browsers may issue requests with different user agent fields. In the following example, a user with the Microsoft Internet Explorer browser

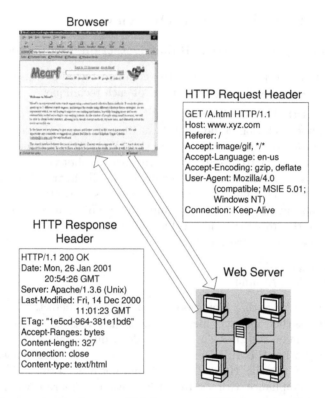

Fig. 9.2. Communication between Web browser and Web server via HTTP protocol

attempts to download a PDF document from our Website. Two HTTP request messages are generated as a result of this action, each having its own user agent field (i.e., MSIE 5.01 and contype). These requests are recorded in the Web server logs as:

    203.94.250.186 - - [01/Jan/2001:15:18:04 -0600] "GET /grad-info/finapp.pdf
    HTTP/1.1" 200 3993 "http://www.cs.umn.edu/grad-info/"
    "Mozilla/4.0 (compatible; MSIE 5.01; Windows 98; bplnet-100)"

    203.94.250.186 - - [01/Jan/2001:15:18:08 -0600] "GET /grad-info/finapp.pdf
    HTTP/1.1" 200 3993 "-" "contype"

The robot detection problem becomes more complicated when robot designers attempt to hide their identities by using the same user agent field as standard Web browsers. In this situation, it is impossible to detect these robots using their user agent fields alone. Similarly, some anonymizer Websites allow Web users to camouflage their accesses by transforming the user agent fields of their browsers into robot-like (i.e., non-Mozilla) values such as SilentSurf and Turing Machine [2].

3. **IP Address Check**

Another way to detect robots is by matching the IP address of a Web client against those of known robots. Even though there are many Websites that provide a list of IP addresses for known Web robots[3], it is difficult to maintain an up-to-date database of all the robots. This is because new robots keep cropping up as the World Wide Web continues to expand. Another problem with this approach is that the same IP address could be used by Web users for surfing the Web and by robots to automatically download some files from a Website. The IP address check approach is applicable only if the robot has been previously identified. One way to detect new robots is by examining the top visiting IP addresses of the Web clients and manually verifying the origin of each client. Unfortunately, this technique is time consuming, and often discovers robots that are already well-known.

4. **Count of HEAD requests and HTTP requests with unassigned referrers**

The guidelines for Web robot designers also suggest that ethical robots should (1) moderate their rate of information acquisition, (2) operate only when the server is lightly loaded (e.g., at night), and (3) use the HEAD request method whenever possible. The request method (e.g., GET, HEAD, and POST) of an HTTP request message determines what type of action the Web server should perform on the resource requested by the Web client. For example, a Web server responds to a GET request by sending a message, consisting of some header information along with a message body, which contains the requested file. However, the response to a HEAD request contains only the message header, thus incurring less communication overhead. This is the reason why Web robots are encouraged to use the HEAD request method. In principle, one can examine sessions with a large number of HEAD requests to discover ethical Web robots. In addition, one should also look for sessions that have a large number of requests with unassigned referrer fields. The referrer field is provided by the HTTP protocol to allow a Web client (particularly, a Web browser) to specify the address of the Web page that contains the link the client followed in order to reach the current requested page. For example, whenever a user requests the page http://www.xyz.com/A.html by clicking on a hyperlink found at http://www.xyz.com, the user's browser will generate an HTTP request message with its referrer field assigned to http://www.xyz.com. Since most robots do not assign any value to

---

[2] The anonymizer Website SilentSurf is located at http://www.noproxy.com, while the Turing Machine Website is located at http://anonymizer.com

[3] For example, the Web Robot Database located at http://info.webcrawler.com/mak/projects/robots/active/

their referrer fields, these values appear as "-" in the Web server logs. Both of these heuristics are not entirely reliable because non-robots can sometimes generate HEAD request messages (for example, when proxy servers attempt to validate their cache contents) and HTTP messages with unassigned referrer values (for example, when a user clicks on a bookmarked page or types in a new URI in the address window).

9.2.2 Motivation for Proposed Robot Detection Technique

The previous discussion suggests that a more robust technique is needed to identify visits by camouflaging and previously unknown Web robots. Before we describe our proposed technique, it is essential to understand what are the different types of robots that are available today (cf. Table 9.2). This is because each type of robot may exhibit different characteristics depending on their navigational goals. Knowing the characteristics and navigational goals of these robots can help us to identify the set of relevant features for predicting Web robot sessions.

Table 9.2. Typical characteristics and navigational goals of Web Robots

Client's Type	Examples	Navigational Goals	Characteristics
Search Engine	T-Rex (Lycos),	maximize coverage	breadth first search,
	Scooter (Altavista)	of a Website	unassigned referrer.
Offline Browser	Teleport Pro,	download Website	varied behavior.
	Offline Explorer	to local disk	
Email Collector	EmailDigger,	maximize coverage	unassigned referrer,
	Extractor Pro	of home pages	ignores image files.
Link Checker	LinkScan,	check for broken	HEAD request,
	Xenu's Link Sleuth	links	unassigned referrer.

For example, consider a search engine robot. The ultimate goal of such robot is to retrieve as many documents as possible in a short period of time. As a result, these robots often use a breadth-first retrieval strategy to maximize their coverage and parallel retrieval strategy in order to speed up their operations.

Another type of Web robot is called a link checker, which is a utility program used by Website administrators to check for broken hyperlinks and missing pages. Many link checkers would use the HEAD request message to check if a hyperlink still exists. A Web server would respond to the HEAD request by sending a message header containing a status code that indicates whether the request has succeeded or failed.

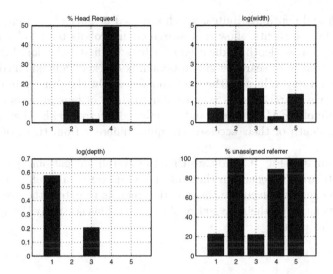

Fig. 9.3. Comparison between the navigational patterns of several known Web clients. The horizontal axis represents each type of client: (1) users from the Computer Science department at University of Minnesota, (2) search engine robots, (3) offline browsers, (4) link checkers, and (5) email collectors, while the vertical axis represents the various session characteristics

Web robots are also designed for various other purposes. For example, email collectors are robots that automatically collect email addresses posted on the Web. These robots tend to retrieve only HTML pages, while ignoring image and other file formats. Offline browsers are either stand-alone browsers or add-on utilities that allow a Web user to download the entire Website (or a portion of it) to a local directory for offline viewing. The characteristics of offline browsers may vary, depending on their navigational goals. For instance, offline browsers that download an entire Website behave very similar to search engine robots, while those that download only a small portion of the Website (for pre-caching purposes) resemble human users. Other types of Web robots include personal browsing assistants [9.1, 9.15], shopbots [9.3, 9.10], resume hunters, etc.

The main assumption of this paper is that the navigational patterns of Web robots are distinct from those of human users. These patterns can be characterized in terms of the types of pages being requested, the length of the session, the interval between successive HTML requests, the coverage of the Website, etc. A more in-depth discussion about the navigational characteristics is presented in Sect. 9.3.2.

Figure 9.3 shows a comparison between the navigational patterns of various Web clients. Note the consistency between the observed patterns and the navigational goals of these robots, given in Table 9.2. For example, the HEAD method accounts for almost half of the requests made by link checkers.

The width and depth attributes, which are used to infer the search strategy employed by the clients, show that search engine robots tend to have broad width but shallow depth, indicative of a breadth-first behavior. Figure 9.3 also shows that most search engine robots, link checkers, and email collectors do not assign any values to their referrer fields. On the other hand, offline browsers have characteristics very similar to those of human users in terms of their rate of HEAD requests and unassigned referrer fields. However, the width and depth of their accesses are quite different from those of human users.

The results of Fig. 9.3 indicate that the overall navigational patterns of Web robots are different from those of human users. As a result, one should be able to construct accurate classification models to detect the presence of Web robots based on their navigational patterns. This is exactly the approach taken by this paper.

9.3 Methodology

The goal of our classification task is to build models that will automatically map each session into one of the two predefined classes: robot versus non-robot. This requires (1) a preprocessing step to transform the Web data into individual sessions, (2) extraction of features that best describe the properties of each class, (3) a procedure to the determine the class label of each session, (4) a technique to assess how good the labeling scheme is, (5) an algorithm to build classification models, and (6) a metric to evaluate the performance of the models. The first two requirements are described in Sects. 9.3.1–9.3.2, respectively. The third requirement is needed to provide the training data for the classification algorithm, and is described in Sect. 9.3.3. The fourth requirement is described in Sect. 9.3.5, while the last two requirements are discussed in Sect. 9.3.4.

9.3.1 Data Source and Preprocessing

Web server logs are used as the data source for our experiments. A typical Web server log contains information such as the IP address and user agent of the Web client, the URI of the requested and referred pages, the request method and protocol used, the timestamp and status code of each request, and the size of the file that was transferred. Entries in the Web server logs are ordered according to their timestamps.

The first preprocessing step is to transform the raw clickstream data into server sessions. Without client-side tracking, cookies, or embedded session identifiers, it is extremely difficult to identify the individual sessions in the Web server logs reliably. A standard way to do this is by grouping together the Web log entries according to their IP address and user agent fields [9.5, 9.16].

This approach may not work well for two reasons. First, each IP address/user agent pair may contain more than one session, e.g., sessions created by Web users who share the same proxy server and use the same type of browser. Secondly, a session that contains multiple IP addresses or user agents will be fragmented into multiple sessions.

Cooley [9.4] has addressed the first problem by modifying his session identification heuristic to fragment each IP address/user agent pair into smaller sessions whenever the session contiguity condition is violated. This condition simply states that the referrer field of the current request of a session must match one of the pages previously retrieved by the session (unless the referrer value is not assigned).

Developing heuristics for solving the second problem is essential for detecting robot activities, since many search engine robots tend to parallelize their Web retrieval operations. These heuristics are also useful for identifying user sessions if client tracking technologies are unavailable. In this paper, we have proposed a new session identification algorithm for handling the second problem.

In order to match a log entry l_j to its corresponding session, we partition the list of currently active sessions H into 4 groups: candidateSet[1], candidateSet[2], candidateSet[3], and candidateSet[4]. The first group contains sessions that have the same IP address and user agent as l_j. The second group, candidateSet[2], contains sessions that have the same user agent as l_j but different a IP address. However, these sessions share a common domain name with l_j (e.g., crawler1.googlebot.com and crawler2.googlebot.com both have the same domain name, googlebot.com). We use a reverse DNS lookup program to resolve the hostname of each IP address. The third group, candidateSet[3], contains sessions that have the same user agent and prefix IP address as l_j (e.g., both 160.94.178.151 and 160.94.178.153 share a common class C address, 160.94.178). This group is needed because the hostnames for some sessions can not be resolved by the reverse DNS lookup program due to server timeout, non-existent host/domain error, etc. The last group, candidateSet[4], accounts for sessions that have the same IP address as but user agents different from l_j.

Table 9.3 summarizes the key steps of our session identification algorithm. For each log entry, l_j, we first find the candidate sessions that may potentially contain l_j. The getCandidates function (Table 9.4) will generate the four groups of candidate sessions mentioned above. Next, the bestCandidate function (Table 9.4) will choose the most likely session to contain l_j. In this function, candidateSet[1] is scanned first, since it is the one that will most likely contain l_j. If no matching session is found, then candidateSet[2] is scanned, followed by candidateSet[3], and finally candidateSet[4]. We use a 30-minute session timeout and the session contiguity conditions to determine whether l_j belongs to a given session.

Table 9.3. Modified Session Identification Heuristic

type logEntry {		type session {	
ip : string,	*request* : URI	*count* : integer	
time : seconds,	*referrer* : URI	*list* : array of logEntry	
agent : string,	*method* : string	*Class* : integer }	
status : string	*protocol* : string }		

1. Let H denotes the set of active sessions.
2. Let L denotes the time-ordered Web log entries.
3. Let T denote the session timeout.
4. for each $l_j \in L$ do
5.   for each $s_j \in H$ do
6.   if $(s_j.list[s_j.count].time - l_j.time > T)$ then
7.   close session s_j
8.   end;
9.   candidateSet = getCandidates(H, l_j)
10.   if (candidateSet is NULL) then
11.   create new session s'
12.   add l_j to $s'.list$ and increment $s'.count$
13.   add s' to H
14.   else
15.   assign = bestCandidate(candidateSet, l_j)
16.   if (assign is NULL) then
17.   create new session s'
18.   add l_j to $s'.list$ and increment $s'.count$
19.   add s' to H
20.   else
21.   add l_j to assign.$list$
22.   increment assign.$count$
23. end;

One potential pitfall of our approach is that it may inadvertantly group together sessions that belong to different users. For example, requests that come from different machines but with the same domain name (e.g., lnx02.cs.umn.edu and lnx03.cs.umn.edu) could be grouped together even though they belong to different users. This is because our session identification heuristics are capable of distinguishing between the two sessions as long as they do not overlap each other in terms of their set of requested pages or their request times. Unfortunately, we found that both conditions can be violated by sessions due to users from our university. We have attempted to correct this problem by ignoring candidateSet[2] and candidateSet[3] for clients with hostnames that end with umn.edu or with other known non-proxy hostnames. We verify this step by checking all the sessions that still contain multiple IP adddresses.

9.3.2 Feature Vector Construction

After the server sessions have been created, the next step is to derive the properties of each session. Each session is broken up into several episodes,

Table 9.4. The getCandidates and bestCandidate function.

---

function getCandidates(H: set of session, l_j: logEntry)
1. Let candidateSet be a two-dimensional array of sessions
2. for each $s_j \in H$ do
3.   if (containsAgent(s_j, $l_j.agent$)) then
4.   if (containsIP(s_j, $l_j.ip$)) then
5.   add s_j to candidateSet[1]
6.   else if (sameDomain(s_j, $l_j.ip$)) then
7.   add s_j to candidateSet[2]
8.   else if (sameAddressClass(s_j, $l_j.ip$)) then
9.   add s_j to candidateSet[3]
10.   else
11.   if (containsIP(s_j, $l_j.ip$)) then
12.   add s_j to candidateSet[4]
13. end;
14. return candidateSet

function bestCandidate(C: two dimensional array of sessions, l_j: logEntry)
1. assign = NULL
2. if ($l_j.referrer$ is a local page) then
3.   for i=1 to 4 do
4.   assign = find $s_k \in C[i]$ such that $(l_j.time - s_k.list[s_k.count].time)$
            is minimum and $l_j.referrer \in$ requestSet(s_k)
5,   if assign is not NULL then return assign
6.   end;
7. else referrer is an external page or unassigned
8.   for i=1 to 4 do
9.   assign = find $s_k \in C[i]$ such that $(l_j.time - s_k.list[s_k.count].time)$
            is minimum and $l_j.referrer \in$ referrerSet(s_k)
10   if assign is not NULL then return assign
11.   end;
12. return NULL

---

where each episode corresponds to a request for an HTML file. (We will use the terms episode and request interchangeably throughout this paper.) We associate each episode with a tuple, (p_i, p_j), where p_i and p_j are respectively the requested and the referred HTML pages. A session that does not contain any request for HTML files will have a single episode, $(-, -)$, associated with its last log entry.

Table 9.5 presents a summary of attributes that can be derived from the server sessions. The values of these attributes may change as the session length increases. The computation of temporal attributes such as *totalTime*, *avgTime*, and *stdevTime* is illustrated in Fig. 9.4. *totalTime* is approximated by the interval between the first and the last log entries of the session, while the *avgTime* and *stdevTime* attributes are computed based on the intervals between two successive HTML requests in the session.

The *width* and *depth* attributes are computed from the graph representing all the HTML requests in a particular session. The *width* attribute measures the number of leaf nodes generated in the graph while the *depth* attribute measures the maximum depth of the tree(s) within the graph. For example, if a session contains the requests {(/A,-), (/A/B,/A), (/A/B/C,/A/B)}, then the session's *width* is 1 while its *depth* is 3. As another example, a session that contains requests for {(/A,-) (/A/B,/A), (/C,-), (/D,-) } will have a *width* of 3 and a *depth* of 2.

MultiIP and *MultiAgent* are two binary features that are used to indicate whether a session contains multiple IP addresses or user agents. The rest of the attributes in Table 9.5 are self-explanatory.

Table 9.5. Summary of attributes derived from the Web server sessions. These attributes are used for class labeling (denoted as Classify) and representing the access features of a session (denoted as Feature)

Id	Attribute Name	Remark	Purpose
1	totalPages	total number of pages requested.	Feature
2	% Image	% of image pages (.gif/.jpg) requested.	Feature
3	% Binary Doc	% of binary documents (.ps/.pdf) requested.	Feature
4	% Binary Exec	% binary program files (.cgi/.exe) requested.	Feature
5	robots.txt	indicates whether robots.txt file is requested.	Classify
6	% HTML	% of HTML pages requested.	Feature
7	% Ascii	% of Ascii files (.txt/.c/.java) requested.	Feature
8	% Zip	% of compressed files (.zip/.gz) requested.	Feature
9	% Multimedia	% of multimedia files (.wav/.mpg) requested.	Feature
10	% Other	% of other file formats requested.	Feature
11	*totalTime*	approximated total time of the session.	Feature
12	*avgTime*	average time between two HTML requests.	Feature
13	*stdevTime*	standard deviation of time between requests.	Feature
14	*Night*	for requests made between 12am to 7am.	Feature
15	*Repeated*	fraction of repeated requests.	Feature
16	*Error*	% of requests with status $\geq$ 400.	Feature
17	GET	% of requests made with GET method.	Feature
18	POST	% of requests made with POST method.	Feature
19	HEAD	% of page requests made with HEAD method.	Classify
20	OTHER	% of requests made with other methods.	Feature
21	*width*	width of the traversal (in the URL space).	Feature
22	*depth*	depth of the traversal (in the URL space).	Feature
23	*length*	session length (total no of HTML requests).	Ignore
24	referrer = "-"	% of requests with unassigned referrer.	Classify
25	*MultiIP*	indicates whether session contains multiple IP.	Feature
26	*MultiAgent*	indicates whether session contains multiple agents.	Feature

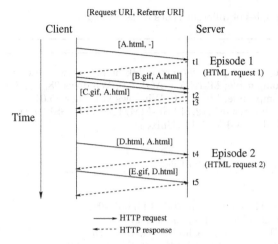

Fig. 9.4. There are two requests in this session. $t1$, $t2$, $t3$, $t4$, and $t5$ are the timestamps recorded in the server logs. The total time of the session is $t5 - t1$, while the period between the two requests is $t4 - t1$

9.3.3 Session Labeling

We use the common robot detection techniques described in Sect. 9.2.1 to determine the class label of each session. The session labeling algorithm is summarized in Table 9.7:

1. If a session s contains a request for the robots.txt file, then the session is declared to be a robot session (denoted as $s.Class = 1$).
2. We label the rest of the sessions according to their user agents. In order to do this, we have divided the user agents into four categories: Type 1 (known robots), Type 2 (known browsers), Type 3 (possible robots), and Type 4 (possible browsers). This categorization is done semi-automatically. Type 3 agents contain agent names that would suggest that they are likely to be robots, while Type 4 agents are possible browsers or helper applications invoked by the browsers. Table 9.6 shows some examples of the different types of agents. In our experiments, we label sessions with Type 3 agents as robots ($s.Class = 1$) while those with Type 4 agents as non-robots ($s.Class = 0$).
3. For sessions that contain multiple user agents, we use a labeling scheme that favors non-robot, as opposed to robot, sessions. There are two main reasons for doing this. Firstly, we observed that the majority of these multi-agent sessions contain combinations of Type 2 and Type 3 agents or Type 3 and Type 4 agents. Further analysis showed that these sessions are due to Web users who invoke a helper application while surfing the Web. In the following example, Go!Zilla is a download manager program used by the Web user to retrieve a zipped file:

Table 9.6. Examples of different types of user agents

User agent	Type
ArchitextSpider	Type 1
Mozilla/4.0 (compatible; MuscatFerret/2.0; http://www.webtop.com/)	Type 1
Mozilla/4.0 (compatible; MSIE 5.0; AOL 6.0; Windows 98; DigExt)	Type 2
Mozilla/4.0 (compatible; MSIE 5.0; Windows 98) Opera 5.01 [en]	Type 2
Lynx/2.8.3rel.1 libwww-FM/2.14 SSL-MM/1.4.1 OpenSSL/0.9.6	Type 2
www4mail/2.4 libwww-FM/2.14 (Unix; I)	Type 3
unknown/1.0	Type 3
contype	Type 4
Windows-Media-Player/7.00.00.1956	Type 4

155.239.194.112 - - [01/Jan/2001:14:38:37 -0600]
"GET / mein/blender/plugins/ HTTP/1.1"
200 1562 "http://www.rash.f2s.com/links.htm"
"Mozilla/4.0 (compatible; MSIE 5.0; Windows 98)"

155.239.194.112 - - [01/Jan/2001:14:43:34 -0600]
"GET / mein/blender/plugins/plugins.zip HTTP/1.1"
206 626775 "http://www-users.cs.umn.edu/ mein/blender/plugins/"
"Go!Zilla 3.5 (www.gozilla.com)"

Secondly, there are several multi-agent type sessions with combinations of Type 1 and Type 2 agents. A typical example of a Type 1 user agent found in this combination is "Java1.1"[4]. Upon further investigation, we found that these multi-agent sessions are created by Web browsers that are accessing the HTML pages containing Java applets. In fact, our session labeling algorithm will classify all multi-agent sessions as non-robots except for those with combinations of Type 1 and Type 3 agents.

4. There are other heuristics we can use to supplement the above labeling scheme. For example, if all the requests are made using the HEAD method, then the session is most likely created by a link checker robot. Another heuristic could be based on the referrer field of the session. If a Web client does not assign a referrer value to any of its requests, then there is a strong possibility that the client is a Web robot, as long as the number of requests is large. Otherwise, the session is more likely created by a Web browser. This is because a Web browser can generate requests with unassigned referrer values when a user submits a URI from the address window or clicks on a bookmark entry (these are known as user-input clicks). When the number of requests is large, the probability that a Web browser generates only user-input clicks is low. Thus, by selecting an appropriate threshold t on the minimum number of requests, one can potentially identify new robot sessions.

---

[4] This user agent is often associated with the various Java-based agents crawling our Website. This is why it is initially categorized as a Type 1 agent.

Table 9.7. Session Labeling Algorithm

Labeling Algorithm (H: array of sessions, t: length threshold) {
1. for each session $s \in H$ do
2.   if s contains a request for robots.txt then $s.Class = 1$
3.   Let $Agents =$ getUserAgent(s)
4.   Let $AgentTypes =$ getAgentTypes($Agents$)
5.   if $|Agents| = 1$ then
6.   if $Type1 \in AgentTypes$ or $Type3 \in AgentTypes$
7.   then $s.Class = 1$
8.   else $s.Class = 0$
9.   else   session contains multiple agent types
10.   if ($Type2 \in AgentTypes$ or $Type4 \in AgentTypes$)
11.   then $s.Class = 0$
12.   else $s.Class = 1$
13.   if $s.Class = 0$ and $|s.list| > t$ then
14.   if HEAD(s) = 100% or UnassignReferrer(s) = 100%
15.   then $s.Class = 1$
16. end;

9.3.4 Classification

After deriving the session features, classification models are built using the C4.5 decision tree algorithm [9.17]. There are two main objectives we would like to achieve: (1) to find a good model for predicting Web robot sessions based upon their access features, and (2) to be able to detect robotic activities as soon as possible, with reasonably high accuracy.

There are various metrics we can use to evaluate the performance of a classifier. Accuracy is a reasonable metric, as long as the data set remains evenly distributed (between robots and non-robots). Otherwise, we need to compensate the imbalanced class distribution via stratification, or by using techniques such as bagging and boosting. This issue is described further in Sect. 9.4.1. In the area of information retrieval, recall and precision are two popular metrics for evaluating the performance of binary classifiers:

$$\text{recall, } r = \frac{\text{no. of robot sessions found correctly}}{\text{total no. of actual robot sessions}} \text{ and} \tag{9.1}$$

$$\text{precision, } p = \frac{\text{no. of robot sessions found correctly}}{\text{total no. of predicted robot sessions}}. \tag{9.2}$$

A classifier that assigns the value 1 to every session will have perfect recall but poor precision. In practice, the two metrics are often summarized into a single value, called the F_1-measure [9.19].

Given the objective of detecting Web robot activities as early as possible, it is necessary to build the classification models incrementally (after each request) and determine the minimum number of requests needed to identify robot sessions accurately. In order to do this, we need to generate the data sets for each request. For example, the data set for one request is generated from all the sessions because each session has at least one episode. If a session contains only a single request, we use all the information within the session to create its feature vector. However, if the session contains more than one request, we should use only the session information up to its first HTML request. As another example, for the data set with two requests, we ignore all the sessions that have only one request, and consider only those sessions with at least two HTML requests. Sessions having more than two requests are then truncated to ensure that their feature vectors contain only the session information up to their second HTML request. This procedure is repeated until there are very few sessions left to create a new data set (we have created data sets up to seven requests in our experiments).

9.3.5 Identifying Mislabeled Sessions

The overall performance of a classifier depends primarily on the choice of the learning algorithm and the quality of the training data. Evaluation of the various learning algorithms on this data set will not be discussed here because it is not the main focus of this paper. Instead, we turn our attention to the problem of identifying mislabeled sessions in order to improve the quality of the training data, as well as to assess the effectiveness of our session labeling scheme. In particular, we are interested in knowing the main reasons for the sample mislabeling – are they due to poor choice of labeling heuristics, insufficient information in the data, or misleading information in the data? In this section, we describe a technique for identifying mislabeled sessions using an ensemble filtering approach [9.2]. The idea here is to build multiple classifiers from the training instances and use their misclassification errors to identify the mislabeled samples. In our approach, the misclassification errors are also weighted according to the classifiers' accuracies.

Our technique uses the classification models built from all the attributes given in Table 9.5. This includes the attributes that are used to determine the class label of the session (e.g., robots.txt and % HEAD request). Since the C4.5 algorithm prunes the decision trees during model building in order to avoid overfitting, the leaf nodes of the trees contain a probability distribution for each class. We denote these probabilities as $P(0|X)$ and $P(1|X)$, where $P(i|X)$ is the probability that the sample X belongs to class i.

Suppose there is an ensemble of k classifiers, $C_1, C_2, \cdots, C_k$, built from the training data. Let $t(X)$ be the true class of sample X according to the session labeling heuristics, while $c_m(X)$ is the class label assigned to X by the classifier C_m. Furthermore, let A_m denotes the accuracy of classifier C_m.

Using the above definitions, we define the false positive $FP_m(X)$ and false negative $FN_m(X)$ scores for each sample X, and classifier C_m, according to the following formulas:

$$FP_m(X) = \begin{cases} 0 & if \ t(X) = c_m(X) \\ A_m \times |P(c_m(X)|X) - P(t(X)|X)| & if \ t(X) \neq c_m(X) \\ & and \ t(X) = 0 \end{cases} \quad (9.3)$$

and

$$FN_m(X) = \begin{cases} 0 & if \ t(X) = c_m(X) \\ A_m \times |P(c_m(X)|X) - P(t(X)|X)| & if \ t(X) \neq c_m(X) \\ & and \ t(X) = 1 \end{cases} \quad (9.4)$$

The overall false positive or false negative score of a sample X is given by $FP(X) = \sum_{m=1}^{k} FP_m(X)$ and $FN(X) = \sum_{m=1}^{k} FN_m(X)$. A large $FP(X)$ score indicates that the session is currently assigned to be a non-robot even though the classification models suggest that it is actually a robot. If we examine the log entries for these false positive sessions, we can verify whether they are indeed non-robots or are being mislabeled. Later, we will show that many of the sessions with high false positive scores are indeed mislabeled due to the presence of camouflaging and previously unknown robots.

Meanwhile, sessions with large $FN(X)$ scores are robot sessions that are often misclassified as non-robots. Since our session labeling algorithm is stricter towards robots than non-robots, there is less opportunity for a non-robot to be mislabeled as a robot. Thus, the false negative sessions are more likely due to misclassification by the learning algorithm rather than due to mislabeling of the training samples.

9.4 Experimental Evaluation

9.4.1 Experimental Data Set

The experiments described in this section were performed on Web server logs recorded from January 1 through January 31, 2001 at the University of Minnesota Computer Science department. We have consolidated the logs for the two Computer Science department Web servers at http://www.cs.umn.edu and http://www-users.cs.umn.edu. Log entries that correspond to redirection requests from one server to the other are removed to prevent duplication. The consolidated Web logs contain a total of 1,639,119 entries. After preprocessing, 180,602 sessions are created, and the distribution of robot and non-robot sessions are shown in Table 9.8.

The overall data set is then partitioned into two groups: G1, which contains all the training instances of known (Type 1 and Type 2) user agents, and G2, which contains all the instances of Type 3 and Type 4 user agents. Classification models are built using samples from G1 only (i.e., clean data, denoted as E0), or a mixture of G1 and G2 (i.e., noisy data, E1). We repeat the sampling and model building procedure 10 times for each data set. A summary of the dataset description is given in Table 9.9.

In order to account for the unequal sizes of robot and non-robot sessions, we weighted the training and test sets to ensure that robot and non-robot

Table 9.8. Session distribution for various number of requests

# Request	# Type 1 Agents	# Type 2 Agents	# Type 3 Agents	# Type 4 Agents
1	8040	165835	2805	3922
2	3075	49439	550	1201
3	2050	30277	247	803
4	1627	21436	159	639
5	1415	15618	100	523
6	1091	12108	74	424
7	908	9775	59	379

Table 9.9. Summary of the data set description

Experiment	Description
E0	Both training and test data sets contain only G1 samples.
E1	Both training and test data sets contain G1 and G2 samples.

where
G1 : contains Type 1 (known Robot) and Type 2 (known Browser) agents.
G2 : contains Type 3 (possible Robot) and Type 4 (possible Browser) agents

sessions have equal representation. For instance, suppose a (training or test) data set contains 100 robots and 1,000 non-robots, then each robot session is duplicated 10 times to ensure that both classes have equal distribution during model building. This procedure is called stratification by oversampling. We have also conducted similar experiments using the full unweighted data set, along with other stratification strategies (e.g., stratification by undersampling). The results for the stratification by undersampling are quite similar to the ones presented in this paper, while the results for the unweighted data are less conclusive[5]. We have omitted the discussion of these results due to lack of space.

9.4.2 Correlation Analysis

Our initial task is to determine whether there exists an attribute that correlates highly with sessions due to Web robots. Figure 9.5 shows the correlation between each attribute and the robot class label as the number of requests increases from 1 to 4:

1. As expected, the attributes used to determine the class labels (i.e., robots.txt, % HEAD, and % unassigned referrer) have very strong positive correlation with the robot sessions. Although their correlations are relatively large, none of them exceeds 0.5. This suggests that using any of these attributes alone is insufficient to detect Web robots effectively.

---

[5] A full discussion of these results is reported in our technical report.

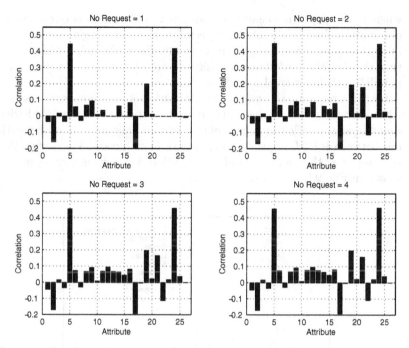

Fig. 9.5. Correlation between each access attribute and the robot class label for various numbers of requests. The horizontal axis corresponds to the attribute IDs given in Table 9.5

More importantly, the values of these attributes can be easily manipulated by robot designers to hide their true identities. For example, a robot designer can easily create a Web robot that does not use the HEAD request and does not access the robots.txt file, but assigns a referrer value to each of its request messages.

2. After one request, the best predictors for robots, beside the attributes used for class labeling, are % image (attribute 2) and % GET request (attribute 17). These attributes have strong anti-correlation with robot sessions, agreeing with our intuition that most robots tend to ignore image files while most browsers often use the GET request method to retrieve their files. Another HTTP request method called POST (attribute 18) has a small negative correlation because it is often used by browsers to submit HTML forms. Attributes 8 (% Zip) and 9 (% Multimedia) are positively correlated due to sessions with Type 3 agents (which are mostly download utility robots). Attributes such as *avgTime, stdevTime, width, depth*, and *Repeated* do not play a significant role for the data set with one request because their values are either all zeros or all ones.

3. After two requests, the *avgTime, width*, and *depth* attributes become more significant. The width is positively correlated with robot sessions,

whereas the depth is negatively correlated. This confirms our previous observation that many robots, especially the search engine ones, use a breadth-first search strategy to retrieve files from a Website. Also, notice that the *MultiIP* attribute is positively correlated, due to robots that parallelize their retrieval operations.

4. After three requests, the *stdevTime* attribute becomes non-zero. A somewhat surprising fact is that this attribute is positively correlated with robot sessions, indicating that robots tend to have more irregular periods between HTML requests, compared to human users. We verified this by comparing the average standard deviations for various user agents, as shown in Fig. 9.6.

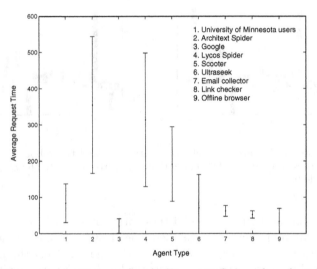

Fig. 9.6. The mean *avgTime* and *stdevTime* attribute values for various Web clients

5. The *Night* attribute has a positive correlation with robot sessions. Figure 9.7 illustrates the hourly traffic at our Web server, after filtering out the anomalous linbot session of Fig. 9.1. Notice that the number of page requests due to Web robots are almost uniformly distributed throughout the day, while the number of page requests due to non-robot sessions peak at normal business hours[6]. Thus, it is surprising that the *Night* attribute is positively correlated with robot sessions. Upon closer examination, we found that this is because most robot sessions are long, spanning into the 12am to 7am time window (which was used to determine the value of the *Night* attribute). Out of the 10,845 robot sessions, 3,127 (28.8%)

---

[6] The observed traffic pattern is very similar to the e-commerce traffic observed by Rosenstein in [9.18].

of them are night crawlers, compared to 30,661 (18.1%) out of 169,757 non-robot sessions that have $Night = 1$.

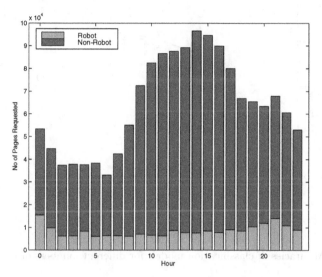

Fig. 9.7. Summary of hourly Web traffic at the University of Minnesota Computer Science department Web server

9.4.3 Classifier Performance

Figure 9.8 shows the classification accuracies of the various models induced from our data sets. The results show that after four requests, we can obtain models with an average accuracy close to 90%. Also, their precision and recall (Fig. 9.9) consistently stay above 82 % and 95%, respectively, after more than three requests. While the addition of noisy data (for E1) does not degrade the classifier precision by much, the recall decreases by as much as 5%. The small difference between E0 and E1 curves for a large number of requests can be explained by the small number of long sessions with Type 3 and Type 4 agents in the data sets (Table 9.8).

There is a dramatic improvement in all three performance measures when the number of requests increases from one to two. This is due to the fact that attributes such as *avgTime*, *width*, *depth*, and *Repeated* become meaningful after one request.

C4.5 also provides an auxiliary program to generate classification rules from the decision trees. Table 9.10 presents some of the high-confidence rules for the robot classes. Indeed, most of the rules seem to agree with our initial correlation analysis. For the data set with one request, the best rules for predicting robot sessions tend to have many attributes in their antecedent.

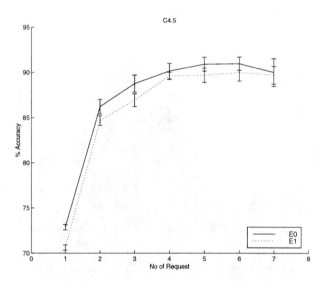

Fig. 9.8. Accuracies of classification models for different numbers of requests.

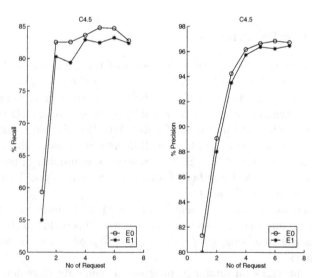

Fig. 9.9. Recall and Precision of classification models for different numbers of requests

These rules tend to overfit the data, which explains the low recall of the results. Additionally, most of these rules predict robots based on the absence of requests for image files.

For the data set with two requests, the *avgTime*, *width*, and *depth* attributes help to improve the accuracy of predicting non-robot sessions. In the example rule given in Table 9.10, robots are classified by sessions that access the server at night, with average request time between 32 and 737 seconds, have low traversal depth, and retrieve very few image and binary executable files. Also, as the number of requests increases, shorter rules involving the *width* attribute begin to emerge.

Table 9.10. Some of the high confidence decision rules produced for different numbers of requests

# Requests	Induced rules [Confidence]
1	$Night = 1$, $totalPages > 3$, % Image ≤ 0.0026, % Binary Exec ≤ 0.9285, % HTML ≤ 0.0526, % Ascii ≤ 0.9, $MultiAgent = 0$, % Other $\leq 0.0312 \longrightarrow$ **CLASS 1** [97.1%]
2	$avgTime \leq 737$, $Night = 1$, % Image ≤ 0.2592, % Binary Exec ≤ 0.3636, $avgTime > 32$, $depth \leq 1$, $multiIP = 1$ $\longrightarrow$ **CLASS 1** [98.2%]
3	% Binary Doc ≤ 0.0096, % Binary Exec ≤ 0, $totalTime > 1861$, $Error > 0.1$, $width > 2 \longrightarrow$ **CLASS 1** [99.6%]
4	$totalPages > 4$, % Image ≤ 0.1, % HTML > 0.6666, $width > 2$, $MultiAgent = 1$, $GET > 0.9473 \longrightarrow$ **CLASS 1** [98.5%] $Night = 1$, $width \leq 1$, $height > 1 \longrightarrow$ **CLASS 0** [99.7%]

9.4.4 Finding Mislabeled Data

We have also conducted a series of experiments to identify sessions that are often misclassified by the classification models. We use the technique described in Sect. 9.3.5 to identify sessions with high false positive and false negative scores. Table 9.11 shows a summary of the log entries for some of the sessions with the highest false positive scores (i.e., sessions that are predicted to be robots but labeled as non-robots). We observe that most of these sessions are indeed mislabeled due to camouflaging or previously unknown robots.

1. The first session contains a user agent that is very similar to a conventional browser. However, the session seems to cover a significant portion of the Website without focusing on any specific topic, which is why there is a strong possibility that this session belongs to a Web robot. Upon resolving the hostname of the session, our suspicion is confirmed because NEC Research, which owns the domain name, is known to have a Scientific Bibliography Search Engine.

Table 9.11. Sessions identified as having large false positive scores

IP Address/Hostname	Time	Requested Page
kablam.nj.nec.com	14:47:23	/~jcui
kablam.nj.nec.com	14:48:32	/~hsieh/misc.html
kablam.nj.nec.com	14:49:06	/Research/dmc/html/abstracts.html
kablam.nj.nec.com	14:49:15	/~kencham/./abstracts/VCR-Ops.html
kablam.nj.nec.com	15:14:03	/~gini/motion.html
kablam.nj.nec.com	15:36:13	/~wijesek/research/qusMetrics.html
64.3.57.99	5:06:42	/employment
64.3.57.99	5:06:42	/grad-info
64.3.57.99	5:06:42	/reg-info/csMinor.html
64.3.57.99	5:06:42	/industry.html
tpa1.hire.com	13:59:42	/~hngo/vnsa/may27-jul24/msg01844.html
tpa1.hire.com	14:01:20	/~ssparikn/resume/shwetal_resume.html
tpa1.hire.com	14:12:27	/~whalen/resume.html
tpa1.hire.com	4:31:38	/~steinmet/pages/steinmetzresume.html
rfx-64-6-194-38.users.reflexcom.com	14:51:00	/~myers/resume.html
rfx-64-6-194-38.users.reflexcom.com	14:58:25	/~tjiang/resume.html
rfx-64-6-194-38.users.reflexcom.com	15:03:45	/~littau/resume.html
rfx-64-6-194-38.users.reflexcom.com	15:11:17	/~tnnguyen/resume

User Agent
Mozilla/4.05 [en] (Win95; U)
Mozilla/4.05 [en] (Win95; U)
Mozilla/4.05 [en] (Win95; U)
Mozilla/4.05 [en] (Win95; U)
Mozilla/4.05 [en] (Win95; U)
Mozilla/4.05 [en] (Win95; U)
Microsoft Internet Explorer/4.40.426 (Windows 95)
Microsoft Internet Explorer/4.40.426 (Windows 95)
Microsoft Internet Explorer/4.40.426 (Windows 95)
Microsoft Internet Explorer/4.40.426 (Windows 95)
Mozilla/4.75 [en] (X11; U; Linux 2.2.16-3 i686)
Mozilla/4.75 [en] (X11; U; Linux 2.2.16-3 i686)
Mozilla/4.75 [en] (X11; U; Linux 2.2.16-3 i686)
Mozilla/4.75 [en] (X11; U; Linux 2.2.16-3 i686)
Mozilla/4.0 (compatible; MSIE 5.01; Window 98; DigExt)
Mozilla/4.0 (compatible; MSIE 5.01; Window 98; DigExt)
Mozilla/4.0 (compatible; MSIE 5.01; Window 98; DigExt)
Mozilla/4.0 (compatible; MSIE 5.01; Window 98; DigExt)

2. The second example also looks highly suspicious despite having a user agent declared as Microsoft Internet Explorer. It is very unlikely that a user will be able to view all four HTML pages in one second. Unfortunately, we were unable to resolve the IP address to determine the origin of the client.

3. The third example is interesting because the pages retrieved by the session are mostly resume files. Our classifiers were able to detect the large width of the traversal to infer that this session is very likely due to accesses by a robot. This observation is confirmed after resolving the IP address of the session (i.e., hire.com).

4. The fourth example is another session we believe is due to a resume hunter robot. However, it is interesting to note that the domain name of the client belongs to a broadband Internet Service Provider. Thus, the traditional techniques of inspecting the user agent and IP address field of a Web client may not be able to detect this robot.

On the other hand, there are very few mislabeled non-robot sessions with high false negative scores (cf. Table 9.12).

Only one of the four sessions in this figure is clearly mislabeled. SilentSurf (i.e., the fourth session) was initially thought to be a Type 3 robot. However, many of our classification models have identified it to be a non-robot. Upon further analysis, we found that SilentSurf is in fact an anonymizer Website that changes the user agent of a browser into a robot-like value. The rest of the false negative sessions in this figure could be due to misclassifications by our models. These sessions contain mostly robots whose navigational patterns are very similar to human users (e.g., offline browsers) and download utility programs that have very short session lengths (which makes it difficult to predict correctly their class labels).

9.5 Conclusions

While our initial correlation analysis suggests that many of the attributes used to label the Web sessions (e.g., robots.txt, % HEAD requests, and % unassigned referrers) are good robot predictors, these attributes alone are often insufficient to detect Web robots. Instead, we have strived to build accurate predictive models for Web robots by using access features derived from the navigational patterns of the various Web clients. The results of our experiments on the Computer Science department Web server logs suggest that Web robots can be detected with more than 90% accuracy after four requests. These classification models can be used to detect camouflaging robots that have very similar navigational patterns as other known robots.

Throughout this paper, we assume that the navigational patterns of Web robots are significantly different from those of human users. In fact, we have observed that Web robots traversing a Website with the same information

Table 9.12. Sessions identified as having large false positive scores

IP Address/Hostname	Time	Requested Page
ns.mof.go.jp	18:42:16	/~subraman/cgi-bin/art.cgi
ns.mof.go.jp	18:48:13	/~subraman/cgi-bin/art.cgi
ns.mof.go.jp	18:48:20	/~subraman/arts/main.html
ns.mof.go.jp	18:48:20	/~subraman/arts
cip123.studcs.uni-sb.de	7:06:13	/~mobasher/webminer/wurvey/survey.html
cip123.studcs.uni-sb.de	7:20:51	/~mobasher/webminer/wurvey/survey.html
cip123.studcs.uni-sb.de	7:28:20	/~mobasher/webminer/wurvey/survey.html
212.160.318.34	8:16:31	/~hougen
212.160.318.34	8:21:06	/departmental
212.160.318.34	8:21:06	/Research/airvl
63.87.224.21	4:45:02	/~ssparikh
63.87.224.21	4:45:05	/~ssparikh/images/headsil.jpg
63.87.224.21	4:45:06	/~ssparikh/images/back/ivy.gif

User Agent
-
-
-
-
java1.1.8
java1.1.8
java1.1.8
Offline Explorer/1.3
Offline Explorer/1.3
Offline Explorer/1.3
SilentSurf/1.1x [en] (X11; 1; $MyVersion)
SilentSurf/1.1x [en] (X11; 1; $MyVersion)
SilentSurf/1.1x [en] (X11; 1; $MyVersion)

need will exhibit similar navigational patterns. This suggests that perhaps we may need to construct different classification models for each type of robot in order to achieve better predictive accuracy. We hope to explore this possibility in our future work.

Our session labeling procedure may introduce some errors into the classification models. We have observed that the labeling errors are more prevalent for robot sessions compared to non-robot sessions, since our session labeling algorithm is biased toward non-robots. Another future direction of this research will be to consider some of the existing analytical models for quantifying the errors due to mislabeled examples [9.2].

We also need to refine the features used for building the classifiers. For example, we can incorporate the Web content and structure information into the session features. Further investigation is also needed to study the effect of other types of navigational patterns not captured by our data.

Acknowledgements

This article was reprinted from Data Mining and Knowledge Discovery (Vol. 6, No. 1, Jan. 2002) with permission of Kluwer Academic Publishers. This work was partially supported by NSF grant #ACI–9982274 and by Army High Performance Computing Research Center contract number DAAH04–95–C–0008. The content of this work does not necessarily reflect the position or policy of the government, and no official endorsement should be inferred. Access to computing facilities was provided by AHPCRC and the Minnesota Supercomputing Institute.

References

9.1 M. Balabanovic, Y. Shoham: Learning information retrieval agents: Experiments with automated Web browsing. In: *On-line Working Notes of the AAAI Spring Symposium Series on Information Gathering from Distributed, Heterogeneous Environments, 1995*

9.2 C. Brodley, M. Friedl: Identifying mislabeled training data. Journal of Artificial Intelligence Research, 11, 131-167 (1999)

9.3 D. Clark: Shopbots Become Agents for Business Change. IEEE Computer, 33(2), 18-21 (2000)

9.4 R. Cooley: Web Usage Mining: Discovery and Application of Interesting Patterns from Web Data. PhD thesis, University of Minnesota, 1999

9.5 R. Cooley, B. Mobasher, J. Srivastava: Data preparation for mining World Wide Web browsing patterns. Knowledge and Information Systems, 1(1) (1999)

9.6 D. Eichmann: Ethical Web agents. Computer Networks and ISDN Systems, 28(1) (1995)

9.7 L. Graham: Keep your bots to yourself. IEEE Software, 17 (6), 106-107 (2000)

9.8 M. Gray: Measuring the growth of the Web.
http://www.mit.edu/people/mkgray/growth/ (1993)

9.9 S. Jackson: Building a better spider trap.
http://www.spiderhunter.com/spidertrap (1998)

9.10 J. Kephart, A. Greenwald: Shopbot Economics. In: *Agents, 1999* pp. 378-379

9.11 C. Kolar, J. Leavitt, M. Mauldin: Robot exclusion standard revisited.
http://www.kollar.com/robots.html (1996)

9.12 M. Koster: Guidelines for robot writers.
http://info.webcrawler.com/mak/projects/robots/ guidelines.html (1994)

9.13 M. Koster: A standard for robot exclusion.
http://info.webcrawler.com/mak/projects/robots/ norobots.html (1994)

9.14 M. Koster: Robots in the Web: threat or treat. ConneXions, 9(4) (1995)

9.15 H. Lieberman: Letizia: An agent that assists Web browsing. In: *Proc. of the 1995 International Joint Conference on Artificial Intelligence, Montreal, Canada 1995*

9.16 P. Pirolli, J. Pitkow, R. Rao: Silk from a sow's ear: Extracting usable structures from the Web. In: *CHI-96, Vancouver, Canada 1996*

9.17 J. Quinlan: *C4.5: Programs for Machine Learning* (Morgan Kaufmann, 1993)

9.18 M. Rosenstein: What is actually taking place on Web sites: E-commerce lessons from Web Server logs. In: *ACM Conference on Electronic Commerce, Minneapolis, MN 2000*

9.19 C.J.V. Rijsbergen: *Information Retrieval* (Butterworths, London, 1979)
9.20 M. Yoon: Web robot detection.
   http://www.arsdigita.com/doc/robot-detection (2000)

10. Web Ontology Learning and Engineering: An Integrated Approach

Roberto Navigli[1], Paola Velardi[1], and Michele Missikoff[2]

[1] Università di Roma "La Sapienza", Italy
[2] IASI-CNR, Italy

Abstract

The importance of domain ontologies is widely recognized, particularly in relationship to the expected advent of the so-called *Semantic Web*. An ontology is an infrastructure able to provide a precise account of the concepts characterizing a given application domain. It represents a shared understanding of a given reality, thus fostering better communication and cooperation among users, and interoperability among systems. Despite the significant amount of work in the field, ontologies are still scarcely used in Web-based applications. One of the main problems is the difficulty in defining the *content*, i.e., the identification and definition of relevant concepts in the domain. The solution proposed in this chapter starts from the idea that the corpus of documents produced by a community is the most representative (although implicit) repository of concepts. We present a method and a tool, OntoLearn – aimed at the extraction of knowledge from Web-sites and, more generally, from documents shared among the members of virtual organizations – to support the construction of a domain ontology.

10.1 Introduction

The development of the *Semantic Web* [10.2, 10.25], aimed at improving the "semantic awareness" of computers connected via the Internet, requires a systematic, computer-oriented representation of the world. Such a world model is often referred to as an ontology.

The goal of a domain ontology is to reduce (or eliminate) the conceptual and terminological confusion among the members of a virtual community of users (e.g., tourist operators, commercial enterprises, medical practitioners) who need to share electronic documents and information of various kinds. This is achieved by identifying and properly defining a set of relevant concepts that characterize a given application domain. An ontology is therefore a *shared understanding* of some domain of interest [10.14]. It contains a set of concepts (e.g., entities, attributes, and processes), together with their definitions and their interrelationships. In other words, an ontology is an explicit, agreed specification about a shared conceptualization. The construction of a shared understanding, i.e., a unifying conceptual framework, fosters:

- *Communication* and *cooperation* among people;
- Better enterprise organization;
- *Interoperability* among systems;
- *System engineering benefits* (reusability, reliability, and specification).

Ontologies may have different degrees of formality but they necessarily include a vocabulary of terms with their meanings (definitions) and their relationships.

The construction of an ontology requires a thorough domain analysis that is accomplished by[1]:

1. Examining *the vocabulary* that is used to describe the relevant objects and processes of the domain;
2. Developing *rigorous definitions* about the terms (concepts) in that vocabulary;
3. Characterizing the *conceptual relationships* among those terms.

Creating ontologies is a difficult process that involves specialists from several fields. Philosophical ontologists and Artificial Intelligence (AI) logicists are usually involved in the task of defining the basic kinds and structures of concepts that are applicable in every possible domain. The issue of identifying these very few "basic" principles, referred to as the *Top Ontology* (TO), is not a purely philosophical one, since there is a clear practical need for a model which has as much generality as possible to ensure reusability across different domains [10.13].

Domain modelers and knowledge engineers are involved in the task of identifying the key domain conceptualizations, and describing them according to the organizational *backbones* established by the Top Ontology. The result of this effort is referred to as the *Upper Domain Ontology* (UDO), which usually includes a few hundred application domain concepts.

While many ontology projects eventually succeed in the task of defining an Upper Domain Ontology[2], populating the third level, what we call the *Specific Domain Ontology* (SDO), is the actual barrier that very few projects can overcome (e.g., Wordnet [10.3, 10.28], Cyc [10.5], and EDR [10.18]), at the price of inconsistencies and limitations. Figure 10.1 shows the three levels of generality of a Domain Ontology.

It turns out that, although domain ontologies are recognized as crucial resources for the Semantic Web [10.2], they are in practice not available, and when available they are *not used* outside specific research environments[3].

---

[1] IDEF5 Ontology Description Capture Method Overview. See http://www.idef.com/overviews/idef5.htm.

[2] In fact, many ontologies are already available on the Internet including a few hundred more or less extensively defined concepts.

[3] For example, Wordnet is widely used in the Computational Linguistics research community, but large scale IT applications based on WordNet are not available.

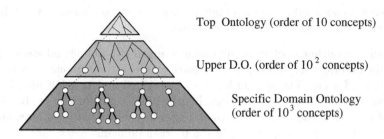

Top Ontology (order of 10 concepts)

Upper D.O. (order of 10^2 concepts)

Specific Domain Ontology
(order of 10^3 concepts)

Fig. 10.1. The three levels of generality of a Domain Ontology

The features most needed to build *usable* ontologies are as follows:

Coverage. the domain concepts *must be there*: the SDO must (for application purposes) be sufficiently populated. Tools are needed to extensively support the task of identifying the relevant concepts and the relationships among them.

Consensus. decision making is a difficult activity for one person and it gets even harder when there is a group of people who must reach a consensus on a given issue and when, in addition, the group is geographically dispersed. When a group of enterprises decide to cooperate in a given domain, they have to first agree on many basic issues, i.e., they must reach a consensus on the business domain. Such a common view must be reflected by the domain ontology.

Accessibility. the ontology must be easily *accessible*: tools are needed to easily integrate the ontology within an application that may clearly show its decisive contribution, e.g., by improving the ability to share and exchange information through the Web.

There has been a growing awareness on the importance of ontologies in information systems. Despite the significant amount of work carried out in recent years, ontologies are still scarcely applied and used. Research has mainly addressed the basic principles, such as knowledge representation formalisms, though limited attention has been devoted to more practical issues, such as techniques and tools aimed at the actual construction of an ontology (i.e., its actual *content*). Though, recently, a number of contributors proposed methods to extract terminology and word relationships from domain data and Websites [10.7, 10.8, 10.12, 10.17], there are two major drawbacks in the reported approaches:

– First, what is learned from available documents is mainly a list of terms and term relationships. The definition (i.e., the *semantic interpretation*) of these terms is still left to the ontology engineer;
– Second, ontology learning [10.20, 10.24] has been investigated in isolation from ontology *engineering* (as also remarked in [10.7]) and ontology *validation* issues. This is a serious drawback, since an ontology where the relevant

concepts do not conform with a domain view of a given community will be scarcely used, or may even be disregarded.

We shortly outline a software environment aimed at building and assessing a domain ontology for intelligent information integration among a virtual community of users. The system has been developed and is being tested in the context of two European projects, Fetish [10.22] and Harmonise [10.23], where it is used as the basis of a semantic interoperability platform for small and medium enterprises operating in the tourism domain.

The main focus of this chapter is the description of a tool, *OntoLearn*, aimed at extracting knowledge from electronic documents to support the rapid construction of a domain ontology.

Figure 10.2 reports the proposed ontology engineering method, i.e., the sequence of steps and the intermediate output that are produced in building a domain ontology.

As shown in Fig. 10.2, ontology engineering is an iterative process involving machine concept *learning* (*OntoLearn*), machine-supported concept *validation* (*Consys*), and *management* (*SymOntoX*).

OntoLearn explores available documents and related Websites to learn domain concepts, and to detect taxonomic relationships among them. Initially, concept learning is based on the use of external, generic knowledge sources (specifically WordNet and SemCor, illustrated later); then, in the next cycles, the domain ontology progressively comes in use when it starts to be adequately populated (as indicated by the *self learning* cycle in the figure). The subsequent processing step in Fig. 10.2 is *ontology validation*. This is a continuous process supported by a Web-based *groupware* aimed at consensus building, *Consys* [10.10], to achieve thorough validation with representatives of the communities that are active in the application domain.

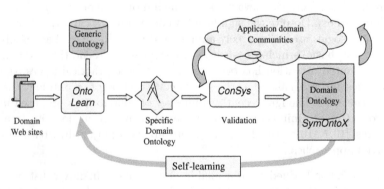

Fig. 10.2. The Ontology Engineering Chain

Ontology learning, validation, and management are the modules of an integrated ontology engineering architecture. Several cycles are necessary:

the learning cycle highlights the progressively growing role of the domain ontology as a background knowledge for learning new concepts; the validation cycle highlights the many interactions that are necessary among knowledge engineers and domain experts in order to assess the information in the domain ontology.

OntoLearn operates in connection with SymOntoX [10.27], an *ontology management* system used by the ontology engineers to define concepts and their mutual relationships, thus allowing a semantic net to be constructed. SymOntoX uses a knowledge representation method, referred to as *OPAL* (Object, Process, Actor modeling Language) [10.11], that is an extension of XML-based formalisms such as RDF [10.29] and OIL [10.19]. The ontology engineers use the environment provided by SymOntoX to attach automatically learned concept sub-trees under the appropriate nodes of the upper domain ontology, to enrich concepts with additional information, and to perform consistency checks.

The rest of the chapter is organized as follows: in Sect. 10.2 we describe in more detail the OntoLearn system, which is the main focus of this chapter: Sect. 10.2.1 describes the method to extract terminology from Websites, Sect. 10.2.2 presents the knowledge-based semantic interpretation method, along with a summary of the knowledge representation scheme, Sect. 10.2.3 describes the creation of a Specific Domain Ontology, and Sect. 10.2.4 provides a way for creating a domain ontology from WordNet. Finally, Sect. 10.3 presents the results of the first year of experiments. Further research and expected outcomes are discussed in Sect. 10.4, the Conclusion.

10.2 The OntoLearn System

Figure 10.3 shows the architecture of the OntoLearn system. There are three main phases. First, a domain terminology is *extracted* from available texts in the application domain (specialized Websites or documents exchanged among members of a virtual community) and *filtered* using statistical techniques and documents from different domains for comparative analysis. Terminology is at the surface of relevant domain concepts.

Second, terms are *semantically interpreted*, i.e., we associate unambiguous *concept* names with the extracted terms.

Third, taxonomic (i.e., generalization/specialization) and similarity relationships among concepts are detected, and a *Specific Domain Ontology* (hereafter SDO) is generated. Ontology matching (i.e., the integration of SDO with the existing upper ontology) is performed in connection with SymOntoX.

Initially, we assume that only a small *upper domain ontology* is available (a realistic assumption); therefore, semantic interpretation is based on external (non domain-specific) knowledge sources, such as *WordNet* [10.3, 10.28] and the semantically tagged corpus *SemCor* [10.26]. WordNet is a large lexical

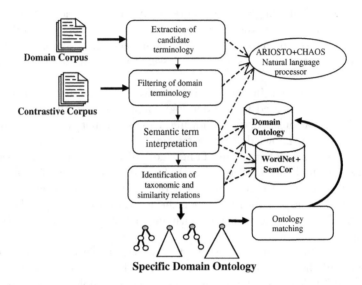

Fig. 10.3. Architecture of OntoLearn

knowledge base, whose popularity has recently been growing even outside the computational linguistic community. SemCor is a corpus of semantically annotated sentences, i.e., every word is annotated with a sense tag, selected from the WordNet sense inventory for that word.

As soon as the ontology engineering and validation processes result in a sufficiently rich *domain ontology*, the role of the latter in automatic concept learning progressively overshadows that of WordNet. Eventually, new terms are semantically disambiguated and taxonomically organized using only the information already stored in the domain ontology.

10.2.1 Identification of Relevant Domain Terminology

The objective of this phase is to extract a domain terminology from the available documents. The documents are retrieved by browsing Websites with an initial set of domain terms[4], and then progressively specializing the search when new terms are learned.

We use a linguistic processor, ARIOSTO+CHAOS [10.1], to extract from the domain documents a list of syntactically plausible terminological patterns, e.g., compounds (*credit card*), prepositional phrases (*board of directors*), and adjective-noun relationships (*manorial house*). Then, two measures based on information theory are used to filter out non-terminological (e.g.,

---

[4] In our application domain, an initial upper ontology of about 300 terms was available.

last week) and non-domain specific (e.g., *net income* in a Tourism domain) terms.

The first measure, called *Domain Relevance*, computes the conditional probability of occurrence of a candidate term in the application domain (e.g., Tourism) relative to other corpora that we use for a comparative analysis (e.g., Medicine, Economy, Novels, etc.).

More precisely, given a set of n domains $\{D_1, \dots, D_n\}$, the domain relevance of a term t in the domain D_k is computed as

$$DR_{t,k} = \frac{P(t|D_k)}{\max_{1 \leq j \leq n} P(t|D_j)} \, , \tag{10.1}$$

where $P(t|D_k)$ is estimated by

$$E(P(t|D_k)) = \frac{f_{t,k}}{\sum_{t' \in D_k} f_{t',k}} \, , \tag{10.2}$$

where $f_{t,k}$ is the frequency of term t in the domain D_k.

The second measure, called *Domain Consensus* (DC), computes the *entropy* of the probability distribution of a term across the documents of the application domain. The underlying idea is that only terms that are *frequently* and *consistently* referred to in the available domain documents reflect some *consensus* on the use of that term. DC is an entropy, defined as

$$DC_{t,k} = \sum_{d \in D_k} \left(P_t(d) log \frac{1}{P_t(d)} \right) \, , \tag{10.3}$$

where $P_t(d)$ is the probability that a document d includes t. These two measures have been extensively evaluated in [10.15] and [10.16].

Terminology filtering is obtained through a linear combination of the two filters:

$$DW_{t,k} = \alpha DR_{t,k} + (1 - \alpha)DC_{t,k}^{norm} \, , \tag{10.4}$$

where $DC_{t,k}^{norm}$ is a normalized entropy and $\alpha \in (0,1)$.

Let T be the terminology extracted after the filtering phase. Using simple string inclusion, we generate a *forest* of *lexicalized trees*, $\mathcal{F}$. Figure 10.4 is an example of a lexicalized tree $\mathcal{T}$ extracted from our Tourism corpus.

Clearly, lexicalized trees do not capture many taxonomic relationships between terms, for example, between *public transport service* and *bus service* in Fig. 10.4.

Terminology is often considered the surface realization of relevant domain concepts. However, in the absence of semantic interpretation, it is not possible to fully capture the conceptual relationships between concepts, for example, the is-a relationship between "bus service" and "public transport service" in Fig. 10.4.

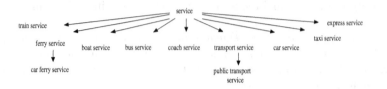

Fig. 10.4. A lexicalized tree

10.2.2 Semantic Interpretation of Terms

The process of *semantic interpretation* is one that associates with each term $t = w_n \cdot \ldots \cdot w_2 \cdot w_1$ (where w_i is an atomic word) the appropriate concept name. The idea is that, though the string t is not usually included in the start-up ontology, there should be a conceptual entry for each possible sense of w_i ($i = 1, \ldots, n$); for example, though there are no concepts associated with *room service*, we may find concept descriptions for *room* and *service* individually. Therefore, it should be possible to compositionally create a definition for t, selecting the appropriate concept definition for w_i ($i = 1, \ldots, n$), given the string context t in which it appears.

As mentioned previously, we use WordNet as a start-up ontology, since the upper domain ontology is initially quite poor. In WordNet, a word sense is uniquely identified by a set of terms called *synset* (e.g., the sense #3 of *transport* is defined by transportation#4, shipping#1, and transport#3), and a textual definition called *gloss* (e.g., "*the commercial enterprise of transporting goods and materials*"). Synsets are taxonomically structured in a lattice, with a number of "root" concepts called *unique beginners* (e.g., entity#1, something#1). WordNet includes over 120,000 words (and over 170,000 synsets), but very few domain terms; for example, *transport* and *company* are individually included, but not *transport company* as a unique term.

Formally, a semantic interpretation is defined as follows: let $t = w_n \cdot \ldots \cdot w_2 \cdot w_1$ be a valid term belonging to a lexicalized tree $\mathcal{T}$. The process of *semantic interpretation* is one that associates to each word w_k in t the appropriate WordNet synset S_i^k, where i is the ith synset ($i = 1, \ldots, n$) associated with w_k. The *sense* of t is hence defined as

$$S(t) = \bigcup_k \{S^k\}, S^k \in Synsets(w_k) \wedge w_k \in t \tag{10.5}$$

where $Synsets(w_k)$ is the set of synsets representing a distinct sense of the word w_k. For instance,

$$S(\text{``transport company''}) = \{\{\text{transportation\#4, shipping\#1,}$$
$$\text{transport\#3\}, company\#1\}\}$$

corresponding to sense #1 of *company* ("*an institution created to conduct business*") and sense #3 of *transport*, previously reported. Semantic interpretation is achieved by intersecting semantic information associated with each alternative sense of the words in t with the others, and then selecting the "best" intersection. Semantic information is extracted from WordNet and represented in a diagrammatic form, according to a representation scheme described in the next subsection.

Semantic Representation of Concepts. For each sense of a word, several other types of semantic relationships are supplied in WordNet, though they are not systematically and formally defined. As a first effort, we tried to establish a connection between semantic relationships in WordNet and the concept representation scheme adopted in OPAL.

According to OPAL an ontology is a *semantic net*, constructed by supplying a set of concepts and their semantic relationships[5]. In each description, a reference is made to the linguistic counterpart in WordNet, and a graphic symbol is refereed. The symbol will be used in constructing the diagrams (semantic nets) presented in the next subsections[6].

Generalization - This is an asymmetric relationship, often indicated as an *is-a* relationship, that links a concept to its more general concepts (e.g., *Hotel Is-A Accomodation*). Its inverse is called *specialization*. In the linguistic realm, this relationship, defined between *synsets*, is called *hyperonymy* ($\overset{@}{\rightarrow}$), and its inverse is called *hyponymy* ($\overset{\sim}{\rightarrow}$).

Aggregation - This is an asymmetric relationship that connects a concept representing a whole to another representing a component. It is often indicated as a *has-a* relationship (e.g., *Reception Has-A Hotel*).

Its inverse is called *decomposition*. In the linguistic realm, this relationship, defined between *synsets*, is called *meronymy* ($\overset{\#}{\rightarrow}$), and *holonymy* ($\overset{\%}{\rightarrow}$) is its inverse.

Similarity - This is a symmetric relationship that links two concepts that are considered similar in the given domain. A similarity degree is often indicated (e.g., *Hotel SimilarTo[0.8] Motel*).

In the linguistic realm this relationship, defined between synsets, is called synonymy when the similarity degree is 1[7], while *similarity* ($\overset{\&}{\rightarrow}$) and *correlation* ($\overset{\wedge}{\rightarrow}$) are used to indicate progressively weaker levels of similarity. In WordNet there is also a *dissimilarity* relationship, called *antonymy* ($\overset{!}{\rightarrow}$), applying, for example, to *liberal* and *conservative*, indicating a degree of similarity equal to zero. Furthermore, *pertainymy* ($\overset{\backslash}{\rightarrow}$) relates the nominal and adjectival realization of a concept (e.g., *mother* and *maternal*).

---

[5] See [10.11] for a formal description of the OPAL knowledge representation scheme.

[6] In the rest of this section, $S_1 \overset{R_1,R_2}{\rightarrow} S_2$ stands for $S_1 \overset{R_1}{\rightarrow} S_2 \vee S_1 \overset{R_2}{\rightarrow} S_2$, while $S_1 \overset{R}{\rightarrow}^{\leq 3} S_2$ stands for $S_1 \underbrace{\overset{R}{\rightarrow} \ldots \overset{R}{\rightarrow}}_{\leq 3} S_2$.

[7] Strict synonyms are those belonging to the same synset.

Relatedness - This is a semantic relationship that connects two symmetrically related concepts in the given domain. This relationship assumes specific, domain dependent interpretations. For example, in *Hotel RelatedTo Airport* the relationship subsumes *physical proximity*. This weakly defined relationship does not have a counterpart in WordNet, but it can be induced from concept definitions and from the semantically annotated sentences in the SemCor corpus. Parsing the definition (i.e., the gloss in WordNet) of a given concept, and the semantically annotated sentences including that concept, we generate a linguistic counterpart of "relatedness", represented by the *gloss* relationship ($\overset{gloss}{\rightarrow}$) and the *topic* relationship ($\overset{topic}{\rightarrow}$). The idea is that, if a concept c_2 appears in the definition of another concept c_1, or if c_2 appears in the near proximity of c_1 in an annotated sentence including c_1, then c_1 and c_2 are "related", i.e., $c_1 \overset{gloss}{\rightarrow} c_2$ or $c_1 \overset{topic}{\rightarrow} c_2$, respectively. For example: "*The room(#1)s were very small but they had a nice view(#2)*" produces: $room(\#1) \overset{topic}{\rightarrow} view(\#2)$.

Concept disambiguation. In order to disambiguate the words in a term $t = w_n \cdot \ldots \cdot w_2 \cdot w_1$ we proceed as follows:

a) If t is the first analyzed element of $\mathcal{T}$, disambiguate the root node (w_1 if t is compound) of $\mathcal{T}$ either in a manual or an automatic way[8].

b) For any $w_k \in t$ and any synset S_i^k of w_k, create a *semantic net* (SN). Semantic nets are automatically created using the semantic relationships described in the previous subsection, extracted from WordNet and Sem-Cor (and, possibly, from the Domain Upper Ontology). To reduce the size of an SN, concepts, S_i^k, at a distance greater than three edges from the SN centre, are excluded. Figure 10.5a is an example of an SN generated for sense #1 of *room*. Then, let $SN(S_i^k)$ be the semantic network for sense i of word w_k.

c) Starting from the "head" w_1 of t, for any pair of words w_{k+1} and w_k ($k = 1, \ldots, n-1$) belonging to t, intersect alternative pairs of the SNs. Let $I = SN(S_i^{k+1}) \cap SN(S_j^k)$ be such an intersection for sense i of word w_{k+1} and sense j of word w_k. Note that in each step k, the word w_k is already disambiguated, either as an effect of root disambiguation (for $k = 1$) or as a result of step $k - 1$.

d) For each alternative intersection, identify common *semantic patterns* in I in order to select the sense pairs (i, j) producing the "strongest" intersection (occasionally more than one pair). To this end, given two arbitrary synsets S_1 and S_2, we use the following heuristics[9]:

---

[8] Automatic root disambiguation can be performed in quite the same way as the procedure presented here by intersecting each root with its descendants and with all the other roots. However, a fully automatic procedure is a very delicate matter because choosing the wrong sense, that is, the wrong collocation for the root term in the hierarchy, would affect all its descendants in $\mathcal{T}$.

[9] Some of these heuristics have been inspired by the work presented in [10.4] and [10.9].

1. *colour*, if S_1 is in the same adjectival cluster as *chromatic#3* and S_2 is a hyponym of a concept that can assume a color like *physical object#1*, *food#1*, etc. (e.g., $S_1 \equiv yellow\#1$ and $S_2 \equiv wall\#1$);

2. *domain*, if the gloss of S_1 contains one or more domain labels and S_2 is a hyponym of those labels (for example, *white#3* is defined as "*(of wine) almost colorless*", and is therefore the best candidate for *wine#1* to disambiguate the term *white wine*);

3. *synonymy*, if

$$(a) \ S_1 \equiv S_2 \ \text{or} \ (b) \ \exists N \in Synset_{WN} : S_1 \xrightarrow{\ \ \searrow\ \ } N \equiv S_2$$

(for example, in the term *open air*, both the words belong to synset $\{open\#8, \ air\#2, \ \ldots, \ outdoors\#1\}$);

4. *hyperonymy/meronymy* path, if

$$\exists M \in Synset_{WN} : S_1 \xrightarrow{@,\#^{\leq 3}} M \xleftarrow{\sim,\%^{\leq 3}} S_2$$

(for instance, *mountain#1* $\xrightarrow{\#}$ *mountain peak#1* $\xrightarrow{@}$ *top#3* provides the right sense for each word of *mountain top*);

5. *hyponymy/holonymy* path, if

$$\exists M \in Synset_{WN} : S_1 \xrightarrow{\sim,\%^{\leq 3}} M \xleftarrow{@,\#^{\leq 3}} S_2$$

(for example, in *sand beach*, *sand#1* $\xrightarrow{\%}$ *beach#1*);

6. *parallelism*, if

$$\exists M \in Synset_{WN} : S_1 \xrightarrow{@^{\leq 3}} M \xleftarrow{@^{\leq 3}} S_2$$

(for instance, in *enterprise company*, *organization#1* is a common ancestor of both *enterprise#2* and *company#1*);

7. *gloss*, if

$$(a) \ S_1 \xrightarrow{gloss} S_2 \ \text{or} \ (b) \ S_1 \xleftarrow{gloss} S_2$$

(for instance, in *picturesque village*, WordNet provides the example "*a picturesque village*" for sense 1 of *picturesque*; in *Website*, the gloss of *web#5* contains the word *site*; in *waiter service*, the gloss of *restaurant attendant#1* and the hyperonym of *waiter#1* contain the word *service*);

8. *topic*, if $S_1 \xrightarrow{topic} S_2$ (like for the term *archeological site*, where both words are tagged with sense 1 in a SemCor file; notice that WordNet provides no mutual information about them);

9. *gloss+hyperonymy/meronymy* path, if

$$\exists G, M \in Synset_{WN} : S_1 \xrightarrow{gloss} G \xrightarrow{@,\#^{\leq 3}} M \xleftarrow{\sim,\%^{\leq 3}} S_2$$
$$\lor S_1 \xrightarrow{gloss} G \xrightarrow{\sim,\%^{\leq 3}} M \xleftarrow{@,\#^{\leq 3}} S_2$$

(for instance, in *railway company*, the gloss of *railway#1* contains the word *organization* and *company#1* $\xrightarrow{@^2}$ *organization#1*);

10. *gloss+parallelism*, if

$$\exists G, M \in Synset_{WN} : S_1 \xrightarrow{gloss} G \xrightarrow{@^{\leq 3}} M \xleftarrow{@^{\leq 3}} S_2$$

(for instance, in *transport company*, the gloss of *transport*#3 contains the word *enterprise* and *organization*#1 is a common ancestor of both *enterprise*#2 and *company*#1);

11. *gloss+gloss*, if

$$\exists G \in Synset_{WN} : S_1 \overset{gloss}{\to} G \overset{gloss}{\leftarrow} S_2$$

(for example, in *mountain range*, *mountain*#1 and *range*#5 both contain the word *hill* so that the right senses can be chosen);

12. *hyperonymy/meronymy+gloss* path, if

$$\exists G, M \in Synset_{WN} : S_1 \overset{@,\#^{\leq 3}}{\to} M \overset{\sim,\%^{\leq 3}}{\leftarrow} G \overset{gloss}{\leftarrow} S_2$$
$$\lor S_1 \overset{\sim,\%^{\leq 3}}{\to} M \overset{@,\#^{\leq 3}}{\leftarrow} G \overset{gloss}{\leftarrow} S_2$$

13. *parallelism+gloss*, if

$$\exists G, M \in Synset_{WN} : S_1 \overset{@^{\leq 3}}{\to} M \overset{@^{\leq 3}}{\leftarrow} G \overset{gloss}{\leftarrow} S_2.$$

Figure 10.5b shows a strong intersection between *transport*#3 and *company*#1. The bold arrows identify a pattern matching the "gloss+parallelism" heuristics (rule 10 above):

$$\text{transport\#3} \overset{gloss}{\to} \text{enterprise\#2} \overset{@}{\to}^1 \text{organization\#1} \overset{@}{\leftarrow}^2 \text{company\#1}$$

e) Finally, for each intersection I, a vector is created measuring the number and weight of matching semantic patterns, as sketched in Fig. 10.6. That is, while disambiguating the subterm $w_{k+1} \cdot w_k$, given the sense S_j^{k+1} for word w_{k+1} and all possible n senses of w_k, each intersection $SN(S_j^{k+1}) \cap SN(S_1^k), \ldots, SN(S_j^{k+1}) \cap SN(S_n^k)$ is evaluated as a vector, and the sum represents the "score" vector for S_j^{k+1}. If no mutual information is retrieved (that is, the sum is $\underline{0}$), the process is repeated between w_{k+1} and w_i ($i = k-1, \ldots, 1$) until a positive score is calculated. The best "score" vector (according to a lexicographic ordering) determines the sense for w_{k+1}. The process does not take into account the sense chosen for w_k in the previous iteration, because of a well acknowledged polysemy of words coded in WordNet [10.6] (in fact other senses may bring important information to the semantic interpretation process).

Semantic nets are generated for each possible sense of the words in a terminological string, like "room service", or "transport company". Semantic nets are automatically built using available ontological information (e.g., WordNet), textual definitions, and sentence contexts. Semantic nets are then intersected to find common semantic patterns.

10.2.3 Creating a Specific Domain Ontology

Initially, all the terms in a tree $\mathcal{T}$ are independently disambiguated. Subsequently, taxonomic information in WordNet (or in the upper domain ontology) is used to detect *is-a* relationships between *concepts*, e.g., *ferry service* $\overset{@}{\to}$ *boat service*.

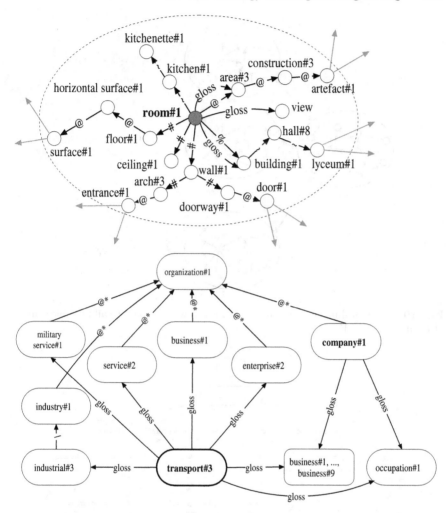

Fig. 10.5. a) example of semantic net for the concept *room*#1; b) example of intersecting semantic patterns for *transport*#3 and *company*#1

In this phase, since all the elements in $\mathcal{T}$ are jointly considered, some interpretation error produced in the previous disambiguation step is corrected. In addition, certain concepts are fused in a unique concept name, called "semantic domain", on the basis of pertainymy, similarity, and synonymy relationships (e.g., respectively, *manor house* and *manorial house*, *expert guide* and *skilled guide*, *bus service* and *coach service*). Notice again that we detect semantic relationships between *concepts*, not words. For example, *bus*#1 and *coach*#5 are synonyms, but this relationship does not hold for other senses of these two words. Each lexicalized tree $\mathcal{T}$ is finally transformed into a *domain concept* tree $\mathcal{T}$. Figure 10.7 shows the concept tree obtained from the lexi-

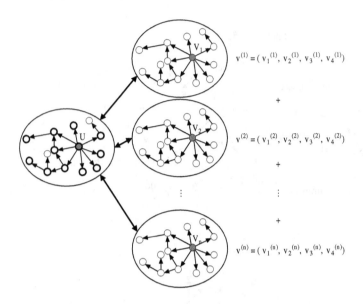

$$v^{(1)} = (v_1^{(1)}, v_2^{(1)}, v_3^{(1)}, v_4^{(1)})$$

$$+$$

$$v^{(2)} = (v_1^{(2)}, v_2^{(2)}, v_3^{(2)}, v_4^{(2)})$$

$$+$$

$$\vdots \qquad \vdots$$

$$+$$

$$v^{(n)} = (v_1^{(n)}, v_2^{(n)}, v_3^{(n)}, v_4^{(n)})$$

Fig. 10.6. The evaluation of a sense U of term u with respect to all possible senses of term v $(V_1, \ldots, V_n)$

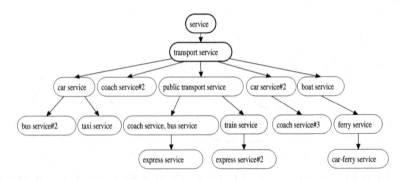

Fig. 10.7. A Domain Concept Tree

calized tree of Fig. 10.4. For clarity, in Fig. 10.7 concepts are labelled with the associated terms (rather than with synsets), and numbers are shown only when more than one semantic interpretation holds for a term, as for *coach service* and *bus service* (e.g., sense #3 of *bus* refers to "old cars").

Taxonomic relationships are detected between concepts, not terms. "Coach" and "bus" denote the same concept when "coach" means "a vehicle for many passengers, used for public transportation", while "coach" is a kind of "car" when it means "passenger car, carriage".

10.2.4 Creating a Domain Ontology from WordNet

In case it is not available, an Upper Domain Ontology can be extracted from WordNet through proper pruning and trimming, accomplished as follows:

- After the Domain Concept Trees are attached under the appropriate nodes in WordNet in either a manual or an automatic manner, as discussed in Sect. 10.2.2, all branches not containing a domain node can be removed from the WordNet hierarchy;
- An intermediate node in the remaining WordNet hierarchy is pruned whenever the following conditions hold together[10]:
    i. it has no "brother" nodes;
    ii. it has only one direct hyponym;
    iii. it is not the root of a Domain Concept Tree;
    iv. it is not at a distance ≤ 2 from a WordNet *unique beginner* (this is to preserve a "minimal" top ontology).

Condition (i) prevents flattening of the hierarchy (the nodes in light grey in Fig. 10.8 would be deleted, thus losing important semantic information). Condition (ii) must hold because a node with more than one hyponym is surely valuable, as it collocates at least two nodes under the same concept; conversely, a node with only one hyponym gives no additional information and provides no further classification. Condition (iii) is trivial: no domain node can be deleted. Condition (iv) is also quite intuitive: nodes very high in the hierarchy represent the essential core of abstract concepts that cannot be deleted. Figure 10.8 shows an example of pruning the nodes located over the Domain Concept tree with root *wine#1*.

10.3 Evaluation of the OntoLearn System

OntoLearn is a knowledge extraction system aimed at improving human productivity in the time consuming task of building a domain ontology. Though a complete field evaluation is still in progress within the Harmonise project using the Consys groupware, some crude facts indicate validity of our method[11]. Our experience in building a tourism ontology for the European project Harmonise reveals that, after one year of ontology engineering activities, the tourism experts were able to release the most general layer of the tourism ontology, manually identifying about 300 concepts. Then, we decided to speed up the process developing the OntoLearn system, aimed at supporting the

---

[10] The pruning step is performed from the bottom (the root domain node's hyperonyms) to the top of the hierarchy (the *unique beginners*).

[11] Although a formal evaluation is in progress, the semantic richness of the ontology is evident.

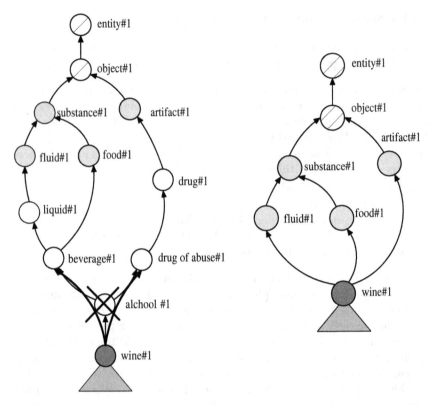

Fig. 10.8. An intermediate step and the final pruning step over the Domain Concept Tree for *wine#1* (in dark grey)

ontology engineering tasks. This produced a significant acceleration in ontology building, since in the next six months[12] the tourism ontology reached 3,000 concepts (although a comparison is not really feasible).

The OntoLearn system has also been evaluated independently from the ontology engineering process. We extracted from a one million word corpus of travel descriptions (downloaded from tourism websites) a terminology of 3,840 terms, manually evaluated[13] by domain experts participating in the Harmonise project.

Two usual evaluation measures have been employed: *recall* and *precision*. Recall is defined as follows:

---

[12] The time span also includes the effort needed to test and tune OntoLearn. Manual verification of automatically acquired domain concepts actually required few days.

[13] Here manual evaluation is simply deciding whether an extracted term is relevant, or not for the tourism domain.

$$recall = \frac{\#correct\ terms}{\#domain\ relevant\ terms} \cdot 100 \ , \tag{10.6}$$

thus providing a percentage of the domain extracted terms with respect to the overall number of terms in the considered domain (which can, of course only be estimated).

Precision gives the percentage of domain terms with respect to all the retrieved terms. It is defined as

$$precision = \frac{\#correct\ terms}{\#retrieved\ terms} \cdot 100 \ . \tag{10.7}$$

We obtained a precision ranging from 72.9% to about 80% and a recall of 52.74%. The precision shift is motivated by the well-known fact that the intuition of experts may significantly differ[14]. The recall has been estimated by submitting a list of about 6,000 syntactic candidates (first step of Sect. 10.2.1) to the experts, requiring them to mark truly terminological entries, and then comparing this list with that obtained by our filtering method described in step 2 of Sect. 10.2.1.

The authors personally evaluated the semantic disambiguation algorithm, described in Sect. 10.2, using a test bed of about 650 extracted terms that have been manually assigned to the appropriate WordNet concepts. These terms altogether contributed to the creation of 90 syntactic trees. The entire process of semantic disambiguation and the creation of domain trees has been evaluated, leading to an overall 84.5% precision (see Fig. 10.9). The precision grows to about 89% for highly structured sub-trees, as those in Fig. 10.7. In fact, the phase described in Sect. 10.2.3 significantly contributes to eliminating disambiguation errors (on average, a 5% improvement).

Besides, the analysis highlighted that some heuristics contribute more than others. In particular, rules making use of glosses ((7), (9), ..., (13)) bring precise semantic information for term disambiguation, thus showing the importance of using rich resources like WordNet.

Variations on the structure of semantic nets have also been considered, both by including the information conveyed by certain kinds of relationships (pertainymy, attribute, similarity) and by applying some cuts on the quantity of hyponyms and on the higher part of WordNet's name hierarchy. The best result was reached by including all kinds of semantic relationships and by applying reasonable cuts.

10.4 Conclusions

In this chapter, a method and the main lines of the OntoLearn tool, aimed at supporting the ontology engineering process, have been presented. The

---

[14] This fact stresses the need of a consensus building groupware, such as Consys.

method allows one to extract domain concepts and to detect semantic relationships among them. As mentioned in the previous section, the use of OntoLearn within the Harmonise project produced a remarkable increase in productivity in ontology building.

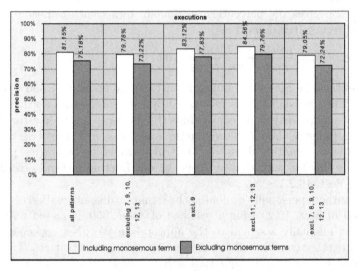

Fig. 10.9. Precision obtained excluding some of the heuristics discussed in Sect. 10.2.2

Letting aside the outcomes of a specific application (though real-world test beds do matter), we envisage several general advantages of OntoLearn with respect to ontology learning methods described in the literature:

- Many methods have been proposed to extract domain terminology or word associations from texts and to use this information to build or enrich an ontology. *Terms*, however, are invariably identified with domain *concepts*, while we propose an actual *semantic interpretation* of terms.
- Thanks to semantic interpretation, we are able to detect not only taxonomic, but also other types of relationships (e.g., similarity, pertainymy). The amount of extracted semantic relationships is being extended in our ongoing work, exploiting the information obtained from the intersections of semantic nets (see Fig. 10.5b).
- Though WordNet is not an ontological standard for the Semantic Web, it is *de facto* one of the most widely used general purpose lexical databases, as also witnessed by the considerable funding by the European Community to its extension (the EuroWordNet project [10.21]). An explicit relationship between a domain ontology and WordNet may favour interoperability and harmonization between different ontologies.

– Ontology learning issues have been considered in strict connection with ontology engineering and validation issues. We conceived a general architecture, that we described only in a short form, for the sake of brevity. The interested reader may consult to the references for details.

Our ongoing research follows two directions: the first concerns the extension of the type of ontological knowledge that can be learned from documents. OntoLearn extracts mainly taxonomic knowledge and similarity relationships, but semantic net intersections offer information for learning other types of concept descriptors. The second direction concerns the design of a more objective ontology evaluation experiment, in the context of some interesting Web-based application. One such evaluation is in progress within a document retrieval and classification task.

References

10.1 R. Basili, M.T. Pazienza, P. Velardi: An Empirical Symbolic Approach to Natural Language Processing. Artificial Intelligence, 85, pp. 59-99 (1996)

10.2 T. Berners-Lee: *Weaving the Web* (Harper, San Francisco, 1999)

10.3 C. Fellbaum: *WordNet: an electronic lexical database* (Cambridge, MIT press, 1995)

10.4 S. Harabagiu, D. Moldovan: Enriching the WordNet Taxonomy with Contextual Knowledge Acquired from Text. AAAI/MIT Press, 1999

10.5 D.B. Lenat: CYC: a large scale investment in knowledge infrastructure. Communication of the ACM, 3(11) (1993)

10.6 R. Krovetz: Homonymy and polysemy in Information Retrieval. *Proceedings of ACL/EACL 1997*

10.7 A. Maedche, S. Staab: Semi-automatic Engineering of Ontologies from Text. *Proceedings of the Twelfth International Conference on Software Engineering and Knowledge Engineering (SEKE'2000)*

10.8 A. Maedche, S. Staab: Learning Ontologies for the Semantic Web. *Proceedings of the 2nd International Workshop on the Semantic Web, Hongkong, China, May 2001*

10.9 R. Milhalcea, D. Moldovan: eXtended WordNet: progress report. *NAACL 2001 Workshop on WordNet and Other Lexical Resources, Pittsbourgh, June 2001*

10.10 M. Missikoff, X.F. Wang: Consys-A Group Decision-Making Support System For Collaborative Ontology Building. In: *Proc. of Group Decision and Negotiation 2001 Conference, La Rochelle, France, 2001*

10.11 M. Missikoff: *OPAL-A Knolwedge-Based Approach for the Analysis of Complex Business Systems* (LEKS, IASI-CNR, Rome, 2000)

10.12 E. Morin: Automatic Acquisition of semantic relationships between terms from technical corpora. *Proc. of 5th International Congress on Terminology and Knowledge extraction, TKE-99, 1999*

10.13 B. Smith, C. Welty: *Ontology: towards a new synthesis, Formal Ontology in Information Systems* (ACM Press, 2001)

10.14 M. Uschold, M. Gruninger: Ontologies: Principles, Methods and Applications. The Knowledge Engineering Review, 11(2) (1996)

10.15 P. Velardi, M. Missikoff, R. Basili: Identification of relevant terms to support the construction of Domain Ontologies. *ACL-EACL Workshop on Human Language Technologies, Toulouse, France, July 2001*

10.16 P. Velardi, M. Missikoff, P. Fabriani: Using Text Processing Techniques to Automatically enrich a Domain Ontology. *Proc. of ACM Conf. On Formal Ontologies and Information Systems, ACM FOIS, Ogunquit, Maine, October 2001*

10.17 P. Vossen: Extending, Trimming and Fusing WordNet for Technical Documents. *NAACL 2001 workshop on WordNet and Other Lexical Resources, Pittsbourgh, July 2001*

10.18 T. Yokoi: The EDR electronic dictionary. Communications of the ACM, 38(11) (1993)

10.19 DAML+OIL. http://www.daml.org/2001/03/daml+oil-index

10.20 ECAI-2000 1st Workshop on Ontology Learning. http://ol2000.aifb.uni-karlsruhe.de/

10.21 EuroWordNet. http://www.hum.uva.nl/~ewn/

10.22 Fetish EC project ITS-13015. http://fetish.singladura.com/index.php

10.23 Harmonise EC project IST-2000-29329. http://dbs.cordis.lu

10.24 IJCAI-2001 2nd Workshop on Ontology Learning. http://ol2001.aifb.uni-karlsruhe.de/

10.25 Semantic Web Community Portal. http://www.semanticweb.org/index.html

10.26 SemCor The semantic concordance corpus. http://mind.princeton.edu/wordnet/doc/man/semcor.htm

10.27 SymOntos, a symbolic ontology management system. http://www.symontos.org

10.28 WordNet 1.6. http://www.cogsci.princeton.edu/~wn/w3wn.html

10.29 RDF. http://www.w3c.org/RDF/

11. Browsing Semi-Structured Texts on the Web Using Formal Concept Analysis

Richard Cole[1], Florence Amardeilh[1], and Peter Eklund[2]

[1] University of Queensland, Australia
[2] University of Wollongong, Australia

Abstract

Browsing unstructured Web texts using Formal Concept Analysis (FCA) confronts two problems. Firstly, online Web data is sometimes unstructured and any FCA system must include additional mechanisms to discover the structure of input sources. Secondly, many online collections are large and dynamic, so a Web robot must be used to automatically extract data when it is required. These issues are addressed in this chapter, which reports a case study involving the construction of a Web-based FCA system used for browsing classified advertisements for real estate properties[1]. Real estate advertisements were chosen because they represent a typical semi-structured information source accessible on the Web. Furthermore, data is relevant only for a short period of time. Moreover, the analysis of real estate data is a classic example used in introductory courses on FCA. However, unlike the classic FCA real estate example, whose input is a structured relational database, we mine Web-based texts for their implicit structure. The issues encountered when mining these texts, and their subsequent presentation to the FCA system, are examined in this chapter. Our method uses a handcrafted parser for extracting structured information from the real estate advertisements, which are then browsed via a Web-based front-end employing rudimentary FCA system features. The user is able to quickly determine the trade-offs between different attributes of real estate properties and to alter the constraints of the search in order to locate good candidate properties. Interaction with the system is characterized as a mixed initiative process in which the user guides the computer in the satisfaction of constraints. These constraints are not specified apriori, but rather drawn from the data exploration process. Further, the chapter shows how the Conceptual Email Manager, a prototype FCA text information retrieval tool, can be adapted to the problem.

11.1 Introduction: Information Extraction and the Web

Since the creation of the DARPA Message Understanding Conferences (MUC) in 1987, Information Extraction (IE) has become an independent new field of research at the crossroad of Natural Language Processing (NLP), Text Mining, and Knowledge Discovery in Databases (KDD). For this reason, the methods and techniques of IE are strongly influenced by developments in

---

[1] In Formal Concept Analysis the term *property* has a special meaning, similar to *attribute*. In this chapter, *property* is used only with the meaning of *real estate property*, e.g., a house or apartment.

these related research topics. Moreover, IE can be useful for any collection of documents from which one would want to extract facts, and the World Wide Web is such a collection.

11.1.1 Definitions – Information Extraction

The objective of IE [11.13] is to locate and identify specific information from a natural language document. The key element of IE systems is the set of extraction rules, or extraction patterns, that identify the target information according to a scenario. Once an extraction pattern is identified, the IE system reduces extracted information to a more structured form, such as a database table. Each record in the table must also have a link back to the original document [11.26]. As a result, tools for visual representation, fact comparison, and automatic pattern analysis play an important role in the resulting presentation of data derived from IE systems.

Through the case study presented in this chapter, "rental accommodation" classifieds, we define the various terms used in the IE field. First, the "rental classified" scenario itself represents a way to format the target information, e.g., "the location, the renting price, the number of bedrooms, and the phone number". Second, each scenario is defined by a list of patterns that describes possible ways to talk about one of its facets, such as the pattern "for rent". Third, each pattern includes a set of extraction rules defining how to retrieve this pattern in the text. Fourth, each rule is composed of several fields, either constants or variables, representing a particular element of the information to extract. For example, concerning the pattern "for rent", we might employ (or learn) a pattern such as the following:

```
for rent <cr>
<variable location> - phone <variable phone_number> <cr>
$<variable rental_price> <cr>
<variable num_bedrooms> Bedrm.
```

11.1.2 Web Documents and Text Diversity

Some approaches to information extraction on the Web assume that all Web pages are semi-structured, since they contain HTML tags. However, Hsu [11.15] provides a fine-grained categorization for Web documents as follows: *structured Web pages* provide itemized information and each element can be correctly extracted based on some uniform syntactic clues, such as delimiters or the order of elements. *Semi-Structured Web pages* may contain missing elements, multiple value elements, permutations, and exceptions. Finally, *unstructured Web pages* require linguistic knowledge to correctly extract elements. It seems, therefore, that when it comes to extracting information from Web pages, the same sorts of problems and features facing information extraction on natural language documents also apply to the Web domain;

When IE systems for structured text perform well, the information can be easily extracted using format descriptions. However, IE systems for unstructured text need several additional processing steps beyond the construction of extraction rules. These additional steps are typically based upon patterns that involve syntactic relations between words or classes of semantic words. They generally use NLP techniques and cannot be compared to the work of a human expert (although they also provide useful results). Likewise, IE systems for semi-structured text cannot limit themselves to rigid extraction rules, more suited to structured text, but must be able to switch context to apply NLP techniques for free text. Nevertheless, systems for semi-structured texts do use delimiters, such as HTML tags, in order to construct extraction rules and patterns. Thus, a profitable approach to IE from semi-structured texts on the Web is a hybrid of the two.

Moreover, information on the Web is also highly dynamic. Website structure and the presence of hyperlinks are also important facets not present in traditional natural language documents. It may, for instance, be necessary to follow hyperlinks to obtain all the pertinent information from online databases. Web documents are both stylistically different from natural language texts and may be globally distributed over multiple sites and platforms. Hence, the Web IE problem represents a special challenge for the field because of the nature of medium.

11.1.3 Architecture and Components

The first step of the basic IE process is to extract each relevant element of the text through a local analysis, i.e., the system examines each field to decide if it is a new element to add to the pattern or if it relates to an existing element. Next, the system interlinks those elements and produces larger and/or new elements. Finally, only the pertinent elements regarding the patterns are translated into the output format, e.g., the scenario. Moreover, information to be extracted can be potentially in any part of the document, as is often the case with unstructured texts. In these situations, the elements are extracted (as above) and a second process is required to link together all elements dealing with the same scenario.

This IE process is implemented differently if the system is based either on a knowledge engineering approach combined with natural processing language methods, or on a statistics-based automatic training approach. In the first, experts examine sample texts and manually construct the extraction rules with an appropriate level of generality to produce the best performance. This "training" is effective but time consuming. In the second approach, once a training corpus has been annotated, a classifier is run so that the system learns how to analyze new texts. This is faster, but requires sufficient volume of training data to achieve reasonable outcomes [11.2]. Most IE systems compromise by using rules that are manually created and classifier components that are automatically generated.

To elaborate, IE systems use part or all of the following components: First, *segmentation* divides the document into segments, e.g., sentences, and other components, such as images and tables. Second, *lexical analysis* tags parts of speech, disambiguates words, and identifies regular expressions such as names, numbers, and dates. This gives some information about the words, their position in the text and/or sentence, their type, and, sometimes, their meaning. Lexical analysis generally uses dictionaries (and/or ontologies). Third, *syntactic analysis* identifies and tags nominal phrases, verbal phrases, and other relevant structures as a *partial analysis*, or, alternatively, each of the individual elements, i.e., nouns, verbs, determinants, prepositions, conjunctions, etc., as a *complete analysis*. Fourth, *information extraction* creates rules to identify pertinent elements, to retrieve suitable patterns, and to store them according to a predefined format corresponding to the information extraction scenario. This last phase also examines co-reference relations, not often explicit, such as the use of pronouns to qualify a person, a company, or an event. It is the only component specific to the domain [11.1].

Finally, Lawrence and Giles [11.20] claim that 80% of the Web is stored in the hidden Web (for example, in pages dynamically generated from some database), using XML/XSL to generate pages based on specific user requests to a database. This implies a special need for tools that can extract information from such pages. Thus, Information Extraction from Web sites is often performed using *wrappers*. Wrapper generation has evolved independent of the IE field, deploying techniques that are less dependent on grammatical sentences than NLP-based techniques. A wrapper, in the Web environment, converts information implicitly stored as an HTML document into information explicitly stored as a data structure for further processing. Wrappers can be constructed manually, by writing the code to extract information, or automatically, by specifying the Web page structure through a grammar, and translating it into code. In either case, wrapper creation is somewhat tedious, and as Web pages change or new pages appear, new wrappers must be created. Consequently, Web Information Extraction often involves the study of semiautomatic and automatic techniques for wrapper creation. *Wrapper induction* [11.19] is a method for automatic wrapper generation using inductive machine learning techniques. Using this technique, the task is to compute from a set of examples a generalization that explains the observations as an inductive learning problem.

11.1.4 Related Work

The IE field has developed over the last decade due to two factors: firstly, the exponential growth in digital document collections, and secondly, through the organization of the Message Understanding Conferences (MUC), held from 1987 to 1998 and sponsored by DARPA. The MUC conferences coordinated multiple research groups in order to stimulate research by evaluating various IE systems. Each participating site had six months to build an IE system for a predetermined domain. The IE systems were evaluated on the same domain and corpus, allowing direct comparison. The results were scored by an official scoring program, using the standard information retrieval measures. The MUC conferences demonstrated that fully automatic IE systems can be built with state-of-the-art technology, and that, for some selected tasks, their performance is as good as the performance of human experts [11.27]. Despite these outcomes, building IE systems still requires a substantial investment in time and expertise and remains somewhat of an art.

Some of the systems developed during the MUC period were applied, or can be applied, to the Web Information Extraction problem. On the one hand, both FASTUS [11.3, 11.14] and HASTEN [11.17] based their approaches on NLP techniques and developed the IE architecture mentioned above. They are operational systems but are still time- and resource-consuming in their scenario setup. On the other hand, automatic training systems are based on either unsupervised algorithms, combined with a bottom-up approach when extracting the rules, such as CRYSTAL [11.21]; or a supervised algorithm along with a top-bottom approach, such as WHISK [11.22] and SRV [11.11]. Interestingly, with respect to this chapter, WHISK used real estate classified ads as its document collection. Finally, another system named PROTEUS [11.13], uses dictionaries along with a set of regular expressions to mine documents in a top-bottom approach.

Simultaneously with these developments, the wrapper generation communities also developed some IE systems using machine learning algorithms to generate extraction patterns for online information sources. SHOPBOT WIEN, SOFTMEALY, and STALKIER belong to a group of systems that generate wrappers for fairly structured Web pages using delimiter-based extraction patterns [11.9, 11.19].

To conclude, search engines are not powerful enough for all the tasks associated with IE systems. They return a collection of documents, but they cannot extract relevant information from these documents. Thus, the Web Information Extraction field will continue to be an active area of research. As information systems will need to automate the process as far as possible to cope with the large amount of dynamic data found on the Web, IE systems will keep using machine learning techniques, rendering them beyond the scope of generalist search indexes. Nevertheless, a combination of different approaches, achieving hybrid and domain-specific search indexes, is believed to be a promising direction for IE [11.18, 11.22].

11.1.5 The Interaction Paradigm and Learning Context

Mixed initiative [11.16] is a process from human-computer interaction involving humans and machines sharing tasks best suited to their individual abilities. The computer performs computationally intensive tasks and prompts human clients to intervene either when the machine is unable to make a decision or when resource limitations demand intervention. Mixed initiative requires that the client determine trade-offs between different attributes, and alter search constraints to locate objects that satisfy an information requirement. This process is well suited to data analysis using an unsupervised symbolic machine learning technique called Formal Concept Analysis (FCA), an approach demonstrated in our previous work [11.5, 11.6, 11.7, 11.10] and inspired by the work of Carpineto and Romano [11.4].

This chapter reinforces these ideas by reusing the real estate browsing domain, a tutorial exercise in the introductory FCA literature. The browsing program for real estate advertisements (RFCA) is more primitive than the Conceptual Email Manager CEM [11.5, 11.6, 11.7, 11.10], which uses concept lattices to browse email and other text documents. Unlike CEM, RFCA is a Web-based program, creating a different set of engineering and technical issues in its implementation. However, RFCA is limited, and when the analysis of the rental advertising requires nested line diagrams (and other more sophisticated FCA system features), we reuse CEM to show how that program can produce nested line diagrams for the real estate data imported from the Web. Other related work demonstrates mixed initiative extensions by using concept lattice animation, notably the algorithms used in CERNATO, and joint work in the open source GoDA collaboration[2].

This article is structured as follows. Section 11.2 describes practical FCA systems and their coupling with relational database management systems (RDBMS). This highlights the necessity of structured input when using FCA, and therefore highlights the nature of the structure discovery problem. Section 11.3 describes the Web robot used to mine structure from real estate advertisements. This section details the methods required to extract structured data from unstructured Web collections and measures their success in terms of precision and recall. Section 14.3 shows the Web-based interface for browsing structured real estate advertisements. Section 11.5 demonstrates how real estate data can be exported and the CEM program re-used to deploy nested line diagrams and zooming [11.23]. Figure 11.1 shows the Homes on-line homepage that acts as the source of unstructured text for our experiment.

11.2 Formal Concept Analysis and RDBMSs

FCA [11.12] has a long history as a technique for data analysis. Two software tools, TOSCANA [11.24] and ANACONDA, embody a standard methodology for data analysis based on FCA. Following this methodology, data is organized as a table in an RDBMS (see Fig. 11.2) and is modeled mathematically as a *many-valued context*, (G, M, W, I), where G is a set of objects, M is a set of attributes, W is a set of attribute values, and I is a relation between G, M, and W such that if (g, m, w_1) and (g, m, w_2) then $w_1 = w_2$. We define

---

[2] See http://toscanaj.sf.net, and the framework project http://tockit.sf.net.

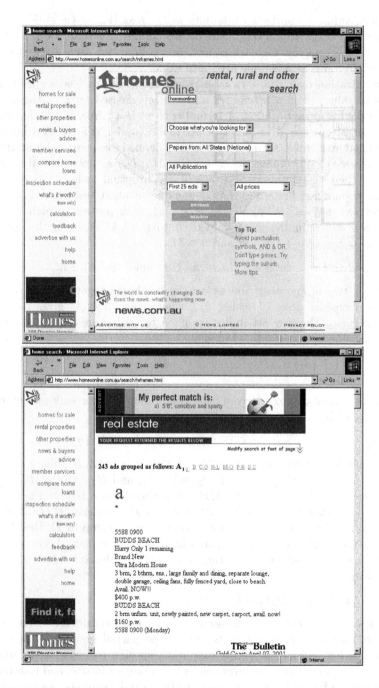

Fig. 11.1. The Homes on-line home-page. The site acts as the source of unstructured texts for our experiment

the set of values taken by an attribute $m \in M$ as $W_m = \{w \in W \mid \exists g \in G : (g, m, w) \in I\}$. An interpretation of this definition is that in the RDBMS table there is one row for each object, one column for each attribute, and each cell contains at most one attribute value.

Organization over the data is achieved via conceptual scales that map attribute values to new attributes and are represented by a mathematical entity called a *formal context*. A formal context is a triple, (G, M, I), where G is a set of objects, M is a set of attributes, and I is a relation between objects and attributes. A *conceptual scale* is defined for a particular attribute of the many-valued context: if $\mathbb{S}_m = (G_m, M_m, I_m)$ is a conceptual scale of $m \in M$, then we require $W_m \subseteq G_m$. The conceptual scale can be used to produce a summary of data in the many-valued context as a derived context. The context derived by $\mathbb{S}_m = (G_m, M_m, I_m)$, with respect to plain scaling from data stored in the many-valued context (G, M, W, I), is the context (G, M_m, J_m), where, for $g \in G$ and $n \in M_m$,

$$(g, n) \in J \Leftrightarrow: \exists w \in W : (g, m, w) \in I \quad \text{and} \quad (w, n) \in I_m .$$

Scales for two or more attributes can be combined together in a derived context. Consider a set of scales, S_m, where each $m \in M$ gives rise to a different scale. The new attributes supplied by each scale can be combined together using a special type of union:

$$N := \bigcup_{m \in M} \{m\} \times M_m .$$

Then, the formal context derived from combining all these scales together is (G, N, J), with

$$(g, (m, n)) \in J \Leftrightarrow: \exists w \in W : (g, m, w) \in I \quad \text{and} \quad (w, n) \in I_m .$$

A concept of a formal context (G, M, I) is a pair (A, B) where $A \subseteq G$, $B \subseteq M$, $A = \{g \in G \mid \forall m \in B : (g, m) \in I\}$, and $B = \{m \in M \mid \forall g \in A : (g, m) \in I\}$. For a concept (A, B), A is called the *extent* and is the set of all objects that have all of the attributes in B. Similarly, B is called the *intent* and is the set of all attributes possessed in common by all the objects in A. As the number of attributes in B increases, the concept becomes more specific, i.e., a specialization ordering is defined over the concepts of a formal context by

$$(A_1, B_1) \leq (A_2, B_2) :\Leftrightarrow B_2 \subseteq B_1 .$$

More specific concepts have larger intents and are considered "less than" ($<$) concepts with smaller intents. The same partial ordering is achieved by considering extents, in which case more specific concepts have smaller extents. The partial ordering over concepts is always a lattice, and commonly drawn using a Hasse diagram. Concept lattices can be exploited to achieve an efficient labeling; each attribute has a single maximal concept (with respect to the specialization ordering) possessing that attribute. If attribute labels are

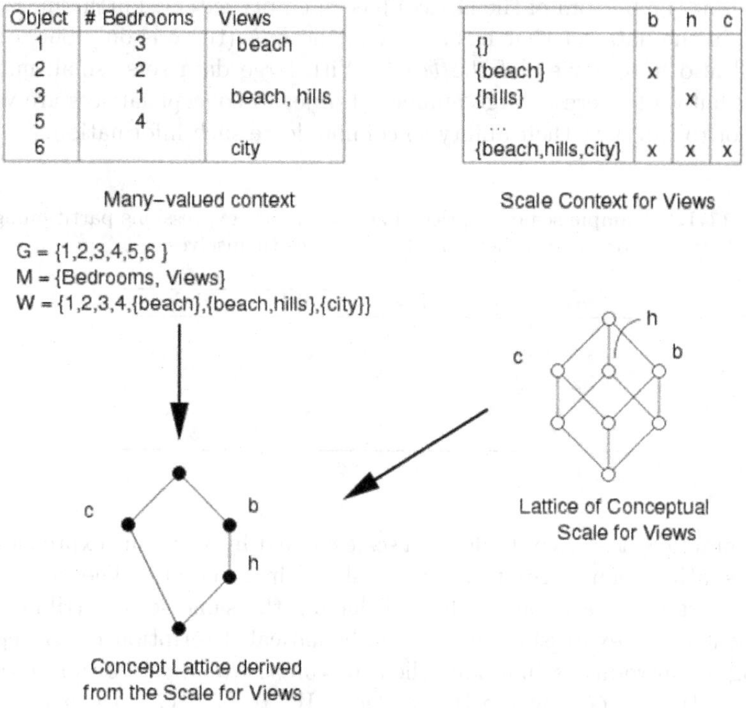

Object	# Bedrooms	Views
1	3	beach
2	2	
3	1	beach, hills
5	4	
6		city

	b	h	c
{}			
{beach}	x		
{hills}		x	
...			
{beach,hills,city}	x	x	x

Many–valued context   Scale Context for Views

G = {1,2,3,4,5,6 }
M = {Bedrooms, Views}
W = {1,2,3,4,{beach},{beach,hills},{city}}

Lattice of Conceptual
Scale for Views

Concept Lattice derived
from the Scale for Views

Fig. 11.2. Example showing the process of generating a derived concept lattice from a many-context and a conceptual scale for the attribute *Views*

attached only to their maximal concepts, then the intent of a concept can be determined by collecting labels from all greater concepts. A similar situation is achieved for objects. Each object has a minimal concept to which its label is attached, and the extent of a concept can be determined by collecting labels from all lesser concepts. Attribute and object labels are disambiguated by attaching object labels from below and attribute labels from above.

Consider Fig. 11.2. An RDMS table contains a list of real estate properties (objects 1–6), the number of bedrooms, and the types of views the properties afford. The many-valued context has two attributes: *#Bedrooms* and *Views*. *Views* is organized by the scale context shown on the top-right of Fig. 11.2. The scale context has all possible combinations of *beach, hills,* and *city* views as objects and introduces three new attributes: *b, h,* and *c*. The set of scale objects must contain all the attribute values taken on by objects for the attribute being scaled. The scale is applied to the many-valued context to produce a derived context giving rise to the derived concept lattice shown in Fig. 11.2 (lower left). This lattice reveals that there are no objects having views of both the *hills* and the *city*, since the most specific concept (the concept at the bottom of the lattice) has an empty extent. Furthermore, any object in the data set that has a view of the *hills* (there is only one, object 3) will also have a view of the *beach*. With large data sets (small number of attributes of interest, large number of objects) concept lattices are vastly superior to tables in their ability to communicate such information.

Table 11.1. Example scale for price. The objects are expressions partitioning the attribute values for price rather than being values themselves.

price	low	mid	high
<$150	×		
$150-$200	×	×	
$200-$250		×	
$250-$300		×	×
>$300			×

In practice, it is easier to define a scale context by attaching expressions to objects rather than attribute values, as shown in Table 11.1. The expressions denote a range of attribute values all having the same scale attributes. To represent these expressions in the mathematical description of conceptual scaling, we introduce a function called the *composition operator* for attribute m, $\alpha_m : W_m \to G_m$, where $W_m = \{w \in W \mid \exists g \in G : (g, m, w) \in I\}$. This maps attribute values to scale objects. The derived scale then becomes (G, N, J) with

$$(g, (m, n)) \in J \Leftrightarrow: \exists w \in W : (g, m, w) \in I \quad \text{and} \quad (\alpha_m(w), n) \in I_m \ .$$

The main purpose of this summary of FCA is to reinforce that in practice FCA works with structured object-attribute data in RDBMS form, in conjunction with a collection of conceptual scales. Furthermore, this section describes the mechanism by which FCA is applied to data.

11.3 Web-Robot for Extracting Structured Data from Unstructured Sources

The initial purpose of selecting the real-estate classified domain was that it conforms to a classic introductory example in FCA. Our initial intention was to obtain access to Newslimited's structured classifieds database for student experiments introducing FCA [11.8]. However, Newslimited did not have the resources to directly assist our efforts so we constructed a purpose built script and interface to determine the query parameters from the Newslimited Website. This interface is shown in Fig. 11.3 and the extraction Dataflow system diagram is shown in Fig. 11.4.

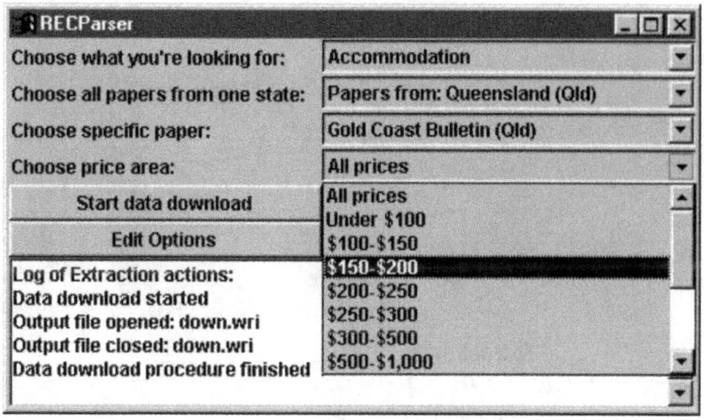

Fig. 11.3. Interface to WebRobot that "extracts" the advertisements from the Newclassifieds Website

We began with a sequence of real estate advertisements in an HTML file rather than with the ideal RDBMS export format. The first task is to separate the advertisements from the surrounding HTML markup, and to segment the advertisements into self-contained objects, one for each property. This was done using a string processing algorithm. An example advertisement is shown in Fig. 11.5. The text refers to six different properties, three with a rental price of $250 per week and three with a price of $300. All properties are located in the suburb of *Arundel*. The format of the advertisements presents three main challenges: (i) the information about properties overlaps, i.e., the single instance of the word *Arundel* indicates that all six properties are in Arundel; (ii) there are many aliases for the same basic attribute, e.g., double garage and dble garage; and (iii) some information is very specific, e.g., 1.up garage or near golf course.

An LL(1) parser was constructed using the Metamata Java Compiler Compiler (JavaCC)[3] to parse advertisements of this type. The parser is able to handle the first two of these challenges with reasonable success. The parser recognizes predefined attributes and discards all unrecognized information.

The initial segmentation was able to extract 89% of the advertisements. The remaining 11% were of low quality and omitted, they did not include a rental price and were therefore not meaningful. The parser recognized 64 attributes of which 53 were single valued, i.e., true or false. The remaining 11 attributes, including rental price, number of bedrooms, and car park type, were multi-valued. To assess the accuracy of the parser, precision and recall were measured for each attribute and then aggregated. A summary of the most common and most important attributes for 53 rental properties is given in Table 11.2.

---

[3] See http://www.metamata.com/JavaCC/

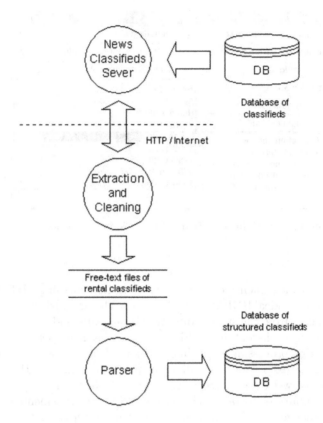

Fig. 11.4. Extraction DataFlow System Diagram

N_A is the set of identified attribute values, and N_B the set of correct attribute values. The *precision* of multi-valued attributes is calculated as the number of correctly identified attribute values ($|N_A \cap N_B|$) as a proportion of the number of identified attributes values ($|N_A|$). The *recall* is the number of correctly identified attribute values ($|N_A \cap N_B|$) as a proportion of the number of correct attribute values ($|N_B|$).

Averaging the most important attributes – *Location, Price, Bedroom, Furnished, and Car parking* – weighted by their frequency yields a precision of 100% and a recall of 95%, while the inclusion of the *Other* attribute reduces the recall to <70%. All real estate advertisements leave out some information about the property they advertise, presumably because of the cost per word for advertising. As a result, we would expect the recall of actual information about the property being advertised to be much less with respect to the actual property.

```
1 FOR RENT - ARUNDEL - Phone 55948184
2 $300
3 4 Bedrm, in-grnd pool, dble garage, near shops and school
4 3 bedrm, tripple garage, immac. presented, close to transport
5 Exec. 3 Bedrm + study, pool, dble garage, all ammen. close to
school
6 $250
7 Leafy 3 bedrm, double garage, avail. Aug.
8 3 bedrm townhouse, resort fac. 1.up garage, 2 bathroom and on-
suite.
9 Townhouse, 2 bedroom, resort fac. garage, near golf course and
transport.
```

Fig. 11.5. A rental classified advertisement illustrating multiple aliases for attributes (as in abbreviations such as Bedrm=bedroom), multiple objects (as rental properties described on lines 3, 4, 5, 7, 8 and 9) in a single advertisement (all lines) clustered on an primary key attribute: in this case the two prices $300 and $250

Table 11.2. Recall and Precision for 53 unseen real estate advertisements.

	Location	Price	Bedroom	Furnished	Car Park	Other
Frequency	100%	100%	100%	26.4%	50.9%	88.7%
Precision	100%	100%	100%	100%	100%	100%
Recall	94.3%	100%	98.1%	71.4%	96.3%	68.1%

One of the strengths of FCA is that it allows the user to compose views of the data that separate objects at different levels of detail. For example, the user may have a coarse distinction based on price, but a fine distinction based on proximity to facilities. Table 11.2 shows poor recall for attributes in the group *Other*. When combined with the knowledge that the advertisements contain only partial descriptions of the data, this places a practical limit on the level of detail that can be usefully explored. This limit can be extended if the initial data source is a database or XML file containing more extensive information about the features of properties for rent.

The LL(1) parser was very fast, building the relational database and storing the multi-valued context in under eight seconds on a Pentium-III 300 MHz computer for an entire week of advertisements, approximately 3,400 properties listed in the local newspaper.

11.4 RFCA – the Web-Based FCA Interface

The Web-based user interface presents a Web page with a scale selector as shown in Fig. 11.6. The client selects a suitable scale to browse through the classified advertisements. The scales are predefined. The classifieds are now

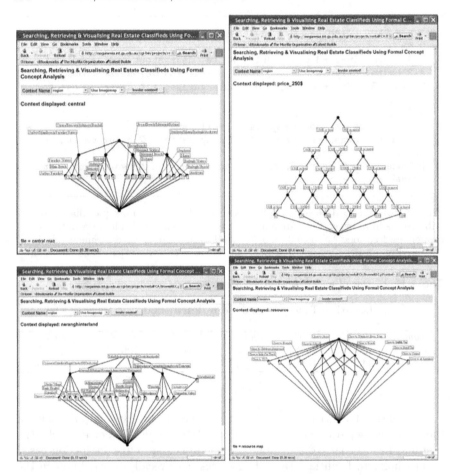

Fig. 11.6. The RENTAL-FCA prototype: Scales are predefined and selected from the "Context Name" menu (top-left). This figure shows the conceptual scales for various geographic regions (top) and price and resource scales (lower)

in a structured database format after the parsing described in the previous section. When the user selects a scale, a new Web page is loaded containing the scale image. This image now contains all the resulting extent numbers from the scales, interrogation of the database. The number of objects in the extents are displayed over each vertex in the usual way. The same scale selector is also available on the Web page displaying the scale image. This allows the user to select a new scale without having to go back to the previous page. In other words, the same scale selection should be present on each of the pages displaying a selected scale.

A process that reproduces the Web page dynamically with different scales and extent numbers was implemented. This program creates the scale images

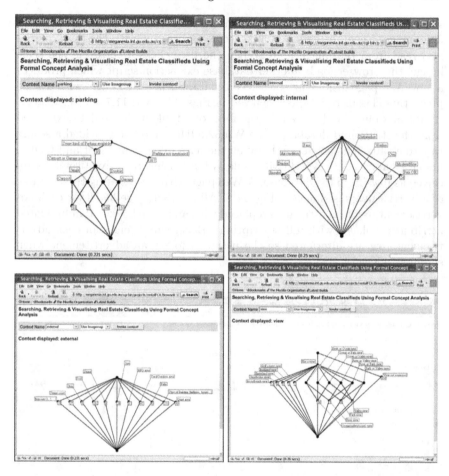

Fig. 11.7. This figure shows the conceptual scales for car parking and fixtures (top) and facilities and views (lower)

after each selection by the user. A database connection and support for reading the scale files from the server are implemented.

The Web pages with extent numbers do not exist as files but are generated on demand. When a scale is selected, the script calls the graph drawing program as a system command with the new scale name as parameter. This drawing program draws a concept lattice corresponding to the context. The result is stored as a PNG file representing the scale image and an image map representing the coordinates for the vertices in the graph. The image map also contains SQL queries extracted from the current context file. Queries in the image map corresponding to vertices in the PNG file are used to interrogate the database. After executing the graph drawing program the script starts to build the client-side image map. All the vertex coordinates are read in sequence from the image map and transformed to "hot" regions in the clickable image. Each hot region is linked to a CGI script with the SQL queries also read from the image map. When the user clicks a vertex in the scale image, the browser loads another database extraction script which produces a new Web page displaying the selected data. Such a Web page, with a scale presenting classified data, is displayed in Figs. 11.6 and 11.7.

Results must be displayed in the form of a table with the data extracted from the structured database. The Win32::ODBC module provided a secure way to establish a connection between the data extraction script and ODBC under Windows NT. An HTML table is built using the advertisements received as rows from the database. A Web page with the resulting HTML table of advertisements is shown in Fig. 11.8. All attributes are listed for each advertisement, boolean attributes replaced with an image hook and abbreviated attributes replaced with full descriptions. Background colors for each advertisement row are alternated so the user can follow an advertisement when scrolling sideways.

Searching, Retrieving & Visualising Real Estate Classifieds Using Formal Concept Analysis

Result of query:

Id	Region	Location	Price	No. of Bedrooms	Furnished	Type of Carpark	No. of Carparks	Tennis	Pool	Yard	Patio	BBQ Area	Spa	Sauna	Gym	Duplex
house0006	Northern	Labrador	150.0000	2	No	Garage	1			✓						✓
house0020	Northern	Labrador	165.0000	3	No	Garage	1									
house0031	Northern	Labrador	165.0000	3	No	Garage	1									
house0049	Northern	Labrador	135.0000	2	No	Carport	1	✓								
house0071	Northern	Arundel	250.0000	4	No	Garage	2									
house0073	Northern	Southport	205.0000	3	No	Garage	2									
house0076	Northern	Paradise Point	160.0000	2	No	No	1									
house0080	Northern	Southport	220.0000	3	No	No	0	✓								

Fig. 11.8. From the scales view shown in Fig. 11.6 the user can navigate to the objects which are displayed in the structured extracted form as a database table

Sometimes, the original advertisement contains attributes not included in the database that can be of additional interest. The first column in the table is a running number that uniquely identifies the advertisement. This number is inserted when parsing the advertisements. The table contains a column named *Id*. *Id* contains the number of the section from where the advertisement was parsed. So, if the advertisement was originally located in the third section in the free text of the rental classifieds file, the column has the value 3. Using this number, we can create a link from the database advertisements to the originals in the downloaded text file. An example of a resulting Web page displaying the original advertisements is shown in Fig. 11.9.

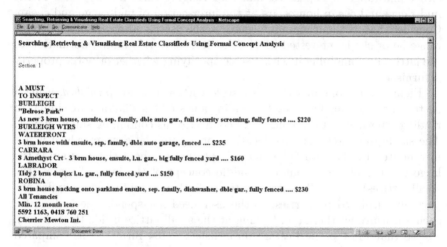

Fig. 11.9. By navigating by the *Id* in Fig. 11.8 the user can recover the text of the original unstructured text. This text can be dynamically generated by a query against the Newclassified Website if copyright is a concern

11.5 Reusing CEM for Nesting and Zooming

Nesting and zooming [11.23, 11.24] are two well established techniques used in FCA. Together, these techniques allow a user to wander around a conceptual landscape [11.25], attempting to find concepts that satisfy their constraints. When searching for a real estate property, there will obviously be compromises between location, price, and other factors. By using concept lattices to show how constraints can be satisfied, users are able to adapt their search to areas more likely to bear fruit. We re-used the CEM program to reinforce these ideas, although the same approach could be implemented with some effort in the RENTAL-FCA prototype.

This contrasts with current online real estate systems which ask the user to provide a specification for the type of property they are interested in and then (in most cases) provide either an empty list or a very long list of candidate properties. Using nesting and zooming in FCA allows questions like, "what are the possibilities for a mid-range house close to the city with a view, and maybe close to a park, shops, or transport," as opposed to questions like "list all mid-range houses that are close to the city, have a view, are close to a park, close to shops, and close to transport."

Consider a person who is new to a city and looking for accommodation. A good place to start is with a decision about price. Figure 11.10 shows a conceptual scale defined for price. The scale shows that most properties are either mid-range or expensive, and that roughly three-fifths of each of the mid-range and cheap houses are at the intersection of mid-range and cheap. Consider that, without more information, the user is uncertain of what price range he or she is interested in. He/she decides to add more information to the lattice by combining it with a scale specifying whether or not a property is furnished.

Figure 11.10 combines the *price* scale with a scale for *furnished*, using a nested line diagram. The rules for reading a nested line diagram are similar to reading a normal lattice. Thick lines connect ovals containing small lattices. The small grey circles show a location for a potential concept which is not instantiated by the data. The first thing to notice about the diagram is the large number of times that the middle concept of the inner diagram (the small lattices) is grey. The grey vertex indicates no mid-range or expensive partially furnished properties, so the user need not spend time looking for such a feature. Furthermore, looking at the small lattice inside the top oval, we see that most properties are unfurnished – 104 furnished as compared with 752 unfurnished.

The user may have an interest in investigating fully furnished mid-range properties and is able to zoom into this concept by selecting it and selecting the zoom operation. He/she could have been more specific and selected a property at the intersection of mid-range, cheap, and fully furnished, since there is such a concept; but for now suppose that he/she selected mid-range and fully furnished. The zoom operation restricts the objects shown in the lattice to only those in the extension of the selected concept.

Figure 11.11 shows a scale that has been zoomed. A small panel in the lower left hand corner shows the zooming history. The two arrow buttons in the tool bar allow moving backward and forward with respect to the zooming operation in a manner similar to navigating in Web browsers. The concept lattice now contains only 69 real estate properties, since the zooming operation has restricted the object set to the extent of the concept for fully furnished and mid-range. The conceptual structure in this lattice is different from that of the general picture without zooming. In the 69 properties shown, proximity to shops implies proximity to water (*Close Water*), and it is impossible to satisfy a desire to be close to University and close to shops in this restricted set of properties.

The user is now able to make a decision between different criteria, perhaps zooming further into the concept labeled *Close Water*, or alternatively retrieving all four properties that are close to shops. Similarly, the user is free to go back and make different zooming choices or include another scale, with still more criteria, to the current scale.

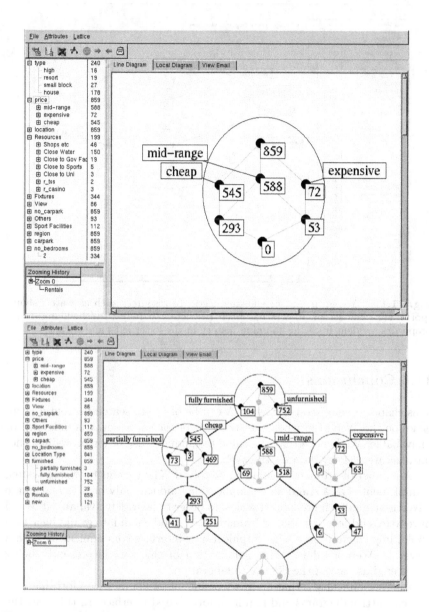

Fig. 11.10. (Top) The derived concept lattice showing how the properties are distributed with respect to three linguistic variables (scale attributes): *cheap, mid-range,* and *expensive.* (Lower) A combination of the scales for *price* and *furniture* using a nested line-diagram

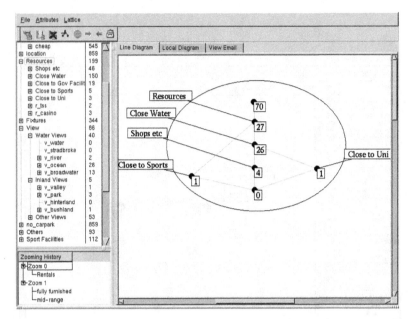

Fig. 11.11. A concept lattice showing access to resources such as water, shops, sports, etc. The set of objects has been restricted to fully furnished, mid-range properties, evident from the zooming history in the lower left-hand corner

11.6 Conclusions

This chapter demonstrates how FCA can be of use to search for rental properties on the Web even when the structure of the source data is unknown or unavailable. We believe that the same technique is useful for browsing other unstructured legacy data on the Web.

A number of problems remain to be solved. The current browsing system is implemented as a stand-alone application, and can only browse real estate advertisements with predefined scales. In order to be widely available it would have to extend to a distributed framework. A good candidate would be a Java applet implementation of the graphical user interface communicating with a server. A Web-based FCA implementation of this sort is presently being engineered as part of the GoDa collaboration.

Another difficulty is that many users are unfamiliar or uncomfortable with concept lattice diagrams and require a form-based interface. In this way, the process and interpretation of the diagrams can be taught to a user using the tool for the first time. The advantage of FCA, even without the concept lattice, is that feedback can be given on the volume of data satisfying search constraints.

The system we implemented obtained its data by parsing small textual descriptions of objects. The increasing use of the Internet is encouraging the storage of more structured information, and thus, in the future, we expect the difficult task of constructing one-off IE parsers to suit specific textual descriptions to disappear as data is directly entered with structure. In other words, browsing XML data using FCA on the Web is significantly more simple than what has been described here, although the techniques for mining structure from unstructured textual sources will be of value in various intelligence applications.

In addition to the interface described in this program, a prototype Web-based interface allowing the construction of derived concept lattices and retrieval of concept extents is available at: http://www.kvocentral.com/software/rentalfca.html

Acknowledgment

The GoDa project is supported by the Australian Research Council (ARC) and the DFG. This research also benefits from the support of the Distributed Systems Technology Research Centre (DSTC Pty Ltd) which operates as part of the Australian Government's CRC program. The authors acknowledge the input of Åge Strand and Peter Becker.

References

11.1 J. Y. Chai: Learning and Generalisation in the Creation of Information Extraction Systems. PhD Thesis, Department of Computer Science, Duke University (1998)

11.2 D. Appelt, D. Israel: Introduction to Information Extraction Technology, Tutorial for IJCAI-99, Stockholm (August 1999)

11.3 D. Applet et al: SRI International: Description of the FASTUS System used for MUC-6. In: *Proceedings of the Sixth Message Understanding Conference (MUC-6), November 1995* pp. 237-248

11.4 C. Carpineto, G. Romano: A Lattice Conceptual Clustering System and its Application to Browsing Retrieval. Machine Learning, 24, pp. 95-122 (1996)

11.5 R. Cole, P. Eklund: Analyzing an Email Collection using Formal Concept Analysis. *European Conf. on Knowledge and Data Discovery, PKDD'99, LNAI 1704, 1999* (Springer Verlag, 1999) pp. 309-315

11.6 R. Cole, G. Stumme: CEM: A Conceptual Email Manager *Proceeding of the 8th International Conf. on Conceptual Structures, ICCS'00, LNAI 1867* (Springer Verlag, 2000) pp. 438-452,

11.7 R. Cole, P. Eklund, G. Stumme: CEM-A Program for Visualization and Discovery in Email. In: D.A. Zighed, J. Komorowski, J. Zytkow (eds.), *Proc. of PKDD'00, LNAI 1910* (Springer-Verlag, 2000) pp. 367-374,

11.8 R. Cole, P. Eklund: Browsing Semi-Structured Web Texts using Formal Concept Analysis. *9th International Conference on Conceptual Structures* (Springer Verlag, LNAI 2120, ICCS'2001, August, 2001) pp. 319-332

11.9 L. Eikil: Information Extraction from the World Wide Web, Norwegian Computer Center, Oslo, pp.12-22 (July, 1999)

11.10 P. Eklund, R. Cole: Structured Ontology and Information Retrieval for Email Search and Discovery. In: M. Hacid, Z. Ras, D.A. Zighed, Y. Kodratof (eds.), *Foundations of Intelligent Systems* (13th International Symposium ISMIS 2002, LNAI 2366, Springer-Verlag, 2002) pp. 75-84

11.11 D. Freitag: Information Extraction from HTML: Application of a general Machine Learning Approach, Carnegie Mellon University (1998)

11.12 B Ganter, R. Wille: Formal Concept Analysis: Mathematical Foundations Springer Verlag (1999)

11.13 R. Grishman: Information Extraction: Techniques and Challenges, New York University, 18 (1997)

11.14 JR. Hobbs: FASTUS: A cascade finite-State Transducer for Extracting Information from Natural-Language Text. In: *Proceedings of the DARPA workshop on Human Language Technology, 1993* pp. 25-35

11.15 CH. Hsu, MT. Dung: Generating Finite-state Transducers for semi-structured Data Extraction From the Web. Information Systems, 23 (8), pp. 521-538 (1998)

11.16 E. Horvitz: Uncertainty, Action and Interaction: In pursuit of Mixed-initiative Computing Intelligent Systems (IEEE, September, 1999) pp. 17-20. http://research.microsoft.com/~horvitz/mixedinit.HTM

11.17 G. Krupka: Description of the SRA system as used for MUC-6. In: *Proceedings of the Sixth Message Understanding Conference (MUC-6), 1995* pp. 221-235

11.18 N. Kushmerick: Gleaning the Web. IEEE Intelligent Systems, 14(2) (March/April 1999)

11.19 N. Kushmerick: Weld DS. and Doorenbos R., Wrapper Induction for Information Extraction. In: *International Conference on Artificial Intelligence (IJCAI-97), 7 (1997)*

11.20 S. Lawrence, CL. Giles: Searching the World Wide Web. Science magazine, 280, pp. 98-100 (April 1998)

11.21 S. Soderland: Learning to extract Text-based Information from the World Wide Web. In: *Proceedings of Third International Conference on Knowledge Discovery and Data Mining (KDD-97) (1997)*

11.22 S. Soderland: Learning Information Extraction rules for Semi-structured and Free Text. Machine Learning, University of Washington, 44 (1999)

11.23 F.C. Vogt: Data Analysis based on a Conceptual File, Classification, Data Analysis and Knowledge Organization, Hans-Hermann Bock, W. Lenski, P. Ihm (eds.), pp. 131-140 (Springer Verlag, Berlin, 1991)

11.24 F. Vogt, R. Wille: TOSCANA: A Graphical Tool for Analyzing and Exploring Data. In: R. Tamassia, I.G. Tollis (eds.), *Graph Drawing '94, LNCS 894* pp. 226-233 (1995)

11.25 R. Wille: Conceptual Landscapes of Knowledge: A Pragmatic Paradigm for Knowledge Processing. In: W. Gaul, H. Locarek-Junge (eds.), *Classification in the Information Age* (Springer, Heidelberg, 1999)

11.26 R. Yangarber, R. Grishman, P. Tapanainen, S. Huttunen: *Unsupervised Discovery of Scenario-Level Patterns for Information Extraction* (New York University & Helsinki University, 8, 2000)

11.27 K. Zechner: A literature survey on Information Extraction and Text Summarization. Term paper, Carnegie Mellon University (1997)

12. Graph Discovery and Visualization from Textual Data

Vincent Dubois and Mohamed Quafafou

IAAI – University of Avignon, France

Abstract

This chapter tackles the problem of knowledge discovery in text collections and the dynamic display of the discovered knowledge. We claim that these two problems are deeply interleaved, and should be considered together. The contribution of this chapter is fourfold: (1) the description of the properties needed for a high-level representation of concept relations in text; (2) a stochastic measure for a fast evaluation of dependencies between concepts; (3) a visualization algorithm to display dynamic structures; and (4) a deep integration of discovery and knowledge visualization, i.e., the placement of nodes and edges automatically guide the discovery of knowledge to be displayed. The resulting program has been tested using two specific data sets based on the specific domains of molecular biology and WWW how-tos. We have also integrated the proposed discovery and visualization methods in a more refined Web mining process.

12.1 Introduction

Extracting and displaying graph structures from text collections is usually done in two steps: first, the graph is extracted; then, it is displayed. These two steps arise whenever a new document is added to the data collection. Moreover, the end user faces a new graph display that is not related to the previous one. In this chapter, we mix the extraction step and the visualization step. As a result, the graph display evolves smoothly when new documents are added. Results are displayed continuously during the discovery process. This is, of course, more user-friendly than waiting blindly for the whole process to complete. To achieve this goal, we present a semantic based on dependencies between concepts for the graph to be extracted, a relevant metric to evaluate these dependencies, an efficient way to compute this measure, a visualization method, and the integration of all these aspects. The resulting implementation is validated on two data sets and an application to Web mining is discussed. The study of results highlights various properties of our method.

In Sect. 12.2, we present the graph semantic and the discovery process. In Sect. 12.3, we present the visualization process and its contribution in the guiding of the mining process. In Sect. 12.4, we give experimental results, and discuss the dynamic and incremental aspects of the Method. In Sect. 12.5, we present a Web content mining application. Section 12.6 concludes this work and shows possible extensions.

12.2 Graph Structure Discovery

12.2.1 Graph Structure Semantic

We chose to represent relationships between concepts using a graph. Each node in this graph is a concept, and each edge represents a dependency between two concepts. For the sake of simplicity, we consider each word as a concept.

Binary mutual entropy. Dependency between two concepts a and b is a high-level and abstract relation. A possible measure is the mutual entropy $H(a; b)$, which is defined using entropy $H(a)$ and conditional entropy $H(a|b)$ as follow: $H(a; b) = H(a) - H(a|b) = H(b) - H(b|a)$.

This use of mutual entropy measure succeeds in describing binary relations between concepts, but it has several flaws: firstly, mutual entropy measure is symmetric, but dependency relation is not necessarily symmetric; secondly, if a concept depends on a set of concepts, it does not necessarily depend on each one separately (e.g., a concept may be created as the result of an exclusive OR on other concepts). This kind of relationship can not be detected by handling only pair of concepts. The main advantage of binary use of the mutual entropy measure is is not being time consuming and easy to use, but it lacks the required description finesse.

Bayesian networks. Bayesian networks have already been successfully used for text categorization [12.1, 12.2] and term disambiguation [12.3]. The Bayesian networks semantic is deep enough to describe the dependency relation and perform model inference [12.4], but related algorithms search the smallest graph describing the data with good enough accuracy. As a consequence, a link between two nodes is always a dependency, but not all dependencies are described in a Bayesian network. Moreover, Bayesian networks are required to be directed acyclic graphs, whereas dependency cycles are common (if a depends on b, it is common to find out that b depends on a).

Thus, we need a formalism with a local description of the dependency (no general property on the graph such as being acyclic), that involves more than two concepts at once (or otherwise it misses some large dependency).

N-ary mutual entropy. Mutual entropy measure can be used with more than two concepts at once. We can express that concept a depends on a set $B = \{b_1, \ldots, b_n\}$ of n concepts by using it as follows: $H(a; B) = H(a) - H(a|b_1, \ldots, b_n)$

But this leads to hyper-graph, and it not easy to think of it or to display it. We propose to flatten it by the following methods: The concept a depends on b iff it exists B minimal such that a depends on B and $b \in B$. Thus, given a concept a, we search all subsets B such that both $H(a; B)$ and $|B|$ are small.

Size penalty. One conditional entropy property is that $H(A|B) \leq H(A|B \cup \{b\})$. So, we need to complete this measure by adding a size penalty. If we code B using A information, the message length (per instance) is given by the conditional entropy $H(B|A)$. It is logical to add to this length the encoding length of the function used to predict B values based on A values. This requires us to code $2^{|B|}$ binary values (one for each possible combination of the concepts in B). If we have l instances to code, then the structure cost per instance is $2^{|B|}/l$. The resulting evaluation function is:

$$f(A, B) = H(A) - H(A|B) - 2^{|B|}/l . \tag{12.1}$$

The optimization process searches for best sets of concepts B for each A. Thus B is considered minimal, because if a concept in B adds no information, then a subset of B would have the same conditional entropy as B, but a lower structural cost, and get a better score.

Definitions. Given a concept A, we call *partial parent sets* concept sets B with a positive measure $f(A, B)$. For each concept A, we accept at most n partial parent sets. We call *parent set* of concept A the union of all (accepted) partial parent sets. We say that a concept B is one of the *parents* of the concept A when B belongs to the parent set of A. We call *score* the $f(A, B)$ value. The graph score is the sum over the graph of all partial parent set scores.

12.2.2 Network Evaluation

Entropy Approximation. N-ary entropy computation is costly, so we use a stochastic approximation of it in order to avoid full computation. The approximation is based on differences between instances (i.e., the set of concepts that are present in one, and only one, of the pair of instances).

Fano's inequality gives:

$$H(P_e) + P_e . \log(|\xi| - 1) \geq H(X|Y) , \tag{12.2}$$

where P_e is the probability of error while predicting X using Y, and $|\xi|$ is the number of different possible values for X. In our context, X is a binary attribute (concept is either present or absent); so, $\log(|\xi| - 1) = 0$ and

$$H(P_e) \geq H(X|Y) . \tag{12.3}$$

Consequently, the conditional entropy is approximated by the entropy of the error of the prediction. We now express this error by using sets of equal attributes[1] (see annex for proof):

$$P_e = 1/2 - 1/2 \sum_Y P(Y)\sqrt{2P(X_=|Y) - 1)} \tag{12.4}$$

---

[1] We define $X_=$ as the event where two instances of variable X share the same value. For example, if X has two possible values, 0 and 1, then $P(X_=) = P(X = 1)P(X = 1) + P(X = 0)P(X = 0)$.

The $\sum$ term describes the mean value of $\sqrt{2P(X_=|Y)-1}$. We approximate it by the square root of the mean value:

$$P_e \simeq 1/2 - 1/2\sqrt{2P(X_=|Y_=)-1} \, . \tag{12.5}$$

We note that $P(X_=|Y_=) = P((X \cup Y)_=)/P(Y_=)$. Thus, the information required to compute our approximation of conditional entropy $H(X|Y)$ is the count of identical attribute sets a in pair of instances. This count is much easier to compute than any entropy measure, and can be stored in a more efficient way. Moreover, only a small sampling of all possible pairs is necessary. More information on this method can be found in [12.5].

12.2.3 Methods and Algorithms

The general method consists of searching among the space of possible parent sets for the best one for each node. It is described in Algorithm 14.

Algorithm 14 Graph structure extraction

Require: C, set of all concepts c_i
Ensure: G is the best graph according to our metric
    $G(w_i)$ is the list of n best partial parent sets for c_i
1: **for all** node w_i **do**
2:    **for all** possible parent set $P \subset C_{\backslash c_i}$ **do**
3:        **if** P is a good enough parent set **then**
4:            Add P to $G(w_i)$
5:            **if** $G(w_i)$ size exceed n **then**
6:                Remove worst partial parent set in $G(w_i)$
7:            **end if**
8:        **end if**
9:    **end for**
10: **end for**

This crude algorithm searches the whole space to find the best possible partial parent sets. Such an exhaustive search of best partial parent sets among all possible subsets is not realistic. A naive implementation of Algorithm 14 leads to exponential complexity. Hopefully, it is easy to improve this loop, by using some pruning heuristics. The score we compute has two parts: the first one is the ability of the model to predict data, the second is the size cost of the structure. While the first part $(H(a) - H(a|B))$ is bounded (by $H(a)$), the second grows exponentially $(2^{|B|}/l)$. As the metric is required to be greater than 0, there is a size limit for B, given by $2^{|B|} < lH(a)$. If the size of a partial parent set is bigger than this limit, it will have a negative score, even if it defines a functional dependency $(H(a|B) = 0 = P_e)$. This property provides an efficient way to ensure that the complexity is not beyond a selected degree by adjusting the structural cost l.

Considering high dimensional data sets, the previous pruning method is not sufficient: complexity remains low at the cost of the sensibility of the metric. The relative weight of the structure cost is forced high. It is not possible to perform an exhaustive search in very high dimensions. The only way to restrict the search space size is to limit the number of possible parents we will consider, using some heuristic [12.6]. Many heuristics are possible, and each one has its own bias. We choose to build a heuristic that produces graphs easy to visualize.

12.3 Visualization Constrains Discovery

12.3.1 Graph Structure Visualization

Graph structure visualization is a complex problem. We will consider two different approaches. The first one is to define an appearance criterion, and to try to maximize it. The main advantage of this approach is that we know which properties our structure has. The main drawback is that such metric computation of appearance criteria is costly. The second approach is to use construction methods that implicitly imply some nice visualization features. We have no control on the appearance criteria, but we do not need to compute them. To display our graph, we chose the second approach: a specific SOM (Self-Organizing Map) that produces the position of each node.

Self Organizing Maps (SOMs) were introduced by Kohonen ([12.7]). They consist of an unsupervised discrete approximation method of a distribution. They are mainly used to display data items to perform classification and clustering. Let us define a simple SOM where I and O are the input and output spaces. The SOM consists of an array of reference vectors $m_i \in I$, with the index i describing O. The reference vector array is usually two-dimensional.

Learning the distribution is performed using the following Algorithm 15:

Algorithm 15 SOM algorithm

1: **repeat**
2:     get $x(t) \in I$ according to the distribution
3:     $c \leftarrow \mathrm{argmin}_i \|x(t) - m_i\|$
4:     $m_i(t + 1) = m_i(t) + h_{c,i}(t)[x(t) - m_i(t)]$
5: **until** convergence

An SOM variant is characterized by $h_{c,i}(t)$, which is called the neighborhood function. It is necessary that $\liminf h_{c,i} = 0$ to ensure the convergence of the network (otherwise, $m_i(t)$ diverges.). Usually, $h_{c,i}$ decreases as m_c and m_i get closer in the array.

SOMs have been used widely on text data to perform unsupervised text classification. WebSOM [12.8] is a typical use of such a method. Using SOMs, it performs word clustering and text classification according to these clusters.

In this process, SOMs are applied directly on the text, and the final result is a text classifier [12.1, 12.2]: the SOM tends to approach the original text distribution. These kind of methods are well-suited for text classification, but are not intended for text analysis.

The approach we chose here is quite different: knowledge is extracted and represented as a network, and the SOM approach is used to display it. SOMs have already been successfully used to display graphs [12.9]. In our case, we use the graph instead of the array. I represents the visualization space, and O the index of the nodes. The neighborhood function is built using the distance between nodes in the graph (in hop count). The SOM then computes the position at each node in the visualization space. This method relies on the following SOM properties to ensure some appearance criteria:

Distribution approximation: A reference vector is a discrete approximation of the distribution. We choose a uniform distribution in the visualization space. Then, the reference vectors (edges) tend to be distributed uniformly in the visualization space.

Metric preservation: Input and output space metrics tend to be compatible. In other words, if two nodes are close in the graph, they will be close in the visualization space.

The distribution approximation property ensures that the distance between any two neighboring edges is approximately the same, and that edges tend to use all the available space (avoiding hole creation). Neighbors in the graph will likely be neighbors in the visualization space, thanks to the metric preservation property. Thus, it will reduce the number of crossing links in the graph. Using both properties, we have enough clues to believe that the resulting graphs will look nice. This method does not require computing any appearance metric; it relies only on the construction properties. However, we do not think that this method is efficient enough. In particular, it requires having the final graph to display. It is not possible to view a result before the full graph has been extracted.

SOM graph visualization methods suppose that the graph to display is given, and find optimal node positions. What happens if we do not freeze it? Let us make the assumption that in each iteration, the graph may be slightly different (i.e., the structure is dynamic). Unchanged parts of the graph are not directly affected by the modifications, and converge to better positions. As soon as no more changes occur, normal convergence starts. But the unchanged parts have already started the convergence process. Thus, it is reasonable to begin iterating before the exact structure is known. In the worst case, only the recently touched parts of the graph need a full iteration number. Let us make an additional assumption: the mean number of links grows. It is easier to display a sparse graph. At the beginning, a few iteration steps produce reasonable positions. As the graph grows, convergence is slower. It is reasonable to think that the initial step leads to an approximative structure, and is more effective than randomly placing nodes in the following steps.

The question now is how to implement the dynamic structure. The structure is not directly involved in Algorithm 15. Only the neighborhood function $h_{c,i}$ is affected by a structural change. Let us rewrite the neighborhood function to extract the structure-dependent part $s(c,i)$: $h_{c,i}(t) = s(c,i)h'_{c,i}(t)$. The term $h'_{c,i}(t)$ does not depend on the structure. We assume $s(c,i)$ is bounded (as a function applied on two finite variables, s and i, it has a finite number of possible values). Then the dynamic structure does not affect convergence.

Information on the structure and node position is useful for graph extraction as a heuristic. It gives a reasonable bias for partial parent selection. Use of distance as a heuristic results in a nice graph.

Algorithm 16 Visualization guided graph extraction

1: **repeat**
2:    get $x(t) \in [0,1] \times [0,1]$
3:    $j \leftarrow \text{argmin}_i \|x(t) - m_i\|$
4:    update C_j partial parent set list using m_j neighbor as candidates
5:    move C_j and its neighbor toward $x(t)$
6: **until** convergence

Algorithm 16 shows how to use node positions as a heuristic for possible parent selection, and how the current graph is considered when placing nodes in the visualization space.

12.3.2 Incremental and Dynamic Properties

One interesting property of our approach is the ability to provide a graphical result in any iteration step. Thus, the current state of extraction is always available to the user. The progress of extraction is displayed on the screen in real time. This allows direct parameter tuning. Although the parameters to be altered during the extraction process require some changes in Algorithm 16, and lead to Algorithm 17.

Algorithm 17 Visualization guided graph extraction (dynamic version)

1: **repeat**
2:    get $x(t) \in [0,1] \times [0,1]$
3:    $j \leftarrow \text{argmin}_i \|x(t) - m_i\|$
4:    update C_j partial parent sets score
5:    update C_j partial parent set list using C_j neighbor as candidates
6:    move C_j and its neighbor toward $x(t)$
7: **until** convergence

In fact, the only difference is that we introduce an additional step to update already computed scores (i.e., mutual entropy approximation for each partial parent set). Note that to avoid computing every score at each parameter modification, updating is done just before using the scores. It means that just after an update, most of the scores are erroneous. Of course, this does not affect computation a lot, as we correct them before use. Another key advantage of the dynamic result display is that the exploration of extracted knowledge is not delayed.

Data used to compute the graph score may be easily upgraded. New data usually becomes available during treatment. It is necessary to have an incremental property to handle data on the fly.

If new data arise, they are handled as follow:

1. Incoming textual data are parsed and added to the data repository.
2. Relevant statistical results are upgraded according to the new data.
3. Pairs of instance difference occurrence counts are updated.
4. If required, new nodes are added to the graph structure as concepts arise, and placed randomly in the visualization space.
5. The SOM neighborhood function rises (this is equivalent to temperature increase in simulated annealing methods) in order to absorb the new data faster.

No additional action is needed. As an iteration occurs, partial parent set scores are updated according to the new metric. The already computed positions are reused, avoiding restarting from scratch. Thus, the algorithm presented here is incremental [12.10].

12.4 Experimental Results

12.4.1 Results on Corpus

First, we apply our approach to discover knowledge and to visualize it from two different texts. In this first step, we do not consider any dynamic aspect of the data. We assume files are full and available. The main objective of these two tests is to determine what kind of knowledge it is possible to extract from the text, using unsupervised probabilistic network structure learning, and to display the extracted knowledge. Before looking at the graphs, we have two important remarks:

- The task is performed here without any prior domain knowledge, or any linguistic tool. It is almost language independent, and could have been realized on most text documents available on the Web. However, to interpret results, human beings require language and domain knowledge.
- The graphs we produce are quite large (we could generate even larger graphs). To present them in this article, we had to reduce their scale. The resulting graph may not be easy to explore. We tried to "zoom" in on some interesting parts, but it is easier to zoom in a digital/electronic version.

Moreover, it is not possible to show in a book format how a graph evolves and reacts to parameters tuning.

Let us now present the text data we focused on.

WWW-HOWTO. This how-to is included in many Linux distributions. It deals with WWW access and servicing. It is a typical technical document, using specific words. It would not be easy to analyze it using natural language text processing, because it would lack a specialized thesaurus. According to a word counting utility, its length is more than 14,000 words. This is enough to perform statistical analysis.

In Fig. 12.5, we show the result of our unsupervised analyze method. By following the links, it is possible to recognize familiar computer domain words, such as *ftp, downloaded, network, man, hosts, virtual, machine, etc.*. We can also distinguish some connected components such as *computer, domain, enable, restart, message, and send*. Thus, it is possible to extract reduced lexical fields automatically.

DNA Corpus. At the opposite end from the WWW-HOWTO file, stands the DNA Corpus. This corpus has been presented and studied in the International Journal of Corpus Linguistics [12.11]. Initial study was based on a linguistic collocational network, in order to extract the emerging use of new patterns of words in the scientific sublanguage associated with molecular biology applied to parasitic plants. Automated probabilistic network construction may be useful to linguists, even if the relation between collocational networks and probabilistic ones is not yet clear.

Fig. 12.1 shows the result we get for the corpus DNA. We have no particular knowledge of molecular biology, and choose from among the extracted patterns some understandable samples. It appears that some of the associations are lexical fields (e.g., deletion, evolutionary) while others (hybridize, intact) reveal the application domain. Even if links are different from collocational links, distance in the PN we produce (in hop counts) between collocate is low (e.g., *gene* and *plant* are collocate, and they are linked by *remains* .

deletion ← gene contains kb similar evolutionary substitutions
evolutionary ← gene plant involved necessary deletion remains phylogenetic organisms hemiparasite properly introns
hybridizes ← chloroplast intact genes codons families stop lack leaf relate altered
intact ← functional species genes templates hybridizes leaf relate barley altered
remains ← gene plant plastids living involved hemiparasite reduction
sites ← protein site gel size homology sequencing
study ← trnas trna sequence isolated single content tests

Characterization of the linguistic relation between a word and its parents requires an expert of the domain. Being able to get any relation without

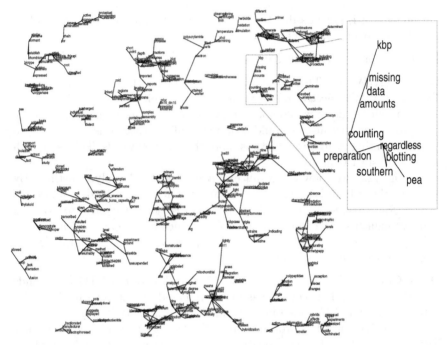

Fig. 12.1. DNA Corpus

knowing its nature was one of our requirements in this study. Interpretation of the relation is an other research topic, and a possible extension of this work.

12.4.2 Convergence Problem

In Fig. 12.2, we draw the score augmentation per 1,000 iteration steps on the WWW-HOWTO file. As an experimental result, we see that the curve may be bounded by a decreasing exponential. This result is in accordance with the convergence property of the presented algorithm: the score is strictly growing, because a partial parent set may be replaced only by a better partial parent set. The score is bounded because there is only a finite number of possible graphs. The growing and bounded score therefore converges, and difference between consecutive scores tends to 0, as suggested by Fig. 12.2. We ensure graph score convergence, but what about graph convergence? The score converges and depends only on a finite number of variables. For a given iteration step, the score is constant. Structure changes only if a better partial parent set has been found, but it is not possible if the score is constant. Then, no better partial parent set may be found, and structure does not evolve.

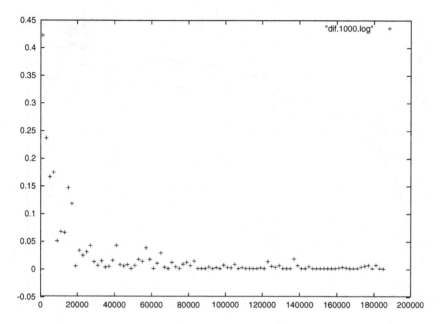

Fig. 12.2. score augmentation per 1,000 iteration steps (WWW-HOWTO)

12.4.3 Effect of Initial Position on Result

The objective of this data set (Mail-HOWTO) is to study the dependency between initial position and graph score after numerous (100,000) iteration steps. We run our program using the same data set and three different initial positions of nodes in the visualization space. Results are shown in Fig. 12.3, where each curve represents the score evolution for a given initialization of word positions in the visualization space. Although the behavior of individual curves is somewhat stochastic, it appears that the distance between them is always less than one unit. As the values are around 20, the relative error is 5%. Thus, the dependency of the curve evolution to the initial position is low.

12.4.4 Effect of New Data Insertion

In this experiment, the initial data set has been cut into four segments. Every 50,000 iteration steps, a new segment has been transmitted to the program. The resulting graph curve is given in Fig. 12.4. The most important effect of each new addition is to invalidate the previous evaluation function. This has two immediate effects:

− the current structure is evaluated using the new notation function. Thus, score may increase or decrease very fast. It decreases if the graph is less

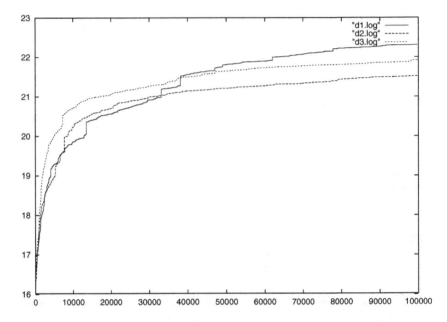

Fig. 12.3. score augmentation per 1000 iteration steps (WWW-HOWTO)

useful given the new data set, and increases if the new data fit the model already built. One interpretation of score decrease is that the method learned on a restricted set of data, and was too close to these data, and not able to generalize any more. This problem is called overfitting.
– the convergence process speed grows: unless new data set may be predicted using current model, the model quality may increase by taking new data into account.

In order to evaluate the change in the model before and after the addition of data, we split the WWW-HOWTO into two parts. We captured the graph before and after insertion of the second part. The result is shown in Figs. 12.5 and 12.6. No parameter was changed between these two snapshots. However, the second graph is sparser than the first one. This is a direct consequence of the graph metric change: on the one hand some links no more holds; on the other hand, some new links appear.

12.4.5 Influence of Data Order

This data set (Mail-HOWTO) has been built by cutting the original file (Mail-HOWTO) into tree segments: A_1, A_2, and A_3. We present these data to our program in two different orders:

order 1: segment A_1, A_2, and then A_3.

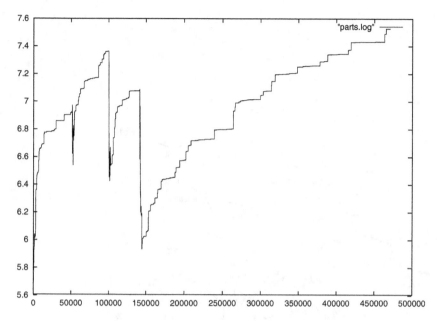

Fig. 12.4. 4-parts segmented text (WWW-HOWTO), one part every 50,000 iteration steps

order 2: segment A_3,A_2, and then A_1.

In both experiments, we sent a new data segment every 10,000 iteration steps. Figure 12.7 presents the score curves at each step for order 1 and order 2.

In the first period (0-10,000), the order 1 discovered graph score is lower than the score of the one extracted considering order 2. We deduce from this observation that segment A_1 is harder to model than segment A_3. In the second period, the curves associated with both order 1 and order 2 grow rapidly. Consequently, segment A_2 seems to be the easiest to model. Networks that were well fitted for segment A_1 and A_3 get even better when A_2 is added. A_2's structure certainly encompasses A_1 and A_3. In the third period, order 1 becomes better than order 2, but the difference between the two curves is roughly one unit. As we find that such a difference on curves occurs only in the initial node positions, we cannot say that, in this case, data order had a great influence on the final result.

12.5 Application: Web Content Mining

12.5.1 Introduction

Searching for information on the Web is becoming more and more complex, and there are many factors determining the success of processes seeking such

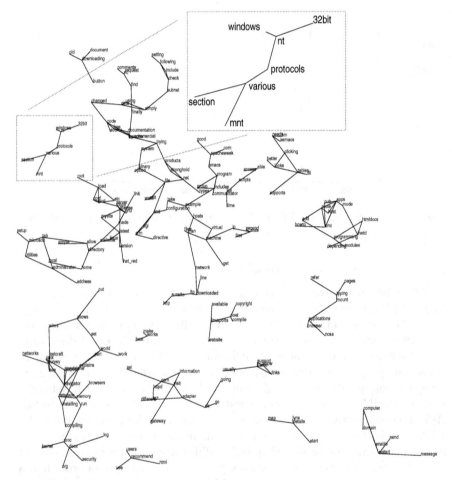

Fig. 12.5. WWW-HOWTO analysis (1)

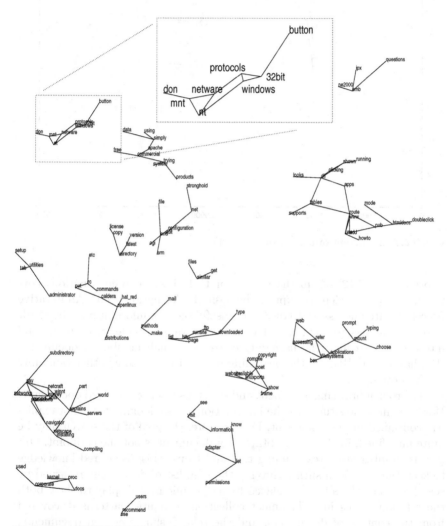

Fig. 12.6. WWW-HOWTO analysis (evolution)

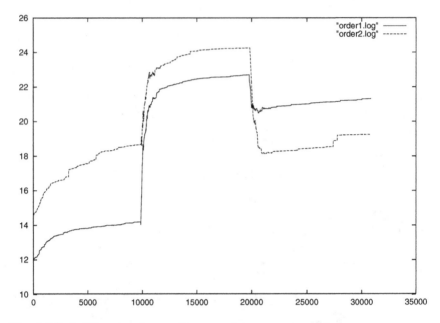

Fig. 12.7. 2 different orders on the same data

information [12.12]. From the user's point of view, in some situations keywords are not sufficient to express the desired information needs. An iterative process is often necessary to discover a useful piece of information. The result of search engines are long lists that are not always suitable for browsing and interactive searching: the user only consults a small number of documents that have their URLs at the beginning of the list, and some other randomly selected ones.

The role of browsing is to focus and select a subset of retrieved documents. However, in some situations the information we are looking for is not explicitly contained in one document. Furthermore, the goal of the search may be quite undefined; in this case, obtaining a document is not the endpoint. Our goal is to integrate a text mining tool and to use this discovered knowledge to keep the user from sifting through the long list of documents returned by search engines. This list is replaced by a dynamic map display to show both the relevant terms in a document collection, semantic patterns discovered from the content of documents, and the relationships between documents. Such tools have been used for Web data; see the Memex project [12.13] for more information on this aspect. In conclusion, we can say that the third search engine generation will certainly be smart integrated engines that enrich data Web crawling and searching features with a collection of interactive, visual, mining, and more generally intelligent functions to make both the search and exploration processes more dynamic, fruitful, understandable, and collaborative.

This work contributes to achieve this dream by integrating different components to meet the following requirements:

- Integration of incremental and focused dynamic crawling with meta-search;
- Freeing the user from sifting through the long list of documents returned by the search engines;
- Extracting comprehensive patterns and useful knowledge from the documents;
- Visual-based support to browsing dynamic document collections;
- Multi-level querying.

12.5.2 General Architecture

We present here a dynamic search and exploration system to help find useful information and to understand events and phenomena. To achieve this goal, we encapsulate the basic architecture of search engines and offer a set of methods to support the following main functions:

- Focused Crawling: we use one or several search engines, for instance AltaVista, Yahoo, Google, etc. The set of URLs resulting from these engines is the starting point of our advanced search and exploration. Firstly, priority queues are used to filter all discovered URLs; the most promising ones are crawled while the less important ones are discarded. This filtering is a heuristic-based method to avoid the crawling of irrelevant directories. Secondly, a term-based approach is used to create a vector representation of the crawled documents. This representation is used in the pruning of useless documents. The documents kept after this filter/focus process are the input for the miner to extract patterns.
- Probabilistic network-based mining: the mining function is useful, as it allows the acquisition of new knowledge that can support the user's understanding, decision, and query refinement. We have integrated a method based on probabilistic networks that leads to the identification of relations between the keywords of a collection of documents. The discovered relations are maintained dynamically as new documents arrive. Our discovery method is closely dependent on the visualization one, and vice-versa.
- Dynamic visualization: a map-based approach is adopted to visualize keywords by dynamically changing their positions in the map according to the new search and/or crawled information. Both discovered relations and documents are visualized on the same map. The position of objects on the map is useful information that is iteratively exploited by the miner. The results of the miner, i.e., discovered knowledge, is then used to adjust the positions of words.

– Multi-query processing: generally, a user's query is processed by the search engine which returns the list of URLs related to a query. This approach is considered here as one-level querying. In contrast to the previous situation, where the user's query is processed only by a search engine, multi-level querying allows users to send queries to different types of engines, which support other functions like crawling, mining, visualizing, etc.

The system presented here integrates the following main functions, i.e., dynamic crawling, mining, visualization, and multi-level querying. The user has the ability to formulate his query using only words, and to process it by the engine that supports the appropriate function: searching, mining, or visualization. When the query is for a search, the user can choose one or more search engines. The following section gives more details on the introduced functions.

12.5.3 Related Work

The methods presented here overlaps different work developed by different communities. Thus, it is important to discuss both the common and different aspects. The first dimension of our work is related to dynamic search and incremental crawling. Coupling dynamic search with conventional static search is an idea has already been defended by different authors [12.14] to complete the ordered search list with possibly new and/or dynamically generated documents. Of course, we share the motivations of this augmenting method; however, we consider a deeper integration between crawling and searching. In fact, crawled pages are analyzed, and knowledge is expressed as a probabilistic network simultaneously learned. This induced knowledge can support user refinements of the queries and accordingly adjust the crawling strategy. Another aspect is efficient crawling. Several papers investigate how to construct efficient crawling methods. Our method overlaps work developed by Cho and Molina. [12.15], as we use a priority queue and two main modules for ranking URLs and updating information. One main contribution is the integration of probabilistic discovery with a map-based visualization method. It is well known that the learning cost of probabilistic networks, such as Bayesian networks, is very high. This cost becomes acute when the learning data is large, which is the case with textual data. Of course, our method is inspired by work developed to learn Bayesian networks from data. The main differences are that our graph is not necessarily acyclic, we only approximate the node, and we use a stochastic method. Our work overlaps with [12.6], which also exploits the idea of restricting the parent of a node to reduce the search cost. Considering the visualization problem, we have chosen a map-based approach because of its interest in managing dynamic aspects of visualization. [12.16] experimentally studies the interest in map-based approaches for search and visualization. Our method overlaps with this work, and more generally, with the self-organizing map algorithm. The main difference with the well-known method based on Kohonen's approach [12.7] is related to the notion of topological neighborhood: neighbors are always the same (static), which is not the case in our method, where the set of neighbors of a node can change (dynamic). Our work proposes a new paradigm where discovery and visualization methods are dependent on each other.

12.5.4 Dynamic Search and Crawling

Search engines are designed to process user's queries that are described in a specific language and to return a long list of URLs. A query is generally reduced to a single or a small set of words and each URL in the returned list is referred to as a hit. In order to return such a list in reasonable time, search engines generally use a local index and/or a local collection of Web pages. Such indexes and collections are built by using a crawler to automatically create an index and collect Web pages that are stored in huge databases. However, this approach faces different problems essentially related to:

- The huge size of the Web [12.17] and the storage of its complete image;
- The rate at which pages change is the second crucial problem to deal with. Predictive Web changing models [12.18] are at the basis of the major strategies for re-indexing the expected modified pages;
- The Web is updated by millions of users in an uncontrolled way. Important existing pages may be deleted and links to authority pages may be broken, which increases users, frustration during Web browsing.

The objective of dynamic searching and crawling is to resolve these problems. The difference in dynamic searching, when compared to static searching, is that it does not limit its results to those found in an outdated index, but explores the Web in its current state.

Dynamic searching is limited by the efficiency cost, essentially the number of irrelevant downloaded pages of a crawler that retrieves and analyzes pages on the fly. This type of crawling is often referred to as focused crawling in that it is guided by a given query or topic of interest. Building such efficient crawlers is a challenge for dynamic Web exploration [12.15]. The main problem is to find different heuristics allowing to limit the number of downloads of irrelevant documents without losing too many pertinent ones.

A focused crawler can be characterized by the strategy it uses to determine which pages it will or will not download. Defining a strategy consists of setting up different filters at different stages of the crawling process. A crawling process can basically be split into two main subparts: (1) downloading a page and (2) extracting the links of a page. These two components communicate by respectively sending to each other the newly downloaded pages and the last extracted links. In this scheme, the strategy can be defined in terms of filters place before and after each of these parts. Three main types of filters can be distinguished:

- structural filters: based on the form of the URL to download
- content filters: used to filter the links of an off-topic page
- contextual filters: used to filter links based on the context in which they were found

The work presented here only makes use of two types of filters, which are sufficient to evaluate the idea we put forth. The first, a contextual filter, uses the nearness of a link to prune links related to servers which already gave bad links The second, a content filter, uses statistical methods to calculate a representative document vector and rejects any document whose vector does not contain one of the keywords. Of course, only the main words are selected according to both document-dependent and document collection-dependent measures, based on term importance as used in automatic text analysis [12.19].

Putting together static and dynamic searching is an interesting approach, as it avoids the static and incomplete information stored in the databases of search engines while still having a set of *a priori* relevant documents to initiate the dynamic crawling. The usefulness of this hybrid approach is detailed in [12.14]. In this case, the seed URLs used to start crawling are the result of a static search: the list of URLs returned by a query on a given subject to a classic search engine.

12.5.5 Integration and Implementation Issues

We have described three main components, dedicated to crawling, mining, and visualization. One major issue is improving the efficiency of the methods supported by each component. For this reason we have used different filters to avoid the analysis of pages not useful. Furthermore, the mining and visualization algorithms are closely related. In fact, the miner uses the visualization space to focus the research while the visualizer takes into account the new extracted knowledge and adapts the dynamic map display.

Unfortunately, the efficiency of the general process depends on different operations:

- Crawling new pages, only the most promising ones being analyzed.
- Update of the graph structure.
- Update of the visualization.

We have introduced different optimization methods to tackle the problems of incrementality (dealing with new documents), score computation, and local search:

- Partial result caching
  When searching for the best parents for a given node (local optimization), score evaluation is the most frequent and costly operation. We need efficient evaluation methods. In particular, full data analysis on each score

evaluation is not a realistic approach for a large data set. This is the reason we approximate the entropy. We now explain how it helps avoiding constant rescan of data.

Difference counts and per-attribute statistics carry all the information required to compute. These partial results are computed only once. This way, no access to the data is done when computing score. If using exact entropy, full joint statistics could not be computed and cached, due to their size. Our cached results require more complex treatment than joint statistics, but they are computed once and for all. When computing score, no original data access is done.

- Stochastic approach

Exhaustive instance pair difference evaluation is a costly task. In order to reduce computational cost, we chose to perform sampling on instance pairs. Information in instance pairs is highly redundant. Thus, even if only a small part of the instance pairs is analyzed, no significant loss of information is observed. Moreover, we rely on a robust score (MML-based). An error that could invalidate a functional dependency only affects a parent set score. It as been observed that given i instances, $o(i)$ pairs are sufficient. It is easy to understand, given differences between instance i and $i + 1$ and the first instance, that anybody could find original instance values. Thus, i differences between instances carry roughly the same information as the data does. This way, we get a fast data preprocessing step to compute MML-based score approximations.

- Local search

Searching our dependency graph is a local task. For each node, we search the best parent sets according to our score. The parents of the node are given by the union of the n best parent sets. Exhaustive search among all possible parent sets is not affordable, for their number is exponential (in terms of number of attributes in the data set). Pruning methods efficiently reduce the search space, using the MML score property; due to structural cost, maximal size of promising parents set is bounded. Moreover, as soon as n good parents have been found, this bound diminishes, because the score to beat (initially 0) increases. But even these methods are not enough for high dimensional data (and textual data are very high dimensional). To limit search space, we need potent heuristics. In the next section, we will present such a heuristic.

12.5.6 Experimental Results

An important task is the global multi-level structure exploration. We consider about 440 documents crawled and analyzed. All the information contained is analyzed, and a probabilistic network is learned by structuring the information contained in all crawled documents. Of course, the graph shown in the next screen changes dynamically, and the user observes these evolutions. As we have said before, the dynamic modification of this structure is guided by two criteria. The first one is the search for the graph optimizing a scoring function, whereas the second is related to its visualization. The graph with the best appearance is kept. Multi-level explorations can be achieved by our proposed system. In fact, the user may explore the global structure of the whole graph or focus his attention only on a local part of the structure represented by a connected component.

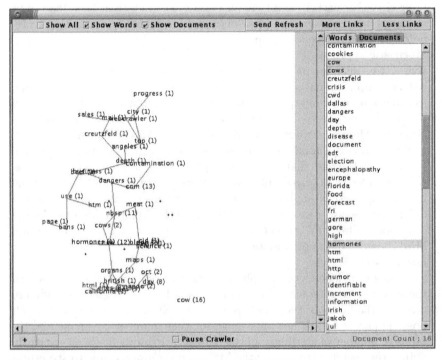

Fig. 12.8. Focused explorations of the mad cow problem

Figure 12.8 shows an example resulting from the query "mad cow". The list of the words used to construct the graph is given in the right part of the screen. Thus, the user can focus his exploration considering only a set of the extracted words. Figure 12.8 shows an example where we have selected the words "cow", "cow", and "hormones". Only connected components containing at least one word from the selected ones are drawn. Only one component is drawn in Fig. 12.8 and involves other words like "British", "organs", "nbsp", "danger", and "contamination". This information and the structure can play an important role in querying reformulation and textual data analysis [12.20].

12.6 Conclusions

In this chapter, we have presented an original approach to deal with text mining in document collection. The resulting algorithms and methods have be implemented and produce dynamic and incremental graph structure extraction and visualization. These properties have been validated on two different and specific domains. Using visualization data (i.e., node position) for guiding the extraction process has proved fruitful, and could be applied to other domains. Mixing mining and visualization is valuable for both aspects, as exploration needs heuristics, and dynamic display quality benefits from these particular heuristics. This work can be extended to handle the problem of Web mining. For example, mining search results (such as those provided by search engines like Google, Lycos, etc.) provide specific domain knowledge, which is useful to reformulate the query.

Proofs We want to establish that Eq. (12.4) holds. We have

$$P_e = \sum_Y P(Y) \min(P(X = 0|Y), P(X = 1|Y))) . \tag{12.6}$$

We use $\min(p, 1 - p) = (1 - |2p - 1|)/2$ and $2p(1 - p) = q \Leftrightarrow p = 1/2 \pm 1/2\sqrt{1 - 2q}$.

$P_e = 1/2 \sum_Y P(Y)(1 - |2P(X = 0|Y) - 1|)$, or

$P_e = 1/2 \sum_Y P(Y)(1 - |1 \pm \sqrt{1 - 22P(X = 0|Y)P(X = 1|Y)} - 1|)$, or

$P_e = 1/2 \sum_Y P(Y)(1 - \sqrt{1 - 2P(X \neq |Y)})$, or

$P_e = 1/2 - 1/2 \sum_Y P(Y)\sqrt{2P(X_= |Y) - 1)}$.

References

12.1 A. McCallum, K. Nigam: A comparison of event models for naive bayes text classification (1998)

12.2 K. Nigam, A. McCallum, S. Thrun, T. Mitchell: Text classification from labeled and unlabeled documents using em (1999)

12.3 G. Escudero, L. arquez, G. Rigau: Naive bayes and exemplar-based approaches to word sense disambiguation revisited (2000)

12.4 L.E. Sucar, J. Ruiz-Suárez: Learning structure from data and its application to ozone prediction. Applied Intelligence, 7, 327-338 (1997)

12.5 V. Dubois, M. Quafafou: Discovering graph structures in high dimensional spaces. In: Data Mining II (2000)

12.6 N. Friedman, I. Nachman, D. Peter: Learning bayesian network structure from massive datasets: The sparse candidate algorithm. In: *Proceedings of the Fifteenth Annual Conference on Uncertainty in Artificial Intelligence (UAI-99)* (San Francisco, CA, Morgan Kaufmann Publishers 1999) pp. 196-205

12.7 T. Kohonen: *Self-Organizing Maps* (Springer-Verlag 1994)

12.8 T. Honkela, S. Kaski, K. Lagus, T. Kohonen: Websom- self-organizing maps of document collections (1997)

12.9 B. Meyer: Self-organizing graphs-a neural network perspective of graph layout (1998)

12.10 V. Dubois, M. Quafafou: Incremental and dynamic text mining. In: M.S. Hacid, Z.W. Ras, D.A. Zighed, Y. Kodratoff (eds.), *Foundations of Intelligent Systems, 13th International Symposium, ISMIS 2002, Lyon, France, June 27-29, 2002, Proceedings. Volume 2366 of Lecture Notes in Computer Science* (Springer, 2002) pp. 265-273

12.11 G.C. Williams: Collocational networks: Interlocking patterns of lexis in a corpus of plant biology research articles. International Journal of Corpus Linguistic, 3, 151-171 (1998)

12.12 D. Hawking, N. Craswell, D. Harman: Results and challenges in web search evaluation. In: www8, Toronto (1999)

12.13 S. Chakrabarti, S. Srivastava, M. Subramanyam, M. Tiwari: Using memex to archive and mine community web browsing experience. In: www9 (2000)

12.14 I.B. Schaul, M. Herscovici, M. Jacovi, Y.S. Maarek, D. Pelleg, M. Shtalhaim, V. Soroka, S. Ur: Adding support for dynamic and focused search with fetuccino. In: WWW8 (1999)

12.15 J. Cho, H.G. Molina: The evolution of the web and implications for an incremental crawler. In: *Proceedings of 26th International Conference on Very Large Databases, VLDB, 2000*

12.16 L. X.: Searching and browsing on map displays. In: *Proceedings of ASIS'95 (1995)*

12.17 A. Brodera, R. Kumar, F. Maghoul, P. Raghavan, S. Rajagopalan, A.T.R. Stata, J. Wiener: Graph structure in the web. In: WWW9. (2000)

12.18 B.E. Brewington, G. Cybenko: How dynamic is the web? In: WWW9 (2000)

12.19 G. Salton, C.S. Yang, C.T. Yu: A theory of term importance in automatic text analysis. Journal of the American Society for Information Science, 26 (1975)

12.20 V. Dubois, M. Quafafou, B. Habegger: Mining crawled data and visualizing discovered knowledge. In: N. Zhong, Y.Y. Yao, J. Liu, S. Ohsuga (eds.), *Web Intelligence: Research and Development, First Asia-Pacific Conference, WI 2001, Maebashi City, Japan, October 23-26, 2001, Proceedings Volume 2198 of Lecture Notes in Computer Science* (Springer, 2001) pp. 493-497

Part III

Emerging Agent Technology

13. Agent Networks: Topological and Clustering Characterization

Xiaolong Jin and Jiming Liu

Hong Kong Baptist University, Hong Kong

Abstract

In this chapter, we will introduce the notion of an agent network and its topology. Further, we will show how to model the computational complexity of distributed problem solving by means of studying the topological and clustering characteristics of agent networks formed by multi-agent systems. Specifically, we will examine two multi-agent-based representations for satisfiability problems (SATs), i.e., agents representing clauses and agents representing variables, and experimentally characterize the topologies of the resulting agent networks. Based on the experimental results, we will study how different topologies reflect the computational complexities of agent networks, and further discuss how to balance complexities in intra- and inter-agent computations. In the above studies, random agent networks are used to compare with real-world agent networks. Particularly, the clustering characteristic is one of the properties used to differentiate between real-world agent networks and random agent networks. In order to provide a quantitative measurement for characterizing random agent networks, we will study their clustering characteristics, focusing on the average value and lower bounds of clustering coefficients.

13.1 Introduction

13.1.1 Small World Phenomena

In [13.17], Milgram firstly proposed the notion of *small world*. Later, Watts and Strogatz mathematically formulated a small world topology based on *characteristic path length* and *clustering coefficient* [13.24]. In Sect. 13.2.1, we will give a formal definition of small world topology. Small world phenomena have been extensively found in natural systems (e.g., human society [13.17], food Web in ecology [13.18]) as well as in man-made systems (e.g., the World Wide Web [13.2]). Walsh observed such phenomena in search problems [13.23], such as *graph coloring*, *time tabling*, and *quasigroup problems*. He further experimentally proved that a small world topology would make a search process very difficult.

13.1.2 Agent Networks

Definition 13.1.1. *An **agent network** is a virtual graph corresponding to a multi-agent system, where vertices are agents, and edges, also called links,* *are the implicit or explicit relationships of cooperation, coordination, or competition among agents*[1].

The topology of an agent network can be defined as follows:

Definition 13.1.2. *The* **topology** *of an agent network is its geometric property, which reflects the connectivity of the network.*

Multi-agent systems have been widely used in distributed problem solving tasks, such as *constraint satisfaction problems* (CSPs) [13.14, 13.25] and *satisfiability problems* (SATs) [13.7, 13.10, 13.13]. In order to solve a given task, agents in a multi-agent system will implicitly or explicitly form an agent network. As an agent usually cooperates, coordinates, or competes with some agents, but not all agents, the resulting agent network will not be fully connected. Hence, a straightforward question is: what is the topology of the network formed by agents in a multi-agent system? Answering this question will help us understand the performance of the system. It will also guide us in designing more reasonable methods for solving problems.

13.1.3 Satisfiability Problems

Satisfiability problems (SATs) have attracted much attention from the communities of artificial intelligence as well as other related fields of computer science. This is mainly because SATs are NP-complete problems with the simplest form. Furthermore, researchers have found that many practical problems can be formulated into SATs.

A SAT is to determine whether or not a propositional formula has a solution, i.e., an assignment to its variables that make all clauses satisfied. A propositional formula (i.e., a SAT) F is usually represented in conjunctive normal form (CNF):

$$F = \wedge_{i=1}^{m} Cl_i,$$

where Cl_i is a disjunction of several literals, namely, variables or their negations.

Many methods have been proposed to solve SATs. The conventional methods can be categorized into two types. The first type is based on the DPLL procedure [13.4, 13.5], for example, GRASP [13.15], REL_SAT [13.11], SATZ [13.12], and TABLEAU [13.3]. The second type is based on *local search* [13.6, 13.22], such as GSAT [13.22], WalkSAT [13.16, 13.21], SDF [13.19, 13.20], and UnitWalk [13.8, 13.9]. All these conventional methods work sequentially. Recently, two distributed multi-agent-based approaches have been proposed to solve SATs, namely, MASSAT [13.10] and Multi-DB [13.7]. In both distributed approaches, the authors employed a multi-agent system where agents were used to represent one or more variables.

---

[1] The word 'virtual' implies that there may not exist physical links among agents in a multi-agent system.

13.1.4 Problem Statements

MAASAT [13.10] is an autonomy oriented satisfiability solving approach. When solving a SAT, MASSAT implicitly creates an agent network. In this chapter, without loss of generality, we will take MASSAT as a case study to demonstrate how to characterize agent networks and their corresponding complexities. Specifically, we will address the following questions:

1. If agents in a multi-agent system are used to represent clauses in a SAT problem, what topology can be observed in the resulting agent network?
2. In MASSAT, agents are employed to represent variables of a SAT problem. Does an agent network formed in MASSAT show a small world topology?
3. If the topology of an agent network varies according to specific problem representations, how does it reflect the computational complexity of the given SAT problem?

To study the topologies of agent networks, random agent networks will be used to compare with other real-world agent networks. In particular, the clustering coefficient [13.24] will be used as a property to characterize random agent networks and other agent networks. In order to offer a quantitative means for characterizing random agent networks, we will also study the clustering characteristics of random agent networks. In particular, we will address the average value and lower bounds for clustering coefficients.

13.1.5 Organization of This Chapter

In this section, we introduced some basic notions, i.e., small world phenomena, agent networks, and SATs, which will be used throughout later discussions. Further, we stated the problems to be addressed. The remainder of this chapter is organized as follows. Sect. 13.2 examines two specific distributed multi-agent-based representations for SATs, and characterizes the topologies of the resulting agent networks. We will discuss the effects of different topologies on computational complexity, and balanced computational complexities in intra- and inter-agent organizations in Sect. 13.3. In Sects. 13.4 and 13.5, we will study the clustering characteristics of random agent networks. Sect. 13.6 concludes this chapter.

13.2 Topologies of Agent Networks

When using a multi-agent system to solve a SAT, there exist different representations. What we are interested in here is the topologies of agent networks under different representations. In this section, in order to study the computational complexities of multi-agents systems, we will examine two typical distributed agent representations: (1) agents representing clauses and (2) agents representing variables.

13.2.1 Representation I

In this subsection, we will experimentally study the multi-agent representation of a SAT problem where agents are used to represent clauses. Based on the experimental results, we attempt to answer the first question in Sect. 13.1.4.

Clause-based representation and the corresponding agent network.
In representation I, agents represent clauses of a SAT problem. A clause is a disjunction of several literals, i.e., variables or their negations. To satisfy a clause is to assign values to variables in the clause such that at least one literal is true. Since the agent is employed to represent a clause, the agent should make at least one literal true in order to satisfy this clause. Because a variable can appear in multiple clauses simultaneously, the agents that have common variables should cooperate to assign their variables with consistent values as well as to satisfy their respective clauses. In this case, an agent acts as a vertex of an agent network. If two agents have a common variable, there will exist a link between the corresponding two vertices.

Figure 13.1 presents a schematic diagram. In this figure, agents a_1 and a_2 have the same variable A, and agents a_1 and a_3 have the same variable B; hence there are two links: between agents a_1 and a_2 and between agents a_1 and a_3. Because agents a_2 and a_3 have no identical variable, there is no link between them.

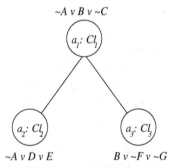

Fig. 13.1. A schematic diagram to encode three clauses into an agent network, where each vertex denotes an agent, and each line denotes a link between two corresponding agents. $a_i : Cl_i$ means agent a_i represents clause Cl_i

Experimental testing. Based on the above representational encoding scheme, we conducted some experiments using benchmark SAT problems: Uniform-3-SAT problems and Flat Graph Coloring problems [13.1]. We randomly selected 10% of the instances from each testset. As two important properties of an agent network, we calculated for each testset its average *characteristic path length*, L_G, and *clustering coefficient*, C_G.

Table 13.1 shows our experimental results. Here, characteristic path length L_G and clustering coefficient C_G of agent network G are defined as follows [13.24]:

Given an agent network $G = \langle A, B \rangle$, where $A = \{a_1, a_2, \cdots, a_n\}$, i.e., the set of agents and $B = \{b_1, b_2, \cdots, b_m\}$, i.e., the set of links between agents in A. Then,

1. *Charactertistic path length:*

$$L_G = \frac{2}{n \cdot (n-1)} \sum_{i,j \in \{1,\cdots,n\}} d_{i,j}, \tag{13.1}$$

where $d_{i,j}$ is the shortest distance between two agents a_i and a_j.

2. *Clustering coefficient:*

$$C_G = \frac{1}{n} \sum_{i=1}^{n} c_{a_i}, \tag{13.2}$$

where c_{a_i} is called the *clustering ratio* (also called *clustering* in short) of agent a_i. Assume $d(a_i)$ is the degree of a_i (i.e., the number of neighboring agents of a_i), and $e(a_i)$ is the number of existing links between agent a_i and its neighbors. Therefore,

$$c_{a_i} = \frac{e(a_i)}{\frac{(d(a_i)+1)\cdot d(a_i)}{2}} = \frac{2 \cdot e(a_i)}{(d(a_i)+1) \cdot d(a_i)} \tag{13.3}$$

In general, L_G represents a global property of agent network G that indicates the connectivity of G. C_G, on the other hand, represents a local property that reflects the average connectivity of cliques in G. In essence, C_G denotes the possibility that two agents are neighbors that have a common neighboring agent.

Small world topology. Watts and Strogatz defined graph G with n vertices and m edges has a small world topology if and only if

$$L_G \approx L_{random} \quad \text{and} \quad C_G \gg C_{random}, \tag{13.4}$$

where L_{random} and C_{random} are the characteristic path length and clustering coefficient of random graphs that also have n vertices and m edges, respectively [13.24].

Note that in the above definition, graph G is required to be connected. If not, L_G will not make sense. In this sense, it requires $k \gg ln(n)$, where $k = \frac{2m}{n}$ is the average degree of vertices in graph G. Note also that there are a large number of random graphs with n vertices and m edges. Furthermore, their characteristic path lengths L_{random} and clustering coefficients C_{random} are usually different, therefore the L_{random} and C_{random} used in determining the small world topology of a graph should be the average values.

In Sect. 13.5, we will present a method for calculating the average clustering coefficient. Generally speaking, connected random agent networks with n agents and m links have the average clustering coefficient as follows:

Table 13.1. Experimental results of representation I on benchmark SATs: Uniform-3-SAT and Flat Graph Coloring. In these experiments, we randomly selected 10% of the instances from each testset and calculated the average L_G, C_G, and μ values on the selected instances. For each instance, we generated 10 random agent networks with the same number of agents and links. We calculated the average L_G of these random agent networks in order to compare with the corresponding instance. C_G was calculated according to Eq. (13.5) (see Sect. 13.2.1)

TestSet	Agents	Links	L_G	L_{random}	C_G	C_{random}	μ
Uf50	218	4071.7	1.8361	1.8293	0.4443	0.2153	2.071
Uf75	325	6124.6	1.9231	1.8950	0.4173	0.1620	2.614
Uf100	430	8112.5	2.0065	1.9453	0.4066	0.1351	3.106
Uf125	538	10240.4	2.0769	1.9916	0.3992	0.1184	3.514
Uf150	645	12275.8	2.1460	2.0401	0.3931	0.1073	3.854
Uf175	753	14366	2.2051	2.0861	0.3894	0.0992	4.148
Uf200	860	16433.8	2.2598	2.1300	0.3873	0.0932	4.408
Uf225	960	18193.8	2.3089	2.1754	0.3842	0.0889	4.587
Uf250	1065	20168.2	2.3522	2.2150	0.3826	0.0852	4.768
Flat30	300	1882.8	3.1955	2.5328	0.5759	0.1834	3.962
Flat50	545	3825.75	3.4331	2.6699	0.5671	0.1554	4.694
Flat75	840	6092.4	3.6729	2.7913	0.5642	0.1441	5.154
Flat100	1117	8071.2	3.8749	2.8929	0.5639	0.1407	5.368
Flat125	1403	10257.6	4.0185	2.9651	0.5648	0.1371	5.583
Flat150	1680	12159.75	4.1551	3.0454	0.5632	0.1367	5.619

$$C_G = \frac{2n^2 \cdot (n-1) + 2m \cdot (2m-n)}{n \cdot (n-1) \cdot (2m+n)}. \tag{13.5}$$

The above definition of a small world is qualitative. In order to measure the "small worldiness" of a graph, Walsh provides a quantitative measurement, *proximity ratio* μ [13.23]:

$$\mu = \frac{\frac{C_G}{L_G}}{\frac{C_{random}}{L_{random}}} = \frac{C_G \cdot L_{random}}{C_{random} \cdot L_G}. \tag{13.6}$$

A small world topology requires $\mu \gg 1$. The larger the μ, the more "small worldy" the graph (i.e., the graph has more clusters).

To test if there exist small world topologies, we calculated the average L_{random} and C_{random} as well as μ for each testset (see Table 13.1). For each instance in a testset, we generated 10 random agent networks with the same size and calculated their C_{random} and L_{random}, where C_{random} is calculated using Eq. (13.5). The results show that in all testsets, $L_G \approx L_{random}$, $C_G \gg C_{random}$, and $\mu > 2$. This means with representation I, the agent networks show small world topologies.

13.2.2 Representation II

The previous subsection answers the first question in Sect. 13.1.4, i.e., an agent network created with representation I has a small world topology. Next, we will try to answer the second question.

MASSAT. Given a SAT problem, MASSAT divides all variables into groups. Each group contains one or more variables. Then, MASSAT uses an agent to represent each variable group. The Cartesian product of a variable group serves as the local space, where the corresponding agent can reside and move around. An element in the product corresponds to a position where an agent can stay. Each element indicates a value combination of the variables. Therefore, if an agent stays at a certain position, it means the variables are assigned the values corresponding to this position.

After grouping, MASSAT randomly places all agents over their local spaces. From now on, agents will cooperatively explore the positions in their respective local spaces based on a certain evaluation criterion, e.g., the total number of satisfied clauses.

In Multi-DB [13.7], an agent cooperates with its neighbors by sending and receiving *Ok?* or *improve* messages. In MASSAT, there is no explicit agent-to-agent communication. Agents cooperate through sharing their position information. That is, there is a shared blackboard where an agent can write its own position information and read that of other agents.

By having the position information of other agents, an agent knows the values of variables represented by other agents. The agent can then evaluate its current position based on the given criterion. At each step, the agent will probabilistically select a 'better' position or one of the 'best' positions to move to. The agent may also randomly select a position. No matter what evaluation criterion is employed, the final goal is to satisfy all clauses. If all clauses are satisfied, the given SAT is solved and MASSAT stops. For more details about MASSAT, readers are referred to [13.10].

Variable-based representation and the corresponding agent network. In SATs, each clause normally contains several literals (i.e., variables or their negations). In order to satisfy a clause, the related variables should be assigned compatible values. In other words, they should guarantee at least one literal is true. In this sense, there are constraints (i.e., clauses) among variables. Since MASSAT represents variable groups with agents, the constraints among variables are implicitly transformed into constraints among agents. In order to satisfy a constraint (i.e., a clause), agents that represent the variables in the constraint will restrain each other. If we assume each agent represents only one variable, the agent network can be encoded as follows:

A vertex denotes an agent. There exists a link between two agents if and only if two corresponding variables appear in a certain clause simultaneously.

Figure 13.2 provides a schematic diagram where variables A, B, and C are in the same clause $\neg A \vee B \vee \neg C$. There is a link between each pair of agents that represent the variables.

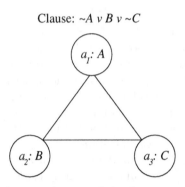

Clause: ~A v B v ~C

Fig. 13.2. A schematic diagram that encodes a clause into an agent network, where each agent denotes one variable. $a_i : X$ means agent a_i represents variable X

Experimental testing. Based on the above encoding scheme, we encoded some benchmark SAT problems, Uniform-3-SAT problems, and Flat Graph Coloring problems [13.1] into agent networks. By calculating their characteristic path lengths and clustering coefficients, we examined if there exist small world topologies. Table 13.2 presents our experimental results.

Non-small-world topology. *From Table 13.2, we note that for all testsets of Uniform-3-SAT, $L_G \approx L_{random}$, $C_G \approx C_{random}$, and $1.1 < \mu < 1.3$. This indicates small world topologies do not exist. For all testsets of Flat Graph Coloring, $L_G \not\approx L_{random}$, $C_G \not\approx C_{random}$, and μ is around 1.65. They do not show small world topologies either. Thus, we can assert that with representation II, there is no small world topology in an agent network.*

In the above encoding scheme, each agent represents only one variable. An agent can also represent several variables. In this case, two agents will be connected by a link if and only if two variables, represented respectively by these agents, appear in a common clause. Further, the agent network in this case is equivalent to the network that is derived by collapsing some of the original agents into a single one. Note that each original agent represents only one variable.

Obviously, the former network is denser than the latter one. A small world topology normally exists in a connected but 'sparse' network. Since the latter agent network does not have a small world topology, the former will not either. This point has been demonstrated in our other experiments. In our results, μ is still less than 2.0. This indicates there does not exist a small world topology.

Table 13.2. Experimental results of representation II on benchmark SATs: Uniform-3-SAT and Flat Graph Coloring. In these experiments, we randomly selected 10% of the instances from each testset, and calculated the average L_G, C_G, and μ values on the selected instances. For each instance, we generated 10 random agent networks with the same number of agents and links. We calculated the average L_G of these random agent networks in order to compare them with the corresponding instances. C_G was calculated using Eq. (13.5)

TestSet	Agents	Links	L_G	L_{random}	C_G	C_{random}	μ
Uf50	50	506.8	1.6931	1.5865	0.5010	0.4688	1.140
Uf75	75	831.2	1.7575	1.7016	0.4025	0.3600	1.154
Uf100	100	1130.6	1.8212	1.7760	0.3348	0.2937	1.168
Uf125	125	1457.2	1.8615	1.8221	0.2968	0.2548	1.190
Uf150	150	1780.4	1.8999	1.8598	0.2712	0.2272	1.219
Uf175	175	2095	1.9394	1.8946	0.2501	0.2067	1.238
Uf200	200	2414.4	1.9684	1.9253	0.2370	0.1912	1.267
Uf225	225	2727.4	2.0003	1.9568	0.2234	0.1788	1.276
Uf250	250	3036.6	2.0314	1.9866	0.2143	0.1689	1.297
Flat30	90	270	3.6618	2.6774	0.3874	0.3338	1.585
Flat50	150	495	3.9350	2.8358	0.3500	0.2957	1.641
Flat75	225	765	4.1855	3.0247	0.3369	0.2789	1.671
Flat100	300	1017	4.3747	3.1787	0.3303	0.2739	1.659
Flat125	375	1278	4.5321	3.2892	0.3318	0.2694	1.697
Flat150	450	1530	4.6465	3.3897	0.3274	0.2676	1.676
Flat175	525	1776	4.7350	3.4791	0.3286	0.2671	1.674
Flat200	600	2037	4.8491	3.5386	0.3253	0.2651	1.681

13.3 Discussions

The previous section experimentally examines the first two questions in Sect. 13.1.4. In this section, we will answer the last one. By doing so, we will further achieve a guiding principle in designing a multi-agent system for solving problems.

13.3.1 Complexities in Different Representations

From the previous section, we note that given a SAT problem, different representation approaches can lead to different agent networks. In the two cases mentioned, the first one shows small world topologies in its resulting agent networks. The second one, i.e., MASSAT, does not generate small world topologies.

To measure computational complexity, Walsh has studied the relationship between topologies (in particular, small world topologies) and complexities. In [13.23], Walsh empirically proved that a small world topology can increase the computational complexity of an algorithm that involves certain heuristics. This is because, first, heuristics normally guide a search process locally, and second, in a small world network, local information cannot predict global properties well.

We have also experimentally validated this assertion based on the previous two representations of SAT problems. Using the same SAT problems, we examined a clause-based representation as opposed to the variable-based representation used in MASSAT. In our examinations, each agent represents a clause. A clause acts as a local space where the corresponding agent can reside and move around. A literal in a clause is a position where an agent can stay. If an agent stays at a certain position, it indicates that the corresponding literal is true. If two agents stay at two positions whose literals are negations of each other, a conflict occurs. For the whole system, to solve a given SAT is to eliminate conflicts. Our experimental results suggest that in such a problem representation, it is normally hard (in term of flips) to solve a problem, but it is relatively easy to solve it with MASSAT.

13.3.2 Balanced Complexities in Intra- and Inter-Agent Computations

In MASSAT, an agent can represent one or more variables. Section 13.2.2 has experimentally shown that, in both cases, the obtained agent networks do not have small world topologies. But, which case is better? Experiments have suggested that as the number of variables represented by an agent in MASSAT increases, the resulting agent networks will be less 'small worldy', because the networks become smaller and denser. As an extreme situation, we can use only one agent to represent all variables. In this case, the network collapses to an isolated vertex. Is this the best situation?

The answer is no. In fact, if an agent represents multiple variables, the agent should assign values to all its variables. Because the variables of an agent are not independent, as the number of variables represented by an agent increases, the intra-agent computational complexity to assign values to its variables becomes harder. Therefore, a good design should balance the intra- and inter-agent computational complexities to achieve the lowest possible total computational cost.

With respect to this point, our experiments have suggested that when an agent represents four or five variables, the total computational cost, in terms of agent movements and variable flips, becomes the lowest.

13.3.3 A Guiding Principle

Based on the above discussions, we can arrive at a guiding principle for designing a multi-agent-based approach to solving a problem:

- It should avoid involving a small world topology in a resulting agent network;
- It should maintain balanced intra- and inter-agent computational complexities so as to achieve the lowest total computational cost.

In the previous sections, we studied the topological characteristics of agent networks. We note from these sections that random agent networks are compared with real-world agent networks. In particular, the clustering coefficient is one of the key properties used to characterize random agent networks and other agent networks. In the following two sections, we will study the clustering characteristics of random agent networks, namely, the average value and lower bounds of the clustering coefficient in random agent networks.

13.4 Average Value of the Clustering Coefficient

If an agent network has n agents, there can exist at most $\frac{n \cdot (n-1)}{2}$ possible links in it. To generate a random agent network with n agents and m links, one can uniformly select m from $\frac{n \cdot (n-1)}{2}$ possible links at random. In this case, each possible link can be selected with a probability of $\frac{2m}{n \cdot (n-1)}$. If random agent network G has n agents and m links, the average degree of agents in G is $\frac{2m}{n}$. This implies that an agent is on average connected to $\frac{2m}{n}$ different agents. As a consequence, there can exist at most

$$\frac{\left(\frac{2m}{n} + 1\right) \cdot \frac{2m}{n}}{2} \tag{13.7}$$

possible links between this agent and its neighbors. But, since this agent (on average) has $\frac{2m}{n}$ neighbors, there are at least $\frac{2m}{n}$ links existing. As to other possible

$$\frac{\left(\frac{2m}{n} + 1\right) \cdot \frac{2m}{n}}{2} - \frac{2m}{n} \tag{13.8}$$

links between this agent and its neighbors, because each link has a probability of $\frac{2m}{n \cdot (n-1)}$ of appearing in this agent network, there will on average exist

$$\left[\frac{\left(\frac{2m}{n} + 1\right) \cdot \frac{2m}{n}}{2} - \frac{2m}{n}\right] \cdot \frac{2m}{n \cdot (n-1)} \tag{13.9}$$

links that also appear in agent network G. Hence, the total number of links existing between this agent and its neighbors is

$$\frac{2m}{n} + \left[\frac{\left(\frac{2m}{n} + 1\right) \cdot \frac{2m}{n}}{2} - \frac{2m}{n}\right] \cdot \frac{2m}{n \cdot (n-1)}. \tag{13.10}$$

Therefore, for any agent a_i in G, its clustering c_{a_i} is :

$$
\begin{aligned}
c_{a_i} &= \frac{\frac{2m}{n} + \left[\frac{\left(\frac{2m}{n}+1\right) \cdot \frac{2m}{n}}{2} - \frac{2m}{n}\right] \cdot \frac{2m}{n \cdot (n-1)}}{\frac{(2m+n) \cdot m}{n^2}} \\
&= \frac{2n^2 \cdot (n-1) + 2m \cdot (2m-n)}{n \cdot (n-1) \cdot (2m+n)}.
\end{aligned}
\tag{13.11}
$$

Hence, the clustering coefficient C_G of random agent network G is

$$C_G = \frac{1}{n} \cdot \sum_{a_i \in G} c_{a_i} = \frac{2n^2 \cdot (n-1) + 2m \cdot (2m-n)}{n \cdot (n-1) \cdot (2m+n)}. \tag{13.12}$$

But, we should note that in Eq. (13.9), it should be guaranteed that

$$\frac{(\frac{2m}{n}+1) \cdot \frac{2m}{n}}{2} - \frac{2m}{n} \geq 0, \tag{13.13}$$

i.e., the total number of possible links between agent a_i and its neighbors should be larger than or equal to the number of existing links. Simplifying Eq. (13.13), we get $m \geq \frac{n}{2}$. Thus, Eq. (13.12) is applicable under the condition $m \geq \frac{n}{2}$.

What, then, is the average value of the clustering coefficient in the case of $m < \frac{n}{2}$? We can understand that if $m \ll \frac{n}{2}$, i.e., there are only a few links that exist in agent network G, the existing links will normally not be connected to the same agent. In this case, the agent with a link has clustering 1. The isolated agent has clustering 0. Since there are $2m$ agents whose clusterings are 1, we can readily get clustering coefficient C_G:

$$C_G = \frac{2m}{n}. \tag{13.14}$$

Altogether, in a general case, we have the following average value for the clustering coefficient in random agent networks:

$$C_G = \begin{cases} \frac{2n^2 \cdot (n-1) + 2m \cdot (2m-n)}{n \cdot (n-1) \cdot (2m+n)} & if\ m \geq \frac{n}{2} \\ \frac{2m}{n} & if\ m < \frac{n}{2} \end{cases} \tag{13.15}$$

Theorem 13.4.1. *Let $\{G(n,m)\}$ be a set of all possible random agent networks with n agents and m links, and $C(n,m)$ be the average value of the clustering coefficient in $\{G(n,m)\}$. Then,*

$$C(n,m) = \begin{cases} \frac{2n^2 \cdot (n-1) + 2m \cdot (2m-n)}{n \cdot (n-1) \cdot (2m+n)} & if\ m \geq \frac{n}{2} \\ \frac{2m}{n} & if\ m < \frac{n}{2} \end{cases} \tag{13.16}$$

In Fig. 13.3, we show both theoretical and experimental results on an example of agent networks with 100 agents. We can see from Fig. 13.3 that the experimental result is very similar to the theoretical one.

In agent network $G(n,m)$, let k denote the average degree of agents. Then, we can get

$$m = \frac{n \cdot k}{2}. \tag{13.17}$$

Substituting the above equation into Eq. (13.16) in the case of $m \geq \frac{n}{2}$, we get

$$C(n,k) = \frac{2}{k+1} + \frac{k \cdot (k-1)}{(n-1) \cdot (k+1)}. \tag{13.18}$$

We conclude that for a set of random agent networks with $n \gg 1$ and $n \gg k$, their average clustering coefficient value is approximately equal to $\frac{2}{k+1}$.

13.5 Lower Bounds of the Clustering Coefficient

In the previous section, we have presented the average value of the clustering coefficient in random agent networks with n agents and m links. In this section, let us deduce a lower bound for the clustering coefficient in an agent network. Given an agent network, G, which has n agents and m links, for $\forall a_i$ ($i \in 1, 2, ..., n$), let its degree be $d(a_i)$. There can be at most $\frac{(d(a_i)+1) \cdot d(a_i)}{2}$ links between a_i and its neighbors. Further, we assume that there actually exist $e(a_i)$ links. The clustering ratio c_{a_i} of a_i and clustering coefficient C_G of G are

$$c_{a_i} = \frac{2 \cdot e(a_i)}{(d(a_i) + 1) \cdot d(a_i)} \tag{13.19}$$

and

$$C_G = \frac{1}{n} \cdot \sum_{i=1}^{n} \frac{2 \cdot e(a_i)}{(d(a_i) + 1) \cdot d(a_i)}, \tag{13.20}$$

respectively. In any agent network, it is known that $\forall a_i$, $e(a_i) \geq d(a_i)$. We have

$$C_G = \frac{1}{n} \cdot \sum_{i=1}^{n} \frac{2 \cdot e(a_i)}{(d(a_i) + 1) \cdot d(a_i)} \geq \frac{1}{n} \cdot \sum_{i=1}^{n} \frac{2 \cdot d(a_i)}{(d(a_i) + 1) \cdot d(a_i)}$$

$$= \frac{1}{n} \cdot \sum_{i=1}^{n} \frac{2}{d(a_i) + 1}. \tag{13.21}$$

It is also known that in any agent network, there is a relationship between the total number of agent degrees and the number of links, i.e.,

$$\sum_{i=1}^{n} d(a_i) = 2m. \tag{13.22}$$

Then, the problem of finding a lower bound for the clustering coefficient C in agent network G becomes an optimization problem:

$$\min \sum_{i=1}^{n} \frac{1}{d(a_i) + 1} \quad s.t. \quad \sum_{i=1}^{n} d(a_i) = 2m. \tag{13.23}$$

Let $x_i = d(a_i) + 1$; the above problem can be rewritten as

$$\min \sum_{i=1}^{n} \frac{1}{x_i} \quad s.t. \quad \sum_{i=1}^{n} x_i = 2m + n \tag{13.24}$$

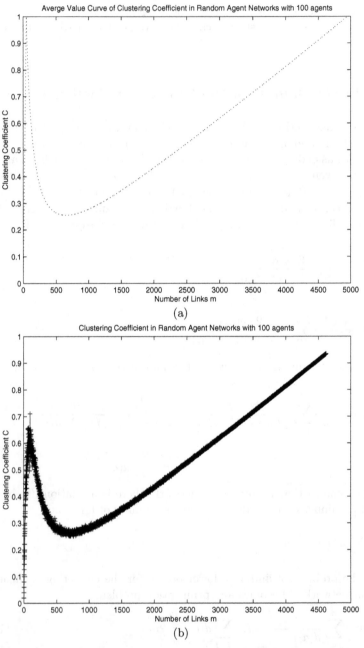

Fig. 13.3. (a) The average value curve of the clustering coefficient in agent networks with 100 agents. (b) An experimental result of the clustering coefficient in agent networks, also with 100 agents. For each $m \in [1, 4950]$, we generated 100 random agent networks and calculated the average value of the clustering coefficient on these agent networks

Further, let $z = 2m + n$ and $f(x_1, x_2, ..., x_n) = \sum_{i=1}^{n} \frac{1}{x_i}$. From Eq. (13.24), we have

$$x_n = z - \sum_{i=1}^{n-1} x_i. \tag{13.25}$$

Substituting x_n in $f(x_1, x_2, ..., x_n)$ with this equation, we get

$$f(x_1, x_2, ..., x_{n-1}) = \sum_{i=1}^{n-1} \frac{1}{x_i} + \frac{1}{z - \sum_{i=1}^{n-1} x_i}. \tag{13.26}$$

To find the extremal value of the function $f(x_1, x_2, ..., x_{n-1})$, we derivate its partial differential coefficient for x_i $(i = 1, ..., n-1)$:

$$\begin{cases} \frac{\partial f}{\partial x_1} = \frac{1}{(z - \sum_{i=1}^{n-1} x_i)^2} - \frac{1}{x_1^2} \\ \frac{\partial f}{\partial x_2} = \frac{1}{(z - \sum_{i=1}^{n-1} x_i)^2} - \frac{1}{x_2^2} \\ \vdots \qquad \vdots \\ \frac{\partial f}{\partial x_{n-1}} = \frac{1}{(z - \sum_{i=1}^{n-1} x_i)^2} - \frac{1}{x_{n-1}^2} \end{cases} \tag{13.27}$$

Then, letting $\frac{\partial f}{\partial x_1} = \frac{\partial f}{\partial x_2} = \cdots = \frac{\partial f}{\partial x_{n-1}} = 0$, we get the following equation system:

$$\begin{cases} \frac{1}{(z - \sum_{i=1}^{n-1} x_i)^2} - \frac{1}{x_1^2} = 0 \\ \frac{1}{(z - \sum_{i=1}^{n-1} x_i)^2} - \frac{1}{x_2^2} = 0 \\ \vdots \qquad \vdots \\ \frac{1}{(z - \sum_{i=1}^{n-1} x_i)^2} - \frac{1}{x_{n-1}^2} = 0 \end{cases} \tag{13.28}$$

Since $x_i > 0$ and $z - \sum_{i=1}^{n-1} x_i > 0$, we simplify Eq. (13.28) and get

$$\begin{cases} x_1 + \sum_{i=1}^{n-1} x_i = z \\ x_2 + \sum_{i=1}^{n-1} x_i = z \\ \vdots \qquad \vdots \\ x_{n-1} + \sum_{i=1}^{n-1} x_i = z \end{cases} \tag{13.29}$$

Solving the above Eq. (13.29), we easily get $x_1 = x_2 = \cdots = x_{n-1} = \frac{z}{n}$. Substituting in Eq. (13.25), we get $x_n = \frac{z}{n}$. So,

$$\sum_{i=1}^{n} \frac{1}{x_n} = n \cdot \frac{n}{z} = \frac{n^2}{2m+n}. \tag{13.30}$$

Therefore,

$$C \geq \frac{1}{n} \cdot \sum_{i=1}^{n} \frac{2}{d(a_i) + 1} \geq \frac{2}{n} \cdot \frac{n^2}{2m+n} = \frac{2n}{2m+n}, \tag{13.31}$$

i.e., $\frac{2n}{2m+n}$ is a lower bound for the clustering coefficient in an agent network.

Theorem 13.5.1. *Let $\{G(n,m)\}$ be a set of all possible random agent networks with n agents and m links, and $C_i(n,m)$ be the clustering coefficient of $G_i(n,m) \in \{G(n,m)\}$. Then,*

$$C_i(n,m) \geq \frac{2n}{2m+n}. \tag{13.32}$$

In Fig. 13.4, we plot the average value curve and lower bound curve of the clustering coefficient in agent networks with 100 agents. From this figure, we can note that the lower bound curve is a monotonously decreasing curve. It is close to the average value curve in the region where m is close to n. Thus, in this region, this lower bound is good. But, as the number of links m increases, the lower bound curve gradually moves away from the average value curve. This indicates that the lower bound becomes worse. Here, we make a conjecture:

Conjecture 13.5.1. For an agent network $G(n,m)$ with n agents and m links,

$$f(n,m) = \frac{2m}{n \cdot (n-1)} \tag{13.33}$$

is a lower bound for its clustering coefficient.

In Fig. 13.5, we plot this conjectured lower bound and compare it with the proven lower bound as well as the average value. We note that as $m \longrightarrow \frac{n\cdot(n-1)}{2}$, $f(n,m)$ curve gets very close to the average value curve. This means $f(n,m)$ becomes a better lower bound as $m \longrightarrow \frac{n\cdot(n-1)}{2}$.

13.6 Conclusions

Small world phenomena have been widely observed in both natural systems and man-made systems. A small world topology will make a search process difficult, which is based on certain heuristics. When using a distributed multi-agent system to solve a SAT problem, it will implicitly or explicitly form an agent network. In this chapter, we first defined agent networks and their topologies. Next, in order to characterize the computational complexities of agent networks formed in solving SATs with multi-agent systems, we studied their topological characteristics. Specifically, we provided two distributed agent representations of SATs, where each agent represents one clause or one variable, respectively.

We experimentally showed that different representations can make agent networks manifest different topological characteristics. The agent network created in MASSAT does not have a small world topology. On the contrary, the agent network created by clause-based representation exhibits a small world topology. This finding can, to some extent, reveal that MASSAT is better than the approach where agents represents clauses.

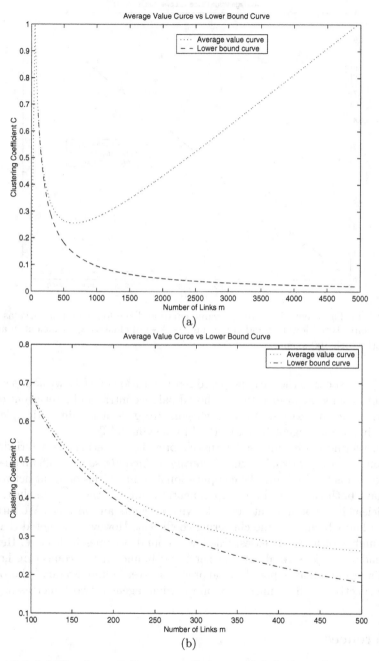

Fig. 13.4. (a) The theoretical average value curve and the lower bound curve of clustering coefficient in agent networks with 100 agents. (b) A zoom-in figure of the curves in (a) where $m \in [n, 500]$

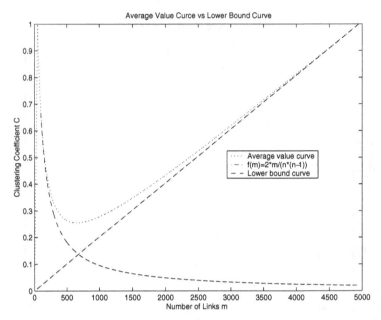

Fig. 13.5. The theoretical average value curve and the lower bound curve as well as the conjectured lower bound curve (i.e., $\frac{2m}{n\cdot(n-1)}$) of clustering coefficient in agent networks with 100 agents

With respect to the variable-based representation of SATs without a small world topology, we argued that it should balance intra- and inter-agent computational complexities. As a conclusion, we gave a guiding principle for designing a multi-agent-based method for solving SATs.

Furthermore, we addressed in this chapter the clustering characteristics of random agent networks, i.e., the *clustering coefficients*, so as to offer a quantitative means for studying the computational complexities of agent networks. We gave both theoretical and experimental results of the average clustering coefficient in random agent networks with n agents and m links. We further gave a lower bound for the clustering coefficient. However, we noted that as the number of links increases, this lower bound becomes ineffective. Hence, we made a conjecture about another lower bound. If the conjecture holds, this lower bound, as opposed to the previous lower bound, gradually becomes more effective as the connectivity of a random agent network increases.

References

13.1 http://www.intellektik.informatik.tu-darmstadt.de/satlib/
13.2 L. A. Adamic: The small world Web. In: *Proceedings of European Conference on Digital Libraries (ECDL'99), volume 1696 of LNCS* (Springer, 1999) pp. 443-452

13.3 J. M. Crawford, L. D. Auton: Experimental results on the crossover point in random 3SAT. Artificial Intelligence, 81(1-2), 31-57 (1996)

13.4 M. Davis, G. Logemann, D. Loveland: A machine program for theorem proving. Communications of the ACM, 5, 394-397 (1962)

13.5 M. Davis, H. Putname: A computing procedure for quantification theory. Journal of the ACM, pp. 201-15 (1960)

13.6 J. Gu: Efficient local search for very large-scale satisfiability problem. *SIGART Bulletin*, 3, 8-12 (1992)

13.7 K. Hirayama, M. Yokoo: Local search for distributed SAT with complex local problems. In: *Proceedings of the First International Joint Conference on Autonomous Agent and Multiagent Systems (AAMAS 2002), Bologna, Italy, 7 2002* pp. 1199-1206

13.8 E. A. Hirsch, A. Kojevnikov: Solving boolean satisfiability using local search guided by unit clause elimination. In: *Proceedings of the Seventh International Conference on Principles and Practice of Constraint Programming (CP'01), 2001*

13.9 E. A. Hirsch, A. Kojevnikov: Unitwalk: A new SAT solver that uses local search guided by unit clause elimination. In: *Proceedings of the Fifth International Symposium on the Theory and Applications of Satisfiability Testing (SAT 2002)* pp. 35-42 (2002)

13.10 X. Jin, J. Liu: Multiagent SAT (MASSAT): Autonomous pattern search in constrained domains. In: *Proceedings of the Third International Conference on Intelligent Data Engineering and Automated Learning (IDEAL'02), 2412 of LNCS* pp. 318-328 (2002)

13.11 B.J.B. Jr., R.C. Schrag: Using CSP look-back techniques to solve real-world SAT instances. In: *Proceedings of the fourteenth National Conference on Artificial Intelligence (AAAI-97)* pp. 203-208 (1997)

13.12 C. M. Li, Anbulagan: Look-ahead versus look-back for satisfiability problems. In: *Proceedings of the Third International Conference on Principles and Practice of Constraint Programming (CP'97)* pp. 341-345 (1997)

13.13 J. Liu, X. Jin, J. Han: Distributed problem solving without communication - An examination of computationally hard satisfiability problems. International Journal of Pattern Recognition and Artificial Intelligence, 16(8) (2002)

13.14 J. Liu, H. Jing, Y.Y. Tang: Multi-agent oriented constraint satisfaction. Artificial Intelligence, 136, 101-144 (2002)

13.15 J. P. Marques-Silva, K. A. Sakallah: Grasp - A new search algorithm for satisfiability. In: *Proceedings of IEEE/ACM International Conference on Computer-Aided Design, 1996*

13.16 D. McAllester, B. Selman, H. Levesque: Evidence for invariants in local search. In: *Proceedings of the Fourteenth National Conference on Artificial Intelligence (AAAI'97)* pp. 321-326 (1997)

13.17 S. Milgram: The small world problem. Psychology Today, 2, 60-67, (1967)

13.18 J.M. Montoya, R.V. Sole: Small world patterns in food webs (2000)

13.19 D. Schuurmans, F. Southey: Local search characteristics of incomplete SAT procedures. In: *Proceedings of the Seventeenth National Conference on Artificial Intelligence (AAAI-2000)* pp. 297-302 (2000)

13.20 D. Schuurmans, F. Southey: Local search characteristics of incomplete SAT procedures. Artificial Intelligence, 132(2), 121-150 (2001)

13.21 B. Selman, H. A. Kautz, B. Cohen: Noise strategies for improving local search. In: *Proceedings of the Twelfth National Conference on Artificial Intelligence (AAAI'94)* pp. 337-343, 1994

13.22 B. Selman, H. Levesque, D. Mitchell: A new method for solving hard satisfiability problems. In: *Proceedings of the Tenth National Conference on Artificial Intelligence (AAAI'92)* pp. 440-446 (1992)

13.23 T. Walsh: Search in a small world. In: *Proceedings of the Sixteenth International Joint Conference on Artificial Intelligence (IJCAI'99), 1999*

13.24 D.J. Watts, S.H. Strogatz: Collective dynamics of small world networks. Nature, 393, 440-442 (1998)

13.25 M. Yokoo, K. Hirayama: Algorithms for distributed constraint satisfaction: A review. *Autonomous Agents and Multi-Agent Systems*, 3(2), 198-212 (2000)

14. Finding the Best Agents for Cooperation

Francesco Buccafurri, Luigi Palopoli, Domenico Rosaci, and
Giuseppe M.L. Sarnè

DIMET, Università "Mediterranea" di Reggio Calabria, Italy

Abstract

One of the basic issues to face in multi-agent systems is effectively implementing cooperation among agents. To this aim, a non-trivial problem has to be solved: among a (possibly large) number of agents, how do we detect which agents are promising candidates for cooperation? In this chapter, a new approach for dealing with this problem is proposed. We define a formal model for representing agents, and a number of semantic properties exploited for detecting fruitful cooperation. On the basis of this model we design a multi-agent system, called SPY, capable of managing and supporting cooperation in the agent community. The system learns semantic properties by monitoring the user behavior in such a way that it adapts its response to user expectation.

14.1 Introduction

Cooperation is often considered one of the key concepts in multi-agent systems (often denoted by MAS) [14.4, 14.10, 14.11, 14.12, 14.14, 14.17, 14.18, 14.34]. Indeed, each agent, in an agent community, does not have to learn only by its own discovery, but also by exploiting knowledge coming from other agents. Cooperation is usually implemented by integrating multiple (even heterogeneous) knowledge sources [14.2, 14.3, 14.5, 14.21, 14.24, 14.25, 14.30]. But, a basic problem concerns the coordination of agent behavior in order to satisfy both the knowledge of the interest domain and the individual requirements. For this purpose, learning and adaptation are considered essential by many researchers in this field [14.7, 14.27, 14.28, 14.31, 14.33]. In order to realize such a cooperation, some techniques developed in the field of machine learning have been introduced in various multi-agent systems [14.8, 14.9, 14.19, 14.20, 14.32].

This chapter makes a contribution in this context. Basically, we study the following problem: Consider a network of agents, each supporting a user, and let the individual knowledge of each agent be an ontology also embedding knowledge about the user behavior; when a user contacts other agents in order to integrate their ontologies with that of his agents or, in other words, when a user asks for the cooperation of other agents, how can he select, from the universe of possible agents, the most appropriate ones?

For giving a solution to the above problem, we propose a formal framework in which we represent by quantitative information a number of semantic properties that we consider important for detecting fruitful cooperation. Such properties take into account structural similarities among agent ontologies, user perception, and the attraction power of agents in the community. Potentially appropriate agents for cooperating with an agent, say a, of the network, are obtained by solving a linear system. The solution of such a system provides the user of a with a number of agent lists, each containing the *most appropriate* agents for cooperation, from which the user can choose agents he wants to contact for supporting his activity. The multiplicity of such choice lists depends on the multiplicity of the properties that can be used as preference criteria. Users are free to use the suggested lists even partially, or to ignore them. In any case, a user's behavior induces a modification of some coefficients (describing reactive properties) in such a way that lists suggested in the future are (hopefully) closer to the user's real needs. Therefore, the system *learns* from the user's behavior about how to provide him with suggestions that meet, as much as possible, his expectations.

On the basis of this model, we design a multi-agent system with a client/server architecture: client agents, possibly cooperating with one another, support user activity, while the server coordinates cooperation. Among all possible applications, we have considered the case of agent cooperation for helping the user in retrieving information. The user is provided with a set of recommendations that are generated both by his agent and by agents the system has detected as promising for cooperation.

This chapter is organized as follows. Section 14.2 compares our work with other proposals. Section 14.3 illustrates how individual knowledge of agents is represented in our model. Section 14.4 is the core of the chapter, since it describes which semantic properties we represent and how we model and solve the problem of cooperation. Section 14.5 describes how cooperation between two agents is implemented, that is, the integration of the two respective knowledge bases. In Sect. 14.6, the system SPY, implementing the cooperation model, is presented. Section 14.7 gives a practical description of a system execution session, and some discussion about the system behavior. Finally, Sect. 14.8 gives the conclusions.

14.2 Related Work

In the context of machine learning approaches, Weiß [14.33] illustrates the progress made in the available work on learning and adaptation in multi-agent systems, and provides a general survey of multi-agent systems using adaptation and learning. In [14.31], a demonstration of how reinforcement learning agents can learn cooperative behavior in a simulated social environment is provided, specifying that if cooperation is done intelligently, each agent can benefit from other agents' instantaneous information, episodic experience, and learned knowledge. Byrne and Edwards [14.7] concern themselves with how a group of intelligent agents can work together in order to solve a problem or achieve a common goal by using machine learning techniques to refine their knowledge. An example of a practical situation that needs to be modeled in a group of agents is presented in [14.27], where a probabilistic reciprocity mechanism is introduced to generate stable and cooperative behavior among a group of self-interested agents. In [14.29], authors identify procedures and environments under which self-interested agents may find it beneficial to help others, and point out that the sharing of experiences among reciprocative agents will limit the exploitative gains of selfish agents.

A large numbers of multi-agent systems using learning techniques have been proposed in the literature. Among these, we cite some significant proposals:

- In [14.25], a learning system called COLLAGE, which endows the agents with the capability to learn how to choose the most appropriate coordination strategy from a set of available coordination strategies, is presented.
- *Amalthaea* [14.20], which is an evolving multi-agent ecosystem for personalized filtering, discovery, and monitoring of information sites.
- Choi and Yoo [14.9] present an information retrieval system where a multi-agent learning approach to information retrieval on the Web, in which each agent learns its environment from the user's relevance feedback by using a neural network mechanism, is proposed.
- In [14.19], a system of collaborative agents is proposed where the collaboration among agents assisting different users is introduced in order to improve the efficiency of the local learning.
- The system *Challenger* [14.8], consists of agents which individually manage local resources and communicate with one another to share their resources, in the attempt of utilizing them more efficiently, in order to obtain desirable global system objectives.
- Wang et al. [14.32] present a multi-agent architecture applied to cooperative system engineering, useful in modeling activities and in providing support to cooperative tasks.
- In [14.6], the authors introduce a principle, called *Win or Learn Fast (WOLF)*, for varying a learning rate in a new reinforcement learning technique. They examine this technique theoretically and also present empirical results on a variety of more general stochastic games.
- Schillo and Funk [14.26] aim to establish a mechanism that enables agents to cope with environments that contain both selfish and cooperative entities, where the mixture and the behavior of these entities is previously unknown to all agents. Authors achieve this by enabling agents to evaluate trust in others, based upon the observations they gather.

Such techniques open, on the one hand, the possibility of integrating individual agent knowledge for acquiring an enhanced knowledge of the environment. On the other hand, they consider the problem of determining which agents are promising candidates for suitable knowledge integration. But, differently from our approach, none of them proposes automatic techniques for solving such a problem.

We point out that in the above approaches the knowledge involved in the cooperative exchange is not stored in a complex data structure, but it generally consists of unstructured elementary information about both the environment and the various actions performed by the agents. But an important issue that recently emerged in the MAS field deals with the necessity of organizing the available knowledge in *ontologies* [14.15, 14.16, 14.13, 14.22], which are sophisticated content-oriented data structures.

14.3 The Knowledge Bases

Throughout the chapter, we refer to a given set of agents Λ of cardinality n, and we assume that all agents in Λ can cooperate with each other. Thus, we can see the set Λ as a undirected complete graph of agents whose arcs represent possible cooperation. Without loss of generality, we identify agents in Λ by the cardinal numbers $\{1, \ldots, n\}$.

14.3.1 An Ontology for Describing the Domain of Interest

Since we consider only homogeneous agents, we suppose that a unique environment is associated with our agent network. We represent such an environment in our model by a set of objects. For the rest of the section, we consider a set of objects O as given.

A domain D on O is a set of classes of objects. We suppose that a suitable semantics underlying the classification provided in a domain is given. The notion of domain is formally defined next.

Definition 14.3.1. *A domain on O* , denoted by D , is a set $D \subseteq 2^O$ such that (1) $\forall o \in O$, $\{o\} \in D$ and (2) there exists an element r of D, called *the root* such that, $\forall o \in O$, $o \in r$. Elements of D are called *classes*.

In other words, a domain is a set of object classes containing, at least, a root class collecting all the objects of O and, for each object o of O, the singleton class $\{o\}$. Throughout the rest of the section, we assume a domain D on O as given.

Set containment induces a partial ordering among elements of a domain D. A domain D plus this ordering is called *ontology*.

Definition 14.3.2. An *ontology* on D, denoted by O_D, is a partially ordered set $\langle D, \subseteq \rangle$. The *ontology graph of* O_D is a directed graph $G(O_D)$ with a set of nodes coinciding with D and set of arcs A defined as the binary relation obtained as the transitive reduction of the relation $\subseteq$ of O_D[1]. The node r of $G(O_D)$, where r is the root of D, is called *root of* $G(O_D)$.

---

[1] (A, B) is in the transitive reduction of $\subseteq$ iff $A \subseteq B$ and $\nexists C$ such that $A \subseteq C$ and $C \subseteq B$.

Note that, as a consequence of condition (2) of Definition 14.3.1, each non-root node is reachable from the root r of $G(O_D)$. Furthermore, as a result of condition(1) of Definition 14.3.1, nodes of $G(O_D)$ without degree 0 coincide with singleton classes of D. An ontology based on a generalization hierarchy is suitable for representing many real-world situations, like topics in Web engines and Websites, items in e-commerce, staff hierarchy of an organization, and so on. It is worth noting that this is not the only possible choice for representing the environment of agents. Indeed, in different contexts, as semi-structured data in Web documents, other kinds of ontologies may be better suited (for example, *OEM-graphs* [14.1], *SDR-networks* [14.23], etc.).

14.3.2 The Local Knowledge Base

The ontology represents common knowledge about the environment in which the agents work. However, each agent may have a partial view of the ontology representing the portion of the world the user monitored by the agent selects by his activity. Inside this portion of the ontology, different priorities for the classes can be inferred by exploiting user behaviour. This is encoded in the notion of the *Local Knowledge Base* (*LKB* for short), defined next.

Definition 14.3.3. Given an ontology O_D on D and an agent a, a *Local Knowledge Base* LKB^a (of a on O_D) is a pair $\langle G^a, \rho^a \rangle$, such that (i) $G^a = \langle N^a, A^a \rangle$ is a sub-graph of $G(O_D)$ (i.e., $N^a \subseteq D$, $A^a \subseteq A$) containing the root r of $G(O_D)$ and each $n \in N^a$ is reachable in G^a from r, and (ii) ρ is a function, called the *priority function*, defining a real weight ranging from 0 to 1 associated with each arc (i, j) of G^a such that

$$\rho(i,j) = \frac{c_{ij}}{\sum_{k \in Adj(i)} c_{ik}} \tag{14.1}$$

where $Adj(i)$ is the set of nodes adjacent to i, and for each $k \in Adj(i)$, c_{ik} counts how many times the user of a has selected an object (that is, a leaf node) through a path selection including the arc (i, k). Note that the coefficients c_{ij} in a path $\langle r, i_1, ... i_s \rangle$ are updated only when the leaf node i_s, corresponding to a single object of the domain, is selected. The root r of $G(O_D)$ is also called the *root of LKB^a*.

A Local Knowledge Base, representing the local view of the agent, is then obtained by extracting a sub-graph from the ontology graph including all the classes accessed by the user (and, thus, at least the root node). Moreover, arcs of the graph so obtained are weighted for assigning highest priority to most accessed classes.

14.4 Extraction of the Semantic Properties

Besides his local agent, each user looks at the other agents on the network as a source of potentially interesting information in order to enrich the support to his activity. Interest in agents can be defined by considering some semantic properties. Such properties, useful for driving users' choices, are of two types: (i) *local properties*, taking into account information stored in the LKBs, and (ii) *global properties*, merging local properties with external knowledge extracted from the general context. An important feature of the model is that the merge performed in the construction of global properties is based on an adaptive learning technique involving some parameters by which the user behavior is taken into account. In other words, global properties exploit an important kind of property (encoded in a number of parameters) directly reflecting reactions of users to system advice. We call such additional properties *reactive properties*. Next, we describe the set of properties used in the model.

14.4.1 Local Properties: Similarity

The only local property we consider is the property we call *similarity* between two agents i and j, representing a measure of the similarity of their corresponding LKBs. Such a coefficient is a real value ranging from 0 to 1.

Definition 14.4.1. Let i and j be two agents. Let $G^i = \langle N^i, A^i \rangle$ and $G^j = \langle N^j, A^j \rangle$ be the two graphs of their LKBs. Let ρ^i and ρ^j be their corresponding priority functions. We define the *similarity* S_{ij} between i and j as

$$S_{ij} = 1 - \frac{1}{|A^i \cup A^j|} \sum_{(h,k) \in A^i \cup A^j} \gamma_{hk},$$

where

$$\gamma_{hk} = \begin{cases} |\rho^i(h,k) - \rho^j(h,k)| & \text{if } (h,k) \in A^i \cap A^j \\ 1 & \text{otherwise.} \end{cases}$$

Observe that the term $\frac{1}{|A^i \cup A^j|} \sum_{(h,k) \in A^i \cup A^j} \gamma_{hk}$ in the expression defining S_{ij} (for two agents i and j) represents a *dissimilarity* between agents i and j. This is defined as the mean of a number of contributions γ_{hk}, each corresponding to an arc (h,k) belonging to the set $A^i \cup A^j$. For the common arcs of the two LKBs, that is, the arcs belonging to the intersection between A^i and A^j, γ_{hk} is the difference (in absolute value) between their respective priority functions (note that such a difference is a real value ranging from 0 to 1). In other words, common arcs can be view as "homologous" arcs, and their dissimilarity measures how much these arcs differ in terms of weight. For the remaining arcs $(h,k) \notin A^i \cap A^j$, we assign the value 1 to the coefficient γ_{hk}. Indeed, an arc belonging to A^i but not belonging to A^j is not a "homologous" arc in the LKB graph of the agent j (and vice versa), and thus this is the case of maximum dissimilarity, leading to a contribution (to the overall dissimilarity) saturated at the value 1.

14.4.2 Global Properties: Interest and Attractiveness

Recall that global properties merge local properties with knowledge extracted from the context. In this section, we introduce the notion of *interest coefficient*, representing a measure of the global properties of a given agent as perceived by another agent. Hence, for a pair of agents i and j, the interest coefficient, besides the similarity between i and j, must take into account also knowledge extracted from the context. But, *which kind of contextual knowledge has to be considered as meaningful?* We make the following assessment in our model: the knowledge extracted from the context, used by the agent i for defining the interest coefficient I_{ij} with respect to another agent j, is a measure of the global interest of all the other agents (different from i) with respect to the agent j, that is, a measure of a sort of *attractiveness* of the agent j as perceived by the agent i. Recalling that the interest, besides the contextual knowledge, must take into account also the local knowledge (that is, the similarity), the above definition of contextual knowledge requires that, for each $i \in \Lambda \setminus \{j\}$,

$$I_{ij} = \phi_{ij}(S_{ij}, \mu_{ij}(\{I_{kj} \mid k \neq i, j\})), \qquad (14.2)$$

where μ_{ij} and ϕ_{ij} are suitable functions yielding real values from 0 to 1. In particular, μ_{ij} returns a measure of the attractiveness of the agent j detected by the agent i from the value of the interest coefficients of all the agents (different from i) w.r.t j, while ϕ_{ij} combines such a measure with the similarity S_{ij}. Clearly, the function ϕ_{ij} also plays the role of weighing the importance, for agent i, of the local knowledge with respect to the contextual one.

For μ_{ij} and ϕ_{ij} (where i and j are two agents), we define in our model the following: (i) μ_{ij} is a function returning the average of the interest coefficients of all the other agents different from j, and (ii) ϕ_{ij} is a function computing a linear combination of the similarity coefficient between i and j, and the attractiveness of j w.r.t i. Applying the above definitions for μ_{ij} and ϕ_{ij}, (14.2) becomes the following linear system:

$$(\forall i \in \Lambda \setminus \{j\}) \qquad (14.3)$$
$$(I_{ij} = \psi_{ij} \cdot (P_i \cdot S_{ij} + (1 - P_i) \cdot S_{ij} \cdot \tfrac{1}{n-2} \textstyle\sum_{k \in \Lambda \setminus \{i,j\}} I_{kj})),$$

where ψ_{ij} and P_i, for each $i \in \Lambda \setminus \{j\}$, are adaptive parameters ranging from 0 to 1, and representing a measure of *reactive* properties that we suppose to be learned from the user behavior. ψ_{ij} plays the role of a reducing factor, filtering the advice of the system on the basis of the user behavior, while P_i measures the importance that the user gives to the local knowledge (similarity) with respect to the contextual one. Note that both ψ_{ij} and P_i can be estimated once the reactive properties are defined. We deal with this issue in the next section. Thus, given an agent j, any value assignment to the interest coefficients of all the other agents with respect to j must satisfy (14.4). The next theorem shows that, for every value of the parameters occurring in (14.4), there exists a unique solution of the linear system (14.4), that is, a value assignment to the interest coefficients satisfying (14.4). Obviously, such a solution can be polynomially computed.

Lemma 14.4.1. *Given an agent $j \in \Lambda$ and a set of 3-tuples of $[0,1]$ real coefficients $\{\langle P_i, \psi_{ij}, S_{ij} \rangle \mid i \in \Lambda \setminus \{j\} \wedge (\forall h \in \Lambda \setminus \{j\}, \ P_h \neq 0 \vee \psi_{ij} \neq 1 \vee S_{ij} \neq 1)\}$, there exists a unique solution of the system (14.4).*

Proof. The lemma is trivially true for the case $0 \leq n \leq 2$. Thus, we have to prove the claim for $n > 2$. To this aim, it is sufficient to show that the rank r of the coefficient matrix H of (14.4) is full[2], i.e., $r = n - 1$. Without loss of generality, just for notation convenience, we suppose $1 < j < n$. The coefficient matrix H is

$$H = \begin{pmatrix} -1 & \psi_{1j} \cdot S_{1j} \cdot \frac{1-P_1}{n-2} & \cdots & \psi_{1j} \cdot S_{1j} \cdot \frac{1-P_1}{n-2} \\ \cdots & \cdots & \cdots & \cdots \\ \psi_{nj} \cdot S_{nj} \cdot \frac{1-P_n}{n-2} & \psi_{nj} \cdot S_{nj} \cdot \frac{1-P_n}{n-2} & \cdots & -1 \end{pmatrix}.$$

We proceed by contradiction, supposing that $r < n - 1$. In such a case, there exists a row i of H that can be expressed as a linear combination of the other rows by means of $n - 2$ coefficients, say a_h, $h = 1..n$, $h \neq i, j$. In particular, for the diagonal element $H(i, i) = -1$, the following holds:

$$H(i,i) = \sum_{h=1..n, h \neq i,j} \psi_{hj} \cdot S_{hj} \cdot \frac{1 - P_h}{n - 2} \cdot a_h = -1. \tag{14.4}$$

For the other elements $H(i, t)$ of the row i, where $t = 1..n$, $t \neq i, j$, we obtain

$$H(i,t) = \psi_{tj} \cdot S_{tj} \cdot \frac{1 - P_t}{n - 2} = -a_t + \sum_{h=1..n, h \neq i,j,t} \psi_{hj} \cdot S_{hj} \cdot \frac{1 - P_h}{n - 2} \cdot a_h, \tag{14.5}$$

that is,

$$\psi_{tj} \cdot S_{tj} \cdot \frac{1-P_t}{n-2} = -a_t \left(1 + \psi_{tj} \cdot S_{tj} \cdot \frac{1-P_t}{n-2} \right) + \sum_{h=1..n, h \neq i,j} \psi_{hj} \cdot S_{hj} \cdot \frac{1-P_h}{n-2} \cdot a_h. \tag{14.6}$$

using (14.4), (14.6) becomes

$$\psi_{tj} \cdot S_{tj} \cdot \frac{1 - P_t}{n - 2} = -a_t \cdot \left(1 + \psi_{tj} \cdot S_{tj} \cdot \frac{1 - P_t}{n - 2} \right) - 1 \tag{14.7}$$

from which we derive that $a_t = -1$. For symmetry, $a_h = -1$ for each $h = 1..n$, $h \neq i, j$. As a consequence, (14.4) becomes

$$\sum_{h=1..n, h \neq i,j} \psi_{hj} \cdot S_{hj} \cdot \frac{1 - P_h}{n - 2} = 1. \tag{14.8}$$

Hence, we have reached a contradiction, since by hypothesis for each h, $P_h > 0$ or $\psi_{hj} < 1$ or $S_{hj} < 1$. Thus, (14.8) is false, since $\sum_{h=1..n, h \neq i,j} \psi_{hj} \cdot S_{hj} \cdot \frac{1-P_h}{n-2} < 1$.

---

[2] Recall that the size of H is $n - 1$.

Theorem 14.4.1. *Given an agent $j \in \Lambda$ and a set of 3-tuples of $[0,1]$ real coefficients $\{\langle P_i, \psi_{ij}, S_{ij} \rangle \mid i \in \Lambda \backslash \{j\} \wedge (\forall h \in \Lambda \backslash \{j\},\ P_h \neq 0 \vee \psi_{ij} \neq 1 \vee S_{ij} \neq 1)\}$, there exists a unique $(n-1)$-tuple of $[0,1]$ real values $S = \langle I_{1j}, \ldots, I_{(j-1)j}, I_{(j+1)j}, \ldots I_{nj} \rangle$ satisfying (14.4).*

Proof. Existence and uniqueness is ensured by Lemma 14.4.1. Thus, we have only to prove that I_{hj}, for each $h \in \Lambda \backslash \{j\}$, belongs to the interval $[0,1]$. The theorem is trivially true for the case $0 \leq n \leq 2$. Thus, we have to prove the claim for $n > 2$.

$\mathbf{I_{hj} \geq 0}$, **for each** $\mathbf{h \in \Lambda \backslash \{j\}}$. We start by proving that $I_{hj} \geq 0$, for each $h \in \Lambda \backslash \{j\}$. In particular, we show that the set $V_j = \{h \in \Lambda \backslash \{j\} \mid I_{hj} < 0\}$ is empty. We proceed by contradiction, supposing that $V_j \neq \emptyset$. Let $W_j = \{h \in \Lambda \backslash \{j\} \mid I_{hj} \geq 0\}$.

(14.4) can be rewritten as follows:

$$I_{ij} = \psi_{ij} \cdot S_{ij} \cdot \left(P_i + \frac{1 - P_i}{n - 2} \cdot \left(\sum_{r \in V_j} I_{rj} + \sum_{r \in W_j} I_{rj} - I_{ij} \right) \right).$$

Thus,

$$I_{ij} \cdot \left(1 + \frac{\psi_{ij} \cdot S_{ij} \cdot (1 - P_i)}{n - 2} \right) =$$

$$\psi_{ij} \cdot S_{ij} \cdot \left(P_i + \frac{1 - P_i}{n - 2} \cdot \left(\sum_{r \in V_j} I_{rj} + \sum_{r \in W_j} I_{rj} \right) \right).$$

Now, posing $a_{ij} = 1 + \frac{\psi_{ij} \cdot S_{ij} \cdot (1 - P_i)}{n - 2}$, and applying the summation for each $i \in V_j$, we obtain

$$\sum_{i \in V_j} I_{ij} = \sum_{i \in V_j} \frac{\psi_{ij} \cdot S_{ij}}{a_{ij}} \cdot \left(P_i + \frac{1 - P_i}{n - 2} \cdot \left(\sum_{r \in V_j} I_{rj} + \sum_{r \in W_j} I_{rj} \right) \right),$$

from which we derive

$$\sum_{r \in V_j} I_{rj} \cdot \left(1 - \sum_{i \in V_j} \frac{\psi_{ij} \cdot S_{ij} \cdot (1 - P_i)}{a_{ij} \cdot (n - 2)} \right) =$$

$$\sum_{i \in V_j} \frac{\psi_{ij} \cdot S_{ij}}{a_{ij}} \cdot \left(P_i + \frac{1 - P_i}{n - 2} \cdot \sum_{r \in W_j} I_{rj} \right).$$

Posing $b_j = \sum_{i \in V_j} \frac{\psi_{ij} \cdot S_{ij} \cdot (1 - P_i)}{a_{ij} \cdot (n - 2)} = \sum_{i \in V_j} \frac{1}{1 + \frac{n - 2}{C_{ij} \cdot S_{ij} \cdot (1 - P_i)}}$, we obtain

$$\sum_{r \in V_j} I_{rj} = \frac{1}{1 - b_j} \cdot \sum_{i \in V_j} \frac{\psi_{ij} \cdot S_{ij}}{a_{ij}} \cdot \left(P_i + \frac{1 - P_i}{n - 2} \cdot \sum_{r \in W_j} I_{rj} \right). \qquad (14.9)$$

Since V_j is not empty by hypothesis, $\sum_{r \in V_j} I_{rj} < 0$. As a consequence, (14.9) is false, since its right-hand term is greater than or equal to 0. Indeed, as can be easily verified, $b_j < 1$. We have thus reached a contradiction, and hence the set V_j must be empty. It remains to be proved that $I_{hj} \leq 1$, for each $h \in \Lambda \setminus \{j\}$.

$\mathbf{I_{hj} \leq 1}$, for each $\mathbf{h \in \Lambda \setminus \{j\}}$. We shall demonstrate that the set $M_j = \{h \in \Lambda \setminus \{j\} \mid I_{hj} > 1\}$ is empty. We proceed by contradiction, supposing that $M_j \neq \emptyset$. Let $N_j = \{h \in \Lambda \setminus \{j\} \mid I_{hj} \leq 1\}$.

First, observe that M_j cannot be a singleton. Indeed, if in the tuple $S = \langle I_{1j}, \ldots, I_{(j-1)j}, I_{(j+1)j}, \ldots I_{nj} \rangle$ just one element, say I_{hk}, is greater than 1, then (14.4) is not satisfied, as $\psi_{hk} \cdot S_{hk} \cdot (P_h + (1-P_h) \cdot \frac{1}{n-2} \sum_{r \in \Lambda \setminus \{h,k\}} I_{rk}) \leq 1$.

Thus, we have to consider only the case $|M_j| \geq 2$. Denote by s the cardinality of M_j.

First, rewrite (14.4) as follows:

$$I_{ij} = \psi_{ij} \cdot S_{ij} \cdot \left(P_i + \frac{1 - P_i}{n - 2} \cdot \left(\sum_{r \in M_j} I_{rj} + \sum_{r \in N_j} I_{rj} - I_{ij} \right) \right).$$

Thus,

$$I_{ij} \cdot \left(1 + \psi_{ij} \cdot S_{ij} \cdot \frac{(1 - P_i)}{n - 2} \right) =$$
$$\psi_{ij} \cdot S_{ij} \cdot \left(P_i + \frac{1 - P_i}{n - 2} \cdot \left(\sum_{r \in M_j} I_{rj} + \sum_{r \in N_j} I_{rj} \right) \right).$$

Now, posing $a'_{ij} = 1 + \psi_{ij} \cdot S_{ij} \cdot \frac{(1-P_i)}{n-2}$ and applying the summation for all $i \in M_j$, we obtain

$$\sum_{i \in M_j} I_{ij} = \sum_{i \in M_j} \frac{\psi_{ij} \cdot S_{ij}}{a'_{ij}} \cdot \left(P_i + \frac{1 - P_i}{n - 2} \cdot \left(\sum_{r \in M_j} I_{rj} + \sum_{r \in N_j} I_{rj} \right) \right),$$

from which we derive

$$\sum_{i \in M_j} I_{ij} \cdot \left(1 - \sum_{i \in M_j} \frac{\psi_{ij} \cdot S_{ij} \cdot (1 - P_i)}{a'_{ij} \cdot (n - 2)} \right) =$$
$$\sum_{i \in M_j} \frac{\psi_{ij} \cdot S_{ij}}{a'_{ij}} \cdot \left(P_i + \frac{1 - P_i}{n - 2} \sum_{r \in N_j} I_{rj} \right).$$

Posing $b'_j = \sum_{i \in M_j} \frac{\psi_{ij} \cdot S_{ij} \cdot (1 - P_i)}{a'_{ij} \cdot (n-2)} = \sum_{i \in M_j} \frac{1}{1 + \frac{n-2}{\psi_{ij} \cdot S_i \cdot (1 - P_i)}}$, we have

$$\sum_{r \in M_j} I_{rj} = \tag{14.10}$$

$$\frac{1}{1 - b'_j} \cdot \sum_{i \in M_j} \frac{\psi_{ij} \cdot S_{ij}}{a'_{ij}} \cdot \left(P_i + \frac{1 - P_i}{n - 2} \cdot \sum_{r \in N_j} I_{rj} \right)$$

The right hand term of (14.10) is upper bounded by

$$\sum_{i \in M_j} \frac{1}{1 + \frac{1 - P_i}{n - 2}} \cdot \left(P_i + \frac{1 - P_i}{n - 2} \cdot \sum_{r \in N_j} I_{rj} \right).$$

In turn, the second term in the summation above is upper bounded by

$$\sum_{i \in M_j} \frac{1}{1 + \frac{1 - P_i}{n - 2}},$$

since $\sum_{r \in N_j} I_{rj} \le n - s \le n - 2$.

But, $\frac{1}{1 + \frac{1 - P_i}{n - 2}} < 1$, for each $i \in M_j$, and, thus, the right hand term of (14.10) is less than s. As a consequence, (14.10) is false, since $\sum_{r \in M_j} I_{rj} > s$, and thus we have reached a contradiction. This concludes the proof.

The above result allows us to define the *interest coefficients list* of an agent j as the unique solution of (14.4).

Definition 14.4.2. Given an agent $j \in \Lambda$, the *interest coefficient list of* j is the unique $(n - 1)$-tuple of real values $\langle I_{1j}, \ldots, I_{(j-1)j}, I_{(j+1)j}, \ldots I_{nj} \rangle$ satisfying (14.4). Given an agent $i \ne j$, the *interest coefficient of* i w.r.t j is the value I_{ij} occurring in the interest coefficient list of j.

Besides the interest property, from the knowledge of the interest coefficients lists, agents can exploit a second type of property. Indeed, an agent can compare different agents on the basis of their *attractiveness coefficient*, representing the component of the interest capturing only the contextual knowledge.

Definition 14.4.3. Given a pair of agents $i, j \in \Lambda$, the *attractiveness of* j *perceived by* i is the real coefficient A_{ij} (ranging from 0 to 1) defined as $A_{ij} = \frac{1}{n - 2} \sum_{k \in \Lambda \setminus \{i,j\}} I_{kj}$, where $\langle I_{1j}, \ldots, I_{(j-1)j}, I_{(j+1)j}, \ldots I_{nj} \rangle$ is the interest coefficients list of the agent j.

14.4.3 Choice Lists

Suppose the user of an agent i has the intention of contacting other agents in order to establish a cooperation. Suppose the similarities between i and every other agent is known, along with both the interest coefficient of i with respect to every other agent and the attractiveness of all the agents perceived by i. As previously discussed, such values can be effectively computed once a number of parameters are set (actually, they can be suitably initialized and updated by learning from the behavior of the user, as we shall explain later). Thus, three agent lists can be presented to the user i associated with the agent i, each associated with one of the properties: similarity, interest and attractiveness. We denote these lists $L_S(i)$, $L_I(i)$, and $L_A(i)$, respectively. $L_S(i)$ (or $L_I(i)$, or $L_A(i)$) is the list of the $n-1$ agents j (different from i) ordered by decreasing similarity (or interest, or attractiveness) coefficient S_{ij} (or I_{ij}, or A_{ij}). When the user i chooses an agent j from the list $L_S(i)$ (or $L_I(i)$, or $L_A(i)$), it means that he perceived only the property of similarity (or interest, or attractiveness) about the agent j. From the choices of the users, useful knowledge can thus be drawn that is potentially usable as feedback for correcting advice given to them. This issue is discussed in the next section.

14.4.4 Reactive Properties

By reactive properties we mean properties describing reactions of users to the suggestions received from the system at a given time that must be taken into account for adapting future responses of the system. We implement such adaptation of the system to the user behavior by including into the interest coefficient definition (see Sect. 14.4.2) some specific coefficients that are automatically updated when the system is running. In this section, we describe both the role of such coefficients and the rules defining their adaptation to user behavior. Recall that, given a pair of agents i and j, for defining the interest coefficient I_{ij}, two parameters P_i and ψ_{ij} must be set. They are real parameters ranging from 0 to 1. P_i encodes the *preference property* and is called *preference coefficient* of the agent i, while ψ_{ij} is the product $B_{ij} \cdot C_{ij}$ between the *benevolence coefficient* B_{ij} and *consent coefficient* C_{ij} of i with respect to j. Recall that, given an agent i, by $L_S(i)$, $L_I(i)$, and $L_A(i)$ we denote the three choice lists presented by the system to the user of agent i .

The Preference Property. It is described by a real coefficient ranging from 0 to 1, denoted by P_i, and called *preference coefficient*. The property measures how much, for an agent i, the similarity is more important than the attractiveness for defining global properties. It is easily recognizable that, in the definition of interest given in Sect. 14.4.2, the coefficient P_i plays just this role. Now, we define how the coefficient P_i is updated. Suppose that at a given time the user of the agent i makes a selection of agents. Denote by SI_i (or SS_i, or SA_i) the set of the agents that the user has selected from the list $L_I(i)$ (or $L_S(i)$, or $L_A(i)$). Such choices are interpreted in order to define how to update the coefficient P_i. We adopt the following reasoning: the ratio of the number of agents selected according to the similarity property to the total number of selected agents provides us a perception of the importance the user gives to similarity versus attractiveness. Thus, such a ratio could be used for evaluating the *new* value of P_i. How do we *infer* the number of agents selected for their similarity? Certainly all the agents of SS_i are chosen for their similarity? On the contrary, it is reasonably assumed that agents of

SA_i do not give any contribution to the above number, since they has been chosen *only* on the basis of the attractiveness property. What about agents in SI_i? Here, the choice was done on the basis of the interest property, which mixes similarity and attractiveness. But we can use the old value of P_i for inferring which portion of SI_i has been chosen for the similarity. And this is coherent with the semantic we have given to the preference coefficient. Thus, the total number of agents chosen on the basis of similarity can be assumed to be $P_i \cdot |SI_i| + |SS_i|$. Taking into account the above observations, using P_i after a selection step we updated,

$$P_i = \frac{1}{2} \cdot \left(\frac{| SI_i | \cdot P_i + | SS_i |}{| SI_i | + | SS_i | + | SA_i |} + P_i \right),$$

where $| SI_i | + | SS_i | + | SA_i |$ is the total number of selected agents. This updating is obtained by computing the average between the new contribution with the old value of P_i in order to keep memory of the past and to avoid drastically changing the coefficient.

The Benevolence Property. This property measures a sort of availability of the agent j which a user i requires share knowledge. Such a property is used in order to weigh the interest of i with respect to j. For instance, an agent j that recently, and for several times, has denied collaboration in favor of i should become of little interest to i. The parameter encoding such a knowledge is called *benevolence coefficient*, denoted by B_{ij}, and takes real values ranging from 0 to 1. $B_{ij} = 0$ ($B_{ij} = 1$) means the agent j is completely unavailable (available) to fulfill the requests of i. The response of j to requests of i updates the value of B_{ij} according to the following rules:

$$B_{ij} = \begin{cases} min(1, B_{ij} + \delta) & \text{if } j \text{ grants the request of } i \\ max(1, B_{ij} - \delta) & \text{if } j \text{ denies the request of } i \end{cases}$$

where δ is a (reasonably small) positive real value.

The Consent Property. This property describes how much the user of an agent i trusts suggestions of the system regarding another agent j on the basis of the interest property. The coefficient associated with this property is denoted by C_{ij}, and is called *consent coefficient*. The updating rules defining how to adapt the coefficients C_{ij} after a user selection step take into account only the portion of the selection performed on the list $L_I(i)$. Indeed, from this portion of the user selection, we can draw information about the opinion of the user about the suggestions provided by the system. For instance, if the user of i completely trusts the system capability of providing the best suited agent for cooperation by providing the list $L_I(i)$, he will choose only the first k agents appearing in $L_I(i)$, where k is the size of his selection extracted from $L_I(i)$. This is not in general the case; that is, some of the k agents chosen from $L_I(i)$ do not occur in the set of the first k agents of $L_I(i)$. We defined updating rules by taking into account the above observations according to the following idea: every agent h chosen by the user from $L_I(i)$ produces a gain of the consent coefficient C_{ih} if h is a candidate from the system to be selected, and produces an attenuation of C_{ih} otherwise. More formally, given an agent i and a selection S_i (set of agents) extracted by the user of i from $L_I(i)$, for each $h \in S_i$,

$$C_{ih} = \begin{cases} min(1, C_{ih} + \delta) & \text{if } h \text{ appears among the first } |S_i| \text{ elementsof } L_I(i) \\ max(0, C_{ih} - \delta) & \text{otherwise} \end{cases}$$

where δ is a (reasonably small) positive real value.

14.5 Local Knowledge Base Integration

Cooperation between two agents is implemented in our model by the integration of their LKBs. Thus, the user of an agent i, which has selected an agent j from one of the three choice lists, can solicit the cooperation of j by consulting the *Integrated Knowledge Base* LKB^{ij}, obtained by integrating LKB^i with LKB^j. We show next how LKB^{ij} is defined. Once the LKB^{ij} has been computed, the integration of the knowledge of agent j with that of the client agent i is simply implemented by replacing its LKB with the new LKB^{ij}.

Definition 14.5.1. Let i be an agent. Let $L \in \{L_S(i), L_I(i), L_A(i)\}$ be one of the choice lists of i, and j be an agent selected by i from the list L. Let $G^i = \langle N^i, A^i \rangle$ and $G^j = \langle N^j, A^j \rangle$ be the two graphs of their LKBs .

The *Integrated Knowledge Base* of j in i, denoted by IKB^{ij}, is the pair $\langle G^{ij}, \rho^{ij} \rangle$, where $G^{ij} = \langle N^i \cup N^j, A^i \cup A^j \rangle$ and $\rho^{ij}(h, k)$ (i.e., the priority function) is computed on the basis of the coefficients c_{hk} of the arcs leaving h in G^{ij} as defined by Eq. 14.1. Such coefficients, according to the semantics we have given to the priority function, are defined as follows:

$$c_{hk} = \begin{cases} c_{hk}^i & \text{if } (h, k) \in A^i \setminus A^j \\ c_{hk}^j & \text{if } (h, k) \in A^j \setminus A^i \\ c_{hk}^i + c_{hk}^j & \text{if } (h, k) \in A^i \cap A^j \end{cases}$$

where we denote that the source ontology the coefficients refer to by a superscript.

In other words, the coefficient of an arc in the integrated LKB is obtained by copying the corresponding coefficient from the source LKB (say i, in case the arc belongs only to i), or by summing up the corresponding coefficients (in case the arc appears in both LKBs). The integration process is further illustrated by the following example.

Example 14.5.1. Consider two agents a and b in a music application, and let L_A and L_B of Figure 14.1 be their respective LKBs (for brevity, we do not show in the figure the common ontology on which the two LKBs are defined). In Fig. 14.1, the LKB L_{AB} obtained by integrating the LKBs L_A and L_B is

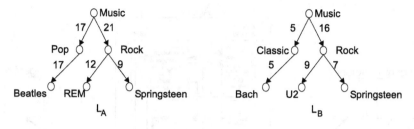

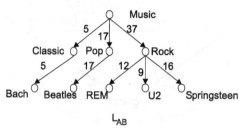

Fig. 14.1. Integration of two LKBs

also shown. Observe that, according to Definition 14.5.1, the coefficients of the arcs (Music, Rock) and (Rock, Springsteen) in L_{AB} have been obtained as sums of the coefficients of the corresponding arcs in L_A and L_B. Indeed, these arcs appear both in L_A and in L_B. Moreover, all the other arcs of L_{AB} keep the coefficients of the LKBs they come from. For instance, the coefficient of the arc (Classic, Bach) is 5, corresponding to the value weighting the arc (Classic, Bach) in the LKB L_B.

Once the integration process is completed and the LKB^{ij} has been computed, we update the LKB of i to be $LKB^i = LKB^{ij}$.

14.6 The SPY System Architecture

The SPY system implements the model described in the previous sections. It is based on a client server architecture as shown in Fig. 14.2. Cooperation of client agents is coordinated and managed from the server side (i.e., the *Agency*). Client agents, possibly cooperating with each other, support user activity. Among all possible approaches, we have considered the case of agent cooperation for helping the user in retrieving information. The user is provided with a set of recommendations that are generated both by his agent and by agents the system has detected as promising for cooperation. The server maintains a copy of the LKB of each agent in the network, updating it each time a change arises. Figure 14.2 zooms in on the (generic) client agent i, showing its architecture.

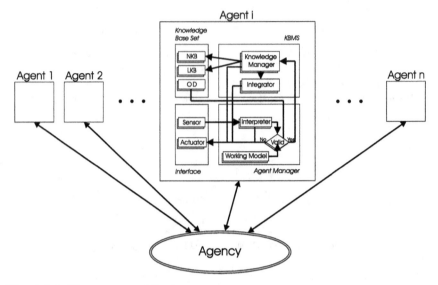

Fig. 14.2. The system architecture

It consists of four main modules that are: (1) the *Interface*, (2) the *Knowledge Base Set*, (3) the *Knowledge Base Management System*, and (4) the *Agent Manager*. We illustrate in detail each of these modules and their function.

1. *Interface.* It is the front-end of the client agent and provides the user with tools for visualizing choice lists (see Sect. 14.4.3), selecting agents from which to ask for cooperation and, consequently, performing the LKB update (by calling the integration procedure). Of course, in general, the interface must also include tools that depend on the application setting. Recall that, in our case, the chosen application setting is information retrieval. As is usual in front-end modules, it is possible to identify two sub-modules, called *Sensor* and *Actuator*, the former including all the functions allowing the user to send requests to the system, and the latter collecting the functions managing the answer of the system to the user's request (see Fig. 14.2). Figure 14.3 reports a snapshot of the interface at run time. The OK button, the search box, the tool for selecting agents in the three list boxes, and the *LKB update* button belong to the sub-module *Sensor*. When the user clicks on the OK button, the system advertises that he would like to contact the selected agents for submitting them the keywords specified in the search box. Furthermore, by clicking on the LKB Update button, the user may require the integration of his local knowledge base and the set of local knowledge bases of the contacted agents, in order to enrich local knowledge with external knowledge he considers useful. The box visualizing the three-choice list and the box

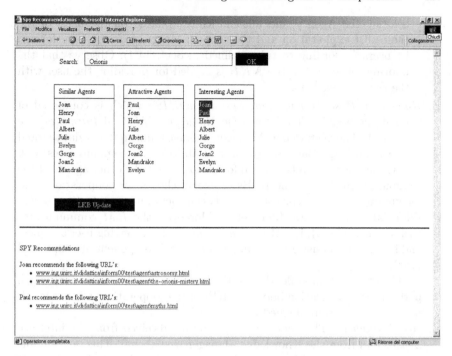

Fig. 14.3. The agent interface module

containing the set of recommendations provided by the system belong to
the sub-module *Actuator*.

2. *Knowledge Base Set.* It contains all the information needed for the agent
   for performing its task. Management and update of such information are
   carried out by module (3) (i.e., the *KBMS*) which is described below.
   As shown in Fig. 14.2, the Knowledge Base Set is composed of the following three sub-modules:
   - the sub-module *OD*, which is the local copy of the Ontology Domain
     (recall that it represents the *apriori* knowledge shared by all the agents
     of the network);
   - the sub-module *LKB*, which is the Local Knowledge Base of the agent
     i (recall that it embeds knowledge about how the user explores the
     Ontology Domain);
   - the sub-module *Network Knowledge Base (NKB)*, which stores the list
     of agents of the community and the lists of similarity, attractiveness,
     and interest coefficients $(S_{ij}, A_{ij}, I_{ij},$ for $j \neq i)$ currently determined
     between the agent i and all the other agents j of the network. Such
     coefficients are computed by the Agency according to rules defined
     in Sect. 14.4, and transmitted to the client agent each time a change
     arises. Recall that the computation of such coefficients needs the knowl-
     edge of preference, benevolence, and consent coefficients. These are

kept by the Agency and automatically updated on the basis of the choice list exploitation and the cooperation grants/denials performed by users (according to rules defined in Sect. 14.4). Observe that the information stored in the NKB is needed for providing the user with the three choice lists.

3. *Knowledge Base Management System* (*KBMS*), which is composed of two sub-modules, called *Knowledge Manager* (*KM*) and *Integrator* (see Fig. 14.2). The Knowledge Manager manages and updates data stored in the Knowledge Base Set (described above), and communicates with the Agency both for retrieving information about the other agents of the community (including the LKBs of agents chosen for cooperation) and for guaranteeing the consistency between local and remote (i.e., server side) data (coefficients, LKB, etc.). Moreover, the KM communicates with the module (4), i.e., the *Agent Manager*, for receiving user's request and sending him answers (the functions of this component are explained next).

    The other sub-module, that is, the Integrator, is the component which performs the integration between LKBs of the cooperating agents according to the procedure defined in Sect. 14.5.

4. *Agent Manager.* This module accepts inputs received from the interface *Sensor* and from the Knowledge Manager (see above). This latter information is transmitted (after suitable processing) to the interface *Actuator* in order to make it available to the user. The Agent Manager is composed of the following sub-modules:

    – the *Interpreter*, which analyzes the Sensor data and, after having checked their correctness (by submitting a request to the sub-module *Validation Component*, which we will describe presently), sends a request to the $KBMS$; the request can be of two types: either a *search request*, or an *update request*. The former is concerned with the supported application, the latter corresponds to the user's request of integrating the LKB of his agent with the LKBs of the agents chosen for cooperation;

    – the *Working Model Component*, which stores the list of the legal actions the user can carry out through the Interface. For instance, the Working Model Component could contain a rule of the type: *the user cannot choose more than 10 agents from the interesting agent list.*

    – the *Validation Component*, whose task is deciding whether the request is consistent with both the Domain Ontology and the Working Model of the agent. An example of failure of such a test is the case where the user submits a keyword not corresponding to any concept appearing in the Ontology Domain. This task is implemented by access to both the Ontology Domain and the Working Model Component. If the request is valid, then it is sent to the $KBMS$; otherwise, a message of *illegal action* is sent to the interface *Actuator*.

14.7 Experiments

In this section, we provide a practical demonstration of the system by tracing its execution through an experimental session. Moreover, we summarize some results obtained from the analysis of a number of experimental execution sessions.

Consider the case of Fig. 14.3. A user, named Frank, has selected two agents from the Interesting Agent list, namely Joan and Paul, and has specified the keyword *Orionis* in the Search box. After Frank clicks on the OK button, the *Sensor* transmits the user specifications to the Agent Manager. This module generates an update request to the $KBMS$ (requiring the update of the preference and the consent coefficients in the NKB), and then verifies that the specified keyword both corresponds to a concept of the Ontology Domain and satisfies the Working Model. Then, the Agent Manager generates a search request for the $KBMS$. As a consequence, the Knowledge Manager of the $KBMS$ solicits from the Agency the transmission of the LKBs of Joan and Paul. After this, the Knowledge Manager looks for the concept *Orionis* in the LKBs of Joan and Paul, and finds it in both of them. Starting from the concept *Orionis* in the LKB of Joan, the Knowledge Manager finds the leaf nodes accessed with a priority greater than a suitable threshold. In our case, such nodes correspond to the URLs www.ing.unirc.it/didattica/inform00/test/agent1/astronomy. html and www.ing.unirc.it/didattica/inform00/test/agent/theorionismistery .html. Similarly, from the analysis of the Paul's LKB, the Knowledge Manager finds the URL www.ing.unirc.it/didattica/inform00/test/agent2/miths.html. The three URLs are then transmitted by the Knowledge Manager to the Actuator, for visualizing them as recommendations. Now, Frank, after having received the recommendations, decides to click on www.ing.unirc.it/didattica/ inform00/test/agent1/astronomy.html. Then, the Sensor notifies the Agent Manager, and this generates a request to the $KBMS$ for updating the priority coefficients of the LKB's nodes involved in the action performed by Frank (note that the action of accessing the node *Orionis* implies the updating of all the ancestor nodes of *Orionis* in Frank, LKB). At this point, Frank, being satisfied with the recommendations provided by Joan, decides to integrate his LKB with Joan's LKB. In order to obtain such an integration, he clicks on the button *LKB Update*. Then, the Sensor transmits a signal to the Agent Manager, which sends an update request to the $KBMS$; as a consequence, the Knowledge Manager of the $KBMS$ uses the Integrator component for carrying out the integration between the user's LKB and Joan's LKB, and updates Frank's LKB with the integrated LKB. Finally, the Knowledge Manager sends the updated LKB to the Agency.

From the analysis of the behavior the system during several experiments, we observe that the semantic properties, encoded by suitable coefficients, drive users to select from the agent network the most promising candidate agents for fruitful cooperation. User choices are exploited, as feedback for adapting coefficients—in such a way that a trade-off among similarity and attractiveness on the one hand, and agent congestion and user dissatisfaction on the other hand—is obtained. To summarize, we report here three meaningful cases tested during experiments: (i) An agent a with high similarity and low attractiveness perceived by another agent b. The user of b decides to contact a less similar, but more attractive, agent c, and this means that the current similarity does not fully satisfy b. Since b has chosen c, as expected it makes choices more similar to those of c than to those of a, and the similarity between a and b decreases, coherently with the dissatisfaction of the user. (ii) An agent a with high interest and low similarity (or low attractiveness) perceived by another agent b. The user of b decides to contact a less interesting, but more similar (or more attractive) agent c. As a consequence, the interest for a perceived by b decreases, due to the decreasing of the consent coefficient C_{ba}. (iii) An agent a with high interest and high attractiveness perceived by another agent b. The user of b knows that high attractiveness probably means a long waiting time for obtaining answers from a, and decides to contact a less interesting agent c. As a consequence, the interest of b for a decreases.

14.8 Conclusions

In this chapter, a framework for representing and managing cooperation among agents in a multi-agent community is provided. The core of the proposal is the definition of a formal model based on several semantic properties and on a linear system involving some coefficients associated with such properties. The solution to such a system allows the user to find the best agents for cooperation in the network, that is, those agents from which the most fruitful cooperation can be expected. Cooperation between two agents is implemented by the integration of their knowledge bases. The main direction for extending this work in our future research concerns the case of heterogeneous domains (in this chapter we assume that agents work in a common environment). In such a case, problems arising from the possible occurrence of semantic heterogeneities have to be faced. Another future work is of implementation type: besides enriching the current prototype, we plan to update the system for incorporating extensions to the formal model.

References

14.1 S. Abiteboul: Querying semi-structured data. *Proceedings of the 6th International Conference on Database Theory (ICDT 97), Delphi, Lecture Notes in Computer Science 1186* (Springer, 1997) pp. 1-18

14.2 S. Adali, R. Emery: A uniform framework for integrating knowledge in heterogeneous knowledge systems. *Proceedings of the Eleventh International Conference on Data Engineering (ICDE 95), Taipei, IEEE Computer Society (1995)* pp. 513-520

14.3 S. Adali, V.S. Subrahmanian: Amalgamating knowledge bases: algorithms, data structures, and query processing. Journal of Logic Programming, 28(1), 45-88 (1996)

14.4 K. Arisha, S. Kraus, R. Ross F. Ozcan, V. Subrahmanian: Impact: the interactive maryland platform for agents collaborating together. IEEE Intelligent Systems Magazine, 14(2), 64-72 (1998)

14.5 G. Boella, L. Lesmo: Norms and cooperation: two sides of social rationality. *Proceedings of the Seventeenth International Joint Conference on Artificial Intelligence (IJCAI 2001), Seattle* (Morgan Kaufmann Publishers, 2001) pp. 4-10

14.6 M. Bowling, M. Veloso: Rational and convergent learning in stochastic games. *Proceedings of the Seventeenth International Joint Conference on Artificial Intelligence (IJCAI 2001), Seattle* (Morgan Kaufmann Publishers, 2001) pp. 1021-1026

14.7 C. Byrne, P. Edwards: Collaborating to refine knowledge. *Proceedings of The Machine Learning '95 Workshop on Agents that Learn from Other Agents, Tahoe City, electronically on WWW (1995)*

14.8 A. Chavez, A. Moukas, P. Maes: Challenger: a multi-agent system for distributed resource allocation. *Proceedings of the First International Conference on Autonomous Agents (Agents 97), Marinadel Rey* (ACM Press, 1997) pp. 323-331

14.9 Y.S. Choi, S.I. Yoo: Multi-agent web information retrieval: neural network based approach. *Proceedings of the Third International Symposium Intelligent Data Analysis (IDA 99), Amsterdam, Lecture Notes in Computer Science 1642* (Springer, 1999) pp. 499-512

14.10 T. Dagaeff, F. Chantemargue, B. Hirsbrunner: Emergence-based cooperation in a multi-agent system. *Proceedings of the Second European Conference on Cognitive Science (ECCS 97)* (Manchester Univ. of Manchester Press, 1997) pp. 91-96

14.11 J.E. Doran, S. Franklin, N.R. Jennings, T.J. Norman: On cooperation in multi-agent systems. The Knowledge Engineering Review, 12(3), 309-314 (1997)

14.12 M. Fisher, J. Muller, M. Schroeder, G. Staniford, G. Wagner: Methodological foundations for agent-based systems. The Knowledge Engineering Review, 12(3), 323-329 (1997)

14.13 F. Gandon: Engineering an ontology for a multi-agents corporate memory system. *Proceedings of the Eighth International Symposium on the Management of Industrial and Corporate Knowledge (ISMICK 2001)* (Compigne, 2001) pp. 209-228

14.14 P.J. Gmytrasiewicz, E.H. Durfee: Rational coordination in multi-agent environments. Autonomous Agents and Multi-Agent Systems, 3(4), 319-350 (2000)

14.15 N.Guarino, C.A. Welty, C. Partridge: Towards ontology-based harmonization of web content standards. *International Conference on Conceptual Modeling ER Workshops (ER 2000), Salt Lake City, Lecture Notes in Computer Science 1920* (Springer, 2000) pp. 1-6

14.16 T. Helmy, S. Amamiya, M. Amamiya: User's ontology-based autonomous interface agents. *Proceedings of The Second International Conference on Intelligent Agent Technology (IAT 2001), Maebashi City* (World Scientific, 2001) pp. 264-273

14.17 C. Iglesias, M. Garijo, J. Centeno-Gonzalez, J.R. Velasco: Analysis and design of multiagent systems using MAS-common KADS. *Proceedings of Agent Theories, Architectures, and Languages (ATAL 97), Providence, Lecture Notes in Computer Science 1365* (Springer, 1997) pp. 313-327

14.18 S. Kraus: Negotiation and cooperation in multi-agent environments. Artificial Intelligence, 94(1-2), 79-97 (1997)

14.19 Y. Lashkari, M. Metral, P. Maes: Collaborative interface agents. *Proceedings of the Twelfth National Conference on Artificial Intelligence (AAAI 94), Seattle* (AAAI Press, 1994) 1, pp. 444-450

14.20 A. Moukas, P. Maes: Amalthaea: An evolving multi-agent information filtering and discovery system for the WWW. Autonomous Agents and Multi-Agent Systems, 1(1), 59-88 (1998)

14.21 M. Mundhe and S. Sen: Evolving agent societies that avoid social dilemmas. *Proceedings of Genetic and Evolutionary Computation Conference (GECCO 2000), Las Vegas* (Academic Press, 2000) pp. 809-816

14.22 M.H. Nodine, J. Fowler, B. Perry: Active information gathering in infosleuth. International Journal of Cooperative Information Systems, 9(1-2), 3-27 (2000)

14.23 L. Palopoli, G. Terracina, D. Ursino: A graph-based approach for extracting terminological properties of elements of XML documents. *Proceedings of the 17th International Conference on Data Engineeringm (ICDE 2001), Heidelberg* (IEEE Computer Society, 2001) pp. 330-337

14.24 L. Peshkin, K.E. Kim, N. Meuleau, L.P. Kaelbling: Learning to cooperate via policy search. *Proceedings of The Sixteenth Conference on Uncertainty in Artificial Intelligence, San Francisco* (Morgan Kaufmann Publishers, 2000) pp. 307-314

14.25 M.V. Nagendra Prasad, V.R. Lesser: Learning situation-specific coordination in cooperative multi-agent systems. Autonomous Agents and Multi-Agent Systems, 2(2), 173-207 (1999)

14.26 M. Schillo, P. Funk: Learning from and about other agents in terms of social metaphors. *Proceedings of the Agents Learning about, from and with other Agents Workshop of the 16th Joint Conference on Artificial Intelligence, Stockholm* (Morgan Kaufmann Publishers, 1999)

14.27 S. Sen, Reciprocity: A foundational principle for promoting cooperative behavior among self-interested agents. *Proceedings of the First International Conference on Multi-Agent Systems (ICMAS 95), Menlo Park* (AAAI Press/MIT Press, 1995) pp. 322-329

14.28 S. Sen: Active information gathering in infosleuth. Trends in Cognitive Science, 1(9), 334-339 (1997)

14.29 S. Sen, A. Biswas, S. Debnath: Believing others: pros and cons. *Proceedings of the Fourth International Conference on Multi-Agent Systems (ICMAS 2000), Boston* (IEEE Computer Society 2000) pp. 279-286

14.30 R. Sun: Individual action and collective function: from sociology to multi-agent learning. Cognitive Systems Research, 2(1), 1-3 (2001)

14.31 M. Tan: Multi-agent reinforcement learning: Independent vs. cooperative agents. *Proceedings of the Tenth International Conference on Machine Learning (ICML 93), Amherst* (Morgan Kaufmann 1993) pp. 330-337

14.32 A. Wang, C. Liu, R. Conradi: A multi-agent architecture for cooperative software engineering. *Proceedings of The Eleventh International Conference on Software Engineering and Knowledge Engineering (SEKE 99), Kaiserslautern* (Knowledge Systems Institute, 1999) pp. 1-22

14.33 G. Weiß: Adaptation and learning in multi-agent systems: Some remarks and a bibliography, Adaptation and Learning in Multi-Agent Systems. Lecture Notes in Computer Science (Springer Verlag, 1996) 1042, pp. 1-21

14.34 M. Wooldridge, N.R. Jennings: The cooperative problem-solving process. Journal of Logic and Computation, 9(4), 563-592 (1999)

15. Constructing Hybrid Intelligent Systems for Data Mining from Agent Perspectives

Zili Zhang[1,2] and Chengqi Zhang[3]

[1] Deakin University, Australia
[2] Southwest China Normal University, China
[3] University of Technology, Sydney, Australia

Abstract

Data mining, the central activity in the process of knowledge discovery in databases, is concerned with finding patterns in data. Many data mining techniques/algorithms that are used to look for such patterns have been developed in domains that range from space exploration to financial analysis. However, a single data mining technique has not been proved appropriate for every domain and data set. Instead, several techniques may need to be integrated into hybrid systems and used cooperatively during a particular data mining operation. That is, hybrid solutions are crucial for the success of data mining. On the other hand, the design and development of hybrid intelligent systems is difficult because they have a large number of parts or components that have many interactions. Existing software development techniques (for example, object-oriented analysis and design) cannot manage these complex interactions efficiently as these interactions may occur at unpredictable times, for unpredictable reasons, between unpredictable components. From a multi-agent perspective, agents are autonomous and can engage in flexible, high-level interactions. They are good at complex, dynamic interactions. To this end, an agent-based framework was proposed to facilitate the construction of hybrid intelligent systems. In this chapter, we will present the agent-based framework first. We then discuss how to apply this framework to construct hybrid intelligent systems for data mining based on a case study. Combining these two cutting-edge technologies, it is expected that more and more difficult real-world problems can be solved. This chapter is a firm step in this direction.

15.1 Introduction

Knowledge Discovery in Databases (KDD) is the non-trivial process of identifying valid, novel, potentially useful, and ultimately understandable structures in data. Data mining is the central step in the KDD process, concerned with applying computational techniques (i.e., data mining algorithms implemented as computer programs) to actually find patterns in the data [15.1]. A number of data mining techniques/algorithms have been developed and used, one at a time, in domains that range from space exploration to financial analysis. Basic algorithms for data mining include linear and multiple regression, top-down induction of decision trees, the covering algorithm for rule induction, finding frequent itemsets and association rules, distance-based prediction and clustering, and learning probabilistic models [15.2]. However, a single data mining technique has not been proven appropriate for every domain and data set. Instead, several techniques may need to be integrated into hybrid systems and used cooperatively during a particular data mining operation [15.4]. That is, hybrid solutions are crucial for the success of data mining.

On the other hand, the design and development of hybrid systems is difficult because they have a large number of parts or components that have many interactions. Existing software development techniques (for example, object-oriented analysis and design) cannot manage these complex interactions efficiently as these interactions may occur at unpredictable times, for unpredictable reasons, between unpredictable components. To this end, an agent-based framework was proposed to facilitate the construction of hybrid intelligent systems [15.5]. This framework has been successfully applied to build a hybrid intelligent system for financial investment planning [15.6]. Many techniques/packages including – fuzzy logic, neural networks, genetic algorithms, expert systems, operations research software package, matrix operation software package, portfolio selection models based on standard probability theory, fuzzy probability theory, and possibility distribution theory – were integrated under the unifying agent framework in the system.

Recently, agent techniques have been applied to distributed data mining. In [15.9] and [15.10], the authors describe a parallel/distributed data mining system PADMA (PArallel Data Mining Agents) that uses software agents for local data accessing and analysis, and a Web-based interface for interactive data visualization. PADMA has been used in medical applications. In [15.11], an agent-based meta-learning system for large-scale data mining applications, which is called JAM (Java Agents for Meta-learning), is described. JAM was empirically evaluated against real credit card transaction data, where the target data mining application was to compute predictive models that detect fraudulent transactions. However, these works are focusing on one of the many steps in data mining.

Kerber, Livezey, and Simoudis reported a hybrid system (called Recon) for data mining [15.4]. In Recon, inductive, clustering, case-based reasoning, and statistical package are integrated and used collaboratively. Recon adopted a typical client/server architecture. However, the adaptability and robustness of Recon do not meet the requirements of real-world applications.

The emphasis of this chapter will be in trying to combine two cutting edge technologies, agent and data mining, by applying the proposed agent-based hybrid framework to construct hybrid intelligent systems for data mining. The work presented in this chapter further verifies the proposed agent framework, and provides an easy way to construct hybrid systems for data mining, which will drive real-world applications. The hybrid systems for data mining based on the framework have two essential characteristics that differentiate our work from existing ones:

- New data mining techniques can be plugged into the system and out-of-date techniques can be dynamically deleted from the system;
- Data mining technique agents can interact at runtime with ease under this framework, while in other non-agent based systems, these interactions must be determined at design-time.

The remainder of this chapter is organized as follows. Section 15.2 explains in detail why hybrid solutions are required for data mining tasks. The suitability of agent perspectives for modeling hybrid intelligent systems is elaborated in Sect. 15.3. Agent-oriented methodologies used to analyze and design agent-based hybrid intelligent systems are outlined in Sect. 15.4. The analysis, design, and implementation of agent-based hybrid systems for data mining are detailed along with a case study in Sect. 15.5. An evaluation of the framework is provided in Sect. 15.6. Finally, Sect. 15.7 has concluding remarks.

15.2 Data Mining Needs Hybrid Solutions

Data mining is concerned with finding patterns in data. It is about exploring the unknown (patterns) in large data sets. Data is a set of facts (e.g., cases in a database). The output of a data mining algorithm is typically a pattern or a set of patterns that are valid in the given data. A pattern is an expression or statement that describes, in a given language, relationships among the facts in a subset of the given data; it is, in some sense, simpler than the enumeration of all facts in the subset. Typical representative patterns are equations, classification and regression trees, classification and regression rules, and positive and negative association rules.

The main data mining tasks include predictive modeling (classification and regression), clustering (grouping similar objects) and summarization (as exemplified by association rule discovery). According to the main tasks, data mining techniques can be divided into five classes [15.3]:

- Predictive modeling. The goal is to predict some field(s) in a database based on other fields. If the field being predicted is a numeric (continuous) variable then the prediction problem is a regression problem. If the field is categorical then it is a classification problem. *There are a wide variety of techniques for classification and regression.*
- Clustering. Unlike classification, we do not know the number of desired "clusters" in clustering. Thus, clustering algorithms typically employ a two-stage search: an outer loop over possible cluster numbers and an inner loop to fit the best possible clustering for a given number of clusters. Given a number, k, of clusters, clustering methods can be divided into three classes: metric-distance-based methods, model-based methods, and partition-based methods.

- Data summarization. One common method in data summarization is taking vertical slices (fields) of the input data. This class of methods is different from those above in that, rather than predicting a specified field (e.g, classification) or grouping cases together (e.g., clustering), its goal is to find relations among fields (e.g., association rules).
- Dependency modeling. Insight into data is often gained by deriving some causal structure within the data. Models of causality can be probabilistic or deterministic, as in derived functional dependencies, between fields in the data. Density estimation methods fall, in general, under this category; so do methods for explicit causal modeling.
- Change and deviation detection. These methods account for sequence information, be it time-series or some other ordering. The distinguishing feature of this class of methods is that ordering of observations is important and must be accounted for.

In a nutshell, there is a variety of methods related to different main tasks in data mining, and outputs of different data mining methods also have different forms. We provide a simple example of a hybrid intelligent system.

The example is about how to identify a set of "promising" securities to be included in an investment portfolio based on the historical fundamental and technical data about the securities. This is a very appropriate domain for data mining for two reasons [15.4]. First, because the number of available securities being traded in the various exchanges is very large, identifying appropriate securities for the goals of a particular portfolio is based on the close examination of the performance of these securities. Without the use of data mining techniques, analysts can only closely examine small amounts of such data. Second, while analysts are able to state criteria for identifying securities that can potentially meet a set of investment goals, they cannot identify all the necessary criteria. Furthermore, even after a set of securities is identified, large volumes of data relating to these securities still has to be examined in order to fine-tune the stated performance criteria, as well as to identify others not previously considered by the analyst. For this simple task, no single data mining technique is adequate. Methods to formulate a pattern (hypothesis) and test its validity on the target databases are needed. Methods to discover other relevant patterns from target databases are also required. Some other methods, including classification methods (to classify each security), inductive learning methods, and visualization techniques, are also helpful for this task. If we construct a computer system to perform this task, it is evident that this system is a hybrid system integrating different techniques.

Once again, data mining is an iterative sequence of many steps, where many techniques may be involved in each step. These techniques need to be integrated into hybrid systems and used cooperatively for data mining tasks.

15.3 Agent Perspectives Are Suitable for Modeling Hybrid Intelligent Systems

As pointed out in Sect. 15.1, the design and development of different hybrid intelligent systems is difficult because they have a large number of parts or components that have many interactions. It is too difficult to determine most of the interactions at design time. Figure 15.1 shows a typical development cycle in the construction of hybrid intelligent systems [15.13]. Most current hybrid intelligent systems are built either from scratch or following this development process.

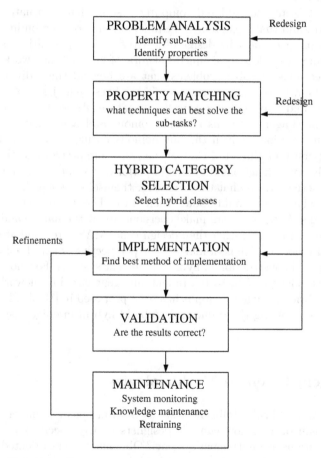

Fig. 15.1. Hybrid intelligent system development cycle

There are some shortcomings to the hybrid intelligent systems developed by following this process. The most obvious one is that the organization of such a hybrid system is not adaptive. Once the techniques are selected in the property matching stage, it is difficult to change or replace them, even though one may find better ones later on. Another difficulty lies in the hybrid category selection phase. In this phase, developers must choose the type of hybrid system required for solving the specific problem. This is not an easy job to do. The hybrid intelligent systems' inherent complexity means that it is impossible to know a priori all potential links or relationships among components constituting a system; interactions will occur at unpredictable times, for unpredictable reasons, between unpredictable components. For this reason, it is futile to try and predict or analyze all the possibilities at design time. At this stage, agent technology is very promising for taking on these responsibilities.

An agent is an encapsulated computer system that is situated in some environment, and that is capable of flexible, autonomous action in that environment in order to meet its design objectives [15.14]. A multi-agent system can be defined as a loosely coupled network of entities that work together to make decisions or solve problems that are beyond the individual capabilities or knowledge of each entity [15.15]. These entities – agents – are autonomous and may be heterogeneous in nature. As we have pointed out, agents in multi-agent systems are autonomous and can engage in flexible, high-level interactions. Given the autonomous nature of agents, they have their own persistent thread of control (i.e., they are active) and they decide for themselves which actions they should perform at what time. The fact that agents are active means that they know for themselves when they should be acting and when they should update their state. The flexible nature of interactions means that agents can make decisions about the nature and scope of interactions at runtime rather than design time. MASs are good at complex, dynamic interactions. Thus, a multi-agent perspective is suitable for modeling, design, and construction of hybrid intelligent systems. For more detailed descriptions, refer to [15.5, 15.16]. In fact, an agent-based framework for constructing hybrid intelligent systems has been proposed in [15.5]. This chapter is focused on applying the framework to build hybrid intelligent systems for data mining.

15.4 Agent-Oriented Methodologies

When building hybrid intelligent systems from an agent point of view, another problem occurs: how can the designers clearly specify and structure their applications as multi-agent systems? Obviously, agent-oriented software engineering methodologies are required.

Existing software development techniques (for example, object-oriented analysis and design) are inadequate for multi-agent system analysis and design [15.17]. There is a fundamental mismatch between the concepts used by object-oriented developers (and, indeed, by other mainstream software

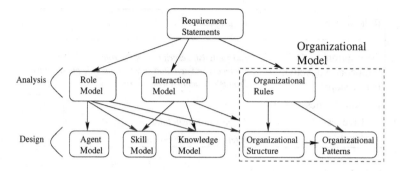

Fig. 15.2. Relationships between models

engineering paradigms) and the agent-oriented perspective. In particular, extant approaches fail to adequately capture an agent's flexible, autonomous problem solving behavior, the richness of an agent's interactions, and the complexity of an agent system's organizational structure. For these reasons, this section outlines a methodology that has been specifically tailored to the analysis and design of agent-based hybrid intelligent systems. The main models used in this methodology, summarized in Fig. 15.2, are mainly based on the Gaia methodology [15.17], but with extensions of skill and knowledge models that are essential for hybrid intelligent systems.

15.4.1 The Role Model

The role model identifies the key roles in the system. Here, a role can be viewed as an abstract description of an entity's expected function. A role is defined by four attributes: *responsibilities, permissions, activities*, and *protocols*.

Responsibilities determine functionality and, as such, are perhaps the key attribute associated with a role. Responsibilities are divided into two types: *liveness properties* and *safety properties*. Liveness properties intuitively state that "something good happens". They describe those states of affairs that an agent must bring about, given certain environmental conditions. In contrast, safety properties are *invariants*. Intuitively, a safety property states that "nothing bad happens" (i.e., that an acceptable state of affairs is maintained across all states of execution).

In order to realize responsibilities, a role has a set of *permissions*. Permissions are the "rights" associated with a role. The permissions of a role thus identify the resources that are available to that role in order to realize its responsibilities. Permissions tend to be *information resources*. For example, a role might have associated with it the ability to read a particular item of information, or to modify another piece of information. A role can also

Role Schema:	*name of role*
Description	*short English description of the role*
Protocols and Activities	*protocols and activities in which the role plays a part*
Permissions	*"rights" associated with the role*
Responsibilities	
Liveness	*liveness responsibilities*
Safety	*safety responsibilities*

Fig. 15.3. Template for role schemata

have the ability to generate information. The *activities* of a role are computations associated with the role that may be carried out by the agent without interacting with other agents.

Finally, a role is also identified with a number of *protocols*, which define the way that it can interact with other roles. A role model is comprised of a set of *role schemata*, one for each role in the system. A role schema draws together the various attributes discussed above into a single place (see Fig. 15.3).

The formal notation for expressing protocols, activities, permissions, and responsibilities adopted by Gaia will be used. To introduce these concepts, the example of a PRICEWATCHER role will be used. The purpose of this role is to monitor whether the trading price of a specific security is exceeded the expected value of the shareholder. The protocols and activities involved in the PRICEWATCHER role include: InformShareholder, GetInitializeInformation, GetPrice, and Compare. The activity names (like Compare) are underlined to distinguish them from protocols.

The following is an illustration of the permissions associated with the role PRICEWATCHER:

    reads supplied *SecurityCode* // Security code used in Share Exchanger
              supplied *ExpectedValue* // The value the shareholder expected
              supplied *TradingPrice* // The current trading price of the security

This specification defines three permissions for PRICEWATCHER: it says that the agent carrying out the role has permissions to access the values of *SecurityCode*, *ExpectedValue*, and *TradingPrice*. The **supplied** keyword is used to indicate that some roles are parameterized by certain values. Another two types of permission are **changes** (read and modify) and **generates** (produce a resource). Note that these permissions relate to the *knowledge* that the agent has.

The liveness responsibilities for the PRICEWATCHER role might be

- whenever the share exchange is not closed, get the trading price of the specific security (indicated by the *SecurityCode*);
- whenever the trading price is exceeded by the expected value, inform the shareholder.

Following the Gaia notation, liveness properties are specified via a *liveness expression*, which defines the "life-cycle" of the role and is a regular expression. The general form of a liveness expression is

$$\textsc{RoleName} = expression,$$

where RoleName is the name of the role whose liveness properties are being defined, and *expression* is the liveness expression defining the liveness properties of RoleName. The atomic components of a liveness expression are either *activities* or *protocols*. The operators for liveness expressions are shown in Table 15.1.

Table 15.1. Operators for liveness expressions

Operator	Interpretation
$x.y$	x followed by y
$x\|y$	x or y occurs
$x*$	x occurs 0 or more times
$x+$	x occurs 1 or more times
x^{ω}	x occurs infinitely often
$[x]$	x is optional
$x\|\|y$	x and y are interleaved

Thus, the liveness responsibilities of the PriceWatcher role can be expressed as

$$\textsc{PriceWatcher} = (\mathsf{GetInitializeInformation}) + .(\mathsf{GetPrice}, \mathsf{Compare}) + .(\mathsf{InformShareholder}) *.$$

This expression says that PriceWatcher consists of executing the protocol GetInitializeInformation, followed by the protocol GetPrice, followed by the activity Compare and the protocol InformShareholder.

Safety requirements are specified by means of a list of predicates. These predicates are typically expressed over the variables listed in a role's permission attribute. By convention, safety expressions are listed as a bulleted list, each item in the list expressing an individual safety responsibility.

When all these are put together, the schema for the PriceWatcher role results (Fig. 15.4).

Role Schema: PRICEWATCHER
Description: This role involves monitoring whether the trading price of a specific security is exceeded the expected value of the shareholder.
Protocols and Activities: InformShareholder, GetInitializeInformation, GetPrice, Compare

Permissions:
reads supplied *SecurityCode* // Security code used in Share Exchanger
         supplied *ExpectedValue* // The value the shareholder expected
         supplied *TradingPrice* // The current trading price of the security

Responsibilities Liveness: PRICEWATCHER = (GetInitializeInformation)+.(GetPrice, Compare)+.(InformShareholder)∗ Safety: • True

Fig. 15.4. Schema for role PRICEWATCHER

15.4.2 The Interaction Model

There are inevitably dependencies and relationships between the various roles
in a multi-agent organization. Indeed, such interplay is central to the way in
which the system functions. Given this fact, interactions obviously need to be
captured and represented in the analysis phase. Such links between roles are
represented in the *interaction model*. This model consists of a set of *protocol
definitions*, one for each type of inter-role interaction. Here, a protocol can
be viewed as an institutionalized pattern of interaction.

A protocol definition consists of the following attributes:

- *purpose*: brief textual description of the nature of the interaction (e.g.,
  "information request", "schedule activity", and "assign task");
- *initiator*: the role(s) responsible for starting the interaction;
- *responder*: the role(s) with which the initiator interacts;
- *inputs*: information used by the role initiator while enacting the protocol;
- *outputs*: information supplied by/to the protocol responder during the
  course of the interaction;
- *processing*: brief textual description of any processing the protocol initiator
  performs during the course of the interaction.

As an illustration, the GetPrice protocol is considered, which forms part
of the PRICEWATCHER role (Fig. 15.5). This states that the protocol Get-
Price is initiated by the role PRICEWATCHER and involves the role SHARE-
EXCHANGER. This protocol involves PRICEWATCHER providing SHAREEX-
CHANGER with the *SecurityCode*, and results in SHAREEXCHANGER return-
ing the value of the *TradingPrice* for security designated by *SecurityCode*.

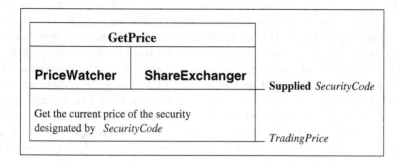

Fig. 15.5. The GetPrice protocol definition

15.4.3 Organizational Rules

Role models precisely describe all the roles that constitute the computational organization in terms of their functionalities, activities, and responsibilities, as well as in terms of their interaction protocols and patterns, which establish the position of each role in the organization. However, such role models cannot be considered as the sole organizational abstraction upon which to base the entire development process. Rather, before the design process actually defines the role model and, consequently, the whole organization, the analysis phase should define how the organization is expected to work, i.e., the organizational rules. These describe the constraints that the actual organization, once defined, will have to respect.

The explicit identification of organizational rules is of particular importance in the context of open agent systems. With the arrival of new and previously unknown agents, the overall organization must somehow enforce its internal coherency despite the dynamic and untrustworthy environment. The identification of global organizational rules allows the hybrid system designer to explicitly define whether, and when, to allow newly arrived agents to enter the organization and, once accepted, what their position in the organization should be.

In summary, the analysis phase is responsible for collecting all the specifications from which the design of the computational organization can start (refer to Fig. 15.2). The output of the analysis phase should be a 3-tuple, $\langle PR, PP, OL \rangle$, where PR are the preliminary roles of the system, PP are the preliminary protocols (which have already been discovered to be necessary for the preliminary roles), and OL are the organizational rules.

15.4.4 The Agent Model

The purpose of the agent model is to document the various *agent types* that will be used in the system under development, and the *agent instances* that will realize these agent types at runtime.

An agent type is best thought of as a set of agent roles. There may, in fact, be a one-to-one correspondence between roles (as identified in the role model) and agent types. However, this need not be the case. A designer can choose to package a number of closely related roles in the same agent type for the purpose of convenience. Efficiency will also be a major concern at this stage: a designer will almost certainly want to optimize the design, and one way of doing this is to aggregate a number of agent roles into a single type.

The agent model is defined using a simple *agent type tree*, in which leaf nodes correspond to roles, and other nodes correspond to agent types. If an agent type t_1 has children t_2 and t_3, then this means that t_1 is composed of the roles that make up t_2 and t_3.

15.4.5 The Skill Model

The aim of the skill model is used to identify the main skills of each agent role. Skills mainly consist of the basic *services* required to be able to perform a role.

A service is defined as a *function* of the agent. For each service that may be performed by an agent, it is necessary to document its properties. Specifically, one must identify the *inputs, outputs, pre-conditions*, and *post-conditions* of each service. Inputs and outputs to services will be derived in an obvious way from the interaction model. Pre- and post-conditions represent constraints on services. These are derived from the safety properties of a role. Note that, by definition, each role will be associated with at least one service.

The services that an agent will perform are derived from the list of protocols, activities, responsibilities, and the liveness properties of a role. The inference mechanisms used by roles are also needed to be identified in this model.

15.4.6 The Knowledge Model

The knowledge model is used to identify the different knowledge levels needed by each agent identified in the agent model. The first level of knowledge is for agent interaction and communication. This involves domain-specific and domain-independent terminologies and their relationships. The domain-specific terms identified, and their relationships, will result in the construction of a domain-dependent ontology for a specific application. The domain-independent terms identified, and their relationships, will result in the customization of a domain-independent ontology from some available general-purpose ontologies.

The second level of knowledge is some domain knowledge related to specific problem solving techniques. This part of knowledge can be represented by typical *if-then* rules. These rules are also domain-specific.

The third level of knowledge is meta knowledge that directs the activities of an agent. This part of knowledge can also be represented by *if-then* rules. These rules are more abstract than those in the second level.

15.4.7 Organizational Structures and Patterns

In the design of a multi-agent system, as well as in the design of any organization, the role model should derive from the organizational structure that is explicitly chosen. Thus, organizational structures should be viewed as first-class abstractions in the design of multi-agent systems.

The definition of the system's overall organizational structure can derive from the specifications collected during the analysis phase, as well as from other factors related to efficiency, simplicity of application design, and organizational theory. In any case, a methodology cannot start the analysis phase by attempting to define a complete role model that implicitly sets the organizational structure. Rather, the definition of the organizational structure is a design choice that should not be anticipated during the analysis phase.

The obvious means by which to specify an organization are the inter-agent relationships that exist within it. There is no universally accepted terminology set for organizational relationships: different types of organizations make use of entirely different organizational concepts.

The aim of organizational patterns is to encourage reuse of predefined components and architectures in order to ease and speed up the work of both designers and developers. With the availability of catalogs of *organizational patterns*, designers can recognize in their multi-agent systems the presence of known patterns, and reuse definitions from the catalog. In addition, designers can also be guided by the catalog in the choice of the most appropriate organizational patterns for their multi-agent system. Of course, for patterns to be properly exploited, the organizational structure must have been explicitly identified in the design phase.

The design phase builds on the output of the analysis phase and produces a complete specification of the multi-agent system. The design stage can now be summarized as follows

- Create an agent model: (1) aggregate roles into agent types, and refine to form an agent type hierarchy; (2) document the instances of each agent type using instance annotations.
- Develop a skill model by examining activities, protocols, and safety and liveness properties of roles.
- Develop knowledge model from the interaction model and agent model.
- Identify organizational structures and organizational patterns that respect the organizational rules.

In the next section, the details of analysis and design of an agent-based hybrid data mining system will be given, which is based on the methodology outlined here.

15.5 Agent-Based Hybrid Systems for Data Mining

As discussed in Sects. 15.1 and 15.2, there are many well established data mining techniques. However, neither of the techniques is a panacea for solving problems involving hundreds of thousands of high dimensional records. A data mining technique can work well in some domains but fail in others [15.7]. To integrate different data mining techniques into hybrid systems and use them cooperatively is of paramount importance. For one data mining task, three techniques may be required to be put together, for another task, five techniques may need to be integrated. Thus, the requirements for such hybrid systems can be identified as follows:

- Different agents based on different data mining techniques can be dynamically tailored to different agent-based hybrid intelligent systems for different data mining tasks;
- Agents based on newly invented data mining techniques can be added to the systems, and agents with out-of-date techniques can be dynamically deleted from the systems. That is, adding agents to the systems or deleting agents from the systems can be done without changing the system design;
- Interactions at runtime are allowed.

15.5.1 Analysis and Design

Based on the methodology described in Sect. 15.4, the first step in the analysis process is to identify the roles in the hybrid system for data mining. Keeping the requirements in mind, the following roles can be identified. To avoid redundancy, only the schema for some of the identified roles are given.

- A role acting as an interface to the users, called USERHANDLER (Fig. 15.6).
- A role overseeing the whole process inside the system (WORKPLANNER, Fig. 15.7).
- To meet the requirement of dynamic tailoring, two roles are needed: one keeping track of the profiles of other roles performing data mining related tasks (CAPABILITYRECORDER, Fig. 15.8) and one checking the profiles (CAPABILITYMATCHER, Fig. 15.9).
- A role to visualize the mined results in some situations (RESULTVISUALIZER).
- The data mining behavior is covered by the ATTRIBUTESELECTOR (Fig. 15.10), FREQUENTITEMSETIDENTIFIER, DATACLEANSER, ASSOCIATIONRULEMINER (Fig. 15.11), DECISIONTREEINDUCER, CLASSIFICATIONRULEGENERATOR, CROSSVALIDATOR, PREDICTIONEVALUATOR, and RESULTACCURACYANALYZER (Fig. 15.12) roles.
- The final role is that of the USER (Fig. 15.13) who requires the data mining results.

Role Schema: UserHandler
Description: Receives request from the user and oversees process to ensure appropriate results are returned.
Protocols and Activities: AwaitCall, InformUser, ProduceAssignments, ReceiveResults
Permissions: **reads supplied** *UserRequirements* // what user wants   *Results* // mined results or nil
Responsibilities Liveness: UserHandler = AwaitCall. ProduceAssignments. ReceiveResults. InformUser Safety: • True

Fig. 15.6. Schema for role UserHandler

Role Schema: WorkPlanner
Description: This role elaborates a work plan for data mining and is in charge of ensuring that such a work plan is fulfilled.
Protocols and Activities: GetUserRequirements, <u>ProducePlan</u>, GetCapabilities, DelegateTasks
Permissions: **reads supplied** *UserRequirements* // detailed service requirements   *Capabilities* // capabilities of other roles **generates** plan // work plan
Responsibilities Liveness: WorkPlanner = GetUserRequirements.GetCapabilities.<u>ProducePlan</u>.DelegateTasks Safety: • *UserRequirements* is clear and available

Fig. 15.7. Schema for role WorkPlanner

Role Schema: CAPABILITYRECORDER
Description: Add capabilities advertised by roles to CapabilityDatabase, or delete capabilities unadvertised by roles from the database.
Protocols and Activities: ReceiveAdvertisement, AddCapability, DeleteCapability
Permissions: **reads supplied** $CapabilityInfo$ // Advertised or unadvertised capability information **changes** $CapabilityDatabase$ // add or delete capability
Responsibilities Liveness: CAPABILITYRECORDER = ReceiveAdvertisement. (AddCapability|DeleteCapability) Safety: • CapabilityDatabase exists and access rights are granted

Fig. 15.8. Schema for role CAPABILITYRECORDER

Role Schema: CAPABILITYMATCHER
Description: Matches capabilities requested by a role with those capabilities in CapabilityDatabase. Informs requester role with the ROLENAME whose capability matched with the requested one.
Protocols and Activities: GetRequestedCapability, AccessCapabilityDatabase, InformRequester, CompareCapability
Permissions: **reads supplied** $RequestedCapability$ // capability requested by a role   **supplied** $CapabilityDatabase$
Responsibilities Liveness: CAPABILITYMATCHER = GetRequestedCapability. (AccessCapabilityDatabase.CompareCapability)+.InformRequester Safety: • True

Fig. 15.9. Schema for role CAPABILITYMATCHER

Role Schema: ATTRIBUTESELECTOR
Description: Access the databases for mining and determine the lists of attributes that are important for a specific task.
Protocols and Activities: ReceiveTask, AccessDatabase, AskforHelp, InformRelevantRole
Permissions: **reads supplied** *Task* // Task delegated by other roles   **supplied** *Databases* // Databases to be mined **generates** *Attributelists* // Important attributes for the mining task
Responsibilities Liveness: ATTRIBUTESELECTOR = ReceiveTask.AccessDatabase.AskforHelp)+.InformRelevantRole Safety: • True

Fig. 15.10. Schema for role ATTRIBUTESELECTOR

Role Schema: ASSOCIATIONRULEMINER
Description: Mines association rules from the itemsets prepared by other roles (e.g., ATTRIBUTESELECTOR).
Protocols and Activities: ReceiveTask, AccessItemsets, SendMinedResults, InformVisualizer, Processing
Permissions: **reads supplied** *Itemsets* // Data prepared by other roles **generates** *MinedResults* // processing the data and   // generate results
Responsibilities Liveness: ASSOCIATIONRULEMINER = RECEIVETASK.AccessItemsets.(Processing)+. (SendMinedResults| INFORMVISUALIZER) Safety: • ASSOCIATIONRULEMINER has the requested capability

Fig. 15.11. Schema for role ASSOCIATIONRULEMINER

Role Schema: RESULTACCURACYANALYZER
Description:
Analyzes the accuracy and performance
of different data mining related roles.
Protocols and Activities:
ReceiveResults, InformUserHandler, TestResults
Permissions:
reads supplied *Results* // Mined results
generates *Measurement*
// Evaluating results of the accuracy and performance, etc.
Responsibilities
Liveness:
RESULTACCURACYANALYZER =
ReceiveResults.(TestResults.InformUserHandler
Safety:
• Mined results are available

Fig. 15.12. Schema for role RESULTACCURACYANALYZER

Role Schema: USER
Description:
Organization or individual requesting a decision.
Protocols and Activities:
MakeCall, GiveRequirements
Permissions:
generates *UserDetails* // Owner of user information
UserRequirements // Owner of user requirements
Responsibilities
Liveness:
USER = (MakeCall.GiveRequirements)+
Safety:
• True

Fig. 15.13. Schema for role USER

With the role definitions in place, the next stage is to define the associated interaction models for these roles. Here, we focus on the interactions associated with the ASSOCIATIONRULEMINER role.

This role interacts with the WORKPLANNER role to obtain the task that it will accomplish (ReceiveTask protocol, Fig. 15.14a). It interacts with the DATAPREPARER role (maybe ATTRIBUTESELECTOR, FREQUENTITEMSETIDENTIFIER, etc.) for where it can access the data for mining (AccessItemsets protocol, Fig. 15.14b). When the ASSOCIATIONRULEMINER role finishes mining association rules, it sends the RESULTACCURACYANALYZER role its mining results for the task (SendMinedResults protocol, Fig. 15.14c). For some roles, but not for ASSOCIATIONRULEMINER, the outputs are also sent to the RESULTVISUALIZER role to visualize and display.

In such a system, the most important organizational rule in the organizational model is that *if a role says it has a capability, then it can perform the tasks corresponding to the capability, and will do so when asked.*

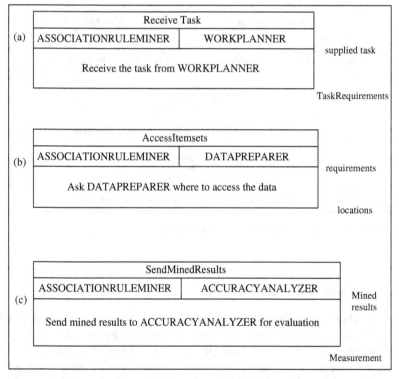

Fig. 15.14. Definition of protocols associated with the ASSOCIATIONRULEMINER role: (a) ReceiveTask, (b) AccessItemsets, and (c) SendMinedResults.

Having completed the analysis of the system, the design phase follows. The most important model to be generated is the agent model (Fig. 15.15). In some cases, there is a one-to-one correspondence between roles and agent types (e.g., USERHANDLER and **InterfaceAgent**, WORKPLANNER and **PlanningAgent**) . In some cases, two or more roles are grouped into a single agent type. For example, the CAPABILITYRECORDER and CAPABILI-TYMATCHER roles are grouped into a **MiddleAgent** because of their high degree of interdependence; the ATTRIBUTESELECTOR, FREQUENTITEMSETI-DENTIFIER, and DATACLEANSER, etc., roles are grouped into **Preprocessin-gAgent**; and so on.

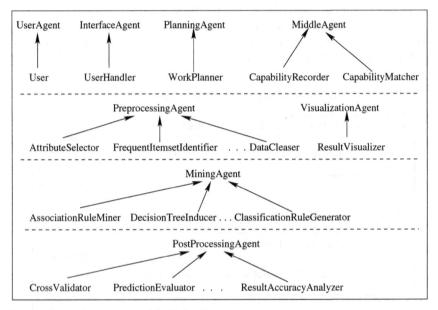

Fig. 15.15. The agent model of the data mining system

15.5.2 Implementation

From the above analysis and design phases, it is clear that there are eight types of agents in the system: user agent, interface agent, planning agent, visualization agent, middle agent, preprocessing agent, mining agent, and post-processing agent. To meet the requirements of such systems, the architecture shown below (Fig. 15.16) is employed to put all these agents together. It is assumed that there are k preprocessing agents, m mining agents, and n post-processing agents. These numbers (k, m, and n) can be dynamically increased or decreased. The behavior of each kind of agent in the system (except the user agent) is briefly described below:

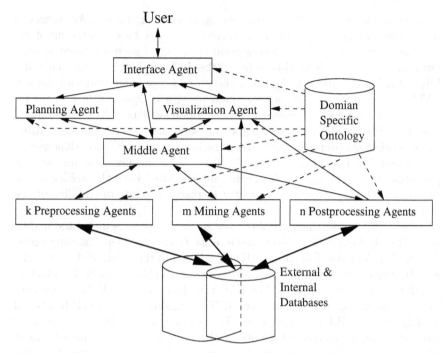

Fig. 15.16. Architecture of agent-based data mining system

- **Interface Agent** This agent interacts with the user (or user agent). It asks the user to provide his requirements, and provides the user with mined results (which may be visualized).
- **Planning Agent** The planning agent is in charge of the activation and synchronization of different agents. It elaborates a work plan and is in charge of ensuring that such a work plan is fulfilled. It receives the assignments from the interface agent.
- **Visualization Agent** It visualizes some mined results and passes them to the interface agent.
- **Middle Agent** It keeps track of the names, ontologies, and capabilities of all registered agents in the system; it can reply to the query of an agent with the name and ontology of an appropriate agent that has the capabilities requested. The introduction of the middle agent in the system ensures flexibility and robustness. How middle agents can bring about flexibility and robustness in the system was discussed in [15.12].
- **Preprocessing Agent** It prepares data for mining.
- **Mining Agent** It mines the data prepared by preprocessing agents and generates relevant patterns.
- **Post-processing Agent** It evaluates the performance and accuracy of mining agents.

The ontology is the foundation for agent communication. All agents in the system interpret the content of received messages based on the ontology.

The most important implementation criterion of such a system is platform independence. With this in mind, the JATLite (Java Agent Template, Lite, *http://java.stanford.edu/*) was chosen to support the implementation. JATLite provides a set of Java templates and a ubiquitous Java agent infrastructure that makes it easy to build common systems. JATLite facilitates especially the construction of agents that send and receive messages using the emerging standard agent communication language KQML (Knowledge Query and Manipulation Language) [15.18]. JATLite does not impose any particular theory of agents. All agents implemented have the ability to exchange KQML messages. This greatly increases the interoperability of the system.

The kernel supporting part of the system is the planning agent and middle agent [15.19]. Figure 15.17 shows the internal structure of the planning agent. The KQML Message Interpreter (KMI) represents the interface between the KQML router and agents. Once an incoming KQML message is detected, it will be passed to the KMI. The KMI transfers incoming KQML messages into a form that agents can understand. The implementation of KMI is based on JATLite KQMLLayer Templates. The planning agent allows the dynamic reorganization of connected agents in the system, while the middle agent allows the agents to be dynamically connected to or disconnected from the system. All these facilitate flexibility and robustness.

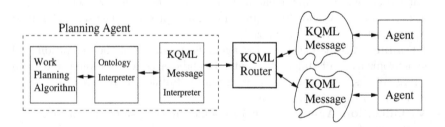

Fig. 15.17. Planning agent structure

Considering there are some legacy data mining (or machine learning) software packages, approaches should be provided to integrate these software packages into the system when needed. This implies that these software packages must be converted into agents in some way. Generally, there are three main approaches to be taken: implementing a *transducer*, implementing a *wrapper*, and *rewriting* the original program [15.20]. Here the second approach, implementing a wrapper, was adopted to wrap the software packages using Java Native Interface [15.21] and JATLite KQML layer templates.

15.5.3 Case Study

The Weka system ([15.8] and *http://www.cs.waikato.ac.nz/ml/weka*) was re-implemented from agent perspectives based on the above discussion. The main focus of Weka is on classifier and filter algorithms. It also includes implementations of algorithms for learning association rules and for clustering data for which no class value is specified.

To re-implement the programs in Weka from agent perspectives, the programs in Weka (written in Java) were compiled into .DLLs (dynamic link libraries) first. The Java Native methods and JATLite KQML layer templates were then employed to wrap these .DLL programs. In this way, all the programs in Weka were equipped with KQML communication capability and were ready to be added to the agent system.

In this agent-based data mining experimental system, in addition to the supporting agents (interface agent, planning agent, middle agent, and so on) there are seven attribute selection-related agents, 25 classifier-related agents, nine filter related agents, and two cluster-related agents.

Figure 15.18 shows the user interface of the system, which can be launched from any Internet Browser or appletviewer.

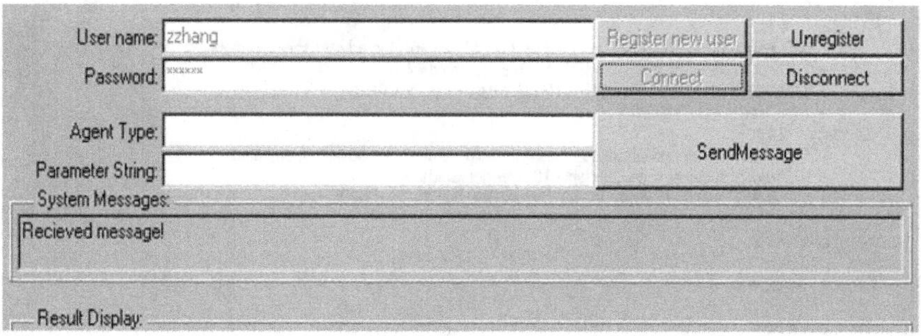

Fig. 15.18. User interface of the system

To use the system, the user needs to type his user name and password in the corresponding fields, and click "register new user" to register for the first time. Thereafter, he just inputs the registered user name and password and clicks "connect". If the user wants to leave the system, he clicks "disconnect" and "unregister".

The system can work in two modes. In one mode, all the data mining related agents can run individually, which is similar to executing the original Weka programs from the command line. In this mode, the user provides the system with the "agent type" and "parameter string" information in the corresponding fields, and then clicks the "SendMessage" button. The system will activate the given agent and display the results in the "result display" window. For example, if we type in "weka.classifiers.m5.M5Prime" in the "agent type" field, and "-t data\cpu.arff" in the "parameter string" field ("data\" was added because all data files in the system are in the "data" subdirectory), the system will display the following results (Fig. 15.19), which are the same as running this program from the command line in Weka:

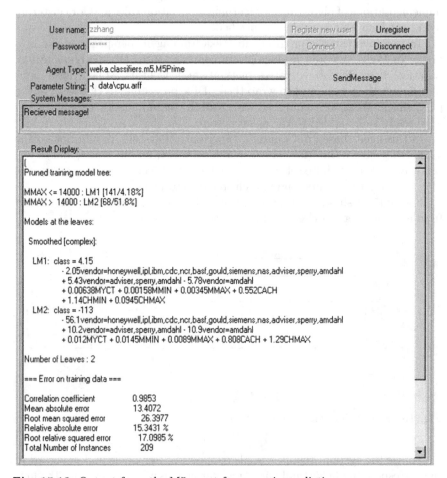

Fig. 15.19. Output from the M5 agent for numeric prediction

In another mode, the planning agent is provided with a work plan. The system then activates different agents based on the work plan. The work plan can be generated automatically, based on meta-knowledge of the task in the planning agent. It can also be edited manually, according to the work plan specification, and loaded into the planning agent. Currently, only the latter is implemented in the experimental system.

It is worth pointing out that agent-based implementation facilitates the integration of different techniques into a system in which the different techniques can work cooperatively to solve complex problems, but it does not directly contribute to the improvement of performance or accuracy of the original algorithms.

15.6 Evaluation

The framework proposed here is suitable for constructing hybrid intelligent systems for different data mining tasks. Here, the evaluation of the framework is given by considering the following:

- *Flexibility.* In the framework, middle agents are used. The presence of the middle agent in the framework allows adaptive system organization. The resulting system, based on the framework, has the ability dynamically to add and delete data mining-related agents (or other kinds of agents), as needed.
- *Robustness.* In the framework, if a particular data mining related agent disappears, a requester agent can find another one with the same or similar capabilities by interrogating the middle agent. Thus, overall system robustness is enabled through the use of the middle agent.
- *Availability of agent capabilities.* As any agent under this framework should register and advertise its capabilities to the middle agent, any other agent can easily access all the capabilities available in the system whenever needed.
- *Interoperability.* One main obstacle of a meaningful interoperation and mediation of services is the syntactic and semantic heterogeneity of data and knowledge that the middle agent accesses and receives from multiple heterogeneous agents and information systems and sources. By using domain-specific ontologies, the semantic heterogeneities are resolved among agents in the framework.
- *Inter-agent interactions.* The interactions among agents are easy, as all agents use the standardized agent communication language KQML.
- *Platform independent.* The prototype was implemented by using Java, and under the support of JATLite. It can execute on any platform with a Java Virtual Machine. Thus, it is platform independent.

From the experimental results, it is clear that all the agents in the experimental system exhibited their behaviors correctly, as specified.

15.7 Conclusions

Data mining involves an iterative process and many different techniques. In this process, different techniques are needed to work cooperatively, rather than individually. This implies that hybrid solutions are needed in data mining. We also noticed that a multi-agent perspective is suitable for modeling, design, and construction of hybrid intelligent systems. To this end, we first proposed an agent-oriented methodology that is especially tailored for the analysis and design of agent-based hybrid intelligent systems. Based on the methodology, we then discussed the analysis and design of agent-based hybrid intelligent systems for data mining. Such systems have the following crucial characteristics that differentiate our work from others:

- new data mining techniques can be added to the system and out-of-date techniques can be deleted from the system dynamically;
- data mining related agents and other agents can interact at runtime with ease under this framework, but in other non-agent based systems these interactions must be determined at design time.

We reimplemented Weka from an agent point-of-view and conducted some experiments under the support of JATLite. Based on the experimental results, it is evident that the proposed framework is applicable to the situations where there are a great number of heterogeneous computing techniques/packages, such as in data mining. More research and experiments are needed for the planning and middle agents.

The agent-based framework proposed in this chapter is not aimed at improving the performance of individual data mining techniques, but at facilitating the construction of hybrid intelligent systems for data mining. By integrating different techniques, many complex problems, which are difficult to tackle using individual techniques alone, can be solved, or better results can be obtained.

References

15.1 U. Fayyad, G.P. Shapiro, P. Smyth: From Data Mining to Knowledge Discovery: An Overview. In: U. Fayyad, G.P. Shapiro, P. Smyth, R. Uthurusamy (eds.), *Advances in Knowledge Discovery and Data Mining* (MIT Press, Cambridge, MA, 1996) pp. 1-34

15.2 S. Dzeroski: Data Mining in a Nutshell. In: S. Dzeroski, N. Lavrac (eds.), *Relational Data Mining* (Springer, 2001) pp. 3-27

15.3 U. Fayyad: Knowledge Discovery in Databases: An Overview. In: S. Dzeroski, N. Lavrac (eds.), *Relational Data Mining* (Springer, 2001) pp. 28-47

15.4 R. Kerber, B. Livezey, E. Simoudis: A Hybrid System for Data Mining. In: S. Goonatilake, S. Khebbal (eds.), *Intelligent Hybrid Systems* (John Wiley & Sons, 1995) pp. 121-142

15.5 Z. Zhang: *An Agent-Based Hybrid Framework for Decision-Making on Complex Problems.* (PhD Thesis, Deakin University, Australia, 2001)

15.6 Z. Zhang, C. Zhang: An Agent-Based Hybrid Intelligent System for Financial Investment Planning. *Proceedings of PRICAI'2002, LNAI 2417* (Springer, August 2002) pp. 355-364

15.7 K. Cios, W. Pedrycz, R. Swiniarski: *Data Mining Motheds for Knowledge Discovery* (Kluwer Academic Publishers, 1998)

15.8 I. Witten, E. Frank: *Data Mining: Practical Machine Learning Tools and Techniques with Java Implementations* (Morgan Kaufmann publishers, 2000)

15.9 H. Kargupta, B. Stafford, I. Hamzaoglu: Web Based Parallel/Distributed Medical Data Mining Using Software Agents. *Proceedings of 1997 Fall Symposium* (American Informatics Association, 1997) http://www.eecs.wsu.edu/~hillol/pubs.html

15.10 H. Kargupta, I. Hamzaoglu, B. Stafford: Scalable, Distributed Data Mining Using an Agent Based Architecture. *Proceedings of Knowledge Discovery and Data Mining* (AAAI Press, 1997) pp. 211-214

15.11 A. Prodromidis, P. Chan, S. Stolfo: Meta-learning in Distributed Data Mining Systems: Issues and Approaches. In: H. Kargupta, P. Chan (eds.), *Advances in Distributed and Parallel Knowledge Discovery* (AAAI/MIT Press, 1999)

15.12 Z. Zhang, C. Zhang: An Improvement to Matchmaking Algorithms for Middle Agents. *Proceedings of the First International Joint Conference on Autonomous Agents and Multi-Agent Systems* (ACM Press, July 2002) pp. 1340-1347

15.13 S. Goonatilake, S. Khebbal (eds.): *Intelligent Hybrid Systems* (Wiley, 1995)

15.14 M. Wooldridge: Agent-Based Software Engineering. IEE Proc. Software Engineering, 144(1), 26-37 (1997)

15.15 G. Weiss (ed.): *Multiagent Systems: A Modern Approach to Distributed Artificial Intelligence* (The MIT Press, 1999)

15.16 Z. Zhang, C. Zhang: Agent-Oriented Approaches to Constructing Hybrid Intelligent Systems. *Proceedings of 7th International Conference on Neural Information Processing 1* (Taejon, Korea, 2000) pp. 258-263

15.17 M. Wooldrige, N. Jennings, D. Kinny: The Gaia Methodology for Agent-Oriented Analysis and Design. Journal of Autonomous Agents and Multi-Agent Systems, 3(3), 285-312 (2000)

15.18 T. Finin, Y. Labrou, J. Mayfield: KQML as an Agent Communication Language. In: J.M. Bradshaw (ed.), *Software Agents* (AAAI Press/ The MIT Press, Menlo Park, CA, 1997) pp. 291-316

15.19 K. Decker, K. Sycara, M. Williamson: Middle Agents for the Internet. *Proceedings of 15th International Joint Conference on Artificial Intelligence (Nogoya, Japan, 1997)* pp. 578-583

15.20 M.R. Genesereth, S.P. Ketchpel: Software Agents. Communications of the ACM, 37(7), 48-53 (1994)

15.21 R. Gordon: *Essential JNI: Java Native Interface* (Prentice-Hall, New Jersey, 1998)

16. Making Agents Acceptable to People

Jeffrey M. Bradshaw[1], Patrick Beautement[2], Maggie R. Breedy[1], Larry Bunch[1], Sergey V. Drakunov[3], Paul J. Feltovich[1], Robert R. Hoffman[1], Renia Jeffers[1], Matthew Johnson[1], Shriniwas Kulkarni[1], James Lott[1], Anil K. Raj[1], Niranjan Suri[1], and Andrzej Uszok[1]

[1] Institute for Human and Machine Cognition
    University of West Florida, USA
[2] QinetiQ, Malvern Technology Centre, UK
[3] Tulane University, USA

Abstract

Because ever more powerful intelligent agents will interact with people in increasingly sophisticated and important ways, greater attention must be given to the technical and social aspects of how to make agents acceptable to people [16.72]. From a technical perspective, we want to help ensure the protection of agent states, the viability of agent communities, and the reliability of the resources on which they depend. To accomplish this, we must guarantee, insofar as is possible, that the autonomy of agents can always be bounded by an explicit enforceable policy that can be continually adjusted to maximize the agents' effectiveness and safety for both human beings and computational environments. From a social perspective, we want agents to be designed to fit well with how people actually work together. Explicit policies governing human-agent interaction, based on careful observation of work practice and an understanding of current research in the social sciences and cognitive engineering, can help assure that effective and natural coordination, appropriate levels and modalities of feedback, and adequate predictability and responsiveness to human control are maintained. These factors are key to providing the reassurance and trust that are the prerequisites to the widespread acceptance of agent technology for non-trivial applications.

16.1 Introduction

Since the beginning of recorded history, people have been fascinated with the idea of non-human agencies.[1] Popular notions about androids, humanoids, robots, cyborgs, and science fiction creatures permeate our culture, forming the backdrop against which software agents are perceived. The word robot, derived from the Czech word for drudgery, entered public discourse following Karel Capek's 1921 play *RUR: Rossum Universal Robots* [16.21] (Fig. 16.1).

---

[1] Works by authors such as Schelde [16.80] and Clute and Nicholls [16.26], who have chronicled the development of popular notions about androids, humanoids, robots, and science fiction creatures, are a useful starting point for agent designers wanting to plumb the cultural context of their creations. Lubar's chapter "Information beyond computers" in [16.64] provides a useful grand tour of the subject. See Ford, Glymour, and Hayes [16.40] for a delightful collection of essays on android epistemology.

While Capek's robots were factory workers, the public has also at times embraced the romantic dream of robots as "digital butlers" who, like the mechanical maid in the animated feature *The Jetsons* would someday putter about the living room performing mundane household tasks (Fig. 16.2).[2] Despite such innocuous beginnings, the dominant public image of artificially intelligent creatures has often been more a nightmare than a dream. Would the awesome power of robots reverse the master-slave relationship with human beings (Fig. 16.3)?[3] Would seeing the world through the eyes of agents lead to dangerously distortions of reality (Fig. 16.4)? Everyday experiences of computer users with the mysteries of ordinary software, riddled with annoying bugs, incomprehensible features, and dangerous viruses reinforce the fear that the software powering autonomous creatures would pose even more problems. The more intelligent the robot, the more capable of pursuing its own self-interest rather than that of its human masters (Fig. 16.5); the more human-like the robot, the more likely it is to exhibit human frailties and eccentricities (Fig. 16.6). Such latent images cannot be ignored in the design of software agents-indeed, there is more than a grain of truth in each of them!

---

[2] It is interesting to note that today's robotic vacuum cleaners have little resemblance to mechanical maids. However, that is true in part because they are conceived as inexpensive single-function appliances and not multi-purpose assistants. Were our current technical prowess sufficient to build cheap, smart, and versatile robotic assistants, there is little doubt that we would prefer models that featured a "good brain and an unspecialized body" [16.68]. See also Sect. 16.4.2 below.

[3] Whether or not such futures are plausible is besides the point – there is no doubt that the fears are real for many people right now. For example, Bill Joy notes the "prophecy" of the Unabomber, asserting that while his "mentality was criminal, his vision is rather realistic": 'What we do suggest is that the human race might easily permit itself to drift into a position of such dependence on the machines that it would have no practical choice but to accept all of the machines' decisions. As society and the problems that face it become more and more complex and machines become more and more intelligent, people will let machines make more of their decisions for them, simply because machine-made decisions will bring better results than man-made ones. Eventually a stage may be reached at which the decisions necessary to keep the system running will be so complex that human beings will be incapable of making them intelligently. At that stage the machines will be in effective control. People won't be able to just turn the machines off, because they will be so dependent on them that turning them off would amount to suicide.' *Theodore Kaczynski* – the criminal Unabomber. On the other hand, just one year ago Stephen Hawking, the noted physicist, suggested using genetic engineering and biomechanical interfaces to computers in order to make possible a direct connection between brain and computers 'so that artificial brains contribute to human intelligence rather than opposing it.' The professor concedes it would be a long process, but important to ensure biological systems remain superior to electronic ones. "In contrast with our intellect, computers double their performance every 18 months," he told Focus magazine. "So the danger is real that they could develop intelligence and take over the world." [16.55].

Fig. 16.1. Scene from Capek's play *Rossum Universal Robots*

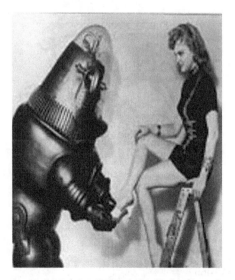

Fig. 16.2. Electro the Robot (aka Robby the Robot) as digital butler to Anne Frances

Fig. 16.3. Powerless in the grasp of a robot (From *Astounding Science Fiction,* October 1953)

Fig. 16.4. Select-O-Vision

Fig. 16.5. A robot thief

Fig. 16.6. In Arthur C. Clarke's *2001: A Space Odyssey*, human beings are compelled to hide from the psychotic computer HAL

"Agents occupy a strange place in the realm of technology," summarizes Don Norman, "leading to much fear, fiction, and extravagant claims" [16.72]. By their ability to operate independently without constant human supervision, they can perform tasks that would be impractical or impossible using traditional software applications. On the other hand, this additional autonomy, if unchecked, also has the potential of effecting severe damage if agents are poorly designed, buggy, or malicious. Because ever more powerful intelligent agents will increasingly differ from software that people are accustomed to, we need to take into account social issues no less than the technical ones if the agents we design and build are to be acceptable to people. Continues Norman:

The technical aspect is to devise a computational structure that guarantees that from the technical standpoint, all is under control. This is not an easy task.

"The social part of acceptability is to provide reassurance that all is working according to plan.... This is also a non-trivial task" [16.72].[4]

This chapter summarizes our efforts to address, through a policy-based approach (Sect. 16.2), some of the technical and social aspects of agent design for increased human acceptability. From a technical perspective, we want to help ensure the protection of agent states, the viability of agent communities, and the reliability of the resources on which they depend. To accomplish this, we must guarantee, insofar as is possible, that the autonomy of agents can always be bounded by explicit enforceable policy that can be continually adjusted to maximize the agents' effectiveness and safety for both human beings and computational environments (Sect. 16.3).

From a social perspective, we want agents to be designed to fit well with how people actually work together. Explicit policies governing human-agent interaction, based on careful observation of work practice and an understanding of current research in the social sciences and cognitive engineering, can help assure that effective and natural coordination, appropriate levels and modalities of feedback, and adequate predictability and responsiveness to human control are maintained (Sect. 16.4). In short, interaction among humans and agents must be graceful and should enhance rather than hinder human work. All these factors are key to providing the reassurance and trust that are the prerequisites to the widespread acceptance of agent technology for non-trivial applications.[5]

---

[4] Similarly, Alan Kay has written: "It will not be an agent's manipulative skills, or even its learning abilities, that will get it accepted, but instead its safety and ability to explain itself in critical situations. At the most basic level the thing we want most to know about an agent is not how powerful it can be, but how trustable it is" [16.56].

[5] A more complete study of many of these topics can be found in [16.11, 16.112]. For an entertaining and informative general characterization of various approaches to human-centered computing, see [16.51].

16.2 Addressing Agent Acceptability Through the Use of Policy

The idea of building strong social laws into intelligent systems can be traced at least as far back as the 1940s to the science fiction writings of Isaac Asimov [16.6]. In his well-known stories of the succeeding decades, he formulated a set of basic laws that were built deeply into the positronic-brain circuitry of each robot so that it was physically prevented from transgressing them. Though the laws were simple and few, the stories attempted to demonstrate just how difficult they were to apply in various real-world situations. In most situations, although the robots usually behaved "logically," they often failed to do the "right" thing, typically because the particular context of application required subtle adjustments of judgments on the part of the robot (e.g., determining which law took priority in a given situation, or what constituted helpful or harmful behavior).[6]

Shoham and Tennenholtz [16.84] introduced the theme of social laws into the agent research community, where investigations have continued under two main headings: *norms* and *policies*. Drawing on precedents in legal theory, social psychology, social philosophy, sociology, and decision theory [16.119], *norm-based* approaches have grown in popularity [16.9, 16.33, 16.62, 16.63]. In the multi-agent system research community, Conte and Castelfranchi [16.32] found that norms were variously described as constraints on behavior, ends or goals, or obligations. For the most part, implementations of norms in multi-agent systems share three basic features:

1. they are designed offline, or
2. they are learned, adopted, and refined through the purposeful deliberation of each agent; and
3. they are enforced by means of incentives and sanctions.

Interest in *policy-based* approaches to multi-agent and distributed systems has also grown considerably in recent years (http://www.policy-workshop.org). While sharing much in common with norm-based approaches, policy-based perspectives differ in subtle ways. Whereas in everyday English the term *norm* denotes a practice, procedure, or custom regarded as typical or widespread,

---

[6] In an insightful essay, Roger Clarke explores some of the implications of Asimov's stories about the laws of robotics for information technologists [16.25]. Weld and Etzioni [16.120] were the first to discuss the implications of Asimov's first law of robotics for agent researchers. Like most norm-based approaches described below (and unlike most policy-based approaches), the safety conditions are taken into account as part of the agents' own learning and planning processes rather than as part of the infrastructure. In an important response to Weld and Etzioni's "call to arms," Pynadath and Tambe [16.76] develop a hybrid approach that marries the agents' probabilistic reasoning about adjustable autonomy with hard safety constraints to generate "policies" governing the actions of agents. The approach assumes a set of homogeneous agents who are motivated to cooperate and follow optimally-generated policies.

a *policy* is defined by the American Heritage Online dictionary as a "course of action, guiding principle, or procedure considered expedient, prudent, or advantageous." Thus, in contrast to the relatively descriptive basis and self-chosen adoption (or rejection) of norms, policies tend to be seen as prescriptive and externally-imposed entities. Whereas norms in everyday life emerge gradually from group conventions and recurrent patterns of interaction, policies are consciously designed and put into and out of force at arbitrary times by virtue of an explicitly-recognized authority.[7] These differences are generally reflected in the way most policy-based approaches differ from norm-based ones with respect to the three features mentioned above. Policy-based approaches

1. support dynamic runtime policy changes, and not merely static configurations determined in advance;
2. work involuntarily with respect to the agents, that is, without requiring the agents to consent or even be aware of the policies being enforced, thus aiming to guarantee that even the simplest of agents comply with policy; and
3. wherever possible, are enforced preemptively, preventing in advance buggy, poorly designed, unsophisticated, or malicious agents from doing harm, rather than rewarding them or imposing sanctions on them after the fact.

In the following subsections, we define policy in the sense that it is used in this section 16.2.1 and distinguish it from related concepts 16.2.2. We then offer definitions of the two major types of policy 16.2.3, describe the relationship between autonomy and policy 16.2.4, discuss both traditional focus areas and new challenges for policy management 16.2.5, and outline the most important aspects and benefits 16.2.6.

16.2.1 What Is Policy?

In agent and distributed computing contexts, policy can be defined as *an enforceable, well-specified constraint on the performance of a machine-executable action by a subject in a given situation.*

- *enforceable:* In principle, an action controlled by policy must be of the sort that it can be prevented, monitored, or enabled by the system infrastructure;
- *well-specified:* Policies are well-defined declarative descriptions;
- *constraint on the performance:* The objective of policy is to ensure, with or without the knowledge or cooperation of the entity being governed, that the policy administrator's intent is carried out with respect to whether or not the specified policy governed action takes place;

---

[7] While it is true that, over time, norms can be formalized into laws, policies are explicit and formal at the outset by their very nature.

- *machine-executable action:* In addition to purely machine-executable actions, we include situations where a person is responsible for completing an action and then somehow signaling that fact to the machine;
- *subject:* The subject is either a human being or a hardware or software component, or a group of such entities;
- *situation:* Policy applicability may be determined by a variety of preconditions and contextual factors.

16.2.2 Distinguishing Policy from Related Concepts

It is evident that not every constraint in an agent system should be managed as an element of policy. Nor should policy in the sense we are discussing it here be confused with other related concepts. For example, the term *policy* is often used to describe what we will call a "Big P" policy, referring to the sorts of high-level declarations of objectives or preferences that one finds in discussions of strategic policy, public policy, or foreign policy. While it is true that every policy of the sort we are concerned with (call them "little p") is motivated by some higher level objective, "Big P" policies comprise a diversity of elements, some of which involve real world considerations that go far beyond distributed computing issues. Resolving the ambiguities and contradictions of complex and "soft" goals, guidelines, and tradeoffs at the "Big P" level is more the stuff of human deliberation and automated planning than of policy management frameworks which are best suited to analysis and implementation of well-understood constraints *after* the difficult preliminary framing has been done.[8]

Policy management also should not be confused with planning or workflow management, which are related but separate functions. Planning mechanisms are generally *deliberative* (i.e., they reason deeply and actively about activities in support of complex goals) whereas policy mechanisms tend to be *reactive* (i.e., concerned with simple actions triggered by some environmental event) [16.43]. Whereas plans are a unified roadmap for accomplishing some coherent set of objectives, bodies of policy collected to govern some sphere of activity are made up of diverse constraints imposed by multiple potentially – disjoint stakeholders and enforced by mechanisms that are more or less independent from the ones directly involved in planning. Plans tend to be strategic and comprehensive, while policies, in our sense, are by nature tactical and piecemeal. In short, we might say that while policies constitute the "rules of the road" – providing the stop signs, speed limits, and lane markers that serve to coordinate traffic and minimize mishaps – they are not sufficient to address the problem of "route planning."[9]

---

[8] The relationship between policy at human and computational levels is a subject we are currently investigating.

[9] For an example of how planning and policy management capabilities can complement on another, see [16.113]. Planning can also be used to help assure successful execution of obligation policies.

Policies should not be mistaken for business rules, for while motivations for business rules sometimes overlap with those for policy-based approaches, these two different attempts to enforce regularities on complex systems have usually maintained a different focus. In a manner similar to the world of policies, we can distinguish between "Big B" and "little b" business rules. A "Big B" business rule "pertains to any of the constraints that apply to the behavior of people in the enterprise, from restrictions on smoking to procedures for filling out a purchase order" [16.58]. On the other hand, "little b" business rules pertain "to the facts which are recorded as data and constraints on changes to the values of those facts. That is, the concern is what data may or may not be recorded in the information system" [16.58]. Like "Big P" policies, "Big B" business rules have a much broader scope than "little p" policies. The "little b" rules, on the other hand, are certainly narrower than "little p" policies, to the extent that the former are restricted to governing the kinds of actions that can be performed on a particular instance of a business database rather than to a broader concept of action in general.

Finally, it should be realized that unwanted circumstances cannot be prevented, nor required events be made to happen, by policy management mechanisms alone. A variety of potential failures must be considered and counteracted in the design of safe and effective agent systems, including extreme events; hardware failure; human error; incorrect system design, specification, or implementation; and inconsistency, redundancy, inaccuracy, or incompleteness of agent knowledge and system information [16.43].

16.2.3 Types of Policy

Drawing on their long history of policy research, Sloman et al.[1.34] define the two major types of policy, *authorizations* and *obligations:*

- "A positive authorization policy defines the actions that a subject is permitted to perform on a target. A negative authorization policy specifies the actions that a subject is forbidden to perform on a target".
- "Obligation policies specify the action that a subject must perform on a set of target objects when an event occurs. Obligation policies are always triggered by events, since the subject[10] must know when to perform the specified action".[11]

---

[10] In the KAoS policy management framework (see Sect. 16.3), a type of enforcer called an *enabler* can be defined to assist subjects in fulfilling obligations, thus reducing, or ideally eliminating, the need for the agent itself to fully understand the policy and to know when and how to undertake its responsibilities [16.16]. Enablers can also be defined for some types of authorization policies.

[11] Some systems differentiate a second class of obligations that requires a given desired state to be continuously maintained by an unspecified action (e.g., Agent A must maintain at least 10 widgets in the bin) in contrast to normal obligations that require a specific action to be performed in response to a trigger (e.g., IF

16.2.4 Autonomy and Policy

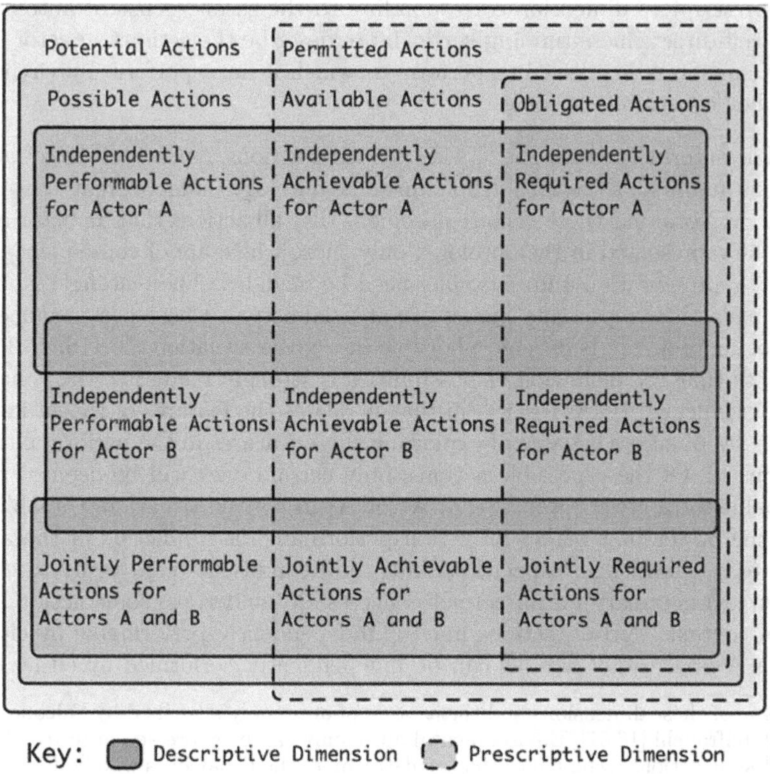

Key: ■ Descriptive Dimension   ⊡ Prescriptive Dimension

Fig. 16.7. Dimensions of autonomy

Some important dimensions relating to autonomy can be straightforwardly characterized by reference to Fig. 16.7 [12] Note that there are two basic dimensions:

---

the number of widgets <= 10 THEN Agent must fill the bin with widgets). For example, Pynadath and Tambe [16.76] distinguish between four classes of *safety constraints:* forbidden actions, forbidden states, required actions, and required states. In KAoS (see Sect. 16.3), forbidden actions correspond to negative authorization policies, while required actions and states map to positive obligation policies. Since many states of the world are outside of system control and cannot be forbidden *a priori,* they can best be handled by representing a forbidden state (ideally with some safety margin) as a trigger to a positive obligation policy that requires the agent to achieve a permissible state.

[12] These dimensions are explained more fully in [16.112]. Note that in this section we emphasize those dimensions that are most pertinent to our discussion of adjustable autonomy; see elsewhere for examples of other possible dimensions (e.g., self-impositions, norms). We can make a rough comparison between

- a descriptive dimension-corresponding to the sense of autonomy as self-sufficiency-that stretches horizontally to describe the actions an actor in a given context is capable of performing; and
- a prescriptive dimension-corresponding to the second sense of autonomy as self-directedness-running vertically to describe the actions an actor in a given context is allowed to perform or which it must perform by virtue of policy constraints in force.

The outermost rectangle, labeled potential actions, represents the set of all actions across all situations defined in some ontology under current consideration. [13] Note that there is no requirement that all actions that an actor may take be represented in the ontology; only those which are of consequence for policy representation and reasoning need be included. The rectangle labeled possible actions represents the set of potential actions whose performance by one or more actors is deemed plausible in a given situation [16.116, 16.117]. [14] Note that the definition of possibilities is strongly related to the concept of affordances [16.118, 16.119], in that it relates the features of the situation to classes of actors capable of exploiting these features in the performance of actions. [15] Of these possible actions, only certain ones will be deemed performable for a given actor [16] (e.g., Actor A) in a given situation. Capability, i.e., the power that makes an action performable, is a function of the abilities (e.g., knowledge, capacities, skills) and conditions (e.g., ready-to-hand resources) necessary for an actor to successfully undertake some action in a given context. Certain actions may be independently performable by either Actor A or B; other actions can be independently performed by either one

---

some of these dimensions and the aspects of autonomy described by Falcone and Castelfranchi [16.37]. Environmental autonomy can be expressed in terms of the possible actions available to the agent-the more the behavior is wholly deterministic in the presence of a fixed set of environmental inputs, the smaller the range of possible actions available to the agent. The aspect of self-sufficiency in social autonomy relates to the ranges of what can be achieved independently vs. in concert with others; deontic autonomy corresponds to the range of permissions and obligations that govern the agent's choice among actions.

[13] The term ontology is borrowed from the philosophical literature, where it describes a theory of what exists. Such an account would typically include terms and definitions only for the very basic and necessary categories of existence. However, the common usage of ontology in the knowledge representation community is as a vocabulary of representational terms and their definitions at any level of generality. A computational system's "ontology" defines what exists for the program-in other words, what can be represented by it.

[14] The evaluation of possibility admits varying degrees of confidence-for example, one can distinguish mere plausibility of an action from a more studied feasibility. These nuances of possibility are not discussed in this chapter.

[15] As expressed by Norman: "Affordances reflect the possible relationships among actors and objects: they are properties of the world" [16.119].

[16] For discussion purposes, we use the term actor to refer to either a biological entity (e.g., human, animal) or an artificial agent (e.g., software agent, robotic agent).

or the other uniquely. [17] Yet other actions are jointly performable by a set of actors. [18]

Along the prescriptive dimension, declarative policies may specify various permissions and obligations [16.34]. An actor is free to the extent that its actions are not limited by permissions or obligations. Authorities may impose or remove involuntary policy constraints on the actions of actors. Alternatively, actors may voluntarily enter into agreements that mutually bind them to some set of policies for the duration of the agreement. The effectivity of an individual policy specifies when it is in or out of force. The set of permitted actions is determined by authorization policies that specify which actions an actor or set of actors is allowed (positive authorizations or A+ policies) or not allowed (negative authorizations or A- policies) to perform in a given context. [19] The intersection of what is possible and what is permitted delimits the set of available actions. Of those actions that are available to a given actor or set of actors, some subset may be judged to be independently achievable in the current context. Some actions, on the other hand, would be judged to be only jointly achievable.

Finally, the set of obligated actions is determined by obligation policies that specify actions that an actor or set of actors is required to perform (positive obligations or O+ policies) or for which such a requirement is waived (negative obligations or O- policies). Jointly obligated actions are those that two or more actors are explicitly required to perform.

A major challenge in the design of intelligent systems is to ensure that the degree of autonomy is continuously and transparently adjusted in order to meet whatever performance expectations have been imposed by the system designer and the humans and agents with which the system interacts [16.110, 16.112, 16.120]. We note that is not the case that "more" autonomy is always better: [20] as with a child left unsupervised in city streets during rush hour, an unsophisticated actor insufficiently monitored and recklessly endowed with unbounded freedom may pose a danger both to itself and to others. On the

---

[17] Note that Figure 16.7 does not show every possible configuration of the dimensions, but rather exemplifies a particular set of relations holding for the actions of a particular set of actors in a given situation. For example, although we show A and B sharing the same set of possible actions, this need not always be the case. Also, note that the range of jointly achievable actions has overlap only with Actor B and not Actor A.

[18] Authority relationships may be, at the one extreme, static and fixed in advance and, at the other, determined by negotiation and persuasion as the course of action unfolds.

[19] We note that some permissions (e.g., network bandwidth reservations) involve allocation of finite and/or consumable resources, whereas others do not (e.g., access control permissions). We note that obligations typically require allocation of finite abilities and resources; when obligations are no longer in effect, these abilities and resources may become free for other purposes.

[20] In fact, the multidimensional nature of autonomy argues against even the effort of mapping the concept of "more" and "less" to a single continuum.

other hand, a capable actor shackled with too many constraints will never realize its full potential.

Thus, a primary purpose of adjustable autonomy is to maintain the system being governed at a sweet spot between convenience (i.e., being able to delegate every bit of an actor's work to the system) and comfort (i.e., the desire to not delegate to the system what it can't be trusted to perform adequately). [21]

The coupling of autonomy with policy mechanisms gives the agent maximum freedom for local adaptation to unforeseen problems and opportunities while assuring humans that agent behavior will be kept within desired bounds. If successful, adjustable autonomy mechanisms give the added bonus of assuring that the definition of these bounds can be appropriately responsive to unexpected circumstances.

In principle, the actual adjustment of an agent's level of autonomy could be initiated either by a human being, the agent, or some other software component. [22] To the extent we can adjust agent autonomy with reasonable dynamism (ideally allowing handoffs of control among team members to occur anytime) and with a sufficiently fine-grained range of levels, teamwork mechanisms can flexibly re-negotiate roles and tasks among humans and agents as needed when new opportunities arise or when breakdowns occur. Such adjustments can also be anticipatory when agents are capable of predicting the relevant events [16.104, 16.105]. Research in adaptive function allocation-the dynamic assignment of tasks among humans and machines-provides some useful lessons for implementations of adjustable autonomy in intelligent systems [16.105].

When evaluating options for adaptively reallocating tasks among team members, it must be remembered that dynamic role adjustment comes at a cost-in both computational and human terms. Measures of expected util-

---

[21] We note that reluctance to delegate can also be due to other reasons. For example, some kinds of work may be enjoyable to people-such as skilled drivers who may prefer a manual to an automatic transmission.

[22] Cohen and Fleming [16.105] draw a line between those approaches in which the agent itself wholly determines the mode of interaction with human beings (mixedinitiative) and those where this determination is imposed externally (adjustable autonomy). Additionally, mixed-initiative systems are considered by Cohen and Fleming to generally consist of a single user and a single agent. However, it is clear that these two approaches are not mutually exclusive and that, in an ideal world, agents would be capable of both reasoning about when and how to initiate interaction with human beings and subjecting themselves to the external direction of whatever set of explicit authorization and obligation policies were currently in force to govern that interaction. Additionally, there is no reason to limit the notion of "mixed initiative" systems to the single agent-single human case. Hence, we prefer to think of mixed-initiative systems as being systems that are capable of making context-sensitive adjustments to their level of social autonomy (i.e., their level or mode of engagement with human beings), whether a given adjustment is made as a result of reasoning internal to the agent or due to externally imposed policy-based constraints.

ity can be used to evaluate the tradeoffs involved in potentially interrupting the ongoing activities of agents and humans in such situations to communicate, coordinate, and reallocate responsibilities [16.105, 16.108, 16.109]. It is also important to note that the need for adjustments may cascade in complex fashion: interaction may be spread across many potentially distributed agents and humans who act in multiply connected interaction loops. For this reason, adjustable autonomy may involve not merely a simple shift in roles among a human-agent pair, but rather the distribution of dynamic demands across many coordinated actors. [23] Defining explicit policies for the transfer of control among team members and for the resultant modifications required to coordination constraints can prove useful in managing such complexity [16.111]. Whereas goal adoption and the commitment to join and interact in a prescribed manner with a team sometimes occurred as part of a single act in early teamwork formulations, researchers are increasingly realizing the advantages of allowing the acts of goal adoption, commitment to work jointly with a team, and the choice of specific task execution strategies to be handled with some degree of independence [16.7, 16.110].

16.2.5 Benefits of Policy Management

A policy-based approach has many benefits:

Explicit license for autonomous behavior. Policy representations that allow the description of entities and actions at abstract levels (e.g., ontologies) can beneficially underspecify the constraints of policy, giving human stakeholders as much leeway as they require to shape the limits of agent behavior across an arbitrarily large scope of action, while leaving every unmentioned detail completely in the hands of the agents that are closest to the problem. Thus, the coupling of policy with autonomy enables human organizations to *think globally while acting locally*. In short, rather than mistakenly thinking of policy only as a restrictive nuisance, we might more productively think of it as the explicit license by which agents are authorized to make specific decisions and adaptations autonomously in response to novel problems and opportunities as they arise – without violating the constraints imposed by those who are responsible for their behavior.

Reusability. Policies encode sets of useful constraints on agent or component behavior, packaging them in a form in which they can easily be reused as occasions require. By reusing policies when they apply, we reap the lessons learned from previous analysis and experience while saving the time it would

---

[23] As Hancock and Scallen [16.105] rightfully observe, the problem of adaptive function allocation is not merely one of technical elegance. Economic factors (e.g., can the task be more inexpensively performed by humans, by agents, or by some combination?), political and cultural factors (e.g., is it acceptable for agents to perform tasks traditionally assigned to humans?), or personal and moral factors (e.g., is a given task enjoyable and challenging vs. boring and mind-numbing for the human?) are also essential considerations.

have taken to reinvent them from scratch. Policy libraries can package sets of policies that have been pre-approved for particular situations. For example, military applications may have different policy sets defined that come into play for various levels of threat conditions.

Efficiency. In addition to lightening the application developers' workload, well-defined policy management mechanisms can sometimes increase runtime efficiency [16.76]. For example, to the extent that policy conflict resolution can be performed offline in advance, and policies can be converted to an efficient runtime representation, overall performance can be increased [16.16, 16.117].

Extensibility. A well designed policy management capability provides a layer of basic representations and services that can be straightforwardly extended to diverse and evolving platforms and to sets of operational capabilities that are often subject to rapid rates of technology refresh. Ideally, these modifications could be made without extensive manual markup or duplication of information stored elsewhere in the organization.

Context-sensitivity. Explicit policy representation improves the ability of agents, components, and platforms to be responsive to changing conditions without changing their code. In mature policy management systems, such changes to policy can be either made manually through convenient distributed administration capabilities or triggered programmatically by events.

Verifiability. By representing policies in an explicitly declarative form instead of burying them in the implementation code, we can better support important types of policy analysis [16.43]. First – and this is absolutely critical for security policies – we can externally validate whether or not the policies are sufficient for the application's tasks, and we can bring both automated theorem provers and human expertise to this task. Second, there are methods to ensure that agent behavior which follows the policy will also satisfy many of the important properties of reactive systems: liveness, recurrence, safety invariants, and so forth.

Support for simple as well as sophisticated agents. By putting the burden for policy analysis and enforcement on the infrastructure, rather than having to build such knowledge into each of the agents themselves, we ensure that all agents operate within the bounds of policy constraints [16.15]. In this way, even one agent shall not be lost due to policy violations, no matter how simple or sophisticated the agent's design, and the task of agent developers is thereby reduced in complexity [16.48].

Protection from poorly-designed, buggy, or malicious agents. Intelligent systems functioning in complex environments cannot rely on design-time techniques to completely eliminate the possibility of unwanted events occurring during operations.[24] Moreover, even if it could be guaranteed that agents designed by a given group would always function correctly, the fact

---

[24] As Fox and Das [16.43] wisely observe, "the nature of a hazard will frequently be unknown until it actually arrives. In some circumstances, ensuring that a system reliably does what the designers intended – and only what they intended – may be exactly the wrong thing to do!"

is that, as long as reliance on open systems continues to increase, the possibility of buggy or malicious agents designed by others cannot be completely ignored. Various forms of policy-based barriers that can control the actions of such agents through monitoring, analysis, inference, adjustable autonomy, and enforcement methods that are infrastructure-based and independent of the agents' own reasoning, appear to be the most effective ways to reduce the risk of these serious problems [16.59].

Reasoning about agent behavior. As permitted by disclosure policies [16.83], sophisticated agents can reason about the implications of the policies that govern their behavior and the behavior of other agents. To the extent that behavior can be predicted from policy, making accurate and consistent models of agents becomes more feasible.

16.2.6 Applications of Policy Using KAoS and Nomads

At the Institute for Human and Machine Cognition (IHMC), we have developed KAoS and Nomads to support a wide range of policy and domain services. KAoS a collection of componentized policy and domain management services compatible with several popular agent frameworks, including Nomads, the DARPA CoABS Grid, the DARPA ALP/Ultra*Log Cougaar framework (http://www.cougaar.net), CORBA (http://www.omg.org), Voyager (http://www.recursionsw.com/osi.asp), Brahms (www.agentisolutions. com), TRIPS [16.2, 16.3, 16.38], and SFX (http://crasar.eng.usf.edu /research/publications.htm). While initially oriented to the dynamic and complex requirements of software agent applications, KAoS services are also being adapted to general-purpose grid computing (http://www.gridforum.org) and Web Services (http://www.w3.org/2002/ws/) environments as well. A comparison between KAoS, Rei, and Ponder for policy specification, representation, reasoning, and enforcement is given in [16.121]. More complete descriptions of KAoS and Nomads can be found in [16.112, 16.113].

To help motivate a later discussion of different kinds of policy, and to give some idea of the wide range of problems to which policy-based approaches can be applied, we briefly describe some applications.

The DARPA CoABS-sponsored Coalition Operations Experiment (CoAX) (http://www.aiai.ed.ac.uk/project/coax/) is a large international cooperation that models military coalition operations and implements agent- based systems to mirror coalition structures, policies, and doctrines. CoAX aims to show that the agent-based computing paradigm offers a promising new approach to dealing with issues such as the interoperability of new and legacy systems, the implicit nature of coalition policies, security, and recovery from attack, system failure, or service withdrawal [16.4]. The most recent CoAX-related work also investigates issues in composition of semantic web services consistent with negotiated policy constraints [16.114]. KAoS provides mechanisms for overall management of coalition organizational structures represented as domains and policies, while Nomads provides strong mobility, resource management, and protection from denial-of-service attacks to untrusted agents that run in its environment.

Within the DARPA Ultra*Log program (http://www.ultralog.net), we are collaborating with CougaarSoft to extend and apply KAoS policy and domain services to assure the scalability, robustness, and survivability of logistics functionality in the face of information warfare attacks or severely constrained or compromised computing and network resources. In agent societies of over 1,000 agents and hundreds of policies, dynamic policy updates can be committed and distributed across multiple hosts in a matter of seconds, and responses to policy authorization queries average less than 1 ms [16.122].

As part of the Army Research Lab Advanced Decision Architectures Consortium, we have been investigating the use of KAoS and Nomads technologies to enable soldiers in the field to use agents from handheld devices to perform functions such as dynamically tasking sensors and customizing information retrieval. Suri has developed an agile computing platform that provides a foundation for this work [16.90, 16.91, 16.92, 16.94]. We have also commenced an investigation of requirements for policy-based information access and analysis within intelligence applications.

An application focused more on the social aspects of agent policy is within the NASA Cross-Enterprise and Intelligent Systems Programs [16.23], where we are investigating the integration of Brahms, an agent-based design toolkit that can be used to model and simulate realistic work situations in space, with KAoS policy-based models and Nomads's strong mobility and resource control capabilities to drive human-robotic teamwork and adjustable autonomy for highly-interactive autonomous systems, such as the Personal Satellite Assistant (PSA). The PSA is a softball-sized flying robot that is being developed to operate onboard spacecraft in pressurized micro-gravity environments [16.44]. The same approach has also being generalized for use in mobile robots for planetary surface exploration [16.85]. The Office of Naval Research (ONR) is supporting research to extend this work on effective human-agent interaction to unmanned vehicles and other autonomous systems that involve close, continuous interaction with people. As one part of this research, IHMC and the University of South Florida are developing a new robotic platform with carangiform (fish-like) locomotion, specialized robotic behaviors for humanitarian demining, human-agent teamwork, agile computing, and mixed-initiative human control.

We are investigating issues in adjustable autonomy and mixed-initiative behavior for software assistants under funding from the DARPA EPCA (CALO) program [16.112, 16.120]. Under funding from DARPA's Augmented Cognition Program, we are also taking the challenge of effective human-agent interaction one step further as we investigate whether a general policy-based approach to the development of cognitive prostheses can be formulated, in which human-agent teaming could be so natural and transparent that robotic and software agents could appear to function as direct extensions of human cognitive, kinetic, and sensory capabilities (see Sect. 16.4.2).

16.3 Technical Aspects of Agent Acceptability

Norman suggests that the technical considerations include such things as ensuring robustness against technical failures, guarding against error and maliciousness, and protecting privacy [16.72]. We touch on each of these considerations in some way in this section. Later on in the chapter (Sect. 16.4.1), we present examples of policies relating to social aspects of agent behavior. Admittedly the distinction between the two kinds of examples is not always clearcut.

Examples of the kinds of basic infrastructure that will be required to support the technical aspects of agent acceptability are becoming more available. Designed from the ground up to exploit next-generation Internet and Web-Services capabilities, grid-based approaches, for example, aim to provide a universal source of dynamically pluggable, pervasive, and dependable computing power, while guaranteeing levels of security and quality of service that will make new classes of applications possible ([16.42]; http://www.gridforum.org). By the time these sorts of approaches become mainstream for large-scale applications, they will also have migrated to ad hoc local networks of very small devices [16.45, 16.92].

This being said, however, we must go far beyond these current efforts to enable the vision of long-lived agent communities performing critical tasks (Fig. 16.8). Current infrastructure implementations typically provide only very simple forms of resource guarantees and no incentives for agents and other components to look beyond their own selfish interests. At a minimum, future infrastructure must go beyond the bare essentials to provide pervasive *life support services* (relying on mechanisms such as orthogonal persistence and strong mobility [16.88, 16.89]) that help ensure the survival of agents that are designed to live for many years. Beyond the basics of individual agent protection, long-lived agent communities will depend on *legal services*, based on explicit policies, to ensure that individual and societal rights and obligations are monitored and enforced. Benevolent *social services* might also be provided to proactively adjust autonomy to avoid problems and help agents fulfill their obligations [16.112]. Although some of these elements exist in embryonic stage within specific agent systems, their scope and effectiveness has been limited by the lack of underlying support at both platform and application levels.

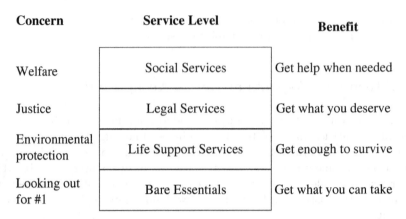

Fig. 16.8. Required elements of future infrastructure for software agents

16.3.1 Examples of Policy Types Relating to Technical Aspects of Agent Acceptability

To better describe the nature of policy as it relates to the technical aspects of agent acceptability, we now discuss several examples. These examples are intended not to be comprehensive but illustrative. Some of them are related to actual policies that we have used in various applications of KAoS; others reflect cases we have anticipated but not yet implemented.

For clarity, we will present example policies in ordinary English rather than in OWL (Web Ontology Language, used to represent KAoS policies). For brevity, the policies will be presented in an incomplete, abbreviated form. Each example is preceded by A+, A-, O+, or O- to indicate whether it is, respectively, a positive authorization, a negative authorization, a positive obligation, or a negative obligation. Note that although we present many of the policies in "IF ... THEN" form for convenient exposition, such conditional information is actually represented in KAoS in the form of OWL property restrictions on action classes rather than in rules. We will look at six categories of technical policy: authentication, data and resource access and protection, communication, resource control, monitoring and response, and mobility.

Authentication policies.

O+: *IF KPAT is launched*
   THEN that instance of KPAT is required to successfully complete a strong authentication process within time T
   PRECEDENCE: A-: no one can use this instance of KPAT
   ELSE O+: this instance of KPAT must terminate.

In this example, which is typical of some of the policies developed within our DARPA Ultra*Log research, the policy assures that strong authentication will be performed each time an effort is made to launch KPAT. Strong authentication is an abstract action that can represent any number of more specific strong authentication methods in the ontology that are available to the system. The authentication might be performed by KPAT itself or delegated to an enabler. One could argue that this policy should be hard wired into the code rather than represented explicitly. That, however, would reduce flexibility in ways that may not be desirable. For example, KPAT administrators at times may want to take this policy out of force in an emergency situation.

In OWL, we represent the precedence conditions as one or more policies. In this case, a negative authorization policy forbids any use of KPAT until the conditions of the obligation policy are fulfilled. Roles are represented straightforwardly as merely one kind of domain or group in which human or agent actors belong. It is recommended to use time, or some more general state indicator, as one of the conditions of obligation fulfillment in order to minimize the risk of the agent getting "stuck" indefinitely. Consequences of non-fulfillment of the obligation (the "ELSE" clause) are also represented as policies. In this case, KPAT is obliged to terminate if the obligation is not successfully fulfilled. In subsequent examples, we will not always list the precedence, conditions of fulfillment, or consequences of non-fulfillment explicitly.

A-: *A user is forbidden from taking any action with account A*
   IF the user has login_failure_count >= n and time since failure <= T

This negative authorization policy, again representative of our Ultra*Log work, deals with authentication failure. After a given number of login failures, the user is locked out of the account until some period of time elapses.

O+: *IF the space station crew member has issued a voice command*
   THEN the Personal Satellite Assistant (PSA) is required to authenticate the crew member's voice within time T
   PRECEDENCE: A-: PSA is forbidden to perform the action corresponding to the command
   ELSE O+: PSA notifies crew member appropriately

This example is drawn from our NASA human-agent teamwork research. Since authorization for some PSA action may depend on who commanded it, authentication of the crewmember's voice is required before the action is performed. The "ELSE" clause embeds a notification policy.

Data and resource access and protection policies.

A-:*Agent X is forbidden from saving data that is unsigned and/or unencrypted*

This data protection policy example specifies that agent X must sign and encrypt all data that it saves. In our work with Ultra*Log, the encryption would be performed by an enabler. In other words, each time X saves data, the policy is enforced through the enabler transparently doing the proper sort of encryption on its behalf.

A-: *All actors in Role R are forbidden from performing any action on servlet*
S

This resource access policy prevents any unauthorized use of Java servlet S by actors (i.e., agents or humans) who are in role R. As in a previous example, the power of abstract specification in the ontology is highlighted: note that this policy can be specified without having to know in advance the particular actions that can be performed on the servlet.[25] Additional Ultra*Log examples of this kind include policies governing actions such as Java JAR file verification, limiting access to private keys, and predicate-based access restriction to blackboard information.

A+: *Users in Role CA Administrator are permitted to perform the revoke certificate operation on the CA Service*

Users or agents in a given role or with a given privilege are authorized here to revoke certificates.

An important part of our current investigations on policy-based information access for intelligence applications concerns disclosure policies. These sorts of policies control the kinds of intelligent responses that can be given as part of queries about which policies are relevant to a given user's analysis or decision-making context [16.114]. In a related application, an agent may want to know about the policies of a given domain before it registers to join. Disclosure policies would determine what kind of policy information could be given to that agent without compromising confidentiality. We are drawing on the work of [16.83, 16.122] to develop more complex strategies for policy disclosure and automated trust negotiation in a variety of circumstances.

Communication policies. Communication has proven to be the most important application of policy within our CoAX research [16.4]. Typically, the domains are configured to be in the "tyrannical" mode, blocking communication among different countries, organizations, or functional groups unless otherwise specified. For example, administrators from the fictional country of Arabello decided on the following restrictive default policy for the actors in their domain:

A-: *Agents in the Arabello domain are forbidden from sending messages to any agent outside the Arabello domain*

However administrators from the Arabello contingent wanted to enable the Arabello Intel agent to be able to send a subset of its reports to the coalition. They specified the following policy, which was assigned a higher precedence for policy conflict resolution purposes:

---

[25] Capability-based access is a term used by Suri to describe an additional level of protection, where all of the details of service implementation are hidden from the client for confidentiality purposes.

A+: *Arabello Intel Agent is permitted to send messages about enemy diesel submarines to any member of the Binni-Coalition domain (sharing messages about any other topic is still forbidden)*

Communication blocking based on message content as illustrated in this example is facilitated by the use of a custom editor within KPAT that allows the administrator to specify the kinds of messages that are to be permitted based on OWL-typing of various message fields [16.90].

A+: *MAD Sensor Agent is permitted to send reports with image resolution X:Y to any member of the Arabello domain*

As part of CoAX, as well as in follow-on Army research, we have also addressed requirements for filtering and transformation of data [16.90, 16.94]. For example the provider of a Magnetic Anomaly Detector (MAD) sensor was willing to share its reports with Arabello, but only on condition that the sensor's signal could be appropriately downgraded in order to prevent Arabello from knowing the full extent of the sensor's capabilities. The policy enforcer-enabler used in this application could be configured by policy to allow three different types of data transformation: a) changes in image resolution, b) changes in frame rate and c) introduction of time lags to prevent transmission of a real time video feed.

Many other types of policy-based transformations could be envisioned for sensor data feeds. A policy enforcer-enabler could, for instance, be implemented to hide sensitive targets or classified infrastructures from the image. This would be used to prevent the release of unnecessary details to the requesting agent by blurring or editing the image appropriately. Another example is an agent that reduces the precision of coordinate values embedded in message content. More generally, such filtering and transformation techniques can be used for sources and methods protection, and as part of the management of information pedigrees and digital rights protection.

In the Ultra*Log application, policies are required to block both sending and receiving of certain kinds of messages. The fact that KAoS policies can specify whether the site of enforcement is to be associated with the subject or the target is useful for this purpose: both the sending and the receiving can be blocked at either the subject or the target side as convenience dictates. *Policy templates* developed for Ultra*Log allow users to specify a composite set of multiple policies more simply as if it were a single policy. To take a simple example, the details of blocking of both sending and receiving messages are accomplished through a simple user interface that presents policy specification options in terms of the more general concept of "communication blocking." As additional examples of communication policies, Ultra*Log also requires that administrators be able to specify which cryptographic modes, transport types, and message formats are allowable in a given context. It also requires limits on message size and system resources in message delivery.

Resource control policies. Whereas resource access policies govern whether or not a resource is made available, resource control policies go a step further to control the amount and rate of resource usage (e.g., CPU, memory, network, hard disk, screen space). For example as part of one of the CoAX scenarios the country of Gao requests permission to host one of its agents on a sensor platform. Because its intentions are unclear and it is distrusted, it is required to run on top of the Aroma VM. Because the Aroma VM is Java-compatible, Gao is not aware of this restriction. Later, when Gao's agent launches a denial-of-service attack which floods the network and begins consuming inordinate amounts of CPU and disk resources, the pattern of misuse is noticed by a Guard, which has been previously authorized to automatically lower the resource limits enforced by the Aroma VM in such situations by one or more policies, such as the following:

O+: *IF a Guard notices a pattern of resource misuse by an agent*
   THEN that Guard must notify its administrator appropriately
   PRECEDENCE A-: The agent is forbidden from using more than 25% of the resource
   ELSE A-: The agent is forbidden from using more than 10% of the resource

The policy requires the Guard to notify the administrator, who can determine whether this is a false alarm (in which case the agent's resources can be restored by a new policy setting) or whether this is a real attack (in which case the administrator may choose to further lower A's resource usage). If the effort to notify the administrator fails, the Guard is authorized to reduce resource usage to 10% on its own. In this case, transparently reducing resource usage is better than peremptorily terminating the agent because in the former case the agent will be unaware that it's misuse has been detected.

The requirement for the Guard to be able to act autonomously in making an initial response to the attack is akin to the need for a sprinkler system in a building to go off in the presence of smoke before the fire department arrives. Though there is a risk that the signal may have been a false alarm, it is still far better in most cases to have limited the potential damage through prompt action. Moreover, in the case of a malicious agent that is attacking the network, the administrator may not be able to reconfigure a remote sensor until a provisional limit is placed on network resource usage.

A+: *Team A is authorized to use 50% of the CPU*

In order to guarantee a certain quality of service to other agents, Team A is limited to 50% in the amount of CPU resources it is authorized to use. In this example, however, note that the policy says nothing about how the CPU resources should be allocated among members of Team A, so internal resource allocation is left to the particular algorithm used by the enforcer performing this task.

Monitoring and response policies. It may sometimes be desirable to represent obligations, for the system to perform specific monitoring and response actions as policy:

O+: *IF an authorization failure event occurs*
THEN the authorization mechanism must record the pertinent data in the system log
PRECEDENCE A-: the authorization mechanism is forbidden to perform any other action
ELSE O+: the authorization mechanism must notify the administrator appropriately

In this example, the authorization mechanism is required to record pertinent data in the system log if an authorization failure event occurs. In another example from Ultra*Log:

O+: *IF there is a new defense posture*
THEN the policy applicability condition monitor must deploy the M&R component group for the new defense posture and deactivate the M&R component for the previous defense posture
PRECEDENCE A-: the policy applicability condition monitor is forbidden to perform any other action
ELSE O+: the policy applicability condition monitor must notify the administrator appropriately

This policy requires a new set of monitoring and response components to be activated when the defense posture changes (e.g., a change from threatcon alpha to threatcon bravo).

Mobility policies.

A-: *Agents that are members of the trust domain are forbidden from moving to host H*

This example illustrates how the movement of software agents from one host to another can be controlled by policy in the same way that any other action is governed, provided appropriate enforcement mechanisms are in place.

In a more complex example based on research by Knoll et al. [16.57], the trust level of a mobile software agent is determined in part by where it has traveled in the past (i.e., there is greater or lesser possibility that it may have been tampered with by a malicious host). The trust level, in turn, is used to limit the permissions of the agent in the future:

A-: *Agents are forbidden from performing sensitive action X*
if their trust level <= threshold

The following example, pertinent to our NASA work on the PSA, obligates the PSA to move away from danger:

O+: *IF a situation dangerous to a PSA is present in some location*
THEN the PSA must move out of that location[26]

16.4 Social Aspects of Agent Acceptability

Norman suggests that the social aspects of agent acceptability include things
such as providing reassurance that everything is working according to plan,
providing an understandable and controllable level of feedback about agent's
intentions and actions, and accurately conveying the agent's capabilities and
limitations [16.72]. In short, human beings must be informed enough to be
able to easily step in and help when the situation becomes more than the
agents can handle, and agents on their part must be made more competent in
conveying the appropriate information to humans and acting in partnership
with them. Speaking of the central problems of conventional automation,
Norman writes:

> "The problem ... is that automation is at an intermediate level of
> intelligence, powerful enough to take over control that used to be
> done by people, but not powerful enough to handle all abnormalities.
> Moreover, its level is insufficient to provide the continual, appropriate
> feedback that occurs naturally among human operators. To solve this
> problem, the automation should either be made less intelligent or
> more so, but the current level is quite inappropriate ... . Problems
> result from inappropriate application, not overautomation" [16.70].

Teamwork has become the most widely accepted metaphor for describing
the nature of cooperation in multi-agent systems. Whereas early research on
agent teamwork focused mainly on agent-agent interaction [16.28, 16.115],
teamwork principles are now being formulated in the context of human-agent
interaction [16.11, 16.14]. Unlike autonomous systems, designed primarily to
take humans out of the loop, many new efforts are specifically motivated by
the need to support close continuous multimodal human-agent interaction
[16.22, 16.27, 16.53, 16.61, 16.73, 16.114].

The KAoS policy-based teamwork model defines what constitutes a team,
and the nature of many of its collaborative activities. Elsewhere, we have
outlined a preliminary perspective on the basic principles and pitfalls of adjustable autonomy and human-centered teamwork gleaned from the literature [16.14]. The set of policies we are designing for human-robotic interaction goes beyond the traditional policy concerns about security and safety in
significant ways. As one example, consider how policy can be used to ensure
effective communication among team members. Previous research on generic

---

[26] Consistent with Asimov's laws; however, the PSA might be obliged by a higher-
level policy to stay if its presence was needed to help a human being.

teamwork models has explored this issue to a limited degree within the context of communication required to form, maintain, and abandon joint goals. However, more research is needed to address the complexities of maintaining mutual awareness in human-agent, as opposed to agent-agent, interaction.

With previous research in agent teamwork, we share the assumption that, to the extent possible, teamwork knowledge should be modeled explicitly and separately from the problem-solving domain knowledge. Policies for agent safety and security, as well as context- and culturally-sensitive teamwork behavior, can be represented as KAoS policies that enable many aspects of the nature and timing of the agent's interaction with people to be appropriate, without requiring each agent to individually encode that knowledge. Agent designers can concentrate on developing unique agent capabilities, while assuming that many of the basic rules of effective human-agent coordination will be built into the environment as part of the policy infrastructure.

16.4.1 Examples of Policy Types Relating to Social Aspects of Agent Acceptability

In contrast to the examples of technical policies in Section 16.3.1 above, our work to begin encoding these social issues in policy is relatively recent and is likely to evolve considerably in the near future. Some of this will require the resolution of difficult research issues; we are beginning implementation with those policies that are most straightforward, and will then continue to progress incrementally to more complex ones. We will give examples from six categories of policy: organization, notification, conversation, nonverbal expression, collaboration, and adjustable autonomy.

Organization policies. Some policy management systems, in part as an artifact of their mode of policy representation, require many or all of what we call organization policies to be represented as "meta level," "higher order," or some other sort of special policy. In KAoS, many of these can be specified uniformly, in the same way that other kinds of policies are represented.

A+:*Individuals of the class Domain Manager are permitted to approve policies*

The KAoS actor ontology distinguishes between people and various classes of agents. Most agents can only perform *ordinary actions*, however various components that are part of the infrastructure (e.g., domain manager, guard), as well as authorized human users, may variously be permitted or obligated to perform *policy actions,* such as policy approval and enforcement.

A+:*Any person in the Manager Role is permitted to authorize check payment IF the same person is not also the check issuer*

The example specifies a dynamic separation of duty, where the issuer of the check is not allowed to also be the one who authorizes payment on that check.

A-: *An agent is forbidden to register to domain D IF it is already registered to any individual of the class domain*

A-: *An agent is forbidden to register to any individuals of the class domain IF it is already registered to domain D*

The pair of policies above specifies that an agent cannot simultaneously be registered as a member of both the domain D and some other domain.

Notification Policies. Building on the work of [16.81, 16.97], we are developing KAoS notification policies in the context of our NASA applications. The vision of future human-agent interaction is that of loosely coordinated groups of humans and agents. As capabilities and opportunities for autonomous operation grow in the future, agents will perform their tasks for increasingly long periods of time with only intermittent supervision. Most of the time routine operation is managed by the agents while the human crews perform other tasks. Occasionally, however, when unexpected problems or novel opportunities arise, humans must assist the agents. Because of the loose nature of these groups, such communication and collaboration must proceed asynchronously and in a mixed-initiative manner. Humans must quickly come up to speed on situations with which they may have had little involvement for hours or days. Then, they must cooperate effectively and naturally with the agents as true team members.[27] Hence the challenge of managing notification and situation awareness for the crewmembers.

Various ontologies supporting notification (e.g., basic concepts for categories of events, roles, notifications, latency, focus of attention, and presence) form the foundation of this work. In conjunction with these ontologies, notification policies and their parameter settings are created, as in the example below:

O+: *IF new notification = true AND utility >= notifyThreshold AND utility < doItThreshold*
THEN notify the space station crew member appropriately

Human attention is a scarce resource. When an important event is signaled, the utility of various alternatives (e.g., notify the crew member, perform some required action without interrupting the person, or do nothing) is evaluated. If a notification is required and the current task is well-defined, the KAoS-Brahms infrastructure will take into account the task and other contextual factors to perform the notification in a manner that is context-sensitive to modality, urgency, and the location of the human being. Because

---

[27] Actually, this vision points to two major opportunities for policy-based help: 1. the use of policy to assure that unsupervised autonomous agent behavior is kept within safe and secure limits of prescribed bounds, even in the face of buggy or malicious agent code; and 2. the use of policy to assure effective and natural human-agent team interaction, without individual agents having to be specifically programmed with the knowledge to do so.

such knowledge resides in the infrastructure, rather than as part of the knowledge of each agent, agent development is simplified.

Conversation Policies. Explicit conversation policies simplify the work of both the agent and the agent designer [16.12, 16.13, 16.47, 16.48]. In comparison to unrestricted agent dialogue, conversation policies reduce the agents' inferential burden by limiting the space of alternative conversational productions and parameters that they need to consider, both in generating messages and in interpreting messages received from other agents. Because a significant measure of conversational planning for routine interactions can be encoded in conversation policies offline and in advance, the agents can devote more of their computational power at runtime to other things.

O+: *IF response lag of conversation X > M minutes*
    THEN the agent must terminate conversation X

This conversation policy requires the agent to unilaterally terminate a conversation when a lag of M minutes has elapsed in waiting for a response, preventing conversations from staying open indefinitely.

A-:*Agents are forbidden to send a message of any type other than Reply*
    IF the type of the conversation is Request-Reply AND the previous message
    type of the conversation is Request

This conversation policy enforces a sequence of messages in the Request-Reply conversation type, such that a message of type *request* must always be followed by a message of type *reply*.

A+:*Agents are permitted to send TRANSCOM messages with return receipts*

This conversation policy example from the Ultra*Log application, allows the sending of TRANSCOM messages that require return receipts. Note that this policy would be appropriate only in a tyrannical domain that prohibited all messages that were not explicitly permitted.

More complex sorts of policies, dealing, for example, with Clark's concept of common ground [16.24] or improvisational approaches to conversation [16.78], will also be important in effective human-agent interaction. Though such policies go beyond what is possible in the current version of KAoS, we have started to address them as part of a collaboration with Allen et al. [16.2, 16.3] in the near future [16.112, 16.120].

Nonverbal Expression Policies. Where possible, agents usually take advantage of explicit verbal channels for communication in order to reduce the need for relying on current primitive robotic vision and auditory sensing capabilities [16.69]. On the other hand, animals and human beings often rely on visual and auditory signals instead of explicit verbal communication for many aspects of coordinated activity. As part of our work on human-robotic interaction for NASA, the Army, and the Navy, we are developing policies to govern various nonverbal forms of expression in hardware and software agents. These nonverbal behaviors will be designed to express not only the current state of the agent but, importantly, also to provide rough clues about what it is going to do next. In this way, people can be better enabled to participate with the agent in coordination, support, avoidance, and so forth. In this sense, nonverbal expressions are an important ingredient in enabling human-agent teamwork.

Maes and her colleagues were among the first to explore this possibility of software agents that continuously communicate their internal state to the user via facial expressions (e.g., thinking, working, suggesting, unsure, pleased, and confused) [16.66]. Breazeal and Scassellati [16.18] have taken inspiration from research in child psychology [16.97] to develop robot displays that reflect four basic classes of preverbal social responses: affective (changing facial expressions), exploratory (visual searching, maintaining mutual regard with a human being), protective (turning away head), and regulatory (expressing feedback to gain caregiver attention, cyclical waxing and waning of internal states, habituating, and signalling internal motivation). Books on human etiquette [16.118] contain many descriptions of appropriate behavior in a wide variety of social settings. Finally, in addition to this previous work, we think that behavior displayed among human beings [16.68] and groups of animals will be one of the most fruitful sources of policy for effective nonverbal expression in agents. Our initial study indicates that there are useful agent equivalents for each of Smith's [16.86] ten categories of widespread vertebrate animal cooperation and coordination displays [16.115].

O+: *IF the current task of the PSA is of type uninterruptible*
THEN the PSA must blink a red light until the current task is finished
PRECEDENCE: A-: The PSA is forbidden from performing any tasks but the current one

This policy requires the PSA to blink a red light while it is busy performing an uninterruptible task. During this time, it is also forbidden from performing any task but the current one. Related messages it may want to give with a similar signal might include: "I am unable to make contact with anybody," "Do not attempt to communicate with me (for whatever reason, e.g., 'my line is bugged')." On the positive side, various uses of a green light might signal messages such as: "I am open for calls," "I need to talk to someone," or "May I interject something into this conversation? " Displays in this general interactional category clearly have benefits for coordination among groups of agents by providing information about which agents are (or are not) in a position to interact with others, in what ways, when, and so forth.

O+: *IF a conversation has been initiated with someone*
THEN the PSA must face the one with whom it is conversing, as long as it is in sight, until the conversation has finished

This policy implements a kind of display associated with maintaining a previously established association. This display might be especially useful when the PSA is moving around the room and needs to let someone else know that it is still attending to an ongoing conversation.

O+: *IF the current task of the PSA is to move some distance greater than D*
THEN the PSA must signal its intention to move for S seconds
PRECEDENCE: A-: The PSA is forbidden from executing its move

It's no fun being hit in the head by a flying robot that suddenly decides to go on the move. This policy prevents the PSA from moving until it has first signaled its intention to move for some number of seconds. Besides the pre-move signaling, some kind of signaling could also take place during the move itself. In addition to this movement signaling policy, other policies should be put in place to require the PSA to stay at a safe and comfortable distance from humans, other robotic agents, and space station structures and equipment.

Collaboration Policies.

O+: *IF an agent becomes aware that a team goal has been achieved, or has become unachievable or irrelevant*
THEN the agent must notify the other team members in an appropriate manner
PRECEDENCE: A-: The team member is forbidden from actions that are performed only in order to achieve the former team goal

A similar version of this policy is one of the centerpieces of the classic theory of teamwork originally proposed by Cohen and Levesque [16.28]. Though there is potentially a lot of complexity buried in the machinery that determines whether the condition is true, the policy imperative that results from this condition is relatively simple and can be represented straightforwardly in KAoS. All the foundational ontologies and mechanisms developed to support other kinds of notification policies can also be brought to bear in this context. In this sense, the example can be seen as just a special kind of notification policy.

O+: *IF an agent suspends work on a current task in order to attend to a new higher priority task*
THEN the agent must notify the other actors involved in an appropriate manner
PRECEDENCE: A-: The agent is forbidden from executing its new task

Just as a sudden physical move might surprise other actors unless it is appropriately signaled in advance, so also an unexpected change in current task might be jarring to others unless it is heralded in some fashion. Note that this policy presupposes that additional actors, beyond the team members themselves, that share a joint goal may require notification.

392   J.M. Bradshaw et al.

Adjustable Autonomy Policies. Humans and agents may play mutual roles that vary according to the relative degree of initiative appropriate for a given situation (Fig. 16.9). [28] At the one extreme, traditional systems are designed to carry out the explicit commands of humans with no ability to ignore orders (i.e., executive autonomy), generate their own goals (i.e., goal autonomy), or otherwise act independently of environmental stimuli (i.e., environmental autonomy). Such systems cannot, in any significant sense, act; they can only be acted upon. At the other end of the spectrum is an imagined extreme in which agents would control the actions of humans. [29] Between these two extremes is the domain of today's agent systems, with most agents typically playing fixed roles as servants, assistants, associates, or guides. Such autonomous systems are designed with fixed assumptions about what degree of initiative is appropriate for their tasks. They execute their instructions without considering that the optimal level of autonomy may vary by task and over time, or that unforeseen events may prompt a need for either the human being or the agent to take more control. At the limit of this extreme are strong, silent systems with only two modes: fully automatic and fully manual [16.77]. In practice this can lead to situations of human "underload," with the human being having very little to do when things are going along as planned, followed by situations of human "overload," when extreme demands may be placed on the human in the case of agent failure.

Although in practice many mixed-initiative system do not live up to their billing, their design goal is to allow agents to dynamically and flexibly assume a range of roles depending on the task to be performed and on the current situation. Research in adjustable autonomy supports this goal through the development of an understanding about how to ensure that, in a given context, the agents are operating at an optimal boundary between the initiative of the human being and that of the agents. People want to maintain that boundary at a sweet spot in the tradeoff curve that minimizes their need to attend to interaction with the agent while providing them with a sufficiently comfortable level of assurance that nothing will go wrong.

O+: *IF elapsed time since last report > time T*

---

[28] For a more fine-grained presentation of a continuum of control between humans and machines, see Hancock and Scallen's [16.50] summary of Sheridan's ten-level formulation. Robert Taylor (personal communication) surmises from his experience that ten may be far more levels than are useful in practice. Barber et al. differentiate three kinds of relationships among agents: command-driven (i.e., the agent is fully subordinated to some other agent), true consensus (i.e., decision-making control is shared equally with other agents), and locally autonomous/master (i.e., the agent makes decisions without consulting other agents and may be allowed to command subordinates).

[29] Of course, in real systems, the relative degree of initiative that could be reasonably taken by an agent or human would not be a global property, but rather relative to particular functions that one or the other was currently assuming in some context of joint work.

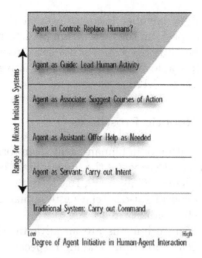

Agent in Control: Replace Humans?

Agent as Guide: Lead Human Activity

Agent as Associate: Suggest Courses of Action

Agent as Assistant: Offer Help as Needed

Agent as Servant: Carry out Intent

Traditional System: Carry out Command

Range for Mixed Initiative Systems

Low   High
Degree of Agent Initiative in Human-Agent Interaction

Fig. 16.9. Spectrum of agent roles in human-agent interaction

THEN the agent must notify the human supervisor about its status in the most appropriate manner.

This simple policy sets the duration of autonomous operation for an agent, requiring it to notify a human supervisor about its status at predetermined intervals.

A+: *Space station crewmembers in the Trusted Operator Role are permitted to override PSA non-critical negative authorizations*

Sometimes, it is critical for authorized human operators to be able to immediately countermand some negative authorization of an agent (i.e., allowing it to do things which it normally is not authorized to do). While this could be done by modifying the policy in the usual way, it is sometimes more practical to do this directly on a one-time temporary basis by overriding a prohibition. However, overriding certain operations (e.g., flying the PSA into a wall of the space station) may require consent of both the space station commander and an authorized person at mission control.

O+: *IF no crewmember is monitoring the environment in space station module X*

THEN PSA must monitor the environment in module X

PRECEDENCE: A+: PSA is permitted to monitor the environment in module X

Sometimes, the PSA may be required to temporarily take upon itself functions that human crewmembers would normally provide. Here, the PSA is given both permission and an obligation to monitor the environment in module X if a crewmember is not currently doing so. Similar policies could come into play when a crew member becomes overloaded or injured, such that he no longer is able to perform the task within predetermined criteria. In such cases, agents could be authorized and/or obligated to assist. For a more complete discussion of adjustable autonomy, mixed-initiative interaction, and the role of policy, see [16.112, 16.120].

16.4.2 Cognitive and Robotic Prostheses

For some researchers, the ultimate in human-agent teamwork is the notion of agents that can function as extensions of the human brain cognitive prostheses and body robotic prostheses [16.39, 16.49, 16.52]. In this section we briefly suggest some preliminary considerations relating to human-agent interaction in the development of such capabilities.

At the outset, we recognize humans are an advantaged lot, each of us having been endowed with a "good brain and an unspecialized body" [16.68], which means that we are in a better position than any other creature to make and use a variety of tools. Moreover, bipedal locomotion has always had the beneficial side effect of freeing one hand to explore the environment and the other to wield those tools. Ford et al. argue that the accumulated tools of human history can all profitably be regarded as prostheses, not in the sense that they compensate for the specific disabilities of any given individual ([16.31]), but rather because they enable us to overcome the biological limitations shared by all of us: with reading and writing anyone can transcend the finite capacity of human memory; "with a power screwdriver anyone can drive the hardest screw; with a calculator, anyone can get the numbers right; with an aircraft anyone can fly to Paris; and with Deep Blue, anyone can beat the world chess champion" [16.39].

The prosthetic perspective can be contrasted with the traditional focus of Artificial Intelligence (AI) on standalone machine competence and its resulting preoccupation with the Turing Test as its measure of success [16.41]. Instead, argues Ford, we should start from a human-centered perspective. This implies that we must shift our goal "from making artificial superhumans who can replace us to making superhumanly intelligent artifacts that can amplify and support our own cognitive abilities" [16.49]. We don't need to jettison the acronym of AI, so long as we now take it to refer to the human's *Augmented* Intelligence.[30]

---

[30] The 1962 report of Engelbart entitled *Augmenting Human Intellect* presciently stressed the theme of "improving the intellectual effectiveness of the individual human being... through extensions of means developed and used in the past to help man apply his native sensory, mental, and motor capabilities. [Like] most systems its performance can best be improved by considering the whole as a set of interacting components rather than by considering the components in isolation" [16.36].

Eyeglasses, a well-known example of an ocular prosthesis,[31] provide a particularly useful example of three foundational concepts that are important to an understanding of cognitive and robotic prostheses:

1. *Transparency.* "Eyeglasses leverage and extend our ability to see, but in no way model our eyes: They don't look or act like them and wouldn't pass a Turing test for being an eye" [16.49]. A key feature of eyeglasses is that they can be used more or less transparently – by our forgetting they are present – just as humans with myopia don't think constantly about the wearing of the contact lenses but rather about the fact that they are seeing more effectively *through* them.[32]

2. *Unity.* Since our goal is not making smart eyeglasses but, rather, augmenting the human's ability to see, the minimum unit of discussion for the design of a prosthesis includes the device, the human being, and the environment in which the human will use the device. This mode of analysis necessarily blurs the line between humans and technology.[33]

3. *Fit.* Your eyeglasses won't fit me; neither will mine do you any good. Prostheses must fit the human and technological components together in ways that synergistically exploit their mutual strengths and mitigate their respective limitations. This implies a requirement for rich knowledge of how humans function.[34]

---

[31] The notion of augmenting sight through eyeglasses was "first mentioned by Roger Bacon in 1268. In the 1665 preface to Micrographia, Robert Hooke goes further, suggesting the addition 'of artificial Organs to the natural... to improve our other senses of hearing, smelling, tasting, and touching" [16.87].

[32] The manner in which perception operates during the use of good tools was insightfully described many years ago by Polanyi: "When we use a hammer to drive a nail, we attend to both nail and hammer, but in a different way. We watch the effect of our strokes on the nail and try to wield the hammer so as to hit the nail most effectively. When we bring down the hammer we do not feel that its handle has struck our palm but that its head has struck the nail. Yet in a sense we are certainly alert to the feelings in our palm and the fingers that hold the hammer. They guide us in handling it effectively, and the degree of attention that was given to the nail is given to the same extent but in a different way to these feelings. The difference may be stated by saying that the latter are not, like the nail, objects of our attention, but instruments of it. They are not watched in themselves; we watch something else while keeping intensely aware of them. I have a subsidiary awareness of the feeling in the palm of my hand which is merged into my focal awareness of my driving in the nail" [16.75].

[33] In 1960, Licklider [16.60] introduced the concept of *man-computer symbiosis:* "the hope is that, in not too many years, human brains and computing machines will be coupled together very tightly and that the resulting partnership will think as no human brain has ever thought and process data in a way not approached by the information-handling machines we know today."

[34] A good example of this is the OZ cockpit display [16.49]. Through a groundbreaking study on the limits of human central and peripheral vision, IHMC's David Still discovered that peripheral vision can pick up 10 times the amount of detail than previously thought. Using this finding, he tailored the design of stimuli in a cockpit display to exploit the human sensory system's natural filter-

The elaboration of foundational concepts that are important to an understanding of cognitive and robotic prostheses, and the study of human functions in particular environments, happily dovetail with progress in the miniaturization of computing devices and the formulation of design principles for wearable computing [16.74]. Mann [16.67] was among the first to elucidate some of the necessary criteria for devices to be successfully subsumed into the human being's eudaemonic space (i.e., where the device seems to be part of the person).[35] He describes three required operational modes for wearable computing:

- **Constancy:** The computer runs continuously, and is *always ready* to interact with the user. Unlike a handheld device, laptop computer, or PDA, it does not need to be opened up and turned on prior to use. The signal flow from human to computer, and computer to human, ... runs continuously to provide a constant user interface.
- **Augmentation:** Traditional computing paradigms are based on the notion that computing is the primary task. Wearable computing, however, is based on the notion that computing is *not* the primary task. The assumption of wearable computing is that the user will be doing something else at the same time as doing the computing. Thus the computer should serve to augment the intellect or augment the senses ...
- **Mediation:** Unlike [traditional computers], the wearable computer can encapsulate us. It doesn't necessarily need to completely enclose us. There are two aspects to this encapsulation:
  i. **Solitude:** It can function as an information filter, and allow us to block out material we might not wish to experience, ... [or] it may simply allow us to alter our perception of reality.
  ii. **Privacy:** Mediation allows us to block or modify information leaving our encapsulated space. In the same way that ordinary clothing prevents others from seeing our naked bodies, the wearable computer may, for example, serve as an intermediary for interacting with untrusted systems.

Because of its ability to encapsulate us, ... [wearable computing devices] may also be able to make measurements of various physiological quantities.

Besides these operational modes, Mann describes six attributes of wearable systems:

---

ing and processing capabilities and to manipulate the data so it provides exactly what the pilot needs to know at any particular time. Stunningly, the OZ cockpit display is completely void of dials and gauges of ordinary cockpits, yet it is easier to learn, more straightforward to control, and more robust to temporary visual system impairment.

[35] Mann's formulations have evolved over time. Here, we discuss the version that can be found at http://www.eyetap.org/defs/glossary/wearcomp. See also Thad Starner's thorough survey of the field in his dissertation on Wearable Computing [16.87].

- **Unmonopolizing** of the user's attention, ... [though it may] mediate (augment, alter, or deliberately diminish) the sensory capabilities.
- **Unrestrictive** to the user: ... 'you can do other things while using it' ...
- **Observable** by the user: it can get your attention continuously if you want it to ...
- **Controllable** by the user ...
- **Attentive** to the environment: it is environmentally aware, multimodal, multi-sensory ... [thus increasing] the user's [situation] awareness ...
- **Communicative** to others: it can be used as a communications medium ...

DARPA's Augmented Cognition (AugCog) Program (http://www.darpa. mil/ito/research/ac/) is an example of an early effort focused on appropriately exploiting and integrating all available channels of communication from agents to humans (e.g., visual, auditory, tactile), and conversely sensing and interpreting a wide range of physiological measures of the human being in real-time so they can be used to tune agent behavior, and thus enhance joint human-machine performance.[36] For example, in IHMC's Adaptive Multi-Sensory Integration (AMI), augmented cognition prototype sets of system sensor agents (e.g., joystick), human sensor agents (e.g., EEG, pupil tracking, arousal meter), human display agents (e.g., visual, auditory, tactile), and adaptive automation agents (e.g., performing specific flight tasks) could work together with a pilot to promote stable and safe flight, sharing and adjusting aspects of control among the human and virtual crew member agents while taking system failures and human attention and stress loads into account [16.112].

While it is still too early to gauge the success of efforts such as AugCog, let alone to prescribe detailed principles for making cognitive and robotic prostheses acceptable to humans, it is clear that such modes of interaction will require new ways of thinking about human-agent interaction. In an insightful essay called *The Teddy* [16.71], Norman discusses some of the issues and implications of the widespread long-term habitual use of such technologies:

- Would we get so dependent that we would become disoriented without them?
- If they are constantly recording every event, should we allow them to be turned off? To protect civil liberties, you must be able to, and an indicator must show if someone's device is listening to you.

---

[36] A related program focused on similar issues with a robotics emphasis is NSF's Robotics and Human Augmentation (http://www.interact.nsf.gov/cise/descriptions.nsf/ 5b8c6c912ebf7f9b8525662c00723201/5e8661fa698fe674852565d9005985ef?Open Document). See also DARPA's Mobile Autonomous Robot Software (MARS) Robotic Vision 2020 Program (http://www.darpa.mil/ito/solicitations/FBO_02-15.html).

- Should it be programmed to always be supportive and encouraging (thus removing us from reality), or to give criticism and correction (thus resembling a nagging parent)? Getting the right balance is difficult in human relationships, how can we expect technology designers to do better?
- If we are never alone, when would we think? Would this accelerate the already tuned-out tendencies of headphone wearers?

16.5 Conclusions

In this chapter, we have outlined some of the technical and social challenges in the problem of making agents acceptable to people and have given examples and explanations of how a policy-based approach might be used to address some of those challenges. We hope that these initial efforts will inspire others to devote greater attention to reusable models and tools to assure the security, safety, naturalness, and effectiveness of human-agent teams.

Acknowledgements

The authors gratefully acknowledge the support of the DARPA CoABS, EPCA(CALO), Augmented Cognition, DAML, and Ultra*Log Programs, the NASA Cross-Enterprise and Intelligent Systems Programs, the Army Research Lab, the Office of Naval Research, the National Technology Alliance, and Fujitsu Labs while preparing this paper. We are also grateful for the contributions of James Allen, Alessandro Acquisti, Mike Bennett, Guy Boy, Kathleen Bradshaw, Mark Burstein, Murray Burke, Alberto Canas, Nate Chambers, Bill Clancey, Rob Cranfill, Cranfill, Grit Denker, Gary Edwards, Rich Feiertag, Ken Ford, Lucian Galescu, Yuri Gawdiak, Mike Goodrich, Mark Greaves, David Gunning, Jack Hansen, Pat Hayes, Mark Hoffman, Wayne Jansen, Hyuckchul Jung, Jim Just, Mike Kerstetter, Mike Kirton, Shri Kulkarni, Henry Lieberman, James Lott, Frank McCabe, Cindy Martin, Robin Murphy, Nicola Muscettola, Jerry Pratt, Debbie Prescott, Timothy Redmond, Sue Rho, Sebastien Rosset, Dylan Schmorrow, Debbie Schrekenghost, Kent Seamons, Mike Shafto, Maarten Sierhuis, Milind Tambe, Austin Tate, William Taysom, Ron Van Hoof, and Tim Wright.

References

16.1 A. Acquisti, M. Sierhuis, W. J. Clancey, J.M. Bradshaw: Agent-based modeling of collaboration and work practices onboard the International Space Station. *Proceedings of the Eleventh Conference on Computer-Generated Forces and Behavior Representation* (Orlando, FL 2002)

16.2 J. Allen, D. K. Byron, M. Dzikovska, G. Ferguson, L. Galescu, A. Stent: An architecture for a generic dialogue shell. Journal of Natural Language Engineering, 6(3), 1-16 (2000)

16.3 J.F. Allen, D.K. Byron, M. Dzikovska, G. Ferguson, L. Galescu, A. Stent: Towards conversational human-computer interaction. AI Magazine, 22(4), 27-35 (2001)

16.4 D. Allsopp, P. Beautement, J.M. Bradshaw, E. Durfee, M. Kirton, C. Knoblock, N. Suri, A. Tate, C. Thompson: Coalition Agents eXperiement (CoAX): Multi-agent cooperation in an international coalition setting. In: A. Tate, J. Bradshaw, M. Pechoucek (eds.), *Special issue of IEEE Intelligent Systems* 17(3), 26-35 (2002)

16.5 R.Ambrose, C. Culbert, F. Rehnmark: An experimental investigation of dexterous robots using Eva tools and interfaces. *AIAA, 4593* (2001)

16.6 I. Asimov: Runaround. In: I. Asimov (ed.), *I, Robot*. pp. 33-51. London, England: Grafton Books. Originally published in *Astounding Science Fiction, 1942* pp. 94-103 (1942/1968)

16.7 K.S. Barber, M. Gamba, C. E. Martin: Representing and analyzing adaptive decision-making frameworks. In: H. Hexmoor, C. Castelfranchi, R. Falcone (eds.), *Agent Autonomy* (Dordrecht, The Netherlands, Kluwer, 2002) pp. 23-42

16.8 K.S. Barber, C.E. Martin: Agent autonomy: Specification, measurement, and dynamic adjustment. *Proceedings of the Workshop on Autonomy Control Software, International Conference on Autonomous Agents* (Seattle, WA, 1999)

16.9 M. Boman: Norms in artificial decision-making. Artificial Intelligence and Law, 7, 17-35 (1999)

16.10 G. Boy: Human-centered design of artificial agents: A cognitive function analysis approach. In: J.M. Bradshaw (ed.), *Handbook of Software Agents*. (Cambridge, MA, AAAI Press/The MIT Press, 2002) (in press)

16.11 J.M. Bradshaw, G. Boy, E. Durfee, M. Gruninger, H. Hexmoor, N. Suri, M. Tambe, M. Uschold, J. Vitek (eds.): *Software Agents for the Warfighter. ITAC Consortium Report* (Cambridge, MA, AAAI Press/The MIT Press 2002)

16.12 J.M. Bradshaw, S. Dutfield, P. Benoit, J.D. Woolley: KAoS: Toward an industrial-strength generic agent architecture. In: J.M. Bradshaw (ed.), *Software Agents* (Cambridge, MA, AAAI Press/The MIT Press 1997) pp. 375-418

16.13 J.M. Bradshaw, M. Greaves, H. Holmback, W. Jansen, T. Karygiannis, B. Silverman, N. Suri, A. Wong: Agents for the masses: Is it possible to make development of sophisticated agents simple enough to be practical? IEEE Intelligent Systems (March-April), 53-63 (1999)

16.14 J.M. Bradshaw, M. Sierhuis, A. Acquisti, P. Feltovich, R. Hoffman, R. Jeffers, D. Prescott, N. Suri, A. Uszok, R. Van Hoof: Adjustable autonomy and human-agent teamwork in practice: An interim report on space applications. In: H. Hexmoor, R. Falcone, C. Castelfranchi (eds.), *Agent Autonomy* pp. 243-280

16.15 J.M. Bradshaw, N. Suri, M.R. Breedy, A. Canas, R. Davis, K.M. Ford, R. Hoffman, R. Jeffers, S. Kulkarni, J. Lott, T. Reichherzer, A. Uszok: Terraforming cyberspace. In: D.C. Marinescu, C. Lee (eds.), *Process Coordination and Ubiquitous Computing* pp. 165-185. Boca Raton, FL: CRC Press. Updated and expanded version of an article that originally appeared in IEEE Intelligent Systems, July 2001, pp. 49-56

16.16 J.M. Bradshaw, A. Uszok, R. Jeffers, N. Suri, P. Hayes, M.H. Burstein, A. Acquisti, B. Benyo, M.R. Breedy, M. Carvalho, D.Diller, M. Johnson, S. Kulkarni, J. Lott, M. Sierhuis, R. Van Hoof: Representation and reasoning for DAML-based policy and domain services in KAoS and Nomads. *Proceedings of the Autonomous Agents and Multi-Agent Systems Conference (AAMAS 2003)* (Melbourne, Australia, New York, NY: ACM Press)

16.17 S.Brainov, H. Hexmoor: Quantifying autonomy. In: H. Hexmoor, C. Castel-franchi, R. Falcone (eds.), *Agent Autonomy*. (Dordrecht, The Netherlands, Kluwer) pp. 43-56

16.18 C. Breazeal, B. Scassellati: How to build robots that make friends and influence people. *IROS* (Kyonjiu, Korea)

16.19 R.R. Burridge, J. Graham, K. Shillcutt, R. Hirsh, D. Kortenkamp: Experiments with an EVA assistant robot. *Proceedings of the Seventh International Symposium on Artificial Intelligence, Robotics and Automation in Space (i-SAIRAS)* (Nara, Japan)

16.20 M.H. Burstein, D. V. McDermott: Issues in the development of human-computer mixed-initiative planning. In: B. Gorayska J.L. Mey (eds.), *Cognitive Technology: In Search of a Humane Interface.* (Elsevier Science)

16.21 K. Capek: R. U. R. (Rossum's Universal Robots). In: P. Kussi (ed.), *Toward the Radical Center: A Karel Capek Reader* (North Haven, CT: Catbird Press)

16.22 H. Chalupsky, Y. Gil, C. A. Knoblock, K. Lerman, J. Oh, D.V. Pynadath, T.A. Russ, M. Tambe: Electric Elves: Agent technology for supporting human organizations. AI Magazine, 2, pp. 11-24

16.23 W.J. Clancey: Simulating activities: Relating motives, deliberation, and attentive coordination. *Cognitive Systems Review, special issue on Situated and Embodied Cognition*

16.24 H.H. Clark: *Arenas of Language Use* (Chicago, IL: University of Chicago Press)

16.25 R. Clarke: Asimov's laws of robotics: Implications for information technology, Parts 1 and 2. IEEE Computer, pp. 53-66.

16.26 J. Clute, P. Nicholls: Grolier Science Fiction: The Multimedia Encyclopedia of Science Fiction (CD-ROM). In: Danbury, CT: Grolier Electronic Publishing

16.27 P. Cohen, R. Coulston, K. Krout: Multimodal interaction during multiparty dialogues: Initial results. *Proceedings of the Fourth IEEE International Conference on Multimodal Interfaces* (Pittsburgh, PA)

16.28 P.R. Cohen, H.J. Levesque: *Teamwork.* Technote 504. Menlo Park, CA, SRI International, March

16.29 R. Cohen, C. Allaby, C. Cumbaa, M. Fitzgerald, K. Ho, B. Hui, C. Latulipe, F. Lu, N. Moussa, D. Pooley, A. Qian, S. Siddiqi: What is initiative? *User Modeling and User-Adapted Interaction* 8(3-4), 171-214

16.30 R. Cohen, M. Fleming: Adjusting the autonomy in mixed-initiative systems by reasoning about interaction. In: H. Hexmoor, C. Castelfranchi, R. Falcone (eds.), *Agent Autonomy* (Dordrecht, The Netherlands: Kluwer) pp. 105-122

16.31 E. Cole, P. Dehdashti: Computer-based cognitive prosthetics: Assistive technology for the treatment of cognitive disabilities. *Proceedings of the Third International ACM Conference on Assistive Technologies (ACM SIGCAPH - Computers and the Physically Handicapped)* (Marina del Rey, CA)

16.32 R. Conte, C. Castelfranchi: *Cognitive and social action* (London, England: UCL Press)

16.33 M. d'Inverno, M. Luck: *Understanding Agent Systems* (Berlin, Germany, Springer-Verlag)

16.34 N. Damianou, N. Dulay, E.C. Lupu, M.S. Sloman: *Ponder: A Language for Specifying Security and Management Policies for Distributed Systems, Version 2.3.* Imperial College of Science, Technology and Medicine, Department of Computing, 20 October 2000

16.35 G. Dorais, R.P. Bonasso, D. Kortenkamp, B. Pell, D. Schrekenghost: Adjustable autonomy for human-centered autonomous systems on Mars. *Proceedings of the AAAI Spring Symposium on Agents with Adjustable Autonomy.*

AAAI Technical Report SS-99-06 (Menlo Park, CA, Menlo Park, CA, AAAI Press)

16.36 D. C. Engelbart: *Augmenting Human Intellect: A Conceptual Framework.* Air Force Office of Scientific Research, AFOSR-3233 Summary Report, SRI Project 3578 (Stanford Research Institute, October)

16.37 R. Falcone, C. Castelfranchi: From automaticity to autonomy: The frontier of artificial agents. In: H. Hexmoor, C. Castelfranchi, R. Falcone (eds.), *Agent Autonomy.* (Dordrecht, The Netherlands, Kluwer) pp. 79-103

16.38 G. Ferguson, J. Allen, B. Miller: TRAINS-95: Towards a mixed-initiative planning assistant. *Proceedings of the Third Conference on Artificial Intelligence Planning Systems (AIPS-96)* (Edinburgh, Scotland) pp. 70-77

16.39 K.M. Ford, C. Glymour, P. Hayes: Cognitive prostheses. AI Magazine, 18(3), 104

16.40 K.M. Ford, C. Glymour, P.J. Hayes (eds.): *Android Epistemology* (Menlo Park, CA: AAAI Press / The MIT Press)

16.41 K.M. Ford, P. Hayes: On computational wings: Rethinking the goals of Artificial Intelligence. *Scientific American. Special issue on Exploring Intelligence* 9 (4), 78-83

16.42 I. Foster, C. Kesselman (eds.): *The Grid: Blueprint for a New Computing Infrastructure.* (San Francisco, CA, Morgan Kaufmann)

16.43 J. Fox, S. Das: *Safe and Sound: Artificial Intelligence in Hazardous Applications* (Menlo Park, CA: AAAI Press/The MIT Press)

16.44 Y. Gawdiak, J.M. Bradshaw, B. Williams, H. Thomas: R2D2 in a softball: The Personal Satellite Assistant. In: H. Lieberman (ed.), *Proceedings of the ACM Conference on Intelligent User Interfaces (IUI 2000)*(New Orleans, LA, New York: ACM Press) pp. 125-128

16.45 N.A. Gershenfeld: *When Things Start to Think* (New York, Henry Holt and Company)

16.46 M.A.Goodrich, D.R. Olsen Jr., J.W. Crandall, T.J. Palmer: Experiments in adjustable autonomy. *Proceedings of the IJCAI_01 Workshop on Autonomy, Delegation, and Control: Interacting with Autonomous Agents.* Seattle, WA, Menlo Park, CA, AAAI Press

16.47 M. Greaves, H. Holmback, J.M. Bradshaw: What is a conversation policy? M. Greaves, J.M. Bradshaw (eds.), *Proceedings of the Autonomous Agents '99 Workshop on Specifying and Implementing Conversation Policies* (Seattle, WA) pp. 1-9

16.48 M. Greaves, H. Holmback, J.M. Bradshaw: Agent conversation policies. In: J.M. Bradshaw (ed.), *Handbook of Agent Technology.* (pp. In: preparation) (Cambridge, MA, AAAI Press/The MIT Press)

16.49 S. Hamilton: Thinking outside the box at IHMC. IEEE Computer, 61-71

16.50 P.A. Hancock, S.F. Scallen: Allocating functions in human-machine systems. In: R. Hoffman, M.F. Sherrick, J.S. Warm (eds.), *Viewing Psychology as a Whole* (Washington, D.C., American Psychological Association) pp. 509-540

16.51 R. Hoffman, P. Feltovich, K.M. Ford, D.D. Woods, G. Klein, A. Feltovich: A rose by any other name? would probably be given an acronym. IEEE Intelligent Systems, July-August, pp. 72-80

16.52 R. R. Hoffman, K.M. Ford, P.J. Hayes, J.M. Bradshaw: The Borg hypothesis. IEEE Intelligent Systems (in press)

16.53 E. Horvitz: Principles of mixed-initiative user interfaces. *Proceedings of the ACM SIGCHI Conference on Human Factors in Computing Systems (CHI'99)* (Pittsburgh, PA, New York: ACM Press)

16.54 E. Horvitz, A. Jacobs, D. Hovel: Attention-sensitive alerting. *Proceedings of the Conference on Uncertainty and Artificial Intelligence (UAI'99)* (Stockholm, Sweden) pp. 305-313

16.55 B. Joy: Why the future doesn't need us. *Wired*, 4

16.56 A. Kay: User interface: A personal view. In: B. Laurel (ed.), *The Art of Human-Computer Interface Design* (Reading, MA, Addison-Wesley) pp. 191-208

16.57 G. Knoll, N. Suri, J.M. Bradshaw: Path-based security for mobile agents. *Proceedings of the First International Workshop onthe Security of Mobile Multi-Agent Systems (SEMAS-2001) at the Fifth International Conference on Autonomous Agents (Agents 2001)* (Montreal, CA, New York: ACM Press)pp. 54-60

16.58 A. Kolber: *Defining Business Rules: What Are They Really?* Revision 1.3, Final Report. Business Rules Group (formerly the GUIDE Business Rules Project, July)

16.59 N.G. Leveson: *Safeware: System Safety and Computers: A Guide to Preventing Accidents and Losses Caused by Technology* (Boston, MA, Addison-Wesley)

16.60 J.C.R. Licklider: Man-computer symbiosis. *IRE Transactions in Electronics. New York: Institute of Radio Engineers.*, 4-11

16.61 H. Lieberman (ed.): *Your Wish is My Command: Programming By Example* (San Francisco, CA: Morgan Kaufmann)

16.62 F. Lopez y Lopez, M. Luck, M. d'Inverno: A framework for norm-based inter-agent dependence. *Proceedings of the Third Mexican Internation Conference on Computer Science*

16.63 F. Lopez, Y. Lopez, M. Luck, M. d'Inverno: Constraining autonomy through norms. *Proceedings of the Conference on Autonomous Agents and Multi-Agent Systems* (Bologna, Italy) pp. 674-681

16.64 S. Lubar: *InfoCulture: The Smithsonian Book of Information and Inventions* (Boston, MA, Houghton Mifflin Company)

16.65 M. Luck, M. D'Inverno, S. Munroe: Autonomy: Variable and generative. In: H. Hexmoor, C. Castelfranchi, R. Falcone (ed.), *Agent Autonomy* (Dordrecht, The Netherlands, Kluwer) pp. 9-22

16.66 P. Maes: Agents that reduce work and information overload. In: J.M. Bradshaw (ed.), *Software Agents* (Cambridge, MA, AAAI Press/The MIT Press 1997) pp. 145-164

16.67 S. Mann: Wearable computing: A first step toward personal imaging. IEEE Computer, 30(2), 25-32 (1997)

16.68 D. Morris: *Peoplewatching* (London, England, Vintage 2002)

16.69 R.R. Murphy: *Introduction to AI Robotics* (Cambridge, MA, The MIT Press 2000)

16.70 D.A. Norman: The 'problem' with automation: Inappropriate feedback and interaction, not over-automation. In: D.E. Broadbent, J. Reason, A. Baddeley (eds.), *Human Factors in Hazardous Situations* (Oxford, England: Clarendon Press 1990) pp. 137-145

16.71 D.A. Norman: *Turn Signals are the Facial Expressions of Automobiles* (Reading, MA, Addison-Wesley 1992)

16.72 D.A. Norman: How might people interact with agents? In: J.M. Bradshaw (ed.), *Software Agents*. (Cambridge, MA, The AAAI Press/The MIT Press 1997) pp. 49-55

16.73 S.L. Oviatt, P. Cohen, L. Wu, J. Vergo, L. Duncan, B. Suhm, J. Bers, T. Holzman, T. Winograd, J. Landay, J. Larson, D. Ferro: Designing the user interface for multimodal speech and pen-based gesture applications: State-of-the-art systems and future research directions. *Human Computer Interaction,* 15(4), 263-322 (2000)

16.74 A.P. Pentland: Wearable intelligence. *Scientific American. Special issue on Exploring Intelligence,* 9(4), pp. 90-95 (1998)

16.75 M. Polanyi: *Personal Knowledge: Toward a Post-Critical Philosophy* (Chicago, IL, The University of Chicago Press 1962)

16.76 D. Pynadath, M. Tambe: Revisiting Asimov's first law: A response to the call to arms. *Proceedings of ATAL 01* (2001)

16.77 N. Sarter, D.D. Woods, C.E. Billings: Automation surprises. In: G. Salvendy (ed.), *Handbook of Human factors/Ergonomics, 2nd Edition* (New York, NY: John Wiley 1997)

16.78 R.K. Sawyer: *Creating Conversations: Improvisation in Everyday Discourse* (Cresskill, NJ, Hampton Press 2001)

16.79 P. Scerri, D. Pynadath, M. Tambe: Adjustable autonomy for the real world. In: R. Falcone (ed.), *Agent Autonomy* (Dordrecht, The Netherlands: Kluwer 2002) pp. 163-190

16.80 P. Schelde: *Androids, Humanoids, and Other Science Fiction Monsters* (New York, New York University Press 1993)

16.81 D. Schreckenghost, C. Martin, P. Bonasso, D. Kortenkamp, T. Milam, C. Thronesbery: Supporting group interaction among humans and autonomous agents. *Submitted for publication* (2003)

16.82 D. Schreckenghost, C. Martin, C. Thronesbery: Specifying organizational policies and individual preferences for human-software interaction. *Submitted for publication* (2003)

16.83 K.E. Seamons, M. Winslet, T. Yu: Limiting the disclosure of access control policies during automated trust negotiation. *Proceedings of the Network and Distributed Systems Symposium* (2001)

16.84 Y. Shoham, M. Tennenholtz: On the synthesis of useful social laws for artificial agent societies. *Proceedings of the Tenth National Conference on Artificial Intelligence* (San Jose, CA 1992) pp. 276-281

16.85 M. Sierhuis, J.M. Bradshaw, A. Acquisti, R. Van Hoof, R. Jeffers, A. Uszok: Human-agent teamwork and adjustable autonomy in practice. *Proceedings of the Seventh International Symposium on Artificial Intelligence, Robotics and Automation in Space (i-SAIRAS)* (Nara, Japan 2003)

16.86 W.J. Smith: *The Behavior of Communicating* (Cambridge, MA, Harvard University Press 1977)

16.87 T.E. Starner: *Wearable Computing and Contextual Awareness.* Doctor of Philosophy, Massachusetts Institute of Technology (1999)

16.88 N. Suri, J.M. Bradshaw, M.R. Breedy, P.T. Groth, G.A. Hill, R. Jeffers: Strong Mobility and Fine-Grained Resource Control in NOMADS. *Proceedings of the 2nd International Symposium on Agents Systems and Applications and the 4th International Symposium on Mobile Agents (ASA/MA 2000)* (Zurich, Switzerland, Berlin: Springer-Verlag 2000)

16.89 N. Suri, J.M. Bradshaw, M.R. Breedy, P.T. Groth, G.A. Hill, R. Jeffers, T.R. Mitrovich, B.R. Pouliot, D.S. Smith: NOMADS: Toward an environment for strong and safe agent mobility. *Proceedings of Autonomous Agents 2000* (Barcelona, Spain, New York: ACM Press 2000)

16.90 N. Suri, J.M. Bradshaw, M.H. Burstein, A. Uszok, B. Benyo, M.R. Breedy, M. Carvalho, D. Diller, P.T. Groth, R. Jeffers, M. Johnson, S. Kulkarni, J. Lott: OWL-based policy enforcement for semantic data transformation and filtering in multi-agent systems. *Proceedings of the Autonomous Agents and*

Multi-Agent Systems Conference (AAMAS 2003) (Melbourne, Australia, New York, NY, ACM Press 2003)

16.91 N. Suri, J.M. Bradshaw, M. Carvalho, M.R. Breedy, T.B. Cowin, R. Saavendra, S. Kulkarni: Applying agile computing to support efficient and policy-controlled sensor information feeds in the Army Future Combat Systems environment. *Proceedings of the Annual U.S. Army Collaborative Technology Alliance (CTA) Symposium* (2003)

16.92 N. Suri, J.M. Bradshaw, M. Carvalho, T.B. Cowin, M.R. Breedy, P. T. Groth, R. Saavendra: Agile computing: Bridging the gap between grid computing and ad-hoc peer-to-peer resource sharing. In: O.F. Rana (ed.), *Proceedings of the Third International Workshop on Agent-Based Cluster and Grid Computing* (Tokyo, Japan 2003)

16.93 N. Suri, J.M. Bradshaw, M.R. Breedy, P.T. Groth, G.A. Hill, R. Jeffers, T.R. Mitrovich, B.R. Pouliot, D.S Smith: NOMADS: Toward an environment for strong and safe agent mobility. *Proceedings of Autonomous Agents 2000* (Barcelona, Spain, ACM Press, NY 2000)

16.94 N. Suri, M. Carvalho, J.M. Bradshaw, M.R. Breedy, T.B. Cowin, P.T. Groth, R. Saavendra, A. Uszok: Mobile code for policy enforcement. *Policy 2003* (Como, Italy 2003)

16.95 M. Tambe, D. Pynadath, C. Chauvat, A. Das, G. Kaminka: Adaptive agent architectures for heterogeneous team members. *Proceedings of the International Conference on Multi-Agent Systems (ICMAS 2000)*

16.96 M. Tambe, W. Shen, M. Mataric, D.V. Pynadath, D. Goldberg, P.J. Modi, Z. Qiu, B. Salemi: Teamwork in cyberspace: Using TEAMCORE to make agents team-ready. *Proceedings of the AAAI Spring Symposium on Agents in Cyberspace* (Menlo Park, CA, Menlo Park, CA, The AAAI Press 1999)

16.97 C. Trevarthen: Communication and cooperation in early infancy: A description of primary intersubjectivity. In: M. Bullowa (ed.), *Before Speech* (Cambridge, England: Cambridge University Press 1979) pp. 321-348

16.98 A. Uszok, J.M. Bradshaw, P. Hayes, R. Jeffers, M. Johnson, S. Kulkarni, M.R. Breedy, J. Lott, L. Bunch: DAML reality check: A case study of KAoS domain and policy services. *Submitted to the Second International Semantic Web Conference (ISWC 03)* (Sanibel Island, Florida, October 2003)

16.99 A. Vanderbilt: *Amy Vanderbilt's New Complete Book of Etiquette: The Guide to Gracious Living* (Garden City, NY, Doubleday and Company 1952/1963)

16.100 H. Verhagen: Norms and artificial agents. *Sixth Meeting of the Special Interest Group on Agent-Based Social Simulation, ESPRIT Network of Excellence on Agent-Based Computing* (Amsterdam, Holland) http://abss.cfpm.org/amsterdam-01/abssnorms.pdf (2001)

16.101 D. Weld, O. Etzioni: The firsts law of robotics: A call to arms. *Proceedings of the National Conference on Artificial Intelligence (AAAI 94)* pp. 1042-1047 (1994)

16.102 M. Winslett, T. Yu, K.E. Seamons, A. Hess, J. Jacobson, R. Jarvis , B. Smith, L. Yu: Negotiating trust on the Web. IEEE Internet Computing, 30-37 (2002)

16.103 K.S. Barber, M. Gamba, C.E. Martin: Representing and analyzing adaptive decision-making frameworks. In: H. Hexmoor, C. Castelfranchi, R. Falcone (eds.), Agent Autonomy. (Dordrecht, The Netherlands, Kluwer) pp. 23-42 (2002)

16.104 G. Boella: Obligations and cooperation: Two sides of social rationality. In: H. Hexmoor, C. Castelfranchi, R. Falcone (eds.), Agent Autonomy. (Dordrecht, The Netherlands, Kluwer) pp. 57-78 (2002)

16.105 R. Cohen, M. Fleming: Adjusting the autonomy in mixed-initiative systems by reasoning about interaction. In: H. Hexmoor, C. Castelfranchi, R. Falcone (eds.), Agent Autonomy. (Dordrecht, The Netherlands, Kluwer) pp. 105-122 (2002)

16.106 R. Falcone, C. Castelfranchi: From automaticity to autonomy: The frontier of artificial agents. In: H. Hexmoor, C. Castelfranchi, R. Falcone (eds.), Agent Autonomy. (Dordrecht, The Netherlands, Kluwer) pp. 79-103 (2002)

16.107 P.A. Hancock, S.F. Scallen: Allocating functions in human-machine systems. In: R. Hoffman, M.F. Sherrick, J.S. Warm (eds.), Viewing Psychology as a Whole. (Washington, D.C., American Psychological Association) pp. 509-540 (1998)

16.108 E. Horvitz: Principles of mixed-initiative user interfaces. *Proceedings of the ACM SIGCHI Conference on Human Factors in Computing Systems (CHI'99)* (Pittsburgh, PA, New York, ACM Press 1999)

16.109 E. Horvitz, A. Jacobs, D. Hovel: Attention-sensitive alerting. *Proceedings of the Conference on Uncertainty and Artificial Intelligence (UAI'99)* (Stockholm, Sweden) pp. 305-313 (1999)

16.110 K. Myers, D. Morley: Directing agents. In: H. Hexmoor, C. Castelfranchi, R. Falcone (eds.), Agent Autonomy (Dordrecht, The Netherlands: Kluwer) pp. 143-162 (2003)

16.111 P. Scerri, D. Pynadath, M. Tambe: Adjustable autonomy for the real world. In: R. Falcone (ed.), Agent Autonomy (Dordrecht, The Netherlands: Kluwer) pp. 163-190 (2002)

16.112 J.M. Bradshaw, P. Feltovich, H. Jung, S. Kulkarni, W. Taysom, A. Uszok, (eds.): Dimensions of adjustable autonomy and mixed-initiative interaction. In: M. Klusch, G. Weiss, M. Rovatsos (eds.), Computational Autonomy (Berlin, Springer-Verlag 2004) (in press)

16.113 A. Uszok, J.M. Bradshaw, R. Jeffers, M. Johnson, A. Tate, J. Dalton, S. Aitken: Policy and contract management for semantic web services. *Proceedings of the AAAI Spring Symposium*, Standford, CA, AAAI Press, March 22-24 (in press)

16.114 A. Uszok, J.M. Bradshaw, R. Jeffers: KAoS: A policy and domain services framework for grid computing and grid computing and semantic web services. In: C. Jensen, S. Prslad, T. Dimitrakos (eds.), Trust Management: Second International Conference (iTrust 2004) Proceedings, Oxford, UK, March/April, Lecture Notes in Computer Science 2995, Berlin, Springer, pp. 16-26

16.115 P. Feltovich, J.M. Bradshaw, R. Jeffers, N. Suri, A. Uszok: Social order and adaptability in animal and human cultures as analogues for agent communities: Toward a policy-based approach. In: A. Omicin, P. Petta, J. Pitt (eds.), Engineering Societies in the Agents World IV. Lecture Notes in Computer Science Series. Berlin, Germany, Springer-Verlag (in press)

16.116 J. Barwise, J. Perry: Situations and Attitudes. Cambridge, MA: MIT Press (1983)

16.117 K. Devlin: Logic and Information. Cambridge, England: Cambridge University Press (1991)

16.118 J.J. Gibson: The Ecological Approach to Visual Perception. Boston, MA: Houghton Mifflin (1979)

16.119 D.A. Norman: Cognitive artifacts. In: J.M. Carroll (ed.), Designing Interaction: Psychology at the Human-Computer Interface. (pp. 17-38). Cambridge: Cambridge University Press (1992)

16.120 J.M. Bradshaw, H. Jung, S. Kulkarni, J. Allen, L. Bunch, N. Chambers, P. Feltovich, L. Galescu, R. Jeffers, M. Johnson, R. Taysom, A. Uszok: Toward trustworthy adjustable autonomy and mixed-initiative interaction in KAoS. Submitted for publication (2004)

16.121 G. Tonti, J.M. Bradshaw, R. Jeffers, R. Montanari, N. Suri, A. Uszok: Semantic Web languages for policy representation and reasoning: A comparison of KAoS, Rei, and Ponder. In: D. Fensel, K. Sycara, J. Mylopoulos (eds.), The Semantic WebISWC 2003. Proceedings of the Second International Semantic Web Conference, Sanibel Island, Florida, USA, October 2003, LNCS 2870. (pp. 419-437). Berlin: Springer (2003)

16.122 J. Lott, J.M. Bradshaw, A. Uszok, R. Jeffers: KAoS policy management for control of security mechanisms in a large-scale distributed system (submitted for publication, 2004)

Part IV

Emerging Soft Computing Technology

17. Constraint-Based Neural Network Learning for Time Series Predictions

Benjamin W. Wah and Minglun Qian

University of Illinois, Urbana-Champaign, USA

Abstract

In this chapter, we have briefly surveyed previous work in predicting noise-free piecewise chaotic time series and noisy time series with high frequency random noise. For noise-free time series, we have proposed a constrained formulation for neural network learning that incorporates the error of each learning pattern as a constraint, a new cross-validation scheme that allows multiple validations sets to be considered in learning, a recurrent FIR neural network architecture that combines a recurrent structure and a memory-based FIR structure, and a violation-guided back propagation algorithm for searching in the constrained space of the formulation. For noisy time series, we have studied systematically the edge effect due to low-pass filtering of noisy time series and have developed an approach that incorporates constraints on predicting low-pass data in the lag period. The new constraints enable active training in the lag period that greatly improves the prediction accuracy in the lag period. Finally, experimental results show significant improvements in prediction accuracy on standard benchmarks and stock price time series.

17.1 Introduction

A *time series* is an ordered sequence of observations made through time, whereas a *time series prediction* problem is the prediction of future data $R(t_0 + h)$ at horizon $h > 0$, given historical data $R(t), t = 1, \cdots, t_0$, in the form of a vector or a scalar. Time series predictions have been used in many areas of science, industry, and commercial and financial activities, such as financial forecasts on stock prices and currency exchange rates, product sale and demand forecasting, population growth, and earthquake activities.

In general, a time series may exhibit nonlinearity, non-stationarity, and possibly periodic behavior such as seasonality. More often than not, observations were contaminated by noise that makes a time series noisy. Figure 17.1 illustrates a noisy nonstationary periodic time series. These four characteristics are described as follows.

1. *Linearity.* A time series is linear if $R(t + h)$ can be expressed as a linear function of some or all of its historical values $f(R(t), R(t-1), \cdots)$; otherwise, it is nonlinear. In this chapter, we are interested in developing general models that can represent nonlinear time series with a continuous function f.

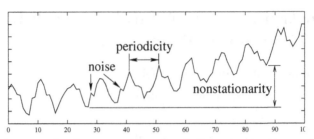

Fig. 17.1. An example of a nonstationary, periodic and noisy time series

2. *Stationarity.* A time series is stationary if it has constant mean and variance; otherwise, it is nonstationary. Nonstationarity is hard to model, as its future behavior may be unpredictable. In this chapter, we are interested in stationary time series as well as a special class of nonstationary time series called piecewise chaotic time series. A time series is *piecewise chaotic* if it consists of several regimes in which each regime corresponds to a chaotic process, and the overall time series is a collection of multiple chaotic regimes.

3. *Periodicity.* A time series with dominant periodic components will exhibit regular periodic variations. Such behavior is displayed, for example, in annual electricity consumption and merchandise sales. Since periodicity is a well studied property and can be eliminated effectively by differencing techniques [17.7], we do not consider it in this chapter.

4. *Random noise* can be present in the entire or some parts of the frequency spectrum of a time series. Since random noise cannot be predicted, we only study time series with high frequency noise and the prediction of their low frequency components.

In the next section, we survey briefly existing work for handling nonlinearity, piecewise chaos and noise in time series. Section 17.3 discusses our newly proposed constrained formulations, along with our violation-guided back-propagation developed to solve the constrained formulations. Section 17.4 presents methods to handle noisy time series using constrained formulations, and illustrates them on the prediction of the low-frequency components of stock prices. Finally, conclusions are drawn in Sect. 17.5.

17.2 Previous Work in Time Series Modeling

A variety of time series models have been proposed and studied in the last four decades. In this section, we review briefly some existing models and present their potential problems when applied to handle nonlinear piecewise chaotic time series with and without noise.

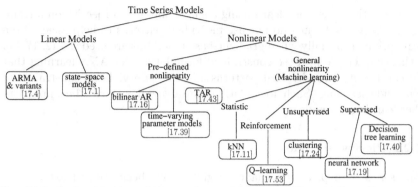

Fig. 17.2. A classification of time-series models for handling nonlinearity

17.2.1 Linearity

Figure 17.2 classifies existing time-series models into linear and nonlinear [17.8].

Linear models work well for linear time series but may fail otherwise. There are three types of linear models.

1. Box-Jenkins *ARIMA* [17.4] and its variations, such as autoregression (AR), moving average (MA), and autoregressive moving average (ARMA), describe future data as a linear combination of historical data and some random processes.
2. *Exponential smoothing* [17.6] models smoothed data $S(t)$ as a function of raw data $R(t)$ by

$$S(t+1) = \alpha R(t) + (1-\alpha)S(t), \qquad 0 < \alpha \leq 1, \tag{17.1}$$

   where α is the only parameter in the model.
3. *State-space models* [17.1] represent inputs as a linear combination of a set of state vectors that evolve over time according to some linear equations. Such vectors and their dimensions are usually hard to choose in practice [17.8].

Nonlinear models can be classified into models with predefined nonlinearity assumptions and general models. The first class includes *bilinear autoregression* [17.16], *time-varying parameter models* [17.39], and *threshold autoregressive models* [17.43]. They are not effective for modeling time series with unknown nonlinear behavior.

Machine learning can handle nonlinear time series because it learns a model without nonlinearity assumptions. Specific methods that can model temporal sequences include statistic learning (such as k-nearest-neighbors (kNN) [17.11]), reinforcement learning (such as Q-learning [17.53]), unsupervised learning (such as clustering methods [17.24]), and supervised learning (such as decision trees [17.40] and artificial neural networks (ANNs) [17.19]).

In general, these methods learn using a single nonlinear objective on a training set. As a result, they do not use individual patterns to help escape from local optima, especially when gradient-based methods are used [17.42, 17.45]. In this chapter, we propose constrained formulations for ANN learning that add constraints on individual patterns and that use violated constraints to help guide learning. Such formulations are general and can be applied to other learning methods.

17.2.2 Piecewise Chaos

Time series with piecewise chaotic regimes have been studied extensively. One approach is to identify local regimes first, before performing learning/predictions on each identified regime. Models using this approach include regime switching models [17.10] and hidden Markov Models [17.26, 17.27]. These approaches are limited because they do not work well unless the changeover points can be correctly identified, and the prediction of changeover points may be as hard as the prediction problem itself.

Without separating the process into two steps, machine learning can learn regime changes by reserving patterns in each regime change to be verified in a cross-validation set. However, traditional learning approaches using a single objective may have difficulties in handling cross validations for multiple regime changes because the single objective containing the sum of the errors in all the cross-validation sets does not provide guidance for refinements when it exceeds a preset threshold. To address this issue, a formulation can be used to constrain the error in each validation set to be satisfied during learning. We show such an approach later, applied to ANN learning and its successful prediction of regime changes in testing.

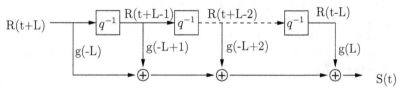

Fig. 17.3. A symmetric FIR filter with $2L$ taps

17.2.3 Random Noise

Random noise is uncorrelated, has zero mean, and is not predictable due to its uncorrelated nature. As its presence in a time series distracts a model from learning useful clean information, especially when the signal-to-noise ratio is relatively low, it is always desirable to eliminate noise before learning.

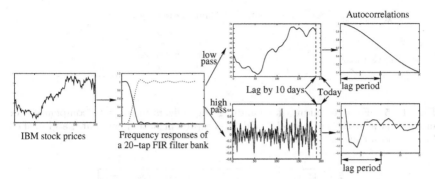

Fig. 17.4. An illustration of a filtering process on a time series of noisy IBM daily closing prices using a 20-tap symmetric low-pass FIR filter to de-noise the time series. Both the low-pass and high-pass data have a 10-day lag. The right two panels show the autocorrelation plots for both filtered time series

In the literature, de-noising is usually done by low-pass filtering or wavelet transforms [17.33, 17.58]. Figure 17.3 illustrates the use of a symmetric FIR filter to generate de-noised data $S(t)$:

$$S(t) = \sum_{j=-L}^{L} R(t+l)g(j), \qquad (17.2)$$

where $g(j)$ is the jth filter coefficient, $2L$ is the number of filter taps, and $R(t)$ is the raw data in the noisy time series.

A symmetric FIR filter is a non-causal filter because its current filtered output depends on future inputs. For example, the filtered output of a $2L$-tap symmetric filter ends at $t_0 - L$ because it depends on raw data that ends at t_0. Such dependencies on future data lead to a lag (sometimes called an edge effect) in the filtered data. Figure 17.4 shows a 10-day lag in both the low-pass and high-pass data of the closing stock prices of IBM, when filtered by a 20-tap symmetric FIR filter.

The edge effect is not a unique artifact of non-causal filters but also occurs when causal filters are used. Although the outputs of causal filters do not depend on future inputs, they reflect a delayed behavior of the original time series and amount to a lag similar to that in non-causal filters.

To overcome edge effects in a time series, a predictor has to first predict missing filtered data in the lag period before predicting into the future. In a time series with high frequency random noise, such predictions will be limited to those of low-pass data, as the autocorrelations between high frequency samples at distances longer than the lag period will be low (Fig. 17.4). As a result, we focus on the predictions of low-pass data in this chapter.

Existing approaches on predicting missing low-pass data in a lag period typically impose some assumptions on the future raw data. Figure 17.5 show some example approaches [17.33, 17.38, 17.57]. Here, flat extension assumes

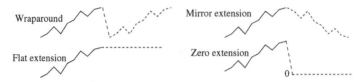

Fig. 17.5. Four techniques for handling edge effects in order to compensate for missing data in the lag period. Solid lines represent actual raw data, and dashed lines stand for extended data

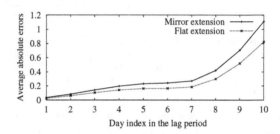

Fig. 17.6. The average absolute errors diverge quickly when predicting missing low-pass data in the lag period of ten days

that future raw data $R(t_0+j)$, $j = 1, 2, \ldots$, is the same as the latest observed raw data $R(t_0)$; mirror extension assumes that future raw data is a mirror image of history data, that is, $R(t_0+j) = R(t_0-j+1)$; wrap-around assumes that after a period of T, the time series repeats itself; and zero extension assumes future data $R(t_0+j)$ to be zero. Using the extended raw data, low-pass filtering is then applied to obtain the de-noised data in the lag period. In this chapter, we do not consider wrap-around and zero extension, as they are applicable only when the time series is stationary and has zero mean.

Figure 17.6 shows the mean absolute errors between the true low-pass data of the closing stock prices of IBM and its corresponding predicted low-pass data using flat and mirror extensions. Although flat extension performs slightly better than mirror extension in this case, both show that the average errors of low-pass data in one part of the lag period (specifically, the last three days) are considerably larger than those in the rest of the lag period. As the low-pass values in the first seven days of the lag period are quite accurate, they can be used as training patterns as if they were true low-pass values. In our approach described in Sect. 17.4.2, we use de-noised data in the first part of the lag period as training patterns, and predict patterns in the latter part of the lag period and beyond. We further use constraints on the raw data in the lag period in order to have more accurate predictions in the lag period.

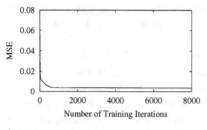
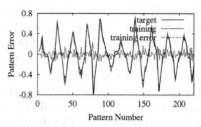

a) Progress of mean squared errors

b) Training errors on individual patterns after 2000 iterations

Fig. 17.7. *Sunspots* time series trained by backpropagation using an unconstrained formulation of Eq. (17.3)

17.2.4 Artificial Neural Networks

It is well known that ANNs are universal function approximators and do not require knowledge of the process under consideration. ANNs for modeling time series generally have special structures that store temporal information either explicitly, using time-delayed structures, or implicitly, using feedback structures. Examples of the first class include time-delayed neural networks (TDNNs) [17.29] and FIR neural networks (FIR-NN) [17.51, 17.52], whereas examples of the latter include recurrent neural networks (RNNs) [17.13, 17.55]. Other architectures, such as radial-basis function network (RBF) [17.36] and supporting vector machines [17.37], store approximate history information in either radial-basis functions or the so-called supporting vectors.

Time-series predictions using ANNs have traditionally been formulated as unconstrained optimization problems that minimize the mean squared errors (MSE) defined as follows:

$$\min_{w} \ \mathcal{E}(w) = \sum_{t=1}^{n} \sum_{i=1}^{N_o} (o_i(t) - d_i(t))^2 \qquad (17.3)$$

where N_o is the number of output nodes in the ANN, $o(t)$ and $d(t)$ are, respectively, the actual and desired outputs of the ANN at time t, w is a vector of all the weights, and the training data consists of patterns observed at $t = 1, \cdots, n$.

Extensive past research has been conducted on designing learning algorithms using an unconstrained formulation in order to lead to ANNs with a small number of weights that can generalize well. However, such learning algorithms have limited success because little guidance is provided in an unconstrained formulation when a search is stuck in a local minimum of the weight space. In this case, the unconstrained objective in Eq. (17.3) does not indicate which patterns are violated and the best direction for the trajectory to move.

Figure 17.7 illustrates the lack of guidance when an unconstrained formulation is used. In this example, an ANN was trained by backpropagation to predict the *Sunspot* time series. Figure 17.7a shows that the MSE in training decreased quickly in the first 1,000 iterations but had little improvement after 2,000 iterations. Further examination of the weights shows that they were almost frozen after 2,000 iterations, and the gradients in all the iterations thereafter were very small. Yet the pattern errors in Fig. 17.7b shows that there were still considerably large errors for some patterns, and that these violated patterns were not identified in an unconstrained formulation. To this end, we propose in the next section a constrained formulation with a constraint on each pattern, and an efficient algorithm for searching in constrained space.

Besides minimizing training errors, cross validations have been used to prevent overfitting in ANN training. Traditional learning involving cross validations generally divides the available historical data into two disjoint training and validation sets and uses the MSE of the validation set as the sole objective. The reason for using only one validation set is due to the limitation of unconstrained formulations that can handle only one objective function.

A problem faced in traditional validations is in choosing a proper cross-validation set. Although there is no defined way on how long and where the validation set should be, one prefers to reserve the last portion of the historical data for validation in order for the ANN to generalize well into the future. Since the training and validations sets are disjoint, the use of the last portion of patterns as a validation set prevents them from being used as training patterns. As a result, the ANN learned does not have access to the most recent patterns in a time series, which are usually the most important for predicting into the future. This is a dilemma in traditional cross validations used in time series predictions.

Another problem faced in traditional validations is related to piecewise chaotic time series. Since piecewise chaotic time series behave differently at changeover points, these points need to be learned specifically in order for the learned system to generalize well. For example, the *Laser* time series [17.54] in Fig. 17.8 is a piecewise chaotic time series with two changeover points at 180 and 600, respectively. To learn these changeover points, we like to have one validation set at the segment around 600 and another at the segment around 1,000, right before the end of the training set. Such multi-objective learning cannot be handled by traditional single-objective formulations, but can be modeled in a constrained formulation that considers the error of each cross-validation set as an additional constraint.

In the rest of this chapter, we describe ANN solutions for predicting time series. We first present, in the next section, constrained formulations and learning algorithms for accurately predicting stationary and piecewise chaotic nonlinear time series. We then present, in Sect. 17.4, the design of efficient

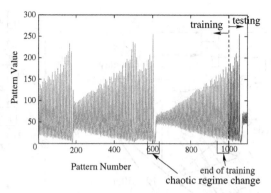

Fig. 17.8. *Laser* is a piecewise chaotic time series that requires at least two validation sets: one for the regime changeover section and another at the end of training

learning algorithms for predicting missing low-pass data of noisy time series in the lag period and beyond.

17.3 Predictions of Noise-Free Stationary and Piecewise Chaotic Time Series

The predictions of a noise-free time series by an ANN are generally carried out by *single-step* (*or iterative*) predictions in which the inputs to the ANN in each prediction step are the true observed (*or* predicted) outputs of the previous prediction step. In both approaches, we adopt a widely used evaluation metric, the *normalized mean square error (nMSE)*, defined as follows:

$$nMSE = \frac{1}{\sigma^2 N} \sum_{t=t_0}^{t_1} (o(t) - d(t))^2, \tag{17.4}$$

where σ^2 is the variance of the true time series in period $[t_0, t_1]$, N is the number of patterns tested, and $o(t)$ and $d(t)$, respectively, are the actual and desired outputs at time t.

In this section, we describe the recurrent FIR ANN architecture used, the constraints developed, the learning algorithm, and our experimental results.

17.3.1 Recurrent FIR Neural Networks

Although a variety of ANN architectures for modeling time series have been studied in the past, there was no consensus on the best architecture to use. For example, Horne and Giles concluded that "recurrent networks usually did better than TDNNs except on the finite memory machine problem" [17.23]. Yet, Hallas and Dorffner stated that "recurrent neural networks do not seem

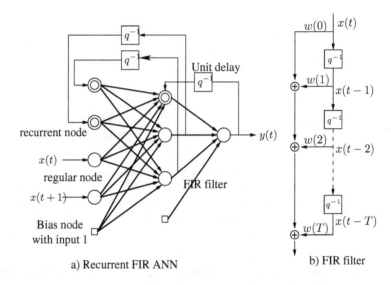

a) Recurrent FIR ANN   b) FIR filter

Fig. 17.9. Structure of a three-layer RFIR: Double concentric circles in (a) indicate recurrent nodes; other circles are non-recurrent nodes; and small boxes are bias nodes with constant input 1. q^{-1} is a one unit time delay

to be able to do so (prediction) under the given conditions" and "a simple feedforward network significantly performs best for most of the nonlinear time series"[17.18]. However, many do agree that the best architecture is problem dependent and that "the efficiency of the learning algorithm is more important than the network model used" [17.28].

Since neither recurrent nor non-recurrent ANNs are found to be superior, we propose a hybrid recurrent ANN called *recurrent FIR neural network* (RFIR). In RFIR, each feedback link has a non-zero delay, and the link between any two nodes has an explicit memory modeled by a multi-tap FIR filter (Fig. 17.9b) for storing history information. Figure 17.9a shows a simple three-layer RFIR in which each feedback link has an associated unit delay. The advantage of this architecture is that it can store more historical information than both RNN and FIR-NN. The concept of RFIR can be generalized easily to existing recurrent ANNs, such as Elman's neural network [17.13], fully recurrent neural networks (FRNN) [17.19], and nonlinear autoregressive networks with exogenous inputs (NARX) [17.9].

17.3.2 Constrained Formulations for ANN Learning

A constrained formulation consists of a learning objective, similar to Eq. (17.3) used in a traditional unconstrained formulation, and multiple constraints, each enforcing a learning criterion. We describe two types of constraints considered, although others may be added as needed.

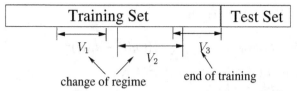

Fig. 17.10. Three validation sets V_1, V_2 and V_3 defined in a training set. Validation sets V_1 and V_2 can be used to cover regime changes in a piecewise chaotic time series

Constraints on individual patterns. To address the lack of guidance in search, we have recently proposed new constraints that account for the error on each training pattern in a constraint [17.45, 17.46, 17.47, 17.48, 17.49]. The resulting optimization problem is as follows:

$$\min_w \quad \mathcal{E}(w) = \sum_{t=1}^{n} \sum_{i=1}^{N_o} \phi((o_i(t) - d_i(t))^2 - \tau) \tag{17.5}$$
$$\text{subject to } h_i(t) = (o_i(t) - d_i(t))^2 \leq \tau \text{ for all } i \text{ and } t,$$

where $\phi(x) = \max\{0, x\}$. Constraint $h_i(t) \leq \tau$ prescribes that the error of the ith output unit on the tth training pattern be less than τ, a predefined small positive value on the learning accuracy to be achieved. Note that $\mathcal{E}(w)$ is modified from Eq. (17.3) in such a way that when all the constraints are satisfied, the objective will be zero and learning stops.

A constrained formulation is beneficial in difficult training scenarios because violated constraints provide additional guidance during a search, leading a trajectory in a direction that reduces overall constraint violations. It can overcome the lack of guidance in Fig. 17.7 when the gradient of the objective function in an unconstrained formulation is very small. A search algorithm solving a constrained formulation has information on all pattern errors and on exactly which patterns have large errors. Hence, it may intentionally modify its search direction by increasing gradient contributions from patterns with large errors.

Constraints on cross validations. Our constrained formulation leads to a new method for cross validations [17.46], shown in Fig. 17.10, that defines multiple validation sets in learning and that includes the error from each validation set in a new constraint. The validation error for the ith output unit from the kth, $k = 1, \cdots, v$, validation set is accounted for by $nMSE$ in both single-step and iterative validations. The advantage of this new approach is that training can be performed on all historical data available, without excluding validation patterns from training as in previous approaches. It considers new constraints on errors in iterative validations, thereby avoiding overfitting of the ANN learned. Further, it allows a new validation set to be defined for each regime transition region in the training data of a piecewise chaotic time series, leading to more accurate predictions of regime changes in the future.

Denote the iterative (*or* single-step) validation-constraint function for the ith output node in the kth validation set as $h_{k,i}^I$ (*or* $h_{k,i}^S$). The constrained formulation of Eq. (17.5) becomes

$$\min_w \mathcal{E}_{av}(t_0, t_1) = \sum_{t=t_0}^{t_1} \sum_{i=1}^{N_o} \phi((o_i(t) - d_i(t))^2 - \tau)$$

subject to $h_i(t) = (o_i(t) - d_i(t))^2 \leq \tau,$ (17.6)

$$h_{k,i}^I = e_{k,i}^I \leq \tau_{k,i}^I,$$
$$h_{k,i}^S = e_{k,i}^S \leq \tau_{k,i}^S,$$

where e^I (or e^S) is the $nMSE$ of the iterative (or single-step) validation error, and $\tau_{k,i}^I$ and $\tau_{k,i}^S$ are predefined small positive constants.

Eq. (17.6) is a constrained nonlinear programming problem (NLP) with non-differentiable function $h_{k,i}^I$. Hence, existing Lagrangian methods that require the differentiability of functions cannot be applied. Existing methods based on penalty formulations have difficulties in convergence when penalties are not chosen properly. Yet, algorithms that sample in real space and that apply the recently developed theory of Lagrange multipliers for discrete constrained optimization [17.50, 17.56] are too inefficient for solving large problem instances. To this end, we describe next an efficient learning algorithm called violation-guided backpropagation (VGBP) [17.47, 17.48].

17.3.3 Violation-Guided Backpropagation Algorithm

Assuming that w is discretized, we first transform Eq. (17.6) into an augmented Lagrangian function:

$$L(w, \lambda) = \mathcal{E}(w) + \sum_{t=1}^{n} \sum_{i=1}^{N_o} \left(\lambda_i(t)\phi(h_i(t) - \tau) + \frac{1}{2}\phi^2(h_i(t) - \tau) \right) \quad (17.7)$$

$$+ \sum_{j=I,S} \sum_{k=1}^{v} \sum_{i=1}^{N_o} \left(\lambda_{k,i}^j \phi(h_{k,i}^j - \tau_{k,i}^j) + \frac{1}{2}\phi^2(h_{k,i}^j - \tau_{k,i}^j) \right).$$

The discretization of w will result in an approximate solution to Eq. (17.6) because a discretized w that minimizes Eq. (17.6) may not be the same as the true optimal w. With fixed error thresholds, Eq. (17.6) may not be feasible with a discretized w but feasible with a continuous w. This is not an issue in ANN learning using Eq. (17.6) because error thresholds in learning are adjusted from loose to tight, and there always exist error thresholds that will lead to a feasible discretized w.

According to the theory of Lagrange multipliers for discrete constrained optimization [17.50], a constrained local minimum in the discrete w space of Eq. (17.6) is equivalent to a saddle point in the discrete w space of Eq. (17.7), which is a local minimum in the w subspace and a local maximum in the λ subspace. To look for the saddle points of Eq. (17.7) by performing descents in the w subspace and ascents in the λ subspace [17.44], we have proposed in Fig. 17.11 the violation guided back-propagation algorithm [17.47, 17.48].

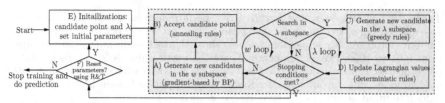

Fig. 17.11. VGBP: an iterative search procedure employing relax-and-tighten (R&T) to solve a discrete constrained formulation for ANN time-series predictions. The shaded box represents the routine to look for saddle points in discretized w subspace and continuous λ subspace

The λ loop carries out ascents in the λ subspace by generating candidates in that subspace in Box (C) and by accepting them using deterministic rules in Box (D). Box (C) increases λ if the violation of the corresponding constraint is found to be larger than a certain threshold. In this way, a training pattern with a large error will have a larger Lagrange multiplier λ.

For a learning problem with a large number of weights and/or training patterns, it is essential that the points generated in Box (A) be likely candidates to be accepted in Box (B). Since Eq. (17.7) is not differentiable, we compute an approximate gradient direction by ignoring single-step and iterative validation errors in Eq. (17.6) and use BP [17.48] to find the gradient of the approximate Lagrangian function $\sum_{t=1}^{n} \sum_{i=1}^{N_o} (1 + \lambda_i(t))(h_i(t) - \tau)$. We then generate a trial point using the approximate gradient and step size η. In this way, a training pattern with a large error will contribute more in the overall gradient direction because of its larger Lagrange multiplier, leading to an effective suppression of constraint violations.

Since the gradient direction computed by BP is based on a heuristic step size and does not consider constraints due to cross validations, a search based on that direction may not always lead to a reduced Lagrangian function value and may get stuck in infeasible local minima of the w subspace. Although restarts can help escape from those local minima, our experimental results have shown that uncontrolled restarts from those points may lead to the loss of valuable local information collected during a search. To address this issue, we propose an annealing strategy in Box (B) that decides whether to go from current point (w, λ) to (w', λ) according to the Metropolis probability [17.47, 17.48]

$$A_T(\mathbf{w}', \mathbf{w})|_\lambda = exp\left\{\frac{(L(\mathbf{w}) - L(\mathbf{w}'))^+}{T}\right\}, \qquad (17.8)$$

where $x^+ = \min\{0, x\}$, and T is introduced to control the acceptance probability.

Tolerances τ, τ^I and τ^S are set by the *relax-and-tighten* strategy proposed in [17.47]. This strategy is based on the observation that looser constraints are easier to satisfy, while achieving larger violations at convergence, and that tighter constraints are slower to satisfy, while achieving smaller violations at

Table 17.1. Single-step and iterative test performance of VGBP in $nMSE$ on *Laser* as compared to other published results. The test set consists of patterns from 1,001 to 1,100. As a comparison, we also show the test performance on patterns from 1001 to 1050. Boxed numbers indicate the best results; N/A stands for data not available. Both runs of VGBP were done on an RFIR with one input node, 20 hidden nodes, one output node, feedbacks from output to hidden layers and from hidden to input layers, but without RFIR on each link

Method	# of weights	Training 100-1000	Single-step predictions 1001-1050	1001-1100	Iterative predictions 1001-1050	1001-1100
FIR-NN [17.52]	1105	0.00044	0.00061	0.023	0.0032	0.0434
ScaleNet [17.15]	N/A	0.00074	0.00437	0.0035	N/A	N/A
VGBP (Run 1)	461	0.00036	0.00043	0.0034	0.0054	0.0194
VGBP (Run 2)	461	0.00107	0.00030	0.00276	0.0030	0.0294

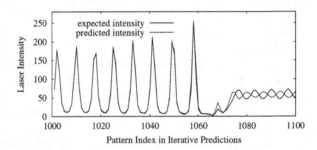

Fig. 17.12. Iterative predictions starting at Pattern 1001 made by an ANN trained by VGBP on a constrained formulation (Run 1 in Table 17.1)

convergence. By using loose constraints at the beginning and by tightening the constraints gradually, learning converges faster with tighter tolerances.

17.3.4 Experimental Results

We have tested our formulations and algorithms on several benchmarks, including *Laser*, *Sunspots*, and five chaotic time series. We present only partial results here [17.46, 17.47, 17.48].

Laser is taken from the Sante Fe time series competition [17.54], in which Wan's FIR-NN [17.52] won the first place on this time series. Table 17.1 shows that VGBP improves Wan's result in terms of both single-step and iterative

Table 17.2. A comparison of single step-prediction performance in $nMSE$ on five methods: Carbon copy (CC), linear [17.52], FIR-NN [17.52], DRNN [17.2], and VGBP. Carbon copy simply predicts the next time-series data to be the same as the proceeding data ($x(t+1) = x(t)$). The training (*resp.* testing) set shows the patterns used in learning (*resp.* testing). *Lorenz attractor* has two data streams labeled by x and z, respectively, whereas *Ikeda attractor* has two streams – real ($Re(x)$) and imaginary ($Im(x)$) parts of a plane wave

Bench-Mark	Training Set	Testing Set	Performance Metrics	CC	Linear	FIR-NN	DRNN	VGBP
MG17	1-500	501-2000	$nMSE$	0.6686	0.320	0.00985	0.00947	0.000057
			# of weights	0	N/A	196	197	121
MG30	1-500	501-2000	$nMSE$	0.3702	0.375	0.0279	0.0144	0.000374
			# of weights	0	N/A	196	197	121
Henon	1-5000	5001-10000	$nMSE$	1.633	0.874	0.0017	0.0012	0.000034
			# of weights	0	N/A	385	261	209
Lorenz	1-4000	4001-5500	$nMSE$ x	0.0768	0.036	0.0070	0.0055	0.000034
			z	0.2086	0.090	0.0095	0.0078	0.000039
			# of weights	0	N/A	1070	542	527
Ikeda	1-10000	10001-11500	$nMSE$ $Re(x)$	2.175	0.640	0.0080	0.0063	0.00023
			$Im(x)$	1.747	0.715	0.0150	0.0134	0.00022
			# of weights	0	N/A	2227	587	574

predictions, using less than half of the weights. Figure 17.12 further shows that VGBP gives accurate iterative predictions for over 60 steps, predicting precisely the regime changeover point and some phase shift afterwards.

We have also tested five more nonlinear chaotic time series: Mackey-Glass 17 (MG17), Mackey-Glass 30 (MG30), Henon, Lorenz, and Ikeda. Table 17.2 compares our results with four other predictors and shows that our ANNs trained by VGBP on constrained formulations use less weights and achieve $nMSE$s that are one to two orders of magnitude smaller.

17.4 Predictions of Noisy Time Series with High Frequency Random Noise

As most financial time series have been found to be noisy and behave like random walks [17.12, 17.58, 17.22], we use daily closing stock prices as examples of noisy time series to illustrate how such series can be learned using constrained formulations. As mentioned earlier, a de-noised low-pass filtered time series lags behind the original series, and a predictor needs to predict the missing low-pass data in the lag period before predicting into the future.

17.4.1 Review on Financial Time Series Predictions

Models for financial time-series forecasting can be classified into linear, nonlinear, and expert system-based. Linear models are still popular in financial time-series predictions, the most popular of which are exponential smoothing, ARIMA models [17.31, 17.32, 17.35, 17.25], and GARCH models (linear in mean and nonlinear in variance) [17.5, 17.34]. Nonlinear models consist mainly of the k-nearest-neighbor methods and ANN methods [17.3, 17.34, 17.41]. Expert systems employ a collection of models (including exponential smoothing methods, ARIMA methods, and moving average), develop a set of rules for selecting a specific method [17.14, 17.35], and select one method for activation when certain conditions are satisfied.

In the past twenty years, there were three Makridakis competitions (also known as M-Competitions [17.30], M2-Competitions [17.31], and M3-Competitions [17.32]) held to test forecasting accuracy, including financial and economic time series. A variety of models, including those aforementioned, were tested. The conclusions reached were rather consistent. We list below some related conclusions drawn from these competitions and the literature [17.17, 17.21, 17.34].

- No single method is clearly superior to other methods in most time series tested.
- Existing methods do not outperform random walk models significantly and statistically in terms of both prediction accuracy and prediction directions (up/down trends). In some cases, they are even worse than random walks.
- Prediction quality is measurement dependent.

17.4.2 Constraint in the Lag Period

In our approach, we first apply flat extensions to generate future raw time series data, obtain low-pass data in the first part of the lag period using the extended raw data, and train an RFIR by VGBP, while including the low-pass data in the first part of the lag period as learning patterns and as additional constraints on raw data . After the RFIR is trained, we use it to predict low-pass data in the latter part of the lag period and into the future.

In order to improve prediction accuracy, we include a special constraint in the lag period by utilizing the available raw data. Let $R(t)$ be the raw data, $S(t)$ be the low-pass data, and $\hat{S}(t)$ be the ANN output at time t (where t can occur during learning or during prediction). Since the low-pass curve is a smoothed version of the raw data curve, the raw data in the lag period generally centers around the true low-pass curve and, consequently, the curve of the ANN outputs (Fig. 17.13). This observation motivates us to add a new constraint in Eq. (17.6) on the difference between the raw data and the ANN outputs in the lag period:

$$h^{lag} = \sum_{t=t_0-m+1}^{t_0} \hat{S}(t) - R(t) \leq \tau^{lag}. \tag{17.9}$$

The new constrained formulation for ANN learning of noisy time series can now be solved by VGBP.

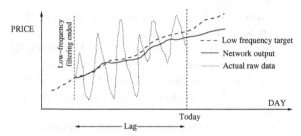

Fig. 17.13. Illustration of an additional constraint in the lag period. The raw data should center around the low-pass data as well as the ANN outputs

17.4.3 Experimental Results

We have conducted experiments by low-pass filtering of time series using a 20-tap filter that incurs a 10-unit lag. After extending the raw data by flat extensions and low-pass filtering of the extended data, we have found experimentally that the last three low-pass values in the lag period have considerably large errors (Fig. 17.6). Figure 17.14 shows the use of the first seven low-pass values in the lag period in training RFIR and the iterative predictions from $t_0 - 2$. True predictions start from $t_0 + 1$.

To test whether our approach will lead to more accurate predictions on future data, we have used the daily closing prices of IBM (symbol IBM), Citigroup (symbol C) and Exxon-Mobil (symbol XOM) between April 1, 1997 and March 31, 2002.

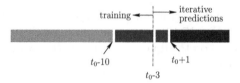

Fig. 17.14. Predictions beyond the lag period using a 20-tap filter

We have also constructed the following five predictors:

CC: Carbon copy that simply copies the most recently available data;

AR(30): Autoregression with an order of 30, using the *TISEAN* implementation [17.20];

NN: Unconstrained formulation trained by BP on RFIR with one input node, two hidden nodes, one output node, feedbacks from hidden to input layers and from output to input layers, and one-tap FIR structure on each feedback link;

LNN: Constrained formulation, with constraints on the lag period and using seven predicted data values in the lag period, trained by VGBP using the same RFIR structure as NN;

IP: Ideal predictor, same as LNN but using seven true values in the lag period.

IP establishes an approximate upper bound for prediction accuracy, as it uses seven error-free low-pass data in the lag period that are not available otherwise.

To compare prediction accuracy, we use $nMSE$, as well as the widely used metric called *hit rate* in financial time series predictions. Let $D(t+h) = \text{sign}(S(t+h) - S(t))$ be the actual direction of change for $S(t)$, and let $\hat{D}(t+h) = \text{sign}(\hat{S}(t+h) - \hat{S}(t))$ be the predicted direction change. We call a prediction for horizon h a *hit* if and only if $\hat{D}(t+h) \times D(t+h) > 0$, and we define the hit rate $H(h)$ as follows:

$$H(h) = \frac{\left|\{t \text{ where } D(t+h)\hat{D}(t+h) > 0\}\right|}{\left|\{t \text{ where } D(t+h)\hat{D}(t+h) \neq 0\}\right|}, \quad t = 1, \cdots, n, \quad (17.10)$$

where $|E|$ represents the number of elements in set E.

Figure 17.15a shows the $nMSE$s of the five predictors for the period between April 1, 1997, and March 31, 2002. (Predictions did not start on April 1, 1997, as an initial window of 200 days were used for training before the first prediction was made.) AR(30) and NN do not perform well because they need to predict iteratively in the 10 day lag period before predicting into the future. LNN improves significantly over CC and AR(30), especially

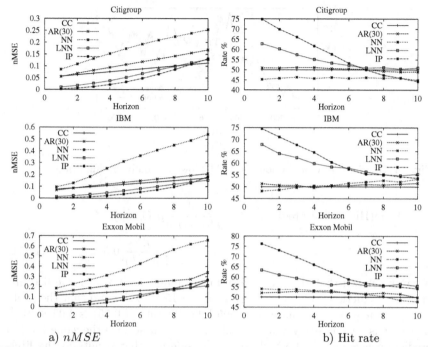

a) $nMSE$ b) Hit rate

Fig. 17.15. Performance of five predictors tested on the daily closing prices of Citigroup, IBM and Exxon-Mobil

for small horizons, and outperforms traditional NN over all horizons. It also shows that LNN has errors closest to those of IP over small prediction horizons, and achieves slightly better $nMSE$ at longer horizons. (Note that IP only gives an *approximate* upper bound on accuracy.)

Figure 17.15b plots the hit rates for the five predictors. It shows that CC, AR(30) and NN behave like random walks over the 10 day horizons, as they always have around a 50% chance to predict the correct direction of price changes. On the other hand, LNN can achieve hit rates significantly higher than 50% over small horizons (one to five days). Finally, IP performs better than LNN for small horizons (one to five days), but performs statistically the same as LNN for large horizons (six days and beyond).

Figure 17.16 plots the predictions of LNN on the next day closing prices of IBM over 1,100 days, as compared to the actual low-pass data. The results show that the predictions track well with the actual low-pass data.

Finally, we analyze the probability that a random walk can achieve the same level of prediction accuracy as LNN, and conclude that it is very unlikely that LNN is a random walk. Since a random walk has a probability of $p = 0.5$ to achieve a hit for each prediction, the probability for it to achieve k hits out of n predictions is governed by a binomial probability:

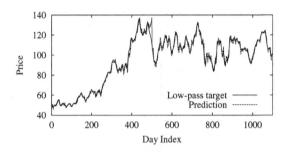

Fig. 17.16. Next-day predictions on IBM's low-pass daily closing prices

$$\text{Prob}(\text{Hits} = k | n \text{ predictions}) = \frac{n!}{k!(n-k)!} p^k (1-p)^{n-k}$$

$$= \frac{n!}{k!(n-k)!} 0.5^n. \tag{17.11}$$

It can be shown that the probability for a random walk to achieve 660 hits in 1,100 predictions (hit rate of 0.6) is only 1.15×10^{-11}, and that the probability for it to achieve more than 605 hits (hit rate of 0.55) is 4.05×10^{-4}. Hence, it is very unlikely that a random walk can achieve the same level of prediction accuracy as LNN. Further, considering that results in the literature on financial time series forecasting achieve next day hit rates below 55% [17.17, 17.21, 17.34], our results are very competitive.

17.5 Conclusions

We have studied the predictions of noise-free as well as noisy time series using ANNs.

For noise-free and possibly piecewise chaotic time series, we have proposed new constrained formulations for ANN learning that allow multiple learning criteria and prior knowledge to be included. Such criteria include testing errors on multiple validation sets that can model regime changes in piecewise chaotic time series, and the error of the ANN learned when validated on the objective used in testing. We have further proposed a new RFIR architecture that combines a recurrent structure and a memory-based FIR structure, and incorporates a violation-guided backpropagation algorithm based on the theory of Lagrange multipliers for discrete constrained optimization.

For noisy time series with high frequency random noise, we have studied systematically the edge effect due to low-pass filtering of noisy time series. To predict missing low-pass data in the lag period, we have developed an approach that estimates low-pass values in the lag period using raw data extended by flat extensions, and that incorporates new constraints on predicted low-pass data in the lag period. The new constraints enable active training in the lag period that greatly improves the prediction accuracy in that period.

References

17.1 M. Aoki: *State Space Modeling of Time Series* (Springer-Verlag, Nerlin, 1987)

17.2 A. Aussem: Dynamical recurrent neural networks towards prediction and modeling of dynamical systems. Neurocomputing, 28, 207-232 (1999)

17.3 S.D. Balkin, J.K. Ord: Automatic neural network modeling for univariate time series. Int'l J. of Forecasting, 16, 509-515 (2000)

17.4 G.E.P. Box, G.M. Jenkins: *Time Series Analysis: Forecasting and Control,* *2nd ed.* (Holden-Day, San Francisco, 1976)

17.5 C. Brooks, S.P. Burke, G. Persand: Benchmarks and the accuracy of GARCH model estimation. Int'l J. of Forecasting, 17, 45-56 (2001)

17.6 R.G. Brown: *Smoothing, Forecasting and Prediction* (Prentice Hall, Englewood Cliffs, NJ, 1963)

17.7 C. Chatfield: *The analysis of time series-an introduction* (Chapman & Hall, London, 5 edition, 1996)

17.8 C. Chatfield: *Time-series forecasting* (Chapman & Hall/CRC, Boca Raton, Florida, 2001)

17.9 S. Chen, S. Billings, P. Grant: Non-linear system identification using neural networks. Int'l J. of Control, 51, 1191-1214 (1990)

17.10 D. Drossu, Z. Obradovic: Regime signaling techniques for non-stationary time series forecasting. In: *Proc. 30th Hwaii Int'l Conf. on System Sciences* (Wailea, HI, USA, 1997) 5, pp. 530-538

17.11 R.O. Duda, P.E. Hart: *Pattern Classification and Scene Analysis* (John Wiley and Sons, 1973)

17.12 R.D. Edwards, J. Magee: *Technical Analysis of Stock Trends* (John Magee, Springfield, MA, 5 edition, 1966)

17.13 J.L. Elman: Finding structure in time. Cognitive Science, 14, 179-211 (1990)

17.14 B.E. Flores, S.L. Pearce: The use of an expert system in the M3 competition. Int'l J. of Forecasting, 16, 485-496 (2000)

17.15 A.B. Geva: ScaleNet - multiscale neural-network architecture for time series prediction. IEEE Trans. on Neural Networks, 9 (5), 1471-1482 (1998)

17.16 C.W.J. Granger, A.P. Andersen: *Introduction to Bilinear Time Series Models* (Vandenhoeck & Ruprect, Göittingen, 1978)

17.17 S. Gutjahr, M. Riedmiller, J. Klingemann: Daily prediction of the foreign exchange rate between the us dollar and the german mark using neural networks. In: *Proc. of SPICES* pp. 492-498 (1997)

17.18 M. Hallas, G. Dorffner: A comparative study on feedforward and recurrent neural networks in time series prediction using gradient descent learning. In: *Proc. of 14th European Meeting on Cybernetics and Systems Research 2,* pp. 644-647 (1998)

17.19 S. Haykin: *Neural Networks: A Comprehensive Foundation* (Prentice Hall, NJ, 2 edition, 1999)

17.20 R. Hegger, T. Schreiber: The TISEAN software package. http://www.mpipks-dresden.mpg.de/ tisean (2002)

17.21 T. Hellstrm: Predicting a rank measure for stock returns. Theory of Stochastic Processes, 6(20), 64-83 (2000)

17.22 T. Hellstrom, K. Holmstrom: *Predicting the Stock Market.* Technical Report Series IMa-TOM-1997-07, Malardalen University, Vasteras, Sweden, 1997

17.23 B.G. Horne, C.L. Giles: An experimental comparison of recurrent neural networks. In: G. Tesauro, D. Touretzky, T. Leen (eds.), *Neural Information Processing Systems* (MIT Press, Cambridge, MA, 1995) pp. 697-704

17.24 A.K. Jain, M.N. Murty, P.J. Flynn: Data clustering: A review. ACM Computing Surveys, 31 (3), 264-323 (1999)

17.25 E.S. Gardner Jr., E.A. Anderson-Fletcher, A.M. Wicks: Further results on focus forecasting vs. exponential smoothing. Int'l J. of Forecasting, 17, 287-293 (2001)

17.26 B.H. Juang, L.R. Rabiner: Hidden Markov models for speech recognition. Technometrics, 33, 251-272 (1991)

17.27 J. Kohlmorgen, K.R. Müller, K. Pawelzik: Analysis of drifting dynamics with neural network hidden markov models. Advances in Neural Information Processing Systems, 10 (1998)

17.28 T. Koskela, M. Lehtokangas, J. Saarinen, K. Kaski: Time series prediction with multilayer perceptron, FIR and Elman neural networks. In: *Proc. of the World Congress on Neural Networks* pp. 491-496 (1996)

17.29 K.J. Lang, G.E. Hinton: *The development of the time-delayed neural network architecture for speech recognition.* Technical Report #CMU-CS-88-152, Carnegie-Mellon University, Pittsburgh, PA, 1988

17.30 S. Makridakis, A. Andersen, R. Carbone, R. Fildes, M. Hibon, R. Lewandowski, J. Newton, E. Parzen, R. Winkler: The accuracy of extrapolation (time series) methods: results of a forecasting competition. Int'l J. of Forecasting, 1, 111-153 (1982)

17.31 S. Makridakis, C. Chatfield, M. Hibon, M. Lawrence, T. Mills, K. Ord, L.F. Simmons: The M2-Competition: a real-time judgementally based forecasting study. Int'l J. of Forecasting, 9, 5-23 (1993)

17.32 S. Makridakis, M. Hibon: The M3-Competition: results, conclusions and implications. Int'l J. of Forecasting, 16, 451-476 (2000)

17.33 T. Masters: *Neural, Novel and Hybrid Algorithms for Time Series Prediction* (John Wiley & Sons, Inc., NY, 1995)

17.34 N. Meade: A comparison of the accuracy of short term foreign exchange forecasting methods. Int'l J. of Forecasting, 18, 67-83 (2002)

17.35 G. Melard, J.M. Pasteels: Automatic arima modeling including interventions, using time series expert software. Int'l J. of Forecasting, 16, 497-508 (2000)

17.36 J. Moody, C. Darken: Fast learning in networks of locally-tuned processing units. Neural Computation, 1(2), 281-294 (1989)

17.37 K. Müller, A. Smola, G. Rätsch, B. Schölkopf, J. Kohlmorgen, and V. Vapnik: Predicting time series with support vector machines. In: *ICANN* pp. 999-1004 (1997)

17.38 F. Murtagh, A. Aussem: Using the wavelet transform for multivariate data analysis and time series forecasting. In: C. Hayashi, H.H. Bock, K. Yajima, Y. Tanaka, N. Ohsumi, Y. Baba, editors, *Data Science, Classification and Related Methods* pp. 617-624 (Springer-Verlag, 1998)

17.39 D.F. Nicholls, A.R. Pagan: Varying coefficient regression. In: E.J. Hannan, P.R. Krishnaiah, M.M. Rao (eds.), *Handbook of Statistics* (North-Holland, Amsterdam, 1985) pp. 413-449

17.40 J.R. Quinlan: Induction of decision trees. Machine Learning, 1, 81-106 (1986)

17.41 S. Ramaswamy: One-step prediction of financial time series, BIS Working Paper No. 57. Technical report, Bank for Interal Settlements, Basle, Switzerland, 1998

17.42 Y. Shang, B.W. Wah: Global optimization for neural network training. IEEE Computer, 29, 45-54 (March 1996)

17.43 H. Tong: *Nonlinear Time Series: A Dynamical System Approach* (Oxford University Press, Oxford, 1990)

17.44 B.W. Wah, Y.X. Chen: Constrained genetic algorithms and their applications in nonlinear constrained optimization. In: *Proc. Int'l Conf. on Tools with Artificial Intelligence* (IEEE, November 2000) pp. 286-293

17.45 B.W. Wah, M.L. Qian: Constrained formulations for neural network training and their applications to solve the two-spiral problem. In: *Proc. Fifth Int'l Conf. on Computer Science and Informatics* 1, pp. 598-601 (February 2000)

17.46 B.W. Wah, M.L. Qian: Time-series predictions using constrained formulations for neural-network training and cross validation. In: *Proc. Int'l Conf. on Intelligent Information Processing, 16th IFIP World Computer Congress* (Kluwer Academic Press, August 2000) pp. 220-226

17.47 B.W. Wah, M.L. Qian: Violation-guided learning for constrained formulations in neural network time series prediction. In: *Proc. Int'l Joint Conference on Artificial Intelligence* (IJCAI, Aug. 2001) pp. 771-776

17.48 B.W. Wah, M.L. Qian: Violation guided neural-network learning fo constrained formulations in time-series predictions. Int'l Journal on Computational Intelligence and Applications, 1(4), 383-398 (December 2001)

17.49 B.W. Wah, M.L. Qian: Constrained formulations and algorithms for stock price predictions using recurrent FIR neural networks. In: *Proc. 2002 National Conf. on Artificial Intelligence* (AAAI, 2002)(accepted to appear)

17.50 B.W. Wah, Z. Wu: The theory of discrete Lagrange multipliers for nonlinear discrete optimization. Principles and Practice of Constraint Programming, pp. 28-42 (October 1999)

17.51 E.A. Wan: Temporal backpropagation for FIR neural networks. IEEE Int'l Joint Conf. on Neural Networks, 1, pp. 575-580 (San Diego, CA., 1990)

17.52 E.A. Wan: *Finite Impulse Response Neural Networks with Applications in Time Series Prediction*. Ph.D. Thesis, Standford University, 1993

17.53 C.J. Watkins: *Models of Delayed Reinforcement Learning*. Ph.D. thesis, Cambridge University (Cambridge, UK, 1989)

17.54 A.S. Weigend, N.A. Gershenfeld (eds.): *Time Series Prediction: Forecasting the future and understanding the past* (Addison-Wesley, 1994)

17.55 R.J. Williams, D. Zipser: A learning algorithm for continually running fully recurrent neural networks. Neural Computation, 1, 270-280 (1989)

17.56 Z. Wu: *The Theory and Applications of Nonlinear Constrained Optimization using Lagrange Multipliers*. Ph.D. Thesis, Dept. of Computer Science, Univ. of Illinois, Urbana, IL (May 2001)

17.57 B.L. Zhang, R. Coggins, M.A. Jabri, D. Dersch, B. Flower: Multiresolution forecasting for future trading using wavelet decompositions. IEEE Trans. on Neural Networks, 12, 766-775 (2001)

17.58 G. Zheng, J.L. Starck, J.G. Campbell, F. Murtagh: Multiscale transforms for filtering financial data streams. J. of Computational Intelligence in Finance, 7, 18-35 (1999)

18. Approximate Reasoning in Distributed Environments

Andrzej Skowron

Warsaw University, Poland

Abstract

Information sources provide us with granules of information that must be transformed, analyzed, and built into structures that support problem solving. Lotfi A. Zadeh has recently pointed out the need to develop a new research branch called Computing with Words and Perceptions (CWP). One way to achieve CWP is through Granular Computing (GC). The main concepts of GC are related to information granule calculi. One of the main goals of information granule calculi is to develop algorithmic methods for construction of complex information granules from elementary ones by means of available operations and inclusion (closeness) measures. These constructions can also be interpreted as approximate schemes of reasoning (AR schemes). The constructed complex granules represent a form of information fusion. Such constructed granules should satisfy some constraints like quality criteria and/or degrees of granule inclusion in (closeness to) a given information granule. In the chapter, we discuss the idea of the rough-neural computing paradigm for inducing AR schemes based on rough sets and, in particular, on rough mereology. Information granule decomposition methods are important components of methods for AR schemes induced from data and background knowledge. We report some recent results on information granule decomposition.

18.1 Introduction

Lotfi A. Zadeh has recently pointed out to a necessity for developing a new research branch called Computing with Words and Perceptions (CWP) (see [18.58, 18.59, 18.60]). The goal of this new research direction is to build foundations for future intelligent computers and information systems performing computations on words rather than on numbers. In this new paradigm, different soft computing tools like neural networks, fuzzy sets, rough sets, or genetic algorithms should work in a complementary, a not competitive, fashion. A great challenge is to develop the foundations for this new computing paradigm and to show that they can help to demonstrate new applications.

Information granulation belongs to intensively studied topics in soft computing (see [18.17, 18.25, 18.36, 18.54, 18.58, 18.59, 18.60]). One of the recently emerging approaches to deal with information granulation is based on information granule calculi (see [18.24, 18.35, 18.40, 18.41, 18.45]). The development of such calculi is important for making progress in many areas such as object identification by autonomous systems (see [18.4, 18.56]), Web

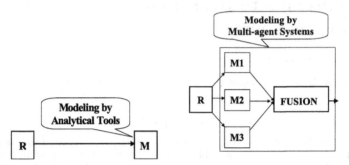

Fig. 18.1. Traditional analytical vs multiagent systems modeling (R is for reality, M,M1,M2 denote models)

mining (see [18.13]) and spatial reasoning (see [18.6, 18.7, 18.55]). In particular, reasoning methods using background knowledge, as well as knowledge extracted from experimental data (e.g., sensor measurements) represented by concept approximations [18.4], are important for making progress in such areas. Moreover, such calculi are important for the development of sensor fusion strategies (see [18.3, 18.8, 18.29, 18.30]). One should take into account that modeling complex phenomena entails the use of local models (captured by local agents), which should next be fused. This process involves the negotiation between agents [18.12] to resolve contradictions and conflicts in local modeling. This kind of modeling will become more and more important in solving the complex real-life problems we are unable to model using traditional analytical approaches. The latter approaches lead to exact models; however, the necessary assumptions used to create them cause the resulting solutions to be *too far* from reality to be accepted (see Fig. 18.1).

One way to achieve CWP is through Granular Computing (GC). The main concepts of GC are related to information granulation and, in particular, to information granules. Information granules, due to Zadeh [18.58], are clumps of objects (points) that are drawn together by indistinguishability, similarity, or functionality. Several approaches concerning formulation of the information granule concept have been proposed.

Any approach to information granulation should make it possible to define complex information granules, e.g., in spatio-temporal reasoning, one should be able to determine if the situation on a road is safe (see Fig. 18.2) on the basis of sensor measurements [18.55] or be able to classify situations in complex games, like soccer [18.50]. These complex information granules consitute a form of information fusion. Any calculus of complex information granules should allow one to (i) deal with the vagueness of information granules; (ii) develop strategies for inducing multi-layered schemes of complex granule construction; (iii) construct robust information granules with respect to deviations of the granules from which they are constructed; and

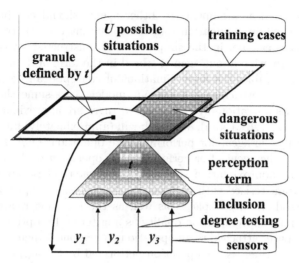

Fig. 18.2. Classification of situations

(iv) develop adaptive strategies for reconstructing induced schemes of complex information granule synthesis.

To deal with vagueness, one can adopt fuzzy set theory [18.57] or rough set theory [18.27], either separately or in combination [18.22]. The second requirement is related to the problem of understanding reasoning from measurements to perception (see [18.59]), to concept approximation learning in layered learning [18.50], and to fusion of information from different sources (see [18.58, 18.59, 18.60]). The importance of searching for *Approximate Reasoning Schemes* (AR schemes) as schemes for new information granule construction, is stressed also in rough mereology. In general, this leads to hierarchical schemes of new information granule construction. This is closely related to ideas of cooperation and conflict resolution in multi-agent systems [18.12]. Among important topics studied are methods for specifying operations on information granules, in particular, for their construction from data and background knowledge, and methods for inducing the hierarchical schemes of information granule construction. One of the possible approaches is to learn such schemes using evolutionary strategies.

Robustness of the scheme means that any scheme produces a higher order information granule that is a clump (e.g., a set) of close information granules rather than a single information granule. Such a clump is constructed, e.g., by means of the scheme from the Cartesian product of input clumps (e.g., clusters) satisfying some constraints. The input clumps are defined by deviations (up to acceptable degree) of input information granules. Using multiagent terminology, we also observe that local agents perform operations on information granules from granule sets "understandable" to them. Hence, granules submitted as arguments by other agents should be approximated by properly tuned approximation spaces creating interfaces between agents. These interfaces can be, in the simplest case, constructed on the basis of exchanged information about agents, stored in the form of decision data tables. From these tables the approximations of concepts can be constructed using rough set approach [18.44]. In our model, we assume that, for any agent ag and its operation $o(ag)$ of arity n, there are approximation spaces $AS_1(o(ag), in), \ldots, AS_n(o(ag), in)$ which will (approximately) filter the granules received by the agent for performing the operation $o(ag)$. In turn, the granule sent by the agent after performing the operation is filtered (approximated) by the approximation space $AS(o(ag), out)$. These approximation spaces are parameterized with parameters allowing one to optimize the size of neighborhoods in these spaces, as well as the inclusion relation [18.36], using the granule approximation quality as a criterion for optimization. Approximation spaces attached to an operation correspond to neuron weights in neural networks, whereas the operation performed by the agent corresponds to the operation realized by the neuron on the vector of real numbers. The generalized scheme of agents returns an output granule in response to input information granules. Such an output granule can, for example, be a cluster of elementary granules. Hence, our schemes (being extensions of schemes for synthesis of complex objects, or granules, developed in [18.35] and [18.33]) realize much more general computations than neural networks operating on vectors of real numbers. The question, if such schemes can be efficiently simulated by classical neural networks, is open. We would like to call extended AR schemes for complex object construction *rough-neural schemes*. The stability of such schemes corresponds to the resistance to noise of classical neural networks.

The methods of inducing AR schemes transforming information granules into new information granules, studied using rough set methods in hybridization with other soft computing approaches, create a core for Rough-Neural Computing (RNC) (see [18.23, 18.24, 18.37, 18.48]).

Another important problem concerns relationships between information granules and words (linguistic terms) in a natural language, and also the possibility of inducing AR schemes as schemes approximating reasoning in natural language. This can strengthen the links between RNC and CWP. It is of great importance for many applications. For example, in case of Web mining (see Fig. 18.3) one is interested in extracting documents relevant to the user's query or to dialog with the user. Hence, the problem of how to construct information granules describing a clump of documents most relevant to the user's query.

Among the approaches to information granulation, RNC attempts to define information granules using rough sets and rough mereology (introduced to deal with vague concepts) in hybridization with methods for constructing more complex information granules by schemes analogous to neural networks.

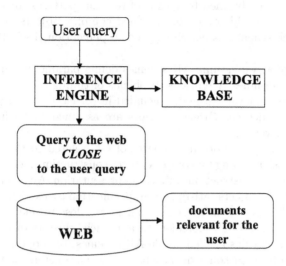

Fig. 18.3. Web mining

In RNC, computations are performed on information granules. Building the foundations of RNC requires the theory of artificial neural networks, rough set theory [18.27], and its extensions like rough mereology (see [18.31, 18.32, 18.33, 18.36, 18.37, 18.38]), as well as some tools created by hybridization with other soft computing approaches, in particular with fuzzy set theory [18.22, 18.57, 18.60] and evolutionary programming [18.15, 18.23].

In the developing of RNC, special roles have different hybrid methods using soft computing tools used to induce the robust AR schemes for complex information granule construction and object classification, as well as methods based on integration of rough sets with neural network techniques, because they are crucial for developing the theory of RNC through synthesis of approximate schemes of reasoning.

We outline a rough-neural computing model as a basis for GC. Our approach is based on rough sets, rough mereology, and information granule calculus.

Rough Mereology [18.32, 18.35] is a paradigm allowing a synthesis of the main ideas of two paradigms for reasoning under uncertainty: fuzzy set theory and rough set theory. We present applications of rough mereology to the important theoretical idea put forth by Lotfi Zadeh [18.58, 18.59], i.e., granularity of knowledge, by presenting the rough-neural computing paradigm.

Information granule decomposition methods are important components of methods for inducing AR schemes from data and background knowledge. In the chapter, we discuss some information granule decomposition methods.

AR schemes are obtained by means of relevant patterns for a given task decomposition of the identified or classified complex objects. The problem of deriving such schemes is closely related to perception [18.2, 18.10, 18.59, 18.53, 18.39].

In the chapter, we assume AR schemes define parameterized operations on information granules. We discuss problems of tuning these parameters to derive from them relevant granules included in (or close to) target concepts to a satisfactory degree. Target concepts are assumed to be incompletely specified and/or vague.

We emphasize an important property of GC related to the necessity of lossless compression tuning for complex object constructions. This means that we map a cluster of constructions into one representation. Any construction in the cluster delivers objects satisfying the specification to a satisfactory degree only if input objects to synthesize are sufficiently close to selected standards (prototypes). In the rough mereological approach, clusters of constructions are represented by the so-called stable AR schemes (of cooperating agents), i.e., schemes robust to some deviations in the parameters of transformed granules. Consequently, the stable AR schemes are able to return objects satisfying to a satisfactory degree the specification not only from standard (prototype) objects, but also from objects sufficiently close to them [18.32, 18.33]. In this way, any stable scheme of complex object construction is a representation of a cluster of similar constructions from clusters of elementary objects.

One can distinguish two kinds of parts (represented by sub-formulas and by sub-terms) of AR schemes. Parts of the first type are represented by expressions from a language, called the *domestic* language L_d that has known semantics (consider, for example, a given information system [18.27]). Representations of parts of the second type of AR scheme are from a language, called *foreign* language L_f (e.g., natural language) that has semantics definable only in an approximate way (e.g., by means of patterns extracted using rough, fuzzy, rough-fuzzy or other approaches). For example, parts of the second kind of scheme can be interpreted as soft properties of sensor measurements [18.4].

For a given expression e representing a given scheme that consists of sub-expressions from L_f, we propose to search for relevant approximations in L_d of the foreign parts from L_f, and to next derive global patterns from the whole expression after replacing the foreign parts by their approximations. This can be a multilevel process, i.e., we are facing problems of discovered pattern propagation through several domestic-foreign layers.

Let us consider some strategies for pattern construction from schemes. The first strategy entails searching for relevant approximations of parts using a rough set approach. This means that each part from L_f can be replaced by its lower or upper approximation with respect to a set B of attributes. The approximation is constructed on the basis of the relevant data table [18.4].

With the second strategy, parts from L_f are partitioned into a number of sub-parts corresponding to cuts (or to the set theoretical differences between cuts) of fuzzy sets [18.57], representing vague concepts, and each sub-part is approximated by means of rough set methods. The third strategy is based on searching for patterns sufficiently included in foreign parts. In all cases, the extracted approximations replace foreign parts in the scheme, and candidates for global patterns are derived from the scheme obtained after the replacement. Searching for relevant global patterns is a complex task because many parameters, e.g., the set of relevant features used in approximation, relevant approximation operators, the number and distribution of objects from the universe of objects among different cuts, and so on, should be tuned. One can use evolutionary techniques for relevant pattern searching to obtain optimal parameters with respect to the quality of synthesized patterns.

We propose an approach for extracting, from data, patterns relevant to a given target concept. The approach is based on information granule decomposition strategies. It is shown that the strategies discussed can be based on the rough set methods developed for decision rules generation and Boolean reasoning [18.14]. We discuss, in particular, methods for decomposition which can be based on background knowledge. In [18.20, 18.45], the reader can find another approach to decomposition.

The chapter is structured as follows. Section 18.2 includes examples of information granules and operations on granules. In particular, in Sect. 18.2.4 we discuss parameterized rough and fuzzy information granules. Sections 18.3 and 18.4 are dedicated to the rough-neural computing paradigm. In Sect. 18.5, methods for decomposition of information granules are outlined.

The chapter summarizes and extends some results presented in our previous works, presented in particular in [18.31, 18.32, 18.33, 18.34, 18.35, 18.36, 18.37, 18.38, 18.43, 18.44, 18.45, 18.46, 18.47].

18.2 Information Granules

18.2.1 Rough Sets and Approximation Spaces

We recall the general definition of approximation space [18.43].

A *parameterized approximation space* (with parameters # and \$) is a system

$$AS_{\#,\$} = (U, I_\#, \nu_\$), \tag{18.1}$$

where

- U is a non-empty set of objects;
- $I_\# : U \to P(U)$, where $P(U)$ denotes the powerset of U, is an *uncertainty function*;
- $\nu_\$: P(U) \times P(U) \to [0, 1]$ is a rough *inclusion function*.

We write I, ν instead of $I_\#, \nu_\$$ to simplify notation. If $X = \{x\}$ and $Y = \{y\}$, we also write $I(x)$ and $\nu(x,y)$ instead of $I(X)$ and $\nu(X,Y)$, respectively. If $p \in [0,1]$, then $\nu_p(x,y)$ $(\underline{\nu}_p(x,y))$ denotes that the condition $\nu(x,y) \geq p$ $(\nu(x,y) \leq p)$ holds.

The uncertainty function defines for every object x a set of similarly described objects, i.e., the *neighborhood* $I_\#(x)$ of x. A constructive definition of the uncertainty function can be based on the assumption that some metrics (distances) are given on attribute values. By ν_{SRI} we denote the *standard rough inclusion* function defined by

$$\nu_{SRI}(X,Y) = \begin{cases} \frac{card(X \cap Y)}{card(X)} & \text{if } X \neq \emptyset \\ 1 & \text{if } X = \emptyset. \end{cases}$$

A set $X \subseteq U$ is *definable in* $AS_{\#,\$}$, if it is a union of some values of the uncertainty function.

The inclusion function defines the degree of inclusion between two subsets of U [18.43]. The inclusion function definition has been generalized in rough mereology to *rough inclusion*.

The lower and upper approximations of subsets of U are defined as follows: For a parameterized approximation space $AS_{\#,\$} = (U, I_\#, \nu_\$)$ and any subset $X \subseteq U$, the *lower* and the *upper approximations* are

$$\begin{aligned} &LOW(AS_{\#,\$}, X) = \{x \in U : \nu_\$(I_\#(x), X) = 1\} \text{ and} \\ &UPP(AS_{\#,\$}, X) = \{x \in U : \nu_\$(I_\#(x), X) > 0\}, \text{ respectively.} \end{aligned} \quad (18.2)$$

The set approximations can be defined in a soft way by allowing one to classify an object x to the lower approximation of X if the neighborhood $I_\#(x)$ is included in X at least to a given a priori satisfactory degree $p \in (0,1]$, and of the complement of X if $I_\#(x)$ is included in X to at most a degree $q < p$, where q is a given a priori degree. Moreover, one can observe that the neighborhood $I_\#(x)$ can be usually defined as a collection of close objects, i.e., it can be defined using rough inclusion. Sets of objects that are definable in a given language can be treated as examples of information granules. We can conclude, then, that primitive notions are *information granules and inclusion* (*closeness*) relations between them. This has been the starting point for investigating rough mereology.

Approximations of concepts (sets) are constructed on the basis of data and background knowledge. Obviously, concepts are also related to new (unseen) objects. Hence, it is very useful to define parameterized approximations, with parameters tuned for approximations of concepts in the searching process. This idea is crucial for methods of construction of concept approximations, in particular for rough set methods. In our notation $\#, \$$ denote vectors of parameters which can be tuned in the process of concept approximation. For a discussion on rough sets in inductive reasoning, the reader is referred to [18.42, 18.48].

Let us now recall some basic definitions [18.14]. If $IS = (U, A)$ is an *information system*, then (a, v) denotes a *descriptor* defined by the attribute

a and its value v, α denotes a Boolean combination of descriptors, and $[\alpha]_{IS}$ (or $[\alpha]_A$) denotes its meaning in IS, i.e., the set of all objects from U satisfying α. The *A-lower* and *A-upper approximations* of $X \subseteq U$ with respect to A are denoted by $\underline{A}X$ and $\overline{A}X$, respectively. By $Inf_A^{IS}(u)$ (or by $Inf_A(u)$), we denote the *signature* of x in IS (i.e., the set $\{(a, a(x)) : a \in A\}$), and by $INF^{IS}(A)$ we denote the set $\{Inf_A(u) : u \in U\}$. If $DT = (U, A, d)$ is a *decision table*, then we assume the set V_d of values of the *decision d* to be equal to $\{1, \ldots, r(d)\}$ for some positive integer $r(d)$ called *the range of d*. The decision d determines a partition $\{C_1, \ldots, C_{r(d)}\}$ of the universe U, where $C_k = \{x \in U : d(x) = k\}$ for $1 \leq k \leq r(d)$. The set C_k is called the *kth decision class* of DT.

18.2.2 Syntax and Semantics of Information Granules

Usually, together with an approximation space, there is also specified a set of formulas Φ expressing properties of objects. Hence, we assume that, together with the approximation space $AS_{\#,\$}$, there are given

- a set of formulas Φ over some language;
- semantics Sem of formulas from Φ, i.e., a function from Φ into the power set $P(U)$.

Let us consider an example [18.27]. We define a language L_{IS} used for elementary granule description, where $IS = (U, A)$ is an information system. The syntax of L_{IS} is defined recursively by

$$
\begin{aligned}
&(a \in V) \in L_{IS}, \text{ for any } a \in A \text{ and } V \subseteq V_a \\
&\text{if } \alpha \in L_{IS} \text{ then } \neg\alpha \in L_{IS} \\
&\text{if } \alpha, \beta \in L_{IS} \text{ then } \alpha \wedge \beta \in L_{IS} \\
&\text{if } \alpha, \beta \in L_{IS} \text{ then } \alpha \vee \beta \in L_{IS}
\end{aligned} \tag{18.3}
$$

The semantics of formulas from L_{IS} with respect to an information system IS is defined recursively by

$$
\begin{aligned}
&Sem_{IS}(a \in V) = \{x \in U : a(x) \in V\} \\
&Sem_{IS}(\neg\alpha) = U - Sem_{IS}(\alpha) \\
&Sem_{IS}(\alpha \wedge \beta) = Sem_{IS}(\alpha) \cap Sem_{IS}(\beta) \\
&Sem_{IS}(\alpha \vee \beta) = Sem_{IS}(\alpha) \cup Sem_{IS}(\beta).
\end{aligned} \tag{18.4}
$$

A typical method used by the rough set approach [18.27] for constructive definition of the uncertainty function is the following: for any object $x \in U$, there is given information $Inf_A(x)$ (information signature of x in A) that can be interpreted as a conjunction $EF_B(x)$ of selectors $a = a(x)$ for $a \in A$ and the set $I_\#(x)$ is equal to

$$
Sem_{IS}(EF_B(x)) = Sem_{IS}\left(\bigwedge_{a \in A} a = a(x)\right). \tag{18.5}
$$

One can consider a more general case, taking as possible values of $I_\#(x)$ any set $\|\alpha\|_{IS}$ containing x. Next, from the family of such sets, the resulting neighborhood $I_\#(x)$ can be selected. One can also use another approach by considering more general approximation spaces, in which $I_\#(x)$ is a family of subsets of U (see [18.16]).

We present now the syntax and the semantics of examples of information granules. These granules are constructed by taking collections of already specified granules. They are parameterized by parameters which can be tuned in applications. In the following sections, we discuss some other kinds of operations on granules, as well as the inclusion and closeness relations for such granules.

Let us note that any information granule g can formally be defined by a pair

$$(Syn(g), Sem(g)) \tag{18.6}$$

consisting of the granules syntax $Syn(g)$ and semantics $Sem(g)$. However, for simplicity of notation, we often use only one component of the information granules to denote it. One can consider another model, in which these components are treated as separate granules, and their fusion produces the above pair of information granules.

Elementary granules. In an information system $IS = (U, A)$, *elementary granules* are defined by $EF_B(x)$, where EF_B is a conjunction of selectors (descriptors) of the form $a = a(x)$, $B \subseteq A$, and $x \in U$. For example, the meaning of an elementary granule $a = 1 \wedge b = 1$ is defined by $Sem_{IS}(a = 1 \wedge b = 1) = \{x \in U : a(x) = 1 \ \& \ b(x) = 1\}$. The number of conjuncts in the granule can be taken as one of a technique parameters to be tuned, the well known in machine learning as the drooping condition [18.18].

One can extend the set of elementary granules, assuming that if α is any Boolean combination of descriptors over A, then $(\overline{B}\alpha)$ and $(\underline{B}\alpha)$ also define, for any $B \subseteq A$., the syntax of elementary granules. The reader can find more details on granules defined by rough set approximations in [18.46, 18.48].

Sequences of granules. Let us assume that S is a sequence of granules and the semantics $Sem_{IS}(\bullet)$ in IS of its elements have been defined. We extend $Sem_{IS}(\bullet)$ on S by

$$Sem_{IS}(S) = \{Sem_{IS}(g)\}_{g \in S}. \tag{18.7}$$

Example 18.2.1. Granules defined by rules in information systems are examples of sequences of granules. Let IS be an information system and let (α, β) be a new information granule received from the rule **if** α **then** β, where α, β are elementary granules of IS. The semantics $Sem_{IS}((\alpha, \beta))$ of (α, β) is the pair of sets $(Sem_{IS}(\alpha), Sem_{IS}(\beta))$. If the right hand sides of rules represent decision classes, then is among parameters to be tuned in classification the number of conjuncts on the left hand sides of rules. A typical goal is to search

for a minimal (or less than minimal) number of such conjuncts (corresponding to the largest generalization) that still guarantee a satisfactory degree of
inclusion in a decision class [18.14, 18.18, 18.42].

Sets of granules. Let us assume that a set G of granules, and the semantics $Sem_{IS}(\bullet)$ in IS for granules from G, have been defined. We extend
$Sem_{IS}(\bullet)$, on the family of sets $H \subseteq G$, by $Sem_{IS}(H) = \{Sem_{IS}(g) : g \in H\}$.
One can consider as a parameter of any such granule its cardinality or its
size (e.g., the length of the granule representation). In the first case, a typical
problem is to search a given family of granules for a granule of the smallest
cardinality sufficiently close to a given one.

Example 18.2.2. One can consider granules defined by sets of rules [18.27,
18.14]. Assume that there is a set of rules $Rule\_Set = \{(\alpha_i, \beta_i) : i = 1, \ldots, k\}$.
The semantics of $Rule\_Set$ is defined by

$$Sem_{IS}(Rule\_Set) = \{Sem_{IS}((\alpha_i, \beta_i)) : i = 1, \ldots, k\}. \tag{18.8}$$

The above-mentioned searching problem for a set of granules corresponds in
the case of rule sets to searching for the simplest representation of a given
rule collection by another set of rules (or a single rule) sufficiently close to
the collection [18.5, 18.52].

Example 18.2.3. Let us consider a set G of elementary information granules
– describing possible *situations* together – with DT_α representing the decision
table for any situation $\alpha \in G$. Assume $Rule\_Set(DT_\alpha)$ to be a set of decision
rules generated from decision table DT_α (e.g., in the minimal form) [18.14].
Now, let us consider a new granule

$$\{(\alpha, Rule\_Set(DT_\alpha)) : \alpha \in G\} \tag{18.9}$$

with semantics defined by

$$\begin{aligned}
\{Sem_{DT}((\alpha, Rule\_Set(DT_\alpha))) : \alpha \in G\} = \\
\{(Sem_{IS}(\alpha), Sem_{DT}(Rule\_Set(DT_\alpha))) : \alpha \in G\}.
\end{aligned} \tag{18.10}$$

An example of a parameter to be tuned is the number of situations represented in such a granule. A typical task is to search for a granule with
the minimal number of situations, creating, together with the rule sets corresponding to them, a granule sufficiently close to the original one.

Extension of granules defined by tolerance relation. We now present
examples of granules obtained by application of a tolerance relation (i.e., a
reflexive and symmetric relation; for more information see [18.43], and, for
clustering methods based on similarity, see [18.11]).

Example 18.2.4. One can consider extensions of elementary granules defined
by a tolerance relation. Let $IS = (U, A)$ be an information system, and let

τ be a tolerance relation on elementary granules of IS. Any pair $(\tau : \alpha)$ is called a τ-*elementary granule*. The semantics $Sem_{IS}((\tau : \alpha))$ of $(\tau : \alpha)$ is the family $\{Sem_{IS}(\beta) : (\beta, \alpha) \in \tau\}$. Parameters to be tuned in searching for relevant tolerance granule can be its support (represented by the number of supporting it objects) and the degree of its inclusion (or closeness) in some other granules, as well as parameters specifying the tolerance relation.

Example 18.2.5. Let us consider granules defined by rules of *tolerance information systems* [18.43]. Let $IS = (U, A)$ be an information system, and let τ be a tolerance relation on elementary granules of IS. If **if** α **then** β is a rule in IS, then the semantics of a new information granule $(\tau : \alpha, \beta)$ is defined by $Sem_{IS}((\tau : \alpha, \beta)) = Sem_{IS}((\alpha, \tau)) \times Sem_{IS}((\beta, \tau))$. Parameters to be tuned are the same as in the case of granules that are sets of more elementary granules, as well as parameters of the tolerance relation.

Clustering of decision and association rules is an important problem in data mining. The reader is referred for measures of closeness of such rules to [18.5, 18.52].

Example 18.2.6. We consider granules defined by sets of decision rules corresponding to evidence given in tolerance decision tables. Let $DT = (U, A, d)$ be a decision table, and let τ be a tolerance on elementary granules of $IS = (U, A)$. Now, any granule $(\alpha, Rule\_Set\,(DT_\alpha))$ can be considered representative of the information granule cluster

$$(\tau : (\alpha, Rule\_Set\,(DT_\alpha))) \tag{18.11}$$

with the semantics

$$\begin{aligned} Sem_{DT}\,((\tau : (\alpha, Rule\_Set\,(DT_\alpha)))) = \\ \{Sem_{DT}\,((\beta, Rule\_Set\,(DT_\beta))) : (\beta, \alpha) \in \tau\}\,. \end{aligned} \tag{18.12}$$

One can see that the case considered is a special case of information granules from Example 18.2.3, with G defined by a tolerance relation.

Dynamic granules. An elementary granule α of the information system IS is non-empty if $\|\alpha\|_{IS} \neq \emptyset$. A non-empty elementary granule β of IS is an extension of α if $\beta = \alpha \wedge \gamma$, where γ is an elementary granule. Let us consider *dynamic granules* defined by some subsets of

$$\{(\beta, Rule\_Set\,(DT_\beta)) : \beta \text{ is an extension of } \alpha\}\,. \tag{18.13}$$

The semantics of these new granules is defined as in the case of sets of granules. Any set G of elementary granules, and granule α, specify new granules

$$\{(\beta, Rule\_Set\,(DT_\beta)) : \beta \text{ is an extension of } \alpha \text{ and } \beta \in G\} \tag{18.14}$$

important for decision making in dynamically changing environments. Let us consider an example. A DT-path is any sequence $\pi = ((\alpha_1, R_1), \ldots, (\alpha_k, R_k))$ such that α_i is an elementary non-empty granule of IS, $R_i = Rule\_Set(DT_{\alpha_i})$

for $i = 1, \ldots, k$, and $\alpha_i = \alpha_{i-1} \wedge \gamma_{i-1}$ for some elementary atomic granule γ_{i-1} (e.g., selector $a = v$) with attributes not appearing in α_{i-1} for $i = 2, \ldots, k$. A granule α_{i-1} is called a *guard* of π if R_{i-1} is not sufficiently close to R_i (what we denote by $non(cl_p(R_{i-1}, R_i))$, where p is the closeness degree). By $Guard(\pi)$ we denote the subsequence of $\alpha_1, \ldots, \alpha_k$ consisting all guards of π. In applications, it is important to search for a minimal (in cardinality) set of granules G satisfying the following condition: for any maximal DT-path π of extensions of α, all guards β from $Guard(\pi)$ (i.e., all points in which it is sufficient to change the decision algorithm represented by the set of decision rules) are from G.

One can also consider dynamic granules with tolerance relations. Let $DT = (U, A, d)$ be a decision table and let τ be a tolerance relation on the elementary granules of $IS = (U, A)$. Two DT-paths $\pi = ((\alpha_1, R_1), \ldots, (\alpha_k, R_k))$ and $\pi' = ((\beta_1, R_1'), \ldots, (\beta_l, R_l'))$ are τ-*similar* if and only if $(\alpha_{i_s}, \beta_{j_s}) \in \tau$ for $s = 1, \ldots, r$, where $Guard(\pi) = (\alpha_{i_1}, \ldots, \alpha_{i_r})$ and $Guard(\pi') = (\beta_{j_1}, \ldots, \beta_{j_r})$. Let us assume τ has the following property:

> **if** $(\beta, \alpha) \in \tau$
> $$\text{then the granules } Rule\_Set\,(DT_\alpha) \qquad (18.15)$$
> $$\text{and } Rule\_Set\,(DT_\beta) \text{ are sufficiently close.}$$

Having such tolerance relations one can search for a set G of guards of a smaller size than before. To specify the task is enough to change the condition for the maximal path in the above formulated problem to the following one: for any maximal path π of extensions of α, there exists a τ-similar path π' to π such that all guards β from $Guard(\pi')$ (i.e., all points where it is sufficient to change the decision algorithm represented by the set of decision rules) are from G.

Labeled graph granules. We discuss *graph granules* and *labeled graph* granules to extend previously introduced granules defined by tolerance relations and dynamic granules.

Example 18.2.7. Let us consider granules defined by pairs (G, E), where G is a finite set of granules with semantics in a given information system $IS = (U, A)$ and $E \subseteq G \times G$. The semantics of a new information granule (G, E) is defined by

$$Sem_{IS}\,((G, E)) = (Sem_{IS}\,(G), Sem_{IS}\,(E)) \qquad (18.16)$$

where

$$Sem(G)_{IS} = \{Sem(g)_{IS} : g \in G\} \text{ and} \qquad (18.17)$$
$$Sem(E)_{IS} = \{(Sem(g), Sem(g')) : (g, g') \in E\}.$$

Example 18.2.8. Let G be a set of granules with semantics over a given information system IS. Labeled graph granules over G are defined by (X, E, f, h), where $E \subseteq X \times X$, $f : X \to G$ and $h : E \to P(G \times G)$. We also assume one additional condition:

$$\text{if } (x, y) \in E \text{ then } (f(x), f(y)) \in h(x, y). \tag{18.18}$$

The semantics of the labeled graph granule (X, E, f, h) is defined by

$$\{(Sem(f(x))_{IS}, Sem(h(x, y))_{IS}, Sem(f(y))_{IS}) : (x, y) \in E\}. \tag{18.19}$$

Let us summarize the considerations presented above. One can define the set of granules G as the least set containing a given set of elementary granules G_0, and closed with respect to operations from a given set of operations on information granules.

We have the following examples of granule construction rules:

$$\frac{\alpha_1, \ldots, \alpha_k\text{-elementary granules}}{\{\alpha_1, \ldots, \alpha_k\}\text{-granule}} \tag{18.20}$$

$$\frac{\alpha_1, \alpha_2\text{-elementary granules}}{(\alpha_1, \alpha_2)\text{-granule}} \tag{18.21}$$

$$\frac{\alpha\text{-elementary granule}, \tau}{(\tau : \alpha)\text{-granule}} \tag{18.22}$$

$$\frac{G\text{-a finite set of granules}, E \subseteq G \times G}{(G, E)\text{-granule}}, \tag{18.23}$$

where τ is a tolerance relation on elementary granules.

Let us observe that in the case of granules constructed with the application of tolerance relations, we have the rule restricted to elementary granules. To obtain a more general rule like

$$\frac{\alpha\text{-graph granule}, \tau}{(\tau : \alpha)\text{-granule}}, \tag{18.24}$$

where τ is a tolerance relation on elementary granules, it is necessary to extend the tolerance (similarity, closeness) relation on more complex objects.

One more interesting class of information granules creates classifiers. This example will be discussed in one of the following sections. Parameters to be tuned are voting strategies and matching strategies of objects against rules, as well as other parameters discussed above, like the closeness of granule in the target granule.

In the examples presented, we have discussed parameterized information granules. We have pointed out that the process of parameter tuning is used to induce relevant (for a given task) information granules. In particular, the process of parameter tuning is performed to obtain a satisfactory degree of inclusion (closeness) for information granules.

In the following section, we discuss inclusion and closeness for information granules.

18.2.3 Granule Inclusion and Closeness

In this section, we will discuss inclusion and closeness of the different information granules introduced in the previous section. Inclusion and closeness are
the basic concepts related to information granules [18.35, 18.45]. Using them,
one can measure the closeness of the constructed granule to the target granule, and the robustness of the construction scheme with respect to deviations
of information granules being components of the construction. For details
and examples of closeness relations, the reader is referred to [18.35, 18.45].

Let us mention that the choice of inclusion or closeness definition depends
very much on the application and the data analyzed. This is the reason that
we have decided to introduce a separate section with this more subjective (or
task-oriented) part of granule semantics.

The *inclusion relation* between granules G and G' to degree at least p,
i.e., $\nu(G, G') \geq p$, will be denoted by $\nu_p(G, G')$. By $\underline{\nu}_p(G, G')$, we denote
the inclusion of G in G' to degree at most p, i.e., that $\nu(G, G') \leq p$ holds.
Similarly, the *closeness relation* between granules G and G' to degree at least
p will be denoted by $cl_p(G, G')$. By p, we denote a vector of parameters (e.g.,
from the interval $[0,1]$ of real numbers). Usually, the set of degrees is assumed
to be a lattice with null (0) and unit (1) elements.

A general scheme for the construction of *hierarchical granules* and their
closeness can be described by the following recursive meta-rule: if granules of
order $\leq k$ and their closeness have been defined, then the closeness $cl_p(G, G')$
(at least to degree p) between granules G and G' of order $k+1$ can be defined
by applying an appropriate operator F to closeness values of components of G
and G', respectively. Certainly, the same scheme can be applied to inclusion
measures.

Elementary granules. We have introduced the simplest case of granules
in information system $IS = (U, A)$. They are defined by $EF_B(x)$, where
EF_B is a conjunction of selectors of the form $a = a(x)$, $B \subseteq A$ and $x \in U$.
Let $G_{IS} = \{EF_B(x) : B \subseteq A \text{ and } x \in U\}$. In [18.27], elementary granules
describe indiscernibility classes with respect to some subsets of attributes. In
a more general setting see [18.43], tolerance (similarity) classes are described.

The crisp inclusion of α in β, where $\alpha, \beta \in \{EF_B(x) : B \subseteq A \ \& \ x \in U\}$
is defined by $Sem_{IS}(\alpha) \subseteq Sem_{IS}(\beta)$, where $Sem_{IS}(\alpha)$ and $Sem_{IS}(\beta)$ are
sets of objects from IS satisfying α and β, respectively. The non-crisp inclusion, known in KDD [18.1], is defined for association rules by means of two
thresholds, t and t':

$$support_{IS}(\alpha, \beta) = card(Sem_{IS}(\alpha \wedge \beta)) \geq t \times card(U) \tag{18.25}$$

$$confidence_{IS}(\alpha, \beta) = \frac{support_{IS}(\alpha, \beta)}{card(Sem_{IS}(\alpha))} \geq t'. \tag{18.26}$$

Elementary granule inclusion in a given information system IS can be defined
using different schemes, e.g., by

$$\nu_t^{IS}(\alpha, \beta) \text{ if and only if } confidence_{IS}(\alpha, \beta) \geq t. \qquad (18.27)$$

The closeness of granules can be defined by

$$cl_{t,t'}^{IS}(\alpha, \beta) \text{ if and only if } \nu_{t,t'}^{IS}(\alpha, \beta) \text{ and } \nu_{t,t'}^{IS}(\beta, \alpha) \text{ hold}. \qquad (18.28)$$

Decision rules as granules. One can define inclusion and closeness of granules corresponding to rules of the form **if** α **then** β using the accuracy coefficients. Given granules $g = (\alpha, \beta)$ and $g' = (\alpha', \beta')$, one can define inclusion and closeness of g and g' by

$$\nu_{t,t'}^{IS}(g, g') \text{ if and only if } \nu_{t,t'}^{IS}(\alpha, \alpha') \text{ and } \nu_{t,t'}^{IS}(\beta, \beta'). \qquad (18.29)$$

The closeness can be defined by

$$cl_{t,t'}^{IS}(g, g') \text{ if and only if } \nu_{t,t'}^{IS}(g, g') \text{ and } \nu_{t,t'}^{IS}(g', g). \qquad (18.30)$$

Another way of defining inclusion of granules corresponding to decision rules is as follows:

$$\nu_t^{IS}((\alpha, \beta), (\alpha', \beta')) \text{ if and only if}$$
$$\nu_{t_1, t_2}^{IS}(\alpha, \alpha') \text{ and } \nu_{t_1, t_2}^{IS}(\beta, \beta') \text{ and } t = w_1 \cdot t_1 + w_2 \cdot t_2, \qquad (18.31)$$

where w_1 and w_2 are some given weights satisfying $w_1 + w_2 = 1$ and $w_1, w_2 \geq 0$.

Measures of closeness of rules are discussed in [18.5, 18.52].

Extensions of elementary granules by tolerance relation. For extensions of elementary granules defined by a similarity (tolerance) relation, i.e., granules of the form $(\tau : \alpha)$ and $(\tau : \beta)$, one can consider the inclusion measure

$$\nu_{t,t'}^{IS}((\tau : \alpha)(\tau : \beta)) \text{ if and only if } \nu_{t,t'}^{IS}(\alpha', \beta') \text{ for any } \alpha', \beta' \qquad (18.32)$$
$$\text{such that } (\alpha, \alpha') \in \tau \text{ and } (\beta, \beta') \in \tau$$

and the closeness measure

$$cl_{t,t'}^{IS}((\tau : \alpha)(\tau : \beta)) \text{ if and only if } \nu_{t,t'}^{IS}((\alpha, \tau)(\beta, \tau))$$
$$\text{and } \nu_{t,t'}^{IS}((\beta, \tau)(\alpha, \tau)). \qquad (18.33)$$

It can be important for some applications to define closeness of an elementary granule α and the granule $(\tau : \alpha)$. A definition reflecting an intuition that α should be a representation of $(\tau : \alpha)$ sufficiently close to this granule is as follows:

$$cl_{t,t'}^{IS}(\alpha, (\tau : \alpha)) \text{ if and only if } cl_{t,t'}^{IS}(\alpha, \beta) \text{ for any } (\alpha, \beta) \in \tau. \qquad (18.34)$$

Sets of rules. An important problem related to association rules is that the number of such rules generated even from a simple data table can be large. Hence, one should search for methods for aggregating close association rules. We suggest that this be defined as searching for some close information granules.

Let us consider two finite sets, $Rule\_Set$ and $Rule\_Set'$, of association rules, defined by

$$Rule\_Set = \{(\alpha_i, \beta_i) : i = 1, \ldots, k\} \tag{18.35}$$

$$Rule\_Set' = \{(\alpha_i', \beta_i') : i = 1, \ldots, k'\}. \tag{18.36}$$

One can treat them as higher order information granules. These new granules, $Rule\_Set$ and $Rule\_Set'$, can be treated as close to a degree at least t (in IS) if and only if there exists a relation rel between sets of rules $Rule\_Set$ and $Rule\_Set'$ such that:

1. For any $Rule \in Rule\_Set$, there is $Rule' \in Rule\_Set'$ such that $(Rule, Rule') \in rel$ and $Rule$ is close to $Rule'$ (in IS) to degree at least t.
2. For any $Rule' \in Rule\_Set'$, there is $Rule \in Rule\_Set$ such that $(Rule, Rule') \in rel$ and $Rule$ is close to $Rule'$ (in IS) to degree at least t.

Another way of defining closeness of two granules, G_1 and G_2, represented by sets of rules can be described as follows:

Let us consider again two granules, $Rule\_Set$ and $Rule\_Set'$, corresponding to two decision algorithms. By $I(\beta_i')$, we denote the set $\{j : cl_p^{IS}(\beta_j', \beta_i')\}$ for any $i = 1, \ldots, k'$.

Now, we assume $\nu_p^{IS}(Rule\_Set, Rule\_Set')$ if and only if for any $i \in \{1, \ldots, k'\}$ there exists a set $J \subseteq \{1, \ldots, k\}$ such that

$$cl_p^{IS}(\bigvee_{j \in I(\beta_i')} \beta_j', \bigvee_{j \in J} \beta_j) \text{ and } cl_p^{IS}(\bigvee_{j \in I(\beta_i')} \alpha_j', \bigvee_{j \in J} \alpha_j), \tag{18.37}$$

and, for closeness, we assume

$$cl_p^{IS}(Rule\_Set, Rule\_Set') \text{ if and only if}$$
$$\nu_p^{IS}(Rule\_Set, Rule\_Set') \text{ and } \nu_p^{IS}(Rule\_Set', Rule\_Set). \tag{18.38}$$

For example, if the granule G_1 consists of the rules: if α_1 then $d = 1$; if α_2 then $d = 1$; if α_3 then $d = 1$; if β_1 then $d = 0$; and if β_2 then $d = 0$; and the granule G_2 consists of the rules if γ_1 then $d = 1$ and if γ_2 then $d = 0$, then

$$cl_p(G_1, G_2) \text{ if and only if } cl_p(\alpha_1 \vee \alpha_2 \vee \alpha_3, \gamma_1)$$
$$\text{and } cl_p(\beta_1 \vee \beta_2, \gamma_2). \tag{18.39}$$

One can consider a searching problem for a granule $Rule\_Set'$ of minimal size such that $Rule\_Set$ and $Rule\_Set'$ are close. Certainly, the above discussed example is only a simple example of closeness measure between rule sets, and, for a given real-life application, one should induce relevant closeness measures.

Granules defined by sets of granules. The methods previously discussed for inclusion and closeness definitions can be easily adopted for granules defined by sets of granules already defined. Let G and H be sets of granules.

The inclusion of G in H can be defined by

$$\nu_{t,t'}^{IS}(G,H) \text{ if and only if for any}$$
$$g \in G \text{ there is } h \in H \text{ for which } \nu_{t,t'}^{IS}(g,h), \tag{18.40}$$

and their closeness can be defined by

$$cl_{t,t'}^{IS}(G,H) \text{ if and only if } \nu_{t,t'}^{IS}(G,H) \text{ and } \nu_{t,t'}^{IS}(H,G). \tag{18.41}$$

Let G be a set of granules and let φ be a property of sets of granules from G (e.g., $\varphi(X)$ if and only if X is a tolerance class of a given tolerance $\tau \subseteq G \times G$). Then $P_\varphi(G) = \{X \subseteq G : \varphi(X) \text{ holds}\}$. Closeness of granules $X, Y \in P_\varphi(G)$ can be defined by

$$cl_t(X,Y) \text{ if and only if } cl_t(g,g') \text{ for any } g \in X \text{ and } g' \in Y. \tag{18.42}$$

We have the following examples of inclusion and closeness propagation rules:

$$\frac{\text{for any } \alpha \in G, \text{ there is } \alpha' \in H \text{ such that } \nu_p(\alpha,\alpha')}{\nu_p(G,H)} \tag{18.43}$$

$$\frac{cl_p(\alpha,\alpha'), cl_p(\beta,\beta')}{cl_p((\alpha,\beta),(\alpha',\beta'))} \tag{18.44}$$

$$\frac{\text{for any } \alpha' \in \tau(\alpha), \text{ there is } \beta' \in \tau(\beta) \text{ such that } \nu_p(\alpha',\beta')}{\nu_p((\tau:\alpha),(\tau:\beta))} \tag{18.45}$$

$$\frac{cl_p(G,G'), cl_p(E,E')}{cl_p((G,E),(G',E'))}, \tag{18.46}$$

where $\alpha, \alpha', \beta,$ and β' are elementary granules, and G and G' are finite sets of elementary granules.

The exemplary rules have a general form, i.e., they are true in any IS (under the chosen definitions of inclusion and closeness). Some of them are derivable from others. We will see in the next part of their chapter that there are also operations for new granule construction specific to a given information system. In this case, one should extract inference rules from given data.

Information granules defined by inclusion and closeness measures. Let us observe that inclusion (closeness) measures can be used to define new granules that are approximations or generalizations of existing ones. Assume that g and h are given information granules, and ν_p is an inclusion measure (where $p \in [0,1]$). A (h,p)-approximation of g is an information granule $\nu_{h,p}$ represented by a set $\{h' : \nu_1(h',h) \wedge \nu_p(h',g)\}$. Now, the lower and upper approximations of given information granules can be easily defined [18.43].

18.2.4 Rough–Fuzzy Granules

In this section, we will briefly discuss approximation schemes of granules and methods for extracting from them relevant patterns when they include fuzzy concepts as foreign parts. We propose to use a rough set approach to define approximations of fuzzy concepts in a constructive way [18.46, 18.47]. The rough set approximations of fuzzy cuts are used in searching for constructive definitions of approximations of fuzzy sets. We use the cut approximations to derive patterns relevant to the target concept approximation. In the process of searching for high quality patterns, evolutionary techniques can be used.

Let $DT = (U, A, d)$ be a decision table, where the decision is the restriction of the fuzzy membership function $\mu : U \to [0, 1]$ to objects from U. Consider real numbers $0 < c_1 < \ldots < c_k$, where $c_i \in (0, 1]$ for $i = 1, \ldots, k$. Any c_i defines c_i-cut by $X_i = \{x \in U : \mu(x) \geq c_i\}$. Assume, $X_0 = U$ and $X_{k+1} = X_{k+2} = \emptyset$.

A *rough-fuzzy granule* (*rf-granule*, for short) corresponding to $(DT, c_1, \ldots, c_k)$ is any granule $g = (g_0, \ldots, g_k)$ such that, for some $B \subseteq A$,

$$Sem_B(g_i) = (\underline{B}(X_i - X_{i+1}), \overline{B}(X_i - X_{i+1})), \text{ for } i = 0, \ldots, k \qquad (18.47)$$
$$\overline{B}(X_i - X_{i+1}) \subseteq (X_{i-1} - X_{i+2}), \text{ for } i = 1, \ldots, k. \qquad (18.48)$$

Any function $\mu^* : U \to [0, 1]$ satisfying the conditions

$$\mu^*(x) = 0 \text{ if } x \in U - \overline{B}X_1 \qquad (18.49)$$
$$\mu^*(x) = 1 \text{ if } x \in \underline{B}X_k \qquad (18.50)$$
$$\mu^*(x) = c_{i-1} \text{ if } x \in \underline{B}(X_{i-1} - X_i), \text{ for } i = 2, \ldots, k - 1 \qquad (18.51)$$

$$c_{i-1} < \mu^*(x) < c_i \text{ if } x \in (\overline{B}X_i - \underline{B}X_i),$$
$$\text{for } i = 1, \ldots, k \text{ and } c_0 = 0 \qquad (18.52)$$

is called a B-*approximation* of μ.

Now, one can choose the lower or upper approximations of parts, i.e., the set theoretical differences between successive cuts, and propagate them along the scheme in searching for relevant patterns. Another strategy is to propagate the global approximation of foreign fuzzy concepts through the scheme describing a target concept.

This problem is of great importance in the classifying situations by autonomous systems on the basis of sensor measurements [18.56].

18.2.5 Classifiers as Information Granules

An important class of information granules creates classifiers. One can observe that sets of decision rules generated from a given decision table $DT = (U, A, d)$ (see [18.40]) can be interpreted as information granules. The *classifier* construction from DT can be described as follows:

1. First, one can construct granules G_j, corresponding to each particular decision $j = 1, \ldots, r$, by taking a collection $\{g_{ij} : i = 1, \ldots, k_j\}$ of left hand sides of decision rules for a given decision.

2. Let E be a set of elementary granules (e.g., defined by the conjunction of descriptors) over $IS = (U, A)$. We can now consider a granule, denoted by

$$Match(e, G_1, \ldots, G_r) \qquad (18.53)$$

for any $e \in E$, as being a collection of coefficients ε_{ij}, where $\varepsilon_{ij} = 1$ if the set of objects defined by e in IS is included in the meaning of g_{ij} in IS, i.e., $Sem_{IS}(e) \subseteq Sem_{IS}(g_{ij})$, and zero otherwise. Hence, the coefficient ε_{ij} is equal to one if and only if granule e matches the granule g_{ij} in IS.

3. Let us now denote by $Conflict\_res$ an operation (resolving the conflict between decision rules recognizing elementary granules) defined on granules of the form

$$Match(e, G_1, \ldots, G_r)$$

with values in the set of possible decisions $1, \ldots, r$. Hence,

$$Conflict\_res(Match(e, G_1, \ldots, G_r)) \qquad (18.54)$$

is equal to the decision predicted by the classifier

$$Conflict\_res(Match(\bullet, G_1, \ldots, G_r)) \qquad (18.55)$$

on the input granule e.

Hence, classifiers are special cases of information granules. Parameters to be tuned include voting strategies, matching strategies of objects against rules, as well as others, like the closeness of granules in the target granule.

Classifier construction is illustrated in Fig. 18.4, where three sets of decision rules are presented for the decision values $1, 2$, and 3, respectively. Hence, $r = 3$. In the figure, we write α_i instead of g_{i1}, β_i instead of g_{i2}, and γ_i instead of g_{i3}, respectively. Moreover, $\varepsilon_1, \varepsilon_2, \varepsilon_3$, denote $\varepsilon_{1,1}, \varepsilon_{2,1}, \varepsilon_{3,1}$; $\varepsilon_4, \varepsilon_5, \varepsilon_6, \varepsilon_7$ denote $\varepsilon_{1,2}, \varepsilon_{2,2}, \varepsilon_{3,2}, \varepsilon_{4,2}$; and $\varepsilon_8, \varepsilon_9$ denote $\varepsilon_{1,3}, \varepsilon_{2,3}$, respectively.

The reader can now easily describe more complex classifiers by means of information granules. For example, one can consider soft instead of crisp inclusion between elementary information granules representing classified objects and the left hand sides of decision rules, or soft matching between the left hand sides of decision rules and decision classes.

Observe that any classifier realizes a kind of $make\_granule$ operation, transforming collections of granules into granules representing decisions.

18.3 Rough-Neural Computing: Weights Defined by Approximation Spaces

In this section, we will discuss the rough-neural computing paradigm using the model for information granule construction introduced in [18.44, 18.45].

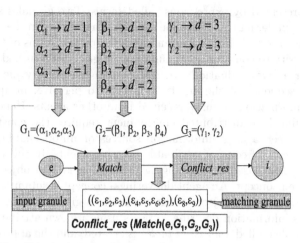

Fig. 18.4. Classifiers as information granules

First, we elaborate a general scheme for information granule construction in distributed systems. Such schemes are parameterized, in particular by local parameterized approximation spaces. These parameterized approximation spaces can be treated as analogous to neural network weights. The parameters should be learned to induce the relevant information granules (see Fig. 18.5).

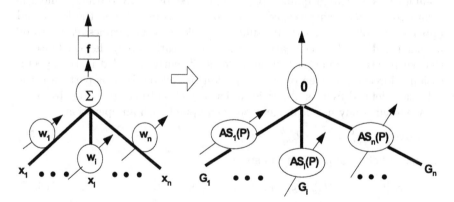

Fig. 18.5. Rough neuron

We use terminology from multiagent literature to explain the basic constructions [18.12].

Teams of agents are organized, e.g., along the schemes of decompositions of complex objects (e.g., representing situations on the road) into trees. The trees are represented by expressions called *terms*. Two granules are defined for any term t under a given valuation val of leaf agents of t in the set of input granules. They are called the lower and upper approximations of t under val.

The necessity to consider approximations of granules returned by a given term t under a given valuation val, rather than the exact value of t under val, is a consequence of the ability of agents to perceive, in approximate sense only, information granules received from other agents. Hence, approximate reasoning in a distributed environment requires the construction of interfaces between agents (information sources or units) enabling effective learning by agents of concepts definable by other agents. In this chapter, we suggest a solution based on exchanging views of agents on objects with respect to a given concept. An agent delivering a concept is submitting positive and negative examples (objects) with respect to a given concept. The agent receiving this information describes objects using its own attributes. In this way a data table (called a *decision table*) is created, and the approximate description of a concept can be extracted by the receiving agent. Our solution is based on the rough set approach. We propose to use parameterized approximation spaces for appropriate tuning of concept perceptions by agents (see Figs. 18.6 and 18.7).

One can consider different problems related to the synthesis of AR schemes defined by terms. For example, one can look for a strategy returning, for a given specification granule, a term t and its valuation val such that the granules defined by the lower and upper values of t under val are sufficiently included in the soft specification granule. Moreover, one can require these granules to be of high quality (e.g., supported by many objects), and the term t to be *robust* with respect to the deviations of val, i.e., the lower and upper values of t under val' are sufficiently close to such values at val to be included in the soft specification granule to a satisfactory degree. Observe that such terms define pattern granules sufficiently included in the specification. Moreover, a given object (situation) is covered by this pattern if the valuation defined by this object is sufficiently close to val (see Fig. 18.2).

We assume any non leaf-agent ag is equipped with an operation

$$o\,(ag) : U_{ag}^{(1)} \times \ldots \times U_{ag}^{(k)} \to U_{ag}^{(0)} \tag{18.56}$$

and has different approximation spaces

$$AS_{ag}^{(i)} = \left(U_{ag}^{(i)}, I_{ag}^{(i)}, \nu_{SRI} \right), \text{ where } i = 0, \ldots, k. \tag{18.57}$$

We assume that the agent ag is perceiving objects by measuring values of some available attributes. Hence, some objects can become indiscernible [18.27]. This influences the specification of any operation $o\,(ag)$. We consider the case where arguments and values of operations are represented by attribute value vectors. Hence, instead of the operation $o\,(ag)$ we have its inexact specification $o^*\,(ag)$, taking as arguments $I_{ag}^{(1)}\,(x_1), \ldots, I_{ag}^{(k)}\,(x_k)$ for some $x_1 \in U_{ag}^{(1)}, \ldots, x_k \in U_{ag}^{(k)}$ and returning the value $I_{ag}^{(0)}\,(o(ag)(x_1, \ldots, x_k))$

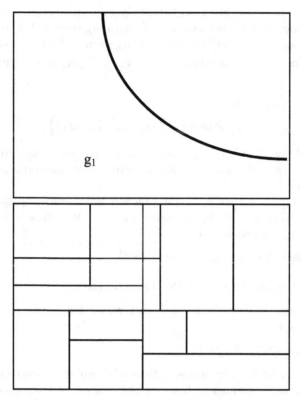

Fig. 18.6. Concept g_1 – information granule of $ag_1 \in Ag$ (top) and communication interface defined by data table (bottom)

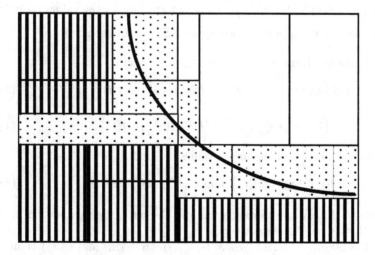

Fig. 18.7. Lower and upper approximation of g_1 by $ag \in Ag$

if $o(ag)(x_1, \ldots, x_k)$ is defined, and the empty set otherwise. This operation can be extended to the operation $o^*(ag)$, with arguments that are definable sets (in approximation spaces attached to arguments), and with values in the family of all non-empty subsets of $U_{ag}^{(0)}$. Let $X_1, \ldots, X_k$ be definable sets. We define

$$o^* (ag) (X_1, \ldots, X_k) =$$
$$\bigcup_{x_1 \in X_1, \ldots, x_k \in X_k} o^* (ag) \left(I_{ag}^{(1)} (x_1), \ldots, I_{ag}^{(k)} (x_k) \right). \tag{18.58}$$

In the sequel, for simplicity of notation, we write $o(ag)$ instead of $o^*(ag)$.

This idea can be formalized as follows. First, we define terms representing agent schemes.

Let $X_{ag}, Y_{ag}, \ldots$ be agent variables for any leaf-agent $ag \in Ag$. Let $o(ag)$ denote a function of arity k. We have mentioned that this is an operation from the Cartesian product of

$$Def\_Sets(AS_{ag}^{(1)}), \ldots, Def\_Sets(AS_{ag}^{(k)}) \tag{18.59}$$

into $P\left(U_{ag}^{(0)}\right)$, where $Def\_Sets(AS_{ag}^{(i)})$ denotes the family of sets definable in $AS_{ag}^{(i)}$. Using the above variables and functors, we define terms in a standard way, for example,

$$t = o(ag) (X_{ag_1}, X_{ag_2}). \tag{18.60}$$

Such terms can be treated as descriptions of complex information granules. By a valuation, we mean any function val defined on the agent variables, with values being definable sets satisfying $val(X_{ag}) \subseteq U_{ag}$ for any leaf-agent $ag \in Ag$. Now, we can define the lower value $val\left(LOW, AS_{ag}^{(i)}\right)(t)$ and the upper value $val\left(UPP, AS_{ag}^{(i)}\right)(t)$ of any term t under the valuation val with respect to a given approximation space $AS_{ag}^{(i)}$ of an agent ag.

1. If t is of the form X_{ag_i} and $val(t) \subseteq U_{ag}^{(i)}$, then

$$val\left(LOW, AS_{ag}^{(i)}\right)(t) = LOW\left(AS_{ag}^{(i)}, val(t)\right) \tag{18.61}$$

$$val\left(UPP, AS_{ag}^{(i)}\right)(t) = UPP\left(AS_{ag}^{(i)}, val(t)\right); \tag{18.62}$$

else, the lower and the upper values are undefined.

$$LOW\left(AS_{ag}^{(i)}, val(t)\right) \text{ and } UPP\left(AS_{ag}^{(i)}, val(t)\right) \tag{18.63}$$

denote the lower approximation and the upper approximation in $AS_{ag}^{(i)}$ of the set $val(t)$, respectively.

2. If $t = o(ag)(t_1, \ldots, t_k)$, where $t_1, \ldots, t_k$ are terms and $o(ag)$ is an operation of arity k, then

a) if $val\left(LOW, AS_{ag}^{(i)}\right)(t_i)$ is defined for $i = 1, \ldots, k$, then

$$val(LOW, AS_{ag}^{(0)})(t) = LOW(AS_{ag}^{(0)}, o(ag) \\ (val(LOW, AS_{ag}^{(1)})(t_1), \ldots, val(LOW, AS_{ag}^{(k)})(t_k))) \; ; \qquad (18.64)$$

else, $val\left(LOW, AS_{ag}^{(0)}\right)(t)$ is undefined;

b) if $val\left(UPP, AS_{ag}^{(i)}\right)(t_i)$ is defined for $i = 1, \ldots, k$, then

$$val(UPP, AS_{ag}^{(0)})(t) = UPP(AS_{ag}^{(0)}, o(ag) \\ (val(UPP, AS_{ag}^{(1)})(t_1), \ldots, val(UPP, AS_{ag}^{(k)})(t_k))) \; ; \qquad (18.65)$$

else, $val(UPP, AS_{ag}^{(0)})(t)$ is undefined.

For illustrative examples of computations of the lower and upper approximations of terms, the reader is referred to [18.45].

Let us observe that the set

$$val(UPP, AS_{ag}^{(0)})(t) - val(LOW, AS_{ag}^{(0)})(t) \qquad (18.66)$$

can be treated as the boundary region of t under val. Moreover, in the process of term construction we have additional parameters to be tuned for obtaining sufficiently high support and confidence, namely, the approximation operations.

A concept X specified by the customer agent is *sufficiently close to t under a given set Val of valuations* if X is included in the upper approximation of t under any $val \in Val$, and X includes the lower approximation of t under any $val \in Val$ as well as the size of the boundary region of t under Val, i.e.,

$$card(\bigcap_{val \in Val} val(UPP, AS_{ag}^{(0)})(t) - \bigcup_{val \in Val} val(LOW, AS_{ag}^{(0)})(t)) \qquad (18.67)$$

is sufficiently small relative to

$$\bigcap_{val \in Val} val(UPP, AS_{ag}^{(0)})(t). \qquad (18.68)$$

18.4 Rough-Neural Computing: Rough Mereological Approach

We now present a conceptual scheme for an adaptive calculus of granules aimed at synthesizing solutions to problems posed under uncertainty. This exposition is based on our earlier analysis presented in [18.32, 18.35, 18.37]. For recent developments, the reader is referred to [18.24, 18.25, 18.48]. We construct a scheme of agents which communicate by relating their respective granules of knowledge by means of transfer functions induced by rough mereological connectives extracted from their respective information systems. Such schemes can be treated as AR schemes. We assume the notation of [18.35] where the reader will find all the necessary information.

In the previous section, some ideas concerning rough-neural computing based on approximation spaces have been discussed. Now, we will present an approach to calculi of granules based on the rough mereological approach. We will concentrate on some ideas. The formal details of rough mereology can be found in [18.31, 18.36, 18.38].

We now formally define the ingredients of our scheme of agents.

18.4.1 Distributed Systems of Agents

We assume that a pair (Inv, Ag) is given where Inv is an *inventory of elementary objects* and Ag is a set of intelligent computing units called *agents*.

We consider an agent $ag \in Ag$. The agent ag is endowed with tools for reasoning about objects in its scope; these tools are defined by components of the agent label. The *label of the agent* ag is the tuple

$$lab(ag) = (\mathcal{A}(ag), M(ag), L(ag), Link(ag), AP\_O(ag), St(ag), \quad (18.69)$$
$$Unc\_rel(ag), H(ag), Unc\_rule(ag), Dec\_rule(ag)) ,$$

where

1. $\mathcal{A}(ag) = (U(ag), A(ag))$ is an information system of the agent ag; we assume as an example that objects (i.e., elements of $U(ag)$) are granules of the form $(\alpha, [\alpha])$, where α is a conjunction of descriptors and $[\alpha]$ denotes the semantics of α in $\mathcal{A}(ag)$ (one may have more complex granules as objects).

2. $M(ag) = (U(ag), [0, 1], \mu_o(ag))$ is a *pre-model* of L_{rm} with a *pre-rough inclusion* $\mu_o(ag)$ on the universe $U(ag)$ [18.31].

3. $L(ag)$ is a set of unary predicates (properties of objects) in a predicate calculus interpreted in the set $U(ag)$; we may assume that formulas of $L(ag)$ are constructed as Boolean combinations of descriptors over B, where $B \subseteq A(ag)$.

4. $St(ag) = \{st(ag)_1, \ldots, st(ag)_n\} \subset U(ag)$ is the set of *standard objects* at ag.

5. $Link(ag)$ is a collection of strings of the form $t = ag_1 ag_2 \ldots ag_k ag$; the intended meaning of a string $ag_1 ag_2 \ldots ag_k ag$ is that $ag_1, ag_2, .., ag_k$ are children of ag in the sense that ag can assemble complex objects (constructs) from simpler objects sent by $ag_1, ag_2, \ldots, ag_k$. In general, we may assume that for some agents ag we may have more than one element in $Link(ag)$, which represents the possibility of re-negotiating the synthesis scheme.

We denote by the symbol $Link$ the union of the family $\{Link(ag) : ag \in Ag\}$.

6. $AP\_O(ag)$ consists of pairs of the form

$$(o(ag, t), ((AS_1(o(ag, t), in), \cdots ,$$
$$AS_n(o(ag, t), in)), AS(o(ag, t), out)) , \quad (18.70)$$

where $o(ag, t) \in O(ag)$, n is the arity of $o(ag, t)$, $t = ag_1 ag_2 \ldots ag_k ag \in Link$, $AS_i(o(ag, t), in)$ is a parameterized approximation space [18.44] corresponding to the ith argument of $o(ag, t)$, and $AS(o(ag, t), out)$ is a parameterized approximation space [18.44] for the output of $o(ag, t)$.

$O(ag)$ is the set of *operations at ag*; any $o(ag, t) \in O(ag)$ is a mapping of the Cartesian product $U(ag) \times U(ag) \times \ldots \times U(ag)$ into the universe $U(ag)$; $o(ag, t)$ is an operation by means of which the agent ag is able to assemble, from objects $x_1 \in U(ag_1), x_2 \in U(ag_2), \ldots, x_k \in U(ag_k)$, the object $z \in U(ag)$, which is an approximation defined by $AS(o(ag, t), out)$ to $o(ag, t)(y_1, y_2, \ldots, y_k) \in U(ag)$, where y_i is the approximation to x_i defined by $AS_i(o(ag, t), in)$. One may choose here either a lower or an upper approximation.

7. $Unc\_rel(ag)$ is the set of *uncertainty relations* unc_rel_i of type

$$(o_i(ag, t), \rho_i(ag), ag_1, \ldots, ag_k, ag, \qquad (18.71)$$

$$\mu_o(ag_1), \mu_o(ag_k), \mu_o(ag), st(ag_1)_i, \ldots, st(ag_k)_i, st(ag)_i) ,$$

where $t = ag_1 ag_2 \ldots ag_k ag \in Link(ag)$, $o_i(ag, t) \in O(ag)$, and

$$\rho_i((x_1, \varepsilon_1), (x_2, \varepsilon_2), \ldots, (x_k, \varepsilon_k), (x, \varepsilon)) \qquad (18.72)$$

holds for $x_1 \in U(ag_1), x_2 \in U(ag_2), \ldots, x_k \in U(ag_k)$ and $\varepsilon, \varepsilon_1, \varepsilon_2, \ldots, \varepsilon_k \in [0, 1]$ if and only if $\mu_o(x, st(ag)_i) \geq \varepsilon_j$, and $\mu_o(x_j, st(ag_j)_i) \geq \varepsilon_j$ for $j = 1, 2, \ldots, k$ for the collection of standards $st(ag_1)_i, st(ag_2)_i, \ldots, st(ag_k)_i, st(ag)_i$ such that

$$o_i(ag, t)(st(ag_1)_i, st(ag_2)_i, \ldots, st(ag_k)_i) = st(ag)_i. \qquad (18.73)$$

The operation o_i performed by ag here is more complex then that of [18.35], as it is composed of three stages: first, approximations to input objects are constructed; next, the operation is performed; and, finally, the approximation to the result is constructed. Relations unc_rel_i provide a global description of this process; in reality, they are compositions of analogous relations corresponding to the three stages. As a result, the unc_rel_i depend on parameters of approximation spaces. This also concerns other constructs discussed here. It follows that, in order to get satisfactory decomposition (or uncertainty and so on) rules, one has to search for satisfactory parameters of approximation spaces (this is analogous to weight tuning in neural computations).

Uncertainty relations express the agents, knowledge about relationships between uncertainty coefficients of the agent ag and uncertainty coefficients of its children. The relational character of these dependencies expresses their intentions.

8. $Unc\_rule(ag)$ is the set of *uncertainty rules* unc_rule_j, of type

$$(o_j(ag, t), f_j, ag_1, ag_2, \ldots, ag_k, ag, \qquad (18.74)$$

$$st(ag_1), st(ag_2), \ldots, st(ag_k), st(ag),$$

$$\mu_o(ag_1), \ldots, \mu_o(ag_k), \mu_o(ag)) ,$$

of the agent, ag where $t = ag_1 ag_2 \ldots ag_k ag \in Link(ag)$ and $f_j : [0, 1]^k \longrightarrow [0, 1]$ is a function which has the property that, for any $x_1 \in U(ag_1), x_2 \in U(ag_2), \ldots, x_k \in U(ag_k)$,

460   A. Skowron

if $o_j(ag,t)(st(ag_1),st(ag_2),\dots,st(ag_k)) = st(ag)$ and   (18.75)

$\mu_o(x_i,st(ag_i)) \geq \varepsilon(ag_i)$ for $i = 1,2,\dots,k$

then $\mu_o(o_j(ag,t)(x_1,x_2,\dots,x_k),st(ag)) \geq$

$f_j(\varepsilon(ag_1),\varepsilon(ag_2),\dots,\varepsilon(ag_k)).$

Figure 18.8 illustrates the idea of uncertainty rules is illustrated.

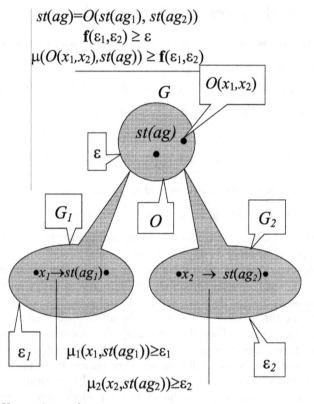

Fig. 18.8. Uncertainty rules

Uncertainty rules provide functional operators, called *rough mereological connectives*, for propagating uncertainty measure values from the children of an agent to the agent; they are applied in negotiation processes, where they inform agents about plausible uncertainty bounds.

9. $H(ag)$ is a strategy which produces uncertainty rules from uncertainty relations (see Sect. 18.2 of [18.35]).

10. $Dec\_rule(ag)$ is a set of *decomposition rules dec_rule$_i$* of type

$(o_i(ag,t),ag_1,ag_2,\dots,ag_k,ag)$ (18.76)

such that

$$(\Phi(ag_1), \Phi(ag_2), .., \Phi(ag_k), \Phi(ag)) \in dec\_rule_i , \qquad (18.77)$$

where

$$\Phi(ag_1) \in L(ag_1), \Phi(ag_2) \in L(ag_2), \ldots,$$
$$\Phi(ag_k) \in L(ag_k), \Phi(ag) \in L(ag) , \qquad (18.78)$$

$$t = ag_1 ag_2 \ldots ag_k ag \in Link(ag) , \qquad (18.79)$$

and there exists a collection of standards $st(ag_1), st(ag_2), \ldots, st(ag_k), st(ag)$ with the properties $o_j(ag,t)(st(ag_1), st(ag_2), ..., st(ag_k)) = st(ag)$, $st(ag_i)$ satisfies $\Phi(ag_i)$, for $i = 1, 2, ..., k$, and $st(ag)$ satisfies $\Phi(ag)$.

Decomposition rules are decomposition schemes in the sense that they describe the standard $st(ag)$ and the standards $st(ag_1), \ldots, st(ag_k)$ from which the standard $st(ag)$ is assembled under o_i in terms of predicates which these standards satisfy.

We may sum up the contents of (1) through (10) by saying that the possible sets of children for any agent ag are specified, and, relative to each team of children, decompositions of standard objects at ag into sets of standard objects at the children, and uncertainty relations as well as uncertainty rules, which relate similarity degrees of objects at the children to their respective standards, and the similarity degree of the object built by ag to the corresponding standard object at ag, are given.

We take rough inclusions of agents as measures of uncertainty in their respective universes. We observe that, because distinct agents use distinct mereological languages, the mereological relation of being a part is not transitive globally over the whole synthesis scheme.

18.4.2 Approximate Synthesis of Complex Objects

The process of synthesis of a complex object (signal, action) of agents by the scheme above defined consists, in our approach, of the two communication stages viz. the top-down communication/negotiation process and the bottom-up communication/assembling process. We outline the AR scheme construction in the language of approximate formulas.

For simplicity of exposition and to avoid unnecessarily tedious notation, we assume that the relation $ag' \leq ag$, which holds for agents $ag', ag \in Ag$ if and only if there exists a string $ag_1 ag_2 \ldots ag_k ag \in Link(ag)$ with $ag' = ag_i$ for some $i \leq k$, that orders the set Ag into a tree. We also assume that $O(ag) = \{o(ag,t)\}$ for $ag \in Ag$, i.e., that each agent has a unique assembling operation for a unique t.

To this end, we build a logic $L(Ag)$ [18.35] in which we can express global properties of the synthesis process.

Elementary formulas of $L(Ag)$ are of the form $\langle st(ag), \Phi(ag), \varepsilon(ag) \rangle$, where $st(ag) \in St(ag), \Phi(ag) \in L(ag), \varepsilon(ag) \in [0,1]$ for any $ag \in Ag$. Formulas of

$L(ag)$ form the smallest extension of the set of elementary formulas closed under the propositional connectives $\vee$, $\wedge$, and $\neg$, and under the modal operators $[\,]$, and $<>$.

To introduce a semantics for the logic $L(ag)$, we first specify the meaning of satisfaction for elementary formulas. The meaning of a formula $\Phi(ag)$ is defined classically as the set $[\Phi(ag)] = \{u \in U(ag) : u$ has the property $\Phi(ag)\}$; we will denote the fact $u \in [\Phi(ag)]$ by the symbol $u \models \Phi(ag)$. We now extend the satisfiability predicate $\models$ to approximate formulas: for $x \in U(ag)$, we say that x satisfies an elementary formula $\langle st(ag), \Phi(ag), \varepsilon(ag)\rangle$; using symbols, $x \models< st(ag), \Phi(ag), \varepsilon(ag) >$ iff

$$st(ag) \models \Phi(ag) \text{ and } \mu_o(ag)(x, st(ag)) \geq \varepsilon(ag). \tag{18.80}$$

We let

$$x \models \neg\langle st(ag), \Phi(ag), \varepsilon(ag)\rangle \text{ iff it is not true that} \tag{18.81}$$
$$x \models \langle st(ag), \Phi(ag), \varepsilon(ag)\rangle$$
$$x \models \langle st(ag)_1, \Phi(ag)_1, \varepsilon(ag)_1\rangle \vee \langle st(ag)_2, \Phi(ag)_2, \varepsilon(ag)_2\rangle \text{ iff}$$
$$x \models \langle st(ag)_1, \Phi(ag)_1, \varepsilon(ag)_1\rangle \text{ or } x \models \langle st(ag)_2, \Phi(ag)_2, \varepsilon(ag)_2\rangle. \tag{18.82}$$

In order to extend the semantics over modalities, we first introduce the notion of a selection: by a *selection* over Ag we mean a function sel which assigns to each agent ag an object $sel(ag) \in U(ag)$.

For two selections sel, sel'(i.e., $sel, sel' : Ag \rightarrow \bigcup U_{ag}$ and $sel(ag), sel'(ag) \in U_{ag}$ for any $ag \in Ag$) we say that sel *induces* sel'; in symbols $sel \rightarrow_{Ag} sel'$ when

$$sel(ag) = sel'(ag) \text{ for any } ag \in Leaf(Ag) \text{ and} \tag{18.83}$$
$$sel'(ag) = o(ag, t)(sel'(ag_1), sel'(ag_2), \ldots, sel'(ag_k))$$
$$\text{for any } t = ag_1 ag_2 \ldots ag_k ag \in Link.$$

We extend the satisfiability predicate $\models$ to selections: for an elementary formula $\langle st(ag), \Phi(ag), \varepsilon(ag)\rangle$, we let

$$sel \models \langle st(ag), \Phi(ag), \varepsilon(ag)\rangle \text{ iff } sel(ag) \models \langle st(ag), \Phi(ag), \varepsilon(ag)\rangle. \tag{18.84}$$

We now let $sel \models<>< st(ag), \Phi(ag), \varepsilon(ag) >$ when there exists a selection sel' satisfying the following conditions:

$$sel \rightarrow_{Ag} sel' \tag{18.85}$$
$$sel' \models \langle st(ag), \Phi(ag), \varepsilon(ag)\rangle. \tag{18.86}$$

In terms of logic $L(Ag)$, it is possible to express the problem of synthesis of an approximate solution to the problem posed to the team Ag. We denote by $head(Ag)$ the root of the tree $(Ag, \leq)$.

In the process of top-down communication, a requirement Ψ received by the scheme from an external source (which may be called a *customer*) is decomposed into approximate specifications of the form $\langle st(ag), \Phi(ag), \varepsilon(ag)\rangle$

for any agent ag of the scheme. The decomposition process is initiated at the agent $head(Ag)$ and propagated down the tree.

We are able now to formulate the synthesis problem.

Synthesis problem

Given a formula

$$\alpha : \langle st(head(Ag)), \Phi(head(Ag)), \varepsilon(head(Ag)) \rangle \tag{18.87}$$

find a selection sel over the tree $(Ag, \leq)$ *with the property sel* $\models<> \alpha$.

A solution to the synthesis problem with a given formula

$$\langle st(head(Ag)), \Phi(head(Ag)), \varepsilon(head(Ag)) \rangle \tag{18.88}$$

is found by negotiations among the agents. Negotiations are based on uncertainty rules of agents, and their successful result can be expressed by a top-down recursion in tree $(Ag, \leq)$ as follows: given a local team $ag_1 ag_2 \ldots ag_k ag$ with the formula $\langle st(ag), \Phi(ag), \varepsilon(ag) \rangle$ already chosen in negotiations on a higher tree level, it is sufficient that each agent ag_i choose a standard $st(ag_i) \in U(ag_i)$, a formula $\Phi(ag_i) \in L(ag_i)$, and a coefficient $\varepsilon(ag_i) \in [0,1]$ such that

$$(\Phi(ag_1), \Phi(ag_2), \ldots, \Phi(ag_k), \Phi(ag)) \in Dec\_rule(ag) \tag{18.89}$$

with standards $st(ag)$, $st(ag_1)$, ..., $st(ag_k)$;

$$f(\varepsilon(ag_1), \ldots, \varepsilon(ag_k)) \geq \varepsilon(ag) \tag{18.90}$$

where f satisfies $unc\_rule(ag)$ with $st(ag)$, $st(ag_1)$, ..., $st(ag_k)$ and $\varepsilon(ag_1)$, ..., $\varepsilon(ag_k)$, $\varepsilon(ag)$.

For a formula

$$\alpha : \langle st(head(Ag)), \Phi(head(Ag)), \varepsilon(head(Ag)) \rangle , \tag{18.91}$$

we call an α-*scheme* an assignment of a formula $\alpha(ag) : \langle st(ag), \Phi(ag), \varepsilon(ag) \rangle$ to each $ag \in Ag$ in such manner that formulas (18.89) and (18.90) are satisfied, and $\alpha(head(Ag))$ is

$$\langle st(head(Ag)), \Phi(head(Ag)), \varepsilon(head(Ag)) \rangle. \tag{18.92}$$

We denote this scheme with the symbol

$$sch(\langle st(head(Ag)), \Phi(head(Ag)), \varepsilon(head(Ag)) \rangle). \tag{18.93}$$

We say that a selection sel is *compatible* with a scheme

$$sch(\langle st(head(Ag)), \Phi(head(Ag)), \varepsilon(head(Ag)) \rangle) \tag{18.94}$$

when $\mu_o(ag, t)(sel(ag), st(ag)) \geq \varepsilon(ag)$ for each leaf agent $ag \in Ag$, where

$$\langle st(ag), \Phi(ag), \varepsilon(ag) \rangle \tag{18.95}$$

is the value of the scheme at ag for any leaf $ag \in Ag$.

Any leaf agent realizes its approximate specification by choosing a construct satisfying the specification from the subset $Inv \cap U(ag)$ of the inventory of primitive constructs.

The goal of negotiations can now be summarized as follows.

Proposition 18.4.1. *(Sufficiency Criterion) For a given a requirement*

$$\langle st(head(Ag)), \Phi(head(Ag)), \varepsilon(head(Ag)) \rangle \tag{18.96}$$

we have
if *a selection sel is compatible with a scheme*

$$sch(\langle st(head(Ag)), \Phi(head(Ag)), \varepsilon(head(Ag)) \rangle) \tag{18.97}$$

then

$$sel \models <> \langle st(head(Ag)), \Phi(head(Ag)), \varepsilon(head(Ag)) \rangle. \tag{18.98}$$

The bottom-up communication consists of agents sending the chosen constructs to their parents. The root agent $root(Ag)$ assembles the final construct.

There is a parallel between the proposed calculi of granules in distributed systems and neural computing. Let us point out some analogies [18.36, 18.38]:

1. Any elementary team of agents may be regarded as a model of a neuron with inputs supplied by agents $ag_1, ag_2, \ldots, ag_k$, the output returned by ag, and a parameterized family of activation functions represented as rough mereological connectives.
2. Values of rough inclusions are counterparts of weights in a traditional neural network. Let us observe that, in our case, the resulting network is a parameterized system of simple networks, indexed by synthesis schemes.
3. Learning in this new kind of a neural network is based also on back-propagation mechanisms in which the incoming signal (a customer specification) is assigned a proper scheme, and a proper set of weights is set in negotiation and cooperation processes among local teams and the agents therein.

These processes of learning would require new algorithms, and one possible way out is to base the process of learning on familiar techniques of neural networks by encoding all the constructs whose activation functions are tractable (e.g., piece-wise differentiable) approximations to rough mereological connectives in a neural network. As a result, we would obtain a closed-loop system providing feedback information from the distributed system to the neural network. The theory and practice of such systems is to come in future.

18.5 Extracting AR-Schemes from Data and Background Knowledge

In this section, we present some methods of information granule decomposition aimed at extracting from data decomposition rules. We restrict our considerations to methods based only on experimental data. This approach can be extended to the case of information granule decomposition methods using background knowledge [18.46, 18.47].

The search methods discussed in this section return local granule decomposition schemes. These local schemes can be composed using techniques discussed in the previous section. The received schemes of granule construction (which can be treated as AR schemes) have the following property: if the input granules are sufficiently close to input concepts (standards), then the output granule is sufficiently included in the target concept (standard), provided this property is preserved locally (see Proposition 18.4.1 in Sect. 18.4.2 and [18.35]).

The above may be formulated in terms of a synthesis grammar [18.38], with productions corresponding to the local decomposition rules. The relevant derivations over a given synthesis grammar represent AR schemes. Note that synthesis grammars reflect processes that arise in a multiagent systems involved in cooperation, negotiation, and conflict-resolving actions when attempting to provide a solution to the specification of a problem posed to the root. Complexities of membership problems for languages generated by synthesis grammars may be taken ex definitione as complexities of the underlying synthesis processes.

18.5.1 Granule Decomposition

In this section, we show that in some cases decomposition can be performed using methods for specific rule generation based on Boolean reasoning [18.14]. Moreover, we present how the decomposition stable with respect to information granule deviations can be obtained.

First, the representation problem for operations on information granules will be discussed. We assume that a (partial) operation $f : G_1 \times \ldots \times G_k \to H$, with arguments from the sets $G_1, \ldots, G_k$ of information granules, and values in the set H of information granules, is partially specified by a data table (information system) [18.27]. In Fig. 18.9, R denotes constraints specifying the domain of f, i.e., arguments of f that can be composed using f if and only if they satisfy constraints from R.

Any row in the data table corresponds to an object that is a tuple $(g_1, \ldots, g_k, f(g_1, \ldots, g_k))$, where $(g_1, \ldots, g_k)$ belongs to the domain of f. The attribute values for a given object consist of

1. values of attributes from sets $A_{G_1}, \ldots, A_{G_k}$ on information granules $g_1, \ldots, g_k$ (attributes are extracted from some preassumed feature languages $L_1, , \ldots, L_k$);
2. values of attributes characterizing relations among information granules $g_1, \ldots, g_k$, specifying conditions under which the tuple $(g_1, \ldots, g_k)$ belongs to a relevant part of the domain of f;
3. values of attributes selected for the information granule $f(g_1, \ldots, g_k)$ description.

In this way, partial information about the function f is given. In our considerations, we assume the objects indiscernible by condition attributes are indiscernible by decision attributes, i.e., the decision table $DT = (U, A, d)$ considered is consistent [18.27]. We also assume that the representation is consistent with a given function on information granules, i.e., any image obtained by f of the Cartesian product of indiscernibility classes defined by condition attributes is included in a decision indiscernibility class.

Now, we explain in what sense the decision table $DT = (U, A, d)$ can be treated as partial information about the function $f : G_1 \times \ldots \times G_k \to H$. For $i = 1, \ldots, k$, let

$$G_i^{DT} = \{g_i \in G_i : \text{ there exists in } DT \text{ an object } (g_1, \ldots, g_i, \ldots, g_k, h)\}.$$

One can define H^{DT} in an analogous way. The decision table DT defines a function

$$f_{DT} : G_1/IND(A_{G_1}) \times \ldots \times G_k/IND(A_{G_k}) \to H^{DT}/IND(d) \quad (18.99)$$

where A_{G_i} is a set of attributes used to describe granules from G_i and $IND(A_{G_i})$ is the indiscernibility relation defined by $A_{G_i} (i = 1, \ldots, k)$. by

$$f_{DT}([g_1]_{IND(A_{G_1})}, \ldots, [g_k]_{IND(A_{G_k})}) = [h]_{IND(d)} \text{ iff} \\ (g_1, \ldots, g_k, h) \text{ is an object of } DT. \quad (18.100)$$

We assume a consistency modeling condition for f is satisfied, namely

$$f([g_1]_{IND(A_{G_1})} \times \ldots \times [g_k]_{IND(A_{G_k})}) = \\ f_{DT}([g_1]_{IND(A_{G_1})}, \ldots, [g_k]_{IND(A_{G_k})}) \quad (18.101)$$

for any $(g_1, \ldots, g_k) \in G_1^{DT} \times \ldots \times G_k^{DT}$.

The function description can be induced from such a data table by interpreting it as a decision table with the decision corresponding to the attributes specifying the values of the function f.

We assume that a family of inclusion relations $\nu_p^i \subseteq G_i \times G_i$, $\nu_p^H \subseteq H \times H$, and a family of closeness relations, $cl_p^1, \ldots, cl_p^k, cl_p^H$ for every $p \in [0, 1]$ and $i = 1, \ldots, k$, are given [18.35]. Let us assume that two thresholds, t and p, are given. We define a relation $Q_{t,p}^{DT}(Pattern_1, \ldots, Pattern_k, \overline{v})$ between granules called patterns $Pattern_1, \ldots, Pattern_k$ from pattern languages $L_1, \ldots, L_k$ for arguments of f, and the target pattern $\overline{v}$ representing the decision value vector, in the following way:

$$Q_{t,p}^{DT}(Pattern_1, \ldots, Pattern_k, \overline{v}) \quad (18.102)$$

if and only if the two conditions

$$\nu_p^H(f(Sem_{DT}(Pattern_1) \times \ldots \times Sem_{DT}(Pattern_k)), [\overline{v}]_{IND(d)}) \quad (18.103)$$

$$card(Sem_{DT}(Pattern_1) \times \ldots \times Sem_{DT}(Pattern_k)) \geq t \quad (18.104)$$

are satisfied.

Let us now consider the following decomposition problem:

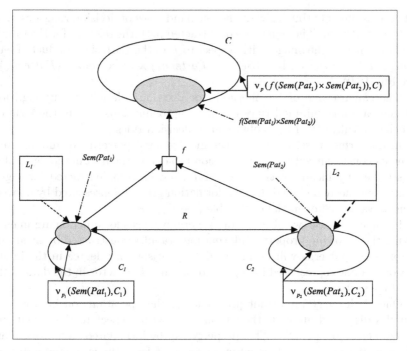

Fig. 18.9. Decomposition of information granule

Granule decomposition problem
Input:

– two thresholds t, p;
– pattern languages $L_1, \ldots, L_k$;
– a decision table $DT = (U, A, d)$ representing an operation $f : G_1 \times \ldots \times G_k \to H$, where $G_1, \ldots, G_k$ and H are given sets of information granules;
– a fixed decision value vector $\overline{v}$ represented by a value vector of decision attributes.

Output:

– a tuple $(Pattern_1, \ldots, Pattern_k) \in L_1 \times \ldots \times L_k$ of patterns such that

$$Q^{DT}_{t,p}(Pattern_1, \ldots, Pattern_k, \overline{v}). \tag{18.105}$$

We consider a description by decision rules extracted from the data table specifying the function f. Any left hand side of a decision rule can be divided into parts corresponding to different arguments of the function f. The ith part, denoted by $Pattern_i$, specifies a condition which should be satisfied by the ith argument of f to obtain the function value specified by the decision attributes. For simplicity, we do not consider conditions specifying the relations between arguments. In this way, the left hand sides of decision rules describe patterns $Pattern_i$. The semantics of extracted patterns relevant for the target can be defined as the image, with respect to f, of the Cartesian product of sets $Sem_{DT}(Pattern_i)$, i.e., by $f(Sem_{DT}(Pattern_1) \times \ldots \times Sem_{DT}(Pattern_k))$ (see Fig. 18.9).

One can use one of the methods for decision rule generating, e.g., for the generation of minimal rules or their approximations (say in the form of association rules) [18.14], to obtain such decision rules.

In the former case, we receive the most general patterns for function arguments consistent with a given decision table, i.e., the information granules constructed by means of the function f from patterns extracted for arguments are exactly included in the information granule represented by a given decision value vector in the data table (see Fig. 18.9).

In the latter case, we obtain more general patterns for function arguments having the following property: information granules constructed by means of f from such patterns will be included to a satisfactory degree in the information granule represented by a given decision value vector in the data table (see Fig. 18.9).

One of the very important properties of the operations on information granules discussed above is their robustness with respect to the deviations of arguments (see [18.38]). This property can be formulated as follows: if an information granule constructed by means of f from the extracted patterns $Pattern_1, \ldots, Pattern_k$ satisfies the target condition, then the information granule constructed from patterns, $Pattern'_1, \ldots, Pattern'_k$ sufficiently close to $Pattern_1, \ldots, Pattern_k$, respectively, satisfies the target condition too. In this way, to obtain the following problem:

Robust decomposition problem (RD problem)
   Input:

- thresholds t, p;
- pattern languages $L_1, \ldots, L_k$;
- a decision table $DT = (U, A, d)$ representing an operation $f : G_1 \times \ldots \times G_k \to H$, where $G_1, \ldots, G_k$ and H are given sets of information granules;
- a fixed decision value vector $\overline{v}$ represented by a value vector of decision attributes.

Output:

- a tuple $(p_1, \ldots, p_k)$ of parameters;
- a tuple $(Pattern_1, \ldots, Pattern_k) \in L_1 \times \ldots \times L_k$ of patterns such that

$$Q_{t,p}^{DT}(Pattern'_1, \ldots, Pattern'_k, \overline{v}) \tag{18.106}$$

if $cl_{p_i}^i(Sem_{DT}(Pattern_i), Sem_{DT}(Pattern'_i))$ for $i = 1, \ldots, k$.

It is possible to search for the solution of the RD problem by modifying the previous approach of decision rule generation. In the process of rule generation, one can impose a stronger discernibility condition by assuming that objects are discernible if their tolerance classes are disjoint. Certainly, one can tune parameters of tolerance relations to obtain rules of satisfactory quality. We would like to stress that efficient heuristics for solving these problems can be based on Boolean reasoning [18.14].

Searching for relevant patterns for information granule decomposition can be based on methods for tuning parameters of rough set approximations of fuzzy cuts or on concepts defined by differences between cuts (see Sect. 18.2.4). In this case, pattern languages consist of parameterized expressions describing the rough set approximations of *parts* of fuzzy concepts as fuzzy cuts or differences between cuts. Hence, an interesting research direction related to the development of new hybrid rough-fuzzy methods arises, aimed at developing algorithmic methods for rough set approximations of parts of fuzzy sets relevant to information granule decomposition.

An approach presented in this section can be extended to local granule decomposition based on background knowledge [18.46, 18.47].

18.6 Conclusions

We have outlined a general scheme for rough-neural computing based on knowledge granulation ideas using rough mereological tools. Important practical problems are the construction of such schemes (networks) for rough-neural computing and of algorithms for parameter tuning. We now foresee two possible approaches: one in which we would rely on new, original decomposition, synthesis, and tuning methods in the manner of those in [18.35], but in the presence of approximation spaces; a second in which a rough-neural computing scheme would be encoded by a neural network in such a way that the optimization of weights in the neural net approach lead to satisfactory solutions for the rough-neural computing scheme (see [18.9] for an attempt in this direction).

We have also discussed an approach for extracting relevant patterns from parameterized schemes of information granule construction, consisting of parts from different information sources. The schemes can also be treated as AR schemes built on the basis of perception by means of information granule calculi. Relevant output patterns (information granules) can be obtained by tuning of the AR scheme parameters. We have emphasized the necessity of approximations (in an accessible language) of information granules that are parts of schemes and are expressed in another language, called the foreign language.

Several research directions are related to the AR schemes and rough neural networks discussed. We list such directions, together with examples of problems.

1. *Developing foundations for information granule systems.* Certainly, more work is needed to develop solid foundations for the synthesizing and analyzing information granule systems. In particular, methods for constructing hierarchical information granule systems and methods for the representation of such systems should be developed.

2. *Algorithmic methods for inducing parameterized productions.* Some methods have already been reported, such as the discovery of rough mereological connectives from data (see [18.32]), and methods based on decomposition (see [18.30, 18.33, 18.40, 18.46]). However, these are only initial steps toward algorithmic methods for inducing parameterized productions from data. One interesting problem is to determine how such productions can be extracted from data and background knowledge.

3. *Algorithmic methods for synthesizing AR schemes.* It was observed (see [18.33, 18.37]) that problems of negotiation and conflict resolution are of great importance for synthesizing AR schemes. Problems arise when we are searching a given set of agents for a granule sufficiently included or close to a given one. These agents, often working with different systems of information granules, can derive different granules, and their fusion will be necessary to obtain the relevant output granule. In the fusion process, negotiations and conflict resolutions are necessary. Much more work should be done in this direction by using the existing results on negotiation and conflict resolution. In particular, Boolean reasoning methods seem to be promising [18.33].

    Another problem is related to the size of production sets. These sets can be large, and it is important to develop learning methods for extracting *small* candidate production sets when extending temporary derivations from huge production sets. For solving this kind of problem, methods for clustering productions should be developed to reduce the size of production sets. Moreover, dialogue and cooperative strategies between agents can help reduce the search space for the necessary extension of temporary derivations.

4. *Algorithmic methods for learning in rough neural networks.* Basic problems in rough neural networks are related to selecting relevant approximation spaces and to parameter tuning. One can examine to what extent existing methods for classical neural methods can be used for learning in rough neural networks. However, it seems that, to deal with real-life applications, new approaches and methods for learning in rough neural networks should be developed. In particular, due to the fact that high quality approximations of concepts can be often obtained only through dialogue and negotiation, processes among agents in which the concept approximation is gradually constructed. Hence, for rough neural network learning methods based on dialogue, negotiation and conflict resolution should be developed. In some cases, one can directly use rough set and Boolean reasoning methods (see [18.45]). However, more advanced cases

need new methods. In particular, hybrid methods based on rough and fuzzy approaches can bring new results [18.22].

5. *Fusion methods in rough neurons.* A basic problem in rough neurons is the fusion of the inputs (information) derived from information granules. This fusion makes it possible to contribute to the construction of new granules. When a the granule constructed by a rough neuron consists of characteristic signal values made by relevant sensors, a step in the direction of solving the fusion problem can be found in [18.28].

6. *Adaptive methods.* Certainly, adaptive methods for discovery of productions, for learning of AR schemes, and rough neural networks should be developed [18.15].

7. *Discovery of multiagent systems relevant to given problems.* Quite often, agents and communication methods among them are not given a priori with the problem specification, and it is a challenge to develop methods for discovering multiagent system structures relevant to given problems, in particular, methods for discovering relevant communication protocols.

8. *Construction of multiagent systems for complex real-life problems.* There are challenging problems related to applying a methodology discussed to real-life problems, such as control of autonomous systems (see page of the WITAS project [18.56]), Web mining problems (see [18.13, 18.40]), sensor fusion (see [18.3, 18.28, 18.29]), and spatial reasoning (see [18.6, 18.7]).

9. *Evolutionary methods.* For all of the above methods, it is necessary to develop evolutionary searching methods for (semi-) optimal solutions [18.15].

10. *Parallel algorithms.* The problems discussed are of high computational complexity. Parallel algorithms searching for AR schemes and methods for their hardware implementation are one important research direction. Moreover, reasoning based on synthesis of AR schemes using DNA computing [18.26] or quantum computing [18.21] should be developed.

Acknowledgements

I would like to thank Professor Lech Polkowski for close cooperation on rough mereology, Professor Jarosław Stepaniuk for cooperation on information granule models, and Professor James F. Peters for cooperation on sensor fusion methods, insightful comments, and a number of clarifying discussions on the chapter presented. Many thanks to Professor Ning Zhong for his kind invitation to write this article.

The research has been supported by the grant 3711C00226 from the Ministry of Scientific Research and Information Technology of the Republic of Poland.

References

18.1 R. Agrawal, H. Mannila, R. Srikant, H. Toivonen, A. Verkano: Fast discovery of association rules. In: U.M. Fayyad, G.P. Shapiro, P. Smyth, R. Uthurusamy (Eds.), *Advances in Knowledge Discovery and Data Mining* (The AAAI Press/The MIT Press) pp. 307-328 (1996)

18.2 L.W. Barsalou: Perceptual symbol systems. Behavioral and Brain Sciences, 22, 577-660 (1999)

18.3 R.R. Brooks, S.S. Iyengar: Multi-sensor Fusion. Upper Saddle River, NJ, Prentice-Hall PTR (1998)

18.4 P. Doherty, W. Łukaszewicz, A. Skowron, A. Szałas: Combining rough and crisp knowledge in deductive databases. In: [18.24], pp. 189-218

18.5 G. Dong, J. Li: Interestigness of discovered association rules in terms of neighborhood based unexpectedness, PAKDD (1998) 72-86

18.6 I. Düntsch (Ed.): Spatial Reasoning (special issue). Fundamenta Informaticae, 45(1-2) (2001)

18.7 M.T. Escrig, F. Toledo: *Qualitative Spatial Reasoning: Theory and Practice* (IOS Press, Amsterdam 1998)

18.8 J. Fraden: *Handbook of Modern Sensors: Physics, Design, and Applications* (Springer-Verlag, Berlin 1996)

18.9 L. Han, J.F. Peters, S. Ramanna, R. Zhai: Classifying faults in high voltage power systems: A rough-fuzzy neural computational approach. *Proceedings of the 7th International Workshop on Rough Sets, Fuzzy Sets, Data Mining and Granular Soft Computing (RSFDGrC'99), Lecture Notes in Artificial Intelligence, 1711* (Springer-Verlag, Berlin 1999) pp. 47-54

18.10 S. Harnad (ed.): *Categorical Perception: The Groundwork of Cognition* (Cambridge University Press, New York 1987)

18.11 T. Hastie, R. Tibshirani, J.H. Friedman: *The Elements of Statistical Learning: Data Mining, Inference, and Prediction* (Springer Series in Statistics, Springer Verlag, Berlin 2001)

18.12 M.N. Huhns, M.P. Singh (eds.): *Readings in Agents* (Morgan Kaufmann, San Mateo 1998)

18.13 H. Kargupta, Ph. Chan: *Advances in Distributed and Parallel Knowledge Discovery* (AIII Press/MIT Press, Cambridge 2001)

18.14 J. Komorowski, P. Pawlak, L. Polkowski, A. Skowron, Rough sets: A tutorial. In: S.K. Pal, A. Skowron (eds.), *Rough-Fuzzy Hybridization: A New Trend in Decision-Making* (Springer-Verlag, Singapore 1998) pp. 3-98

18.15 J.R. Koza: *Genetic Programming II: Automatic Discovery of Reusable Programs* (MIT Press, Cambridge, MA 1994)

18.16 T.Y. Lin: Granular computing on binary relations I. Data mining and neighborhood systems, In: [18.34], 1, pp. 107-121

18.17 T.Y. Lin, Y.Y. Yao, L. Zadeh (eds.): *Data Mining, Rough Sets and Granular Computing* (Physica-Verlag, Heidelberg 2001)

18.18 T.M. Michell: *Machine Learning* (McGraw-Hill, Portland 1997)

18.19 H.S. Nguyen, S.H. Nguyen, A. Skowron: Decomposition of task specification. *Proceedings of the 11th International Symposium on Foundations of Intelligent Systems, June 1999, Warsaw, Poland, Lecture Notes in Computer Science, 1609* (Springer-Verlag, Berlin 1999) pp. 310-318

18.20 H.S. Nguyen, A. Skowron, J. Stepaniuk: Granular computing: A rough set approach. Computational Intelligence, 17(3), 514-544 (2001)

18.21 M.A. Nielsen, I.L. Chuang: *Quantum Computation and Quantum Information* (Cambridge University Press, Cambridge 2000)

18.22 S.K. Pal, A. Skowron (eds.): *Rough-Fuzzy Hybridization: A New Trend in Decision Making* (Springer-Verlag, Singapore 1999)

18.23 S.K. Pal, W. Pedrycz, A. Skowron, R. Swiniarski (eds.): Special Volume on Rough-Neural Computing. Neurocomputing, 36, 1-262 (2001)

18.24 S.K. Pal, L. Polkowski, A. Skowron (eds.): *Rough-Neural Computing: Techniques for Computing with Words* (Springer-Verlag, Berlin 2004)

18.25 S.K. Pal, J.F. Peters, L. Polkowski, A. Skowron: Rough-neuro computing: An introduction. In: [18.24] pp. 16-43

18.26 G. Paun, G. Rozenberg, A. Salomaa: *DNA Computing* (Springer-Verlag, Berlin 1998)

18.27 Z. Pawlak: Rough Sets: *Theoretical Aspects of Reasoning about Data* (Kluwer Academic Publishers, Dordrecht 1991)

18.28 Z. Pawlak, J.F. Peters, A. Skowron, Z. Suraj, S. Ramanna, M. Borkowski: Rough measures: Theory and applications. Bulletin of the International Rough Set Society, 5(1-2), 177-184 (2001)

18.29 J.F. Peters, S. Ramanna, A. Skowron, M. Borkowski, J. Stepaniuk, Z. Suraj: Sensor fusion: A rough granular approach. In: *Proceedings of the Joint 9th International Fuzzy Systems Association (IFSA) World Congress and 20th North American Fuzzy Information Processing Society (NAFIPS) Int. Conf., Vancouver, British Columbia, Canada, 25-28 June 2001* (IEEE 2001) pp. 1367-1371

18.30 J.F. Peters, A. Skowron, J. Stepaniuk: Information granules in spatial reasoning. In: *Proceedings of the Joint 9th International Fuzzy Systems Association (IFSA) World Congress and 20th North American Fuzzy Information Processing Society (NAFIPS) Int. Conf., Vancouver, British Columbia, Canada, 25-28 June 2001* (IEEE 2001) pp. 1355-1361

18.31 L. Polkowski, A. Skowron: Rough mereology: a new paradigm for approximate reasoning. International J. Approximate Reasoning, 15(4), 333-365 (1996)

18.32 L. Polkowski, A. Skowron: Rough mereological approach to knowledge-based distributed AI. In: *J.K. Lee, J. Liebowitz, and J.M. Chae (eds.), Critical Technology, Proceedings of the Third World Congress on Expert Systems, February 5-9, Seoul, Korea* (Cognizant Communication Corporation, New York 1996) pp. 774-781

18.33 L. Polkowski, A. Skowron: Rough mereological foundations for design, analysis, synthesis, and control in distributed systems. Information Sciences An International Journal, 104(1-2), 129-156 (1998)

18.34 L. Polkowski, A. Skowron (eds.): Rough Sets in Knowledge Discovery, 1-2. Studies in Fuzziness and Soft Computing (Physica-Verlag, Heidelberg 1998) pp. 18-19,

18.35 L. Polkowski, A. Skowron: Towards adaptive calculus of granules. In: [18.60], 1, pp. 201-227

18.36 L. Polkowski, A. Skowron: Rough mereology in information systems. A case study: Qualitative spatial reasoning. In: L. Polkowski, T.Y. Lin, S. Tsumoto (eds.), *Rough Sets: New Developmentsin Knowledge Discovery in Information Systems*, Studies in Fuzziness and Soft Computing, 56 (Physica-Verlag, Heidelberg 2000) pp. 89-135

18.37 L. Polkowski, A. Skowron: Rough-neuro computing. *Proceedings of the Second International Conference on Rough Sets and Current Trends in Computing (RSCTC'2000), October 16-19, 2000, Banff, Canada*, Lecture Notes in Artificial Intelligence, 2005 (Springer-Verlag, Berlin 2001) pp. 57-64

18.38 L. Polkowski, A. Skowron: Rough mereological calculi of granules: A rough set approach to computation. Computational Intelligence, 17(3), 472-492 (2001)

18.39 I. Rock: *Indirect Perception* (The AAAI Press/ MIT Press, Cambridge 1996)

18.40 A. Skowron: Toward intelligent systems: Calculi of information granules. Bulletin of the International Rough Set Society, 5(1-2), 9-30, see also: [18.51] 251-260

18.41 A. Skowron: Approximate reasoning by agents in distributed environments. In: [18.61] pp. 28-39

18.42 A. Skowron, J. Komorowski, Z. Pawlak, L. Polkowski: A rough set perspective on data and knowledge. In: W. Kloesgen, J. Żytkow (eds.), Handbook of KDD (Oxford University Press 2002) pp. 134-149

18.43 A. Skowron, J. Stepaniuk: Tolerance approximation spaces. Fundamenta Informaticae, 27(2-3), 245-253 (1996)

18.44 A. Skowron, J. Stepaniuk, S. Tsumoto: Information granules for spatial reasoning. Bulletin of the International Rough Set Society, 3(4), 147-154 (1999)

18.45 A. Skowron, J. Stepaniuk: Information granules: Towards foundations of granular computing. International Journal of Intelligent Systems, 16(1), 57-86 (2001)

18.46 A. Skowron, J. Stepaniuk, J.F. Peters: Extracting patterns using information granules. Bulletin of the International Rough Set Society, 5(1), 135-142 (2001)

18.47 A. Skowron, J. Stepaniuk: Information granule decomposition. Fundamenta Informaticae, 47(3-4), 337-350 (2001)

18.48 A. Skowron, J. Stepaniuk: Information granules and rough-neural computing. In: [18.24] pp. 43-84

18.49 J. Stepaniuk: Knowledge discovery by application of rough set models. In: L. Polkowski, S. Tsumoto, T.Y. Lin (eds.), Rough Set Methods and Applications, New Developments in Knowledge Discovery in Information Systems (Physica-Verlag, Heidelberg 2000) pp. 137-233

18.50 P. Stone: Layered Learning in Multiagent Systems: A Winning Approach to Robotic Soccer (MIT Press, Cambridge 2000)

18.51 T. Terano, T. Nishida, A. Namatame, S. Tsumoto, Y. Ohsawa, T. Washio: New Frontiers in Artificial Intelligence, Lecture Notes in Artificial Intelligence 2253, (Springer-Verlag, Berlin 2001)

18.52 H. Toivonen, M. Klemettinen, P. Ronkainen, K. Hätönen, H. Mannila: Pruning and grouping discovered association rules. In: MLnet Familiarization Workshop on Statistics (ML and KDD 1995) pp. 47-52

18.53 D. Waltz: The importance of importance. AI Magazine, 20(3), 18-26 (1999)

18.54 P.P. Wang (Ed.): Granular Computing (Wiley, New York 2001)

18.55 WWW SPACENET page. http://agora.scs.leeds.ac.uk/spacenet/

18.56 WITAS project Web page. http://www.ida.liu.se/ext/witas/eng.html

18.57 L.A. Zadeh: Fuzzy sets. Information and Control, 8, 333-353 (1965)

18.58 L.A. Zadeh: Fuzzy logic = Computing with words. IEEE Trans. on Fuzzy Systems, 4, 103-111 (1996)

18.59 L.A. Zadeh: A new direction in AI: Toward a computational theory of perceptions. AI Magazine, 2(1), 73-84 (2001)

18.60 L.A. Zadeh, J. Kacprzyk (eds.): Computing with Words in Information/Intelligent Systems, 1-2, Studies in Fuzziness and Soft Computing, 30, (Physica-Verlag, Heidelberg 1999)

18.61 N. Zhong, J. Liu, S. Ohsuga, J. Bradshaw (eds.): Intelligent Agent Technology: Research and Development. Proceedings of the 2nd Asia-Pacific Conference on IAT, Maebashi City, Japan, October 23-26, 2001 (World Scientific, Singapore 2001)

18.62 N. Zhong, A. Skowron, S. Ohsuga (eds.): New Directions in Rough Sets, Data Mining, and Granular Soft Computing. Proceedings of the 7th International Workshop (RSFDGr'99), November 1999, Yamaguchi, Japan, Lecture Notes in Artificial Intelligence, 1711, (Springer-Verlag, Berlin 1999)

19. Soft Computing Pattern Recognition, Data Mining and Web Intelligence

Sankar K. Pal, Sushmita Mitra, and Pabitra Mitra

Indian Statistical Institute, India

Abstract

The relevance of fuzzy logic, artificial neural networks, genetic algorithms, and rough sets to pattern recognition problems is described through examples. Different integrations of these soft computing tools are illustrated. The significance of the soft computing approach in data mining, knowledge discovery, and Web mining is discussed. Various existing algorithms and tools in this regard are reviewed. Finally, some research challenges and the scope of future research are outlined.

19.1 Introduction

Soft computing is a consortium of methodologies which work synergistically and provide in one form or another flexible information processing capabilities for handling real-life ambiguous situations. Its aim is to exploit the tolerance for imprecision, uncertainty, approximate reasoning, and partial truth in order to achieve *tractability, robustness, low cost solutions,* and *close resemblance to human-like decision making.* In other words, it provides the foundation for the conception and design of high MIQ (Machine IQ) systems, and therefore forms the basis of future generation computing systems. At this juncture, Fuzzy Logic (FL), Rough Sets (RS), Artificial Neural Networks (ANN), and Genetic Algorithms (GA) are the principal components, where FL provides algorithms for dealing with imprecision and uncertainty arising from vagueness rather than randomness, RS provides approaches for handling uncertainty arising from limited discernibility of objects, ANN provides the machinery for learning and adaptation, and GA provides techniques for optimization and searching [19.75, 19.112].

Pattern recognition (PR) [19.18, 19.77, 19.103] and machine learning [19.58] form a major area of research and development that encompasses the processing of pictorial and other non-numerical information obtained from interaction between science, technology, and society. The ability to recognize a pattern is the first requirement of any intelligent machine. Pattern recognition is a must component of the so-called Intelligent Control Systems, which involve processing and fusion of data from different sensors and transducers. It is also a necessary function providing failure detection, verification, and diagnosis task. Machine recognition of patterns can be viewed as a two-fold task, consisting of learning the invariant and common properties of a set of samples characterizing a class, and of deciding that a new sample is a possible member of the class by noting that it has properties common to those of the set of samples. Therefore, the task of pattern recognition by a computer can be described as a transformation from the measurement space M to the feature space F and, finally, to the decision space D. Depending on the type of input patterns, one may have speech recognition systems, image recognition or vision systems, medical diagnostic systems, etc.

In recent years, the rapid advances being made in computer technology have ensured that large sections of the world population are able to gain easy access to computers (on account of falling costs), and their use is now commonplace in all walks of life. Government agencies and scientific, business, and commercial organizations are routinely using computers not just for computational purposes but also for storage, in massive databases, of the immense volumes of data that they routinely generate or require from other sources. Large scale computer networking has ensured that such data has become accessible to more and more people. In other words, we are in the midst of an information explosion, and there is urgent need for methodologies that will help us bring some semblance of order into the phenomenal volumes of data that can readily be accessed by us with a few clicks of the keys of our computer keyboard. Traditional statistical data summarization and database management techniques are just not adequate for handling data on this scale, for intelligently extracting information or, rather, knowledge that may be useful for exploring the domain in question or the phenomena responsible for the data and for providing support to decision-making processes. This quest had thrown up some new phrases, for example, *data mining* and *knowledge discovery in databases (KDD)*.

The massive databases that we are talking about are generally characterized by the presence of not just numeric, but also textual, symbolic, pictorial and aural data. They may contain redundancy, errors, imprecision, and so on. KDD is aimed at discovering natural structures within such massive and often heterogeneous data. Therefore, PR plays a significant role in the KDD process. However, KDD is being perceived as being capable of not just knowledge discovery using generalizations and magnifications of existing and new pattern recognition algorithms, but also of the adaptation of these algorithms to enable them to process such data, its storage and access, its preprocessing and cleaning, its interpretation, visualization, and application of results, and its modeling of and support for the overall human-machine interaction. What make KDD feasible today, and will in the future, are the rapidly falling cost of computation and the simultaneous increase in computational power, which together make possible the routine implementation of sophisticated, robust, and efficient methodologies hitherto thought to be too computation-intensive to be useful.

Data mining is that part of knowledge discovery which deals with the process of identifying valid, novel, potentially useful, and ultimately understandable patterns in data, and excludes the knowledge interpretation part of KDD. Therefore, as it stands now, data mining can be viewed as applying PR and machine learning principles in the context of voluminous, possibly heterogeneous, data sets. Furthermore, soft computing-based (i.e., involving fuzzy sets, neural networks, genetic algorithms and rough sets) PR methodologies and machine learning techniques seem to hold great promise for data mining [19.65]. The motivation for this is provided by their ability to handle imprecision, vagueness, uncertainty, approximate reasoning, and partial truth, and they lead to tractability, robustness, and low-cost solutions. In this context, case-based reasoning [19.71], which is a novel Artificial Intelligence (AI) problem-solving paradigm, has a significant role to play, as is evident from the recent book edited by Pal, Dillon, and Yeung [19.71].

Over the last decade, we have witnessed an explosive growth in the information available on the World Wide Web (WWW). Today, Web browsers provide easy access to a myriad sources of text and multimedia data. More than a billion pages are indexed by search engines, and finding the desired information is not an easy task. This profusion of resources has prompted a need for developing automatic mining techniques on the WWW, thereby giving rise to the term Web mining.

To proceed toward Web intelligence, obviating the need for human intervention, we need to incorporate and embed artificial intelligence into Web tools. The necessity of creating server-side and client-side intelligent systems that can effectively mine for knowledge across the Internet, and, in particular, Web localities, is drawing the attention of researchers from the domains of information retrieval, knowledge discovery, machine learning, and AI, among others. However, the problem of developing automated tools in order to find, extract, filter, and evaluate the information desired from unlabelled, distributed, and heterogeneous Web data is far from being solved. To handle these characteristics and overcome some of the limitations of existing methodologies, soft computing seems to be a good candidate; the research area combining the two may be termed as soft Web mining.

This chapter is organized as follows: we first illustrate with examples the relevance of different soft computing tools to pattern recognition problems. Different integrations among them are then described. Section 19.3 describes the basic notions of knowledge discovery in databases and data mining. This is followed in Sect. 19.4 by a survey explaining the role of the aforementioned soft computing tools and their hybridizations, categorized on the basis of the different data mining functions implemented. Next, a brief outline of the applications of different soft computing tools for Web mining is provided. The chapter concludes by indicating some challenges and future research directions.

19.2 Soft Computing Pattern Recognition

19.2.1 Relevance of Fuzzy Set Theory in Pattern Recognition

Fuzzy sets were introduced in 1965 by Zadeh [19.111] as a new way to represent vagueness in everyday life. They are generalizations of conventional (crisp) set theory. Conventional sets contain objects that satisfy precise properties required for membership. Fuzzy sets, on the other hand, contain objects that satisfy imprecisely defined properties to varying degrees. A fuzzy set A of the universe X is defined as a collection of ordered pairs

$$A = \{(\mu_A(x), x), \ \forall x \in X\} \,,$$

where $\mu_A(x)(0 \leq \mu_A(x) \leq 1)$, gives the degree of belonging of the element x to the set A, or the degree of possession of an imprecise property represented by A. Different aspects of fuzzy set theory, including membership functions, basic operations, and uncertainty measures can be found in [19.72, 19.111].

In this section, we explain some of the uncertainties which one often encounters while designing a pattern recognition system, and the relevance of fuzzy set theory in handling them. Let us consider first the case of processing and recognizing a graytone image pattern. Conventional approaches to image analysis and recognition [19.29, 19.89] consist of segmenting the image into meaningful regions, extracting their edges and skeletons, computing various features (e.g., area, perimeter, centroid, etc.) and primitives (e.g., line, corner, curve, etc.) of and relationships among the regions, and, finally, developing decision rules and grammars for describing, interpreting, and/or classifying the image and its sub-regions. In a conventional system, each of these operations involves crisp decisions (e.g., yes or no, black or white, 0 or 1) to make regions, features, primitives, properties, relations, and interpretations clear.

Since the regions in an image are not always clearly defined, uncertainty can arise within every phase of the aforementioned tasks. Any decision made at a particular level will have an impact on all higher level activities. An image recognition system should have sufficient provision for representing and manipulating the uncertainties involved at every processing stage, i.e., in defining image regions, their features, and relations among them, so that the system retains as much of the information content of the data as possible. If this is done, the ultimate output (result) of the system will possess minimal uncertainty (and, unlike conventional systems, it may not be biased or affected as much by lower level decision components).

In short, gray information is expensive and informative. Once it is thrown away, there is no way to get it back. Therefore, one should try to retain this information as long as possible, throughout the decision making tasks, for its full use. When a crisp decision is required to be made at the highest level, one can always through away or ignore this information.

Let us now consider the case of a decision-theoretic approach to pattern classification. With the conventional probabilistic and deterministic classifiers [19.18, 19.103], the features characterizing the input patterns are considered to be quantitative (numeric) in nature. The patterns having imprecise or incomplete information are usually ignored or discarded from their designing and testing processes. The impreciseness (or ambiguity) may arise from various sources. For example, instrumental error or noise corruption in the experiment may lead to only partial or partially reliable information being available on a feature measurement F. In some cases, it may be convenient to use linguistic variables and hedges. In such cases, it is not appropriate to give exact representation to uncertain feature data. Rather, it is reasonable to represent uncertain feature information by fuzzy subsets.

Uncertainty in classification or clustering of patterns may also arise from the overlapping nature of the various classes. This overlapping may result from fuzziness or randomness. In the conventional technique, it is usually assumed that a pattern may belong to only one class, which is not necessarily true in real-life applications. A pattern can and should be allowed to have degrees of membership in more than one class. It is, therefore, necessary to convey this information while classifying a pattern or clustering a data set.

From the aforementioned examples, we see that the concept of fuzzy sets can be used at the *feature level* in representing input data as an array of membership values denoting the degree of possession of certain properties, in representing linguistically phrased input features for their processing, and in weakening the strong commitments for extracting ill-defined image regions, properties, primitives, and relations among them; and at the *classification level* for representing class membership of objects in terms of membership values. *In other words, fuzzy set theory provides a notion of embedding: we find a better solution to a crisp problem by looking in a large space at first, which has different (usually less) constraints and, therefore, allows the algorithm more freedom to avoid errors forced by commission to hard answers in intermediate stages.*

The capability of fuzzy set theory in pattern recognition problems has been reported adequately since the late sixties. A cross-section of the advances with applications is available in [19.6, 19.72, 19.75].

19.2.2 Relevance of Neural Network Approaches

Neural network *(NN)* models [19.50, 19.90] try to emulate the biological neural network and nervous system with electronic circuitry. *NN* models have been studied for many years with the hope of achieving human-like performance (artificially), particularly in the field of pattern recognition, by capturing the key ingredients responsible for the remarkable capabilities of the human nervous system. Note that these models are extreme simplifications of the actual human nervous system.

*NN*s are designated by the network topology, connection strength between pairs of neurons (called weights), node characteristics, and the status updating rules. Node characteristics mainly specify the primitive types of operations it can perform, like summing the weighted inputs coming to it and then amplifying it or doing some fuzzy aggregation operations. The updating rules may be for weights and/or states of the processing elements (neurons). Normally, an objective function is defined; it represents the complete status of the network, and the set of its minima corresponds to the set of stable states of the network. Since there are interactions among the neurons, the collective computational property inherently reduces the computational task and makes the system fault tolerant. Thus, NN models are also suitable for tasks where collective decision making is required. Hardware implementations of neural networks are also attempted.

Neural network-based systems are usually reputed to enjoy the following major characteristics:

- *adaptivity,* for adjusting the connection strengths to new data or information,
- *speed,* due to massively parallel architecture,
- *robustness,* due to missing, confusing, ill-defined or noisy data,
- *ruggedness,* due to failure of components,
- *optimality,* as regards error rates in performance.

For any pattern recognition system, one desires to achieve the above-mentioned characteristics. Moreover, there exists some direct analogy between the working principles of many pattern recognition tasks and neural network models. For example, image processing and analysis in the spatial domain mainly employ simple arithmetic operations at each pixel site in parallel. These operations usually involve information of neighboring pixels (co-operative processing) in order to reduce the local ambiguity and to attain global consistency. An objective measure is required (representing the overall status of the system), the optimum of which represents the desired goal. The system thus involves collective decisions. On the other hand, we notice that neural network models are based on parallel and distributed working principles (all neurons work in parallel and independently). The operations performed at each processor site are simple and independent of the others. The overall status of a neural network can also be measured.

Again, the task of recognition in a real-life problem involves searching a complex decision space. This becomes more complicated when there is no prior information on class distribution. Neural network-based systems use adaptive learning procedures, learn from examples and attempt to find a useful relation between input and output, however complex it may be, for decision-making problems. Neural networks are also reputed to model complex nonlinear boundaries and to discover important underlying regularities in the task domain. These characteristics demand methods for constructing and refining neural network models for various recognition tasks. *In short, neural networks are natural classifiers having resistance to noise, tolerance to distorted images and patterns (i.e., an ability to generalize), superior ability to recognize partially occluded or degraded images or overlapping pattern classes (or classes with highly nonlinear boundaries), and potential for parallel processing.*

19.2.3 Genetic Algorithms for Pattern Recognition

Genetic Algorithms (GAs) [19.28, 19.81] are adaptive computational procedures modeled on the mechanics of natural genetic systems. They express their ability by efficiently exploiting historical information to speculate on new offspring with improved expected performance [19.28]. GAs are executed iteratively on a set of coded solutions, called *population,* with three basic operators: *selection/reproduction, crossover,* and *mutation.* They use only the payoff (objective function) information and probabilistic transition rules for moving to the next iteration. They are different from most of the normal optimization and search procedures in four ways:

– GAs work with the coding of the parameter set, not with the parameters themselves.
– GAs work simultaneously with multiple points, and not with a single point.
– GAs search via sampling (a blind search) using only the payoff information.
– GAs search using stochastic operators, not deterministic rules.

One may note that the methods developed for pattern recognition and image processing are usually problem dependent. Moreover, many tasks involved in the process of analyzing or identifying a pattern need appropriate parameter selection and efficient search in complex spaces in order to obtain optimal solutions. This makes the process not only computationally intensive, but also leads to the possibility of losing the exact solution. Therefore, the application of genetic algorithms for solving certain problems of pattern recognition, which need optimization of computation requirements, and robust, fast, and close approximate solutions, appears to be appropriate and natural [19.81].

19.2.4 Relevance of Rough Sets

The theory of rough sets [19.83] has emerged as a major mathematical tool for managing uncertainty that arises from granularity in the domain of discourse, i.e., from indiscernibility between the objects in a set, and has proved to be useful in a variety of pattern recognition tasks. It offers mathematical tools to discover hidden patterns in data and therefore its importance, as far as data mining is concerned, can in no way be overlooked. A fundamental principle of a rough set-based learning system is to discover redundancies and dependencies between the given features of a problem to be classified. It approximates a given concept from below and from above, using *lower* and

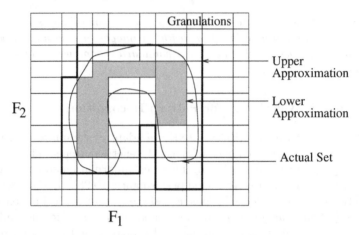

Fig. 19.1. Lower and upper approximations in a rough set

upper approximations. Figure 19.1 provides a schematic diagram of a rough set.

A rough set learning algorithm can be used to obtain a set of rules in IF–THEN form from a *decision table*. The rough set method provides an effective tool for extracting knowledge from databases. Here, one first creates a knowledge base, classifying objects and attributes within the created decision tables. Then, a knowledge discovery process is initiated to remove some undesirable attributes. Finally, the data dependency is analyzed, in the reduced database, to find the minimal subset of attributes called *reduct.*

19.2.5 Integration and Hybrid Systems

Integration of the individual soft computing tools help in designing hybrid systems which are more versatile and efficient compared to stand-alone use of the tools. The most visible integration in the soft computing community is that of neural networks and fuzzy sets [19.75]. Neuro-fuzzy systems have been successfully developed for decision making, pattern recognition, and image processing tasks. The hybridization falls in two major categories: a neural network (termed fuzzy neural network) equipped with the capability of handling fuzzy information to augment its application domain, and a fuzzy system (termed neural-fuzzy systems) augmented by neural networks to enhance some of its characteristics, like flexibility, speed, adaptability, and learning. Both the classes of hybridisation, and their application to various pattern recognition problems, are described in [19.75].

There are some applications where the integration of GAs with fuzzy sets and ANNs is found to be effective. For example GAs are sometimes found essential for overcoming some of the limitations of fuzzy set theory, specifically to reduce the subjective nature of membership functions. Note that the other way of integration, i.e., incorporating the concept of fuzziness into GAs, has not been tried seriously. Synthesis of ANN architectures can be done using GAs as an example of neuro-genetic systems. Such an integration may help in designing an optimum ANN architecture with appropiate parameter sets. Methods for designing neural network architectures using GAs are primarily divided into two parts. In one part, the GA replaces the learning method to find appropiate connection weights of some predefined architecture. In another part, GAs are used to find the architecture itself, and it is then evaluated using some learning algorithms. Literature is also available on integration of fuzzy sets, neural networks, and genetic algorithms [19.71, 19.73, 19.75].

Recently, rough sets have been integrated with both fuzzy sets and neural networks. Several rough-fuzzy hybrid systems are discussed in [19.75]. In the framework of rough-neuro integration [19.78], two broad approaches are available, namely, use of rough sets for encoding weights of knowledge-based networks [19.3], and designing neural network architectures which incorporate roughness at the neuron level. GAs have also been used for the fast generation of rough set reducts from an indiscernibility matrix.

Evolutionary Rough Fuzzy MLP Let us demonstrate an integration (Fig. 19.2) of four soft computing tools with the concept of modularity in order to artificially build more intelligent system [19.76]. The knowledge flow structure of evolutionary rough fuzzy MLP is illustrated in Fig. 19.3. Here, each of the soft computing tools act synergistically to contribute to the final performance of the system as follows. Rough set rules are used for extracting crude domain knowledge, which when encoded in a fuzzy MLP results not only in fast training of the network, but also in automatic determination of the network size. The GA operators are adaptive and use the domain knowledge extracted with rough sets for even faster learning. The fuzziness incorporated at the input and output help in better handling of uncertainties and overlapping classes.

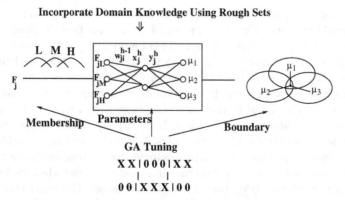

Fig. 19.2. Components of the modular rough-fuzzy Modular Rough-fuzzy MLP

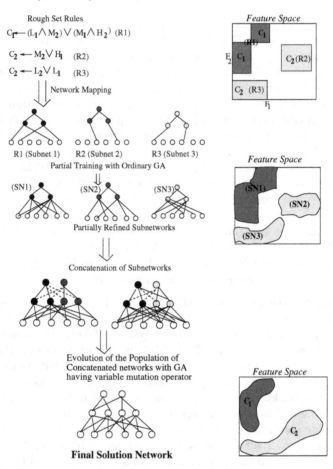

Fig. 19.3. Knowledge flow in modular rough-fuzzy MLP

The evolutionary modular rough fuzzy MLP has been applied to a number of real world problems like speech recognition and medical diagnosis. In case of speech recognition [19.76], the system is found to correctly classify 84% of the samples, while the fuzzy MLP correctly classifies only 78% and the MLP correctly classifies only 59%. The system also improves computation time significantly. For determining the stages of Cervical Cancer [19.60], the system provides results identical to that of medical experts in 83% of the cases. In other cases, the stagings were also close. In addition to the above, performance logical rules were extracted from the trained system. The rules coincided with the guidelines adopted by medical practicioners for staging. In the rough fuzzy MLP, the final network has a structure imposed on the weights. Hence, crisp logical rules can be easily extracted from the networks. This makes the system suitable for knowledge discovery in databases, something to be discussed later on. The rules obtained are found to be superior to those of several popular methods, as measured with some quantitative indices. For example, on the speech recognition data, the rules obtained using the modular rough-fuzzy MLP have an accuracy of 81.02% with 10 rules, while the popular C4.5 rule generation algorithm has an accuracy of 75.00% using 16 rules. The fraction of samples uncovered by the rules obtained by us is only 3.10%, whereas the C4.5 rules have uncovered 7.29% of the samples. The confusion index is also low for the proposed method, compared to C4.5 [19.60].

19.3 Knowledge Discovery and Data Mining

Knowledge discovery in databases (KDD) is defined as *the nontrivial process of identifying valid, novel, potentially useful, and ultimately understandable patterns in data* [19.22, 19.15]. *Data* is a set of facts F, and a *pattern* is an expression E in a language L describing the facts in a subset F_E of F. E is called a pattern if it is simpler than the enumeration of all facts in F_E. A measure of certainty, measuring the *validity* of discovered patterns, is a function C mapping expressions in L to a partially or totally ordered measure space M_C. An expression E in L about a subset $F_E \subset F$ can be assigned a certainty measure $c = C(E, F)$. *Novelty* of patterns can be measured by a function $N(E, F)$ with respect to changes in data or knowledge. Patterns should potentially lead to some *useful* actions, as measured by some utility function $u = U(E, F)$ mapping expressions in L to a partially or totally ordered measure space M_U. The goal of KDD is to make patterns *understandable* to humans. This is measured by a function $s = S(E, F)$ mapping expressions E in L to a partially or totally ordered measure space M_S.

Interestingness of a pattern combines validity, novelty, usefulness, and understandability, and can be expressed as $i = I(E, F, C, N, U, S)$, which maps expressions in L to a measure space M_I. A pattern $E \in L$ is called *knowledge* if for some user-specified threshold $i \in M_I$, $I(E, F, C, N, U, S) > i$ [19.22]. One can select some thresholds $c \in M_C$, $s \in M_S$, and $u \in M_u$, and term a pattern E knowledge

$$iff \quad C(E, F) > c, \quad and \quad S(E, F) > s, \quad and \quad U(E, F) > u. \quad (19.1)$$

19.3.1 Data Mining

KDD refers to the overall process of turning low-level data into high-level knowledge. The overall KDD process is outlined in Fig. 19.4, it is interactive and iterative in nature. An important step in the KDD process is data mining. Data mining is an interdisciplinary field with a general goal of predicting outcomes and uncovering relationships in data. It uses automated tools employing sophisticated algorithms to discover hidden patterns, associations, anomalies, and/or structures from large amounts of data stored in data warehouses or other information repositories. Data mining tasks can

Fig. 19.4. The KDD process

be descriptive, *i.e.*, discovering interesting patterns describing the data, and predictive, *i.e.*, predicting the behavior of the model based on available data.

The more common model functions in current data mining practice include: *classification, regression, clustering, rule generation, discovering association rules, summarization, sequence analysis.*

The rapid growth of interest in data mining is due to (i) the falling cost of large storage devices and increasing ease of collecting data over networks, (ii) the development of robust and efficient machine learning algorithms to process this data, and (iii) the falling cost of computational power, enabling use of computationally intensive methods for data analysis [19.59].

19.4 Soft Computing for Data Mining

Recently, various soft computing methodologies have been applied to handle the different challenges posed by data mining [19.65]. The main constituents of soft computing, at this juncture, include fuzzy logic, neural networks, genetic algorithms, and rough sets. Each of them contributes a distinct methodology for addressing problems in its domain. This is done in a cooperative, rather than a competitive, manner. The result is a system more intelligent and robust than a traditional one, providing a human-interpretable, low cost, approximate solution, as compared to traditional techniques.

Let us first describe the roles and significance of the individual soft computing tools and their hybridizations, followed by the various systems developed for handling the different functional aspects of data mining. A suitable preference criterion is often optimized during mining. It may be mentioned that there is no universally best data mining method; choosing particular soft computing tools, or some combination with traditional methods, is entirely dependent on the particular application, and requires human interaction to determine the suitability of an approach.

19.4.1 Fuzzy Sets

KDD is mainly concerned with identifying interesting patterns and describing them in a concise and meaningful manner [19.22]. Fuzzy models can be said to represent a prudent and user-oriented sifting of data, qualitative observations, and calibration of commonsense rules in an attempt to establish meaningful and useful relationships between system variables [19.86]. Despite the growing versatility of knowledge discovery systems, there is an important component of human interaction that is inherent in any process of knowledge representation, manipulation, and processing. Fuzzy sets are inherently inclined toward coping with linguistic domain knowledge and producing more interpretable solutions.

The notion of *interestingness*, which encompasses several features such as validity, novelty, usefulness, and simplicity, can be quantified through fuzzy sets. Fuzzy dissimilarity of a discovered pattern with a user-defined vocabulary has been used as a measure of this interestingness [19.51]. As an extension to the above methodology, *unexpectedness* can also be defined in terms of a *belief system*, where if a belief b is based on previous evidence ξ, then $d(b|\xi)$ denotes the degree of belief b. In soft belief systems, a weight w_i is attached to each belief b_i. The degree of a belief may be measured with conditional probability, the Dempster-Shafer belief function, or the frequency of the raw data. Here, the interestingness of a pattern E, relative to a belief system B and evidence ξ, may be formally defined as

$$I(E, B, \xi) = \sum_{b_i \in B} w_i |d(b_i|E, \xi) - d(b_i|\xi)|. \tag{19.2}$$

This definition of interestingness measures the amount by which the degrees of belief change as a result of a new pattern E.

The role of fuzzy sets is categorized below, based on the different functions of data mining that are modeled.

Clustering. In data mining, one is typically interested in a focused discovery of structure and an eventual quantification of functional dependencies existing therein. This helps prevent searching for meaningless or trivial patterns in a database. Researchers have developed fuzzy clustering algorithms for this purpose [19.104]. Pedrycz has designed fuzzy clustering algorithms [19.85] using (a) contextual information and (b) induced linguistic space for better focusing of the search procedure in KDD.

Achieving focus is important in data mining because there are too many attributes and values to be considered, and can result in combinatoric explosion. Most unsupervised data mining approaches try to achieve attribute focus by first recognizing the most interesting features. Mazlack [19.56] suggests a converse approach of progressively reducing the data set by partitioning and eliminating the least important attributes to reduce intra-item dissonance within the partitions. A *soft* focus is used to handle both crisp and imprecise data. It works by progressive reduction of cognitive dissonance, leading to an increase in useful information. The objective is to generate cohesive and comprehensible information *nuggets* by sifting out uninteresting attributes. A combined distance metric takes care of different types of attributes simultaneously, thus avoiding any taxonomic structure. Non-crisp values are handled by granularization followed by partitioning.

Mining association rules. Wei and Chen [19.106] have mined generalized association rules with fuzzy taxonomic structures. A crisp taxonomy assumes that a child belongs to its ancestor with degree one. A fuzzy taxonomy is represented as a directed acyclic graph, each of whose edges represents a fuzzy *IS-A* relationship with degree μ ($0 \leq \mu \leq 1$). The partial belonging of an item in a taxonomy is taken into account while computing the degrees of support and confidence.

Au and Chan [19.2] utilize an *adjusted difference* between observed and expected frequency counts of attributes for discovering fuzzy association rules in relational databases. Instead of dividing quantitative attributes into fixed intervals, they employ linguistic terms to represent the regularities and exceptions revealed. Here, no user-supplied thresholds are required, and quantitative values can be directly inferred from the rules. The linguistic representation leads to the discovery of *natural* and more understandable rules. The algorithm allows one to discover both *positive* and *negative* rules, and can deal with fuzzy class boundaries as well as missing values in databases. The use of fuzzy techniques buries the boundaries of adjacent intervals of numeric quantities, resulting in resilience to noises such as the inaccuracies in physical measurements of real-life entities. The effectiveness of the algorithm was demonstrated on a transactional database of a PBX system and on a database concerning industrial enterprises in mainland China.

Analyzing functional dependencies. Fuzzy logic has been used for analyzing inference based on functional dependencies (FDs) between variables in database relations. Fuzzy inference generalizes both imprecise (set-valued) and precise inference. Similarly, fuzzy relational databases generalize their classical and imprecise counterparts by supporting fuzzy information storage and retrieval [19.34]. Inference analysis is performed using a special abstract model which maintains vital links to classical, imprecise, and fuzzy relational database models. These links increase the utility of the inference formalism in practical applications involving catalytic inference analysis, including knowledge discovery and database security. FDs are an interesting notion from a knowledge discovery standpoint, since they allow one to express, in a condensed form, some properties of the real world which are valid on a given database. These properties can then be used in various applications, such as reverse engineering or query optimization. Bosc et al. [19.8] use a data mining algorithm to extract and discover extended FDs, represented by gradual rules composed of linguistic variables.

Data summarization. Summary discovery is one of the major components of knowledge discovery in databases. It provides the user with comprehensive information for grasping the essence of a large amount of information in a database. Fuzzy set theory is also used for data summarization [19.47]. Typically, fuzzy sets are used for an interactive top-down summary discovery process which utilizes fuzzy *IS-A* hierarchies as domain knowledge. Here, generalized tuples are used as a representational form of a database summary including fuzzy concepts. By virtue of fuzzy *IS-A* hierarchies, where fuzzy *IS-A* relationships common in actual domains are naturally expressed, the discovery process comes up with more accurate database summaries.

Linguistic summaries of large sets of data are derived as linguistically quantified propositions with a degree of validity [19.109]. This corresponds to the preference criterion involved in the mining task. The system consists of a summarizer (like *young*), a quantity in agreement (like *most*), and the truth or validity (say, 0.7). Single-attribute simple summarizers often need to be extended for some confluence of attribute values, implying combinatorial problems due to the huge number (all possible combinations) of summaries involved and the determination of the most appropriate or valid one.

Chiang et al. [19.13] have used fuzzy linguistic summary for mining time series data. The system provides human interaction in the form of a graphic display tool to help users pre-mine a database and determine what knowledge could be discovered. The model is used to predict the online utilization ranks of different resources, including CPU and storage.

19.4.2 Neural Networks

Neural networks were earlier thought to be unsuitable for data mining because of their inherent *black-box* nature. No information was available from them in symbolic form, suitable for verification or interpretation by human beings. Recently, there has been widespread activity aimed at redressing this situation, by extracting the embedded knowledge in trained networks in the form of symbolic rules [19.102]. This serves to identify the attributes that, either individually or in a combination, are the most significant determinants of the decision or classification. Unlike fuzzy sets, the main contribution of neural nets to data mining stems from rule extraction and clustering.

Rule extraction. In general, the primary input to a connectionist rule extraction algorithm is a representation of the trained neural network, in terms of its nodes, its links, and, sometimes, its data set. One or more hidden and output units are used to automatically derive the rules, which may later be combined and simplified to arrive at a more comprehensible rule set. These rules can also provide new insights into the application domain. The use of neural nets helps in (i) incorporating parallelism and (ii) tackling optimization problems in the data domain. The models are usually suitable in *data-rich* environments.

Typically, a network is first trained to achieve the required accuracy rate. Redundant connections of the network are then removed using a pruning algorithm. The link weights and activation values of the hidden units in the network are analyzed, and classification rules are generated [19.53, 19.102].

Rule evaluation. Here, we provide some quantitative measures to evaluate the performance of the generated rules [19.76]. They are: *user's accuracy, kappa, fidelity, confusion, coverage, rulebase size, computational complexity,* and *confidence*. They relate to the *preference criteria or goodness of fit* chosen for the rules.

Clustering and self-organization. One of the big challenges of data mining is the organization and retrieval of documents from archives. Kohonen et al. [19.44] have demonstrated the utility of a huge self-organizing map (SOM) with more than one million nodes to partition a little less than seven million patent abstracts, where the documents are represented by 500-dimensional feature vectors. Vesanto et al. [19.105] employ a step-wise strategy by partitioning the data with a SOM, followed by its clustering. Alahakoon et al. [19.1] perform hierarchical clustering of SOMs, based on a spread factor which is independent of the dimensionality of the data.

Shalvi and DeClaris [19.92] have designed a data mining technique, combining Kohonen's self-organizing neural network with data visualization, for clustering a set of pathological data containing information regarding the patients' drugs, topographies (body locations), and morphologies (physiological abnormalities). Koenig [19.42] has combined SOM and Sammon's nonlinear mapping for reducing the dimension of data representation for visualization purposes.

Regression. Neural networks have also been used for a variety of classification and regression tasks [19.14]. Time series prediction has been attempted by Lee and Liu [19.48]. They have employed a neural oscillatory elastic graph matching model with hybrid radial basis functions for tropical cyclone identification and tracking.

19.4.3 Neuro-Fuzzy Computing

Neuro-fuzzy computation [19.75] is one of the most popular hybridizations widely reported in literature. It comprises a judicious integration of the merits of neural and fuzzy approaches, enabling one to build more intelligent decision making systems. This approach incorporates the generic advantages of artificial neural networks, like massive parallelism, robustness, and learning in *data-rich* environments, into the system. The modeling of imprecise and qualitative knowledge in natural or linguistic terms, as well as the transmission of uncertainty, is possible through the use of fuzzy logic. Besides these generic advantages, the neuro-fuzzy approach also provides the application-specific merits highlighted earlier.

The rule generation aspect of neural networks is utilized to extract more *natural* rules from fuzzy neural networks [19.62]. The fuzzy MLP [19.63] and fuzzy Kohonen network [19.64] have been used for linguistic rule generation and inferencing. Here the input, besides being in quantitative, linguistic, or set forms, or a combination of these, can also be incomplete. The components of the input vector consist of membership values to the overlapping partitions of linguistic properties *low*, *medium*, and *high* corresponding to each input feature. Output decision is provided in terms of class membership values.

The models are capable of

- Inferencing, based on complete and/or partial information
- Querying the user for unknown input variables that are key to reaching a decision
- Producing justification for inferences in the form of IF-THEN rules.

The connection weights and node activation values of the trained network are used in the process. A *certainty factor* determines the confidence in an output decision. Note that this certainty refers to the preference criterion for the extracted rules, and is different from the notion of certain patterns of Equation (19.1).

A neuro-fuzzy knowledge-based network by Mitra et al. [19.61] is capable of generating both *positive* and *negative* rules in linguistic form to justify any decision reached. In the absence of positive information regarding the belonging of a pattern to class C_k, the complementary information about the pattern not belonging to class $C_{k'}$ is used for generating the negative rules. The a priori class information and the distribution of pattern points in the feature space are taken into account while encoding the crude *domain knowledge* from the data set of the connection weights. Fuzzy intervals and linguistic sets are used in the process. The network topology is automatically determined, followed by refinement using the growing and/or the pruning of links and nodes. The knowledge-based network converges sooner, resulting in more meaningful rules.

19.4.4 Genetic Algorithms

Knowledge discovery systems have been developed using genetic programming concepts [19.23]. The *MASSON* system [?], where intentional information is extracted for a given set of objects, is popular. The problem addressed is that of finding common characteristics of a set of objects in an object-oriented database. Genetic programming is used to automatically generate, evaluate, and select object-oriented queries. GAs are also used for several other purposes, like fusion of multiple data types in *multimedia* databases, and automated program generation for mining multimedia data [19.101].

However, the literature in the domain of GA-based data mining is not as rich as that of fuzzy sets. We provide below a categorization of a few interesting systems based on the functions modeled.

Regression. Noda et al. [19.69] use GAs to discover *interesting* rules in a dependence modeling task, where different rules can predict different goal attributes. Generally, attributes with high information gain are good predictors of a class when considered individually. However attributes with low information gain could become more relevant when attribute interactions are taken into account. This phenomenon is associated with rule interestingness. The degree of interestingness of the consequent is computed based on the relative frequency of the value being predicted by it. In other words, the rarer the value of a goal attribute, the more interesting a rule it predicts. The authors attempt to discover a few interesting rules (knowledge nuggets) instead of a large set of accurate (but not necessarily interesting) rules.

Association rules. Lopes et al. [19.52] evolve association rules of type IF C THEN P, which provide a high degree of accuracy and coverage. While the *accuracy* of a rule measures its degree of confidence, its *coverage* is interpreted as the comprehensive inclusion of all the records that satisfy the rule. Hence, $accuracy = \frac{|C \cap P|}{|C \cap P| + |C \cap \overline{P}|}$ and $coverage = \frac{|C \cap P|}{|C \cap P| + |\overline{C} \cap P|}$. Note that quantitative measures for rule evaluation have been discussed in Sect. 19.4.2, with reference to neural networks.

19.4.5 Rough Sets

Rough set applications to data mining generally proceed along the following directions.

1. *Decision rule induction from attribute value table* [19.35, 19.67, 19.97]. Most of these methods are based on generation of discernibility matrices and reducts.
2. *Data filtration by template generation* [19.88]. This mainly involves extracting elementary blocks from data based on the equivalence relation. Genetic algorithms are also sometimes used at this stage for searching, so that the methodologies can be used for large data sets.

Besides the above, reduction of memory and computational requirements for rule generation, and working on dynamic databases [19.93], are also considered. Recently, the use of rough fuzzy hybrid systems for case-based reasoning on large data sets have been proposed [19.74]. The system, based on the principle of information granulation, provides fast case retrieval and generation and a significant reduction in the average number of features required to represent a case. Some of the rough set-based systems developed for data mining include (i) the KDD-R system based on VPRS (Variable Precision Rough Set) model [19.114] and (ii) the rule induction system based on LERS (Learning from Examples based on Rough Set Theory) [19.31]. LERS has been extended in [19.32] to handle *missing* attributes using the closest fit.

19.5 Web Mining

The Web is a vast collection of completely uncontrolled heterogeneous documents. Thus, it is huge, diverse, and dynamic, and raises the issues of scalability, heterogeneity, and dynamism, respectively. Due to these characteristics, we are currently drowning in information, but starving for knowledge, thereby making the Web a fertile area of data mining research with the huge amount of information available online. Data mining refers to the non-trivial process of identifying valid, novel, potentially useful, and ultimately understandable patterns in data.

Web mining can be broadly defined as the discovery and analysis of useful information from the World Wide Web (WWW). In Web mining, data can be collected at the server side, at the client side, at proxy servers, or obtained from an organization's database. Depending on the location of the source, the type of data collected differs. It also has extreme variation both in its content (e.g., text, image, audio, symbolic) and in the meta information that might be available. This makes for the techniques used for a particular Web mining task to vary widely. Some of the characteristics of Web data are

(a) unlabelled
(b) distributed
(c) heterogeneous (mixed media)
(d) semi-structured
(e) time varying
(f) high dimensional

Therefore, Web mining basically deals with mining a large and hyper-linked information base having the aforementioned characteristics. Also, being interactive, most Web applications have human interface as a key component. Some of the issues which have come to light, as a result, concern

(a) need for handling context sensitive and imprecise queries,
(b) need for summarization and deduction,
(c) need for personalization and learning.

Thus, Web mining, though considered to be a particular application of data mining, warrants a separate field of research, mainly because of the aforementioned characteristics of the data and because of human-related issues.

19.5.1 Web Mining Components and the Methodologies

Web mining can be viewed as consisting of four tasks, shown in Fig. 19.5, according to Etzioni [19.19]. Each task is described below along with a survey of the existing methodologies and tools for the task.

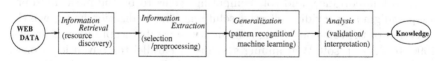

Fig. 19.5. Web mining subtasks

Information Retrieval (Resource Discovery) Resource discovery or information retrieval (IR) deals with automatic retrieval of all relevant documents, while at the same time ensuring that the irrelevant ones are fetched as little as possible. The IR process mainly includes document representation, indexing, and searching for documents.

Information Selection and Extraction and Preprocessing Once the documents have been retrieved, the challenge is to automatically extract knowledge and other required information without human interaction. Information Extraction (IE) is the task of identifying specific fragments of a single document that constitute its core semantic content.

Generalization In this phase, pattern recognition and machine learning techniques are usually used on the extracted information. Most of the machine learning systems deployed on the Web learn more about the user's interests than about the Web itself. A major obstacle when learning about the Web is the labelling problem: data is abundant on the Web, but it is unlabelled. Many data mining techniques require inputs labelled as positive (yes) or negative (no) examples with respect to some concept.

Analysis Analysis is a data driven problem which presumes that there is sufficient data available so that potentially useful information can be extracted and analyzed. Human beings play an important role in the information or knowledge discovery process on the Web since it is an interactive medium. This is especially important for validation and/or interpretation of the mined patterns which take place in this phase.

Based on the aforementioned four phases (Fig. 19.5), *Web mining can be viewed as the use of data mining techniques to automatically retrieve, extract, and evaluate information for knowledge discovery from Web documents and services. Here, evaluation includes both generalization and analysis.*

19.6 Soft Computing for Web Mining

Web data, being inherently unlabelled, imprecise and/or incomplete, heterogeneous, and dynamic, appears to be a very good candidate for its mining in a soft computing framework. Besides, because human interaction is a key component in Web mining, issues such as context-sensitive and approximate queries, summarization and deduction, and personalization and learning are of utmost importance, and soft computing seems to be the most appropriate paradigm for providing effective solutions. This realization has drawn the attention of the soft computing community, since its inception in or around 1996, to develop soft Web mining systems in parallel to conventional ones. In the following sections, we discuss some applications of each of the soft computing tools.

19.6.1 Fuzzy Logic for Web Mining

The application of fuzzy logic so far made to Web mining tasks falls mainly under IR and generalization (clustering, association). These attempts will be described here, along with different commercially available systems. Some of the prospective areas which need immediate attention are also outlined.

Information Retrieval. Yager describes in [19.108] a framework for formulating linguistic and hierarchical queries. It describes an IR language which enables users to specify the interrelationships between desired attributes of documents sought using linguistic quantifiers. Examples of linguistic quantifiers include most, at least, about half. Let Q be a linguistic expression corresponding to a quantifier such as most; then, it is represented as a fuzzy subset Q over $I = [0, 1]$ in which, for any proportion r belonging to I, $Q(r)$ indicates the degree to which r satisfies the concept indicated by the quantifier Q. Gedeon and Koczy [19.27] deal with the problem of automatic indexing and retrieval of documents when it cannot be guaranteed that the user queries include the actual words that occur in the documents that should be retrieved. Fuzzy tolerance and similarity relations are presented, and the notion of hierarchical co-occurrence, which allows the introduction of two or more hierarchical categories of words in the documents, is defined.

As an example of the use of fuzzy set theory to extend boolean information retrieval, we discuss the methodology proposed by Pasi and Bordonga [19.82] for semi-structured document (e.g., HTML) retrieval. It models the concept of graduality of relevance of a document to the user query. Formally, a document is represented as a fuzzy binary relation:

$$R_d = \sum_{(t) \in T} \mu_d(t)/t \,, \tag{19.3}$$

where R_d is the representation of document $d \in D$, the set of archive documents, $t \in T$, the set of index terms, and $\mu_d : D \times T \to [0, 1]$ is the membership function of R_d. μ_d is a dynamic function, with $\mu_d(t, s)$ expressing the significance of the term t in section s of document d. $\mu_d(t, s)$ are based on the semantics of the section s. For example, in sections containing formatted text, like *author* and *keywords*, a single occurrence of a term makes it fully significant ($\mu_d(t, s) = 1$) for that section, and $\mu_d(t, s)$ is a boolean function. On the other hand, for sections containing textual descriptions, $\mu_d(t, s)$ can be computed as a function of the normalized term frequency for that section, for example,

$$\mu_d(t, s) = tf_{dst} IDF_t \,, \tag{19.4}$$

in which IDF_t is the inverse document frequency of the term t, tf_{dst} is the normalized term frequency defined as

$$tf_{dst} = \frac{OCC_{dst}}{MAXOCC_{sd}} \,, \tag{19.5}$$

where OCC_{dst} is the number of occurrences of the term t in section s of document d, and $MAXOCC_{sd}$ is a normalization parameter depending on the section's length. To obtain the overall degree of significance of a term in a document, computed over all the sections, an aggregation scheme, Ordered Weighted Average (OWA), is used:

$$\mu(d,t) = OWA_{lq}(\mu_1(d,t),\dots,\mu_n(d,t)) \ . \tag{19.6}$$

Parameters l and q are determined by user-specified weights for the sections.

A query $< t, w >$ is represented by terms t_i and the corresponding weights w_i. The query is evaluated by an E function for a given document, and then aggregation operators are used. Thus, the result of a query evaluation is represented as a fuzzy subset of the archived documents, given by

$$R_d(t) = \sum_{d \in D} \mu_W(d,t)/d. \tag{19.7}$$

This brings to light the fact that fuzzy boolean IR models are more flexible in representing both document contents and information needs.

Generalization. *Clustering:* Etzioni and Zamir [19.21] has listed the key requirements of Web document clustering as a measure of relevance, browsable summaries, the ability to handle overlapping data, snippet tolerance, speed, and incremental characteristics. In [19.46] fuzzy c medoids (FCNdd) and fuzzy c Trimmed medoids (FCTMdd) are used for clustering of Web documents and snippets (outliers). In [19.36] a fuzzy clustering technique for Web log data mining is described. Here, an algorithm called CARD (Competitive Agglomeration of Relational Data) for clustering user sessions is described; it considers the structure of the site and the of URL for computing the similarity between two user sessions. This approach requires the definition and computation of dissimilarity and similarity between each session pair, and the forming of a similarity or fuzzy relation matrix, prior to clustering. Since the data in a Web session involves an access method (GET or POST), a URL, a transmission protocol (HTTP or FTP), etc, which are all non-numeric, correlation between two user sessions, and hence their clustering, is best handled using the fuzzy set approach. Other techniques for clustering Web data include those using hypergraph-based clustering [19.66].

Association Rule Mining: Some algorithms for mining association rules using fuzzy logic techniques have been suggested in [19.33]. They deal with the problem of mining fuzzy association rules understandable to humans from a database containing both quantitative and categorical attributes. Association rules of the form if X is A then Y is B, where X and Y are attributes and A and B are fuzzy sets, are mined. Nauck [19.68] has developed a learning algorithm that creates *mixed* fuzzy rules involving both categorical and numerical attributes.

Prospective Areas of Application. *Search Engines:* There is immense scope for applying fuzzy logic to improve Web search from the points of view of deduction, matching, and ranking. To add human-like deductive capabilities to search engines, the use of fuzzy logic is not an option; rather, it is a necessity. Regarding matching, a possible approach is to compromise sightly on precision (which is anyway very difficult to achieve, due to millions of Web pages) and retrieve the most relevant documents from an expanded domain. The retrieved documents may then be clustered during or after the search, or filtered at the client side, or both. The concept of linguistic variables and membership functions can be used for keyword matching. Similarly, for page ranking, the degree of closeness of hits in a document can be used for its computation. For example, variables like close, far, nearby may be used to represent the distance between hits in a document for a given query. Similarly, fuzzy variables, like reputation or importance, attached to the URL which is referencing a particular page, can be used in calculating page ranks. For example, in Fig. 19.6, which shows a neuro-fuzzy IR system, match parameters such as proximity and subjectivity in queries can be found using fuzzy sets.

Let us consider the popular Google search engine, which is considered highly effective. In [19.11] a schematic diagram (Figure 3) of the technology behind Google has been explained; in it we can see that the lexicon gives wordIDs to each word of the query, and wordIDs are then matched. If the query contains quantifiers like less, very less, or more, then instead of blindly rejecting or selecting pages based on their absence or presence in the document, a smoother transition, based on their membership value, is a better option. For fuzzy queries, i.e., if the query text includes linguistic variables like almost, somewhat, more or less, or about, we can provide more relevant documents by giving grades of membership to different results. When we consider hits, greater weight should be given to those documents in which query words are closer to each other than to those in which they are far apart.

Similarity Measures: There are certain questions like: What is the distance between two URLs? Which two URL's are always requested together? Which users have common interests and request similar documents? that appear to be better handled in a fuzzy set theoretic framework, since answers to these questions need not always be crisp.

Others: Some other areas where fuzzy logic may be applied include ontology, matching techniques, recognition technology, summarization, e-commerce, content management, database querying, information aggregation and fusion, customization, and profiling.

19.6.2 Neural Networks and Learning Systems for Web Mining

A neural network (NN) can formally be defined as *a massively parallel interconnected network of simple (usually adaptive) processing elements which is intended to interact with objects of the real world in the same way as biological systems do.* NNs are designated by the network topology, connection strength between pairs of neurons (called weights), node characteristics, and status updating rules. Normally, an objective function is defined which represents the complete status of the network and its set of minima corresponds to the set of stable states of the network. Neural network-based systems are usually reputed to enjoy the following major characteristics: generalization capability, adaptivity to new data or information, speed due to massively parallel architecture, robustness to missing, confusing, or ill-defined and/or noisy data, and capability for modelling nonlinear decision boundaries.

Neural networks have been applied, so far, to tasks like IR, IE, and clustering (self-organization) of Web mining, and for personalization. We summarize the existing literature on these lines. Some of the prospective areas which need immediate attention are also discussed.

Information Retrieval. Artificial neural networks provide a convenient method of knowledge representation for IR applications. Also, their learning ability helps to achieve the goal of implementing adaptive systems. Shavlik and Eliassi [19.94] suggest an agent, the WAWA-IE+IR system (Wisconsin Adaptive Web Assistant), using neural networks with reinforcement learning, which uses two network modules, namely, ScorePage and ScoreLink. ScoreLink uses unsupervised learning while ScorePage uses supervised learning in the form of advice from users. The system uses KBNN (Knowledge-based Neural Networks) as its knowledge base to encode the initial knowledge of users, which is then refined. This approach has the following advantages: (a) the agent is able to perform reasonably well initially because it can utilize the users' prior knowledge, and (b) users' prior knowledge does not have to be correct, as it is refined through learning. Information is derived by extracting rules from KBNNs [19.95]. In order to map large sized Web pages into fixed sized neural networks, a sliding window is used. This parses each page by considering three words at a time, and HTML tags like $< p >, < /p >, < br >$ act as window breakers. Using self-generated training examples it can act also as a self-tuning agent. Rules of the type "when *'precondition'* then *'action'* " are extracted, where actions could be of the type "*strength* followed by 'show page' or 'follow link' or 'avoid showing page' ". Here, *strength* could be *weakly, moderately, strongly,* or *definitely,* which are determined by the weight of the links between the layers of the neural network.

In [19.12], Chen et al. have implemented several search methods in Java based on neural networks and genetic search on databases, intranets, and the Internet. Mercure [19.10] is another IR system, based on multi-layered networks, that allows document retrieval, using a spreading activation process, and query optimization, using relevance back-propagation. This model consists of an input layer which represents users, information needs, a term neuron layer, a document neuron layer, and an output layer representing the result of query evaluation.

Lim [19.49] has developed a concept of visual keywords which are abstracted and extracted from visual documents using soft computing techniques. Each visual keyword is represented as a neural network or a soft cluster center. Merkl and Rauber [19.57] have shown how to use hierarchical feature maps to represent the contents of a document archive. After producing a map of the document space using self-organizing maps, the system provides a hierarchical view of the underlying data collection in the form of an atlas. Using such modelling, the user can easily zoom into particular regions of interest while still having general maps for overall orientation. An ANN-based hybrid Web text mining system has been described in [19.26].

Information Extraction. Most IE systems that use learning fall into two groups: one that uses relational learning [19.25, 19.99] to learn extracted patterns, and the another that learns parameters of Hidden Markov Models (HMM) and uses them to extract information [19.7]. In [19.24] wrapper induction techniques are combined with the adaBoost algorithm (Schapire and Singer, 1998) called Boosted Wrapper Induction (BWI), and the system has outperformed many of the relational learners and is competitive with WAWA-IE and HMM. For a brief and comprehensive view of the various learning systems used in Web content mining, one may refer to Table I.

Self Organization (WEBSOM). The emerging field of text mining applies methods of data mining and exploratory data analysis to analyze text collections and to convey information to users in an intuitive manner. Visual map-like displays provide a powerful and fast medium for portraying information about large collections of text. Relationships between text items and collections, such as similarity, clusters, gaps, and outliers, can be communicated naturally using spatial relationships, shading, and colors. In WEBSOM [19.44], the self-organizing map (SOM) algorithm is used to automatically organize very large and high-dimensional collections of text documents onto two-dimensional map displays. The map forms a document landscape where similar documents appear close to each other at different points of the regular map grid. The landscape can be labelled with automatically identified descriptive words that convey properties of each area and also act as landmarks during exploration. With the help of an HTML-based interactive tool, the ordered landscape can be used in browsing the document collection and in performing searches on the map. An organized map offers an overview of an unknown document collection, helping the user familiarize himself with the domain. Map displays that are already familiar can be used as visual frames of reference for conveying properties of unknown text items. For example, thematically arranged static document landscapes provide a meaningful background for dynamic visualizations of time-related properties of the data. The mathematical preliminaries, background, basic ideas, implications, and numerous applications of self-organizing maps are described in a recent book [19.43].

Personalization. Personalization means that the content and search results are tailored as per users, interests and habits. Neural networks may be used for learning user profiles with training data collected from users or systems, as in [19.94]. Since user profiles are highly nonlinear functions, neural networks seem to be an effective tool to learn them. An agent which learns user profiles using Bayesian classifier is "Syskill & Webert" [19.84]. Once the user profiles have been learned, it can be used to determine whether the users would be interested in another page. However, this decision is made by analyzing the HTML source of a page, and it requires the page to be retrieved first. To avoid network delays, we allow the user to prefetch all pages accessible from the index page and store them locally. Once this is done, Syskill & Webert can learn a new profile and make suggestions about pages to visit quickly. Once the HTML is analyzed, it annotates each link on the page with an icon indicating the user's rating, or its prediction of the user's rating, together with the estimated probability that a user would like the page. Note that these ratings and predictions are specific only to one user, and do not reflect on how other users might rate the pages. As described above, the agent is limited to making suggestions about which link to follow from a single page only. This is useful when someone has collected a nearly comprehensive set of links about a topic. A similar system which assists users in browsing software libraries has been built by Drummond et al. [19.17].

Personalized page ranking: As mentioned in section 19.4.1, page ranks are important since human beings find it difficult to scan through the entire list of documents returned by the search engine in response to their queries. Therefore, it is desirable, for convenience, to get the pages ranked with respect to relevance to user queries so that users can get the desired documents by scanning only the first few pages.

Let us consider here again the case of the popular search engine Google [19.11]. It computes the rank of a page a using

$$\Pr(a) = 1 - d + d \sum_{i=1}^{n} \frac{\Pr(T_i)}{C(T_i)} \ , \tag{19.8}$$

where d is the damping factor, $\Pr(a)$ is the rank of page a, which has pages $T_1, T_2, \ldots, T_n$ pointing to it, and $C(a)$ is the number of outgoing links from page a.

Note that it takes into consideration only the popularity of a page (reputation of incoming links) and richness of information content (number of outgoing links), and does not take care of other important factors like

- User preference: Whether the link matches with the preferences of the user, established from his history
- Validity: Whether the link is currently valid or not
- Interestingness: Whether the page is of overall interest to the user or not

These should also be reflected in the computation of page ranks. The learning and generalization capabilities of ANNs can be exploited for determining user preference and interestingness. User preference can be incorporated by training a neural network based on user history. Since ANNs can model nonlinear functions and learn from examples, they appear to be good candidates for computing the interestingness of a page. Self-organizing neural networks can be used to filter out invalid pages dynamically.

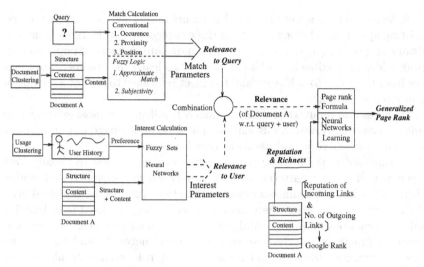

Fig. 19.6. Neuro fuzzy IR system

An ANN can compute the page rank from a combination of each of the parameters hub, authority, reputation, validity, interestingness, user preference, with weights assigned to each that the user can modify, thereby refining the network for his interest. These factors may sometimes also be characterized by fuzzy variables. For example, variables like close, far, nearby may be used to represent the distance between hits in a document for a given query. Similarly, fuzzy variables like reputation and importance can be attached to the URL that is referencing a particular page. This is unlike Google, where these variables are considered to be crisp. This signifies the importance of synergistically integrating ANN with fuzzy logic under the *neuro-fuzzy paradigm* [19.63] for computing page ranks. In the next paragraph, we describe more details of a proposed neuro-fuzzy IR system.

Neuro-fuzzy IR: A schematic diagram of a proposed neuro-fuzzy IR system is shown in Fig. 19.6. It shows that the total relevance of a document is a combination of relevance with respect to a query (match parameters) and relevance with respect to the user (interest parameters). This total relevance, when combined with richness and reputation of document A (currently reflected in the Google rank), give the generalized page rank. The dotted links represent areas not addressed by existing algorithms. Means of computing each of the quantitities, namely, relevance with respect to a query, relevance with respect to the user, and richness and reputation, are mentioned below.

Relevance to a query: Here, fuzzy logic can be used in computing match parameters by handling subjectivity in queries and approximate matching in documents, thereby better modelling the relevance to query. Also, structured documents can be handled more effectively in the framework of fuzzy logic. Literature described in Sect. 19.6.1 addresses many of these tasks.

Relevance with respect to a user: Literature in this area is relatively scarce. Existing approaches belong mainly to three categories: (a) learning from user history or profile, (b) clustering of users into homogeneous groups, and (c) using relevance feedback. ANNs can be used to learn the nonlinear user profiles from their previous history and to reflect relevance to user of a document A in interest parameters.

Richness and reputation: This parameter is reflected in most existing page ranking systems. However, efficient computation of the page rank is an open research issue where neural networks may be used.

Clustering and Classification: Neural networks can be used to classify Web pages as well as user patterns, in both supervised and unsupervised modes. Their ability in model complex nonlinear functions can also be exploited here.

Deduction: Another area where neural networks may be used is in building deductive capabilities in Web mining systems. As mentioned earlier, complex, nonlinear functions may be learned using neural networks and logical rules may be extracted from trained networks using rule extraction algorithms. The logical rules are interpretable, and help in generating deductions.

19.6.3 Genetic Algorithms for Web Mining

Genetic algorithms (GAs), a biologically inspired technology, are randomized search and optimization techniques guided by the principles of evolution and natural genetics. They are efficient, adaptive, and robust search processes, producing near-optimal solutions and having a large amount of implicit parallelism. GAs are executed iteratively on a set of coded solutions (genes), called population, with three basic operators: selection/reproduction, crossover, and mutation. They use only the payoff (fitness function) information and probabilistic transition rules for moving to the next iteration.

The literature explaining the use of genetic algorithms to Web mining seems to be even poorer than that of fuzzy logic and neural networks. Genetic algorithms are used mainly in search, optimization, and description. Here, we describe some of the attempts:

Search and Retrieval: A GA-based search to find other relevant homepages, given some user supplied homepages, has been implemented in G-Search [19.16]. Web document retrieval by genetic learning of importance factors of HTML tags has been described in [19.41]. Here, the method learns the importance of tags from a training text set. GAs have also been applied for feature selection in text mining [19.55].

Query Optimization: In [19.9], Boughanem et al. have developed a query reformulation technique using genetic algorithms in which a GA generates several queries that explore different areas of the document space and determines the optimal one. Yang and Korfhage [19.110] have presented an evolutionary algorithm for query optimization by re-weighting the document indexing without query expansion. Kraft et al. [19.45] apply genetic programming in order to improve weighted boolean query formulation.

Document Representation: Gordon [19.30] adopted a GA to derive better descriptions of documents. Here, each document is assigned N descriptions, where each description is a set of indexing terms. Then, genetic operators and relevance judgements are applied to these descriptions in order to determine the best one in terms of classification performance in response to a specific query.

Distributed Mining: GEMGA (Gene Expression Messy Genetic Algorithm) [19.37] which is a sub-quadratic, highly parallel evolutionary search algorithm, is found specially suitable for distributed data mining applications, including the Web. The foundation of GEMGA is laid on the principles of both decomposing black box search into iterative construction of partial ordering and performing selection operations in the relation, class, and sample spaces. The GEMGA is designed based on an alternate perspective of evolutionary computation, proposed by the SEARCH framework, that emphasizes the role of gene expression or intra-cellular flow of genetic information. The research on the use of GEMGA in distributed data mining is growing fast, and it deals with the problems of finding patterns in data in an environment where both the data and the computational resource are distributed. In [19.38], KarGupta et al., suggest CDM (Collective Data Mining) as a new approach to distributed data mining (DDM) from heterogeneous sites. It is pointed out that naive approaches to distributed data analysis in an heterogeneous environment may lead to ambiguous situations. CDM processes information locally and builds a global model in situations where it is not possible, as in the case of Web, to collect all the data in a common data warehouse to process.

Regarding prospective applications of GAs, let us consider here the case of *Adaptive Web Sites.* These are sites that automatically improve their organization and presentation by learning from visitor access patterns [19.20]. They focus on the problem of *index page synthesis*, where an index page is a page consisting of a set of links that covers a particular topic. Their basic approach is to analyze Web access logs and find groups of pages that occur together often in user visits (and hence represent coherent topics in users' minds) and to convert them into index pages. Here, genetic algorithms may be used for the prediction of user preferences, dynamic optimization, and the evolution of Web pages.

19.6.4 Rough Sets for Web Mining

Rough sets are characterized by their ability for granular computation. In rough set theory, a concept B is described by its lower ($\underline{B}$) and upper ($\bar{B}$) approximations, defined with respect to some indiscernibility relation. The use of rough set theory for knowledge discovery [19.98] and rule mining [19.3, 19.76] is widely acknowledged. However, the current literature on application of rough sets to Web mining, like the genetic approach, is very scanty. Some

Web mining tasks where rough set theory have been used are mentioned below.

Information Retrieval. *Granular Information Retrieval:* An important application of rough sets is in granular information retrieval. In rough set theory, *information granules* refer to homogeneous blocks or clusters, of documents as described by a particular set of features which may vary over the clusters. The approach is efficient in many document retrieval tasks where the documents are clustered into homogeneous groups before they are retrieved. Documents are often represented by a large number of features or words, and dimensionality reduction needs to be performed for clustering. However, instead of representing all the clusters by the same set of features (words), in granular IR, using rough set theory, each cluster is represented by different feature sets. This is closer to reality because different word sets are important for different document classes. Wong et al. [19.107] suggest reducing the dimensionality of terms by constructing a term hierarchy in parallel to a document hierarchy.

Handling Heterogeneous Data: Rough sets, as well as rough-fuzzy hybrid systems [19.79], have been used for handling multimedia data and information fusion. A system where rough sets have been used for retrieval of multimedia objects is described in [19.100].

Association/Clustering. Rough sets have been used for document clustering [19.39] and mining of Web usage patterns [19.54]. Uses of variable precision rough sets [19.54] and tolerance relations are important in this regard.

Some additional areas where rough sets may be applied include :

- Web Structure Mining: Rough mereology [19.87], which has the ability to handle complex objects, is a potential tool for mining multimedia objects as well as complex representations, like Web graphs, and semantic structures.
- Multi-agent systems and collaborative mining.
- Rough-neuro computing (RNC), as a means of computing with words (CWW), is likely to play an important role in natural language query formulation.
- Rough set theory can also be used for the purpose of approximate information retrieval, where the set of relevant documents may be rough and represented by its upper and lower approximations. The lower approximation refers to the set which is definitely relevant and the upper approximation denotes the maximal set which may possibly be relevant. Dynamic and focussed search, exploiting the above concept, may also help in developing efficient IR systems.

19.7 Conclusions

Soft computing methodologies, involving fuzzy sets, neural networks, genetic algorithms, rough sets, and their hybridizations, have widely been used to solve pattern recognition, data mining, and Web mining problems. They strive to provide approximate solutions at low cost, thereby speeding up the process. A categorization has been provided, based on the different soft computing tools and their hybridizations, the mining function implemented, and the preference criterion selected by the model.

Fuzzy sets, which constitute the oldest component of soft computing, are suitable for handling issues related to the understandability of patterns, incomplete and/or noisy data, mixed media information, and human interaction, and can provide approximate solutions faster. They have been mainly used in clustering, discovering association rules and functional dependencies, summarization, time series analysis, Web applications, and image retrieval.

Neural networks are suitable for data-rich environments, and are typically used for extracting embedded knowledge in the form of rules, quantitative evaluation of these rules, clustering, self-organization, classification, and regression. They have an advantage over other types of machine learning algorithms in scaling [19.5].

Neuro-fuzzy hybridization exploits the characteristics of both neural networks and fuzzy sets in generating natural or linguistic rules, handling imprecise and mixed mode data, and modeling highly nonlinear decision boundaries. Domain knowledge, in natural form, can be encoded in the network for improved performance.

Genetic algorithms provide efficient search algorithms for selecting a model, from mixed media data based on some preference criterion or objective function. They have been employed in regression and in discovering association rules. Rough sets are suitable for handling different types of uncertainty in data, and have been mainly utilized for extracting knowledge in the form of rules.

Other hybridizations typically enjoy the generic and application-specific merits of the individual soft computing tools that they integrate. Data mining functions modeled by such systems include rule extraction, data summarization, clustering, incorporation of domain knowledge, and partitioning. It is to be noted that partitioning, i.e., the modular approach, provides an effective direction for scaling up algorithms and speeding up convergence. Case-based reasoning (CBR), a novel AI problem solving paradigm, has recently drawn the attention of both soft computing and data mining communities. A profile of potential applications is available in [19.71].

Some of the challenges in the use of these methodologies include the following:

– Scalability of extremely large heterogeneous databases spread over multiple files, possibly in different disks or across the Web in different geographical locations. Often combining such data into a very large single file may be infeasible.

- Feature evaluation and dimensionality reduction to improve prediction accuracy. Some recent work in this direction is available in [19.4, 19.40, 19.70, 19.96].
- Choice of metrics and evaluation techniques to handle dynamic changes in data.
- Incorporation of domain knowledge and user interaction.
- Quantitative evaluation of performance.
- Efficient integration of soft computing tools. In this connection, the Computational Theory of Perceptions, as explained by Zadeh [19.113], needs attention.

References

19.1 D. Alahakoon, S.K. Halgamuge, B. Srinivasan: Dynamic self organizing maps with controlled growth for knowledge discovery. IEEE Transactions on Neural Networks, 11, 601-614 (2000)

19.2 W.H. Au, K.C.C. Chan: An effective algorithm for discovering fuzzy rules in relational databases. In: *Proceedings of IEEE International Conference on Fuzzy Systems FUZZ IEEE 98* (Alaska, May 1998) pp. 1314-1319

19.3 M. Banerjee, S. Mitra, S.K. Pal: Rough fuzzy MLP: Knowledge encoding and classification. IEEE Transactions on Neural Networks, 9(6), 1203-1216 (1998)

19.4 S. Bengio, Y. Bengio: Taking on the curse of dimensionality in joint distribution using neural networks. IEEE Transactions on Neural Networks, 11, 550-557 (2000)

19.5 Y. Bengio, J.M. Buhmann, M. Embrechts, J.M. Zurada: Introduction to the special issue on neural networks for data mining and knowledge discovery. IEEE Transactions on Neural Networks, 11, 545-549 (2000)

19.6 J.C. Bezdek, S.K. Pal (eds.): *Fuzzy Models for Pattern Recognition: Methods that Search for Structures in Data* (IEEE Press, New York, 1992)

19.7 D. Bikel, R. Schwartz, R. Weischedel: An algorithm that learns what's in a name. Machine learning, 34 (Special issue on Natural Language Learning)(1/3), 211-231 (1999)

19.8 P. Bosc, O. Pivert, L. Ughetto: Database mining for the discovery of extended functional dependencies. In: *Proceedings of NAFIPS 99* (New York, USA, June 1999) pp. 580-584

19.9 M. Boughanem, C. Chrisment, J. Mothe, C.S. Dupuy, L. Tamine: Connectionist and genetic approaches for information retrieval. In: F. Crestani, G. Pasi (eds.), *Soft Computing in Information Retrieval: Techniques and Applications* (Physica Verlag, Heidelberg, 2000) 50, pp. 102-121.

19.10 M. Boughanem, T. Dkaki, J. Mothe, C. Soule-Dupuy: Mercure at trec7. In: *Proceedings of the 7th International Conference on Text Retrieval, TREC7* (Gaithrsburg, MD, 1998)

19.11 S. Brin, L. Page: The anatomy of a large scale hypertextual web search engine. In: *Proceedings of Eighth International WWW Conference* (Brisbane, Australia, April 1998) pp. 107-117

19.12 H. Chen, M. Ramsay, P. Li: The Java search agent workshop. In: F.Crestani,
G.Pasi (eds.), *Soft Computing in Information Retrieval: Techniques and Applications* (Physica Verlag, Heidelberg, 2000) 50, pp. 122-140

19.13 D.A. Chiang, L.R. Chow, Y.F. Wang: Mining time series data by a fuzzy
linguistic summary system. Fuzzy Sets and Systems, 112, 419-432 (2000)

19.14 V. Ciesielski, G. Palstra: Using a hybrid neural/expert system for database
mining in market survey data. In: *Proc. Second International Conference on
Knowledge Discovery and Data Mining (KDD-96)* (Portland, OR, August 2-4,
1996 AAAI Press) pp. 38

19.15 K.J. Cios, W. Pedrycz, R. Swiniarski: *Data Mining Methods for Knowledge
Discovery* (Kluwer, Dordrecht, 1998)

19.16 F. Crestani, G. Pasi, (eds.): *Soft Computing in Information Retrieval: Techniques and Application* (Physica-Verlag, Heidelberg, 2000)

19.17 C. Drummond, D. Ionescu, R. Holte: *A learning agent that assists the browsing of software libraries.* Technical Report TR-95-12 (University of Ottawa,
1995)

19.18 R.O. Duda, P.E. Hart: *Pattern Classification and Scene Analysis* (John Wiley, New York, 1973)

19.19 O. Etzioni: The World Wide Web: Quagmire or gold mine. Communications
of the ACM, 39(11), 65-68 (1996)

19.20 O. Etzioni, M. Perkowitz: Adaptive web sites: An AI challenge. In: *Proceedings of Fifteenth National Conference on Artificial Intelligence* (Madison,
Wisconsin, July 1998)

19.21 O. Etzioni, O. Zamir: Web document clustering: A feasibility demonstration.
In: *Proceedings of the 21st Annual International ACM SIGIR Conference, 1998*
pp. 46-54

19.22 U.M. Fayyad, G. Piatetsky-Shapiro, P. Smyth, R. Uthurusamy (eds.): *Advances in Knowledge Discovery and Data Mining* (AAAI/MIT Press, Menlo
Park, CA, 1996)

19.23 I.W. Flockhart, N.J. Radcliffe: A genetic algorithm-based approach to data
mining. In: *The Second International Conference on Knowledge Discovery and
Data Mining (KDD-96)* (Portland, OR, August 2-4 1996 AAAI Press) pp. 299

19.24 D. Freitag, N. Kushmerick: Boosted wrapper induction. In: *Proceedings of
AAAI, 2000* pp. 577-583

19.25 D. Freitag, A. McCallum: Information extraction from HMM's and shrinkage.
In: *Proceedings of AAAI-99 Workshop on Machine Learning for Information
Extraction* (Orlando, FL, 1999)

19.26 H. Fukuda, E.L.P. Passos, A.M. Pacheco, L.B. Neto, J. Valerio, V.Jr.De
Roberto, E.R. Antonio, L. Chigener: Web text mining using a hybrid system.
In: *Proceedings of the Sixth Brazilian Symposium on Neural Networks, 2000*
pp. 131-136

19.27 T. Gedeon, L. Koczy: A model of intelligent information retrieval using fuzzy
tolerance relations based on hierarchical co-occurrence of words. In: F. Crestani,
G. Pasi (ed.), *Soft Computing in Information Retrieval: Techniques and Applications*, volume50 (Physica Verlag, Heidelberg, 2000) pp. 48-74

19.28 D.E. Goldberg: *Genetic Algorithms in Search, Optimization and Machine
Learning* (Addison-Wesley, Reading, MA, 1989)

19.29 R.C. Gonzalez, P. Wintz: *Digital Image Processing* (Addison-Wesley, Reading, MA, 1987)

19.30 M.D. Gordon: Probabilistic and genetic algorithms for document retrieval.
Communications of the ACM, 31(10), 208-218 (1988)

19.31 J.W. Grzymala-Busse: LERS-A knowledge discovery system. In: L.Polkowski, A.Skowron (eds.), *Rough Sets in Knowledge Discovery 2, Applications, Case Studies and Software Systems* (Physica-Verlag, Heidelberg, 1998) pp. 562-565

19.32 J.W. Grzymala-Busse, W.J. Grzymala-Busse, L.K. Goodwin: A closest fit approach to missing attribute values in preterm birth data. In: *Proceedings of RSFDGrC'99* (Yamaguchi, Japan, November 1999) pp. 405-413

19.33 A. Gyenesei: A fuzzy approach for mining quantitative association rules. TUCS technical reports 336, University of turku, Department of Computer Science, Lemminkisenkatu14, Finland, March 2000

19.34 J. Hale, S. Shenoi: Analyzing FD inference in relational databases. Data and Knowledge Engineering, 18, 167-183 (1996)

19.35 X. Hu, N. Cercone: Mining knowledge rules from databases: A rough set approach. In *Proceedings of the 12th International Conference on Data Engineering* (Washington, February 1996 IEEE Computer Society) pp. 96-105

19.36 A. Joshi, R. Krishnapuram: Robust fuzzy clustering methods to support web mining. In: *Proc Workshop in Data Mining and Knowledge Discovery, SIGMOD, 1998*, 15, pp. 1-8

19.37 H. Kargupta: The gene expression messy genetic algorithm. In: *Proceedings of the IEEE International Conference on Evolutionary Computation* (Nagoya University, Japan, 1996) pp. 631-636

19.38 H. Kargupta, B.H. Park, D. Hershberger, E. Johnson: Collective data mining: A new perspective toward distributed data mining. *Advances in Distributed and Parallel Knowledge Discovery* (MIT/AAAI Press, 1999)

19.39 S. Kawasaki, N.B. Nguyen, T.B. Ho: Hierarchical document clustering based on tolerance rough set model. In: *Proceedings of the Sixth International Conference on Knowledge Discovery and Data Mining (KDD-2000) Workshop on Text Mining Boston, MA* (August 2000)

19.40 R. Kewley, M. Embrechta, C. Breneman: Data strip mining for the virtual design of pharmaceuticals with neural networks. IEEE Transactions on Neural Networks, 11, 668-679 (2000)

19.41 S. Kim, B.T. Zhang: Web document retrieval by genetic learning of importance factors for html tags. In: *Proceedings of the International Workshop on Text and Web mining* (Melbourne, Australia, August 2000) pp. 13-23,

19.42 A. Koenig: Interactive visualization and analysis of hierarchical neural projections for data mining. IEEE Transactions on Neural Networks, 11, 615-624 (2000)

19.43 T. Kohonen: *Self-organising Maps* (Springer, Berlin, Germany, second edition, 1997)

19.44 T. Kohonen, S. Kaski, K. Lagus, J. Salojarvi, J. Honkela, V. Paatero, A. Saarela: Self organization of a massive document collection. IEEE Transactions on Neural Networks, 11, 574-585 (2000)

19.45 D.H. Kraft, F.E. Petry, B.P. Buckles, T. Sadasivan: The use of genetic programming to build queries for information retrieval. In: *Proceedings of the IEEE Symposium on Evolutionary Computation* (Orlando, FL, 1994)

19.46 R. Krishnapuram, A. Joshi, L. Yi: A fuzzy relative of the k-medoids algorithm with application to document and snippet clustering. In: *Proceedings of IEEE Intl. Conf. Fuzzy Systems - FUZZIEEE 99, Korea, 1999*

19.47 D.H. Lee, M.H. Kim: Database summarization using fuzzy ISA hierarchies. IEEE Transactions on Systems Man and Cybernetics. Part B-Cybernetics, 27, 68-78 (1997)

19.48 R.S.T. Lee, J.N.K. Liu: Tropical cyclone identification and tracking system using integrated neural oscillatory leastic graph matching and hybrid RBF network track mining techniques. IEEE Transactions on Neural Networks, 11, 680-689 (2000)

19.49 J.H. Lim: Visual keywords: from text retrieval to multimedia retrieval. In: F.Crestani, G.Pasi (eds.), *Soft Computing in Information Retrieval: Techniques and Applications* (Physica Verlag, Heidelberg, 2000), 50, pp. 77-101

19.50 R.P. Lippmann: Pattern classification using neural networks. IEEE Communications Magazine, pp. 47-64 (1989)

19.51 B. Liu, W. Hsu, L.F. Mun, H.Y. Lee: Finding interesting patterns using user expectation. IEEE Transactions on Knowledge and Data Engineering, 11, 817-832 (1999)

19.52 C. Lopes, M. Pacheco, M. Vellasco, E. Passos: Rule-evolver: An evolutionary approach for data mining. In: *Proceedings of RSFDGrC'99* (Yamaguchi, Japan, November 1999) pp. 458-462

19.53 H.J. Lu, R. Setiono, H. Liu: Effective data mining using neural networks. IEEE Transactions on Knowledge and Data Engineering, 8, 957-961 (1996)

19.54 V.U. Maheswari, A. Siromoney, K.M. Mehata: The variable precision rough set model for web usage mining. In: *Proceedings of the First Asia-Pacific Conference on Web Intelligence (WI-2001)* (Maebashi, Japan, October 2001)

19.55 M.J. Martin-Bautista, M.A. Vila: A survey of genetic feature selection in mining issues. In: *Proceedings of the Congress on Evolutionary Computation (CEC 99), 1999* pp. 13-23

19.56 L.J. Mazlack: Softly focusing on data. In: *Proceedings of NAFIPS 99* (New York, June 1999) pp. 700-704

19.57 D. Merkl, A. Rauber: Document classification with unsupervised artificial neural networks. In: F.Crestani, G.Pasi (eds.), *Soft Computing in Information Retrieval: Techniques and Applications*, volume50, (Physica Verlag, Heidelberg, 2000) pp. 102-121

19.58 T.M. Mitchell: *Machine Learning* (McGraw-Hill, New York, 1997)

19.59 T.M. Mitchell: Machine learning and data mining. Communications of the ACM, 42(11), 1999

19.60 P. Mitra, S. Mitra, S.K. Pal: Staging of cervical cancer with soft computing. IEEE Trans. Biomedical Engineering, 47(7), 934-940 (2000)

19.61 S. Mitra, R.K. De, S.K. Pal: Knowledge-based fuzzy MLP for classification and rule generation. IEEE Transactions on Neural Networks, 8, 1338-1350 (1997)

19.62 S. Mitra, Y. Hayashi: Neuro-fuzzy rule generation: Survey in soft computing framework. IEEE Transactions on Neural Networks, 11, 748-768 (2000)

19.63 S. Mitra, S.K. Pal: Fuzzy multi-layer perceptron, inferencing and rule generation. IEEE Transactions on Neural Networks, 6, 51-63 (1995)

19.64 S. Mitra, S.K. Pal: Fuzzy self organization, inferencing and rule generation. *IEEE Transactions on Systems, Man and Cybernetics, Part A: Systems and Humans*, 26, 608-620, 1996

19.65 S. Mitra, S.K. Pal, P. Mitra: Data mining in soft computing framework: A survey. IEEE Trans. Neural Networks, 13(1), 3-14 (2002)

19.66 B. Mobasher, V. Kumar, E.H. Han: Clustering in a high dimensional space using hypergraph models. Technical Report TR-97-063, University of Minnesota, Minneapolis, 1997.

19.67 T. Mollestad, A. Skowron: A rough set framework for data mining of propositional default rules. *Lecture Notes in Computer Science* 1079, 448-457 (1996)

19.68 D. Nauck: Using symbolic data in neuro-fuzzy classification. In *Proceedings of NAFIPS 99* (New York, June 1999) pp. 536-540

19.69 E. Noda, A.A. Freitas, H.S. Lopes: Discovering interesting prediction rules with a genetic algorithm. In: *Proceedings of IEEE Congress on Evolutionary Computation CEC 99* (Washington DC, July 1999) pp. 1322-1329

19.70 S.K. Pal, R.K. De, J. Basak: Unsupervised feature evaluation : A neuro-fuzzy approach. IEEE Transactions on Neural Networks, 11, 366-376 (2000)

19.71 S.K. Pal, T.S. Dillon, D.S. Yeung: *Soft Computing in Case Based Reasoning* (Springer Verlag, London, 2001)

19.72 S.K. Pal, D. DuttaMajumder: *Fuzzy Mathematical Approach to Pattern Recognition* (John Wiley, Halsted Press, New York, 1986)

19.73 S.K. Pal, A. Ghosh, M.K. Kundu (eds.): *Soft Computing for Image Processing* (Physica Verlag, Heidelberg, 2000)

19.74 S.K. Pal, P. Mitra: Case generation: A rough fuzzy approach. In: *Proc. Intl. Conf. Case Based Reasoning (ICCBR2001)* (Vancouver, Canada, 2001)

19.75 S.K. Pal, S. Mitra: *Neuro-fuzzy Pattern Recognition: Methods in Soft Computing* (John Wiley, New York, 1999)

19.76 S.K. Pal, S .Mitra, P. Mitra: Rough fuzzy MLP: Modular evolution, rule generation and evaluation. IEEE Trans. Knowledge and Data Engineering, 15(1), 14-25 (2003)

19.77 S.K. Pal, A. Pal (eds.): *Pattern Recognition: From classical to modern approaches* (World Scientific, Singapore, 2001)

19.78 S.K. Pal, W. Pedrycz, A. Skowron, R. Swiniarski (eds.): Spl. issue on rough-neuro computing. Neurocomputing, 36(1-4) (2001)

19.79 S.K. Pal, A. Skowron. *Rough Fuzzy Hybridization: A New Trend in Decision Making* (Springer-Verlag, Singapore, 1999)

19.80 S.K. Pal, V. Talwar, P. Mitra: Web mining in soft computing framework: Relevance, state of the art and future direction. IEEE Trans. Neural Networks, 13(5), 1163-1177 (2002)

19.81 S.K. Pal, P.P. Wang (eds.): *Genetic Algorithms for Pattern Recognition* (CRC Press, Boca Raton, 1996)

19.82 G. Pasi, G. Bordonga: Application of fuzzy set theory to extend boolean information retrieval. In: F.Crestani, G.Pasi, (eds.), *Soft Computing in Information Retrieval: Techniques and Applications* (Physica Verlag, Heidelberg, 2000) 50, pp. 21-47

19.83 Z. Pawlak: *Rough Sets, Theoretical Aspects of Reasoning about Data* (Kluwer Academic, Dordrecht, 1991)

19.84 M. Pazzani, J. Muramatsu, D. Billsus: Syskill and webert:identifying interesting web sites. In: *Proceedings of Thirteenth National Conference on AI* pp. 54-61 (1996)

19.85 W. Pedrycz: Conditional fuzzy c-means. Pattern Recognition Letters, 17, 625-632 (1996)

19.86 W. Pedrycz: Fuzzy set technology in knowledge discovery. Fuzzy Sets and Systems, 98, 279-290 (1998)

19.87 L. Polkowski, A. Skowron: Rough mereology: A new paradigm for approximate reasoning. International Journal of Approximate Reasoning, 15(4), 333-365 (1996)

19.88 L. Polkowski, A. Skowron: *Rough Sets in Knowledge Discovery 1 and 2* (Physica-Verlag, Heidelberg, 1998)

19.89 A. Rosenfeld, A.C. Kak: *Digital Picture Processing* (Volume 1-2. Academic Press, New York, 1982)

19.90 D.E. Rumelhart, J.L. McClelland (eds.): *Parallel Distributed Processing: Explorations in the Microstructures of Cognition*, volume1 (MIT Press, Cambridge, MA, 1986)

19.91 T. Ryu, C.F. Eick: MASSON: discovering commonalties in collection of objects using genetic programming. In: *Genetic Programming 1996: Proc. First Annual Conference* (Stanford University, CA, July 28-31 1996 MIT Press) pp. 200-208

19.92 D. Shalvi, N. DeClaris: Unsupervised neural network approach to medical data mining techniques. In: *Proceedings of IEEE International Joint Conference on Neural Networks* (Alaska, May 1998) pp. 171-176

19.93 N. Shan, W. Ziarko: Data-based acquisition and incremental modification of classification rules. Computational Intelligence, 11, 357-370 (1995)

19.94 J. Shavlik, T. Eliassi: A system for building intelligent agents that learn to retrieve and extract information. *International Journal on User Modeling and user adapted interaction*, April 2001 (Spl. issue on User Modeling and Intelligent Agents)

19.95 J. Shavlik, G.G. Towell: Knowledge-based artificial neural networks. Artificial Intelligence, 70(1-2), 119-165 (1994)

19.96 C.K. Shin, S.J. Yu, U.T. Yun, H.K. Kim: A hybrid approach of neural network and memory based learning to data mining. IEEE Transactions on Neural Networks, 11, 637-646 (2000)

19.97 A. Skowron: Extracting laws from decision tables - a rough set approach. Computational Intelligence, 11, 371-388 (1995)

19.98 A. Skowron, L. Polkowski (eds.), *Rough Sets in Knowledge Discovery* (Physica-Verlag, Heidelberg, 1998)

19.99 S. Soderland: Learning information extraction rules for semi-structured and free text. *Machine learning*, 34 (Special issue on Natural Language Learning) 233-272 (1999)

19.100 U. Straccia: A framework for the retrieval of multimedia objects based on four-valued fuzzy description logics. In F.Crestani, G.Pasi, (ed.), *Soft Computing in Information Retrieval: Techniques and Applications* (Physica Verlag, Heidelberg, 2000) 50, pp. 332-357

19.101 A. Teller, M. Veloso: Program evolution for data mining. The International Journal of Expert Systems, 8, 216-236 (1995)

19.102 A.B. Tickle, R. Andrews, M. Golea, J. Diederich: The truth will come to light: Directions and challenges in extracting the knowledge embedded within trained artificial neural networks. IEEE Transactions on Neural Networks, 9, 1057-1068 (1998)

19.103 J.T. Tou, R.C. Gonzalez: *Pattern Recognition Principles* (Addison-Wesley, London, 1974)

19.104 I.B. Turksen: Fuzzy data mining and expert system development. In: *Proceedings of IEEE International Conference on Systems, Man, Cybernetics* (San Diego, CA, October 1998) pp. 2057-2061

19.105 J.Vesanto, E.Alhoniemi: Clustering of the self-organizing map. IEEE Transactions on Neural Networks, 11, 586-600 (2000)

19.106 Q.Wei, G.Chen: Mining generalized association rules with fuzzy taxonomic structures. In: *Proceedings of NAFIPS 99* (New York, June 1999) pp. 477-481

19.107 S.K. Wong, Y.Y. Yao, C.J. Butz: Granular information retrieval. In F.Crestani, G.Pasi (eds.), *Soft Computing in Information Retrieval: Techniques and Applications* (Physica Verlag, Heidelberg, 2000) 50, pp. 317-331.

19.108 R.Yager: A framework for linguistic and hierarchical queries for document retrieval. In: F.Crestani, G.Pasi (eds.), *Soft Computing in Information Retrieval: Techniques and Applications* (Physica Verlag, Heidelberg, 2000) 50, pp. 3-20

19.109 R.R. Yager: On linguistic summaries of data. In: W.Frawley, G.P. Shapiro (eds.), *Knowledge Discovery in Databases* (AAAI/MIT Press, Menlo Park, CA, 1991) pp. 347-363

19.110 J.J. Yang, R.Korfhage: Query modification using genetic algorithms in vector space models. TR LIS045/I592001, Department of IS, University of Pittsburg (1992)

19.111 L.A. Zadeh: Fuzzy sets. Information and Control, 8, 338-353 (1965)

19.112 L.A. Zadeh: Fuzzy logic, neural networks, and soft computing. Communications of the ACM, 37, 77-84 (1994)

19.113 L.A. Zadeh: A new direction in AI: Towards a computational theory of perceptions. AI Magazine, 22, 73-84 (2001)

19.114 W. Ziarko, N. Shan: KDD-R: A comprehensive system for knowledge discovery in databases using rough sets. In: *Proc. Third International Workshop on Rough Sets and Soft Computing (RSSC'94, 1994)* pp. 164-173

20. Dominance-Based Rough Set Approach to Knowledge Discovery (I): General Perspective

Salvatore Greco[1], Benedetto Matarazzo[1], and Roman Slowinski[2,3]

[1] University of Catania, Italy
[2] Poznan University of Technology, Poland
[3] Polish Academy of Sciences, Poland

Abstract

This chapter is devoted to knowledge discovery from data, taking into account prior knowledge about preference semantics in patterns to be discovered. The data concern a set of objects (situations, states, examples) described by a set of attributes (properties, features, characteristics). The attributes are, in general, divided into condition and decision attributes, corresponding to input and output descriptions of an object. The set of objects is partitioned by decision attributes into decision classes. A pattern discovered from the data has a symbolic form of decision rule or decision tree. In many practical problems, some condition attributes are defined on preference ordered scales, and the decision classes are also preference ordered. The known methods of knowledge discovery unfortunately ignore this preference information, risking drawing wrong patterns. To deal with preference-ordered data, we propose to use a new approach called Dominance-based Rough Set Approach (DRSA). Given a set of objects described by at least one condition attribute with preference-ordered scale and partitioned into preference-ordered classes, the new rough set approach is able to approximate this partition by means of dominance relations. The rough approximation of this partition is a starting point for induction of "*if ...*, *then ...* " decision rules. The syntax of these rules is adapted to represent preference orders. The DRSA analyzes only facts present in data, and possible inconsistencies are identified. It preserves the concept of granular computing; however, the granules are dominance cones in evaluation space, and not bounded sets. It is also concordant with the paradigm of computing with words, as it exploits the ordinal, and not necessarily the cardinal, character of data. The basic DRSA and its major extensions are presented in two consecutive parts in this book. In the present part, we give a general perspective of DRSA, explaining its use in the context of multicriteria classification, choice, and ranking. Moreover, we present a variant of DRSA that handles missing values in data sets.

20.1 Introduction: Three Types of Prior Knowledge to Be Included in Knowledge Discovery

We live in an information-rich but knowledge-poor environment. *Knowledge discovery* and *data mining* techniques try to bridge the gap between data generation and data comprehension.

Discovering knowledge from data means being able to find concise classification patterns that say some truth about objects described by the data. They are useful for the explanation of past classifications, and for prediction of future classifications in applications such as medical or technical diagnostics, performance evaluation, and risk assessment. The objects are described by a set of *attributes*, called also properties, or features, or characteristics. The attributes may be on either *the condition* or *the decision* side of the description, corresponding to the input or the output of a classification system. The *objects* may be situations, individuals, states, examples, etc.

The data set in which classification patterns are searched for is called *learning sample*. Learning of patterns from this sample should take into account available *prior knowledge* that may include the following items:

(i) domains of attributes, i.e., sets of values that an attribute may take while being meaningful to a user's perception,

(ii) division of attributes into condition and decision attributes, restricting the range of patterns to functional relations between condition and decision attributes,

(iii) preference order in domains of some attributes and semantic correlation between pairs of these attributes, requiring the patterns to observe the dominance principle.

In fact, item (i) is usually taken into account in knowledge discovery. With this prior knowledge alone, one can discover patterns called *association rules* [20.1], showing strong relationships between values of some attributes, without mandating which attributes will be on the condition side and which ones on the decision side in the rules.

If item (i) is combined with item (ii) in the prior knowledge, then one can consider a partition of the learning sample into decision classes defined by decision attributes. The patterns to be discovered then have the form of *decision trees* or *decision rules* representing functional relations between condition and decision attributes. These patterns are typically discovered by machine learning and data mining methods [20.42]. As there is a direct correspondence between decision tree and rules, we will concentrate further our attention on decision rules.

As item (iii) is crucial for this chapter, let us explain it in more detail. Consider an example of a data set concerning students' achievements in a high school. Suppose that among the attributes describing the students are results in *mathematics (Math)* and *physics (Ph)*, and *general achievement (GA)*. The domains of these attributes are composed of three values: *bad, medium*, and *good*. This information constitutes item (i) of prior knowledge. Item (ii) is also available because, clearly, *Math* and *Ph* are condition attributes while *GA* is a decision attribute. The preference order of the attribute values is obvious: *good* is better than *medium* and *bad*, and *medium* is better than *bad*. It is known, moreover, that *Math* is semantically correlated with *GA*, as well as *Ph* is with *GA*. This is, precisely, item (iii) of the prior knowledge.

Attributes with preference-ordered domains are called *criteria* because they involve an evaluation. We will use the term *regular attributes* for those attributes whose domains are not preference ordered. *Semantic correlation between two criteria* (condition and decision) means that an improvement on one criterion should not worsen evaluation on the second criterion while other attributes and criteria are unchanged. In our example, improvement of a student's score in *Math* or *Ph* should not worsen his or her *GA*. In general, semantic correlation between condition criteria and decision criteria requires that object x dominating object y on all condition criteria (i.e., x having evaluations at least as good as y on all condition criteria) should also dominate y on all decision criteria (i.e., x should have evaluations at least as good as y on all decision criteria). This principle is called *dominance principle* (or Pareto principle) and it is the only objective principle that everybody agrees to observe in multicriteria comparisons of objects.

What classification patterns can be drawn from the student data set? How does item (iii) influence the classification patterns? The answer to the first question is *"if..., then..."* decision rules. Each decision rule is characterized by *a condition profile* and *a decision profile*, corresponding to vectors of threshold values of regular attributes and criteria in the condition and decision parts of the rule, respectively. The answer to the second question is that condition and decision profiles of a decision rule should observe the dominance principle if the rule has at least one pair of semantically correlated criteria spanned over the condition and decision parts. We say that one profile *dominates* another if they both involve the same values of regular attributes, and the values of criteria of the first profile are not worse than the values of criteria of the second profile.

Let us explain the dominance principle with respect to decision rules on the student example. Suppose that two rules induced from the student data set relate *Math* and *Ph* on the condition side, with *GA* on the decision side:

$rule\#1 : if \; Math = medium \; and \;\; Ph = medium, \; then \; GA = good,$
$rule\#2 : if \; Math = good \quad\;\; and \;\; Ph = medium, \; then \; GA = medium.$

The two rules do not observe the dominance principle because the condition profile of rule#2 dominates the condition profile of rule#1, while the decision profile of rule#1 dominates the decision profile of rule#2. Thus, in the sense of the dominance principle, the two rules are inconsistent; that is, they are wrong.

One could say that the above rules are true because they are supported by examples of students from the learning sample, but this would mean that the examples are also inconsistent. The *inconsistency* may come from many sources, e.g.,

− missing attributes (regular ones or criteria) in the description of objects; maybe the data set does not include such attributes as *opinion of a student's tutor* (*OT*), expressed only verbally during the assessment of student's *GA* by the school teachers' council,

– unstable preferences of decision makers; maybe the members of the school
teachers' council changed their view on the influence of *Math* on *GA* during
the assessment.

Handling these inconsistencies is of crucial importance for knowledge discovery. They cannot be simply considered as noise or error to be eliminated
from data, or amalgamated with consistent data by some averaging operators,
but they should be identified and presented as uncertain patterns.

If item (iii) would be ignored in prior knowledge, then the handling of
above-mentioned inconsistencies would be impossible. Indeed, there would be
nothing wrong in rules #1 and #2: they are supported by different examples
discerned by considered attributes.

It has been acknowledged by many authors that *rough set theory* provides
an excellent framework for dealing with inconsistency in knowledge discovery [20.39, 20.48, 20.49, 20.50, 20.51, 20.56, 20.64, 20.74, 20.75]. The paradigm
of rough set theory is that of *granular computing*, because the main concept
of the theory – rough approximation of a set – is built with blocks of objects
indiscernible by a given set of attributes, called *granules of knowledge*. When
the attributes are regular ones, the granules are bounded sets. Decision rules
induced from rough approximation of a classification are also built with such
granules. While taking into account prior knowledge of type (i) and (ii), the
rough approximation and the inherent rule induction ignore, however, prior
knowledge of type (iii). In consequence, the resulting decision rules may be
inconsistent with the dominance principle.

The authors have proposed an extension to the granular computing
paradigm that permits taking into account prior knowledge of type (iii),
in addition to either (i) alone [20.36], or (i) and (ii) together [20.18, 20.20,
20.26, 20.31, 20.32, 20.59, 20.62]. The combination of the new granules with
the idea of rough approximation makes the so-called *Dominance-based Rough
Set Approach* (DRSA).

DRSA and its major extensions are presented in two consecutive parts in
this book. In this part, we present the concept of granules, enabling us to
handle prior knowledge of type (iii); then, we present the basic DRSA with
an illustrative example and, later, DRSA for multicriteria choice and ranking,
and a variant of DRSA that handles missing values in data sets.

20.2 The Influence of Preference Order in Data on Granular Computing

In this section, we answer the following question: how should we define the
granule of knowledge in the attribute space in order to take into account prior
knowledge about preference order in data when searching for decision rules?

As it is usual in knowledge discovery methods, information about objects
is represented in a *data table*, in which rows are labeled by *objects* and contain

the values of attributes for each corresponding object, whereas columns are labeled by *attributes* and contain the values of each corresponding attribute for the objects.

Let U denote a finite set of objects (universe), and Q denote a finite set of attributes divided into set C of *condition attributes* and set D of *decision attributes*; $C \cap D = \emptyset$. Also, let $X_C = \prod_{q=1}^{|C|} X_q$ and $X_D = \prod_{q=1}^{|D|} X_q$ be attribute spaces corresponding to sets of condition and decision attributes, respectively. Elements of X_C and X_D can be interpreted as possible evaluations of objects on attributes from set $C = \{1, \ldots, |C|\}$ and from set $D = \{1, \ldots, |D|\}$, respectively. Therefore, X_q is the set of possible evaluations with respect to attribute q of objects considered. The value of object x on attribute $q \in Q$ is denoted by x_q. Objects x and y are *indiscernible* by $P \subseteq C$ if $x_q = y_q$ for all $q \in P$ and, analogously, objects x and y are indiscernible by $R \subseteq D$ if $x_q = y_q$ for all $q \in R$. Sets of indiscernible objects are equivalence classes of the corresponding *indiscernibility relation* I_P or I_R. Moreover, $I_P(x)$ and $I_R(x)$ denote equivalence classes including object x. I_D makes a partition of U into a finite number of decision classes $\boldsymbol{Cl} = \{Cl_t, t = \{1, \ldots, n\}\}$. Each $x \in U$ belongs to one, and only one, class $Cl_t \in \boldsymbol{Cl}$.

The above definitions take into account prior knowledge of type (i) and (ii). In this case, the **granules of knowledge are bounded sets** in X_P and X_R ($P \subseteq C$ and $R \subseteq D$), defined by partitions of U induced by indiscernibility relations I_P and I_R, respectively. Then, classification patterns to be discovered are functions representing granules $I_R(x)$ by granules $I_P(x)$ in condition attribute space X_P, for any $P \subseteq C$ and any $x \in U$.

If prior knowledge includes item (iii) in addition to items (i) and (ii), then the indiscernibility relation is unable to produce granules in X_C and X_D, taking into account the preference order. To do so, it has to be substituted by the dominance relation in X_P and X_R ($P \subseteq C$ and $R \subseteq D$). Suppose, for simplicity, that all condition attributes in C and all decision attributes in D are criteria, and that C and D are semantically correlated.

Let $\succeq_q$ be a *weak preference relation* on U (often called outranking) representing a preference on the set of objects with respect to criterion $q \in \{C \cup D\}$; $x_q \succeq_q y_q$ means "x_q is at least as good as y_q with respect to criterion q". On the one hand, we say that x *dominates* y with respect to $P \subseteq C$ (shortly, x P-*dominates* y) in condition attribute space X_P (denoted by $xD_P y$) if $x_q \succeq_q y_q$ for all $q \in P$. Assuming, without loss of generality, that the domains of the criteria are numerical, i.e., $X_q \subseteq \boldsymbol{R}$ for any $q \in C$, and that they are ordered such that preference increases with value, one can say that $xD_P y$ is equivalent to $x_q \geq y_q$ for all $q \in P, P \subseteq C$. Observe that, for each $x \in X_P$, $xD_P x$, i.e., P-dominance is reflexive. On the other hand, the analogous definition holds in decision attribute space X_R (denoted by $xD_R y$), $R \subseteq D$.

The dominance relations $xD_P y$ and $xD_R y$ ($P \subseteq C$ and $R \subseteq D$) are directional statements where x is a subject and y is a referent.

If $x \in X_P$ is the referent, then one can define a set of objects $y \in X_P$ dominating x, called *P-dominating set*, $D_P^+(x) = \{y \in U : yD_P x\}$.

If $x \in X_P$ is the subject, then one can define a set of objects $y \in X_P$ dominated by x, called *P-dominated set*, $D_P^-(x) = \{y \in U : xD_P y\}$.

P-dominating sets $D_P^+(x)$ and *P*-dominated sets $D_P^-(x)$ correspond to *positive* and *negative dominance cones* in X_P, with origin x.

As for decision attribute space X_R, $R \subseteq D$, the *R*-dominance relation enables us to define sets

$$Cl_{\overline{R}}^{\geq x} = \{y \in U : yD_R x\}, \quad Cl_{\overline{R}}^{\leq x} = \{y \in U : xD_R y\}.$$

$Cl_{t_q} = \{x \in X_D : x_q = t_q\}$ is a decision class with respect to $q \in D$. $Cl_{\overline{R}}^{\geq x}$ is called *the upward union* of classes, and $Cl_{\overline{R}}^{\leq x}$ is called *the downward union* of classes. If $y \in Cl_{\overline{R}}^{\geq x}$, then y belongs to class Cl_{t_q}, $x_q = t_q$, or better, on each decision attribute $q \in R$; if $y \in Cl_{\overline{R}}^{\leq x}$, then y belongs to class Cl_{t_q}, $x_q = t_q$, or worse, on each decision attribute $q \in R$. The downward and upward unions of classes correspond to *positive* and *negative dominance cones* in X_R, respectively.

In this case, the **granules of knowledge are open sets** in X_P and X_R, defined by dominance cones $D_P^+(x)$ and $D_P^-(x)$ ($P \subseteq C$) and $Cl_{\overline{R}}^{\geq x}$ and $Cl_{\overline{R}}^{\leq x}$ ($R \subseteq D$), respectively. Then, classification patterns to be discovered are functions representing granules $Cl_{\overline{R}}^{\geq x}$ and $Cl_{\overline{R}}^{\leq x}$ by granules $D_P^+(x)$ and $D_P^-(x)$, respectively, in condition attribute space X_P, for any $P \subseteq C$ and $R \subseteq D$ and for any $x \in X_P$.

In both cases above, the functions are sets of decision rules.

20.3 Dominance-Based Rough Set Approach (DRSA)

20.3.1 Granular Computing with Dominance Cones

When discovering classification patterns, the set D of decision attributes is, usually, a singleton, $D = \{d\}$. Let us take this assumption for further presentation, although it is not necessary for DRSA. Decision attribute d makes a partition of U into a finite number of classes $\boldsymbol{Cl} = \{Cl_t, t \in \{1, \ldots, n\}\}$. Each $x \in U$ belongs to one and only one class $Cl_t \in \boldsymbol{Cl}$. The upward and downward unions of classes boil down, respectively, to

$$Cl_t^{\geq} = \bigcup_{s \geq t} Cl_s, \quad Cl_t^{\leq} = \bigcup_{s \leq t} Cl_s, \quad t = 1, \ldots, n.$$

Notice that, for $t = 2, \ldots, n$, we have $Cl_t^{\leq} = U - Cl_{t-1}^{\leq}$, i.e., all the objects not belonging to class Cl_t, or better, belong to class Cl_{t-1}, or worse.

Let us explain how the rough set concept has been generalized to DRSA in order to enable granular computing with dominance cones (for more details, see [20.18, 20.20, 20.26, 20.31, 20.62]).

Given a set of criteria $P \subseteq C$, the inclusion of an object $x \in U$ to the upward union of classes $Cl_t^{\geq}$, $t = 2, \ldots, n$, creates an *inconsistency in the sense of dominance principle* if one of the following conditions holds:

- x belongs to class Cl_t or better, but it is P-dominated by an object y belonging to a class worse than Cl_t, i.e., $x \in Cl_t^{\geq}$ but $D_P^+(x) \cap Cl_{t-1}^{\leq} \neq \emptyset$,
- x belongs to a worse class than Cl_t, but it P-dominates an object y belonging to class Cl_t, or better, i.e., $x \notin Cl_t^{\geq}$, but $D_P^-(x) \cap Cl_t^{\geq} \neq \emptyset$.

If, given a set of criteria $P \subseteq C$, the inclusion of $x \in U$ to $Cl_t^{\geq}$, $t = 2, \ldots,$ n, creates an inconsistency in the sense of dominance principle, we say that x belongs to $Cl_t^{\geq}$ *with some ambiguity*. Thus, x belongs to $Cl_t^{\geq}$ *without any ambiguity* with respect to $P \subseteq C$ if $x \in Cl_t^{\geq}$ and there is no inconsistency in the sense of dominance principle. This means that all objects P-dominating x belong to $Cl_t^{\geq}$, i.e., $D_P^+(x) \subseteq Cl_t^{\geq}$. Geometrically, this corresponds to the inclusion of the complete set of objects contained in the positive dominance cone originating in x, into the positive dominance cone $Cl_t^{\geq}$ originating in Cl_t.

Furthermore, x *possibly belongs to* $Cl_t^{\geq}$ with respect to $P \subseteq C$ if one of the following conditions holds:

- according to decision attribute d, x belongs to $Cl_t^{\geq}$,
- according to decision attribute d, x does not belong to $Cl_t^{\geq}$, but it is inconsistent in the sense of dominance principle with an object y belonging to $Cl_t^{\geq}$.

In terms of ambiguity, x possibly belongs to $Cl_t^{\geq}$ with respect to $P \subseteq C$ if x belongs to $Cl_t^{\geq}$ with or without any ambiguity. Due to reflexivity of the dominance relation D_P, the above conditions can be summarized as follows: x *possibly belongs* to class Cl_t, or better, with respect to $P \subseteq C$ if, among the objects P-dominated by x, there is an object y belonging to class Cl_t, or better, i.e., $D_P^-(x) \cap Cl_t^{\geq} \neq \emptyset$. Geometrically, this corresponds to the nonempty intersection of the set of objects contained in the negative dominance cone originating in x, with the positive dominance cone $Cl_t^{\geq}$ originating in Cl_t.

For $P \subseteq C$, the set of all objects belonging to $Cl_t^{\geq}$ without any ambiguity constitutes the P-*lower approximation* of $Cl_t^{\geq}$, denoted by $\underline{P}(Cl_t^{\geq})$, and the set of all objects that possibly belong to $Cl_t^{\geq}$ constitutes the P-*upper approximation* of $Cl_t^{\geq}$, denoted by $\overline{P}(Cl_t^{\geq})$:

$$\underline{P}(Cl_t^{\geq}) = \{x \in U \colon D_P^+(x) \subseteq Cl_t^{\geq}\}, \quad \overline{P}(Cl_t^{\geq}) = \{x \in U \colon D_P^-(x) \cap Cl_t^{\geq} \neq \emptyset\},$$
for $t = 2, \ldots, n$.

Analogously, one can define the P-*lower approximation* and the P-*upper approximation* of $Cl_t^{\leq}$ as follows:

$\underline{P}(Cl_t^{\leq}) = \{x \in U: D_P^-(x) \subseteq Cl_t^{\leq}\}, \ \overline{P}(Cl_t^{\leq}) = \{x \in U: D_P^+(x) \cap Cl_t^{\leq} \neq \emptyset\},$
for $t = 1, \ldots, n-1$.

The P-lower and P-upper approximations, so defined, satisfy the following *inclusion properties* for all $P \subseteq C$:

$$\underline{P}(Cl_t^{\geq}) \subseteq Cl_t^{\geq} \subseteq \overline{P}(Cl_t^{\geq}), \quad t = 2, \ldots, n,$$

$$\underline{P}(Cl_t^{\leq}) \subseteq Cl_t^{\leq} \subseteq \overline{P}(Cl_t^{\leq}), \quad t = 1, \ldots, n-1.$$

All the objects belonging to $Cl_t^{\geq}$ and $Cl_t^{\leq}$ with some ambiguity constitute the *P-boundary* of $Cl_t^{\geq}$ and $Cl_t^{\leq}$, denoted by $Bn_P(Cl_t^{\geq})$ and $Bn_P(Cl_t^{\leq})$, respectively. They can be represented in terms of upper and lower approximations as follows:

$$Bn_P(Cl_t^{\geq}) = \overline{P}(Cl_t^{\geq}) - \underline{P}(Cl_t^{\geq}), t = 2, \ldots, n,$$
$$Bn_P(Cl_t^{\leq}) = \overline{P}(Cl_t^{\leq}) - \underline{P}(Cl_t^{\leq}), t = 1, \ldots, n-1.$$

P-lower and P-upper approximations of unions of classes $Cl_t^{\geq}$ and $Cl_t^{\leq}$ have an important *complementarity property*. It states that if object x belongs without any ambiguity to class Cl_t, or better, it is impossible that it could belong to class Cl_{t-1}, or worse, i.e., $\underline{P}(Cl_t^{\geq}) = U - \overline{P}(Cl_{t-1}^{\leq}), t = 2, \ldots, n$.

Due to the complementarity property, $Bn_P(Cl_t^{\geq}) = Bn_P(Cl_{t-1}^{\leq})$, for $t = 2, \ldots, n$, which means that if x belongs with ambiguity to class Cl_t, or better, it also belongs with ambiguity to class Cl_{t-1}, or worse.

From the knowledge discovery point of view, P-lower approximations of unions of classes represent *certain knowledge* given by criteria from $P \subseteq C$, while P-upper approximations represent *possible knowledge*, and the P-boundaries contain *doubtful knowledge* given by criteria from $P \subseteq C$.

The above definitions of rough approximations are based on a strict application of the dominance principle. However, when defining non-ambiguous objects, it is reasonable to accept a limited proportion of negative examples, particularly for large data tables. Such an extended version of DRSA is called the Variable-Consistency DRSA model (VC-DRSA) [20.34].

For any $P \subseteq C$, we say that $x \in U$ belongs to $Cl_t^{\geq}$ *with no ambiguity at consistency level* $l \in (0, 1]$ if $x \in Cl_t^{\geq}$ and at least $l*100\%$ of all objects $y \in U$ dominating x with respect to P also belong to $Cl_t^{\geq}$, i.e.,

$$\frac{card(D_P^+(x) \cap Cl_t^{\geq})}{card(D_P^+(x))} \geq l, \quad t = 2, \ldots, n.$$

The level l is called *consistency level* because it controls the degree of consistency between objects qualified as belonging to $Cl_t^{\geq}$ without any ambiguity. In other words, if $l < 1$, then at most $(1-l)*100\%$ of all objects $y \in U$ dominating x with respect to P do not belong to $Cl_t^{\geq}$, and thus contradict the inclusion of x in $Cl_t^{\geq}$.

Analogously, for any $P \subseteq C$, we say that $x \in U$ belongs to $Cl_t^{\leq}$ *with no ambiguity at consistency level* $l \in (0, 1]$ if $x \in Cl_t^{\leq}$ and at least $l*100\%$ of all the objects $y \in U$ dominated by x with respect to P also belong to $Cl_t^{\leq}$, i.e.,

$$\frac{card(D_P^-(x) \cap Cl_t^{\leq})}{card(D_P^-(x))} \geq l, \quad t = 1, \ldots, n-1.$$

Thus, for any $P \subseteq C$, each object $x \in U$ is either ambiguous or non-ambiguous at consistency level l with respect to the upward union $Cl_t^{\geq}$ ($t = 2, \ldots, n$) or with respect to the downward union $Cl_t^{\leq}$ ($t = 1, \ldots, n$-1).

The concept of non-ambiguous objects at some consistency level l leads naturally to the definition of P-lower approximations of the unions of classes $Cl_t^{\geq}$ and $Cl_t^{\leq}$:

$$\underline{P}^l(Cl_t^{\geq}) = \{x \in Cl_t^{\geq} : \frac{card(D_P^+(x) \cap Cl_t^{\geq})}{card(D_P^+(x))} \geq l\}, \quad t = 2, \ldots, n,$$

$$\underline{P}^l(Cl_t^{\leq}) = \{x \in Cl_t^{\leq} : \frac{card(D_P^-(x) \cap Cl_t^{\leq})}{card(D_P^-(x))} \geq l\}, \quad t = 1, \ldots, n-1.$$

Given $P \subseteq C$ and consistency level l, we can define the P-*upper approximations* of $Cl_t^{\geq}$ and $Cl_t^{\leq}$, denoted by $\overline{P}^l(Cl_t^{\geq})$ and $\overline{P}^l(Cl_t^{\leq})$, respectively, by complementation of $\underline{P}^l(Cl_{t-1}^{\leq})$ and $\underline{P}^l(Cl_{t+1}^{\geq})$ with respect to U:

$$\overline{P}^l(Cl_t^{\geq}) = U - \underline{P}^l(Cl_{t-1}^{\leq}), \quad \overline{P}^l(Cl_t^{\leq}) = U - \underline{P}^l(Cl_{t+1}^{\geq}).$$

$\overline{P}^l(Cl_t^{\geq})$ can be interpreted as the set of all the objects belonging to $Cl_t^{\geq}$ *possibly ambiguous* at consistency level l. Analogously, $\overline{P}^l(Cl_t^{\leq})$ can be interpreted as the set of all the objects belonging to $Cl_t^{\leq}$ *possibly ambiguous* at consistency level l. The P-*boundaries* (P-doubtful regions) of $Cl_t^{\geq}$ and $Cl_t^{\leq}$ are defined as

$$Bn_P^l(Cl_t^{\geq}) = \overline{P}^l(Cl_t^{\geq}) - \underline{P}^l(Cl_t^{\geq}), t = 2, \ldots, n,$$
$$Bn_P^l(Cl_t^{\leq}) = \overline{P}^l(Cl_t^{\leq}) - \underline{P}^l(Cl_t^{\leq}), t = 1, \ldots, n-1.$$

The *variable consistency* model of the dominance-based rough set approach provides some degree of flexibility in assigning objects to lower and upper approximations of the unions of decision classes. It can easily be demonstrated that, for $0 < l' < l \leq 1$ and $t = 2, \ldots, n$,

$$\underline{P}^l(Cl_t^{\geq}) \subseteq \underline{P}^{l'}(Cl_t^{\geq}) \quad \text{and} \quad \overline{P}^{l'}(Cl_t^{\geq}) \subseteq \overline{P}^l(Cl_t^{\geq}).$$

The *variable consistency* model is inspired by Ziarko's model of the *variable precision* rough set approach [20.73, 20.74]; however, there is a significant difference in the definition of rough approximations, because $\underline{P}^l(Cl_t^{\geq})$ and $\overline{P}^l(Cl_t^{\geq})$ are composed of non-ambiguous and ambiguous objects at consistency level l, respectively, while Ziarko's $\underline{P}^l(Cl_t)$ and $\overline{P}^l(Cl_t)$ are composed of P-indiscernibility <u>sets</u> such that at least $l*100\%$ of these sets are included in Cl_t or have a non-empty intersection with Cl_t. If one would like to use Ziarko's definition of variable precision rough approximations in the context of multiple criteria classification, then the P-indiscernibility sets should be substituted by P-dominating sets $D_P^+(x)$; however, then the notion of ambiguity that naturally leads to the general definition of rough approximations (see [20.63]) loses its meaning. Moreover, a bad side effect of a direct use of Ziarko's definition is that a lower approximation $\underline{P}^l(Cl_t^{\geq})$ may include objects y assigned to Cl_h, where h is much less than t, if y belongs to the $D_P^+(x)$ that was included in $\underline{P}^l(Cl_t^{\geq})$. When the decision classes are preference ordered, it is reasonable to expect that objects assigned to far worse classes than the considered union are not counted in the lower approximation of this union.

For every $P \subseteq C$, the objects that are consistent in the sense of dominance principle with all upward and downward unions of classes are P-*correctly classified*. For every $P \subseteq C$, the *quality of approximation of classification* $\mathbf{Cl}$ by a set of criteria P is defined as the ratio between the number of P-correctly classified objects and the number of all the objects in the data sample set. Since the objects P-correctly classified are those that do not belong to any P-boundary of unions $Cl_t^{\geq}$ and $Cl_t^{\leq}$, $t = 1, \ldots, n$, the quality of approximation of classification $\mathbf{Cl}$ by a set of criteria P, can be written as

$$\gamma_P(\mathbf{Cl}) = \frac{card((U - (\bigcup_{t=1,\ldots,n-1} Bn_P(Cl_t^{\leq})) \cup (\bigcup_{t=2,\ldots,n} Bn_P(Cl_t^{\geq}))))}{(U)}$$

$$= \frac{card((U - (\bigcup_{t=2,\ldots,n} Bn_P(Cl_t^{\geq}))))}{(U)}.$$

$\gamma_P(\mathbf{Cl})$ can be seen as a measure of the quality of knowledge that can be extracted from the data table, where P is the set of criteria and $\mathbf{Cl}$ is the considered classification.

Each minimal subset $P \subseteq C$ such that $\gamma_P(\mathbf{Cl}) = \gamma_C(\mathbf{Cl})$ is called a *reduct* of $\mathbf{Cl}$, and is denoted by $RED_{\mathbf{Cl}}$. Let us remark that a data sample set can have more than one reduct. The intersection of all reducts is called the *core*, and is denoted by $CORE_{\mathbf{Cl}}$. Criteria in $CORE_{\mathbf{Cl}}$ cannot be removed from the data sample set without deteriorating the quality of the knowledge to be discovered. This means that, in set C, there are three categories of criteria:

- *indispensable* criteria included in the core,
- *exchangeable* criteria included in some reducts, but not in the core,
- *redundant* criteria, neither indispensable nor exchangeable, and thus not included in any reduct.

Let us note that reducts are minimal subsets of attributes and criteria, conveying the relevant knowledge contained in the learning sample; this knowledge is relevant for explanation of patterns in data, but not necessarily for prediction.

It has been shown in [20.17, 20.29] that the quality of classification satisfies the properties of set functions called *fuzzy measures*. For this reason, one can use the quality of classification for calculation of indices measuring the relevance of particular attributes and/or criteria, as well as the strength of interactions between them; the useful indices are the value index and the interaction indices of Shapley and Banzhaf, the interaction indices of Murofushi-Soneda and Roubens, and the Möbius representation. All these indices can help assess the informational dependence among the considered attributes and criteria, and help choose the best reduct.

20.3.2 Induction of Decision Rules

The dominance-based rough approximations of upward and downward unions of classes can serve to induce a generalized description of objects contained in the data table in terms of "*if* ..., *then* ..." decision rules. For a given upward or downward union of classes, $Cl_t^{\geq}$ or $Cl_s^{\leq}$, the decision rules induced under a hypothesis that objects belonging to $\underline{P}(Cl_t^{\geq})$ or $\underline{P}(Cl_s^{\leq})$ are positive examples and all the others are negative, suggest an assignment to "class Cl_t, or better", or to "class Cl_s, or worse", respectively. On the other hand, the decision rules induced under a hypothesis that objects belonging to the intersection $\overline{P}(Cl_s^{\leq}) \cap \overline{P}(Cl_t^{\geq})$ are positive examples, and all the others are negative, suggest an assignment to some classes between Cl_s and Cl_t ($s < t$).

In the case of preference-ordered data, it is meaningful to consider the following five types of decision rules:

1) *certain* $D_{\geq}$-*decision rules*, providing lower profile descriptions for objects belonging to $Cl_t^{\geq}$ without ambiguity: *if* $x_{q1} \succeq_{q1} r_{q1}$ and $x_{q2} \succeq_{q2} r_{q2}$ and $\ldots x_{qp} \succeq_{qp} r_{qp}$, *then* $x \in Cl_t^{\geq}$, where, for each $w_q, z_q \in X_q$, "$w_q \succeq_q z_q$" means "w_q is <u>at least</u> as good as z_q",

2) *possible* $D_{\geq}$-*decision rules*, providing lower profile descriptions for objects belonging to $Cl_t^{\geq}$ with or without ambiguity: *if* $x_{q1} \succeq_{q1} r_{q1}$ and $x_{q2} \succeq_{q2} r_{q2}$ and $\ldots x_{qp} \succeq_{qp} r_{qp}$, *then* x possibly belongs to $Cl_t^{\geq}$,

3) *certain* $D_{\leq}$-*decision rules*, providing upper profile descriptions for objects belonging to $Cl_t^{\leq}$ without ambiguity: *if* $x_{q1} \preceq_{q1} r_{q1}$ and $x_{q2} \preceq_{q2} r_{q2}$ and $\ldots x_{qp} \preceq_{qp} r_{qp}$, *then* $x \in Cl_t^{\leq}$, where, for each $w_q, z_q \in X_q$, "$w_q \preceq_q z_q$" means "w_q is <u>at most</u> as good as z_q",

4) *possible* $D_{\leq}$-*decision rules*, providing upper profile descriptions for objects belonging to $Cl_t^{\leq}$ with or without ambiguity: *if* $x_{q1} \preceq_{q1} r_{q1}$ and $x_{q2} \preceq_{q2} r_{q2}$ and $\ldots x_{qp} \preceq_{qp} r_{qp}$, *then* x possibly belongs to $Cl_t^{\leq}$,

5) *approximate* $D_{\geq \leq}$-*decision rules*, simultaneously providing lower and upper profile descriptions for objects belonging to $Cl_s \cup Cl_{s+1} \cup \ldots \cup Cl_t$, without possibility of discerning to which class: *if* $x_{q1} \succeq_{q1} r_{q1}$ *and* $\ldots x_{qk} \succeq_{qk} r_{qk}$ *and* $x_{qk+1} \preceq_{qk+1} r_{qk+1}$ *and* $\ldots x_{qp} \preceq_{qp} r_{qp}$, *then* $x \in Cl_s \cup Cl_{s+1} \cup \ldots \cup Cl_t$.

In the left hand side of a $D_{\geq}$-decision rule, we can have "$x_q \succeq_q r_q$" and "$x_q \preceq_q r_q'$", where $r_q \leq r_q'$, for the same $q \in C$. Moreover, if $r_q = r_q'$, the two conditions boil down to "$x_q \sim_q r_q$", where for each $w_q, z_q \in X_q$, "$w_q \sim_q z_q$" means "w_q is indifferent to z_q".

Since a decision rule is a kind of implication, by a *minimal* rule we mean an implication such that there is no other implication with the left hand side (LHS) of at least the same weakness (in other words, a rule using a subset of elementary conditions or/and weaker elementary conditions) and the right hand side (RHS) of at least the same strength (in other words, a $D_{\geq}$- or a $D_{\leq}$-decision rule assigning objects to the same union or sub-union of classes, or a $D_{\geq \leq}$-decision rule assigning objects to the same or larger set of classes).

The rules of type 1) and 3) represent certain knowledge extracted from the data table, while the rules of type 2) and 4) represent possible knowledge; the rules of type 5) represent doubtful knowledge because they are supported by ambiguous objects only.

The rules of type 1) and 3) are *exact* if they do not cover negative examples, and they are *probabilistic* otherwise. In the latter case, each rule is characterized by a confidence ratio, representing the probability that an object matching the LHS of the rule also matches its RHS. Probabilistic rules are concordant with the VC-DRSA model mentioned above.

Let us comment on the application of decision rules to the objects described by the criteria from C. When applying $D_{\geq}$-decision rules to object x, it is possible that x either matches the LHS of at least one decision rule or does not match the LHS of any decision rule. In the case of at least one matching, it is reasonable to conclude that x belongs to class Cl_t, the lowest class of the upward union $Cl_t^{\geq}$ resulting from an intersection of the RHSs of all the rules covering x. Precisely, if x matches the LHSs of rules $\rho_1, \rho_2, \ldots, \rho_m$, having RHSs $x \in Cl_{t1}^{\geq}, x \in Cl_{t2}^{\geq}, \ldots, x \in Cl_{tm}^{\geq}$, then x is assigned to class Cl_t, where $t = \max\{t_1, t_2, \ldots, t_m\}$. In the case of no matching, we conclude that x belongs to Cl_1, i.e., to the worst class, since no rule with a RHS suggesting a better classification of x covers this object.

Analogously, when applying $D_{\leq}$-decision rules to object x, we conclude that x belongs either to class Cl_z, the highest class of the downward union $Cl_t^{\leq}$ resulting from the intersection of the RHSs of all the rules covering x, or to class Cl_n, i.e., to the best class, when x is not covered by any rule. Precisely, if x matches the LHSs of rules $\rho_1, \rho_2, \ldots, \rho_m$, having RHSs $x \in Cl_{t1}^{\leq}, x \in Cl_{t2}^{\leq}, \ldots, x \in Cl_{tm}^{\leq}$, then x is assigned to class Cl_t, where $t = \min\{t_1, t_2, \ldots, t_m\}$. In the case of no matching, we conclude that x belongs to the best class Cl_n because no rule with a RHS suggesting a worse classification of x covers this object.

Finally, when applying $D_{\geq\leq}$-decision rules to object x, we conclude that x belongs to the union of all classes suggested in the RHSs of rules covering x.

A set of decision rules is *complete* if it is able to cover all objects from the data table in such a way that consistent objects are re-classified to their original classes, and inconsistent objects are classified to clusters of classes referring to this inconsistency. We call each set of decision rules that is complete and non-redundant *minimal*, i.e., exclusion of any rule from this set makes it non-complete.

In the case of VC-DRSA, the decision rules are induced from P-lower ap-approximations, whose composition is controlled by user-specified consistency level l. In consequence, the value of confidence α for the rule should be constrained from below. It is reasonable to require that the smallest accepted confidence of the rule should not be lower than the currently used consistency level l. Indeed, in the worst case, some objects from the P-lower approximation may create a rule using all criteria from P, thus giving a confidence $\alpha \geq l$.

One of three induction strategies can be adopted to obtain a set of decision rules [20.65]:

- generation of a *minimal* description, i.e., a minimal set of rules,
- generation of an *exhaustive* description, i.e., all rules for a given data table,
- generation of a *characteristic* description, i.e., a set of rules covering relatively many objects, however, not necessarily all objects, from U.

Note that the syntax of decision rules induced from rough approximations, defined using dominance cones, consistently use this type of granule. Each condition profile defines a dominance cone in X_C, and each decision profile defines a dominance cone in X_D. In both cases, the cone is positive for $D_{\geq}$-rules and negative for $D_{\leq}$-rules.

Let also remark that dominance cones corresponding to condition profiles can originate in any point of X_C, without the risk of their being too specific. Thus, contrary to traditional granular computing, the condition attribute space X_C need not be discretized.

Procedures for rule induction from rough approximations have been proposed in [20.29].

In [20.13], a new methodology for the induction of monotonic decision trees from dominance-based rough approximations of preference-ordered decision classes has been proposed.

20.3.3 Illustrative Example

To illustrate the application of DRSA to multicriteria classification, we will use a part of data provided by a Greek industrial bank, ETEVA, which finances industrial and commercial firms in Greece [20.64]. A sample composed of 39 firms has been chosen for the study in cooperation with ETEVA's financial manager. The manager has classified the selected firms into three classes of the bankruptcy risk. The sorting decision is represented by a decision attribute d making a trichotomic partition of the 39 firms:

- $d = $ A means "acceptable",
- $d = $ U means "uncertain",
- $d = $ NA means "non-acceptable".

The partition is denoted by $\boldsymbol{Cl} = \{Cl_A, Cl_U, Cl_{NA}\}$ and, obviously, class Cl_A is better than Cl_U, which is better than Cl_{NA}.

The firms were evaluated using the following twelve criteria ($\uparrow$ means preference increasing with value, and $\downarrow$ means preference decreasing with value):

- $A_1 = $ earnings before interests and taxes/total assets, $\uparrow$
- $A_2 = $ net income/net worth, $\uparrow$
- $A_3 = $ total liabilities/total assets, $\downarrow$
- $A_4 = $ total liabilities/cash flow, $\downarrow$
- $A_5 = $ interest expenses/sales, $\downarrow$
- $A_6 = $ general and administrative expense/sales, $\downarrow$
- $A_7 = $ managers' work experience, $\uparrow$
  (very low = 1, low = 2, medium = 3, high = 4, very high = 5),
- $A_8 = $ firm's market niche/position, $\uparrow$
  (bad = 1, rather bad = 2, medium = 3, good = 4, very good = 5),
- $A_9 = $ technical structure facilities, $\uparrow$
  (bad = 1, rather bad = 2, medium = 3, good = 4, very good = 5),
- $A_{10} = $ organization personnel, $\uparrow$
  (bad = 1, rather bad = 2, medium = 3, good = 4, very good = 5),
- $A_{11} = $ special competitive advantage of firms, $\uparrow$
  (low = 1, medium = 2, high = 3, very high = 4),
- $A_{12} = $ market flexibility, $\uparrow$
  (very low = 1, low = 2, medium = 3, high = 4, very high = 5).

The first six criteria are continuous (financial ratios) and the last six are ordinal. The data matrix is presented in Table 20.1. The main questions to be answered by the knowledge discovery process were the following:

- Is the information contained in Table 20.1 consistent?
- What are the reducts of criteria, ensuring the same quality of approximation of the multicriteria classification as the whole set of criteria?
- What decision rules can be extracted from Table 20.1?
- What are the minimal sets of decision rules?

We will answer these questions using the DRSA.

The first result of the DRSA is a discovery that the financial data matrix is consistent for the complete set of criteria C. Therefore, the C-lower approximation and the C-upper approximation of $Cl^{\leq}_{NA}, Cl^{\leq}_U, Cl^{\geq}_U, and\ Cl^{\geq}_A$ are the same. In other words, the quality of approximation of all upward and

downward unions of classes, as well as the quality of classification, is equal
to 1.

Table 20.1. Financial data matrix

Firm	A_1	A_2	A_3	A_4	A_5	A_6	A_7	A_8	A_9	A_{10}	A_{11}	A_{12}	d
F1	16.4	14.5	59.82	2.5	7.5	5.2	5	3	5	4	2	4	A
F2	35.8	67.0	64.92	1.7	2.1	4.5	5	4	5	5	4	5	A
F3	20.6	61.75	75.71	3.6	3.6	8.0	5	3	5	5	3	5	A
F4	11.5	17.1	57.1	3.8	4.2	3.7	5	2	5	4	3	4	A
F5	22.4	25.1	49.8	2.1	5.0	7.9	5	3	5	5	3	5	A
F6	23.9	34.5	48.9	1.7	2.5	8.0	5	3	4	4	3	4	A
F7	29.9	44.0	57.8	1.8	1.7	2.5	5	4	4	5	3	5	A
F8	8.7	5.4	27.4	3.3	4.5	4.5	5	2	4	4	1	4	A
F9	25.7	29.7	46.8	1.7	4.6	3.7	4	2	4	3	1	3	A
F10	21.2	24.6	64.8	3.7	3.6	8.0	4	2	4	4	1	4	A
F11	18.32	31.6	69.3	4.4	2.8	3.0	4	3	4	4	3	4	A
F12	20.7	19.3	19.7	0.7	2.2	4.0	4	2	4	4	1	3	A
F13	9.9	3.5	53.1	4.5	8.5	5.3	4	2	4	4	1	4	A
F14	10.4	9.3	80.9	9.4	1.4	4.1	4	2	4	4	3	3	A
F15	17.7	19.8	52.8	3.2	7.9	6.1	4	4	4	4	2	5	A
F16	14.8	15.9	27.94	1.3	5.4	1.8	4	2	4	3	2	3	A
F17	16.0	14.7	53.5	3.9	6.8	3.8	4	4	4	4	2	4	A
F18	11.7	10.01	42.1	3.9	12.2	4.3	5	2	4	2	1	3	A
F19	11.0	4.2	60.8	5.8	6.2	4.8	4	2	4	4	2	4	A
F20	15.5	8.5	56.2	6.5	5.5	1.8	4	2	4	4	2	4	A
F21	13.2	9.1	74.1	11.21	6.4	5.0	2	2	4	4	2	3	U
F22	9.1	4.1	44.8	4.2	3.3	10.4	3	4	4	4	3	4	U
F23	12.9	1.9	65.02	6.9	14.01	7.5	4	3	3	2	1	2	U
F24	5.9	-27.7	77.4	-32.2	16.6	12.7	3	2	4	4	2	3	U
F25	16.9	12.4	60.1	5.2	5.6	5.6	3	2	4	4	2	3	U
F26	16.7	13.1	73.5	7.1	11.9	4.1	2	2	4	4	2	3	U
F27	14.6	9.7	59.5	5.8	6.7	5.6	2	2	4	4	2	4	U
F28	5.1	4.9	28.9	4.3	2.5	46.0	2	2	3	3	1	2	U
F29	24.4	22.3	32.8	1.4	3.3	5.0	2	3	4	4	2	3	U
F30	29.7	8.6	41.8	1.6	5.2	6.4	2	3	4	4	2	3	U
F31	7.3	-64.5	67.5	-2.2	30.1	8.7	3	3	4	4	2	3	NA
F32	23.7	31.9	63.6	3.5	12.1	10.2	3	2	3	4	1	3	NA
F33	18.9	13.5	74.5	10.0	12.0	8.4	3	3	3	4	3	4	NA
F34	13.9	3.3	78.7	25.5	14.7	10.1	2	2	3	4	3	4	NA
F35	-13.3	-31.1	63.0	-10.0	21.2	23.1	2	1	4	3	1	2	NA
F36	6.2	-3.2	46.1	5.1	4.8	10.5	2	1	3	3	2	3	NA
F37	4.8	-3.3	71.9	34.6	8.6	11.6	2	2	4	4	2	3	NA
F38	0.1	-9.6	42.5	-20.0	12.9	12.4	1	1	4	3	1	3	NA
F39	13.6	9.1	76.0	11.4	17.1	10.3	1	1	2	1	1	2	NA

The second discovery is a set of 18 reducts of criteria ensuring the same
quality of classification as the whole set of 12 criteria:

$$RED^1_{Cl} = \{A_1, A_4, A_5, A_7\}, \quad RED^2_{Cl} = \{A_2, A_4, A_5, A_7\},$$
$$RED^3_{Cl} = \{A_3, A_4, A_6, A_7\}, \quad RED^4_{Cl} = \{A_4, A_5, A_6, A_7\},$$
$$RED^5_{Cl} = \{A_4, A_5, A_7, A_8\}, \quad RED^6_{Cl} = \{A_2, A_3, A_7, A_9\},$$
$$RED^7_{Cl} = \{A_1, A_3, A_4, A_7, A_9\}, \quad RED^8_{Cl} = \{A_1, A_5, A_7, A_9\},$$
$$RED^9_{Cl} = \{A_2, A_5, A_7, A_9\}, \quad RED^{10}_{Cl} = \{A_4, A_5, A_7, A_9\},$$
$$RED^{11}_{Cl} = \{A_5, A_6, A_7, A_9\}, \quad RED^{12}_{Cl} = \{A_4, A_5, A_7, A_{10}\},$$
$$RED^{13}_{Cl} = \{A_1, A_3, A_4, A_7, A_{11}\}, RED^{14}_{Cl} = \{A_2, A_3, A_4, A_7, A_{11}\},$$
$$RED^{15}_{Cl} = \{A_4, A_5, A_6, A_{12}\}, \quad RED^{16}_{Cl} = \{A_1, A_3, A_5, A_6, A_9, A_{12}\},$$
$$RED^{17}_{Cl} = \{A_3, A_4, A_6, A_{11}A_{12}\}, RED^{18}_{Cl} = \{A_1, A_2, A_3, A_6, A_9, A_{11},$$
$$A_{12}\}.$$

All above subsets of criteria are equally good, and sufficient for perfect approximation of the classification performed by ETEVA's financial manager on the 39 firms. The core of $\textbf{\textit{Cl}}$ is empty ($CORE_{Cl} = \emptyset$), which means that no criterion is indispensable for the approximation. Moreover, all the criteria are exchangeable and no criterion is redundant.

The third discovery is the set of all decision rules. We obtained 74 rules describing $Cl^{\leq}_{NA}$, 51 rules describing $Cl^{\leq}_{U}$, 75 rules describing $Cl^{\geq}_{U}$, and 79 rules describing $Cl^{\geq}_{A}$.

The fourth discovery is the finding of minimal sets of decision rules. Several minimal sets were found; one of them is shown below (in parentheses there is the number of objects supporting the corresponding rule):

1. *if $f(x, A_3) \geq 67.5$ and $f(x, A_4) \geq -2.2$ and $f(x, A_6) \geq 8.7$, then $x \in Cl^{\leq}_{NA}$,* (4),

2. *if $f(x, A_2) \geq 3.3$ and $f(x, A_7) \leq 2$, then $x \in Cl^{\leq}_{NA}$,* (5),

3. *if $f(x, A_3) \geq 63.6$ and $f(x, A_7) \leq 3$ and $f(x, A_9) \leq 3$, then $x \in Cl^{\leq}_{NA}$,* (4),

4. *if $f(x, A_2) \leq 12.4$ and $f(x, A_6) \leq 5.6$, then $x \in Cl^{\leq}_{U}$,* (14),

5. *if $f(x, A_7) \leq 3$, then $x \in Cl^{\leq}_{U}$,* (18),

6. *if $f(x, A_2) \geq 3.5$ and $f(x, A_5) \leq 8.5$, then $x \in Cl^{\geq}_{U}$,* (26),

7. *if $f(x, A_7) \geq 4$, then $x \in Cl^{\geq}_{U}$,* (21),

8. *if $f(x, A_1) \geq 8.7$ and $f(x, A_9) \geq 4$, then $x \in Cl^{\geq}_{U}$,* (27),

9. *if $f(x, A_2) \geq 3.5$ and $f(x, A_7) \geq 4$, then $x \in Cl^{\geq}_{A}$,* (20).

As the minimal set of rules is complete, and composed of $D_{\geq}$-decision rules and $D_{\leq}$-decision rules only, the application of these rules to the 39 firms will result in their exact re-classification to classes of risk.

Minimal sets of decision rules represent the most concise and non-redundant knowledge representations. The above minimal set of nine decision rules uses eight criteria and 18 elementary conditions, i.e., 3.85% of descriptors from the data matrix.

The well-known machine discovery methods cannot deal with multicriteria classification because they do not consider preference orders in the domains of attributes and among the classes. Within multicriteria decision analysis there exist methods for multicriteria classification; however, they do not discover classification patterns from data; they simply apply a preference model, like the utility function in scoring methods (see, e.g., [20.68]), to a set of objects to be classified. In this sense, they are not knowledge discovery methods.

Comparing DRSA to Classical Rough Set Approach (CRSA), one can notice the following differences between the two approaches. CRSA extracts knowledge about a partition of U into classes which are not preference ordered; the granules used for knowledge representation are sets of objects indiscernible by a set of condition attributes.

In case of DRSA and multicriteria classification, the condition attributes are criteria, and classes are preference ordered. The extracted knowledge concerns a collection of upward and downward unions of classes, and the granules used for knowledge representation are sets of objects defined using a dominance relation. This is the main difference between CRSA and DRSA.

There are three remarkable advantages of DRSA over CRSA. The first one is the ability of handling criteria, preference-ordered classes, and inconsistencies in the set of decision examples that CRSA is not able to discover, inconsistencies in the sense of violation of the dominance principle. In consequence, the rough approximations separate the certain part of information from the doubtful one, which is taken into account in rule induction. The second advantage is the analysis of a data matrix without any preprocessing of data; in particular, without any discretization of continuous attributes. The third advantage of DRSA lies in its richer syntax of decision rules induced from rough approximations. The elementary conditions (criterion *rel.* value) of decision rules resulting from DRSA use $rel. \in \{\leq, =, \geq\}$, while those resulting from CRSA use $rel. \in \{ = \}$. The DRSA syntax is more understandable to practitioners and makes the representation of knowledge more synthetic, since its minimal sets of decision rules are smaller than those resulting from CRSA.

20.4 DRSA for Multicriteria Choice and Ranking

One of the first extensions of the DRSA concerned preference-ordered data representing pairwise comparisons (i.e., binary relations) between objects on both, condition and decision attributes [20.19, 20.20, 20.25, 20.28]. Note that, while classification is based on the absolute evaluation of objects, choice and ranking refer to pairwise comparisons of objects. In this case, the patterns (i.e., decision rules) to be discovered from data characterize a comprehensive binary relation in the set of objects. If the comprehensive binary relation is a preference relation, and among the condition attributes there are some semantically correlated criteria, then the data set, serving as a learning sample, can be considered as preferential information of a decision maker (DM) in a multicriteria choice or ranking problem; in consequence, the comprehensive preference relation characterized by decision rules discovered from this data can be considered as a *preference model* of the DM. It may be used to explain a decision policy of the DM and to recommend a good choice or preference ranking with respect to new objects.

Let us consider a finite set A of objects evaluated by a finite set of criteria C. The best choice for the preference ranking in set A is semantically correlated with criteria from set C. The preferential information concerning the multicriteria choice or ranking problem is a data set in the form of a pairwise comparison table (PCT) including pairs of some *reference objects* from a subset $E \subseteq A$, described by preference relations on particular criteria and by a comprehensive preference relation, (for example, a weak preference relation called *outranking relation*). Using the DRSA for the analysis of the PCT, one obtains a rough approximation of the outranking relation by a dominance relation. Decision rules induced from the rough approximation are then applied to the complete set A of objects concerned with the choice or ranking. As a result, one obtains a four-valued outranking relation on this set. In order to obtain a recommendation, it is advisable to use an exploitation procedure based on the net flow score of the objects. Below, we present in greater detail this methodology.

20.4.1 Pairwise Comparison Table (PCT) as a Preferential Information and a Learning Sample

For a set of reference objects that are representative of the decision problem, the DM expresses the preferences with pairwise comparisons. In the following, xSy will denote the presence, while $xS^c y$ will denote the absence of the outranking relation for a pair of objects $(x,y) \in A \times A$.

Let E be a set of reference objects. For each pair $(x,y) \in E \times E$, the DM is asked to select one of the three possibilities:

1. object x is as good as y, i.e., xSy
2. object x is worse then y, i.e., $xS^c y$
3. DM won't compare those objects.

An $m \times (n+1)$ PCT, is then created on the basis of this information. Its first n columns correspond to criteria from set C. The last, $(n+1)$-th, column of PCT represents the binary relation S or S^c. The m rows correspond to pairs of objects being compared by the DM. For each pair in PCT, a difference in criteria values is noted in the corresponding column. If the DM refused to compare two objects, such a pair does not appear in PCT. Let us define PCT more formally.

For any criterion $g_i \in C$, let T_i be a finite set of binary relations defined on A on the basis of the evaluations of objects from A, with respect to the considered criterion g_i, such that, for every $(x,y) \in A \times A$, exactly one binary relation $t \in T_i$ is verified. More precisely, given the domain V_i of $g_i \in C$, if $v_i', v_i'' \in V_i$ are the respective evaluations of $x,y \in A$ by means of g_i and $(x,y) \in t$, with $t \in T_i$, then, for each $w,z \in A$ having the same evaluations v_i', v_i'' by means of g_i, $(w,z) \in t$. Furthermore, let T_d be a set of binary relations defined on set A (comprehensive pairwise comparisons) such that at most one binary relation $t \in T_d$ is verified for every $(x,y) \in A \times A$. The *pairwise comparison table* (PCT) is defined as data table $S_{PCT} = \langle B, C \cup \{d\}, T_C \cup T_d, f \rangle$, where $B \subseteq E \times E$ is a non-empty *set of exemplary pairwise comparisons of reference objects* $T_C = \bigcup_{g_i \in C} T_i$, d is a decision corresponding to the comprehensive pairwise comparison (comprehensive preference relation), and $f: B \times (C \cup \{d\}) \to T_C \cup T_d$ is a total function such that $f[(x,y),g_i] \in T_i$ for every $(x,y) \in A \times A$ and for each $g_i \in C$, and $f[(x,y),d] \in T_d$ for every $(x,y) \in B$. It follows that for any pair $(x,y) \in B$, there is verified one, and only one, binary relation $t \in T_d$. Thus, T_d induces a partition of B. In fact, data table S_{PCT} can be seen as a decision table, since the set of considered criteria C and decision d are different.

We assume that the exemplary pairwise comparisons made by the DM can be represented in terms of *graded preference relations* (for example, "very weak preference", "weak preference", "strict preference", "strong preference", "very strong preference") P_i^h: for each $g_i \in C$ and for every $(x,y) \in A \times A$,

$$T_i = \{P_i^h, h \in H_i\},$$

where H_i is a particular subset of the relative integers and

- $x P_i^h y$, $h > 0$, means that object x is preferred to object y by degree h with respect to the criterion g_i,
- $x P_i^h y$, $h < 0$, means that object x is not preferred to object y by degree h with respect to the criterion g_i,
- $x P_i^0 y$ means that object x is similar (asymmetrically indifferent) to object y with respect to the criterion g_i.

Within the preference context, the similarity relation P_i^0, even if not symmetrical, resembles an indifference relation. Thus, in this case, we call this similarity relation "asymmetric indifference". Of course, for each $g_i \in C$ and for every $(x,y) \in A \times A$,

$$[x P_i^h y, h > 0] \Rightarrow [y P_i^k x, k \leq 0], \quad [x P_i^h y, h < 0] \Rightarrow [y P_i^k x, k \geq 0].$$

The set of binary relations T_d may be defined in a similar way, but $x P_d^h y$ means that object x is comprehensively preferred to object y by degree h. We are considering a PCT where the set T_d is composed of two binary relations defined on A:

- x outranks y (denoted by xSy or $(x, y) \in S$), where $(x, y) \in B$,
- x does not outrank y (denoted by $xS^c y$ or $(x,y) \in S^c$), where $(x,y) \in B$ and $S \cup S^c = B$.

Observe that the binary relation S is reflexive, but not necessarily transitive or complete.

20.4.2 Rough Approximation of Outranking and Non-outranking Relations Specified in PCT

In this subsection we will distinguish between two types of evaluation scales of criteria: *cardinal* and *ordinal*. Let C^N be the set of criteria expressing preferences on a cardinal scale and C^O be the set of criteria expressing preferences on an ordinal scale such that $C^N \cup C^O = C$ and $C^N \cap C^O = \emptyset$. Moreover, for each $P \subseteq C$, we denote by P^O the subset of P composed of criteria expressing preferences on an ordinal scale, i.e., $P^O = P \cap C^O$, and denote by P^N the subset of P composed of criteria expressing preferences on a cardinal scale, i.e., $P^N = P \cap C^N$. Of course, for each $P \subseteq C$, we have $P = P^N \cup P^O$ and $P^N \cap P^O = \emptyset$.

The meaning of the two scales is such that in the case of the cardinal scale one can specify intensity of preference for a given difference of evaluations, while in the case of ordinal scale this is not possible, and one can only establish an order of evaluations.

Multi-graded Dominance. Let $P = P^N$ and $P^O = \emptyset$. Given $P \subseteq C$ $(P \neq \emptyset)$, (x,y), $(w,z) \in A \times A$, the pair of objects (x,y) is said to dominate (w,z), taking into account the criteria from P (denoted by $(x,y)D_P(w,z)$), if x is preferred to y at least as strongly as w is preferred to z with respect to each $g_i \in P$. Precisely, "at least as strongly as" means "by at least the same degree", i.e., $h_i \geq k_i$, where $h_i, k_i \in H_i$, $xP_i^{h_i}y$ and $wP_i^{k_i}z$, for each $g_i \in P$.

Let $D_{\{i\}}$ be the dominance relation confined to the single criterion $g_i \in P$. The binary relation $D_{\{i\}}$ is reflexive $((x, y)D_{\{i\}}(x, y)$, for every $(x, y) \in A \times A)$, transitive $((x, y)D_{\{i\}}(w, z)$ and $(w, z)D_{\{i\}}(u, v)$ imply $(x, y)D_{\{i\}}(u, v)$, for every $(x, y), (w, z), (u, v) \in A \times A)$, and complete $((x, y)D_{\{i\}}(w, z)$ and/or $(w, z)D_{\{i\}}(x, y)$, for all $(x, y), (w, z) \in A \times A)$. Therefore, $D_{\{i\}}$ is a complete preorder on $A \times A$. Since the intersection of complete preorders is a partial preorder, and $D_P = \bigcap_{g_i \in P} D_{\{i\}}, P \subseteq C$, the dominance relation D_P is a partial preorder on $A \times A$. Let $R \subseteq P \subseteq C$ and $(x,y),(u,v) \in A \times A$; then, the following implication holds:

$$(x, y)D_P(u, v) \Rightarrow (x, y)D_R(u, v).$$

Given $P \subseteq C$ and $(x,y) \in B$, we define

- a set of pairs of objects dominating (x, y), called P-*dominating set*,
  $D_P^+(x, y) = \{(w,z) \in B : (w,z) D_P(x,y)\}$,
- a set of pairs of objects dominated by (x,y), called P-*dominated set*,
  $D_P^-(x, y) = \{(w,z) \in B : (x,y) D_P(w,z)\}$.

The P-dominating sets and the P-dominated sets defined on B for all pairs of reference objects from B are "granules of knowledge" that can be used to express P-lower and P-upper approximations of comprehensive outranking relations S and S^c, respectively:

$$\underline{P}(S) = \{(x, y) \in B : D_P^+(x, y) \subseteq S\},$$

$$\overline{P}(S) = \bigcup_{(x,y) \in S} D_P^+(x, y).$$

$$\underline{P}(S^c) = \{(x, y) \in B : D_P^-(x, y) \subseteq S^c\},$$

$$\overline{P}(S^c) = \bigcup_{(x,y) \in S^c} D_P^-(x, y).$$

It has been proved in [20.19] that

$$\underline{P}(S) \subseteq S \subseteq \overline{P}(S), \quad \underline{P}(S^c) \subseteq S^c \subseteq \overline{P}(S^c).$$

Furthermore, the following complementarity properties hold:

$$\underline{P}(S) = B - \overline{P}(S^c), \quad \overline{P}(S) = B - \underline{P}(S^c),$$

$$\underline{P}(S^c) = B - \overline{P}(S), \quad \overline{P}(S^c) = B - \underline{P}(S).$$

The P-boundaries (P-doubtful regions) of S and S^c are defined as

$$Bn_P(S) = \overline{P}(S) - \underline{P}(S), \quad Bn_P(S^c) = \overline{P}(S^c) - \underline{P}(S^c).$$

From the above equations it follows that $Bn_P(S) = Bn_P(S^c)$. The concepts of quality of approximation, reducts, and core can also be extended to the approximation of the outranking relation by multi-graded dominance relations.

In particular, the coefficient

$$\gamma_P = \frac{card(\underline{P}(S) \cup \underline{P}(S^c))}{card(B)}$$

defines the *quality of approximation of S and S^c by $P \subseteq C$*. It expresses the ratio of all pairs of objects $(x,y) \in B$, correctly assigned to S and S^c by the set P of criteria, to all the pairs of objects contained in B. Each minimal subset $P \subseteq C$, such that $\gamma_P = \gamma_C$, is called a *reduct* of C (denoted by $RED_{S_{PCT}}$). Let us note that S_{PCT} can have more than one reduct. The intersection of all B-reducts is called the *core* (denoted by $CORE_{S_{PCT}}$).

It is also possible to use the Variable Consistency Model on S_{PCT} [20.60], allowing some of the pairs in positive or negative dominance sets to belong to the opposite relation, but at least l percent of pairs to belong to the correct one. Then the definition of lower approximations of S and S^c boils down to

$$\underline{P}(S) \ = \{(x,y) \in B : \frac{card(D_P^+(x,y) \cap S)}{card(D_P^+(x,y))} \geq l\},$$
$$\underline{P}(S^c) = \{(x,y) \in B : \frac{card(D_P^-(x,y) \cap S^c)}{card(D_P^-(x,y))} \geq l\}.$$

Dominance Without Degrees of Preference. The degree of graded preference considered above is defined on a cardinal scale of the strength of preference; however, in many real world problems, the existence of such a quantitative scale is rather questionable. This applies to ordinal scales of criteria. In this case, the dominance relation is defined directly on evaluations $g_i(x)$ for all objects $x \in A$. Let us explain this in more detail.

Let $P = P^O$ and $P^N = \emptyset$; then, given $(x,y),(w,z) \in A \times A$, the pair (x,y) is said to dominate the pair (w,z) with respect to P if, for each $g_i \in P$, $g_i(x) \geq g_i(w)$ and $g_i(z) \geq g_i(y)$.

Let D_i be the dominance relation confined to the single criterion $g_i \in P^O$. The binary relation D_i is reflexive, transitive, but non-complete (it is possible that $not\ (x,y)D_i(w,z)$ and $not\ (w,z)D_i(x,y)$ for some $(x,y),(w,z) \in A \times A$). Therefore, D_i is a partial preorder. Since the intersection of partial preorders is also a partial preorder, and $D_P = \bigcap_{g_i \in P} D_{\{i\}}$, $P = P^O$, the dominance relation D_P is also a partial preorder.

If some criteria from $P \subseteq C$ express preferences on a quantitative or a numerical non-quantitative scale, and others express them on an ordinal scale, i.e., if $P^N \neq \emptyset$ and $P^O \neq \emptyset$, then, given $(x,y),(w,z) \in A \times A$, the pair (x,y) is said to dominate the pair (w,z) with respect to criteria from P if (x,y) dominates (w,z) with respect to both P^N and P^O. Since the dominance relation with respect to P^N is a partial preorder on $A \times A$ (because it is a multi-graded dominance) and the dominance with respect to P^O is also a partial preorder on $A \times A$ (as explained above), then the dominance D_P, being the intersection of these two dominance relations, is also a partial preorder. In consequence, all the concepts introduced in the previous point can be restored using this specific definition of dominance relation.

20.4.3 Induction of Decision Rules from Rough Approximations of Outranking and Non-outranking Relations

Using the rough approximations of S and S^c defined in Sect. 20.4.2, it is possible to induce a generalized description of the preferential information contained in a given S_{PCT} in terms of suitable decision rules. The syntax of these rules is based on the concept of *upward cumulated preferences* (denoted by $P_i^{\geq h}$) and *downward cumulated preferences* (denoted by $P_i^{\leq h}$), having the following interpretations:

- $xP_i^{\geq h}y$ means "x is preferred to y with respect to g_i by at least degree h",
- $xP_i^{\leq h}y$ means "x is preferred to y with respect to g_i by at most degree h".

The exact definitions of the cumulated preferences, for each $(x,y) \in A \times A$, $g_i \in C$ and $h \in H_i$, are as follows:

- $xP_i^{\geq h}y$ if $xP_i^k y$, where $k \in H_i$ and $k \geq h$,
- $xP_i^{\leq h}y$ if $xP_i^k y$, where $k \in H_i$ and $k \leq h$.

Let $G_i = \{g_i(x), \; x \in A\}$, $g_i \in C^O$. The decision rules then have the following syntax:

1) **$D_{\geq}$-decision rules**:

$$\text{if } xP_{i1}^{\geq h(i1)}y \text{ and } \ldots \; xP_{ie}^{\geq h(ie)}y \text{ and } g_{ie+1}(x) \geq r_{ie+1} \text{ and}$$
$$g_{ie+1}(y) \leq s_{ie+1} \text{ and } \ldots g_{ip}(x) \geq r_{ip} \text{ and } g_{ip}(y) \leq s_{ip}, \text{ then } xSy,$$

where $P = \{g_{i1}, \ldots, g_{ip}\} \subseteq C$, $P^N = \{g_{i1}, \ldots, g_{ie}\}$, $P^O = \{g_{ie+1}, \ldots, g_{ip}\}$, $(h(i1), \ldots, h(ie)) \in H_{i1} \times \ldots \times H_{ie}$, and $(r_{ie+1}, \ldots, r_{ip})$, $(s_{ie+1}, \ldots, s_{ip}) \in G_{ie+1} \times \ldots \times G_{ip}$; these rules are supported by pairs of objects from the P-lower approximation of S alone;

2) **$D_{\leq}$-decision rules**:

$$\text{if } xP_{i1}^{\leq h(i1)}y \text{ and } \ldots xP_{ip}^{\leq h(ip)}y \text{ and } g_{ie+1}(x) \leq r_{ie+1} \text{ and } g_{ie+1}(y) \geq s_{ie+1}$$
$$\text{and } \ldots g_{ip}(x) \leq r_{ip} \text{ and } g_{ip}(y) \geq s_{ip}, \text{ then } xS^c y,$$

where $P = \{g_{i1}, \ldots, g_{ip}\} \subseteq C$, $P^N = \{g_{i1}, \ldots, g_{ie}\}$, $P^O = \{g_{ie+1}, \ldots, g_{ip}\}$, $(h(i1), \ldots, h(ie)) \in H_{i1} \times \ldots \times H_{ie}$, and $(r_{ie+1}, \ldots, r_{ip})$, $(s_{ie+1}, \ldots, s_{ip}) \in G_{ie+1} \times \ldots \times G_{ip}$; these rules are supported by pairs of objects from the P-lower approximation of S^c alone;

3) **$D_{\geq \leq}$-decision rules**:

$$\text{if } xP_{i1}^{\geq h(i1)}y \text{ and } \ldots xP_{ie}^{\geq h(ie)}y \text{ and } xP_{ie+1}^{\leq h(ie+1)}y \ldots xP_{if}^{\leq h(if)}y \text{ and}$$
$$g_{if+1}(x) \geq r_{if+1} \text{ and } g_{if+1}(y) \leq s_{if+1} \text{ and } \ldots g_{ig}(x) \geq r_{ig} \text{ and}$$
$$g_{ig}(y) \leq s_{ig} \text{ and } g_{ig+1}(x) \leq r_{ig+1} \text{ and } g_{ig+1}(y) \geq s_{ig+1} \text{ and}$$
$$\ldots g_{ip}(x) \leq r_{ip} \text{ and } g_{ip}(y) \geq s_{ip}, \text{ then } xSy \text{ or } xS^c y,$$

where $O' = \{g_{i1}, \ldots, g_{ie}\} \subseteq C$, $O'' = \{g_{ie+1}, \ldots, g_{if}\} \subseteq C$, $P^N = O' \cup O''$, O' and O'' not necessarily disjoint, $P^O = \{g_{if+1}, \ldots, g_{ip}\}$, $(h(i1), \ldots, h(if)) \in H_{i1} \times \ldots \times H_{if}$, and $(r_{if+1}, \ldots, r_{ip})$, $(s_{if+1}, \ldots, s_{ip}) \in G_{if+1} \times \ldots \times G_{ip}$; these rules are supported by pairs of objects from the P-boundary of S and S^c only.

20.4.4 Use of Decision Rules for Decision Support

The decision rules, induced from a given S_{PCT}, describe the comprehensive preference relations S and S^c either exactly (D$_\geq$- and D$_\leq$-decision rules) or approximately (D$_{\geq\leq}$-decision rules). A set of these rules covering all pairs of S_{PCT} represents a preference model of the DM which gives the pairwise comparison of reference objects. Application of these decision rules on a new subset $M \subseteq A$ of objects induces a specific preference structure on M.

In fact, any pair of objects $(u,v) \in M \times M$ can match the decision rules in one of four ways:

- at least one D$_\geq$-decision rule and neither D$_\leq$- nor D$_{\geq\leq}$-decision rules,
- at least one D$_\leq$ -decision rule and neither D$_\geq$- nor D$_{\geq\leq}$-decision rules,
- at least one D$_\geq$-decision rule and at least one D$_\leq$-decision rule, or at least one D$_{\geq\leq}$-decision rule,
- no decision rule.

These four ways correspond, respectively, to the following four situations of outranking:

- uSv and *not* uS^cv, that is *true* outranking (denoted by uS^Tv),
- uS^cv and *not* uSv, that is *false* outranking (denoted by uS^Fv),
- uSv and uS^cv, that is *contradictory* outranking (denoted by uS^Kv),
- *not* uSv and *not* uS^cv, that is *unknown* outranking (denoted by uS^Uv).

The four situations above, which together constitute the so-called *four-valued outranking* [20.37], have been introduced to underline the presence and absence of *positive* and *negative* reasons for the outranking. Moreover, they make it possible to distinguish contradictory situations from unknown ones.

A final *recommendation* (choice or ranking) can be obtained upon a suitable exploitation of this structure, i.e., of the presence and the absence of outranking S and S^c on M. A possible exploitation procedure consists of calculating a specific score, called Net Flow Score, for each object $x \in M$:

$$S_{nf}(x) = S^{++}(x) - S^{+-}(x) + S^{-+}(x) - S^{--}(x),$$

where

$$S^{++}(x) =$$
card($\{y \in M$: there is at least one decision rule which affirms $xSy\}$),

$$S^{+-}(x) =$$
card($\{y \in M$: there is at least one decision rule which affirms $ySx\}$),

$$S^{-+}(x) =$$
card($\{y \in M$: there is at least one decision rule which affirms $yS^cx\}$),

Table 20.2. Decision table with reference objects

Warehouse	A_1	A_2	A_3	d $(ROE \%)$
1	good	medium	good	10.35
2	good	sufficient	good	4.58
3	medium	medium	good	5.15
4	sufficient	medium	medium	-5
5	sufficient	medium	medium	2.42
6	sufficient	sufficient	good	2.98
7	good	medium	good	15
8	good	sufficient	good	-1.55

$$S^{--}(x) =$$

$\mathrm{card}(\{y \in M : \text{there is at least one decision rule which affirms } xS^c y\}).$

The recommendation in ranking problems consists of the total preorder determined by $S_{nf}(x)$ on M; in choice problems, it consists of the object(s) $x^* \in M$ such that $S_{nf}(x^*) = \max_{x \in M}\{S_{nf}(x)\}$.

The procedure described above has been recently characterized with reference to a number of desirable properties [20.37].

20.4.5 Illustrative Example

Let us suppose that a company managing a chain of warehouses wants to buy some new warehouses. To choose the best proposals or to rank them all, the managers of the company decide to analyze first the characteristics of eight warehouses already owned by the company (reference objects). This analysis should give some indication for the choice and ranking of the new proposals. Eight warehouses belonging to the company have been evaluated by the following three criteria: capacity of the sales staff (A_1), perceived quality of goods (A_2), and high traffic location (A_3). The domains (scales) of these attributes are presently composed of three preference-ordered echelons: $V_1 = V_2 = V_3 = \{\text{sufficient, medium, good}\}$. The decision attribute (d) indicates the profitability of warehouses, expressed by the *Return On Equity* (*ROE*) ratio (in %). Table 20.2 presents a decision table with the considered reference objects.

With respect to the set of criteria $C = C^N = \{A_1, A_2, A_3\}$, the following multi-graded preference relations P_i^h, $i = 1,2,3$, were defined:

- $xP_i^0 y$ (and $yP_i^0 x$), meaning that x is *indifferent* to y with respect to A_i if $f(x,A_i) = f(y,A_i)$.
- $xP_i^1 y$ (and $yP_i^{-1}x$), meaning that x is *preferred* to y with respect to A_i if $f(x,A_i) = $ good and $f(y,A_i) = $ medium, or if $f(x,A_i) = $ medium and $f(y,A_i) = $ sufficient,
- $xP_i^2 y$ (and $yP_i^{-2}x$), meaning that x is *strongly preferred* to y with respect to A_i if $f(x,A_i) = $ good and $f(y,A_i) = $ sufficient.

Using the decision attribute, the comprehensive outranking relation was built as follows: warehouse x is at least as good as warehouse y with respect to profitability (xSy) if

$$ROE(x) \geq ROE(y) - 2\% \ .$$

Otherwise, i.e., if $ROE(x) < ROE(y) - 2\%$, warehouse x is *not* at least as good as warehouse y with respect to profitability ($xS^c y$).

The pairwise comparisons of reference objects result in a PCT. Rough set analysis of the PCT leads to the conclusion that the set of decision examples on reference objects is inconsistent. The quality of approximation of S and S^c by all criteria from set C is equal to 0.44. Moreover, $RED_{S_{PCT}} = CORE_{S_{PCT}} = \{A_1, A_2, A_3\}$; this means that no criterion is superfluous.

The C-lower approximations and the C-upper approximations of S and S^c, obtained by means of multi-graded dominance relations, are as follows:
$\underline{C}(S) = \{(1,2),(1,4),(1,5),(1,6),(1,8),(3,2),(3,4),(3,5),(3,6),(3,8),(7,2),(7,4),$
    $(7,5),(7,6),(7,8)\}$,
$\underline{C}(S^c) = \{(2,1),(2,7),(4,1),(4,3),(4,7),(5,1),(5,3),(5,7),(6,1),(6,3),(6,7),(8,1),$
    $(8,7)\}$.

All the remaining 36 pairs of reference objects belong to the C-boundaries of S and S^c, i.e., $Bn_C(S) = Bn_C(S^c)$.

The following minimal $D_{\geq}$-decision rules and $D_{\leq}$-decision rules can be induced from lower approximations of S and S^c, respectively (within parentheses there are the pairs of objects supporting the corresponding rules):

if $xP_1^{\geq 1}y$ *and* $xP_2^{\geq 1}y$, *then* xSy $((1,6),(3,6),(7,6))$,
if $xP_2^{\geq 1}y$ *and* $xP_3^{\geq 0}y$, *then* xSy $((1,2),(1,6),(1,8),(3,2),(3,6),(3,8),$
                $(7,2),(7,6),(7,8))$,
if $xP_2^{\geq 0}y$ *and* $xP_3^{\geq 1}y$, *then* xSy $((1,4),(1,5),(3,4),(3,5),(7,4),(7,5))$,
if $xP_1^{\leq -1}y$ *and* $xP_2^{\leq -1}y$, *then* $xS^c y$ $((6,1),(6,3),(6,7))$,
if $xP_2^{\leq 0}y$ *and* $xP_3^{\leq -1}y$, *then* $xS^c y$ $((4,1),(4,3),(4,7),(5,1),(5,3),(5,7))$,
if $xP_1^{\leq 0}y$ *and* $xP_2^{\leq -1}y$ *and* $xP_3^{\leq 0}y$, *then* $xS^c y$ $((2,1),(2,7),(6,1),(6,3),$
                $(6,7),(8,1),(8,7))$.

Moreover, it was possible to induce five minimal $D_{\geq \leq}$-decision rules from the boundary of approximation of S and S^c:

if $xP_2^{\leq 0}y$ *and* $xP_2^{\geq 0}y$ *and* $xP_3^{\leq 0}y$ *and* $xP_3^{\geq 0}y$, *then* xSy *or* $xS^c y$
        $((1,1),(1,3),(1,7),(2,2),(2,6),(2,8),(3,1),(3,3),(3,7),(4,4),(4,5),$
        $(5,4),(5,5),(6,2),(6,6),(6,8),(7,1),(7,3),(7,7),(8,2),(8,6),(8,8))$,
if $xP_2^{\leq -1}y$ *and* $xP_3^{\geq 1}y$, *then* xSy *or* $xS^c y$ $((2,4),(2,5),(6,4),(6,5),(8,4),(8,5))$,
if $xP_2^{\geq 1}y$ *and* $xP_3^{\leq -1}y$, *then* xSy *or* $xS^c y$ $((4,2),(4,6),(4,8),(5,2),(5,6),(5,8))$,
if $xP_1^{\geq 1}y$ *and* $xP_2^{\leq 0}y$ *and* $xP_3^{\leq 0}y$, *then* xSy *or* $xS^c y$ $((1,3),(2,3),(2,6),(7,3),$
                $(8,3),(8,6))$,
if $xP_1^{\geq 1}y$ *and* $xP_2^{\leq -1}y$, *then* xSy *or* $xS^c y$ $((2,3),(2,4),(2,5),(8,3),(8,4),(8,5))$.

Table 20.3. Ranking of warehouses for sale by decision rules and the Net Flow Score procedure

Warehouse for sale	A_1	A_2	A_3	Net Flow Score	Ranking
1'	good	sufficient	medium	1	5
2'	sufficient	good	good	11	1
3'	sufficient	medium	sufficient	-8	8
4'	sufficient	good	sufficient	0	6
5'	sufficient	sufficient	medium	-4	7
6'	sufficient	good	good	11	1
7'	medium	sufficient	sufficient	-11	9
8'	medium	medium	medium	7	3
9'	medium	good	sufficient	4	4
10'	medium	sufficient	sufficient	-11	9

Using all above decision rules and the Net Flow Score exploitation procedure on ten other warehouses proposed for sale, the managers obtained the result presented in Table 20.3. The dominance-based rough set approach gives a clear recommendation:

- for the **choice problem** it suggests **selecting warehouses 2' and 6'**, having maximum score (9),
- for the **ranking problem** it suggests the **ranking** presented in the last column of Table 20.3, as follows:

$$(2',6') \rightarrow (8') \rightarrow (9') \rightarrow (1') \rightarrow (4') \rightarrow (5') \rightarrow (3') \rightarrow (7',10').$$

20.4.6 Summary

We presented briefly the contribution of the DRSA to multicriteria choice and ranking problems. Let us point out the main features of the described methodology:

- preference information necessary to deal with a multicriteria decision problem, in terms of exemplary decisions, is requested for the DM,
- the rough set analysis of preferential information supplies some useful elements of knowledge about the decision situation; these are the relevance of particular attributes and/or criteria, information about their interaction, minimal subsets of attributes or criteria (reducts) conveying important knowledge contained in the exemplary decisions, and the set of the non-reducible attributes or criteria (core),
- the preference model induced from the preferential information is expressed in a natural and comprehensible language of "*if* ..., *then* ..." decision rules; the decision rules concern pairs of actions and conclude either the presence or the absence of a comprehensive preference relation; conditions

for the presence are expressed in "at least" terms, and conditions for the absence are expressed in "at most" terms,

- the decision rules do not convert ordinal information into numeric information, but keep the ordinal character of input data due to the syntax proposed,
- heterogeneous information (qualitative and quantitative, ordered and nonordered) and scales of preference (ordinal, cardinal) can be processed within the DRSA, while classical methods consider only quantitative ordered evaluations, with rare exceptions,
- no prior discretization of the quantitative domains of criteria is necessary.

20.5 DRSA with Missing Values of Attributes and Criteria

In practical applications, the data table is often not complete because some data are missing. To deal with this case, we proposed in [20.21, 20.22] an extension of the rough set methodology to the analysis of incomplete data tables. The extension concerns both the classical rough set approach (CRSA), based on the use of indiscernibility relations, and the DRSA.

The relations of indiscernibility or dominance between two objects are considered as directional statements where a subject is compared to a referent object. We require that the referent object have no missing data. The two extended rough set approaches boil down to the original approach when there are no missing data. The rules induced from the newly defined rough approximations are either certain or approximate, depending on whether they are supported by consistent objects or not. The way of handling the missing values in the proposed approach seems faithful to available data because the decision rules are *robust*, in the sense of being grounded on at least one object existing in the data set and not based on hypothetical objects created by substituting some possible values for the missing ones.

20.5.1 Generalized Indiscernibility Relation

For any two objects $x, y \in U$, we consider a directional comparison of y to x; object y is called *subject* and object x is called *referent*. We say that subject y is *indiscernible* from referent x, with respect to condition attributes $P \subseteq C$ (denotation yI_Px) if, for every $q \in P$, the following conditions are met:
1) $f(x,q) \neq *$; and 2) $f(x,q) = f(y,q)$ or $f(y,q) = *$, where $*$ denotes a missing value.

This means that the referent object considered for indiscernibility with respect to P should have no missing values on attributes from set P. The binary relation I_P is not necessarily reflexive, and also not necessarily symmetric. However, I_P is transitive.

For each $P \subseteq C$ let us define a set of objects having no missing values on attributes from P:

$$U_P = \{x \in U : f(x,q) \neq * \text{ for each } q \in P\}.$$

For each $x \in U$ and $P \subseteq C$, let $I_P(x) = \{y \in U: yI_Px\}$ denote the class of objects indiscernible from x. Given $X \subseteq U$ and $P \subseteq C$, we define lower and upper approximations of X with respect to P as follows:

$$\underline{I}_P(X) = \{x \in U_P : I_P(x) \subseteq X\},$$

$$\overline{I}_P(X) = \{x \in U_P : I_P(x) \cap X \neq \emptyset\}.$$

Let $X_P = X \cap U_P$. The rough approximation defined as above satisfies the following properties:

- (*Rough inclusion*) For each $X \in U$ and $P \subseteq C$:
$\underline{I}_P(X) \subseteq X_P \subseteq \overline{I}_P(X)$.
- (*Complementarity*) For each $X \in U$ and $P \subseteq C$:
$\underline{I}_P(X) = U_P - \overline{I}_P(U - X)$.

Let us observe that a very useful property of lower approximation within classical rough set theory is that if an object $x \in U$ belongs to the lower approximation of X with respect to $P \subseteq C$, then x also belongs to the lower approximation of X with respect to $R \subseteq C$ when $P \subseteq R$ (this is a kind of monotonicity property). However, the above definition of $\underline{I}_P(X)$ does not satisfy this property of lower approximation, because it is possible that $f(x,q) \neq *$ for all $q \in P$, but $f(x,q) = *$ for some $q \in R - P$. This is quite problematic for some key concepts of the rough set theory, like quality of approximation, reduct, and core.

Therefore, another definition of lower approximation should be considered to corroborate the concepts of quality of approximation, reduct, and core in the case of missing values. Given $X \subseteq U$ and $P \subseteq C$, this definition is as follows:

$$\underline{I}_P^*(X) = \bigcup_{R \subseteq P} \underline{I}_R(X).$$

$\underline{I}_P^*(X)$ is called *cumulative* P-lower approximation of X because it includes all the objects belonging to all R-lower approximations of X, where $R \subseteq P$.

It can be shown that another type of indiscernibility relation, denoted by I_P^*, permits a direct definition of the cumulative P-lower approximation in the usual way. For each $x,y \in U$ and for each $P \subseteq C$, yI_P^*x means that $f(x,q) = f(y,q)$ or $f(x,q) = *$ and/or $f(y,q) = *$, for every $q \in P$. I_P^* is reflexive and symmetric but not transitive. Let $I_P^*(x) = \{y \in U: yI_P^*x\}$ for each $x \in U$ and for each $P \subseteq C$. It has been proved in [20.22] that the above definition of $\underline{I}_P^*(X)$ is equivalent to the following definition: $\underline{I}_P^*(X) =$

$\{x \in U_P^*: I_P^*(x) \subseteq X\}$, where $U_P^* = \{x \in U: f(x,q) \neq *$ for at least one $q \in P\}$.

Using I_P^*, we can give a definition of the P-upper approximation of X, complementary to $\underline{I}_P^*(X)$:

$$\overline{I}_P^*(X) = \{x \in U_P^* : I_p^* \cap X \neq \emptyset\}.$$

For each $X \subseteq U$, let $X_p^* = X \cap U_P^*$. Let us note that $x \in U_P^*$ if and only if, there exists $R \neq \emptyset$ such that $R \subseteq P$ and $x \in U_R^*$.

Rough approximations $\underline{I}_P^*(X)$ and $\overline{I}_P^*(X)$ satisfy the following properties:

- (*Rough inclusion*) For each $X \subseteq U$ and $P \subseteq C$, $\underline{I}_P^*(X) \subseteq X_P^* \subseteq \overline{I}_P^*(X)$;
- (*Complementarity*) For each $X \subseteq U$ and $P \subseteq C$,
$\underline{I}_P^*(X) = U_P^* - \overline{I}_P^*(U - X)$.
- (*Monotonicity of the accuracy of approximation*) For each $X \subseteq U$ and $P,R \subseteq C$, such that $P \subseteq R$, the inclusion $\underline{I}_P^*(X) \subseteq \underline{I}_R^*(X)$ holds. Furthermore, if $U_P^* = U_R^*$, the inclusion $\overline{I}_P^*(X) \supseteq \overline{I}_R^*(X)$ is also true.

Due to the property of monotonicity, when augmenting a set of attributes P, we get a lower approximation of X that is at least of the same cardinality. Thus, we can restore the following key concepts of rough set theory for the case of missing values: accuracy and quality of approximation, reduct, and core.

20.5.2 Illustrative Example

The illustrative example presented at this point will serve us to explain the concepts introduced in Sect. 20.5.1. The director of a school must give a comprehensive evaluation to some students. This evaluation should be based on results of exams in *Mathematics, Physics,* and *Literature*. However, not all the students have passed all three exams and, therefore, there are some missing values. The director gave the examples of evaluation as shown in Table 20.4.

The following lower and upper approximations can be calculated from Table 20.4:

Table 20.4. Student evaluations with missing values

Student	Mathematics	Physics	Literature	Comprehensive evaluation
1	medium	bad	bad	bad
2	good	medium	*	good
3	medium	*	medium	bad
4	*	medium	medium	good
5	*	good	bad	bad
6	good	medium	bad	good

$$\underline{I}_C(\text{good}) = \{6\}, \qquad \underline{I}_C(\text{bad}) = \{1\},$$
$$\overline{I}_C(\text{good}) = \{6\}, \qquad \overline{I}_C(\text{bad}) = \{1\},$$
$$\underline{I}_C^*(\text{good}) = \{2,6\}, \qquad \underline{I}_C^*(\text{bad}) = \{1,5\},$$
$$\overline{I}_C^*(\text{good}) = \{2,3,4,6\}, \; \overline{I}_C^*(\text{bad}) = \{1,3,4,5\}.$$

The quality of approximation of the partition of U using attributes from C is equal to 0.67. There are two reducts: $Red_1 = \{Mathematics, Physics\}$, and $Red_2 = \{Mathematics, Literature\}$. The intersection of Red_1 and Red_2 constitutes the core, i.e., $Core = \{Mathematics\}$.

The following certain rules can be induced from Table 20.4 (within parentheses there are objects supporting the corresponding decision rule):

1) "*if Mathematics* is good *and Physics* is medium, *then* the student is good" (students 2,4,6)
2) "*if Mathematics* is medium *and Literature* is bad, *then* the student is bad" (students 1,5)

It is also possible to induce the following approximate rule from Table 20.4:

3) "*if Mathematics* is medium *and Literature* is medium, *then* the student is bad or good" (students 3, 4)

We claim that decision rules induced from an incomplete data table according to our approach are *robust* in the following sense: among objects supporting a given decision rule, there is at least one object matching exactly all elementary conditions of the rule. Indeed, rule 1) is grounded on student 2 and 6, rule 2) on student 1, and rule 3) on student 3.

20.5.3 Generalized Dominance Relation

Keeping in mind that a comparison of subject y to referent x is directional for any two objects $x,y \in U$, we say that subject y *dominates* referent x with respect to criteria from $P \subseteq C$ (denoted by $y D_P^+ x$) if, for every criterion $q \in P$, the following conditions are met:
(1) $f(x,q) \neq *$, (2) $f(y,q) \geq f(x,q)$ or $f(y,q) = *$.

We also say that subject y is *dominated* by referent x with respect to criteria from $P \subseteq C$ (denoted by $x D_P^- y$) if, for every criterion $q \in P$, the following conditions are met:
(1) $f(x,q) \neq *$, (2) $f(x,q) \leq f(y,q)$ or $f(y,q) = *$.

The above definition means that the referent object considered for dominance D_P^+ and D_P^- should have no missing values on criteria from set P.

The binary relations D_P^+ and D_P^- are not necessarily reflexive. However, D_P^+ and D_P^- are transitive.

For each $P \subseteq C$, we restore the definition of set U_P from Sect. 20.5.1. Given $P \subseteq C$ and $x \in U$, the "granules of knowledge" used for approximation are

- a set of objects dominating x, called P-*dominating set*,
  $D_P^+ = \{y \in U\colon yD_P^+x\}$,
- a set of objects dominated by x, called P-*dominated set*,
  $D_P^- = \{y \in U\colon xD_P^-y\}$.

For any $P \subseteq C$ we say that $x \in U$ belongs to $Cl_t^\geq$ *without any ambiguity* if $x \in Cl_t^\geq$ and, for all the objects $y \in U$ dominating x with respect to P, we have $y \in Cl_t^\geq$, i.e., $D_P^+ \subseteq Cl_t^\geq$. Furthermore, we say that $x \in U$ *could belong* to $Cl_t^\geq$ if there would exist at least one object $y \in Cl_t^\geq$ dominated by x with respect to P, i.e., $y \in D_P^-$.

Thus, with respect to $P \subseteq C$, the set of all objects belonging to $Cl_t^\geq$ without any ambiguity constitutes the P-lower approximation of $Cl_t^\geq$, denoted by $\underline{P}(Cl_t^\geq)$, and the set of all objects that could belong to $Cl_t^\geq$ constitutes the P-upper approximation of $Cl_t^\geq$, denoted by $\overline{P}(Cl_t^\geq)$, for $t = 1, \ldots, n$:

$$\underline{P}(Cl_t^\geq) = \{x \in U_P\colon D_P^+(x) \subseteq Cl_t^\geq\},$$

$$\overline{P}(Cl_t^\geq) = \{x \in U_P : D_P^-(x) \cap Cl_t^\geq \neq \emptyset\}.$$

Analogously, one can define P-lower approximation and P-upper approximation of $Cl_t^\leq$ for $t = 1, \ldots, n$.

Let $(Cl_t^\geq)_P = Cl_t^\geq \cap U_P$ and $(Cl_t^\leq)_P = Cl_t^\leq \cap U_P$, $t = 1, \ldots, n$. The rough approximations defined as above satisfy the following properties:

- (*Rough inclusion*) For each $Cl_t^\geq$ and $Cl_t^\leq$, $t = 1, \ldots, n$, and for each $P \subseteq C$,

  $$\underline{P}(Cl_t^\geq) \subseteq (Cl_t^\geq)_P \subseteq \overline{P}(Cl_t^\geq), \quad \underline{P}(Cl_t^\leq) \subseteq (Cl_t^\leq)_P \subseteq \overline{P}(Cl_t^\leq);$$

- (*Complementarity*) For each $Cl_t^\geq$, $t = 2, \ldots, n$, and $Cl_t^\leq$, $t = 1, \ldots, n\text{-}1$, and for each $P \subseteq C$,

  $$\underline{P}(Cl_t^\geq) = U_P - \overline{P}(Cl_{t-1}^\leq), \quad \underline{P}(Cl_t^\leq) = U_P - \overline{P}(Cl_{t+1}^\geq).$$

To preserve the monotonicity property of the lower approximation (see 20.5.1), it is necessary to use another definition of the approximation for a given $Cl_t^\geq$ and $Cl_t^\leq$, $t = 1, \ldots, n$, and for each $P \subseteq C$:

$$\underline{P}(Cl_t^\geq)^* = \bigcup_{R \subseteq P} \underline{R}(Cl_t^\geq),$$

$$\underline{P}(Cl_t^\leq)^* = \bigcup_{R \subseteq P} \underline{R}(Cl_t^\leq).$$

$\underline{P}(Cl_t^\geq)^*$ and $\underline{P}(Cl_t^\leq)^*$ are called *cumulative* P-lower approximations of unions $Cl_t^\geq$ and $Cl_t^\leq$, respectively, $t = 1, \ldots, n$, because they include all the objects belonging to all R-lower approximations of $Cl_t^\geq$ and $Cl_t^\leq$, where $R \subseteq P$.

It can be shown that another type of dominance relation, denoted by D_P^*, permits a direct definition of the cumulative P-lower approximations in a classical way. For each $x,y \in U$ and for each $P \subseteq C$, yD_P^*x means that $f(y,q) \geq f(x,q)$ or $f(x,q) = *$ and/or $f(y,q) = *$, for every $q \in P$.

Now, given $P \subseteq C$ and $x \in U$, the "granules of knowledge" used for approximation are

- a set of objects dominating x, called P-*dominating set*,
  $$D_P^{+*}(x) = \{y \in U \colon yD_P^*x\},$$
- a set of objects dominated by x, called P-*dominated set*,
  $$D_P^{-*}(x) = \{y \in U \colon xD_P^*y\}.$$

D_P^+ is reflexive but not transitive. It has been proved in [20.22] that the above definitions of cumulative P-lower approximations are equivalent to the following definitions (for U_P^*, see the definition in 20.5.1):

$$\underline{P}(Cl_t^{\geq})^* = \{x \in U_P^* : D_P^{+*}(x) \subseteq Cl_t^{\geq}\},$$

$$\underline{P}(Cl_t^{\leq})^* = \{x \in U_P^* : D_P^{-*}(x) \subseteq Cl_t^{\leq}\}.$$

Using D_P^+ we can give the following definitions of the P-upper approximations of $Cl_t^{\geq}$ and $Cl_t^{\leq}$, complementary to $\underline{P}(Cl_t^{\geq})^*$ and $\underline{P}(Cl_t^{\leq})^*$, respectively:

$$\overline{P}(Cl_t^{\geq})^* = \{x \in U_P^* : D_P^{-*}(x) \cap Cl_t^{\geq} \neq \emptyset\},$$

$$\overline{P}(Cl_t^{\leq})^* = \{x \in U_P^* : D_P^{+*}(x) \cap Cl_t^{\geq} \neq \emptyset\}.$$

For each $Cl_t^{\geq} \subseteq U$ and $Cl_t^{\leq} \subseteq U$, let $(Cl_t^{\geq})^* = Cl_t^{\geq} \cap U_P^*$ and $(Cl_t^{\leq})^* = Cl_t^{\leq} \cap U_P^*$. Let us note that $x \in U_P^*$ if, and only if, there exists $R \neq \emptyset$ such that $R \subseteq P$ and $x \in U_R$.

Rough approximations $\underline{P}(Cl_t^{\geq})^*$, $\underline{P}(Cl_t^{\leq})^*$, $\overline{P}(Cl_t^{\geq})^*$, and $\overline{P}(Cl_t^{\leq})^*$, satisfy the following properties:

- (*Rough inclusion*) For each $Cl_t^{\geq}$ and $Cl_t^{\leq}$, $t = 1, \ldots, n$, and for each $P \subseteq C$,
  $$\underline{P}(Cl_t^{\geq})^* \subseteq (Cl_t^{\geq})^* \subseteq \overline{P}(Cl_t^{\geq})^*, \quad \underline{P}(Cl_t^{\leq})^* \subseteq (Cl_t^{\leq})^* \subseteq \overline{P}(Cl_t^{\leq})^*.$$

- (*Complementarity*) For each $Cl_t^{\geq}$, $t = 2, \ldots, n$, and $Cl_t^{\leq}$, $t = 1, \ldots, n-1$, and $P \subseteq C$,
  $$\underline{P}(Cl_t^{\geq})^* = U_P^* - \overline{P}(Cl_{t-1}^{\leq})^*.$$
  $$\underline{P}(Cl_t^{\leq})^* = U_P^* - \overline{P}(Cl_{t+1}^{\geq})^*.$$

- (*Monotonicity of the accuracy of approximation*) For each $Cl_t^{\geq}$ and $Cl_t^{\leq}$, $t = 1, \ldots, n$, and $P, R \subseteq C$, such that $P \subseteq R$, the following inclusions hold:
  $$\underline{P}(Cl_t^{\geq})^* \subseteq \underline{R}(Cl_t^{\geq})^*, \underline{P}(Cl_t^{\leq})^* \subseteq \underline{R}(Cl_t^{\leq})^*.$$

Furthermore, if $U_P^* = U_R^*$, the following inclusions are also true

$$\overline{P}(Cl_t^{\geq})^* \supseteq \overline{R}(Cl_t^{\geq})^*, \quad \overline{P}(Cl_t^{\leq})^* \supseteq \overline{R}(Cl_t^{\leq})^*.$$

Due to the property of monotonicity, when augmenting a set of attributes P, we get lower approximations of $Cl_t^{\geq}$ and $Cl_t^{\leq}$, $t = 1, \ldots, n$, that are at least of the same cardinality. Thus, we can restore for the case of missing values the key concepts of rough set theory: accuracy and quality of approximation, reduct, and core.

20.5.4 Illustrative Example

Let us consider the example presented in Sect. 20.5.2; however, now, the three attributes are criteria with preference-ordered scales, and the decision classes are also preference-ordered. We shall apply the DRSA with the generalized dominance relation to Table 20.4.

We will approximate first the downward and the upward unions of classes, i.e., the class of students "at most bad" ($Cl_1^{\leq}$) and the class of students "at least good" ($Cl_2^{\geq}$). Since only two classes are considered, these unions coincide with the class of "bad" students (Cl_1) and with the class of "good" students (Cl_2), respectively.

The C-lower approximations, the C-upper approximations, and the C-boundaries of the classes of "good" and "bad" students are equal, respectively, to:

$$
\begin{aligned}
&\underline{C}(good) = \emptyset, &&\underline{C}(bad) = \{1\}, \\
&\overline{C}(good) = \{6\}, &&\overline{C}(bad) = \{1\}, \\
&\underline{C}(good)^* = \emptyset, &&\underline{C}(bad)^* = \{1\}, \\
&\overline{C}(good)^* = \{2,3,4,5,6\}, &&\overline{C}(bad)^* = \{1,2,3,4,5,6\}.
\end{aligned}
$$

Let us note that students 2 and 6 belong to the C-lower approximation of the class of "good" students when this approximation is calculated using the indiscernibility relation (see $\underline{I}_C^*(good) = \{2,6\}$ in Sect. 20.5.2); however, he or she belongs to the boundary of "good" students when this approximation is calculated using the dominance relation. This is true because there is no "bad" student indiscernible with students 2 and 6. Observe, however, that although student 5 has a comprehensive evaluation worse than students 2 and 6 ("bad" vs. "good"), he or she dominates these students with respect to the three criteria. Precisely, student 2 is dominated by student 5 because student 2 has a worse score in *Physics* (medium vs. good), and in *Mathematics* and *Literature* either student 2 or student 5 has a missing score. For this reason, the assignments of students 2 and 5 are inconsistent, and thus they both belong to the C-boundary of the "(at least) good" class constructed using the dominance relation. The inconsistency between student 2 and student 5 cannot be detected using the classical rough set approach based on indiscernibility because these students are discernible with respect to C. A similar explanation holds for students 5 and 6.

The quality of classification using criteria from C is equal to 0.17. There is only one reduct which is also the core; it is composed of one criterion: {*Physics*}.

The following minimal set of decision rules can be obtained from the data table considered (within parentheses there are objects supporting the corresponding decision rules):

1) "*if Physics* $\leq$ bad, *then* student is (at most) bad"
   (students 1,3).
2) "*if Physics* $\geq$ medium *andPhysics* $\leq$ good, *then* student is bad or good",
   i.e., not enough information to assign the student to one class only.
   (students 2,3,4,5,6).

20.6 Conclusions

Knowledge discovery from preference-ordered data differs from usual knowledge discovery since the former involves preference orders in domains of attributes and in the set of decision classes. This requires that a knowledge discovery method applied to preference-ordered data respects the dominance principle. As this is not the case for the well-known methods of data mining and knowledge discovery, they are not able to discover all relevant knowledge contained in the analysed data sample and, worse, they may yield unreasonable discoveries. These deficiencies are addressed by DRSA, based on the concept of rough approximations consistent with the dominance principle. DRSA, moreover, permits to apply the rough set approach to some new fields, like multicriteria decision making and decision under uncertainty. Multiple extensions proposed for DRSA make this approach a useful tool for practical applications.

The rough set analysis of data representing preferential information supplies some useful elements of knowledge about the decision; these are relevance of attributes and/or criteria, information about their interaction (from quality of approximation and its analysis using fuzzy measure theory), minimal subsets of attributes or criteria (reducts) conveying the relevant knowledge contained in the exemplary decisions, and the set of the non-reducible attributes or criteria (core).

In the second part, which follows, we present other useful extensions of DRSA that make of it a yet more general tool for knowledge discovery and decision support.

Acknowledgements

The research of the first two authors has been supported by the Italian Ministry of Education, University and Scientific Research (MIUR). The third author wishes to acknowledge financial support from the State Committee for Scientific Research.

References

20.1 R. Agrawal, H. Mannila, R. Srikant, H. Toivinen, I. Verkamo: Fast discovery of association rules. In: U.M.Fayyad et al. (eds.), *Advances in Knowledge Discovery and Data Mining* (AAAI Press, 1996) pp. 307-328

20.2 M. Allais: The so-called Allais paradox and rational decision under uncertainty. In: M. Allais, O. Hagen (eds.), *Expected Utility Hypotheses and the Allias Paradox* (Reidel, Dordrecht, 1979) pp. 437-681

20.3 E.I. Altman: Financial ratios, discriminant analysis and the prediction of corporate bankruptcy. Journal of Finance, 589-609 (1968)

20.4 T. Bilgic, I.B. Turksen: Measurement-theoretic justification of connectives in fuzzy set theory. Fuzzy Sets and Systems, 76, 289-308 (1995)

20.5 B. Bouchon-Mounier, J. Yao: Linguistic modifiers and gradual membership to a category. International Journal on Intelligent Systems, 7, 26-36 (1992)

20.6 N. Capon: Credit scoring systems: a critical analysis. Journal of Marketing, 46, 32-91 (1982)

20.7 G. Cattaneo: Fuzzy extension of rough sets theory. In: L. Polkowski, A. Skowron (eds.), *Rough Sets and Current Trends in Computing* (LNAI 1424, Springer, Berlin, 1998) pp. 275-282

20.8 K. Dembczynski, S. Greco, R. Slowinski: Methodology of rough-set-based classification and sorting with hierarchical structure of attributes and criteria, Control & Cybernetics, 31 (2002) (to appear)

20.9 D. Dubois, H. Prade: Gradual inference rules in approximate reasoning. Information Sciences, 61, 103-122 (1992)

20.10 D. Dubois, H. Prade: Putting rough sets and fuzzy sets together. In: R. Slowinski (ed.), *Intelligent Decision Support: Handbook of Applications and Advances of the Sets Theory* (Kluwer, Dordrecht, 1992) pp. 203-232

20.11 D. Dubois, H. Prade, R. Yager: A Manifesto: Fuzzy Information Engineering. In: D. Dubois, H. Prade, R. Yager (eds.), *Fuzzy Information Engineering* (J.Wiley, New York, 1997) pp. 1-8

20.12 J. Fodor, M. Roubens: *Fuzzy Preference Modelling and Multicriteria Decision Support* (Kluwer, Dordrecht, 1994)

20.13 J. Giove, S. Greco, B. Matarazzo, R. Slowinski: Variable consistency monotonic decision trees. In: J.J. Alpigini, J.F. Peters, A. Skowron, N. Zhong (eds.), *Rough Sets and Current Trends in Computing* (LNAI 2475, Springer-Verlag, Berlin, 2002) pp. 247-254

20.14 M. Grabisch: Fuzzy integral in multiple-criteria decision making. Fuzzy Sets and Systems, 69, 279-298 (1995)

20.15 S. Greco, M. Inuiguchi, R. Slowinski: Dominance-based rough set approach using possibility and necessity measures. In: J.J. Alpigini, J.F. Peters, A. Skowron, N. Zhong (eds.), *Rough Sets and Current Trends in Computing* (LNAI 2475, Springer-Verlag, Berlin, 2002) pp. 85-92

20.16 S. Greco, M. Inuiguchi, R. Slowinski: A new proposal for fuzzy rough approximations and gradual decision rule representation. In: D. Dubois, J. Grzymala-Busse, M. Inuiguchi, L. Polkowski (eds.), *Rough Fuzzy and Fuzzy Rough Sets* (Springer-Verlag, Berlin, 2003) (to appear)

20.17 S. Greco, B. Matarazzo, R. Slowinski: Fuzzy measures technique for rough set analysis. In: *Proc. 6th European Congress on Intelligent Techniques & Soft Computing* (Aachen, 1998) 1, pp. 99-103

20.18 S. Greco, B. Matarazzo, R. Slowinski: A new rough set approach to evaluation of bankruptcy risk. In: C.Zopounidis (ed.), *Operational Tools in the Management of Financial Risk* (Kluwer Academic Publishers, Boston, 1998) pp. 121-136

20.19 S. Greco, B. Matarazzo, R. Slowinski: Rough approximation of a preference relation by dominance relations. European Journal of Operational Research, 117, 63-83 (1999)

20.20 S. Greco, B. Matarazzo, R. Slowinski: The use of rough sets and fuzzy sets in MCDM. Chapter 14 in: T.Gal, T.Stewart, T.Hanne (eds.), *Advances in Multiple Criteria Decision Making* (Kluwer Academic Publishers, Boston, 1999) pp. 14.1-14.59

20.21 S. Greco, B. Matarazzo, R. Slowinski: Handling missing values in rough set analysis of multi-attribute and multi-criteria decision problems. In: N. Zhong, A. Skowron, S. Ohsuga (eds.), *New Directions in Rough Sets, Data Mining and Granular-Soft Computing* (LNAI 1711, Springer-Verlag, Berlin, 1999) pp. 146-157

20.22 S. Greco, B. Matarazzo, R. Slowinski: Dealing with missing data in rough set analysis of multi-attribute and multi-criteria decision problems. In: S.H. Zanakis, G. Doukidis, C. Zopounidis (eds.), *Decision Making: Recent Developments and Worldwide Applications* (Kluwer Academic Publishers, Boston, 2000) pp. 295-316

20.23 S. Greco, B. Matarazzo, R. Slowinski: Rough set processing of vague information using fuzzy similarity relations. In: C.S. Calude, G. Paun (eds.), *Finite Versus Infinite - Contributions to an Eternal Dilemma*)Springer-Verlag, London, 2000) pp. 149-173

20.24 S. Greco, B. Matarazzo, R. Slowinski: Fuzzy extension of the rough set approach to multicriteria and multiattribute sorting. In: J. Fodor, B. De Baets and P. Perny (eds.), *Preferences and Decisions under Incomplete Knowledge* (Physica-Verlag, Heidelberg, 2000) pp. 131-151

20.25 S. Greco, B. Matarazzo, R. Slowinski: Extension of the rough set approach to multicriteria decision support. INFOR, 38, 161-196 (2000)

20.26 S. Greco, B. Matarazzo, R. Slowinski: Rough sets theory for multicriteria decision analysis. European J. of Operational Research, 129, 1-47 (2001)

20.27 S.Greco, B. Matarazzo, R. Slowinski: Conjoint measurement and rough set approach for multicriteria sorting problems in presence of ordinal criteria. In: A.Colorni, M.Paruccini, B.Roy (eds.), *A-MCD-A: Aide Multi Critère à la Décision - Multiple Criteria Decision Aiding*, European Commission Report, EUR 19808 EN, (Ispra, 2001) pp. 117-144

20.28 S.Greco, B. Matarazzo, R. Slowinski: Rule-based decision support in multicriteria choice and ranking. In: S. Benferhat, Ph. Besnard (eds.), *Symbolic and Quantitative Approaches to Reasoning with Uncertainty* (LNAI 2143, Springer-Verlag, Berlin, 2001) pp. 29-47

20.29 S. Greco, B. Matarazzo, R. Slowinski: Assessment of a value of information using rough sets and fuzzy measures. In: J. Chojcan, J. Leski (eds.), *Fuzzy Sets and their Applications* (Silesian University of Technology Press, Gliwice, 2001) pp. 185-193

20.30 S. Greco, B. Matarazzo, R. Slowinski: Rough set approach to decisions under risk. In: W.Ziarko, Y.Yao (eds.): *Rough Sets and Current Trends in Computing* (LNAI 2005, Springer-Verlag, Berlin, 2001) pp. 160-169

20.31 S. Greco, B. Matarazzo, R. Slowinski: Rough sets methodology for sorting problems in presence of multiple attributes and criteria. European J. of Operational Research, 138, 247-259 (2002)

20.32 S. Greco, B. Matarazzo, R. Slowinski: Multicriteria classification. In: W. Kloesgen, J. Zytkow (eds.), Handbook of Data Mining and Knowledge Discovery (Oxford University Press, New York, 2002, chapter 16.1.9) pp. 318-328

20.33 S. Greco, B. Matarazzo, R. Slowinski: Preference representation by means of conjoint measurement and decision rule model. In: D.Bouyssou, E.Jacquet-Lagrèze, P.Perny, R.Slowinski, D.Vanderpooten, Ph.Vincke (eds.), Aiding Decisions with Multiple Criteria - Essays in Honor of Bernard Roy (Kluwer Academic Publishers, Boston, 2002) pp. 263-313

20.34 S. Greco, B. Matarazzo, R. Slowinski, J. Stefanowski: Variable consistency model of dominance-based rough set approach. In: W.Ziarko, Y.Yao: Rough Sets and Current Trends in Computing (LNAI 2005, Springer-Verlag, Berlin, 2001) pp. 170-181

20.35 S. Greco, B. Matarazzo, R. Slowinski, J. Stefanowski: An algorithm for induction of decision rules consistent with dominance principle. In: W.Ziarko, Y.Yao (eds.): Rough Sets and Current Trends in Computing (LNAI 2005, Springer-Verlag, Berlin, 2001b) pp. 304-313

20.36 S. Greco, B. Matarazzo, R. Slowinski, J. Stefanowski: Mining association rules in preference-ordered data. In: M.-S. Hacid, Z.W. Ras, D.A. Zighed, Y. Kodratoff (eds.), Foundations of Intelligent Systems (LNAI 2366, Springer-Verlag, Berlin, 2002) pp. 442-450

20.37 S. Greco, B. Matarazzo, R. Slowinski, A. Tsoukias: Exploitation of a rough approximation of the outranking relation in multicriteria choice and ranking. In: T.J.Stewart, R.C. van den Honert (eds.), Trends in Multicriteria Decision Making (LNEMS 465, Springer-Verlag, Berlin, 1998) pp. 45-60

20.38 W.M. Goldstein: Decomposable threshold models. Journal of Mathematical Psychology, 35, 64-79 (1991)

20.39 J.W. Grzymala-Busse: LERS - a system for learning from examples based on rough sets. In: R.Slowinski (ed.), Intelligent Decision Support. Handbook of Applications and Advances of the Rough Sets Theor (y. Kluwer, Dordrecht, 1992) pp. 3-18

20.40 M. Inuiguchi, S. Greco, R. Slowinski, T. Tanino: Possibility and necessity measure specification using modifiers for decision making under fuzziness. Fuzzy Sets and Systems, 137, 151-175 (2003)

20.41 M. Inuiguchi, T. Tanino: New fuzzy rough sets based on certainty qualification. In: S. K. Pal, L. Polkowski, A. Skowron (eds.), Rough-Neuro-Computing: Techniques for Computing with Words (Springer-Verlag, Berlin, 2002) pp. 110-126

20.42 R.S. Michalski, I. Bratko, M. Kubat (eds.): Machine Learning and Data Mining - Methods and Applications (Wiley, New York, 1998)

20.43 G.A. Miller: The magical number seven, plus or minus two: some limits on our capacity for information processing. Psychological Review 63, 81-97 (1956)

20.44 F. Modave, M. Grabisch: Preference representation by the Choquet Integral: the commensurability hypothesis. In: Proc. 7th Int. Conference on Information Processing and Management of Uncertainty in Knowledge Based Systems (Paris, La Sorbonne, 1998) pp. 164-171

20.45 A. Nakamura: Applications of fuzzy-rough classification to logics. In: R. Slowinski (ed.),Intelligent Decision Support: Handbook of Applications and Advances of the Sets Theory (Kluwer, Dordrecht, 1992) pp. 233-250

20.46 A. Nakamura, J.M. Gao: A logic for fuzzy data analysis. Fuzzy Sets and Systems, 39, 127-132 (1991)

20.47 E. Orlowska, Introduction: What you always wanted to know about rough sets. In: E. Orlowska (ed.), *Incomplete Information, Rough Set Analysis* (Physica-Verlag, Heidelberg, New York, 1998) pp. 1-20

20.48 Z. Pawlak: *Rough Sets. Theoretical Aspects of Reasoning about Data.* Kluwer Academic Publishers, Dordrecht, 1991

20.49 Z. Pawlak, J.W. Grzymala-Busse, R. Slowinski, W. Ziarko: Rough sets. Communications of the ACM, 38, 89-95 (1995)

20.50 L. Polkowski: *Rough Sets: Mathematical Foundations* (Physica-Verlag, Heidelberg, 2002)

20.51 L. Polkowski, A. Skowron: Calculi of granules based on rough set theory: approximate distributed synthesis and granular semantics for computing with words. In: N.Zhong, A.Skowron, S.Ohsuga (eds.), *New Directions in Rough sets, Data Mining and Soft-Granular Computing* (LNAI 1711, Springer-Verlag, Berlin, 1999) pp. 20-28

20.52 F.S. Roberts: *Measurement theory with applications to decision-making, utility and the social science* (Addison-Wesley Publ., Reading, MA, 1979)

20.53 B. Roy, D. Bouyssou: *Aide Multicritère à la Décision: Méthodes et Cas* (Economica, Paris, 1993)

20.54 T. L. Saaty: *The Analytic Hierarchy Process* (McGraw-Hill, New York, 1980)

20.55 R. Slowinski: A generalization of the indiscernibility relation for rough set analysis of quantitative information. *Rivista di matematica per le scienze economiche e sociali*, 15, 65-78 (1992)

20.56 R. Slowinski (ed.): *Intelligent Decision Support. Handbook of Applications and Advances of the Rough Sets Theory* (Kluwer Academic Publishers, Dordrecht, 1992)

20.57 R. Slowinski: Rough set processing of fuzzy information. In: T.Y.Lin, A.Wildberger (eds.), *Soft Computing: Rough Sets, Fuzzy Logic, Neural Networks, Uncertainty Management, Knowledge Discovery* (Simulation Councils, Inc., San Diego, CA, 1995) pp. 142-145

20.58 R. Slowinski, S. Greco, B. Matarazzo: Rough set analysis of preference-ordered data. In: J.J. Alpigini, J.F. Peters, A. Skowron, N. Zhong (eds.), *Rough Sets and Current Trends in Computing* (LNAI 2475, Springer-Verlag, Berlin, 2002) pp. 44-59

20.59 R. Slowinski, S. Greco, B. Matarazzo: Mining decision-rule preference model from rough approximation of preference relation. In: Proc. 26th IEEE Annual Int. Conference on *Computer Software & Applications (COMPSAC 2002)* (Oxford, England, 2002) pp. 1129-1134

20.60 R. Slowinski, S. Greco, B. Matarazzo: Axiomatization of utility, outranking and decision-rule preference models for multiple-criteria classification problems under partial inconsistency with the dominance principle. Control and Cybernetics, 31 (4), 1005-1035 (2002)

20.61 R. Slowinski, J. Stefanowski: Rough set reasoning about uncertain data. Fundamenta Informaticae, 27, 229-243 (1996)

20.62 R. Slowinski, J. Stefanowski, S. Greco, B. Matarazzo: Rough sets based processing of inconsistent information in decision analysis. Control and Cybernetics, 29, 379-404 (2000)

20.63 R. Slowinski, D. Vanderpooten: A generalised definition of rough approximations. IEEE Transactions on Data and Knowledge Engineering, 12, 331-336 (2000)

20.64 R. Slowinski, C. Zopounidis: Application of the rough set approach to evaluation of bankruptcy risk. Intelligent Systems in Accounting, Finance and Management, 4, 27-41 (1995)

20.65 J. Stefanowski: On rough set based approaches to induction of decision rules. In: L. Polkowski, A. Skowron (eds.), *Rough Sets in Data Mining and Knowledge Discovery* (Physica-Verlag, Heidelberg, 1998) 1, pp. 500-529

20.66 J. Stepaniuk: Knowledge Discovery by Application of Rough Set Models, In: L. Polkowski, S. Tsumoto, T.Y. Lin (eds.): *Rough Set Methods and Application* (Physica Verlag, Heidelberg, 2000) pp. 137-231

20.67 M. Sugeno: *Theory of fuzzy integrals and its applications*. Doctoral Thesis, Tokyo Institute of Technology, 1974

20.68 L.C. Thomas, J.N. Crook, D.B. Edelman (eds.): *Credit Scoring and Credit Control* (Clarendon Press, Oxford, 1992)

20.69 P.P. Wakker, H. Zank: State dependent expected utility for savage's state space. Mathematics of Operations Research, 24, 8-34 (1999)

20.70 L.L. White, A.G. Wilson, D. Wilson (eds.): *Hierarchical Structures* (Elsevier, New York, 1969)

20.71 Y.Y. Yao: Combination of rough and fuzzy sets based on α-level sets. In: T.Y. Lin and N. Cercone (eds.), *Rough Sets and Data Mining: Analysis for Imprecise Data* (Kluwer, Boston, 1997) pp. 301-321

20.72 L.A. Zadeh: A fuzzy set-theoretic interpretation of linguistic hedges. Journal of Cybernetics, 2, 4-34 (1972)

20.73 W. Ziarko: Variable precision rough sets model. Journal of Computer and Systems Sciences, 46, 39-59 (1993)

20.74 W. Ziarko: Rough sets as a methodology for data mining. In: L.Polkowski, A.Skowron (eds.), *Rough Sets in Knowledge Discovery* (Physica-Verlag, Heidelberg, 1998) 1, pp. 554-576

20.75 W. Ziarko, N. Shan: KDD-R, a comprehensive system for knowledge discovery in databases using rough sets. In: T.Y. Lin, A.M. Wildberg (eds.), *Soft Computing: Rough Sets, Fuzzy Logic, Neural Networks, Uncertainty Management, Knowledge Discovery* (Simulation Council Inc., San Diego, 1995) pp. 93-96

20.76 C. Zopounidis, M. Doumpos: A multicriteria decision aid methodology for sorting decision problems: the case of financial distress. Computational Economics, 14, 197-218 (1999)

21. Dominance-Based Rough Set Approach to Knowledge Discovery (II): Extensions and Applications

Salvatore Greco[1], Benedetto Matarazzo[1], and Roman Slowinski[2,3]

[1] University of Catania, Italy
[2] Poznan University of Technology, Poland
[3] Polish Academy of Sciences, Poland

Abstract

While the first part was devoted to the presentation of a general perspective of Dominance-based Rough Set Approach (DRSA) in the context of multicriteria classification, choice, and ranking, this part presents fuzzy set extensions of DRSA, DRSA for decisions under risk, and DRSA for hierarchical decision making. The chapter ends with a section concerning an axiomatic characterization of three preference models inferred from preference ordered data: general utility function, outranking relation, and a set of decision rules resulting from DRSA. It appears that the decision rule model is the most general among the known preference models.

21.1 Introduction

This chapter is a continuation of the first part of presentation of the Dominance-based Rough Set Approach (DRSA) to knowledge discovery from preference ordered data. All the motivations given in the first part are still valid for the second part; however, we wish to present here some further extensions of DRSA, showing its ability to handle a large variety of knowledge discovery tasks.

In the next section, we present fuzzy set extensions of DRSA; then, in the following two sections, DRSA for decisions under risk and DRSA for hierarchical decision making are presented. As a set of decision rules resulting from DRSA can be seen as a preference model in multicriteria decision problems, we present in the last section some results concerning its relations with traditional models of preferences, such as utility functions and outranking relations. This comparison permits us to conclude that the decision rule model is the most general among the known preference models.

21.2 Fuzzy Set Extensions of DRSA

It has been acknowledged by different studies that fuzzy set theory and rough set theory are complementary because they handle different kinds of uncertainty. Fuzzy sets deal with possibilistic uncertainty, connected with imprecision of states, perceptions, and preferences [21.11]. Rough sets deal with uncertainty following from ambiguity of information [21.48]. The two types of uncertainty can be encountered together in real-life problems. For this reason, many approaches have been proposed to combine fuzzy sets with rough sets [21.7, 21.10, 21.20, 21.23, 21.24, 21.41, 21.45, 21.46, 21.50, 21.57, 21.61, 21.71]. The main preoccupation in almost all these studies was related to the fuzzy extension of Pawlak's definitions of lower and upper approximation using fuzzy logical connectives (T-norm, T-conorm, fuzzy implication).

In [21.15, 21.16, 21.20, 21.24], DRSA was extended and characterized by using the *fuzzy dominance relation* in two different ways . These extensions of the rough approximation into a fuzzy context maintain the same desirable properties of the crisp rough approximation of preference-ordered decision classes. These generalizations distinguish all possible cases where either approximating granules in X_C (dominance cones in X_C) are fuzzy, or approximated granules in X_D (preference-ordered decision classes) are fuzzy, or both these granules are fuzzy. These generalizations follow the traditional line of using fuzzy logical connectives in definitions of lower and upper approximation. In fact, there is no rule for the choice of the "right" connective, so this choice is always arbitrary to some extent. For this reason, a new fuzzy rough approximation was proposed in [21.16]. It avoids this choice and exploits the ordinal properties of fuzzy membership degrees only.

Below, we present the core idea of all these approaches to fuzzy rough approximation in three parts: fuzzy DRSA for multicriteria classification, fuzzy DRSA for multicriteria choice and ranking, and fuzzy rough approximation without fuzzy logical connectives, including gradual rules.

21.2.1 Fuzzy DRSA for Multicriteria Classification

When proposing DRSA, it was clear for us that the concept of dominance can be refined by introducing gradedness through the use of fuzzy sets. In this application, a fuzzy set is used in the semantics connected with a degree of preference.

Let $\succeq_q$ be a *fuzzy weak preference relation* (also called fuzzy outranking relation) on U with respect to criterion $q \in C$, i.e., $\succeq_q : U \times U \to [0,1]$, such that $\succeq_q (x,y)$ represents the credibility of the proposition "*x is at least as good as y with respect to criterion q*". Suppose that $\succeq_q$ is a fuzzy partial T-preorder, i.e., it is reflexive ($\succeq_q (x,x) = 1$ for each $x \in U$) and T-transitive ($T(\succeq_q (x,y), \succeq_q (y,z)) \leq \succeq_q (x,z)$ for each $x, y, z \in U$). Using the fuzzy weak preference relations $\succeq_q$, $q \in C$, a *fuzzy dominance relation* on U (denotation $D_P(x,y)$) can be defined for each $P \subseteq C$ as follows:

$$D_P(x,y) = \underset{q \in P}{T} (\succeq_q (x,y)).$$

Given $(x,y) \in U \times U$, $D_P(x,y)$ represents the credibility of the proposition "*x is weakly preferred to y with respect to each criterion q from P*".

Since the fuzzy weak preference relations $\succeq_q$ are supposed to be partial T-preorders, the fuzzy dominance relation D_P is also a partial T-preorder.

Furthermore, let $\boldsymbol{Cl} = \{Cl_t, t \in \{1, \ldots, n\}\}$ be a set of fuzzy classes in U, such that for each $x \in U$, $Cl_t(x)$ represents the membership function of x to Cl_t. As before, we suppose that the classes of $\boldsymbol{Cl}$ are increasingly ordered, i.e., that for all $r,s \in \{1, \ldots, n\}$, such that $r > s$, the elements of Cl_r have a better comprehensive evaluation than the elements of Cl_s.

On the basis of the membership functions of the fuzzy class Cl_t, we can define fuzzy membership functions of two other sets:

1) *fuzzy upward union $Cl_t^{\geq}$*, whose membership function $Cl_t^{\geq}(x)$ represents the credibility of the proposition "*x is at least as good as the objects in Cl_t*",

$$Cl_t^{\geq}(x) = \begin{cases} 1, & \text{if } \exists s \in \{1,\ldots,n\} : Cl_s(x) > 0 \text{ and } s > t \\ Cl_t(x), & \text{otherwise, } i.e., \text{if } \forall s \in \{1,\ldots,n\}, \\ & \text{such that } s > t, Cl_s(x) = 0. \end{cases}$$

2) *fuzzy downward union $Cl_t^{\leq}$*, whose membership function $Cl_t^{\leq}(x)$ represents the credibility of the proposition "*x is at most as good as the objects in Cl_t*",

$$Cl_t^{\leq}(x) = \begin{cases} 1, & \text{if } \exists s \in \{1,\ldots,n\} : Cl_s(x) > 0 \text{ and } s < t \\ Cl_t(x), & \text{otherwise, } i.e., \text{if } \forall s \in \{1,\ldots,n\}, \\ & \text{such that } s < t, Cl_s(x) = 0. \end{cases}$$

The P-lower and the P-upper approximations of $Cl_t^{\geq}$ with respect to $P \subseteq C$ are fuzzy sets in U, whose membership functions (denoted by $\underline{P}\left[Cl_t^{\geq}(x)\right]$ and $\overline{P}\left[Cl_t^{\geq}(x)\right]$) are defined as

$$\underline{P}[Cl_t^{\geq}(x)] = \underset{y \in U}{T}\,(T^*(N(D_P(y,x)), Cl_t^{\geq}(y))),$$

$$\overline{P}[Cl_t^{\geq}(x)] = \underset{y \in U}{T^*}\,(T(D_P(x,y), Cl_t^{\geq}(y))),$$

where T is a T-norm, T^* is a T-conorm [21.12], and N is a strict negation, that is, a strictly decreasing continuous function $N:[0,1] \to [0,1]$ satisfying $N(0) = 1$ $N(1) = 0$.

$\underline{P}[Cl_t^{\geq}(x)]$ represents the credibility of the proposition "*for all $y \in U$, y does not dominate x with respect to criteria from P and/or y belongs to $Cl_t^{\geq}$*", while $\overline{P}[Cl_t^{\geq}(x)]$ represents the credibility of the proposition "*there is at least one $y \in U$ dominated by x with respect to criteria from P which belongs to $Cl_t^{\geq}$*".

The P-lower and P-upper approximations of $Cl_t^{\leq}$ with respect to $P \subseteq C$ (denotation $\underline{P}\left[Cl_t^{\leq}(x)\right]$ and $\overline{P}\left[Cl_t^{\leq}(x)\right]$) can be defined, analogously, as

$$\underline{P}[Cl_t^{\leq}(x)] = \underset{y \in U}{T}\,(T^*(N(D_P(x,y)), Cl_t^{\leq}(y))),$$

$$\overline{P}[Cl_t^{\leq}(x)] = \underset{y \in U}{T^*}(T(D_P(y,x), Cl_t^{\leq}(y))).$$

$\underline{P}[Cl_t^{\leq}(x)]$ represents the credibility of the proposition *"for all $y \in U$, x does not dominate y with respect to criteria from P and/or y belongs to $Cl_t^{\leq}$"*, while $\overline{P}[Cl_t^{\leq}(x)]$ represents the credibility of the proposition *"there is at least one $y \in U$ dominating x with respect to criteria from P which belongs to $Cl_t^{\leq}$"*.

Let us note that using the definition of the T^*-implication, denoted by $I_{T^*,N}^{\rightarrow}$, it is possible to rewrite the definitions of $\underline{P}[Cl_t^{\geq}(x)]$, $\overline{P}[Cl_t^{\geq}(x)]$, $\underline{P}[Cl_t^{\leq}(x)]$, and $\overline{P}[Cl_t^{\leq}(x)]$ as follows:

$$\underline{P}[Cl_t^{\geq}(x)] = \underset{y \in U}{T}(I_{T^*,N}^{\rightarrow}(D_P(y,x), Cl_t^{\geq}(y))),$$

$$\overline{P}[Cl_t^{\geq}(x)] = \underset{y \in U}{T^*}(N(I_{T^*,N}^{\rightarrow}(D_P(x,y)), Cl_t^{\geq}(y))),$$

$$\underline{P}[Cl_t^{\leq}(x)] = \underset{y \in U}{T}(I_{T^*,N}^{\rightarrow}(D_P(x,y), Cl_t^{\leq}(y))),$$

$$\overline{P}[Cl_t^{\leq}(x)] = \underset{y \in U}{T^*}(N(I_{T^*,N}^{\rightarrow}(D_P(y,x)), Cl_t^{\leq}(y))).$$

To introduce the following result, we need the concepts of De Morgan triplet and involutive negation. (T,T^*,N) is a De Morgan triplet if and only if for all $\alpha, \beta \in [0,1], N(T^*(\alpha,\beta)) = T(N(\alpha), N(\beta))$, where N is a strict negation. Negation N is involutive if and only if for all $\alpha \in [0,1], N(N(\alpha)) = \alpha$.

The following results have been proved in [21.24]:

1) $\underline{P}[Cl_t^{\geq}(x)] \leq Cl_t^{\geq}(x) \leq \overline{P}[Cl_t^{\geq}(x)]$
   for each $x \in U$ and for each $t = 2, \ldots, n$,
2) $\underline{P}[Cl_t^{\leq}(x)] \leq Cl_t^{\leq}(x) \leq \overline{P}[Cl_t^{\leq}(x)]$
   for each $x \in U$ and for each $t = 1, \ldots, n-1$,
3) if (T,T^*,N) is a De Morgan triplet, negation N is involutive, and $N(Cl_t^{\geq}(x)) = Cl_{t-1}^{\leq}(x)$ for each $x \in U$ and $t = 2, \ldots, n$, then

   i) $\underline{P}[Cl_t^{\geq}(x)] = N(\overline{P}[Cl_{t-1}^{\leq}(x)]), t = 2, \ldots, n,$
   ii) $\underline{P}[Cl_t^{\leq}(x)] = N(\overline{P}[Cl_{t+1}^{\geq}(x)]), t = 1, \ldots, n-1,$
   iii) $\overline{P}[Cl_t^{\geq}(x)] = N(\underline{P}[Cl_{t-1}^{\leq}(x)]), t = 2, \ldots, n,$
   iv) $\overline{P}[Cl_t^{\leq}(x)] = N(\underline{P}[Cl_{t+1}^{\geq}(x)]), t = 1, \ldots, n-1.$

Results 1) to 3) can be read as the fuzzy counterparts of the following properties of DRSA: 1) (rough inclusion for $Cl_t^{\geq}$) says that $Cl_t^{\geq}$ includes its P-lower approximation and is included in its P-upper approximation; 2) (rough inclusion for $Cl_t^{\leq}$) has an interpretation analogous to 1) with respect to $Cl_t^{\leq}$; 3) (complementarity) says that the P-lower (or P-upper) approximation of $Cl_t^{\geq}$ is the complement of the P-upper (or P-lower) approximation of its complementary set $Cl_{t-1}^{\leq}$ (an analogous property holds for $Cl_t^{\leq}$).

The above definitions of P-lower and P-upper approximations, can be given in terms of *possibility* and *necessity* measures as proposed in [21.15]. Possibility and necessity measures are defined by

$$Poss(B \mid A) = \sup_x K((\mu_A(x), \mu_B(x)) \text{ and}$$

$$Nec(B \mid A) = \inf_x I(\mu_A(x), \mu_B(x)),$$

where K and $I : [0,1] \times [0,1] \rightarrow [0,1]$ are *conjunction* and *implication* functions, such that

K1) $K(0,0) = K(0,1) = K(1,0) = 0$ and $K(1,1) = 1$,

I1) $I(0,0) = I(0,1) = I(1,1) = 1$ and $I(1,0) = 0$.

If the conjunction and implication functions are monotonic, then

K2) $K(a,b) \leq K(c,d)$ if $a \leq c$ and $b \leq d$,

I2) $I(a,b) \leq I(c,d)$ if $a \geq c$ and $c \leq d$.

If K and I satisfy K2) and I2), then we have the following properties, respectively:

$$Poss(B_1 \mid A_1) \leq Poss(B_2 \mid A_2) \text{ and } Nec(B_1 \mid A_2) \leq Poss(B_2 \mid A_1) \text{ if}$$
$$A_1 \subseteq A_2 \text{ and } B_1 \subseteq B_2.$$

Since there exist many conjunction and implication functions, we have many possibility and necessity measures. Thus, there is the question of how to select the possibility and necessity measures. In [21.40], we have answered this question by proposing the level cut conditioning approach. In this approach, one can specify possibility and necessity measures based on the following equivalences:

$$Poss(B \mid A) \leq h \text{ if and only if } Q_h^1(A) \subseteq (Q_h^2(B))^c,$$

$$Nec(B \mid A) \geq h \text{ if and only if } m_h(A) \subseteq M_h(B),$$

where A^c is the complement fuzzy set of A, and inclusion relation $A \subseteq B$ is defined by $\mu_A \leq \mu_B$. Q_h^1, Q_h^2, m_h, and M_h are modifiers varying with parameter $h \in (0,1)$. Examples of Q_h^1, Q_h^2, m_h, and M_h are given in Table 21.1.

When h becomes large, condition $Q_h^1(A) \subseteq (Q_h^2(B))^c$ becomes weak, while condition $m_h(A) \subseteq M_h(B)$ becomes strong.

In Table 21.1, the first equivalence implies that if (most weakly A) $\subseteq$ (most weakly B)^c, then $Poss(B|A) = 0$, but if (most strongly A) $\not\subseteq$ (most strongly B)^c, then $Poss(B|A) = 1$. Similarly, the second equivalence implies that if (most weakly A) $\subseteq$ (most weakly B)^c, then $Nec(B|A) = 1$, but if (most strongly A) $\not\subseteq$ (most strongly B)^c, then $Nec(B|A) = 0$.

In order to treat the above equivalences mathematically, Q_h^1, Q_h^2, m_h, and M_h have been defined by the following membership functions:

Table 21.1. An example of Q_h^1, Q_h^2 and m_h, M_h

Modifier 0		→	h	→	1
Q_h^1	most weakly	more or less	normally	very	most strongly
Q_h^2	most weakly	more or less	normally	very	most strongly
m_h	most strongly	very	normally	more or less	most weakly
M_h	most weakly	more or less	normally	very	most strongly

$$\mu_{Q_h^i(A)}(x) = g_i^Q(\mu_A(x), h), i = 1, 2,$$

$$\mu_{m_h(A)}(x) = g^m(\mu_A(x), h), \quad \mu_{M_h(A)}(x) = g^M(\mu_A(x), h).$$

From the properties of modifiers Q_h^1, Q_h^2, m_h, and M_h, modifier functions g_i^Q, g^m, and g^M should satisfy the following requirements:

q1) $g_i^Q(a, \cdot)$ is lower semi-continuous for all $a \in [0, 1]$,
q2) $g_i^Q(1, h) = 1$ and $g_i^Q(0, h) = 0$ for all $h < 1$,
q3) $g_i^Q(a, 1) = 0$ for all $a \in [0, 1]$,
q4) $g_i^Q(a, \cdot)$ is non-increasing for all $a \in [0, 1]$,
q5) $g_i^Q(\cdot, h)$ is non-decreasing for all $h \in [0, 1]$,
q6) $g_i^Q(a, 0) > 0$ for all $a \in (0, 1)$,
g1) $g^m(a, \cdot)$ and $g^M(a, \cdot)$ are lower and upper semi-continuous for all $a \in [0, 1]$, respectively,
g2) $g^m(1, h) = g^M(1, h) = 1$ and $g^m(0, h) = g^M(0, h) = 0$ for all $h > 0$,
g3) $g^m(a, 0) = 0$ and $g^M(a, 0) = 1$ for all $a \in [0, 1]$,
g4) $g^m(a, \cdot)$ is non-decreasing and $g^M(a, \cdot)$ is non-increasing for all $a \in (0, 1)$,
g5) $g^m(\cdot, h)$ and $g^M(\cdot, h)$ are non-decreasing for all $h \in (0, 1)$,
g6) $g^m(a, 1) > 0$ and $g^M(a, 1) < 1$ for all $a \in (0, 1)$.

Requirements q1) and g1) are technical conditions for existence of possibility and necessity measures satisfying the above equivalence properties. Other requirements follow naturally from the meaning of the modifiers.

Given the modifier functions Q_h^1, Q_h^2, m_h, and M_h, it has been shown in [21.40] that the unique possibility and necessity measures satisfying the above equivalence properties are as follows:

$$\text{Poss}^L(B \mid A) = \inf_h \{h \in [0, 1] : Q_h^1(A) \subseteq (Q_h^2(B))^c\}$$

$$= \sup_x K^L(\mu_A(x), \mu_B(x)), \tag{21.1}$$

$$\mathrm{Nec}^L(B \mid A) = \sup_h \{h \in [0,1] : m_h(A) \subseteq M_h(B)\}$$

$$= \inf_x I^L(\mu_A(x), \mu_B(x)), \qquad (21.2)$$

where the conjunction function K^L and the implication function I^L are defined as

$$K_L(a,b) = \inf_h \{h \in [0,1] : g_1^Q \le N(g_2^Q(b,h))\},$$

$$I_L(a,b) = \sup_h \{h \in [0,1] : g^m \le g^M(b,h)\},$$

where N is an involutive negation.

Conjunction and implication functions K^L and I^L satisfy C2) and I2), respectively. Other properties of K^L and I^L are investigated in [21.40]. It has been shown, moreover, that necessity measures defined by many famous implications, including S-, R-, and reciprocal R-implications with continuous Archimedean T-norms, can be expressed in the above form. The modifiers g^m and g^M associated with these necessity measures are given in [21.40].

The P-lower and the P-upper approximations of $Cl_t^{\ge}$ with respect to $P \subseteq C$, expressed in terms of the above necessity and possibility measures, are fuzzy sets defined as

$$\underline{P}[Cl_t^{\ge}(x)] = Nec^L \left(Cl_t^{\ge} | D_P^+(x) \right) = \inf_{y \in U} I^L \left(D_P(y,x), Cl_t^{\ge}(y) \right),$$

$$\overline{P}[Cl_t^{\ge}(x)] = Poss^L \left(Cl_t^{\ge} | D_P^-(x) \right) = \sup_{y \in U} K^L \left(D_P(x,y), Cl_t^{\ge}(y) \right),$$

where $D_P^+(x)$ is a fuzzy set of objects $y \in U$ dominating x with respect to $P \subseteq C$, and $D_P^-(x)$ is a fuzzy set of objects $y \in U$ dominated by x with respect to $P \subseteq C$. The membership functions of $D_P^+(x)$ and $D_P^-(x)$ are defined as follows:

$$\mu\left(y, D_P^+(x)\right) = D_P(y,x),$$
$$\mu\left(y, D_P^-(x)\right) = D_P(x,y).$$

The P-lower and the P-upper approximations of $Cl_t^{\le}$ with respect to $P \subseteq C$ are

$$\underline{P}[Cl_t^{\le}(x)] = Nec^L(Cl_t^{\le} \mid D_P^-(x)) = \inf_{y \in U} I^L(D_P(x,y), Cl_t^{\le}(y)),$$

$$\overline{P}[Cl_t^{\le}(x)] = Poss^L(Cl_t^{\le} \mid D_P^+(x)) = \sup_{y \in U} K^L(D_P(y,x), Cl_t^{\le}(y)).$$

In [21.15], we proved that the basic properties of rough set theory hold for the above definitions of P-lower and P-upper approximations subject to some conditions on modifiers used in definitions of possibility and necessity measures.

Using the fuzzy dominance relation, one can induce decision rules from fuzzy rough approximations of fuzzy upward and downward unions of decision classes. The syntax of certain and possible decision rules is the same as in DRSA (see 20.3.2); however, the weak preference relations $\succeq_{qi}$ and $\preceq_{qi}$ in the premise are fuzzy. Moreover, with each conclusion of decision rule r, there is associated a *credibility* $\chi(r)$:

– for certain rule r of type D$_\geq$, involving criteria from set P in the premise, and concluding that x belongs to $Cl_t^\geq$, credibility $\chi(r)$ is equal to the highest membership degree $\underline{P}\left[Cl_t^\geq(y)\right]$ of a supporting object y having evaluations equal to the condition profile of r,
– for possible rule r of type D$_\geq$, concluding that x could belong to $Cl_t^\geq$, credibility $\chi(r)$ is equal to the highest membership degree $\overline{C}\left[Cl_t^\geq(y)\right]$ of a supporting object y having evaluations equal to the condition profile of r.

Credibilities for decision rules of type D$_\leq$ are defined analogously.

21.2.2 Fuzzy DRSA for Multicriteria Choice and Ranking

In Sect. 20.4, we presented an extension of DRSA to rough approximation of a comprehensive preference relation, for example, a weak preference relation called outranking relation. This rough approximation has further been extended to a fuzzy context [21.20].

Let S be a *comprehensive fuzzy outranking relation* defined on A. As the knowledge of S is confined to the set $B \subseteq A \times A$ of exemplary pairwise comparisons of reference objects, S is a function, $S{:}B \rightarrow[0,1]$, such that for all $(x,y) \in B$, $S(x,y)$ is the credibility of the statement "x is comprehensively at least as good as y". Of course, $S(x,x) = 1$ for each reference object x (reflexivity). The *comprehensive fuzzy non-outranking relation* S^c is the fuzzy complement of relation S, i.e., $S^c(x,y) = N(S(x,y))$.

As in the crisp case (see Sect. 20.4.1), for each $q \in C$, let H_q be a set of preference degrees with respect to criterion q. Then, for each $q \in C$ and $h \in H_q$, a *fuzzy graded preference relation* on the set of objects A, $P_q^h{:}A \times A \rightarrow[0,1]$, can be introduced, where $P_q^h(x,y)$ is the credibility of the statement "x is preferred to y with respect to criterion q by degree h".

Using a *fuzzy multigraded dominance relation*, built on the basis of fuzzy graded preference relation P_q^h, fuzzy outranking relations S and S^c can be approximated. More precisely, rough approximations of S and S^c are defined using fuzzy positive dominance $D_P^+((x,y),(w,z))$ and fuzzy negative dominance $D_P^-((x,y),(w,z))$, respectively, which are the fuzzy counterparts of sets $D_P^+(x,y)$ and $D_P^-(x,y)$ introduced in Sect. 20.4.2 to approximate crisp relations S and S^c.

Let us present, step by step, the process of building a fuzzy multigraded dominance relation satisfying some desirable properties.

First, on the basis of the fuzzy graded preference relation $P_q^h(x,y)$, $q \in C$, the fuzzy upward cumulated preference (denoted by $P_q^{\geq h}(x,y)$, and the fuzzy downward cumulated preference (denoted by $P_q^{\leq h}(x,y)$) are defined as follows:

1) *fuzzy upward cumulated preference* $P_q^{\geq h}$: $A \times A \to [0,1]$, such that for all $(x,y) \in A \times A$, $P_q^{\geq h}(x,y)$ is the credibility of the statement "*x is preferred to y with respect to q by at least degree h*", i.e.,

$$P_q^{\geq h}(x,y) \;=\; \begin{cases} 1, & if\ \exists k \in H_q : P_q^k(x,y) > 0 \text{ and } k > h \\ P_q^h(x,y), & \text{otherwise, i.e. } if\ \forall k > h,\, P_q^k(x,y) = 0 \end{cases}$$

2) *fuzzy downward cumulated preference* $P_q^{\leq h}$: $A \times A \to [0,1]$, such that for all $(x,y) \in A \times A$, $P_q^{\leq h}(x,y)$ is the credibility of the statement "*x is preferred to y with respect to q by at most degree h*", i.e.,

$$P_q^{\leq h}(x,y) \;=\; \begin{cases} 1, & if\ \exists k \in H_q : P_q^k(x,y) > 0 \text{ and } k < h \\ P_q^h(x,y), & \text{otherwise, i.e. } if\ \forall k < h,\, P_q^k(x,y) = 0 \end{cases}$$

Then, we need a definition for the comparison of degrees of fuzzy preference. This comparison represents the fuzzy counterpart of the statement "*x is preferred to y by at least the same degree as w is preferred to z, with respect to $q{\in}C$*". In formal terms, for $q \in C$, the corresponding crisp statement is "$h{\geq}k$, where $xP_q^h y$ and $wP_q^k z$, h, $k{\in}H_q$" or, in short, "$h{\geq}k$, h, $k{\in}H_q$". From a semantic point of view, the latter comparison is equivalent to the following set of implications: "$k{\geq}r$ implies $h{\geq}r$, for each $r{\in}H_q$". Thus, the statement "$h{\geq}k$, where $xP_q^h y$ and $wP_q^k z$, h, $k{\in}H_q$" is equivalent to "$wP_q^{\geq r} z$ implies $xP_q^{\geq r} y$ for each $r{\in}H_q$". This set of implications is finally expressed in fuzzy terms, and its credibility (denoted by $P_q^+((x,y),(w,z))$) is equal to

$$P_q^+((x,y),(w,z)) \;=\; \underset{r \in H_q}{T}\left(I^{\to}\left(P_q^{\geq r}(w,z), P_q^{\geq r}(x,y)\right)\right).$$

Analogously, given (x,y), $(w,z) \in A \times A$, the comparison statement "*x is preferred to y by at most the same degree as w is preferred to z, with respect to $q{\in}C$*" is equivalent, in crisp terms, to "$wP_q^{\leq r} z$ implies $xP_q^{\leq r} y$ for each $r{\in}H_q$", whose credibility (denoted by $P_q^-((x,y),(w,z))$) is equal to

$$P_q^-((x,y),(w,z)) \;=\; \underset{r \in H_q}{T}\left(I^{\to}\left(P_q^{\leq r}(w,z), P_q^{\leq r}(x,y)\right)\right).$$

The following properties are desirable for $P_q^+((x,y),(w,z))$ and $P_q^-((x,y),(w,z))$, $q \in C$:

1) $[P_q^{\geq r}(x,y) \geq P_q^{\geq r}(w,z)$, for all $r \in H_q] \quad \Leftrightarrow \quad P_q^+((x,y),(w,z)) = 1$,
2) $[P_q^{\leq r}(x,y) \geq P_q^{\leq r}(w,z)$, for all $r \in H_q] \quad \Leftrightarrow \quad P_q^-((x,y),(w,z)) = 1$,
3) $P_q^+((x,y),(w,z)) = P_q^-((w,z),(x,y))$, for all $(x,y),(w,z) \in A \times A$.

The first two properties represent monotonicity of comparisons with respect to cumulated preferences. The third property represents a kind of symmetry of comparisons.

Property 1) says that for each $q \in C$, object x is certainly (with credibility equal to 1) preferred to y by at least the same degree as object w is preferred to z, if and only if for each $r \in H_q$ the fuzzy upward cumulated preference of x over y is not smaller than the fuzzy upward cumulated preference of w over z.

Property 2) has an analogous interpretation, with respect to the fuzzy downward cumulated preference.

Property 3) says that for each $q \in C$, the credibility of the comparison statement "*x is preferred to y by at least the same degree as w is preferred to z*" equals the credibility of the symmetric comparison statement "*w is preferred to z by at most the same degree as x is preferred to y*".

A sufficient condition for properties 1) and 2) is that the fuzzy implication $I^{\rightarrow}$ considered in the definition of $P_q^{+}\left((x,y),(w,z)\right)$ and $P_q^{-}\left((x,y),(w,z)\right)$ satisfies the following requirement for each $a,b \in [0,1]$:

$$I^{\rightarrow}(a,b) = 1 \quad \Leftrightarrow \quad a \le b.$$

The following two conditions are sufficient for property 3):

– for each $h \in H_q-\{1\}$,
$$P_q^{\ge h}(x,y) = N\left(P_q^{\le h-1}(x,y)\right),$$
$$P_q^{\le h-1}(x,y) = N\left(P_q^{\ge h}(x,y)\right),$$
– for all $a,b \in [0,1]$, $I^{\rightarrow}(a,b) = I^{\rightarrow}(N(b),N(a))$.

The implications of Łukasiewicz and Perny-Fodor satisfy the above conditions concerning implications (see [21.12]).

Now, it is possible to define fuzzy positive and fuzzy negative dominance relations. Given $(x,y),(w,z) \in A \times A$ and $P \subseteq C$, let

$$D_P^{+}\left((x,y),(w,z)\right) \quad = \quad \underset{q \in P}{T}\left(P_q^{+}((x,y),(w,z))\right),$$

$$D_P^{-}\left((x,y),(w,z)\right) \quad = \quad \underset{q \in P}{T}\left(P_q^{-}((x,y),(w,z))\right).$$

Let us observe that $D_P^{+}\left((x,y),(w,z)\right)$ represents the credibility of the statement "*x is preferred to y by at least the same degree as w is preferred to z, taking into account all criteria q from P*", while $D_P^{-}\left((x,y),(w,z)\right)$ represents the credibility of the statement "*x is preferred to y by at most the same degree as w is preferred to z, taking into account all criteria q from P*".

Finally, the P-lower and the P-upper approximations of S are fuzzy sets defined on B, whose membership functions (denoted by $\underline{P}[S(x,y)]$ and $\overline{P}[S(x,y)]$) are, respectively,

$$\underline{P}[S(x,y)] = \underset{(w,z) \in B}{T}\left(T^{*}\left(N\left(D_P^{+}\left((w,z),(x,y)\right)\right),S(w,z)\right)\right),$$

$$\overline{P}[S(x,y)] = \underset{(w,z) \in B}{T^{*}}\left(T\left(D_P^{+}\left((x,y),(w,z)\right),S(w,z)\right)\right).$$

$\underline{P}[S(x,y)]$ represents the credibility of the statement *"for all $(w,z)\in B$, (w,z) does not dominate (x,y) and/or w outranks z"*, while $\overline{P}[S(x,y)]$ represents the credibility of the statement *"there is at least one $(w,z)\in B$ such that (x,y) dominates (w,z) and w outranks z"*.

Analogously, the P-lower and the P-upper approximations of S^c are fuzzy sets defined on B, whose membership functions (denoted by $\underline{P}[S^c(x,y)]$ and $\overline{P}[S^c(x,y)]$ are, respectively,

$$\underline{P}[S^c(x,y)] = \mathop{T}_{(w,z)\in B} \left(T^* \left(N \left(D_P^- ((w,z),(x,y)) \right), S^c(w,z) \right) \right),$$

$$\overline{P}[S^c(x,y)] = \mathop{T^*}_{(w,z)\in B} \left(T \left(D_P^- ((x,y),(w,z)), S^c(w,z) \right) \right).$$

Let us note that, using the definition of the T^*-implication, denoted by $I_{\overrightarrow{T^*},N}$, it is possible to rewrite the definitions of $\underline{P}[S(x,y)]$, $\overline{P}[S(x,y)]$, $\underline{P}[S^c(x,y)]$, and $\overline{P}[S^c(x,y)]$ as follows:

$$\underline{P}[S(x,y)] = \mathop{T}_{(w,z)\in B} \left(I_{\overrightarrow{T^*},N} \left(D_P^+ ((w,z),(x,y)), S(w,z) \right) \right),$$

$$\overline{P}[S(x,y)] = \mathop{T^*}_{(w,z)\in B} \left(N \left(I_{\overrightarrow{T^*},N} \left(D_P^+ ((x,y),(w,z)), N(S(w,z)) \right) \right) \right),$$

$$\underline{P}[S^c(x,y)] = \mathop{T}_{(w,z)\in B} \left(I_{\overrightarrow{T^*},N} \left(D_P^- ((w,z),(x,y)), S^c(w,z) \right) \right),$$

$$\overline{P}[S^c(x,y)] = \mathop{T^*}_{(w,z)\in B} \left(N \left(I_{\overrightarrow{T^*},N} \left(D_P^- ((x,y),(w,z)), N(S^c(w,z)) \right) \right) \right).$$

It may be proved that if properties 1) and 2) above hold, then

$$\underline{P}[S(x,y)] \leq S(x,y) \leq \overline{P}[S(x,y)], \ \underline{P}[S^c(x,y)] \leq S^c(x,y) \leq \overline{P}[S^c(x,y)],$$

i.e., in terms of the fuzzy inclusion, S and S^c include their lower approximation and are included in their upper approximation.

If (T, T^*, N) is the De Morgan triplet, N is involutive, and property 3) above is satisfied, then the following complementarity properties hold:

i) $\underline{P}[S(x,y)] = N(\overline{P}[S^c(x,y)]),$
ii) $\overline{P}[S(x,y)] = N(\underline{P}[S^c(x,y)]),$
iii) $\underline{P}[S^c(x,y)] = N(\overline{P}[S(x,y)]),$
iv) $\overline{P}[S^c(x,y)] = N(\underline{P}[S(x,y)]).$

Property i), expressed in terms of the complement of a fuzzy set, means that the P-lower approximation of S is the complement of the P-upper approximation of the complement of S, that is, S^c. Properties ii) to iv) have analogous interpretations.

Remark that the above definitions of P-lower and P-upper approximations can be given in terms of necessity and possibility measures involving modifiers, as in Sect. 21.2.1 for the multicriteria classification problem.

Using the fuzzy positive and fuzzy negative dominance relations, one can induce decision rules from fuzzy rough approximations of fuzzy relations S and S^c defined on B. The syntax of certain and possible decision rules is the same as in DRSA (see Sect. 20.4.2); however, the upward and downward cumulated preferences $P_q^{\geq h}$ and $P_q^{\leq h}$ in the premise are fuzzy. Moreover, with each conclusion of decision rule r, there is associated a *credibility* $\chi(r)$ defined analogously to the credibility of classification rules (see Sect. 21.2.1).

21.2.3 Gradual Rules and Fuzzy Rough Approximations Without Fuzzy Logical Connectives

As mentioned at the beginning of this section, the main preoccupations with almost all fuzzy extensions of Pawlak's definitions of lower and upper approximations were related to the use of fuzzy connectives. In fact, there is no rule for the choice of the "right" connective, so this choice is always arbitrary to some extent.

Another drawback of fuzzy extensions of rough sets involving fuzzy connectives is that they are based on cardinal properties of membership degrees. In consequence, the result of these extensions is sensitive to the order-preserving transformation of membership degrees. For example, consider the T-conorm of Łukasiewicz as a fuzzy connective; it may be used in the definition of both the fuzzy lower approximation (to build fuzzy implication) and the fuzzy upper approximation (as a fuzzy counterpart of a union). The T-conorm of Łukasiewicz is defined as

$$T^*(\alpha,\beta) = \min\{\alpha+\beta, 1\}.$$

Let us consider the following values of arguments: $\alpha = 0.5$, $\beta = 0.3$, $\gamma = 0.2$, and $\delta = 0.1$; and their order-preserving transformations: $\alpha' = 0.4$, $\beta' = 0.3$, $\gamma' = 0.2$, and $\delta' = 0.05$.

The values of the t-conorm in both cases are as follows:

$$T^*(\alpha,\delta) = 0.6, \qquad T^*(\beta,\gamma) = 0.5,$$

$$T^*(\alpha',\delta') = 0.45, \qquad T^*(\beta',\gamma') = 0.5.$$

One can see that the order of the results has changed after the order preserving transformations of the arguments. This means that the Łukasiewicz T-conorm takes into account not only the ordinal properties of the membership degrees, but also their cardinal properties. A natural question arises: is it reasonable to expect from membership degree a cardinal content instead of only an ordinal one? Or, in other words, is it realistic that a human being is able to say in a meaningful way not only that

a) "object x belongs to fuzzy set X more likely than object y",

but even something like

b) "object x belongs to fuzzy set X two times more likely than object y"?

We claim that it is safer to consider information of type a), because information of type b) is rather meaningless for a human being [21.4].

The above doubt about the cardinal content of the fuzzy membership degree shows the need for methodologies which consider the imprecision in perception typical for fuzzy sets, but avoid as much as possible meaningless transformation of information through fuzzy connectives.

A fuzzy extension of rough sets that takes into account the above requirement was proposed in [21.16]. It avoids arbitrary choice of fuzzy connectives and meaningless operations on membership degrees. This approach belongs to the minority of fuzzy extensions of the rough set concept that does not involve fuzzy connectives and cardinal interpretation of membership degrees. Within this minority, it is related to the approach of Nakamura and Gao [21.46], using α-cuts on fuzzy similarity relations between objects; this relationship will be explained in the next section.

This new methodology of fuzzy rough approximation infers the most cautious conclusion from available imprecise information. In particular, we observe that any approximation of knowledge about Y using knowledge about X is based on positive or negative relationships between premises and conclusions, i.e,

i) "the more x is X, the more it is Y" (positive relationship),
ii) "the more x is X, the less it is Y" (negative relationship).

The following simple relationships respectively illustrate i) and ii): "the larger the market share of a company, the larger its profit" (positive relationship) and "the larger the debt of a company, the smaller its profit" (negative relationship). These relationships have been already considered within fuzzy set theory under the name of *gradual rules* [21.9]. The fuzzy rough approach discussed here extends the concept of gradual rules by handling ambiguity of information through fuzzy rough approximations.

Note that the syntax of gradual rules is based on monotonic relationships that can also be found in dominance-based decision rules induced from preference-ordered data. From this point of view, the fuzzy rough approximation proposed in [21.16] is closely related to DRSA.

Next, we define a syntax and a semantics of considered gradual rules; we also show how they represent positive and negative relationships between fuzzy sets corresponding to the premise and the conclusion of a decision rule. In the following point, we are introducing fuzzy rough approximations consistent with the gradual rules considered; the fuzzy rough approximations create a base for induction of decision rules. Then, we present a new scheme of inference with a fuzzy rough *modus ponens* adapted to the considered decision rules. In the last point, we give some remarks on further research directions.

Gradual Decision Rules. We aim to obtain gradual decision rules of the following types:

- *lower-approximation rules with positive relationship* (LP-rule):
  "if condition X has credibility $C(X) \geq \alpha$, then decision Y has credibility $C(Y) \geq \beta$";
- *lower-approximation rules with negative relationship* (LN-rule):
  "if condition X has credibility $C(X) \leq \alpha$, then decision Y has credibility $C(Y) \geq \beta$";
- *upper-approximation rule with positive relationship* (UP-rule):
  "if condition X has credibility $C(X) \leq \alpha$, then decision Y could have credibility $C(Y) \leq \beta$";
- *upper-approximation rule with negative relationship* (UN-rule):
  "if condition X has credibility $C(X) \geq \alpha$, then decision Y could have credibility $C(Y) \leq \beta$".

The above decision rules will be represented by 3-tuples $<X,Y,f^+>$, $<X,Y,f^->$, $<X,Y,g^+>$, and $<X,Y,g^->$, respectively, where X is a given condition, Y is a given decision, and $f^+:[0,1]\rightarrow[0,1]$, $f^-:[0,1]\rightarrow[0,1]$, $g^+:[0,1]\rightarrow[0,1]$, and $g^-:[0,1]\rightarrow[0,1]$ are functions relating the credibility of X with the credibility of Y in lower- and upper-approximation rules. More precisely, functions f^+, f^-, g^+, and g^- permit us to rewrite the above decision rules as follows:

- LP-rule: "if condition X has credibility $C(X) \geq \alpha$, then decision Y has credibility $C(Y) \geq \beta = f^+(\alpha)$";
- LN-rule: "if condition X has credibility $C(X) \leq \alpha$, then decision Y has credibility $C(Y) \geq \beta = f^-(\alpha)$";
- UP-rule: "if condition X has credibility $C(X) \leq \alpha$, then decision Y could have credibility $C(Y) \leq \beta = g^+(\alpha)$".
- UN-rule: "if condition X has credibility $C(X) \geq \alpha$, then decision Y could have credibility $C(Y) \leq \beta = g^-(\alpha)$".

An LP-rule can be regarded as a *gradual rule* [21.9]; it can be interpreted as "the more object x is X, the more it is Y". In this case, the relationship between credibility of premise and conclusion is positive and certain. LN-rule can be interpreted, in turn, as "the less object x is X, the more it is Y", so the relationship is negative and certain. On the other hand, the UP-rule can be interpreted as "the more object x is X, the more it could be Y", so the relationship is positive and possible. Finally, UN-rule can be interpreted as "the less object x is X, the more it could be Y", so the relationship is negative and possible.

Example A. In this example, we shall illustrate the usefulness of the above four types of decision rules.

Let us consider a hypothetical car selection problem in which the maximum speed is used for the evaluation of cars.

We assume that the decision maker evaluates 10 sample cars, listed in Table 21.2, from the viewpoint of acceptability for his purchase. The evaluation

Table 21.2. Decision maker's evaluation of sample cars

Car:	A	B	C	D	E	F	G	H	I	J
Maximum_speed (km/h)	180	200	210	200	190	240	210	220	200	230
Acceptability	0.3	0.4	0.6	0.8	0.2	0.9	0.7	0.9	0.7	1

Table 21.3. Degrees of membership to $X = speedy\_car$, $Y = acceptable\_car$

Car:	A	B	C	D	E	F	G	H	I	J
μ_{speedy_car}	0	0.25	0.5	0.25	0	1	0.5	0.75	0.25	1
$\mu_{acceptable\_car}$	0.3	0.4	0.6	0.8	0.2	0.9	0.7	0.9	0.7	1

score is given as a degree of acceptability between 0 (strong rejection) and 1 (strong acceptance), as shown in the bottom row of Table 21.2.

In the decision maker's opinion, a car whose *maximum_speed* is more than 230 km/h is definitely speedy and a car whose *maximum_speed* is less than 190 km/h is definitely not speedy. Therefore, we may define fuzzy set *speedy_car* for the decision maker by the following membership function:

$$\mu_{speedy\_car}(maximum\_speed(x)) = \begin{cases} 0 & \\ \quad if\ maximum\_speed\,(x) < 190 & \\ \frac{maximum\_speed(x) - 190}{40} & \\ \quad if\, 190 \leq\ maximum\_speed\,(x) \leq 230 & \\ 1 & \\ \quad if\ maximum\_speed\,(x) > 230 & \end{cases}$$

where x is a variable corresponding to a car, and *maximum_speed*$(\cdot)$ is a function which maps from the set of cars to each one's *maximum_speed*.

The values of μ_{speedy_car} are shown in Table 21.3. They correspond to credibility of the statement that car x characterized by some *maximum_speed* is speedy.

From Table 21.3 we can induce the following functions, $f^+(\cdot)$ and $g^+(\cdot)$:

$$f^+(\alpha) = \begin{cases} 0.2 & if\, 0 \leq \alpha < 0.25 \\ 0.4 & if\, 0.25 \leq \alpha < 0.5 \\ 0.6 & if\, 0.5 \leq \alpha < 0.75 \\ 0.9 & if\, 0.75 \leq \alpha \leq 1 \end{cases} \quad and\ g^+(\alpha) = \begin{cases} 0.3 & if\, \alpha = 0 \\ 0.8 & if\, 0 < \alpha \leq 0.5 \\ 0.9 & if\, 0.5 < \alpha \leq 0.75 \\ 1 & if\, 0.75 < \alpha \leq 1 \end{cases}.$$

Using functions $f^+(\cdot)$ and $g^+(\cdot)$, we can represent knowledge in terms of LP- and UP-decision rules having the following syntax:

– LP-rule: "if x is *speedy_car* with credibility $\mu_{speedy\_car}(maximum\_speed(x)) \geq \alpha$, then x is *acceptable_car* with credibility $f^+(\alpha) \geq \beta$";
– UP-rule: "if x is *speedy_car* with credibility $\mu_{speedy\_car}(maximum\_speed(x)) \leq \alpha$, then x is *acceptable_car* with credibility $g^+(\alpha) \leq \beta$".

The particular LP-rules induced from Table 21.3 are

"if $\mu_{speedy\_car}(maximum\_speed(x)) \geq 0$, then $\mu_{acceptable\_car}(x) \geq 0.2$",
"if $\mu_{speedy\_car}(maximum\_speed(x)) \geq 0.25$, then $\mu_{acceptable\_car}(x) \geq 0.4$",
"if $\mu_{speedy\_car}(maximum\_speed(x)) \geq 0.5$, then $\mu_{acceptable\_car}(x) \geq 0.6$",
"if $\mu_{speedy\_car}(maximum\_speed(x)) \geq 0.75$, then $\mu_{acceptable\_car}(x) \geq 0.9$".

The particular UP-rules induced from Table 21.3 are

"if $\mu_{speedy\_car}(maximum\_speed(x)) \leq 0$, then $\mu_{acceptable\_car}(x) \leq 0.3$",
"if $\mu_{speedy\_car}(maximum\_speed(x)) \leq 0.5$, then $\mu_{acceptable\_car}(x) \leq 0.8$",
"if $\mu_{speedy\_car}(maximum\_speed(x)) \leq 0.75$, then $\mu_{acceptable\_car}(x) \leq 0.9$",
"if $\mu_{speedy\_car}(maximum\_speed(x)) \leq 1$, then $\mu_{acceptable\_car}(x) \leq 1$".

Note that the syntax of LP- and UP-rules is similar to the syntax of "at least" and "at most" decision rules induced in DRSA from dominance-based rough approximations of preference-ordered decision classes.

On the other hand, the syntax of LP- and UP-rules is more general than that of usual gradual rules introduced in [21.9]. Indeed, while the usual gradual rules are statements of the type "if $\mu_X(x) \geq \alpha$, then $\mu_Y(x) \geq \alpha$", the LP-rules state "if $\mu_X(x) \geq \alpha$, then $\mu_Y(x) \geq \beta$". Therefore, the latter permits us to consider different credibilities for the premise and the conclusion.

The above decision rules can be modeled using Gaines-Rescher implication as follows:

- an LP-rule "if $\mu_X(x) \geq \alpha$, then $\mu_Y(x) \geq \beta$" is true with the credibility $I(f^+(\mu_X(x)), \mu_Y(x))$, where

$$I(a,b) = \begin{cases} 1 \ if \ a \leq b \\ 0 \ otherwise \end{cases} \text{ is the Gaines-Rescher implication;}$$

- an LN-rule "if $\mu_X(x) \leq \alpha$, then $\mu_Y(x) \geq \beta$" is true with the credibility $I(f^-(\mu_X(x)), \mu_Y(x))$;
- a UP-rule "if $\mu_X(x) \leq \alpha$, then $\mu_Y(x) \leq \beta$" is true with the credibility $I(\mu_Y(x), g^+(\mu_X(x)))$;
- a UN-rule "if $\mu_X(x) \geq \alpha$, then $\mu_Y(x) \leq \beta$" is true with the credibility $I(\mu_Y(x), g^-(\mu_X(x)))$.

For example, an LP-rule says that $\mu_Y(x) \geq f^+(\mu_X(x))$ for any pair of objects x and y. Note that the implication used for modeling the above four gradual rules is based on ordinal properties of its arguments, which are membership degrees of premise and conclusion. Moreover, the output of the implication is crisp, i.e., either 1 (true) or 0 (false).

The usual gradual rules can also be expressed in terms of the Gaines-Rescher implication; however, as these rules involve a single level of credibility α on the premise and on the conclusion, the rule "if $\mu_X(x) \geq \alpha$, then $\mu_Y(x) \geq \alpha$" is true with credibility $I(\mu_X(x), \mu_Y(x))$. When the credibilities of the premise and the conclusion are different, say α and β, then, in the implication $I(\cdot, \cdot)$, we need $f^+(\mu_X(x))$, instead of only $\mu_X(x)$.

Finally, let us observe that functions $f^+(\alpha)$, $f^-(\alpha)$, $g^+(\alpha)$, and $g^-(\alpha)$ can be seen as modifiers [21.72] that give restrictions on the credibility of the conclusion on the basis of the credibility of the premise of a decision rule. Let us also mention that Bouchon-Meunier and Yao [21.5] studied fuzzy rules in relation to general modifiers, and that [21.40] proposed the use of modifiers for specification of the possibility and necessity measures useful for the induction of fuzzy rules (see Sect. 21.2.1).

Fuzzy Rough Approximations As a Basis for Rule Induction. The functions $f^+(\cdot)$, $f^-(\cdot)$, $g^+(\cdot)$, and $g^-(\cdot)$ are related to specific definitions of lower and upper approximations considered within rough set theory. Let us consider a universe of the discourse U and two fuzzy sets, X and Y, defined on U by means of membership functions $\mu_X:U \to [0,1]$ and $\mu_Y:U \to [0,1]$. Suppose that we want to approximate knowledge contained in Y using knowledge about X. Let us also adopt the hypothesis that X is positively related to Y, which means "the more x is in X, the more x is in Y".

Then, the lower approximation of Y, given the information on X, is a fuzzy set $\underline{App}^+(X,Y)$, whose membership function for each $x \in U$, denoted by $\mu[\underline{App}^+(X,Y),x]$, is defined as follows:

$$\mu[\underline{App}^+(X,Y),x] = \inf_{z \in U:\mu_X(z) \geq \mu_X(x)} \{\mu_Y(z)\}.$$

Interpretation of the lower approximation is based on a specific meaning of the concept of ambiguity. According to knowledge about X, the membership of object $w \in U$ in fuzzy set Y is ambiguous at the level of credibility ρ if there exists an object $z \in U$ such that $\mu_X(w) < \mu_X(z)$, however, $\mu_Y(w) \geq \rho$ while $\mu_Y(z) < \rho$. Thus, on the basis of knowledge about X, and taking into account the positive relationship between X and Y, x belongs to fuzzy set Y without ambiguity for any credibility level $\rho \leq \mu[\underline{App}^+(X,Y),x]$.

Observe that the above meaning of ambiguity is concordant with the dominance principle assumed in DRSA.

Analogously, the upper approximation of Y, given the information on X, is a fuzzy set $\overline{App}^+(X,Y)$, whose membership function for each $x \in X$, denoted by $\mu[\overline{App}^+(X,Y),x]$, is defined as follows:

$$\mu[\overline{App}^+(X,Y),x] = \sup_{z \in U:\mu_X(z) \leq \mu_X(x)} \{\mu_Y(z)\}.$$

Interpretation of the upper approximation is based, again, on the above meaning of ambiguity. On the basis of knowledge about X, and taking into account the positive relationship between X and Y, x belongs to fuzzy set Y with some possible ambiguity at credibility level $\rho \leq \omega$, where ω is the greatest membership degree of $z \in U$ in Y, such that ω is exactly equal to $\mu[\overline{App}^+(X,Y),x]$. In other words, even if $\mu_Y(x)$ would be smaller than ω, then due to the fact that there exists $z \in U$ with $\mu_X(z) \leq \mu_X(x)$, but $\mu_Y(z) = \omega > \mu_Y(x)$ (i.e., z and x are ambiguous), another object w, such that $\mu_X(w) = \mu_X(x)$, could belong to Y with credibility ω.

Table 21.4. Rough approximations of the set of cars ($X = $ speedy_car, $Y = $ acceptable_car)

Car:	A	B	C	D	E	F	G	H	I	J
μ_{speedy_car}	0	0.25	0.5	0.25	0	1	0.5	0.75	0.25	1
$\mu_{acceptable\_car}$	0.3	0.4	0.6	0.8	0.2	0.9	0.7	0.9	0.7	1
$\mu[\underline{App}^+(X,Y),x]$	0.2	0.4	0.6	0.4	0.2	0.9	0.6	0.9	0.4	0.9
$\mu[\overline{App}^+(X,Y),x]$	0.8	0.8	0.8	0.8	0.3	1	0.8	0.9	0.8	1

Example A (cont.) According to the above definitions, the lower and upper approximations of the set of cars described in Table 21.2 are presented in Table 21.4.

The concept of ambiguity and its relation with rough approximations can be illustrated on cars from Table 21.4. Let us consider car G. Its acceptability is 0.7. However, there is another car, called C, having membership not smaller than G in speedy_car, but a smaller membership degree than G in acceptable_car. Therefore, acceptability 0.7 for car G is ambiguous. The highest membership degree to acceptable_car for which there is no ambiguity is equal to 0.6. Therefore, this value is the degree of membership of car G in the lower approximation of the set acceptable_car. Considering car G again, let us note that there exists another car, called D, having membership degree not greater than G in speedy_car, but a membership degree greater than G in acceptable_car. Therefore, G is acceptable_car with the membership 0.8, which is possible because of ambiguity between G and D. This is also the highest degree of membership of car G in acceptable_car taking into account the ambiguity; consequently, it is the membership degree of car G in the upper approximation of acceptable_car.

These remarks on the concept of rough approximation can be summarized by saying that for each car in Table 21.4, the degrees of membership to the lower and upper approximations of acceptable_car, respectively, give for each car the maximum degree of membership to acceptable_car without ambiguity and the maximum degree of membership to acceptable_car admitting some possible ambiguity.

The above definitions of fuzzy rough approximations are similar to the definitions of fuzzy lower and upper approximations in terms of α-cuts first proposed by Nakamura and Gao [21.46]. They considered a fuzzy similarity relation on U, and α-cuts on this relation, and defined a membership degree to the lower or upper approximation for each level α and for each object $x \in U$, using the inf and sup operators, respectively, on objects $y \in U$ that are similar to x by at least degree α. Our use of sup and inf in the above definitions is similar; however, we do not consider any fuzzy similarity relation, and the membership degrees obtained for lower or upper approximations depend only on object $x \in U$, and not on the specific level of the α-cut. In this sense, our definitions are non-parametric.

One can remark that the definitions of $App^+(X,Y)$ and $\overline{App}^+(X,Y)$ are useful if X is positively related to Y, i.e., when we suppose that "the more $\mu_Y(x)$, the more $\mu_X(x)$" or "the less $\mu_Y(x)$, the less $\mu_X(x)$". When X is negatively related to Y, then "the more $\mu_Y(x)$, the less $\mu_X(x)$" or "the less $\mu_Y(x)$, the more $\mu_X(x)$". In this case, the definitions of $App^+(X,Y)$ and $\overline{App}^+(X,Y)$ are not useful; and thus we need another type of approximation.

The following definitions of the lower and upper approximations are appropriate when X is negatively related to Y:

$$\mu[\underline{App}^-(X,Y),x] = \inf_{z\in U:\mu_X(z)\leq\mu_X(x)} \{\mu_Y(z)\},$$

$$\mu[\overline{App}^-(X,Y),x] = \sup_{z\in U:\mu_X(z)\geq\mu_X(x)} \{\mu_Y(z)\}.$$

These definitions have a similar interpretation in terms of ambiguity as the previous ones, and they are also concordant with the dominance principle. Of course, the above approximations are not useful when X is positively related to Y. When X is independent of Y, it is not worthwhile approximating Y by X.

It has been proved in [21.16] that the following properties are satisfied:

1) for each fuzzy set X and Y defined on U, and for each $x\in U$

1.1) $\mu[\underline{App}^+(X,Y),x] \leq \mu_Y(x) \leq \mu[\overline{App}^+(X,Y),x]$,

1.2) $\mu[\underline{App}^-(X,Y),x] \leq \mu_Y(x) \leq \mu[\overline{App}^-(X,Y),x]$.

2) for any negation $N(\cdot)$ that is a decreasing function $N:[0,1]\rightarrow[0,1]$ such that $N(1) = 0$ and $N(0) = 1$, for each fuzzy set X and Y defined on U, and for each $x\in U$

2.1) $\mu[\underline{App}^+(X,Y^c),x] = N(\mu[\overline{App}^-(X,Y),x]) = N(\mu[\overline{App}^+(X^c,Y),x]) = \mu[\underline{App}^-(X^c,Y^c),x]$,

2.2) $\mu[\overline{App}^+(X,Y^c),x] = N(\mu[\underline{App}^-(X,Y),x]) = N(\mu[\underline{App}^+(X^c,Y),x]) = \mu[\overline{App}^-(X^c,Y^c),x]$,

2.3) $\mu[\underline{App}^-(X,Y^c),x] = N(\mu[\overline{App}^+(X,Y),x]) = N(\mu[\overline{App}^-(X^c,Y),x]) = \mu[\underline{App}^+(X^c,Y^c),x]$,

2.4) $\mu[\overline{App}^-(X,Y^c),x] = N(\mu[\underline{App}^+(X,Y),x]) = N(\mu[\underline{App}^-(X^c,Y),x]) = \mu[\overline{App}^+(X^c,Y^c),x]$,

where for a given fuzzy set W, the fuzzy set W^c is its complement defined by $\mu_{W^c}(x) = N(\mu_W(x))$.

Results 1) and 2) above can be read as fuzzy counterparts of the following well-known results within the classical rough set theory: 1) says that fuzzy set Y includes its lower approximation and is included in its upper approximation; 2) represents a complementarity property – it says that the lower (upper) approximation of fuzzy set Y^c that is positively related to X is the complement of the upper (lower) approximation of fuzzy set Y that is negatively related to X (see first lines of (2.1) and (2.2)); moreover, the lower (upper) approximation of fuzzy set Y^c that is negatively related to X is the complement of the upper (lower) approximation of fuzzy set Y positively related to X (see the first lines of (2.3) and (2.4)). Result 2) also states other complementarity properties related to the complement of X (see the second lines of (2.1) to (2.4)); this complementarity property has not been considered within the classical rough set theory.

The lower and upper approximations defined above can serve to induce certain and approximate decision rules. Let us note that inferring lower and upper credibility rules is equivalent to finding functions $f^+(\cdot)$, $f^-(\cdot)$, $g^+(\cdot)$, and $g^-(\cdot)$. These functions are defined as follows: for each $\alpha \in [0,1]$,

$$f^+(\alpha) = \begin{cases} \sup\limits_{\mu_X(x) \leq \alpha} \{\mu[\underline{App}^+(X,Y), x]\} & \text{if } \exists x \in U : \mu_x \leq \alpha, \\ 0 & \text{otherwise,} \end{cases}$$

$$f^-(\alpha) = \begin{cases} \sup\limits_{\mu_X(x) \geq \alpha} \{\mu[\underline{App}^-(X,Y), x]\} & \text{if } \exists x \in U : \mu_x \geq \alpha, \\ 0 & \text{otherwise,} \end{cases}$$

$$g^+(\alpha) = \begin{cases} \inf\limits_{\mu_X(x) \geq \alpha} \{\mu[\overline{App}^+(X,Y), x]\} & \text{if } \exists x \in U : \mu_x \geq \alpha, \\ 1 & \text{otherwise,} \end{cases}$$

and

$$g^-(\alpha) = \begin{cases} \inf\limits_{\mu_X(x) \leq \alpha} \{\mu[\overline{App}^-(X,Y), x]\} & \text{if } \exists x \in U : \mu_x \leq \alpha, \\ 1 & \text{otherwise.} \end{cases}$$

Example B. To give a more intuitive explanation of the fuzzy rough approximation, let us consider the following example, in which a continuous relationship between fuzzy set X and fuzzy set Y is assumed. The relationship between the membership degrees of fuzzy sets X and Y is described by the function $h:[0,1] \to [0,1]$ such that $\mu_Y(x) = h[\mu_X(x)]$. Function $h(\cdot)$ is represented in Fig. 21.1.

Let us approximate fuzzy set Y using the knowledge about fuzzy set X. This approximation permits us to define LP- and UP-rules. The result is presented in Fig. 21.1. Let us note that in all its domain, function $h(\cdot)$ is contained between functions $f^+(\cdot)$ and $g^+(\cdot)$. This is concordant with properties (1.1) above, stating that the membership degree of fuzzy set Y (corresponding to a value of function $h(\cdot)$) is always greater than or equal to the membership degree of its lower approximation (represented by a value of function $f^+(\cdot)$), and is smaller than or equal to the membership degree of its upper approximation (represented by a value of function $g^+(\cdot)$).

Figure 21.2 represents the relationship between the membership degrees of two other fuzzy sets, W and Z, described by function $k : [0,1] \to [0,1]$ such that $\mu_Z(x) = k[\mu_W(x)]$.

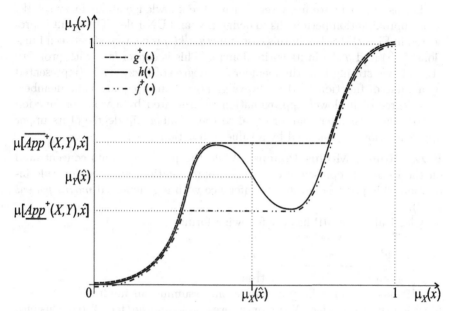

Fig. 21.1. Functions $h(\cdot)$ and $f^+(\cdot)$, $g^+(\cdot)$, representing lower and upper approximations of fuzzy set Y subject to knowledge about fuzzy set X

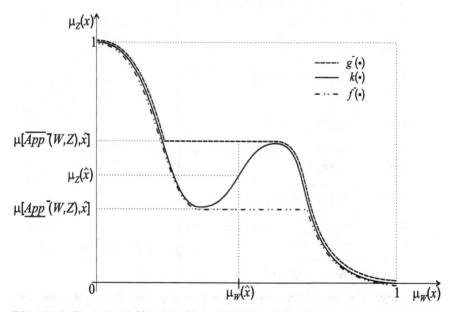

Fig. 21.2. Functions $k(\cdot)$ and $f^-(\cdot)$, $g^-(\cdot)$, representing lower and upper approximations of fuzzy set Z subject to knowledge about fuzzy set W

Let us approximate fuzzy set Z using the knowledge about fuzzy set W. This approximation permits us to define LN- and UN-rules. The result is presented in Fig. 21.2. Also, in this case, function $k(\cdot)$ is contained between functions $f^-(\cdot)$ and $g^-(\cdot)$ in its entire domain. This is concordant with property (1.2) above, stating that the membership degree of fuzzy set Y (represented by a value of function $k(\cdot)$) is always greater than or equal to the membership degree of its lower approximation (represented by a value of function $f^-(\cdot)$), and is smaller than or equal to the membership degree of its upper approximation (represented by a value of function $g^-(\cdot)$).

Fuzzy Rough Modus Ponens. While the previous points concentrated on the issues of representation, rough approximation, and decision rule induction, this point is devoted to inference with a generalized *modus ponens* (MP).

Classically, the MP has the following form:

if	$X \to Y$	is true
and	X	is true
then	Y	is true

MP has the following interpretation: assuming an implication $X \to Y$ (decision rule) and a fact X (premise), we obtain another fact Y (conclusion). If we replace the classical decision rule above by our four kinds of decision rules, then we obtain the following four generalized fuzzy-rough MPs:

$$\left. \begin{array}{ll} \text{if} & \mu_X(x) \geq \alpha \to \mu_Y(x) \geq f^+(\alpha) \\ \text{and} & \mu_X(x) = \alpha' \\ \hline \text{then} & \mu_Y(x) \geq f^+(\alpha') \end{array} \right\} \quad (LP-MP)$$

$$\left. \begin{array}{ll} \text{if} & \mu_X(x) \leq \alpha \to \mu_Y(x) \geq f^-(\alpha) \\ \text{and} & \mu_X(x) = \alpha' \\ \hline \text{then} & \mu_Y(x) \geq f^-(\alpha') \end{array} \right\} \quad (LN-MP)$$

$$\left. \begin{array}{ll} \text{if} & \mu_X(x) \leq \alpha \to \mu_Y(x) \leq g^+(\alpha) \\ \text{and} & \mu_X(x) = \alpha' \\ \hline \text{then} & \mu_Y(x) \leq g^+(\alpha') \end{array} \right\} \quad (UP-MP)$$

$$\left. \begin{array}{ll} \text{if} & \mu_X(x) \geq \alpha \to \mu_Y(x) \leq g^-(\alpha) \\ \text{and} & \mu_X(x) = \alpha' \\ \hline \text{then} & \mu_Y(x) \leq g^-(\alpha') \end{array} \right\} \quad (UN-MP).$$

Summary. The main advantage of this new approach is that it infers the most cautious conclusions from imprecise available information, without using either fuzzy connectives or specific parameters, whose choices are always subjective to some extent. Another advantage of our approach is that it uses only ordinal properties of membership degrees. We noticed that this approach is related to:

- gradual rules, with respect to syntax and semantics of considered decision rules,
- fuzzy rough set approach based on α-cuts, with respect to the specific use of *sup* and *inf* in the definition of lower and upper approximations,
- dominance-based rough set approach, with respect to the idea of a monotonic relationship between the credibilities of premise and conclusion.

We think that this approach gives a new prospect to applications of fuzzy rough approximations in real-world decision problems. More precisely, we envisage the following four extensions of this methodology:

1. Multi-premise generalization: in [21.16], we dealt with relationships between membership degrees of fuzzy sets X and Y; however, in real-world problems, we are interested in relationships between membership degrees in fuzzy sets X_i, $i = 1, \ldots, u$, and Y_j, $j = 1, \ldots, m$, which lead to the consideration of gradual decision rules of the type "*the larger the market share of a company and the smaller its debt, the larger its profit and the better its financial rating*".
2. Variable-precision fuzzy rough approximation.
3. Further generalization of *modus ponens* in order to infer a conclusion from multiple gradual rules.
4. Imprecise input data represented by fuzzy numbers.

21.3 DRSA for Decision Under Risk

In [21.30], we opened a new avenue for applications of the rough set concept. This avenue leads to the classical problem of *decision under risk*. To adapt DRSA to this problem, we substituted the dominance relation with a *stochastic dominance relation* defined on a set of objects meaning acts.[1] We considered the case of traditional additive probability distribution over a set of states of the world; however, the model is rich enough to handle non-additive probability distributions, and even qualitative ordinal distributions. The adapted DRSA gives a representation of DM's preferences under risk in terms of "*if . . . , then. . .* " decision rules induced from rough approximation of preference ordered classification of acts described in terms of outcomes in uncertain states of the world. The preference ordered classification, in this case, constitutes preferential information acquired from the DM.

---

[1] act means action in traditional vocabulary of decision theory, especially when dealing with decision under uncertainty.

21.3.1 DRSA Based on Stochastic Dominance

To apply DRSA to decision under risk, we consider the following basic elements:

- a set $S = \{s_1, s_2, \ldots, s_s\}$ of states of the world, or simply *states*, which are supposed to be mutually exclusive and collectively exhaustive,
- an apriori probability distribution P over the states of the world: more precisely, the probabilities of states $s_1,\ s_2,\ \ldots,\ s_s$ are $p_1,\ p_2,\ \ldots,\ p_s$, respectively ($p_1 + p_2 + \ldots + p_s = 1$, $p_i \geq 0$, $i = 1, \ldots, s$),
- a set $A = \{A_1, A_2, \ldots, A_m\}$ of *acts*,
- a set $X = \{X_1, X_2, \ldots, X_r\}$ of consequences or outcomes that, for the sake of simplicity, are supposed to be expressed in monetary terms, thus $X \subseteq \mathbf{R}$,
- a function $g\colon A \times S \to X$ assigning to each act-state pair $(A_i, s_j) \in A \times S$ a consequence $X_h \in X$,
- a set of classes $\mathbf{Cl} = \{Cl_1, Cl_2, \ldots, Cl_n\}$, such that $Cl_1 \cup Cl_2 \cup \ldots \cup Cl_n = A$, $Cl_p \cap Cl_q = \emptyset$ for each $p, q \in 1, 2 \ldots, n$ with $p \neq q$; the classes of $\mathbf{Cl}$ are preference ordered in increasing order of their indices, in the sense that for each $A_i, A_j \in A$, if $A_i \in Cl_p$ and $A_j \in Cl_q$ with $p > q$, then A_i is preferred to A_j,
- a function $e\colon A \to \mathbf{Cl}$ assigning each act $A_i \in A$ to a class $Cl_t \in \mathbf{Cl}$.

In this context, two different types of dominance relations can be considered:

1) (classical) *dominance*: given $A_i, A_j \in A$, A_i dominates A_j if, for each possible state of the world, act A_i gives an outcome at least as good as act A_j; more formally, $g(A_i, s_k) \geq g(A_j, s_k)$, for each $s_k \in S$,

2) *stochastic dominance*: given $A_i, A_j \in A$, A_i stochastically dominates A_j if, for each outcome $X_h \in X$, act A_i gives an outcome at least as good as X_h with a probability at least as large as act A_j.

Case 1) corresponds to a model in which the utility is state dependent [21.69], while case 2) corresponds to a model of decision under risk proposed by Allais [21.2]. In this chapter, we consider this second case.

On the basis of an apriori probability distribution P, we can assign to each subset of states of the world $W \subseteq S$ ($W \neq \emptyset$) the probability $P(W)$ that one of the states in W is verified, i.e., $P(W) = \sum_{i:s_i \in W} p_i$, and then build up the set Π of all the possible values $P(W)$, i.e.,

$$\Pi = \{\pi \in [0,1]\colon \pi = P(W),\ W \subseteq S\}.$$

We define the function $z\colon A \times S \to \Pi$, assigning to each act-state pair $(A_i, s_j) \in A \times S$ a probability $\pi \in \Pi$, as follows:

$$z(A_i, s_j) = \sum_{r: g(A_i, s_r) \geq g(A_i, s_j)} p_r.$$

Therefore, $z(A_i,s_j)$ represents the probability of obtaining by act A_i an outcome whose value is at least $g(A_i,s_j)$.

On the basis of function $z(A_i,s_j)$, we can define the function ρ: $A \times \Pi \rightarrow X$ as follows:

$$\rho(A_i,\pi) = \min_{j:z(A_i,s_j)\geq\pi} g(A_i,s_j).$$

Thus, $\rho(A_i,\pi) = X_h$ means that by act A_i one can gain *at least* X with a probability greater than or equal to π.

Using function $z(A_i,s_j)$, we can also define function ρ': $A \times \Pi \rightarrow X$ as follows:

$$\rho'(A_i,\pi) = \max_{j:z(A_i,s_j)\leq\pi} g(A_i,s_j).$$

$\rho'(A_i,\pi) = X$ means that by act A_i one can gain *at most* X with a probability smaller than or equal to π.

If the elements of Π, $0 = \pi_{(1)}, \pi_{(2)}, \ldots, \pi_{(w)} = 1$ ($w = card(\Pi)$), are reordered in such a way that $\pi_{(1)} \leq \pi_{(2)} \leq \ldots \leq \pi_{(w)}$, then we have $\rho(A_i,\pi_{(j)}) = \rho'(A_i,1-\pi_{(j-1)})$.

Therefore, $\rho(A_i,\pi_{(j)}) \leq X_h$ is equivalent to $\rho'(A_i,1-\pi_{(j-1)}) \geq X, A_i \in A$, $\pi_{(j)} \in \Pi$, $X \in X$.

Given $A_i, A_j \in A, A_i$ stochastically dominates A_j if and only if $\rho(A_i,\pi) \geq \rho(A_j, \pi)$ for each $\pi \in \Pi$. This is equivalent to saying that, given $A_i, A_j \in A, A_i$ stochastically dominates A_j if and only if $\rho'(A_i,\pi) \leq \rho'(A_j, \pi)$ for each $\pi \in \Pi$.

We can apply DRSA in this context by considering the following correspondence:

- the universe U is the set of acts A,
- the set of condition attributes (criteria) C is the set Π,
- the domain V_π of each criterion $\pi \in \Pi$ is the set X,
- the single decision attribute d specifies the classification of acts from A into classes from Cl,
- the information function f is a function $\boldsymbol{f}$ such that, for all $A_i \in A$ and $\pi \in \Pi$, $\boldsymbol{f}(A_i,\pi) = \rho(A_i,\pi)$ and $\boldsymbol{f}(A_i,d) = e(A_i)$,
- the dominance relation on U is the stochastic dominance relation on A.

The aim of DRSA for decision under risk is to explain the preferences of the DM, represented by his assignments of the acts from A to the classes of Cl, in terms of decision rules involving stochastic dominance on partial profiles corresponding to outcomes X for some probabilities π.

21.3.2 Illustrative Example

The following example illustrates the approach. Let us consider

- set $S = \{s_1,s_2,s_3\}$ of states of the world,

- a priori probability distribution P over the states of the world defined as follows: $p_1 = 0.25$, $p_2 = 0.35$, $p_3 = 0.40$,
- set $A = \{A_1, A_2, A_3, A_4, A_5, A_6\}$ of acts,
- set $X = \{0, 10, 15, 20, 30\}$ of consequences,
- set of classes $Cl = \{Cl_1, Cl_2, Cl_3\}$, where Cl_1 is the set of *bad* acts, Cl_2 is the set of *medium* acts, Cl_3 is the set of *good* acts,
- function $g:A \times S \rightarrow X$, assigning to each act-state pair $(A_i, s_j) \in A \times S$ a consequence $X_h \in X$, and a function $e:A \rightarrow Cl$, assigning each act $A_i \in A$ to a class $Cl_t \in Cl$, as presented in Table 21.5.

Table 21.5. Acts, consequences and assignment to classes from Cl

p_j	A_1	A_2	A_3	A_4	A_5	A_6
s_1 0.25	30	0	15	0	20	10
s_2 0.35	10	20	0	15	10	20
s_3 0.40	10	20	20	20	20	20
d	good	medium	medium	bad	medium	good

Table 21.6. Acts, values of function $\rho(A_i, \pi)$ and assignment to classes from Cl

	A_1	A_2	A_3	A_4	A_5	A_6
0.25	30	20	20	20	20	20
0.35	10	20	20	20	20	20
0.40	10	20	20	20	20	20
0.60	10	20	15	15	20	20
0.65	10	20	15	15	20	20
0.75	10	20	0	15	10	20
1	10	0	0	0	10	10
d	good	medium	medium	bad	medium	good

Table 21.6 shows the values of function $\rho(A_i, \pi)$. It is the data table on which DRSA is applied. Let us explain what the values in the table mean. If we consider the column of an act, say A_3, we see that by act A_3,

- the value 20 in the row corresponding to 0.25 means that the outcome is at least 20 with a probability of at least 0.25,
- the value 15 in the row corresponding to 0.65 means that the outcome is at least 15 with a probability of at least 0.65,
- the value 0 in the row corresponding to 0.75 means that the outcome is at least 0 with a probability of at least 0.75.

If we consider the row corresponding to 0.65, then

- the value 10 relative to A_1 means that, by act A_1, the outcome is at least 10 with a probability of at least 0.65,

- the value 20 relative to A_2 means that, by act A_2, the outcome is at least 20 with a probability of at least 0.65, and so on.

Applying DRSA, we approximate the following *upward* and *downward* *unions* of classes:

- $Cl_2^{\geq} = Cl_2 \cup Cl_3$, i.e., the set of the acts is at least *medium*,
- $Cl_3^{\geq} = Cl_3$, i.e., the set of the acts is (at least) *good*,
- $Cl_1^{\leq} = Cl_1$, i.e., the set of the acts is (at most) *bad*,
- $Cl_2^{\leq} = Cl_1 \cup Cl_2$, i.e., the set of the acts is at most *medium*.

The first result of DRSA is a discovery that the data table (Table 21.6) is not consistent. Indeed, Table 21.6 shows that act A_4 stochastically dominates act A_3; however act A_3 is assigned to a better class (*medium*) than act A_4 (*bad*). Therefore, act A_3 cannot be assigned without doubt to the class of at least *medium* acts, and act A_4 cannot be assigned without doubt to the class of (at most) *bad* acts. In consequence, the lower approximation and the upper approximation of $Cl_2^{\geq}$, $Cl_3^{\geq}$, $Cl_1^{\leq}$, and $Cl_2^{\leq}$ are equal, respectively, to

- $\underline{C}(Cl_2^{\geq}) = \{A_1, A_2, A_5, A_6\} = Cl_2^{\geq} - \{A_3\}$,
  $\overline{C}(Cl_2^{\geq}) = \{A_1, A_2, A_3, A_4, A_5, A_6\} = Cl_2^{\geq} \cup \{A_4\}$,
- $\underline{C}(Cl_3^{\geq}) = \{A_1, A_6\} = Cl_3^{\geq}$, $\quad \overline{C}(Cl_3^{\geq}) = \{A_1, A_6\} = Cl_3^{\geq}$,
- $\underline{C}(Cl_1^{\leq}) = \emptyset = Cl_1^{\leq} - \{A_4\}$, $\quad \overline{C}(Cl_1^{\leq}) = \{A_3, A_4\} = Cl_1^{\leq} \cup \{A_3\}$,
- $\underline{C}(Cl_2^{\leq}) = \{A_2, A_3, A_4, A_5\} = Cl_2^{\leq}$, $\quad \overline{C}(Cl_2^{\leq}) = \{A_2, A_3, A_4, A_5\} = Cl_2^{\leq}$.

Since there are two inconsistent acts on a total of six acts (A_3,A_4), then the quality of approximation (quality of classification) is equal to 4/6.

The second discovery is one reduct of condition attributes (criteria) ensuring the same quality of classification as the whole set Π of probabilities: $RED_{Cl} = \{0.25, 0.75, 1\}$. This means that we can explain the preferences of the DM using the probabilities in RED_{Cl} only. RED_{Cl} is also the core because no probability value can be removed from RED_{Cl} without deteriorating the quality of classification.

The third discovery is sets of decision rules describing the DM's preferences. Below, we present one of minimal sets of decision rules covering all the acts [*we give within brackets a verbal interpretation of the corresponding decision rule*] (we give within parentheses acts supporting the corresponding rule):

1)   if $\rho(A_i, 0.25) \geq 30$, then $A_i \in Cl_3^{\geq}$,
[*if the probability of gaining at least 30 is at least 0.25, then act A_i is (at least) good*] (A_1),

2)   if $\rho(A_i, 0.75) \geq 20$ and $\rho(A_i, 1) \geq 10$, then $A_i \in Cl_3^{\geq}$,
[*if the probability of gaining at least 20 is at least 0.75, and the probability of gaining at least 10 is (at least) 1 (i.e for sure the gain is at least 10), then act A_i is (at least) good*] (A_6),

3)   if $\rho(A_i, 1) \geq 10$, then $A_i \in Cl_2^{\geq}$,
[if the probability of gaining at least 10 is (at least) 1 (i.e., for sure the gain is at least 10), then act A_i is at least medium] $(A_1), (A_5), (A_6)$,

4)   if $\rho(A_i, 0.75) \geq 20$, then $A_i \in Cl_2^{\geq}$,
[if the probability of gaining at least 20 is at least 0.75, then act A_i is at least medium] (A_2, A_6),

5)   if $\rho(A_i, 0.25) \leq 20$ (i.e., $\rho'(A_i, 1) \geq 20$) and $\rho(A_i, 0.75) \leq 15$, (i.e., $\rho'(A_i, 0.35) \geq 15$), then $A_i \in Cl_2^{\leq}$,
[if the probability of gaining at most 20 is (at least) 1 (i.e for sure the gain is at most 20) and the probability of gaining at most 15 is at least 0.35, then act A_i is at most medium] $(A_3), (A_4), (A_5)$,

6)   if $\rho(A_i, 1) \leq 0$, (i.e., $\rho'(A_i, 0.25) \geq 0$), then $A_i \in Cl_1 \cup Cl_2$,
[if the probability of gaining at least 0 is at least 0.25, then act A_i is at most medium] $(A_2), (A_3), (A_4)$,

7)   if $\rho(A_i, 1) \geq 0$ and $\rho(A_i, 1) \leq 0$ (i.e., $\rho'(A_i, 1) = 0$) and $\rho(A_i, 0.75) \leq 15$ (i.e., $\rho'(A_i, 0.35) \geq 10$), then $A_i \in Cl_1 \cup Cl_2$,
[if the probability of gaining at least 0 is 1 (i.e., for sure the gain is at least 0) and the probability of gaining at most 15 is at least 0.35, then act A_i is bad or medium, with no possibility of assigning A_i to only one of the two classes because of ambiguous knowledge] $(A_3), (A_4)$.

Minimal sets of decision rules represent the most concise and non-redundant knowledge contained in Table 21.5 and, consequently, in Table 21.6. The above minimal set of 7 decision rules uses 3 attributes (probabilities 0.25, 0.75, and 1) and 11 elementary conditions, i.e., 26% of descriptors from the original data table (Table 21.6). For larger sets of exemplary acts, the representation in terms of decision rules is even more synthetic (the percentage of descriptors from the original data table is smaller).

Observe that we considered an additive probability distribution; however, an extension to non-additive probability, and even to a qualitative ordinal probability, is straightforward. If the elements of set Π are numerous (like in real applications), a subset $\Pi' \subset \Pi$ of the most significant probability values (e.g., 0, 0.1, 0.2, ..., 0.9, 1) can be considered.

21.4 DRSA for Hierarchical Decision Making

In [21.8], DRSA has been adapted to decision problems, with multiple attributes and criteria, having a hierarchical structure. These problems are often encountered in real life situations, where the process of decision making is decomposable into subproblems; this decomposition may either follow from a naturally hierarchical structure of the evaluation or from a need for the simplification of a complex decision problem. These problems are called *hierarchical classification problems* (HCP).

The hierarchical structure of an HCP has the form of a *tree* whose *nodes* are attributes and/or criteria describing objects. In the root of the tree, there is an overall evaluation, assigning objects to classes; then, in the intermediate nodes, there are subattributes and/or subcriteria, called hierarchical and, finally, in the leafs, there are attributes and/or criteria that do not branch further – they are called flat attributes or criteria. The hierarchy of attributes and criteria seems to be a natural and intuitive concept. For example, when considering the classification of cars, the criterion of fuel consumption may be considered a hierarchical one, consisting of following subcriteria: fuel consumption in urban drive and in highway drive, at 60 km/h and at 120 km/h.

Consideration of hierarchical decision problems is supported by psychological arguments, as pointed out by White, Wilson, and Wilson [21.70]: "the use of hierarchical ordering must be as old as human thought, conscious and unconscious". Psychologists claim that the human brain is limited to about seven (seven, plus or minus two) items in both its short term memory capacity and its discrimination ability [21.43]. Humans have learned how to deal with complexity by hierarchical decomposition. Surprisingly, the hierarchical decomposition of decision problems has gained little attention in scientific decision aiding. The best-known exception is the Analytic Hierarchy Process [21.54].

An example structure of a hierarchical classification problem is shown in Fig. 21.3. The cars are assigned to three classes, acceptable, hardly acceptable, and non-acceptable, on the basis of two attributes (color, country of production) and three criteria (price, max speed, fuel consumption); one of criteria, fuel consumption, is decomposed into four subcriteria.

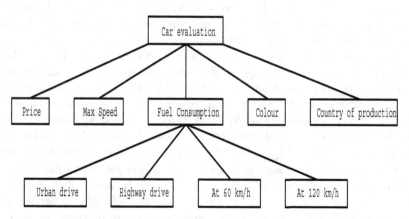

Fig. 21.3. The hierarchy of attributes and criteria for a car classification problem

Subtrees of the hierarchy tree represent subproblems, and each *node* of the tree different from a *leaf* branches into nodes called *direct descendents*. The *root* of the tree refers to an overall decision, each subnode refers to a hierarchical attribute or criterion (subdecision or intermediate decision), and leaves are *flat (ordinary)* attributes and criteria. Note that if the root of a subtree is a hierarchical (regular) attribute, then the corresponding subproblem is a multi-attribute classification problem that can be analysed with CRSA; otherwise, if the root is a hierarchical criterion, then the corresponding subproblem is a multicriteria classification problem that can be analysed with DRSA.

When adapting CRSA and DRSA to HCP, the main difficulty consists of taking into account at each node of the tree the *propagation* of inconsistencies along the tree, i.e., the inconsistent information coming from lower level nodes. In the proposed methodology, the inconsistencies are propagated from the bottom to the top of the tree in the form of *subsets* of possible values of attributes or criteria. Let us note that due to the adaptation of CRSA and DRSA to subsets of possible values of attributes or criteria, one can deal with imprecise evaluations of objects on flat attributes or criteria.

The preference model corresponding to HCP has the form of sets of decision rules for each subproblem. Particular sets of decision rules are generated from rough approximations at each node of the hierarchy. The application of the model consists of progressive classification of new objects at each node of the hierarchy tree.

21.4.1 Rough Set Approach for Attribute Subset Values and Interval Order

Data Representation. The *hierarchical decision table* is presented as a tree TR composed of *subtables*. In some node N_k of TR, there is a subtable ST_k containing the decision attribute or criterion corresponding to node N_k and condition attributes and/or criteria that are direct descendents of N_k.

More formally, the subtable is the 4-tuple $ST_k = \langle U, Q_k, V_k, f \rangle$, where U is a finite set of objects; $Q_k = A_k \cup d_k$ is a finite set of condition and decision attributes such that $A_k = W_k \cup C_k$, where W_k and C_k denote sets of regular attributes and criteria, respectively, corresponding to nodes that are direct descendents of N_k, and d_k is a decision attribute or criterion belonging to the set of condition attributes in the subtable of the *parent node* of N_k; $V_k = \bigcup_{a \in Q_k} V_a$, where V_a is a domain of attribute a; and $f : U \times A_k \to 2^{V_a}$ is a *multiple-valued information function* such that, for each regular attribute $w \in W_k$, $f(x, w) \subseteq V_w$, and, for each criterion $q \in C_k$, $f(x, q) \in [l(x, q), u(x, q)] \subseteq V_q$, where $x \in U$. For each ST_k, a partition of U is considered with respect to the decision attribute or criterion d_k, i.e., classification $Cl(d_k)$. We assume that in each subtable ST_k the objects are assigned to only one class; therefore, the information function for d_k is defined as $f : U \times \{d_k\} \to V_{d_k}$, and $f(x, d_k) \in V_{d_k}$ for each $x \in U$.

Indiscernibility Relation for Subsets of Attribute Values. Two objects, $x, y \in U$, described by subsets of attribute values will be considered as indiscernible on attribute $a \in A_k$ if and only if $f(x,a) \cap f(y,a) \neq \emptyset$. The corresponding *indiscernibility relation* is denoted by $I_a^\cap$. Two objects $x, y \in U$ are indiscernible with respect to the subset of attributes $P \subseteq A_k$, if and only if $xI_a^\cap y$ is true for each $a \in P$. The *P-indiscernibility relation* is denoted by $I_P^\cap$.

The above indiscernibility relation is reflexive and symmetric, but not transitive. This relation satisfies requirements of a tolerance relation. Similar definitions are given, for example, in [21.47, 21.55, 21.66].

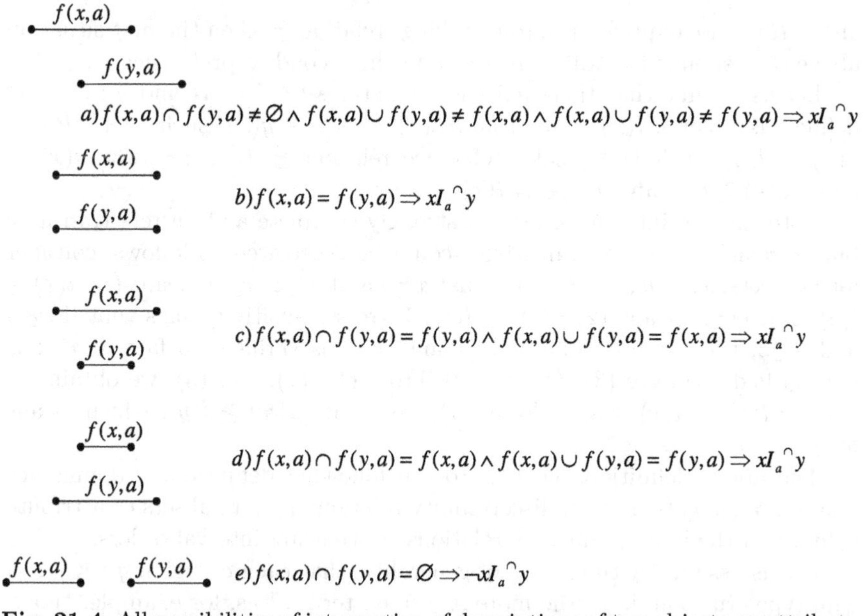

a)$f(x,a) \cap f(y,a) \neq \emptyset \wedge f(x,a) \cup f(y,a) \neq f(x,a) \wedge f(x,a) \cup f(y,a) \neq f(y,a) \Rightarrow xI_a^\cap y$

b)$f(x,a) = f(y,a) \Rightarrow xI_a^\cap y$

c)$f(x,a) \cap f(y,a) = f(y,a) \wedge f(x,a) \cup f(y,a) = f(x,a) \Rightarrow xI_a^\cap y$

d)$f(x,a) \cap f(y,a) = f(x,a) \wedge f(x,a) \cup f(y,a) = f(y,a) \Rightarrow xI_a^\cap y$

e)$f(x,a) \cap f(y,a) = \emptyset \Rightarrow \neg xI_a^\cap y$

Fig. 21.4. All possibilities of intersection of descriptions of two objects on attribute a; Two objects x and y are indiscernible when their description has a common part

Taking into account the indiscernibility relation $I_P^\cap$, all definitions of CRSA may be easily extended. The generalization thus obtained is referred to as Mutiple-Valued CRSA (MV-CRSA). For example, the *P-lower* and the *P-upper approximations* of a class Cl_t, $t \in \{1, \ldots, n\}$, are defined, respectively, as

$$\underline{P}(Cl_t) = \{x \in U : I_P^\cap(x) \subseteq Cl_t\}, \quad \overline{P}(Cl_t) = \bigcup_{x \in Cl_t} I_P^\cap(x),$$

where $I_P^\cap(x) = \{y \in U : xI_P^\cap y\}$ is a *P-elementary* set including x, and $P \subseteq W_k$.

Interval Order. In order to enable imprecise descriptions of objects on criteria, we are required to introduce the definition of interval order, and dominance relation based on it.

The *interval order* is a binary relation R on set U if and only if there exist two functions $g : U \to \mathbf{R}$ (where $\mathbf{R}$ is a set of real numbers) and $q : U \to \mathbf{R}^+$ such that

$$xRy \Leftrightarrow g(x) + q(x) \geq g(y), \text{ and } x, y \in U.$$

Let us observe that

$$xRy \wedge yRx \Leftrightarrow -q(x) \leq g(x) - g(y) \leq q(y) \text{ and}$$
$$xRy \wedge \neg yRx \Leftrightarrow g(x) - g(y) \geq q(y),$$

and if R is a weak preference (outranking) relation $\succeq$, then the first situation above corresponds to indifference $\sim$, and the second to preference $\succ$.

Let us assume that there exist two functions $l : U \to \mathbf{R}$ and $u : U \to \mathbf{R}$ defined as follows: $l(x) = g(x)$ and $u(x) = g(x) + q(x)$, such that $xRy \Leftrightarrow u(x) \geq l(y)$. If R is a weak preference relation $\succeq$, then $x \sim y \Leftrightarrow u(x) \geq l(y) \wedge u(y) \geq l(x)$ and $x \succ y \Leftrightarrow l(x) > u(y)$.

Note that an interval order is a strongly complete and Ferrers transitive binary relation. Ferrers transitivity could be illustrated as follows: consider four objects, $x, y, w, z \in U$, such that $x \succeq y$ and $w \succeq z$; $x \succeq y$ means (1) $u(x) \geq l(y)$, and $w \succeq z$ means (2) $u(w) \geq l(z)$. Ferrers transitivity says that if $x \succeq y$ and $w \succeq z$, then at least one of $x \succeq z$ and $w \succeq y$ is verified. In fact, if $x \succeq z$ is not verified, we have (3) $l(z) > u(x)$. From (1), (2), and (3), we obtain (4) $u(w) \geq l(z) > u(x) \geq l(y)$. From (4), we obtain $u(w) \geq l(y)$, which means $w \succeq y$.

The above definitions lead us to the following definition of dominance relation, aggregating the indiscernibility relations $I_a^\cap$ for subsets of attribute values and the weak preference relations $\succeq_q$ that are interval orders.

Let us assume, without loss of generality, that each criterion $q \in C_k$ is a gain type criterion, i.e., "the more ... the better ..." as, for example, "maximum speed" in the car selection problems." For each $q \in C_k$, we consider a weak preference relation $\succeq_q$ on set U on the basis of values $l(x,q)$ and $u(x,q)$ such that $\succeq_q = \succ_q \cup \sim_q$; thus,

$$x \succeq_q y \Leftrightarrow u(x,q) \geq l(y,q).$$

The *dominance relation* with respect to $P \subseteq A_k$, considering the description of objects by both attributes and criteria, is defined as

$$xD_P^{\langle\rangle}y \Leftrightarrow x \succeq_q y \text{ for all } q \in C_k \text{ and } xI_w^\cap y \text{ for all } w \in W_k.$$

The dominance relation $D_P^{\langle\rangle}$ is reflexive, but it is not Ferrers transitive even if the considered attributes are all interval orders. The following is a counterexample proving this point.

Let us consider four objects, x, y, w, $z \in U$, and two criteria q_1, $q_2 \in C_k$ such that $x \succeq_{q_1} y$, $x \succeq_{q_2} y$, $w \succeq_{q_1} z$, and $w \succeq_{q_2} z$ and, therefore, $x D_P^{\langle\rangle} y$ and $w D_P^{\langle\rangle} z$, with $P = \{q_1, q_2\}$. Let us suppose that

(1) $u(w, q_1) \geq l(z, q_1) > u(x, q_1) \geq l(y, q_1)$ and
(2) $u(x, q_2) \geq l(y, q_2) > u(w, q_2) \geq l(z, q_2)$.

Let us note that (1) and (2) are concordant with (a) $x \succeq_{q_1} y$ (in fact, $u(x,q_1) \geq l(y,q_1)$), (b) $x \succeq_{q_2} y$ (in fact, $u(x,q_2) \geq l(y,q_2)$), (c) $w \succeq_{q_1} z$ (in fact, $u(w,q_1) \geq l(z,q_1)$), and (d) $w \succeq_{q_2} z$ (in fact, $u(w,q_2) \geq l(z, q_2)$). (a) and (b) give $x D_P^{\langle\rangle} y$, while (c) and (d) give $w D_P^{\langle\rangle} z$. However, $x D_P^{\langle\rangle} y$ and $w D_P^{\langle\rangle} z$ derive neither $x D_P^{\langle\rangle} z$ nor $w D_P^{\langle\rangle} y$. In fact, $x D_P^{\langle\rangle} z$ cannot hold because $l(z,q_1) > u(x,q_1)$, while $w D_P^{\langle\rangle} y$ cannot hold because $l(y,q_2) > u(w,q_2)$.

Taking into account the dominance relation $D_P^{\langle\rangle}$, all definitions of DRSA may be easily extended. The generalization thus obtained is referred to as Interval-Valued DRSA (IV-DRSA). Below we update definitions of the granules of knowledge, and of the rough approximation of preference-ordered classes.

The *P-dominated* and the *P-dominating* sets are defined, respectively, as follows:

$$D_P^{\langle\rangle^+}(x) = \left\{ y \in U : y D_P^{\langle\rangle} x \right\}, D_P^{\langle\rangle^-}(x) = \left\{ y \in U : x D_P^{\langle\rangle} y \right\}, \text{ where } P \subseteq C_k.$$

The *P-lower* and the *P-upper approximations* of unions of classes $Cl_t^{\geq}$ and $Cl_t^{\leq}$, for $t \in \{1, \ldots, n\}$, are defined as follows:

$$\underline{P}\left(Cl_t^{\geq}\right) = \left\{ x \in U : D_P^{\langle\rangle^+}(x) \subseteq Cl_t^{\geq} \right\}, \overline{P}\left(Cl_t^{\geq}\right) = \bigcup\nolimits_{x \in Cl_t^{\geq}} D_P^{\langle\rangle^+}(x),$$

$$\underline{P}\left(Cl_t^{\leq}\right) = \left\{ x \in U : D_P^{\langle\rangle^-}(x) \subseteq Cl_t^{\leq} \right\}, \overline{P}\left(Cl_t^{\leq}\right) = \bigcup\nolimits_{x \in Cl_t^{\leq}} D_P^{\langle\rangle^-}(x).$$

Decision Rules. The form of decision rules taking into account the imprecise object description is changed in comparison with the form known from CRSA and DRSA. Moreover, the negative examples are defined differently.

The elementary conditions relative to regular attributes in the premise are formulated as

1) $f(x, a) \cap v \neq \emptyset$, where $a \in W_k$, and $v \subseteq V_a$ is a subset of values of attribute a.

The elementary conditions relative to criteria in the premise are formulated as

2) $u(x, q) \geq r$, where $q \in C_k$, q is of gain type (i.e., "the more ..., the better ...", as, for example, "maximum speed" in the car classification problem), $r \in V_q$, and $u(x, q)$ is an upper bound of interval value of object x on q,
3) $l(x, q) \leq r$, where $q \in C_k$, q is of gain type, $r \in V_q$, and $l(x, q)$ is a lower bound of interval value of object x on q,

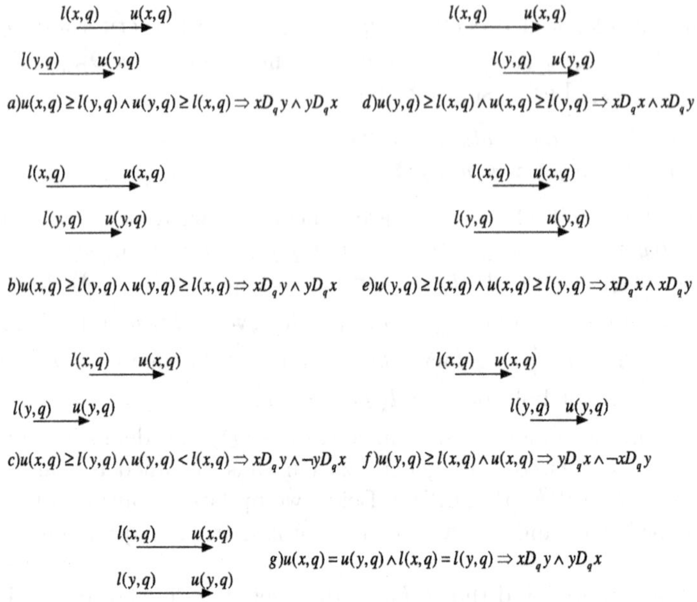

a) $u(x,q) \geq l(y,q) \wedge u(y,q) \geq l(x,q) \Rightarrow xD_q y \wedge yD_q x$ d) $u(y,q) \geq l(x,q) \wedge u(x,q) \geq l(y,q) \Rightarrow xD_q x \wedge xD_q y$

b) $u(x,q) \geq l(y,q) \wedge u(y,q) \geq l(x,q) \Rightarrow xD_q y \wedge yD_q x$ e) $u(y,q) \geq l(x,q) \wedge u(x,q) \geq l(y,q) \Rightarrow xD_q x \wedge xD_q y$

c) $u(x,q) \geq l(y,q) \wedge u(y,q) < l(x,q) \Rightarrow xD_q y \wedge \neg yD_q x$ f) $u(y,q) \geq l(x,q) \wedge u(x,q) \Rightarrow yD_q x \wedge \neg xD_q y$

g) $u(x,q) = u(y,q) \wedge l(x,q) = l(y,q) \Rightarrow xD_q y \wedge yD_q x$

Fig. 21.5. All possible relations between two objects described using intervals on criterion q; object x dominates object y when the upper bound of x is greater than the lower bound of y

4) $l(x,q) \leq r$, where $q \in C_k$, q is of cost type (i.e., "the less ..., the better ...", as, for example, "price" in the car classification problem), $r \in V_q$, and $l(x,q)$ is a lower bound of interval value of object x on q,
5) $u(x,q) \geq r$, where $q \in C_k$, q is of cost type, $r \in V_q$, and $u(x,q)$ is an upper bound of interval value of object x on q.

Decision parts of the rules keep the same form as in CRSA and DRSA. Below, we list the decision rules for multicriteria classification, taking into account the interval object evaluation:

- *certain 'at least'* $D_\geq$-decision rules, where elementary conditions in the premise are in form 1), 2), and 4), and the conclusion is $x \in Cl_t^\geq$,
- *possible 'at least'* $D_\geq$-decision rules, where elementary conditions in the premise are in form 1), 2), and 4), and the conclusion is x could belong to $Cl_t^\geq$,
- *certain 'at most'* $D_\leq$-decision rules, where elementary conditions in the premise are in form 1), 3), and 5), and the conclusion is $x \in Cl_t^\leq$,
- *possible 'at most'* $D_\leq$-decision rules, where elementary conditions in the premise are in form 1), 3), and 5), and the conclusion is that x could belong to $Cl_t^\leq$,

- *approximate* $D_{\geq\leq}$-decision rules, where elementary conditions in the premise are in form 1) to 5), and the conclusion is $x \in Cl_s \cup Cl_{s+1} \cup \ldots \cup Cl_t$, where $s,t \in \{1, \ldots, n\}$ and $s<t$.

Because the generalized indiscernibility and dominance relations are not transitive, when generating any decision rule with a conclusion K, the negative examples with respect to K are those examples which

- are not concordant with K, and
- are either discerned from the positive examples by the indiscernibility relation or are not indifferent with the positive examples in the sense of mutual dominance.

This is explained as follows. Assume that there are three objects, w, y, z $\in U$, described on gain type criterion q only, such that $u(w,q) = 7$, $l(w,q) = 6$, $u(y,q) = 8$, $l(y,q) = 4$, $u(z,q) = 5$, and $l(z,q) = 2$, and w and y belong to class Cl_2, while z belongs to class Cl_1. The lower approximation of $Cl_2^{\geq}$ with respect to $P = \{q\}$ contains only w, because $D_P^{\langle\rangle^+}(w) = \{w,y\}$ and $D_P^{\langle\rangle^+}(y) = \{w,y,z\}$. If y would be considered as a negative example, no certain rule for $Cl_2^{\geq}$ (supported by positive examples from P-lower approximation of $Cl_2^{\geq}$) would be induced. The rule

$$\text{if } (u(x,\ q) \geq 6),\ then\ x \in Cl_2^{\geq}$$

covers $w \in Cl_2^{\geq}$, because $u(w,q) = 7{\geq}6$, but it also covers $y \in Cl_2^{\geq}$, because $u(y,q) = 8{\geq}6$. We accept the above rule, however, even if it covers object y, which is not included in the P-lower approximation of $Cl_2^{\geq}$, because it is concordant with the conclusion of the rule.

21.4.2 Propagation of Inconsistencies and Application of Decision Rules

Let us comment on the *propagation of inconsistencies* along the tree. This is based on the concept of generalized decision. Within CRSA, the *P-generalized decision* of $x \in U$ with respect to the indiscernibility relation is a function $\delta_P(x):U \rightarrow 2^{V_d}$, defined as

$$\delta_P(x) = \{v \in V_d : \exists y \in U, y I_P^{\cap} x \wedge f(y,d) = v\},$$

where $x \in U$ and $P \subseteq W_k$. If an object is not inconsistent with any other object, then $card(\delta_A(x)) = 1$.

Within DRSA, the *P-generalized decision* of object $x \in U$ with respect to the dominance relation is an interval $\delta_P^{\succ}(x) = [l(x,d), u(x,d)]$, where

$$l(x,d) = \min\{v \in V_d : \exists y \in U, y D_P x \wedge f(y,d) = v\},$$

$$u(x,d) = \max\{v \in V_d : \exists y \in U, x D_P y \wedge f(y,d) = v\},$$

and $x,y \in U$ and $P \subseteq C_k$.

In other words, $l(x, d)$ is the lowest decision class of objects dominating x, and $u(x, d)$ is the highest decision class of objects dominated by x. Note that if an object does not cause inconsistency in a decision table, then $l(x, d) = u(x, d)$.

Using the generalized decision, the inconsistencies are propagated from the bottom to the top of the hierarchy. Let ST_k denote a decision subtable corresponding to node N_k of TR, and let ST_l denote a subtable of N_l, a direct descendent of N_k. If object $x \in U$ is inconsistent with other objects in ST_l, then, in the decision subtable $ST_{k,}$, the value of x on $d_l \in C_k$ is as follows:

- $\delta_{C_l}(x)$, if d_l is an attribute,
- $\delta_{C_l}^{\succ}(x)$, if d_l is a criterion,

and d_l corresponds to the decision attribute or criterion in ST_l.

When new objects are submitted to hierarchical classification, the sets of decision rules are to be used progressively, starting from the lowest level of the hierarchy. The decision from each node is propagated up the hierarchy.

Let us comment on the application of decision rules to an object in a particular node. There are two possible cases. First, the node may correspond to an attribute, and then the rules are in the form specified in MV-CRSA. If an object is matched by rules suggesting the same class Cl_t, then the object is assigned to Cl_t without any doubt. Nevertheless, the following doubtful situations can occur:

- object x matches one or more approximate decision rules, e.g.,

  if [conditions] *then* car∈*sport cars* ∪ car∈*family cars*,
  if [conditions] *then* car∈*utility cars* ∪ car∈*family cars*.

- object x matches certain or possible decision rules suggesting different classes, e.g.,

  if [conditions] *then* car∈*utility cars*,
  if [conditions] *then* car∈*family cars*.

In all the above situations, the object is assigned imprecisely to all classes pointed to by the matching rules. When an object does not match the condition part of any decision rule, then it is reasonable to conclude that it may belong to any class.

Second, if a node corresponds to a criterion, then the decision rules are in the form specified in IV-DRSA. When applying $D_\geq$-decision rules to object x, it is possible that x either matches the condition of at least one decision rule, or does not match condition of any decision rule. In the case of at least one matching, it is reasonable to conclude that x belongs to class Cl_t, the lowest class of the upward union $Cl_t^\geq$ resulting from the intersection of all conclusions of rules matching x. In the case of no matching, we conclude that x belongs to Cl_1, i.e., to the worst class.

Analogously, when applying $D_{\leq}$-decision rules to object x, we conclude that x belongs either to class Cl_t, the highest class of the downward union $Cl_t^{\leq}$ resulting from intersection of all conclusions of rules matching x, or to class Cl_n, i.e., to the best class, when x does not match any rule.

However, there may occur four other situations:

- (i) object x matches one or more approximate rules, e.g.,
  if [conditions] *then* car $\in$ *non-acceptable* $\cup$ *hardly acceptable.*
- (ii) object x matches rules with intersecting downward and upward unions of classes, e.g.,
  if [conditions] *then* car $\in$ *hardly acceptable$^{\geq}$* (i.e., hardly acceptable or better),
  if [conditions] *then* car $\in$ *acceptable$^{\leq}$* (i.e., acceptable or worse).
- (iii) object x matches rules with disjoint downward and upward unions of classes, e.g.,
  if [conditions] *then* car $\in$ *acceptable$^{\geq}$* (i.e., acceptable or better),
  if [conditions] *then* car $\in$ *hardly acceptable$^{\leq}$* (i.e., hardly acceptable or worse).
- (iv) object x matches one or more approximate rules and certain rules with intersecting downward and upward unions of classes, e.g.,
  if [conditions] *then* car $\in$ *non-acceptable* $\cup$ *hardly acceptable,*
  if [conditions] *then* car $\in$ *acceptable$^{\geq}$* (i.e., acceptable or better).

The above situations correspond to ambiguous [(i) and (iv)], incomplete [(ii)], and controversial [(iii) and (iv)] knowledge. In all the above situations, object x is assigned imprecisely to an interval of classes. In the first situation, object x is assigned to the interval of classes between the worst and the best class suggested by rules matching x (from non-acceptable to hardly acceptable). In the second and the third situations, object x is assigned to the interval of classes between the worst class of the matching 'at least' rules and the best class of the matching 'at most' rules (in both situations to classes from 'hardly acceptable' to 'acceptable'). In the fourth situation, object x is assigned to the interval of classes that are extreme in the classes suggested by all approximate rules, and the worst class from the intersection of 'at least' rules and/or the best class from the intersection of 'at most' rules matching x.

21.4.3 Illustrative Example

The following illustrative example will serve to explain the concepts introduced in previous points. Let us consider a problem of student qualification to a higher level of study. We are considering students of a Computer Science department choosing their Master's specialization after the third year of studies. In the problem considered, the DM is looking for good students interested in taking a Master's degree in Intelligent Decision Support Systems. The students are assigned to three classes: *desirable, acceptable,* and *unacceptable*. The decision is made on the basis of all evaluations received by students during the Bachelor studies. In the following example, we consider six students described by means of the following two attributes and five (gain type) criteria (see Table 21.7)

- *additional project,*
- *training,*
- *examination score in Statistics,*
- *examination score in Computer Networks,*
- *project score in Computer Networks,*
- *examination score in Databases,*
- *project score in Databases.*

Assume that attributes "*Additional project*" and "*Training*" have the same domain, composed of the following values: *Artificial Intelligence (AI)*, *Statistics*, *Databases*, and *Programming (Progr.)*. The domain of all five criteria is composed of the following evaluations: "*Bad*", "*Sufficient*", and "*Good*". Of course, "*Good*" is better than "*Sufficient*" and "*Sufficient*" is better than "*Bad*".

In the first step of the proposed methodology, we develop a hierarchy of the above evaluation. It seems natural that examination score and project scores from "*Computer Networks*" and "*Databases*" be grouped into two sub-problems. This may also be done with the attributes "*Additional project*" and "*Training*". On the basis of these two attributes, the DM can judge the "*Area of interest*" of a student. The structure of the decomposed problem is presented in Fig. 21.6. The hierarchical criteria "*Computer Network skills*" and "*Database skills*" have the same domain as the component criteria, i.e., "*Bad*", "*Sufficient (Suff.)*", and "*Good*". The hierarchical attribute "*Area of interest*" has the domain composed of the following values: "*Decision Support*","*Knowledge Discovery*", and "*Other*".

The next step of the proposed methodology is the analysis of each sub-problem. In each node, the analysis consists of following phases: evaluation of students according to attributes or criteria considered in the subproblem, analysis of inconsistencies using the rough set approach, and induction of decision rules from rough approximations of decision classes. After this step, the inconsistencies are propagated up the hierarchy tree (if we are not already at the root).

Let us consider the sub-problem with the decision attribute "*Area of interest*". The sub-attributes are "*Additional Project*" and "*Training*". The evaluation of the DM is given in Table 21.8. This sub-problem concerns multi-attribute classification. Students are classified into three classes that are not preference ordered:

- students 1, 4, and 6 belong to class "*Decision Support*" (*DS*),
- students 2 and 5 belong to class "*Knowledge Discovery*" (*KD*),
- student 3 belongs to class "*Other*" (*OT*).

Table 21.7. Evaluations of students during bachelor studies

Student	Statistics (exem)	Comp. Networks (exam)	Comp. Networks (project)	Data-bases (exam)	Data-bases (project)	Addi-tional (project)	Training
1	Suff.	Bad, Suff.	Good	Good	Good	Progr.	AI
2	Good	Suff.	Good	Good	Suff.	Progr.	AI
3	Good	Good	Suff.	Suff.	Good	Data-bases	Data-bases
4	Suff.	Suff.	Suff., Good	Suff.	Suff.	Progr.	Statistics
5	Good	Good	Good	Suff.	Bad	AI	Data-bases, Statistics
6	Bad	Suff.	Bad	Bad	Suff.	AI	Statistics

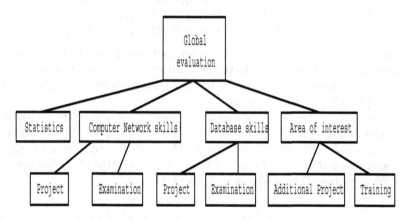

Fig. 21.6. The hierarchy of student qualification problem

Observe that students 1 and 2 have the same description but belong to different classes. A very interesting situation occurs for student 5, whose *Training* included both "*Databases*" and "*Statistics*". According to the generalized indiscernibility relation, student 5 is indiscernible from student 6. The inconsistent students are marked in italics in Table 21.8. Assuming that $P = \{Additional\ project\ (AddProj),\ Training\}$, the rough approximation of classes is as follows:

Table 21.8. The subtree of the *Area of interest* — the multiattribute classification sub-problem

Student	Additional project	Training	Area of interest
1	Progr.	AI	Decision Support
2	Progr.	AI	Knowledge Discovery
3	Databases	Databases	Other
4	Progr.	Statistics	Decision Support
5	AI	Databases, Statistics	Knowledge Discovery
6	AI	Statistics	Decision Support

$$\underline{P}(DS) = \{4\}, \ \overline{P}(DS) = \{1,2,4,5,6\}, \ Bn_P DS) = \{1,2,5,6\}$$
$$\underline{P}(KD) = \emptyset, \quad \overline{P}(KD) = \{1,2,5,6\}, \quad Bn_P(KD) = \{1,2,5,6\}$$
$$\underline{P}(OT) = \{3\}, \ \overline{P}(OT) = \{3\}, \qquad Bn_P(OT) = \emptyset$$

The following decision rules (certain and approximate) have been induced for this subproblem:

if$(f(x, \text{Addproj}) \cap \{\text{Progr}\} \neq \emptyset) \wedge (f(x, \text{Training}) \cap \{\text{Statistics}\} \neq \emptyset)$,
    then $x \in DS$ (4),
if$(f(x, \text{Addproj}) \cap \{\text{Databases}\} \neq \emptyset)$, then $x \in OT$ (3),
if$(f(x, \text{Training}) \cap \{\text{AI}\} \neq \emptyset)$, then $x \in DS \cup KD$ (1,2),
if$(f(x, \text{AddProj}) \cap \{\text{AI}\} \neq \emptyset)$, then $x \in DS \cup KD$ (5,6).

The inconsistencies discovered for students 1 and 2 and students 5 and 6 are propagated and, according to the generalized decision function at the upper level of the tree, the values of the hierarchical attribute *"Area of interest"* are multiple, as shown in Table 21.10.

Let us move to the sub-problem concerning *"Computer Network skills"*. It includes two sub-criteria: *"Project"* and *"Examination"*. This sub-problem is the multicriteria classification problem. The students are assigned to three preference-ordered classes: *"Bad"*, *"Sufficient"*, and *"Good"* . The evaluations given by the DM are shown in Table 21.9.

Table 21.9. The subtree of the *Computer Network skills* – the multicriteria classification subproblem

Student	Examination	Project	Computer Network skills
1	Bad - Sufficient	Good	Good
2	Sufficient	Good	Good
3	Good	Sufficient	Sufficient
4	Sufficient	Sufficient - Good	Sufficient
5	Good	Good	Good
6	Sufficient	Bad	Bad

Observe that, considering interval order evaluations on *"Examination"* and *"Project"*, student 4 is not worse than student 1 and student 2. This means that 4 dominates 1 and 2. Nevertheless, student 4 is assigned to a worse class ("*Sufficient*") than students 1 and 2, ("*Good*"). Therefore, students 1, 2, and 4 are inconsistent with respect to the dominance principle. The inconsistent objects are marked in italics in Table 21.9. The rough approximations of all downward and upward unions of classes (for $P = \{Examination, Project\}$) are as follows:

$$\underline{P}(\text{Sufficient}^{\geq}) = \{1, 2, 3, 4, 5\} \quad \overline{P}(\text{Sufficient}^{\geq}) = \{1, 2, 3, 4, 5\}$$
$$Bn_P(\text{Sufficient}^{\geq}) = \emptyset$$
$$\underline{P}(\text{Good}^{\geq}) = \{5\} \quad \overline{P}(\text{Good}^{\geq}) = \{1, 2, 4, 5\}$$
$$Bn_P(\text{Good}^{\geq}) = \{1, 2, 4\}$$
$$\underline{P}(\text{Sufficient}^{\leq}) = \{3, 6\} \quad \overline{P}(\text{Sufficient}^{\leq}) = \{1, 2, 3, 4, 6\}$$
$$Bn_P(\text{Sufficient}^{\leq}) = \{1, 2, 4\}$$
$$\underline{P}(\text{Bad}^{\leq}) = \{6\} \quad \overline{P}(\text{Bad}^{\leq}) = \{6\}$$
$$Bn_P(\text{Bad}^{\leq}) = \emptyset$$

The following decision rules (certain and approximate) have been induced for this subproblem:

if$(u(x, \text{Project}) \geq \text{Sufficient})$, then $x \in \text{Sufficient}^{\geq}$ $(1, 2, 3, 4, 5)$,
if$(u(x, \text{Examination}) \geq \text{Good})$, then $x \in \text{Sufficient}^{\geq}$ $(3, 5)$,
if$(u(x, \text{Examination}) \geq \text{Good}) \wedge (u(x, \text{Project}) \geq \text{Good})$,
    then $x \in \text{Good}^{\geq}$ (5),
if$(u(x, \text{Project}) \leq \text{Sufficient})$, then $x \in \text{Sufficient}^{\leq}$ $(3, 4, 6)$,
if$(u(x, \text{Project}) \leq \text{Bad})$, then $x \in \text{Bad}^{\leq}$ (6),
if$(u(x, \text{Examination}) \leq \text{Sufficient}) \wedge (u(x, \text{Project}) \geq \text{Good})$,
    then $x \in \text{Sufficient} \cup \text{Good}$ $(1, 2, 4)$.

At the upper level, according to the generalized decision, the criterion "*Computer Network skills*" takes values in intervals, as shown in Table 21.10. We will skip the analysis of the "*Database skills*" subproblem because it is similar to the above.

The final evaluations are shown in Table 21.10. Note that this table is composed of consistent objects. The evaluations modified on the basis of inconsistencies at the lower level are marked in italics.

The decision rules (certain and approximate) for the final decision are presented below:

if$(u(x, \text{Databases}) \geq \text{Good})$, then $x \in \text{Acceptable}^{\geq}$ $(1, 2)$,
if$(u(x, \text{CompNet}) \geq \text{Sufficient}) \wedge (u(x, \text{AreaOfInterest}) \cap \{DS\} \neq \emptyset)$,
    then $x \in \text{Acceptable}^{\geq}$ $(1, 2, 4, 5)$,
if$(u(x, \text{Statistics}) \geq \text{Good}) \wedge (u(x, \text{Databases}) \geq \text{Good})$,
    then $x \in \text{Desirable}^{\geq}$ (2),
if$(u(x, \text{Statistics}) \geq \text{Sufficient})$, then $x \in \text{Acceptable}^{\leq}$ $(1, 2, 6)$,
if$(u(x, \text{Databases}) \geq \text{Sufficient})$, then $x \in \text{Acceptable}^{\leq}$ $(3, 4, 5, 6)$,
if$(u(x, \text{AreaOfInterest}) \cap \{\text{Other}\} \neq \emptyset)$, then $x \in \text{Unacceptable}^{\leq}$ (3),
if$(u(x, \text{CompNet}) \leq \text{Bad})$, then $x \in \text{Unacceptable}^{\leq}$ (6).

594   S. Greco, B. Matarazzo, R. Slowinski

Table 21.10. Global evaluation – multicriteria classification problem

Student	*Statistics* (exam)	*Computer Network skills* (*CompNet*)	*Database skills* (*Databases*)	*Area of interest*	*Global evaluation*
1	Sufficient	Sufficient-Good	Good	Decision Support, Knowledge Discovery	Acceptable
2	Good	Sufficient-Good	Good	Decision Support, Knowledge Discovery	Desirable
3	Good	Sufficient	Sufficient	Other	Unacceptable
4	Sufficient	Sufficient-Good	Sufficient	Decision Support	Acceptable
5	Good	Good	Sufficient	Decision Support, Knowledge Discovery	Acceptable
6	Bad	Bad	Bad	Decision Support, Knowledge Discovery	Unacceptable

Let us observe that analysis of this problem, without partitioning it into smaller subproblems, would be harder. The cognitive effort to evaluate students on the basis of all seven attributes and criteria together is indeed quite large.

Finally, let us show the application of the decision rules for the classification of new students. The two new students are shown in Table 21.11.

The following rules match the first student (1′):

- "*Area of interest*":

  if $(f(x, Training) \cap \{AI\} \neq \emptyset)$, then $x \in DS \cup KD$

- "*Computer Network skills*":

  if $(u(x, Project) \geq Sufficient)$, then $x \in Sufficient^{\geq}$
  if $(l(x, Project) \leq Sufficient)$, then $x \in Sufficient^{\leq}$

The following rules match the second student (2′):

- "*Area of interest*":

Table 21.11. Evaluations of two new students

Student	Statistics (exam)	Comp. Networks (exam)	Comp. Networks (project)	Data-bases (exam)	Data-bases (project)	Addi-tional (project)	Training
1'	Good	Suff.	Suff.	Good	Good	Progr.	AI
2'	Suff.	Good	Bad	Suff.	Good	Data-bases	Progr.

if $(f(x, AddProj) \cap \{Databases\} \neq \emptyset)$, then $x \in OT$

- *"Computer Network skills"*:

if $(u(x, Examination) \geq Good)$, then $x \in Sufficient^{\geq}$
if $(l(x, Project) \leq Bad)$, then $x \in Bad^{\leq}$

On the *"Database skills"* criterion, both students are assigned to the class *"Good"*.

Remark that student 2' is described by controversial information. The evaluations of students on the upper level are presented below.

Table 21.12. Evaluations of new students at the global level

Student	Statistics (exam)	Computer Network skills (CompNet)	Database skills (Databases)	Area of interest	Global evaluation
1'	Good	Sufficient	Good	Decision Support, Knowledge Discovery	?
2'	Sufficient	Bad-Sufficient	Good	Other	?

At the global level the following rules match the first student (1'):

if $(u(x, Databases) \geq Good)$, then $x \in Acceptable^{\geq}$
if $(u(x, CompNet) \geq Sufficient) \wedge (f(x, AreaOfInterest) \cap \{DS\} \neq \emptyset)$,
    then $x \in Acceptable^{\geq}$
if $(u(x, Statistics) \geq Good) \wedge (u(Databases, x) \geq Good)$,
    then $x \in Desirable^{\geq}$

In conclusion, student 1' has been classified as *"Desirable"*.
The following rules match, in turn, the second student (2') at the global level:

if $(u(x, Databases) \geq Good)$, *then* $x \in Acceptable^{\geq}$
if $(u(x, Statistics) \leq Sufficient)$, *then* $x \in Acceptable^{\leq}$
if $(u(x, CompNet) \leq Bad)$, *then* $x \in Unacceptable^{\leq}$
if $(f(x, AreaOfInterest) \cap \{Other\} \neq \emptyset)$, *then* $x \in Unacceptable^{\leq}$

In conclusion, student 2' has been classified between "*Unacceptable*" and "*Acceptable*".

21.5 Comparison of DRSA with Other Paradigms

DRSA aims to give an effective answer to the central problem of any decision-aiding methodology concerning multicriteria and/or multi-attribute classification, that is, the aggregation of the multiple criteria and attributes into a single preference model. In this section, we propose to compare different paradigms used to solve this central problem by different theories (see Table 21.13). In [21.27, 21.33], this comparison was made at the level of axiomatic foundations, which has no precedence in the theoretical research concerning multicriteria classification. The axiomatic approach is interesting for at least three reasons:

- it exhibits differences between preference models and methods,
- it permits the interpretation of methods conceived for one model by another model,
- knowing the basic axioms, one can pass from one method to another with different preference models.

Moreover, in [21.27, 21.33], we have considered aggregation of ordinal criteria, that have been studied much less than cardinal criteria [21.52]. Among several aggregation models, particular interest has been paid recently to an integral proposed by Sugeno [21.67], which is able to deal with ordinal data; it has been considered the most general ordinal aggregation function of the max-min average type. It appears, however, that this function has some unpleasant limitations: the most important is the so-called commensurability [21.44], i.e., the evaluations with respect to each considered criterion should be defined on the same scale. Comparison of the Sugeno integral with the decision rule model at the axiomatic level enables us to show other limitations of the former.

Below, we present the main comparisons concerning the axiomatic foundations of the decision rule model and two traditional models: utility function and outranking relation.

21.5.1 Axiomatic Foundations of Multicriteria Classification Problems and Associated Preference Models

A Representation Theorem. At this point, we consider a finite or denumerable *product space* $X = \prod_{i=1}^{n} X_i$, where X_i is an evaluation scale of

Table 21.13. Different paradigms of aggregation and preference representa

Theory (paradigm)	Main preoccupation (axiomatic basis)	The aggregation result evidences
Social Choice Theory	Voting system or aggregation of rankings	Final ranking
Decision Theory	Definition of preference structures	Relation in A
Measurement Theory	Cancellation property	Function, like in conjoint measurement
Measure Theory, Fuzzy Sets	Capacity or fuzzy measure	Weights or interaction between criteria, like in Choquet integral or Sugeno integral
Machine Learning, Logical Analysis of Data, Rough Sets	Boolean or pseudo-Boolean function, decision rules or decision trees	Knowledge, like in knowledge discovery or data mining

criterion $i = 1, \ldots, n$. With appropriate topological conditions we can also work with infinite non-denumerable space, but in this chapter, for the sake of simplicity, we will skip this possibility. When aggregating multi-criteria evaluations within a preference model, we will exploit an *ordinal* character of the criteria scales only. This means that evaluations on particular criteria are considered as if they were words (linguistic qualifiers, like bad, medium, good, very good) even if an original scale was numerical. Let $(x_i\, z_{-i})$, $x_i \in X_i$ and $z_{-i} \in X_{-i} = \prod_{j=1, j \neq i}^{n} X_j$, denote an element of X equal to z, except for its i-th coordinate being equal to x_i.

The classification decision is generally modeled by one of three models: utility function (e.g., scoring methods [21.6], discriminant analysis [21.3], aggregation-disaggregation methods [21.76]), outranking relation (as in ELEC-TRE TRI [21.53]), or decision rules (as in DRSA).

▷ *Utility function* $f(\cdot)$ gives a real value $f(x)$ to each $x \in X$, and assigns x to $Cl_t^{\geq}$ if $f(x) \geq z_t$, where z_t, $t = 2, \ldots, m$, are $m - 1$ ordered thresholds satisfying

$$z_2 < z_3 < \ldots < z_m.$$

▷ *Outranking relation* S is a binary relation on X such that, for each x, $y \in X$, xSy means "x is (comprehensively) at least as good as y". An outranking relation S on X assigns x to $Cl_t^{\geq} = \bigcup_{s \geq t} Cl_s$ if xSp^t, where p^t, $t = 2, \ldots,$ m, are $m - 1$ reference profiles p^t, such that p^{t+1} dominates p^t (i.e., p^{t+1} is at least as good as p^t with respect to each criterion i, and there is at least one criterion for which p^{t+1} is strictly preferred to p^t), $t = 2, \ldots, m-1$.

▷ A set of "*if* ..., *then*..." *decision rules* is a set of logical implications whose syntax is "*if* x_{i1} is at least as good as r_{i1} *and* x_{i2} is at least as good as r_{i2} *and* ... *and* x_{ih} is at least as good as r_{ih}, *then* $x \in Cl_r^{\geq}$", where $x_{i1}, r_{i1} \in X_{i1}, x_{i2}, r_{i2} \in X_{i2}, \ldots, x_{ih}, r_{ih} \in X_{ih}$, with $\{i1, i2, \ldots, ih\} \subseteq \{1, \ldots, n\}$ and $r = 2, \ldots, m$. These decision rules are called "*at least*" decision rules. Let us consider the case where, for each criterion $i = 1, \ldots, n$, there exists a function $g_i{:}X_i \to \boldsymbol{R}$ such that, for each $x, y \in X$: $g_i(x_i) \geq g_i(y_i) \Leftrightarrow x$ is at least as good as y with respect to criterion i (i.e., x_i is at least as good as y_i). In this case, an "at least" decision rule can also be written as

$$\text{"}if\ g_{i1}(x_{i1}) \geq g_{i1}(r_{i1})\ and\ g_{i2}(x_{i2}) \geq g_{i2}(r_{i2})\ and\ \ldots\ and$$
$$g_{ih}(x_{ih}) \geq g_{ih}(r_{ih}),$$
$$then\ x \in Cl_r^{\geq}\text{"}$$

with $\{i1, i2, \ldots, ih\} \subseteq \{1, \ldots, n\}$ and $r = 2, \ldots, m$. The classification of $x \in X$ with "at least" decision rules is done according to the following procedure:

• $x \in Cl_t$ if and only if there exists a rule matching x that assigns x to $Cl_t^{\geq} = \bigcup_{s \geq t} Cl_s$, and there exists no rule matching x that assigns x to $Cl_s^{\geq}$, where $s > t$;

• $x \in Cl_1$ if and only if there exists no rule matching x.

The following result is a representation theorem for the multiple-criteria classification problem, stating equivalence between a very simple cancellation property, a general utility function, a very general outranking relation, and a set of decision rules. Let us mention that equivalence of the cancellation property considered and the utility function was noted by Goldstein [21.38], within the conjoint measurement approach, for the special case of three classes.

Theorem 1 [21.27] The following four propositions are equivalent:

1) (*cancellation property*) for each $x_i, y_i \in X_i$ and $a_{-i}, b_{-i} \in X_{-i}$, and for each $r, s \in \{1, \ldots, m\}$ (for $i = 1, \ldots, n$),

$$\{(x_i a_{-i}) \in Cl_r\ and\ (y_i b_{-i}) \in Cl_s\} \Rightarrow$$

$$\{(y_i a_{-i}) \in Cl_r^{\geq}\ or\ (x_i b_{-i}) \in Cl_s^{\geq}\}.$$

2) (*utility function*) there exist
   • functions $g_i{:}\ X_i \to \boldsymbol{R}$ for $i = 1, \ldots, n$, called criteria,
   • function $f{:}\ \boldsymbol{R}^n \to \boldsymbol{R}$, non-decreasing in each argument, called the utility function,
   • $m-1$ ordered thresholds z_t, $t = 2, \ldots, m$, satisfying

$$z_2 < z_3 < \ldots < z_m$$

such that for each $x \in X$ and each $t = 2, \ldots, m$,
$$f[g_1(x_1), g_2(x_2), \ldots, g_n(x_n)] \geq z_t \Leftrightarrow x \in Cl_t^{\geq}.$$

3) (*outranking function and relation*) there exist
- functions $g_i\colon X_i \to \boldsymbol{R}$, $i = 1, \ldots, n$, called criteria,
- function $s\colon \boldsymbol{R}^{2n} \to \boldsymbol{R}$, non-decreasing in each odd argument and non-increasing in each even argument, called outranking function,
- m-1 reference profiles p^t, $t = 2, \ldots, m$, satisfying
$$g_i(p^2) \leq g_i(p^3) \leq \ldots \leq g_i(p^m), \text{ for } i = 1, \ldots, n,$$
such that for each $x \in X$ and each $t = 2, \ldots, m$
$$s[g_1(x_1), g_1(p^t), g_2(x_2), g_2(p^t), \ldots, g_n(x_n), g_n(p^t)] \geq 0 \Leftrightarrow x \in Cl_t^{\geq}$$
(N.B. $s[g_1(x_1), g_1(p^t), g_2(x_2), g_2(p^t), \ldots, g_n(x_n), g_n(p^t)] \geq 0 \Leftrightarrow x \ S \ p^t$,
where S is a binary outranking relation).

4) ("*at least*" decision rules) there exist
- functions $g_i\colon X_i \to \boldsymbol{R}$ for each $i = 1, \ldots, n$, called criteria,
- a set of "at least" decision rules whose syntax is
"*if* $g_{i1}(x_{i1}) \geq r_{i1}$ *and* $g_{i2}(x_{i2}) \geq r_{i2}$ *and* $\ldots$ *and* $g_{ih}(x_{ih}) \geq r_{ih}$, *then* $x \in Cl_t^{\geq}$", with $\{i1, i2, \ldots, ih\} \subseteq \{1, \ldots, n\}$, $t = 2, \ldots, m$, such that for each $y \in Cl_t$, $t = 2, \ldots, m$, there is at least one rule implying $y \in Cl_t^{\geq}$ and there is no rule implying $y \in Cl_r^{\geq}$ with $r > t$.

Let us note that the above representation theorem for the multicriteria classification problem starts with a very weak axiomatic condition called cancellation property. Indeed, this property does not require existence of criterion functions g_i, $i = 1, \ldots, n$, or a dominance relation D on X in order to characterize the three preference models. Instead, the meaning of the above cancellation property is as follows, consider the binary weak preference relation $\succeq_i$ defined on X_i, $i = 1, \ldots, n$, such that, for each $x_i, y_i \in X_i$, for each $a_{-i} \in X_{-i}$, and for each $Cl_r \in \boldsymbol{Cl}$:

$$x_i \succeq_i y_i \Leftrightarrow [(y_i a_{-i}) \in Cl_r \Rightarrow (y_i a_{-i}) \in Cl_{\overline{r}}^{\geq}].$$

The cancellation property ensures that the binary weak preference relation $\succeq_i$ on X_i is a complete preorder, that is, strongly complete (for each $x_i, y_i \in X_i$, $x_i \succeq_i y_i$, or $y_i \succeq_i x_i$) and transitive. Consequently, there exists a function $g_i\colon X_i \to \boldsymbol{R}$ such that for each $x_i, y_i \in x_i$,

$$x_i \succeq_i y_i \Leftrightarrow g_i(x_i) \geq g_i(y_i).$$

On the basis of relations $\succeq_i$, $i = 1, \ldots, n$, one can also define a dominance relation D on X as follows: for each $x, y \in X$,

$$xDy \Leftrightarrow x_i \succeq_i y_i \text{ for all } i = 1, \ldots, n.$$

This is, of course, equivalent to

$$xDy \Leftrightarrow g_i(x_i) \geq g_i(y_i) \text{ for all } i = 1, \ldots, n.$$

Cancellation property 1) of Theorem 1 enables us to state the following *condition of coherence between dominance relation D and classification Cl*, for each $x,\ y \in X$,

$$xDy \Rightarrow x \in Cl_r \text{ and } y \in Cl_s, \text{ with } r \geq s.$$

For any subset of criteria $P \subseteq \{1, \ldots, n\}$, and for each pair $x, y \in X$, one can also define a dominance relation D_P on X:

$$xD_P y \Leftrightarrow x_i \succeq_i y_i \text{ for all } i \in P,$$

which is equivalent to

$$xD_P y \Leftrightarrow g_i(x_i) \geq g_i(y_i) \text{ for all } i \in P.$$

Dominance relations D_P, $P \subseteq \{1, \ldots, n\}$, are used in the condition part of decision rules. Being an intersection of complete preorders, binary relations D_P are partial preorders, i.e., they are reflexive and transitive.

Observe, moreover, that Theorem 1 regards a representation of classification Cl in terms of "lower bounds". Theorem 1 can be reformulated in terms of "upper bounds" in such a way that

▷ condition of proposition 2) is expressed as

$$f[g_1(x_1), g_2(x_2), \ldots, g_n(x_n)] \leq w_t \Leftrightarrow x \in Cl_t^{\leq},$$

where w_t, $t = 1, \ldots, m-1$, are $m-1$ suitably ordered thresholds,
▷ condition of proposition 3) is expressed as

$$s[g_1(x_1), g_1(q^t), g_2(x_2), g_2(q^t), \ldots, g_n(x_n), g_n(q^t)] < 0 \Leftrightarrow x \in Cl_t^{\leq},$$

where q^t, $t = 1, \ldots, m-1$, are $m-1$ reference profiles q^t such that q^{t+1} dominates q^t (i.e., q^{t+1} is at least as good as q^t with respect to each criterion $i = 1, \ldots, n$, and there is at least one criterion $j \in \{1, \ldots, n\}$ for which q^{t+1} is strictly preferred to q^t), $t = 1, \ldots, m-2$.
▷ condition of proposition 4) considers a set of decision rules whose syntax is "*if* $g_{i1}(x_{i1}) \leq r_{i1}$ *and* $g_{i2}(x_{i2}) \leq r_{i2}$ *and* ... *and* $g_{ih}(x_{ih}) \leq r_{ih}$, *then* $x \in Cl_t^{\leq}$" with $\{i1, i2, \ldots, ih\} \subseteq \{1, \ldots, n\}$. These decision rules are called "*at most*" decision rules. The classification of $x \in X$ with "at most" decision rules is done according to the following procedure:
  • $x \in Cl_t$ if and only if there exists a rule matching x that assigns x to $Cl_t^{\leq}$, and there exists no rule matching x that assigns x to $Cl_s^{\leq}$, where $s < t$;
  • $x \in Cl_m$ if and only if there exists no rule matching x.

Illustrative Example. Let us consider a multicriteria classification problem inspired by the example of evaluation in a high school proposed by Grabisch [21.14]. Suppose that a high school director wants to assign students to different classes of merit on the basis of their scores in *Mathematics* and *Literature*. The ordinal scale for evaluation in *Mathematics* and *Literature* has been composed of the three grades "*bad*", "*medium*", and "*good*", while the comprehensive evaluation scale has been composed of the two grades "*bad*" and "*good*". The evaluations of student x in *Mathematics* and *Literature* are denoted by x_1 and x_2, respectively. To be criteria, functions $g_1(\cdot)$ and $g_2(\cdot)$ must respect monotonicity, i.e., $g_i(bad) < g_i(medium) < g_i(good)$, $i = 1,2$. For example, a simple way to define $g_1(\cdot)$ and $g_2(\cdot)$ is to set $g_i(bad) = 1$, $g_i(medium) = 2$, $g_i(good) = 3$, $i = 1,2$.

Table 21.14 presents all possible profiles of the students with respect to the two criteria considered, and a classification decision made by the director. Let us note that the classification of students presented in Table 21.14 satisfies proposition 1) of Theorem 1. In fact, each time a student x dominates a student y, student x belongs to the same or higher class than student y.

This can also be seen on the Hasse diagram in Fig. 21.7, where each node corresponds to a profile of evaluations. Profile x corresponding to node α dominates profile y corresponding to node β if α is over β and there is a path from α to β.

The diagram in Fig. 21.8 represents the binary relation R defined on the set of all possible profiles of evaluations $x = \{S1, S2, \ldots, S9\}$ as follows: for each $x,y \in X$,

$$xRy \Leftrightarrow xDy \text{ or } x \in Cl_r \text{ and } y \in Cl_s, \text{ with } r > s.$$

The binary relation R is used in the proof of Theorem 1. In fact, R is reflexive and transitive and, therefore, it is a partial preorder. Thus, there is a function $h \colon X \to \mathbf{R}$ such that for each $x,y \in X$,

$$xRy \Rightarrow h(x) \geq h(y).$$

According to the definition and to the condition of coherence between dominance relation D and classification Cl, we have

$$x \in Cl_r \text{ and } y \in Cl_s, \text{ with } r > s \Leftrightarrow h(x) > h(y). \qquad \text{(i)}$$

On the basis of property (i), it is possible to build function $f \colon \mathbf{R}^n \to \mathbf{R}$ and set of thresholds z_t, $t = 2, \ldots, m$, used in Theorem 1.

In the diagram presented in Fig. 21.8, the arcs representing relation R are drawn from profile x to profile y if and only if x dominates y or x belongs to the class of "*good*" students and y belongs to the class of "*bad*" students. Since relation R is transitive, we are not drawing the arcs between profiles that are already connected by a path, e.g., as profiles "*good-good*" and "*bad-good*" are in relation R; however, there exists a path between them through the profile "*medium-good*", so the direct arc would be redundant.

Table 21.14. Classification of all 9 profiles of possible evaluations

Student	Maths	Literature	Decision	$f[g_1(X), g_2(X)]$	Matching rules
S1	bad	bad	bad	0	# 3
S2	medium	bad	bad	1	# 3
S3	good	bad	good	4	# 1
S4	bad	medium	bad	1	# 3
S5	medium	medium	good	4	# 2
S6	good	medium	good	6	# 1,2
S7	bad	good	bad	2	# 3
S8	medium	good	good	5	# 2
S9	good	good	good	8	# 1,2

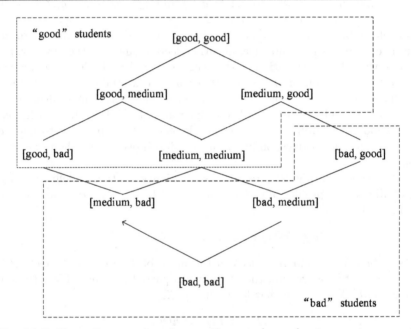

Fig. 21.7. Hasse diagram of student profiles $x = [x_1, x_2]$, where x_1 is a score in *Mathematics* and x_2 is a score in *Literature*

Comparing Figs. 21.7 and 21.8, one can see that in the latter there are two additional arcs. The arc from profile "*medium-medium*" to profile "*bad-good*" does not exist in Fig. 21.7 because profile "medium-medium" does not dominate profile "bad-good"; however, it exists in Fig. 21.8 because, due to the director, the student with profile "*medium-medium*" is classified as better than the student with profile "*bad-good*" (class "*good*" vs. class "*bad*"). For the same reason, in Fig. 21.8 there is an arc from profile "*good-bad*" to profile "*bad-good*", while this arc does not exist in Fig. 21.7.

On the basis of relation R, one can build the function $f[g_1(x_1), g_2(x_2)]$, used in proposition 2) of Theorem 1, in a very simple way: for each profile x

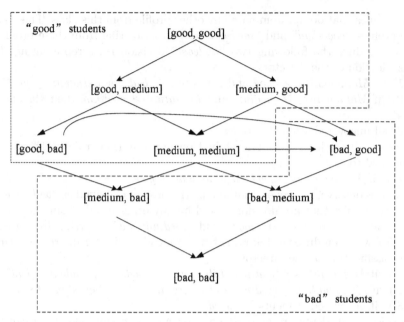

Fig. 21.8. The binary relation R in set of student profiles

$= [x_1, x_2]$ represented by a node α, $f[g_1(x_1), g_2(x_2)]$ is equal to the number of nodes to which there are directed paths starting from α. The function $f[g_1(x_1), g_2(x_2)]$ is represented in Fig. 21.9. It has the property that all the students classified as "*good*" have a greater value of this function than the students classified as "*bad*". Among the "*good*" students, the minimum value of function $f[g_1(x_1), g_2(x_2)]$, equal to 4, is obtained by students with profiles "*good-bad*" or "*medium-medium*". Therefore, the only threshold z_2 is equal to 4; $f[g_1(x_1), g_2(x_2)] \geq 4$ if and only if x is "*good*", and $f[g_1(x_1), g_2(x_2)] < 4$ if and only if x is "*bad*".

An outranking function satisfying conditions present in proposition 3) of Theorem 1 can be built as follows: for each $x, y \in X$ set

$$s[g_1(x_1), g_1(y_1), g_2(x_2), g_2(y_2)] = f[g_1(x_1), g_2(x_2)] - f[g_1(y_1), g_2(y_2)].$$

The only reference profile $p^2 = (p_1^2, p_2^2) \in X$, defined in Theorem 1, can be chosen such that

$$f[g_1(p_1^2), g_2(p_2^2)] = \min_{x \in Class\,\text{"Good"}} \{f[g_1(X_1), g_2(X_2)]\}.$$

It is easy to see that the reference profile p^2 is again one of the profiles "*good-bad*" or "*medium-medium*".

The "*if ...*, *then...*" decision rules specified in proposition 4) of Theorem 1 can be easily built on the basis of the Hasse diagram presented in Fig. 21.7. Let us note that among the profiles classified as "*good*", there are two profiles that do not dominate any other profile from this class. These are the profiles *"good-bad"* and *"medium-medium"*. Starting from these profiles, we can induce the following two *"at least"* decision rules representing the classification of the director:

1) "if *Mathematics* ≥ *good* and *Literature* ≥ *bad*, then *student* ≥ *good*";
2) "if *Mathematics* ≥ *medium* and *Literature* ≥ *medium*, then *student* ≥ *good* ";
3) all uncovered students are *bad*.

Since all students are at least bad on any criterion, rule # 1) can be simplified to

1*) "if *Mathematics* ≥*good*, then *student* ≥ *good*".

Let us observe, moreover, that among the profiles classified as *"bad"*, there are two profiles that are not dominated by any other profile from this class. These are the profiles *"bad-good"* and *"medium-bad"*. Starting from these profiles, we can induce the following two "at most" decision rules representing the classification of the director:

1') "if *Mathematics* ≤ *bad* and *Literature* ≤ *good*, then *student* ≤ *bad*";
2') "if *Mathematics* ≤ *medium* and *Literature* ≤ *bad*, then *student* ≤ *bad*";
3') all uncovered students are *good*.

Since all students are at most good on any criterion, rule # 1') can be simplified to

1'*) "if *Mathematics* ≤ *bad*, then *student* ≤ *bad*".

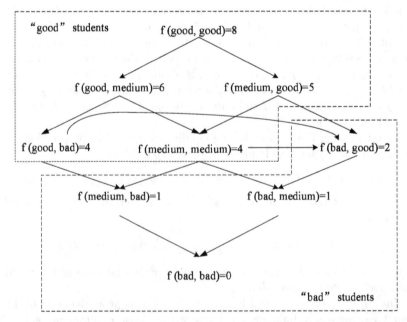

Fig. 21.9. Function $f[g_1(x_1), g_2(x_2)]$ for student profiles $x = [x_1, x_2]$

21.5.2 Conjoint Measurement for Multicriteria Classification Problems with Inconsistencies

The conjoint measurement model presented in Sect. 21.5.1 cannot represent the inconsistency with the dominance principle considered within DRSA. At this point, we present a more general model of conjoint measurement that permits representation of this inconsistency. This model is based on the concepts of dominance-based rough approximation of upward and downward unions of classes $Cl_t^{\geq}$ and $Cl_t^{\leq}$.

The following concepts will be useful: for each $x \in X$, the *lower class* and the *upper class* of x, denoted by $r_*(x)$ and $r^*(x)$, respectively, are defined as follows

$$r_*(x) = \max\{s \in \{1,..,m\}: x \in \underline{C}\left(Cl_s^{\geq}\right)\},$$

$$r^*(x) = \min\{s \in \{1,..,m\}: x \in \underline{C}\left(Cl_s^{\leq}\right)\},$$

where $\underline{C}(Cl_s^{\geq})$ and $\underline{C}(Cl_s^{\leq})$ are the lower approximations of $Cl_s^{\geq}$ and $Cl_s^{\leq}$, respectively, with respect to the set of criteria $C = \{g_1, g_2, \ldots, g_n\}$.

Theorem 2 (see [21.27]) For each set of binary relations $\succeq_i, i = 1, \ldots, n$, that are complete preorders, and for each classification $\boldsymbol{Cl}$, there exist

- functions $g_i: x_i \rightarrow \boldsymbol{R}$, such that $x_i \succeq_i y_i \Leftrightarrow g_i(x_i) \geq g_i(y_i)$, $i = 1, \ldots, n$,
- functions $f^{\geq}: \boldsymbol{R}^n \rightarrow \boldsymbol{R}$ and $f^{\leq}: \boldsymbol{R}^n \rightarrow \boldsymbol{R}$, non-decreasing in each argument, such that

$$f^{\geq}[g_1(x_1), g_2(x_2), \ldots, g_n(x_n)] \leq f^{\leq}[g_1(x_1), g_2(x_2), \ldots, g_n(x_n)]$$

- $m - 1$ ordered thresholds z_t, $t = 2, \ldots, m$,

$$z_2 < z_3 < \ldots < z_m$$

such that for each object $x \in X$, functions $f^{\geq}$ and $f^{\leq}$ assign x to a lower and an upper class, respectively:

$$f^{\geq}[g_1(x_1), g_2(x_2), \ldots, g_n(x_n)] \geq z_t \Leftrightarrow x \in \underline{C}(Cl_t^{\geq}), f^{\leq}[g_1(x_1), g_2(x_2), \ldots, g_n(x_n)] < z_t \Leftrightarrow x \in \underline{C}(Cl_{t-1}^{\leq}),$$

where $C = \{g_1, g_2, \ldots, g_n\}$.

Inconsistency with the dominance principle can also be represented in terms of a set of "*at least*" and "*at most*" decision rules considered together. More formally, a set of "*at least*" and "*at most*" decision rules does not contradict the classification $\boldsymbol{Cl}$ if, for each $x \in Cl_t$, there exists no "*at least*" decision rule for which $x \in Cl_s^{\geq}$, with $s > t$, and there exists no "*at most*" decision rule for which $x \in Cl_s^{\leq}$, with $s < t$. A set of decision rules is complete if, for each $x \in Cl_t^{\geq} = \bigcup_{s \geq t} Cl_s$, there exists a decision rule for which $x \in Cl_s^{\geq}$, with $s \geq t$, and for each $x \in Cl_t^\geq = \bigcup\limits_{s \geq t} Cl_s$ there exists a decision rule for which $x \in Cl_s^\geq$, with $s \leq t$. A set of decision rules represents the classification $\boldsymbol{Cl}$ if it does not contradict $\boldsymbol{Cl}$ and if it is complete.

Theorem 3 (see [21.27]) For each set of binary relations $\succeq_i$, $i = 1, \ldots, n$, that are complete preorders, and for each classification $\boldsymbol{Cl}$, there exists a set of decision rules representing the classification $\boldsymbol{Cl}$.

21.5.3 Summary

The advantage of DRSA over competitor methodologies is the possibility of handling partially inconsistent data that are often encountered in preferential information due to hesitation of decision makers, the unstable character of their preferences, imprecise or incomplete information, and the like. Therefore, we proposed a general model of conjoint measurement that, using the basic concepts of DRSA (lower and upper approximations), is able to represent these inconsistencies by a specific utility function. We showed that these inconsistencies can also be represented in a meaningful way by "*if* ..., *then* ...*" decision rules induced from rough approximations.

As DRSA applied to multiple criteria classification problems and the underlying decision rules exploits only the ordinal properties of the scales of criteria, it is appropriate for the aggregation of ordinal criteria. This challenging problem of multiple criteria decision making has been solved until now by using 'max-min' aggregation functions, with the most general one being the fuzzy integral proposed by Sugeno. In [21.27, 21.33], we have shown that the decision rule model following from DRSA has advantages over the integral of Sugeno; in particular, it can represent some (even consistent) preferences that the Sugeno integral cannot.

The characterization of the decision rule preference model given in this section shows clearly its extraordinary capacity for criteria aggregation in multi-criteria classification problems. The decision rule preference model, apart from its capacity of representation, fulfils the postulates of transparency and interpretability of preference models in decision aiding. The characterization shows that the decision rule preference model is a strong alternative to functional and relational preference models to which it is formally equivalent. Recently, similar benefits of the decision rule model have been proved with respect to multiple criteria choice and ranking problems [21.33].

21.6 Conclusions

In this second part, we presented further extensions of DRSA that let us deal with heterogeneous information: qualitative and quantitative, preference ordered or not, crisp and fuzzy, ordinal and cardinal, without risk or under risk, and having a hierarchical structure.

DRSA gives, moreover, a methodology for building a preference model of a decision maker in terms of decision rules. The decision rules have a special syntax involving partial evaluation profiles and dominance relation on these profiles. The clarity of the rule representation of preferences enables us to see the limits of traditional aggregation functions: utility function and outranking relation. We proposed an axiomatic characterization of these aggregation functions in terms of conjoint measurement and in terms of a set of decision rules. The axioms of "cancellation property" type are the weakest possible. In comparison to other studies on characterization of aggregation functions, our axioms do not require any preliminary assumption about the scales of criteria. A sideresult of these investigations is the corollary that the decision rule aggregation (preference model) is the most general among the known aggregation functions.

Finally, the proposed methodology is based on elementary concepts and mathematical tools (sets and set operations, binary relations), without recourse to any algebraic or analytical structures; the main idea is very natural and even objective, in a certain sense, like the dominance relation.

Acknowledgements

The research of the first two authors has been supported by the Italian Ministry of Education, University and Scientific Research (MIUR). The third author wishes to acknowledge financial support from the State Committee for Scientific Research.

References

21.1 R. Agrawal, H. Mannila, R. Srikant, H. Toivinen, I. Verkamo: Fast discovery of association rules. In: U.M.Fayyad et al. (eds.), *Advances in Knowledge Discovery and Data Mining* (AAAI Press, 1996) pp. 307-328
21.2 M. Allais: The so-called Allais paradox and rational decision under uncertainty. In: M. Allais, O. Hagen (eds.), *Expected Utility Hypotheses and the Allias Paradox* (Reidel, Dordrecht, 1979) pp. 437-681
21.3 E.I. Altman: Financial ratios, discriminant analysis and the prediction of corporate bankruptcy. Journal of Finance, 589-609 (1968)
21.4 T. Bilgic, I.B. Turksen: Measurement-theoretic justification of connectives in fuzzy set theory. Fuzzy Sets and Systems, 76, 289-308 (1995)
21.5 B. Bouchon-Mounier, J. Yao: Linguistic modifiers and gradual membership to a category. International Journal on Intelligent Systems, 7, 26-36 (1992)
21.6 N. Capon: Credit scoring systems: a critical analysis. Journal of Marketing, 46, 32-91 (1982)
21.7 G. Cattaneo: Fuzzy extension of rough sets theory. In: L. Polkowski, A. Skowron (eds.), *Rough Sets and Current Trends in Computing* (LNAI 1424, Springer, Berlin, 1998) pp. 275-282

21.8 K. Dembczynski, S. Greco, R. Slowinski: Methodology of rough-set-based clas-
      sification and sorting with hierarchical structure of attributes and criteria, Con-
      trol & Cybernetics, 31 (2002) (to appear)
21.9 D. Dubois, H. Prade: Gradual inference rules in approximate reasoning. In-
      formation Sciences, 61, 103-122 (1992)
21.10 D. Dubois, H. Prade: Putting rough sets and fuzzy sets together. In: R.
      Slowinski (ed.), *Intelligent Decision Support: Handbook of Applications and Ad-
      vances of the Sets Theory* (Kluwer, Dordrecht, 1992) pp. 203-232
21.11 D. Dubois, H. Prade, R. Yager: A Manifesto: Fuzzy Information Engineer-
      ing. In: D. Dubois, H. Prade, R. Yager (eds.), *Fuzzy Information Engineering*
      (J.Wiley, New York, 1997) pp. 1-8
21.12 J. Fodor, M. Roubens: *Fuzzy Preference Modelling and Multicriteria Decision
      Support* (Kluwer, Dordrecht, 1994)
21.13 S. Giove, S. Greco, B. Matarazzo, R. Slowinski: Variable consistency mono-
      tonic decision trees. In: J.J. Alpigini, J.F. Peters, A. Skowron, N. Zhong (eds.),
      Rough Sets and Current Trends in Computing (LNAI 2475, Springer-Verlag,
      Berlin, 2002) pp. 247-254
21.14 M. Grabisch: Fuzzy integral in multiple-criteria decision making. Fuzzy Sets
      and Systems, 69, 279-298 (1995)
21.15 S. Greco, M. Inuiguchi, R. Slowinski: Dominance-based rough set approach
      using possibility and necessity measures. In: J.J. Alpigini, J.F. Peters, A.
      Skowron, N. Zhong (eds.), *Rough Sets and Current Trends in Computing* (LNAI
      2475, Springer-Verlag, Berlin, 2002) pp. 85-92
21.16 S. Greco, M. Inuiguchi, R. Slowinski: A new proposal for fuzzy rough approx-
      imations and gradual decision rule representation. In: D. Dubois, J. Grzymala-
      Busse, M. Inuiguchi, L. Polkowski (eds.), *Fuzzy Rough Sets: Fuzzy and Rough,
      and Fuzzy along Rough* (Springer-Verlag, Berlin, 2004) (to appear)
21.17 S. Greco, B. Matarazzo, R. Slowinski: Fuzzy measures technique for rough
      set analysis. In: *Proc. 6ᵗʰ European Congress on Intelligent Techniques & Soft
      Computing* (Aachen, 1998) 1, pp. 99-103
21.18 S. Greco, B. Matarazzo, R. Slowinski: A new rough set approach to eval-
      uation of bankruptcy risk. In: C.Zopounidis (ed.), *Operational Tools in the
      Management of Financial Risk* (Kluwer Academic Publishers, Boston, 1998)
      pp. 121-136
21.19 S. Greco, B. Matarazzo, R. Slowinski: Rough approximation of a preference
      relation by dominance relations. European Journal of Operational Research,
      117, 63-83 (1999)
21.20 S. Greco, B. Matarazzo, R. Slowinski: The use of rough sets and fuzzy sets
      in MCDM. Chapter 14 in: T.Gal, T.Stewart, T.Hanne (eds.), *Advances in Mul-
      tiple Criteria Decision Making* (Kluwer Academic Publishers, Boston, 1999)
      pp. 14.1-14.59
21.21 S. Greco, B. Matarazzo, R. Slowinski: Handling missing values in rough set
      analysis of multi-attribute and multi-criteria decision problems. In: N. Zhong,
      A. Skowron, S. Ohsuga (eds.), *New Directions in Rough Sets, Data Mining and
      Granular-Soft Computing* (LNAI 1711, Springer-Verlag, Berlin, 1999) pp. 146-
      157
21.22 S. Greco, B. Matarazzo, R. Slowinski: Dealing with missing data in rough
      set analysis of multi-attribute and multi-criteria decision problems. In: S.H.
      Zanakis, G. Doukidis, C. Zopounidis (eds.), *Decision Making: Recent Devel-
      opments and Worldwide Applications* (Kluwer Academic Publishers, Boston,
      2000) pp. 295-316
21.23 S. Greco, B. Matarazzo, R. Slowinski: Rough set processing of vague infor-
      mation using fuzzy similarity relations. In: C.S. Calude, G. Paun (eds.), *Finite*

Versus Infinite - Contributions to an Eternal Dilemma)Springer-Verlag, London, 2000) pp. 149-173

21.24 S. Greco, B. Matarazzo, R. Slowinski: Fuzzy extension of the rough set approach to multicriteria and multiattribute sorting. In: J. Fodor, B. De Baets and P. Perny (eds.), *Preferences and Decisions under Incomplete Knowledge* (Physica-Verlag, Heidelberg, 2000) pp. 131-151

21.25 S. Greco, B. Matarazzo, R. Slowinski: Extension of the rough set approach to multicriteria decision support. INFOR, 38, 161-196 (2000)

21.26 S. Greco, B. Matarazzo, R. Slowinski: Rough sets theory for multicriteria decision analysis. European J. of Operational Research, 129, 1-47 (2001)

21.27 S.Greco, B. Matarazzo, R. Slowinski: Conjoint measurement and rough set approach for multicriteria sorting problems in presence of ordinal criteria. In: A.Colorni, M.Paruccini, B.Roy (eds.), *A-MCD-A: Aide Multi Critère à la Décision - Multiple Criteria Decision Aiding*, European Commission Report, EUR 19808 EN, (Ispra, 2001) pp. 117-144

21.28 S.Greco, B. Matarazzo, R. Slowinski: Rule-based decision support in multicriteria choice and ranking. In: S. Benferhat, Ph. Besnard (eds.), *Symbolic and Quantitative Approaches to Reasoning with Uncertainty* (LNAI 2143, Springer-Verlag, Berlin, 2001) pp. 29-47

21.29 S. Greco, B. Matarazzo, R. Slowinski: Assessment of a value of information using rough sets and fuzzy measures. In: J. Chojcan, J. Leski (eds.), *Fuzzy Sets and their Applications* (Silesian University of Technology Press, Gliwice, 2001) pp. 185-193

21.30 S. Greco, B. Matarazzo, R. Slowinski: Rough set approach to decisions under risk. In: W.Ziarko, Y.Yao (eds.): *Rough Sets and Current Trends in Computing* (LNAI 2005, Springer-Verlag, Berlin, 2001) pp. 160-169

21.31 S. Greco, B. Matarazzo, R. Slowinski: Rough sets methodology for sorting problems in presence of multiple attributes and criteria. European J. of Operational Research, 138, 247-259 (2002)

21.32 S. Greco, B. Matarazzo, R. Slowinski: Multicriteria classification. In: W. Kloesgen, J. Zytkow (eds.), *Handbook of Data Mining and Knowledge Discovery* (Oxford University Press, New York, 2002, chapter 16.1.9) pp. 318-328

21.33 S. Greco, B. Matarazzo, R. Slowinski: Preference representation by means of conjoint measurement and decision rule model. In: D.Bouyssou, E.Jacquet-Lagrèze, P.Perny, R.Slowinski, D.Vanderpooten, Ph.Vincke (eds.), *Aiding Decisions with Multiple Criteria - Essays in Honor of Bernard Roy* (Kluwer Academic Publishers, Boston, 2002) pp. 263-313

21.34 S. Greco, B. Matarazzo, R. Slowinski, J. Stefanowski: Variable consistency model of dominance-based rough set approach. In: W.Ziarko, Y.Yao: *Rough Sets and Current Trends in Computing* (LNAI 2005, Springer-Verlag, Berlin, 2001) pp. 170-181

21.35 S. Greco, B. Matarazzo, R. Slowinski, J. Stefanowski: An algorithm for induction of decision rules consistent with dominance principle. In: W.Ziarko, Y.Yao (eds.): *Rough Sets and Current Trends in Computing* (LNAI 2005, Springer-Verlag, Berlin, 2001b) pp. 304-313

21.36 S. Greco, B. Matarazzo, R. Slowinski, J. Stefanowski: Mining association rules in preference-ordered data. In: M.-S. Hacid, Z.W. Ras, D.A. Zighed, Y. Kodratoff (eds.), *Foundations of Intelligent Systems* (LNAI 2366, Springer-Verlag, Berlin, 2002) pp. 442-450

21.37 S. Greco, B. Matarazzo, R. Slowinski, A. Tsoukias: Exploitation of a rough approximation of the outranking relation in multicriteria choice and ranking. In: T.J.Stewart, R.C. van den Honert (eds.), *Trends in Multicriteria Decision Making* (LNEMS 465, Springer-Verlag, Berlin, 1998) pp. 45-60

21.38 W.M. Goldstein: Decomposable threshold models. Journal of Mathematical Psychology, 35, 64-79 (1991)

21.39 J.W. Grzymala-Busse: LERS - a system for learning from examples based on rough sets. In: R.Slowinski (ed.), *Intelligent Decision Support. Handbook of Applications and Advances of the Rough Sets Theor* (y. Kluwer, Dordrecht, 1992) pp. 3-18

21.40 M. Inuiguchi, S. Greco, R. Slowinski, T. Tanino: Possibility and necessity measure specification using modifiers for decision making under fuzziness. Fuzzy Sets and Systems, 137, 151-175 (2003)

21.41 M. Inuiguchi, T. Tanino: New fuzzy rough sets based on certainty qualification. In: S. K. Pal, L. Polkowski, A. Skowron (eds.), *Rough-Neuro-Computing: Techniques for Computing with Words* (Springer-Verlag, Berlin, 2002) pp. 110-126

21.42 R.S. Michalski, I. Bratko, M. Kubat (eds.): *Machine Learning and Data Mining - Methods and Applications* (Wiley, New York, 1998)

21.43 G.A. Miller: The magical number seven, plus or minus two: some limits on our capacity for information processing. *Psychological Review* 63, 81-97 (1956)

21.44 F. Modave, M. Grabisch: Preference representation by the Choquet Integral: the commensurability hypothesis. In: Proc. 7th Int. Conference on *Information Processing and Management of Uncertainty in Knowledge Based Systems* (Paris, La Sorbonne, 1998) pp. 164-171

21.45 A. Nakamura: Applications of fuzzy-rough classification to logics. In: R. Slowinski (ed.),*Intelligent Decision Support: Handbook of Applications and Advances of the Sets Theory* (Kluwer, Dordrecht, 1992) pp. 233-250

21.46 A. Nakamura, J.M. Gao: A logic for fuzzy data analysis. Fuzzy Sets and Systems, 39, 127-132 (1991)

21.47 E. Orlowska: Introduction: What you always wanted to know about rough sets. In: E. Orlowska (ed.), *Incomplete Information, Rough Set Analysis* (Physica-Verlag, Heidelberg, New York, 1998) pp. 1-20

21.48 Z. Pawlak: *Rough Sets. Theoretical Aspects of Reasoning about Data.* Kluwer Academic Publishers, Dordrecht (1991)

21.49 Z. Pawlak, J.W. Grzymala-Busse, R. Slowinski, W. Ziarko: Rough sets. Communications of the ACM, 38, 89-95 (1995)

21.50 L. Polkowski: *Rough Sets: Mathematical Foundations* (Physica-Verlag, Heidelberg, 2002)

21.51 L. Polkowski, A. Skowron: Calculi of granules based on rough set theory: approximate distributed synthesis and granular semantics for computing with words. In: N.Zhong, A.Skowron, S.Ohsuga (eds.), *New Directions in Rough sets, Data Mining and Soft-Granular Computing* (LNAI 1711, Springer-Verlag, Berlin, 1999) pp. 20-28

21.52 F.S. Roberts: *Measurement theory with applications to decision-making, utility and the social science* (Addison-Wesley Publ., Reading, MA, 1979)

21.53 B. Roy, D. Bouyssou: *Aide Multicritère à la Décision: Méthodes et Cas* (Economica, Paris, 1993)

21.54 T.L. Saaty: *The Analytic Hierarchy Process* (McGraw-Hill, New York, 1980)

21.55 R. Slowinski: A generalization of the indiscernibility relation for rough set analysis of quantitative information. *Rivista di matematica per le scienze economiche e sociali*, 15, 65-78 (1992)

21.56 R. Slowinski (ed.): *Intelligent Decision Support. Handbook of Applications and Advances of the Rough Sets Theory* (Kluwer Academic Publishers, Dordrecht, 1992)

21.57 R. Slowinski: Rough set processing of fuzzy information. In: T.Y.Lin, A.Wildberger (eds.), *Soft Computing: Rough Sets, Fuzzy Logic, Neural Networks, Uncertainty Management, Knowledge Discovery* (Simulation Councils, Inc., San Diego, CA, 1995) pp. 142-145

21.58 R. Slowinski, S. Greco, B. Matarazzo: Rough set analysis of preference-ordered data. In: J.J. Alpigini, J.F. Peters, A. Skowron, N. Zhong (eds.), *Rough Sets and Current Trends in Computing* (LNAI 2475, Springer-Verlag, Berlin, 2002) pp. 44-59

21.59 R. Slowinski, S. Greco, B. Matarazzo: Mining decision-rule preference model from rough approximation of preference relation. In: Proc. 26th IEEE Annual Int. Conference on *Computer Software & Applications (COMPSAC 2002)* (Oxford, England, 2002) pp. 1129-1134

21.60 R. Slowinski, S. Greco, B. Matarazzo: Axiomatization of utility, outranking and decision-rule preference models for multiple-criteria classification problems under partial inconsistency with the dominance principle. Control and Cybernetics, 31 (4), 1005-1035 (2002)

21.61 R. Slowinski, J. Stefanowski: Rough set reasoning about uncertain data. Fundamenta Informaticae, 27, 229-243 (1996)

21.62 R. Slowinski, J. Stefanowski, S. Greco, B. Matarazzo: Rough sets based processing of inconsistent information in decision analysis. Control and Cybernetics, 29, 379-404 (2000)

21.63 R. Slowinski, D. Vanderpooten: A generalised definition of rough approximations. IEEE Transactions on Data and Knowledge Engineering, 12, 331-336 (2000)

21.64 R. Slowinski, C. Zopounidis: Application of the rough set approach to evaluation of bankruptcy risk. Intelligent Systems in Accounting, Finance and Management, 4, 27-41 (1995)

21.65 J. Stefanowski: On rough set based approaches to induction of decision rules. In: L. Polkowski, A. Skowron (eds.), *Rough Sets in Data Mining and Knowledge Discovery* (Physica-Verlag, Heidelberg, 1998) 1, pp. 500-529

21.66 J. Stepaniuk: Knowledge Discovery by Application of Rough Set Models, In: L. Polkowski, S. Tsumoto, T.Y. Lin (eds.): *Rough Set Methods and Application* (Physica Verlag, Heidelberg, 2000) pp. 137-231

21.67 M. Sugeno: *Theory of fuzzy integrals and its applications.* Doctoral Thesis, Tokyo Institute of Technology (1974)

21.68 L.C. Thomas, J.N. Crook, D.B. Edelman (eds.): *Credit Scoring and Credit Control* (Clarendon Press, Oxford, 1992)

21.69 P.P. Wakker, H. Zank: State dependent expected utility for savage's state space. Mathematics of Operations Research, 24, 8-34 (1999)

21.70 L.L. White, A.G. Wilson, D. Wilson (eds.): *Hierarchical Structures* (Elsevier, New York, 1969)

21.71 Y.Y. Yao: Combination of rough and fuzzy sets based on α-level sets. In: T.Y. Lin and N. Cercone (eds.), *Rough Sets and Data Mining: Analysis for Imprecise Data* (Kluwer, Boston, 1997) pp. 301-321

21.72 L.A. Zadeh: A fuzzy set-theoretic interpretation of linguistic hedges. Journal of Cybernetics, 2, 4-34 (1972)

21.73 W. Ziarko: Variable precision rough sets model. Journal of Computer and Systems Sciences, 46, 39-59 (1993)

21.74 W. Ziarko: Rough sets as a methodology for data mining. In: L.Polkowski, A.Skowron (eds.), *Rough Sets in Knowledge Discovery* (Physica-Verlag, Heidelberg, 1998) 1, pp. 554-576

612   S. Greco, B. Matarazzo, R. Slowinski

21.75 W. Ziarko, N. Shan: KDD-R, a comprehensive system for knowledge discovery in databases using rough sets. In: T.Y. Lin, A.M. Wildberg (eds.), *Soft Computing: Rough Sets, Fuzzy Logic, Neural Networks, Uncertainty Management, Knowledge Discovery* (Simulation Council Inc., San Diego, 1995) pp. 93-96

21.76 C. Zopounidis, M. Doumpos: A multicriteria decision aid methodology for sorting decision problems: the case of financial distress. Computational Economics, 14, 197-218 (1999)

Part V

Statistical Learning Theory

22. Bayesian Ying Yang Learning (I): A Unified Perspective for Statistical Modeling

Lei Xu

Chinese University of Hong Kong, Hong Kong

Abstract

Major dependence structure mining tasks are overviewed from a general statistical learning perspective. Bayesian Ying Yang (BYY) harmony learning has been introduced as a unified framework for mining these dependence structures, with new mechanisms for model selection and regularization on a finite size of samples. Main results are summarized and bibliographic remarks are made. Two typical approaches for implementing learning, namely optimization search and accumulation consensus, are also introduced.

22.1 Introduction: Basic Issues of Statistical Learning

Statistical learning is a process by which an intelligent system $\mathcal{M}$ estimates the underlying distribution and dependence structures among what is to be learned (i.e., among the world $\mathbf{X}$ that $\mathcal{M}$ observes). Specifically, learning is done via a certain medium that is usually a set of samples from the world $\mathbf{X}$. Statistical learning plays an essential role in the literature of neural networks. It also acts as a major part in the field of intelligent data processing, or called *data mining*, with an ever increasing popularity.

According to how a learning medium is used, studies on learning can be further classified into two quite different domains. One is the so-called active learning, with a feature that the strategy of getting samples varies according to the current status of $\mathcal{M}$. The other is the conventional statistical learning, during which samples come from the world according to its underlying distribution in help of a given sampling strategy that is independent of the status of $\mathcal{M}$. This paper focuses on the latter. In this case, learning is characterized by the following basic issues, or ingredients:

– A world $\mathbf{X}$, in the form of an underlying distribution with certain dependence structures.
– A machine $\mathcal{M}$ in an appropriate architecture that is able to accommodate or describe the underlying distribution.
– A finite size set $\mathcal{X} = \{x_t\}_{t=1}^{N}$ of random samples that comes from the underlying distribution.
– A learning principle or theory that coordinates the above three issues in order to get the best estimate by $\mathcal{M}$ of the underlying distribution based on $\mathcal{X}$.
– An efficient algorithm that implements the above learning theory

Though the entire world faced by $\mathcal{M}$ is usually very complicated, we can conduct studies by decomposing a complicated task into a number of learning tasks on much simplified small worlds, as discussed in the next two subsequent sections.

22.2 Dependence Among Samples from One-Object World

We start by considering a simple world $\mathbf{X}$ of only a single object that is observed via a set $\mathcal{X}$ of samples, with each sample $x = [x^{(1)}, \cdots, x^{(d)}]^T$ from the same underlying probability distribution or density $p(x)$. Typically, we can use one or more of the following five ways to describe the dependence structures among $\mathbf{X}$.

(a) **Nonparametric joint density** In general, dependence among variables $x^{(1)}, \cdots, x^{(d)}$ can always be described by a nonstructural or nonparametric joint density. Typical example are the empirical density:

$$p_0(x) = \frac{1}{N} \sum_{t=1}^{N} \delta(x - x_t), \ \ \delta(x) = \begin{cases} \lim_{\delta \to 0} \frac{1}{\delta^d}, & x = 0, \\ 0, & x \neq 0, \end{cases} \tag{22.1}$$

where d is the dimension of x and $\delta > 0$ is a small number. This $p_0(x)$ is equivalent to memorizing and using a given data set $\mathcal{X}$ directly. Another typical example is a non-parametric Parzen window density estimate [22.10]:

$$p_h(x) = \frac{1}{N} \sum_{t=1}^{N} K_h(x, x_t), K_h(x, x_t) \text{ is a kernel located at } x_t, \tag{22.2}$$

which means to use the data set $\mathcal{X}$ after certain smoothing by $K_h(x, x_t)$. In the simplest case, $K_h(x, x_t)$ is a hyper cubic of volume h^d with its center located at x_t, and $p_h(x)$ becomes the widely used histogram estimate that is a smoothed version of Eq. (22.1). The smoothness is controlled by a given parameter $h > 0$ that is called *smoothing parameter*. The other case is $K_h(x, x_t) = G(x|x_t, h^2 I)$ and

$$p_h(x) = \frac{1}{N} \sum_{t=1}^{N} G(x|x_t, h^2 I), \tag{22.3}$$

where, and hereafter in this paper, $G(x|m, \Sigma)$ denotes a Gaussian density with mean vector m and covariance matrix Σ.

Such a nonparametric and non-structural joint density estimate, though conceptually implying all dependence relations among variables, has three major weak points. One is that it is usually a bad estimate when the size N of samples is finite, especially when the dimension d of x is high. The second is that dependence relations are not given directly. The third is the expensive computing cost for each calculation on $p(x)$ at any point $x = x'$.

To improve the weak points, efforts are made along two directions.

(b) **Sample statistics**   The first direction is to describe dependencies among variables $x^{(1)}, \cdots, x^{(d)}$ collectively by sample statistics up to certain orders, instead of a joint density that conceptually implies statistics of all possible orders. The most widely encountered case is statistics up to the second order only, represented in the covariance matrix $\Sigma_x = E(x - m_x)(x - m_x)^T, m_x = Ex$, capturing the linear dependence structures among all the variables subject to the mean square error, where (and throughout this chapter) the notations $E(u) = Eu = E[u]$ denotes the expectation of the random variable u. Equivalently, this case actually assumes that the underlying distribution is a parametric Gaussian density with parameters m_x and Σ_x. Of course, we can also similarly consider higher order statistics. However, the number of such statistics increases exponentially, e.g., in the order of d^m for the m-th order statistics, which brings us to the weak points, similar to those for estimating a nonparametric density.

(c) **Co-occurrence and associations**   The second direction is to focus on the pair-wise dependencies between u and v that denote two or more subsets of variables $x^{(1)}, \cdots, x^{(d)}$. The simplest one is co-occurrence or association. That is, we are interested in finding those events $u = U$ and $v = V$ that will co-occur with high probability, e.g., with the corresponding probability $P(u = U, v = V)$ higher than a pre-specified threshold. We are also interested in an association rule $(u = U) => (v = V)$ that describes how we can infer the occurrence of the event $v = V$ from the occurrence of the event $u = U$, which is measured by its support $P(u = U, v = V)$ and confidence $P(v = V | u = U)$, as defined in *data mining* literature for the market basket analysis (Chap. 7 in [22.17]).

We can see that the task of association rule mining is actually a task simplified from estimating the joint density $p(x)$. It is not necessary, and also expensive, to estimate $p(x)$. What we are interested in are only those events $u = U$ and $v = V$ with $P(u = U, v = V) > s$ and $P(v = V | u = U) > c$ for the pre-specified thresholds $s > 0$ and $c > 0$. In addition to those approaches given in (Chap. 7 in [22.17]), we suggest a stochastic approach with its key idea as follows:

– Find a way such that all the possible values of U and V can be sampled with an equal probability, which is referred as random sampling;
– Make n_s/s such random samples; if a specific pair, $\bar{U}$ and $\bar{V}$, appears more than $s \times n_s/s = n_s$ times, we get $(u = \bar{U}) => (v = \bar{V})$ as a candidate association rule;
– Conditioning on $u = \bar{U}$, make n_c/c random samples on V; if the specific value $\bar{V}$ appears more than $c \times n_c/c = n_c$ times, we take $(u = \bar{U}) => (v = \bar{V})$ as an association rule with its support larger than s and its confidence larger than c.

(d) **Regressions or fittings**   Instead of considering the conditional probability $P(v = V | u = U)$, we may also directly consider the average or expected relation $E(v|u) = \int v P(v|u) dv$, which is non-probabilistic function $v = f(u)$ and usually called regression. The regression can be either linear or nonlinear. Particularly, when v is Gaussian from $G(v|f(u), \sigma_v^2 I)$, the function $f(u)$ fits a set of pairs $\{u_t, v_t\}_{t=1}^N$ in a sense that the average square fitting error $\frac{1}{N} \sum_{t=1}^N \|v_t - f(u_t)\|^2 \approx \sigma_v^2$ is minimized.

(e) **Linear and nonlinear generative structures**   We can also explore dependencies among all the variables $x^{(1)}, \cdots, x^{(d)}$ in help of inner representations via certain hidden or latent structures. One typical structure is the following explicit stochastic linear function:

$$x = Ay + e, \ E(e) = 0, \ e \text{ is independent of } y. \tag{22.4}$$

The earliest effort on this linear model can be traced back to the beginning of the 20th century by Spearman [22.41], and has been followed by various studies in the literature of statistics. In this model, a random sample x of observations is generated via a linear mapping matrix A, from k inner representations or hidden factors in the form $y = [y^{(1)}, \cdots, y^{(k)}]^T$ from a parametric density $q(y)$, disturbed by a noise e. Generally, samples of e are independently and identically distributed (i.i.d.) from a parametric density $q(e)$, and the function forms of $q(y)$ and $q(e)$ are given. The matrix A and the statistics of y_t and e_t are to be learned.

The problem is usually not well defined because there are an infinite number of solutions. To reduce the indeterminacy, we assume that samples of x and e are i.i.d., and, correspondingly, samples of y are also i.i.d. such that dependence among variables of x is equivalently modeled by

$$q(x) = \int q(x - Ay)q(y)dy, \tag{22.5}$$

which implies a parametric density of x, described via the matrix A, the statistics of y and e, and the density forms of $q(y)$ and $q(e)$. Also, Eq. (22.4) implies that all the statistics of x are subject to the constraint. For example, with $\Sigma_x, \Sigma_y, \Sigma_e$ being covariance matrices of x, y, and e, respectively, we have

$$\Sigma_x = A^T \Sigma_y A + \Sigma_e. \tag{22.6}$$

Particularly, when y is Gaussian and uncorrelated with e, e also being uncorrelated with its components, the model Eq. (22.4) is called factor analysis [22.32, 22.39], which was first formulated in [22.3]. In this case, the integral $q(x)$ is analytically solvable, and becomes simply Gaussian.

As discussed in [22.57], this constraint, Eq. (22.6), is not enough to uniquely specify A and Σ_e. If we further impose the constraint

$$A = \phi D, \phi^T \phi = I, \ \Sigma_y = I, \ \Sigma_e = \sigma_e^2 I, \tag{22.7}$$

it follows [22.3, 22.66] that the maximum likelihood learning on $q(x)$ results in ϕ consisting of the k principal eigenvectors of Σ_x and D^2 consisting of the corresponding principal eigenvalues. That is, it becomes equivalent to the so-called principal component analysis (PCA). In recent years, such a special case of factor analysis has also been reiterated in the literature of neural networks under a new name, probabilistic PCA [22.45].

When y is non-Gaussian, the Eq. (22.4) implies not only the constraint Eq. (22.6), but also constraints on higher order statistics such that the indeterminacy on A and Σ_e may be removed. However, the integral $p(x)$ by Eq. (22.5) becomes analytically unsolvable when y is real and non-Gaussian.

Generally, Eq. (22.4) can also be either a nonlinear structure

$$x = g(y, A) + e, \quad q(x) = \int q(x - g(y, A))q(y)dy, \tag{22.8}$$

or a general probabilistic structure

$$q(x) = \int q(x|y)q(y)dy. \tag{22.9}$$

Actually, the above two cases are regarded as being equivalent via the link

$$g(y, A) = E(x|y) - Ee, \quad e = x - g(y, A), \tag{22.10}$$

where e generally relates to y, instead of being independent of y.

(f) **Linear and nonlinear transform structures**   As a dual to generative structure, another typical structure for describing dependence among variables is a forward transform or mapping, $y = f(x, W)$ to inner representations of y.

The simplest case is $y = Wx$ when $Ex = 0$ (otherwise the mean can be subtracted in preprocessing). It specifies a relation $Eyy^T = \Sigma_y = W^T \Sigma_x W$ between the second order statistics of y and x, such that dependence among variables of x are implicitly specified via W and the statistics of y. When Σ_y becomes diagonal, the mapping $y = Wx$ is said to de-correlate the components of x. Moreover, $\max_{W^T W = I} E\|y\|^2$, $E\|y\|^2 = Tr[\Sigma_y]$, leads to the well known principal component analysis (PCA) that extracts the principal second order statistics. The study on PCA can be backtracked to as early as in 1936 [22.19], and has been also widely studied in the literature of statistics, pattern recognition, and neural networks [22.34]. Equivalently, PCA is also reached under the constraint $W^T W = I$ by maximizing the following entropy:

$$\max J(W), \quad J(W) = -\int p(y) \ln p(y) dy, \tag{22.11}$$

when x comes from Gaussian and $y = Wx$ is still Gaussian.

Moreover, when $y = Wx$ satisfies

$$q(y) = \prod_{j=1}^{k} q(y^{(j)}) \tag{22.12}$$

with at most one component being Gaussian, the mapping $y = Wx$ is said to implement independent component analysis (ICA) [22.15, 22.46, 22.7] that extracts statistics of the second order and those higher orders. Also, when each component of y_t is interpreted as a sample of a time series at time t, the ICA solution $\hat{y}_t = Wx_t$ recovers y_t from $x_t = Ay_t$, up to a scaling indeterminacy. That is, the waveform of each component series can be recovered by $\hat{y}_t = Wx_t$, by a process also called blind source separation (BSS) that blindly separates the mixed signal $x_t = Ay_t$, with scaling indeterminacy [22.46].

However, an ICA algorithm with $q(y^{(j)})$ pre-fixed works well only on the cases where either all components of y are super-Gaussians or all components of y are sub-Gaussians [22.2, 22.4], but not on the cases where components of y are partly super-Gaussians and sub-Gaussians. This problem can be solved by learning $q(y^{(j)})$ via a mixture of parametric densities, or equivalently learning the cumulative distribution function (cdf) of $q(y^{(j)})$ by a mixture of parametric cdfs [22.87, 22.72, 22.73]. Efforts have also been made on using a nonlinear mapping $y = f(x, \theta)$ to implement a nonlinear ICA [22.42, 22.66]. However, the satisfaction of Eq. (22.12) will remain unchanged after any component-wise nonlinear transformation on y. Thus, a nonlinear ICA mapping $y = f(x, \theta)$ usually does not retain the feature of performing BBS, since it is no longer able to recover the original y up to scaling indeterminacy, unless in a specific situation with extra constraints imposed [22.42].

Instead of considering nonlinear ICA, we consider in [22.53] a nonlinear mapping $y = f(x, \theta)$ from the perspective of modeling the cumulative distribution function (cdf) of input data x.

22.3 Dependence Among Samples from a Multi-Object World

22.3.1 Dependence Among Samples from a Multi-Object World

We observe a world $\mathbf{X}$ with multiple objects that are either visible or invisible. We start by considering the cases where all the objects are visible, with each label ℓ in a set L denoting a specific object observed via a sample vector $x_\ell = [x_\ell^{(1)}, \cdots, x_\ell^{(d)}]^T$.

We can discover dependence structures within each object $\ell \in L$ individually, in the same ways introduced in the previous subsection. Moreover, dependence structures also exist across different objects. They are described both qualitatively by the topology of L, and quantitatively by dependence structures among variables of x_ℓ across objects.

In those most complicated cases, where there are dependence structures between any pair of objects, the topology of L is a complete graph with every pair of nodes connected. Specifically, if there is no dependence between a pair of objects, the connection between the corresponding two nodes is broken, and can, thus, be removed. Usually, three types of simplified topology are often encountered, described as follows:

(a) *A linear or serial chain*   The simplest case is that the topology is a simple chain, $1, 2, \cdots, \ell - 1, \ell, \ell + 1, \cdots$, with the object ℓ directly connecting to only the object $\ell - 1$ and the object $\ell + 1$. In this case, $x_1, x_2, \cdots, x_{\ell-1}, x_\ell, x_{\ell+1}, \cdots$ form a sequence called a time series, where ℓ

denotes the time. The task of learning is to estimate quantitatively the dependence structures by the joint distribution of the whole sequence in a certain serial/temporal dependence structure. The task is often encountered in signal processing and stochastic process modeling, such as time series prediction and speech, audio, and text processing, via AR and ARMA models and, especially, via the Hidden Markov model [22.36].

(b) *An image and a d-dimensional lattice topology*   We further consider the cases where labels in L are organized in a regular lattice topology, with each object ℓ denoted by a coordinate $(\ell_1, \ell_2, \ldots, \ell_d)$. It reduces to the above chain and time series model when $d = 1$. Moreover, $\{x_\ell : \ell \in L\}$ denotes an image when $d = 2$ and a 3D array or video when $d = 3$, etc., x_ℓ being a pixel. Similarly, the task of learning is to estimate the dependence structures by the joint distribution of all the pixels (usually with the help of certain dependence structures between a pixel and others in its neighborhood), which actually leads to a stochastic field. Such tasks are often encountered in image processing and computer vision.

(c) *A tree topology*   Another typical case is where L has a tree topology, with each object ℓ being a node on the tree, i.e., each node has direct dependence only on its father or children. The task of learning is to estimate the dependence structures by the joint distribution $\{x_\ell : \ell \in L\}$ subject to this tree dependence structure. This is a typical task that has been studied as the popular topic of probabilistic graphical model or Belief networks in the recent literature of AI, statistics, and neural networks; readers are referred to the books [22.35, 22.22].

(d) *A joint temporal topology and spatial topology*   A topology that combines the above case (a) with either of the case (b) and case (c) is also widely studied in the literature. Here, a topology of either of image, d-lattice, and tree describes spatial relations among different objects, while a linear topology describes temporal relations of each object that changes with time, such as encountered in video processing.

22.3.2 Mining Dependence Structure Across Invisible Multi-Object

Next, we further observe a world of invisible multiple objects, with each sample vector x coming from one object $\ell \in L$, but with its label ℓ missing.

Given a set of such samples, we want to discover the dependence structures among the objects, which are again described both qualitatively by the topology of L and quantitatively by the dependence structures among variables of x within and across objects. Still, the tasks of learning can be classified according to what types of topology of L are taken into consideration.

(a) *Finite Mixture*   The simplest case is that we ignore the topology of L, and consider only dependence structures among the variables of x, which depends on the dependence structure between each sample x and each label ℓ. The general form of the dependence is described by a joint distribution $q(x, \ell)$

that describes the joint occurrence of the events that a sample is valued by x and that the sample comes from the object ℓ.

Specifically, $q(x, \ell)$ can be represented via the decomposition $q(x|\ell)q(\ell)$, which describes $q(x, \ell)$ as a casual view on where x comes from. Each of $q(x|\ell), \ell = 1, \cdots, k$, describes the dependence structures within the ℓ-th object, and

$$q(\ell) = \sum_{j=1}^{k} \alpha_j \delta(\ell - j), \text{ with the constraint } \alpha_\ell \geq 0, \sum_{\ell=1}^{k} \alpha_\ell = 1, \quad (22.13)$$

where α_ℓ denotes a priori probability that x comes from the ℓ-th object.

The task of learning is to estimate α_ℓ, and to discover dependence structures within each $q(x|\ell)$, which is equivalent to learning dependence structures in the format

$$q(x) = \sum_{\ell \in L} \alpha_\ell q(x|\ell). \quad (22.14)$$

It is usually called finite mixture in the literature [22.9, 22.38, 22.33]. It is also said to have modular structure in a sense that there are a number of individual modules in an assembly.

As a by-product, we also have the Bayesian rule

$$q(\ell|x) = \alpha_\ell q(x|\ell)/q(x), \quad (22.15)$$

and, thus, another decomposition, $p(x, \ell) = p(\ell|x)p(x)$ with $p(\ell|x)$ describing an inference view on which object the observation x may come from. It results in a partition of a set of samples into different objects, and, thus, is usually called pattern recognition or clustering analysis [22.10]. Particularly, when the dependence structure of $q(x|\ell)$ is simply a Gaussian $q(x|\ell) = G(x|\mu_\ell, \Sigma_\ell)$, Eq. (22.14) becomes the widely used Gaussian mixture [22.9, 22.38, 22.33], and Eq. (22.15) is called Bayesian classifier [22.11]. Moreover, when $\Sigma_\ell = \sigma^2 I$ and $\alpha_\ell = 1/k$, a hard-cut version of Eq. (22.15) leads to the conventional least square clustering [22.77].

(b) *Self-organizing map*   We further consider the cases where L has a given regular d-dimensional lattice topology. Since the label ℓ associated with x is invisible, we are not able to recover the dependence structures among the variables of x across different objects. Alternatively, we re-establish dependence structures according to *a general belief that objects locating topologically in a small neighborhood N_ℓ should be same as or similar to each other*, where a small neighborhood N_ℓ of a knot ℓ usually consists of 2^d knots that are directly connected to ℓ.

The direct placement of all the objects on such a lattice, according to a criterion or measure to judge whether two objects are same or similar, is computationally an NP-hard problem. Instead, this placement can be implemented approximately. Interestingly, a good approximation for this problem is provided by biological brain dynamics of self-organization [22.31], featured by a Mexican hat-type interaction, namely, neurons in near neighborhood excite each other with learning, while neurons far away inhibit each other with de-learning. Computationally, such a dynamic process can be simplified by certain heuristic strategies. Here, we consider two typical ones.

- *One member wins, a family gains*   That is, as long as one member wins in the winner-take-all competition, all the members of a family gain, regardless of whether other members are strong or not. This direction is initialized by a simple and clever technique, i.e., the well known Kohonen self-organizing map [22.23]. In the literature, a great number of studies have been made on extending the Kohonen map. Recently, a general formulation of this strategy has also been proposed [22.54].

- *Stronger members gain and then team together*   That is, a number of stronger members in competition will be picked as winners, who not only gain learning but are also teamed together to become neighbors [22.54]. It can speed up self-organization, especially in the early stages of learning. Also, we can combine the first and the second strategies by using the second in the early stage, and subsequently switching to the first, which is experimentally demonstrated in [22.8]

(c) *Self-organizing graphical topology*   Following a similar line, we can also extend the above studies to a more complicated but given topology of either a tree or a general graph. The only difference is that a small neighborhood N_ℓ of a specific ℓ consists of all the knots that are directly connected to ℓ. and that the number of elements in N_ℓ is not usually 2^d, but equal to the degree of the node ℓ.

(d) *Dynamic self-organizing map*   Another useful but even more complicated case is a combination of a temporal topology (i.e., a line topology) with either of the above cases (a), (b) and (c), which describes not only relations across objects but also how the relations change with time.

22.4 A Systemic View on Various Dependence Structures

Those dependence structures discussed in Sect. 1.2 and Sect. 1.3 can be systematically summarized according to which inner representation is adopted and what kind of inner architectures is used.

Figure 22.1 shows the simplest family F_0, by considering an object as a whole without considering inner representation and inner architecture. The types of dependence here can be classified according to differences in three features. First, focus is put whether on dependencies among all the components of observation as a whole or on the relation between two particular parts of observation. Second, whether temporal dependencies among samples are ignored or considered. Third, whether multiple objects are jointly

The F_0 Family

This F_0 is featured by (0,0) with the 1st '0' denoting a 0-architecture, i.e., an object or distribution is considered as a whole without considering architecture, with the 2nd '0' denoting that no inner representation y is considered.

Different types of dependences in this family come from three features
(*observation, time, topology*)

0 for *observation*: dependence among components of x are considered as a whole
1 for *observation*: dependence between two parts (ξ_t, ς_t) of $x_t = (\xi_t, \varsigma_t)$ is focused
0 for *time*: no dependence between any two samples of x from different times
1 for *time*: dependences among samples from different times are considered
1 for *topology*: only a single object is considered
2 for *topology*: multiple objects without topology are considered
3 for *topology*: multiple objects are considered to be mapped onto a topology

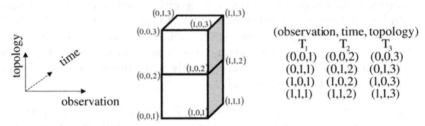

Fig. 22.1. The family F_0

considered without or with a topology. As shown in Fig. 22.1, there are 12 combinatorial types in three groups. T_1 consists of

- $(0, 0, 1)$ (e.g., either a nonparametric density or a simple Gaussian density),
- $(1, 0, 1)$ (e.g., co-occurrence, association, regression and fitting etc.),
- $(0, 1, 1)$ (e.g., with time added, a Gaussian density is extended to a conventional AR, ARMA models),
- $(1, 1, 1)$ (e.g., multi-channel AR, ARMA models).

Moreover, T_2 extends those of T_1 to jointly consider multiple models but ignoring topology, consisting of

- $(0, 0, 2)$ (e.g., finite mixture or Gaussian mixture),
- $(1, 0, 2)$ (e.g., mixture-of-experts [22.12, 22.13, 22.14]),
- $(0, 1, 2)$ (e.g., finite mixture of AR, ARMA models [22.25, 22.48, 22.49, 22.43]),
- $(1, 1, 2)$ (e.g., mixture-of-experts of AR, ARMA models [22.44]).

Furthermore, T_3 extends those of T_1 to jointly consider multiple models and also map them onto a regular topology, with $(0, 0, 3)$ as a self-organizing map (SOM) and other three are further extensions of SOM.

Figure 22.2 shows more sophisticated alternatives of F_0, featured by using what type of five typical inner representations and which of three major architectures, indicated by a code $(i,j), i = 1,2,3, j = 1,2,3,4,5$. With this notation, F_0 can be regarded as a degenerated case with the code $(0,0)$. Different types of inner representations encode different types of dependencies and, thus, perform different pattern recognition and information processing tasks. A same task can be implemented by either of three architectures. As discussed in Sect. 2.1, a forward structure directly implements the mapping $x \rightarrow y$ per sample and indirectly describes data distribution in a differential manner, while a backward structure directly describes data distribution in an integral manner and indirectly implements the mapping $x \rightarrow y$ with an expensive computing cost. A bi-directional architecture trades off the features of the two such that not only modeling data distribution and implementing $x \rightarrow y$ can be both directly made but also each of the two provides a structural constraint to the other as a role of regularization (see Item 3.4 in [22.74]). The least mean square error reconstruction learning [22.94] is a simple example of a BI-architecture that combines a sigmoid post-linear forward structure and a linear backward structure by Eq. (22.4). In fact, this is the key sprit of Bayesian Ying Yang system. The details will be given the subsequent sections.

Types of dependences, under the perspective of BYY system, can be understood by five families featured by a combination of the following two features:

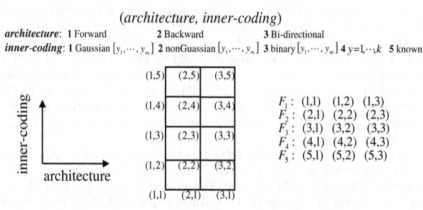

Fig. 22.2. Types of Dependence from BYY System Perspective

As shown in Fig. 22.2, we can further group the 3×5 combinations into five families according to five types of inner representations, with each family consisting of three sub-families featured by each of three architectures. From

F_1 to F_5, constraints on the inner representation are gradually enhanced from de-correlated to becoming independence and then to uniform or known such that indeterminacy of the structures is gradually reduced. The details are briefed as follows:

(a) $F_1 = \{(1,1),(2,1),(3,1)\}$ is featured by a Gaussian vector $y = [y_1, \cdots, y_k]$ with its components de-correlated from each other. Typical examples of $(1,1)$ includes PCA, MCA, and their degenerated case called de-correlated component analysis (DCA) [22.53]. Typical example of $(2,1)$ is factor analysis (FA) by Eq. (22.4), and that of $(3,1)$ is $\min E\|x - AWy\|^2$ that leads to a singular value decomposition (SVD) analysis [22.6]. In this special case, both the linear mapping $x \to y$ and $y \to x$ can be directly obtained. Interestingly, performing PCA is a common special case of three architecture. Similar to Fig. 22.1, they can be further extended to not only temporal PCA (see Eq. (87) in [22.57]; Eq. (163) in [22.53]), temporal FA (see Eqs. (78) and (79) in [22.57]), and temporal SVD by adding in time, as well as hidden layer aided temporal regression by focusing on two part relation, but also to local PCA, local MCA, and local FA [22.54], etc. by jointly considering multiple models but ignoring topology, as well as further to their self-organizing map versions by mapping them on a regular topology.

(b) $F_2 = \{(1,2),(2,2),(3,2)\}$ is featured by a non-Gaussian vector $y = [y_1, \cdots, y_k]$ with components independent from each other such that statistics higher than the second order are also in consideration. A typical example of $(1,2)$ is the ICA discussed at the end of Sect. 2.1. A typical example of $(2,2)$ is non-Gaussian factor analysis (NFA) (see Sect. III in [22.57]), which is superior to ICA via taking observation noise in consideration. An example of $(3,2)$ is an extension of LMSER learning (Sect. 5.4 in [22.53]). Again, it follows from Fig. 22.1 that they can be further extended to temporal versions by adding in time, to local ICA, local NFA, and local LMSER [22.54] by jointly considering multiple models but ignoring topology, as well as further to mapping them on a regular topology.

(c) $F_3 = \{(1,3),(2,3),(3,3)\}$ is featured by a binary vector $y = [y_1, \cdots, y_k]$ with bits independent from each other, which is suitable for problems of encoding data into binary bits. The ICA algorithm made by [22.4] can be regarded as a typical example of $(1,3)$. An example of $(2,3)$ is binary factor analysis (BFA), and that of $(3,3)$ is the LMSER learning [22.94]. Similar to Fig. 22.1, they can be extended to temporal ICA, independent HMM, and its Bi-directional version [22.53, 22.59, 22.57] by adding in time, to local ICA, local BFA, and local LMSER [22.54, 22.55] as well as further to their SOM extensions.

(d) $F_4 = \{(1,4),(2,4),(3,4)\}$ is featured by a discrete label $y = 1, \cdots, k$ that is suitable for making pattern classification via the mapping $x \to y$. The classic perceptron [22.11] and the recent SVM based perceptron [22.47] are both examples of $(1,4)$ for a two classes problem (i.e., $k = 2$). The maximum balanced mapping certainty principle proposed in [22.84] is another example of $(1, 4)$ for $k > 2$. A typical example of $(2, 4)$ is the finite mixture given by Eq. (22.14), and that of $(3, 4)$ is given by BYY learning with a Bi-directional architecture [22.30].

(e) $F_5 = \{(1, 5), (2, 5), (3, 5)\}$ is featured by a uniform distribution for y. It can be regarded as special cases of the above four families with the distribution of y fixed uniformly in a sense that every value of y is taken with an equal probability. As discussed at the end of Sect. 22.2, the mapping $x \to y$ completely describes data distribution in a CDF form [22.53]. Moreover, we may also fix y on a known distribution for specific tasks.

The above discussed perspective of jointly considering forward structure and backward structure, together with typical types of inner representation, provides a systematic view to model various dependence structures. What discussed above is exactly the fundamental sprit of Bayesian Ying Yang system that acts as a unified statistical learning framework. Actually, a quite number of dependence structures mentioned above were initially brought into studies under the guidance of this BYY system. In the rest of this chapter, advances on BYY system and harmony learning will be further overviewed.

22.5 Bayesian Ying Yang System

The world that we consider above can be denoted by $\mathbf{X} = \{X, L\}$ that consists of a number of objects to observe, with L denoting a set of labels with each $\ell \in L$ denoting one object. X is a set of samples with each sample x coming from one of these objects. We have several special cases that correspond to dependence structures previously discussed. The simplest case is that there is only one object in L and X consists of a size N of samples that all come from this object. Thus, L can be ignored. For the cases in Sect. 22.3.1, L has a regular lattice topology and each object ℓ locates at each node n_ℓ. Also, it is already known that each x comes from which node and, thus, can be labeled explicitly, i.e., x_ℓ. Particularly, the topology L for the case (d) can be decomposed into a direct product of a line L_t for time and a regular lattice L_r, i.e., $L = L_t \times L_r$. For the cases in Sect. 22.3.2, the label that indicates where x comes from is actually missing. Through learning, either each sample x is classified into one ℓ of a finite mixture or each sample x is mapped to a node n_ℓ on L such that nodes located nearby describes similar samples. For the case (d) with a topology $L = L_t \times L_r$, x comes at which time is known, and, thus, there is no need to allocate x to a time. Further noticing that the cases of $x = [\xi, \eta]$ are covered as special cases of the above discussed, all the cases in Fig. 22.1 can be summarized by the notation $\mathbf{X} = \{X, L\}$.

Types of dependencies in Fig. 22.2 are further considered via a corresponding representation domain $\mathbf{Y} = \{Y, L\}$ of a BYY system. For simplicity, we start at considering L that consists of a number of isolated objects without any topology. In this case, each $\mathbf{x} = \{x, \ell\}$ denotes a joint event of an observation $x = [x^{(1)}, \cdots, x^{(d)}]^T$ and an object ℓ, subject to a joint underlying distribution $p(\mathbf{x}) = p(x, \ell)$. Corresponding to each $\mathbf{x}$, there is an inner representation $\mathbf{y} = \{y, \ell\}$ in the representation domain $\mathbf{Y} = \{Y, L\}$, subject to a parametric structure of $q(\mathbf{y}) = q(y, \ell)$. Except of the family F_5, representation types of y are rather simple and correspond to the following distributions:

$$q(\mathbf{y}) = q(y, \ell) = q(y|\ell)q(\ell), \quad q(\ell) = \sum_{j=1}^{k} \alpha_j \delta(\ell - j), \quad \alpha_\ell \geq 0, \quad \sum_{\ell=1}^{k} \alpha_\ell = 1,$$

$$q(y|\ell) = \prod_{j=1}^{m_\ell} q(y^{(j)}|\ell), \quad y = [y^{(1)}, \cdots, y^{(m_\ell)}]^T, \tag{22.16}$$

$$q(y^{(j)}|\ell) = \begin{cases} G(y^{(j)}|\mu_\ell^{(j)}, \lambda^{(j)\,2}), & \text{for } F_1, \\ \sum_i \beta_{ji} G(y^{(j)}|\mu_\ell^{(ji)}, \\ \lambda^{(ji)\,2}), \sum_i \beta_{ji} = 1, 0 \leq \beta_{ji} \leq 1, & \text{for } F_2, \\ (q_\ell^{(j)})^{y^{(j)}}(1 - q_\ell^{(j)})^{1-y^{(j)}}, & \text{for } F_3, \\ \delta(y^{(j)} - \mu_\ell^{(j)}), & \text{for } F_4. \end{cases}$$

where k is the number of labels in L, and m_ℓ is the dimension of either a binary or real vector y.

As shown in Fig. 22.3, we consider the joint distribution of $\mathbf{x}$ and $\mathbf{y}$, which can be understood from two complementary perspectives. On the one hand, we can interpret each $\mathbf{x}$ generated from an invisible inner representation $\mathbf{y}$ via a backward path distribution $q(\mathbf{x}|\mathbf{y})$, or a generative model

$$q(\mathbf{x}) = \int q(\mathbf{x}|\mathbf{y})q(\mathbf{y})d\mathbf{y} \tag{22.17}$$

that maps from an inner distribution $q(\mathbf{y})$. On the other hand, we can interpret each $\mathbf{x}$ as being mapped into an invisible inner representation $\mathbf{y}$ via a forward path distribution $p(\mathbf{y}|\mathbf{x})$, or a representative model

$$p(\mathbf{y}) = \int p(\mathbf{y}|\mathbf{x})p(\mathbf{x})d\mathbf{x} \tag{22.18}$$

that matches the inner density $q(\mathbf{y})$.

The two perspectives reflect the two types of Bayesian decomposition of the joint density $q(\mathbf{x}|\mathbf{y})q(\mathbf{y}) = q(\mathbf{x}, \mathbf{y}) = p(\mathbf{x}, \mathbf{y}) = p(\mathbf{x})p(\mathbf{y}|\mathbf{x})$ on $X \times Y$. Without any constraints, the two decompositions should be theoretically identical. Considering real situations, however, the four components $p(\mathbf{y}|\mathbf{x}), p(\mathbf{x}), q(\mathbf{x}|\mathbf{y})$, and $q(\mathbf{y})$ should be subject to certain structural constraints. Thus, we usually have two different but complementary Bayesian representations:

$$p(u) = p(\mathbf{x}, \mathbf{y}) = p(\mathbf{y}|\mathbf{x})p(\mathbf{x}), \quad q(u) = q(\mathbf{x}, \mathbf{y}) = q(\mathbf{x}|\mathbf{y})q(\mathbf{y}), \tag{22.19}$$

As discussed in the original paper [22.88], thanks to the famous Chinese ancient Ying-Yang philosophy, $p(\mathbf{x}, \mathbf{y})$ is called Yang machine, and consists of the observation space (or Yang space) of $p(\mathbf{x})$ and the forward pathway (or Yang pathway) of $p(\mathbf{y}|\mathbf{x})$, and $q(\mathbf{x}, \mathbf{y})$ is called the Ying machine, and consists of the invisible state space (or Ying space) of $q(\mathbf{y})$ and the Ying (or

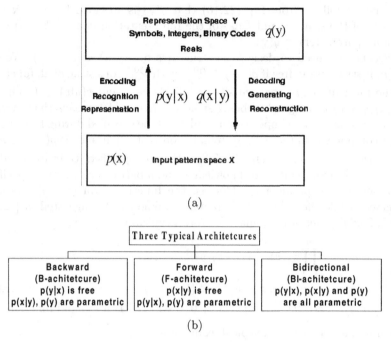

(a)

(b)

Fig. 22.3. Bayesian Ying Yang System and Three Architectures

backward) pathway of $q(\mathbf{x}|\mathbf{y})$. Such a pair of Ying-Yang models is called *the Bayesian Ying Yang (BYY) system.*

This BYY system provides a unified framework for describing various dependence structures in Sect. 22.2 and Sect. 22.3, with details as follows:

– The distribution $p(\mathbf{x})$ is obtained on a set $\mathcal{X}$ of samples from the observed world $\mathbf{X}$, either by the empirical density of Eq. (22.1) or by the nonparametric estimate of Eq. (22.3) with an unknown smoothing parameter h.
– The Ying path, at the special case of only one object in L, covers Eq. (22.9) that includes Eq. (22.5) and Eq. (22.8). Generally, the cases with $k \geq 2$ extend Eq. (22.9) to modular generative structures.
– The Yang path provides the general form of Eq. (22.18) for describing various modular linear and nonlinear transform structures, including those in Sect. 22.2 at a special case of only one object in L.
– All the dependence structures in Sect. 22.3.1 are all covered when L has the topology of a line, a lattice, and a tree, respectively, when the dependence relation between X and L pre-specified. Moreover, they can be further extended to their corresponding self-organizing maps when the dependence relation between X and L is missing.

The task of learning on a BYY system consists of specifying all the aspects of $p(\mathbf{y}|\mathbf{x})$, $q(\mathbf{x}|\mathbf{y})$, and $q(\mathbf{y})$, as well as h (if applicable).

First, it follows from Eq. (22.16) that $q(\mathbf{y})$ is specified by both $\mathbf{k}$ that consists of the scales k and $\{m_\ell\}$ of the representation domain $\mathbf{Y}$ and a set θ_y of all parameters in $q(\mathbf{y})$.

Second, we need to design the structures of $p(\mathbf{y}|\mathbf{x})$ and $q(\mathbf{x}|\mathbf{y})$. We say $p(u|v)$ is structurally free if $p(u|v) \in \mathcal{P}^0_{u|v}$, with $\mathcal{P}^0_{u|v}$ consisting of all functions in the form of $p(u|v)$ that satisfy $\int p(u|v)du = 1$ and $p(u|v) \geq 0$. One of $p(\mathbf{y}|\mathbf{x}), q(\mathbf{x}|\mathbf{y})$ is designed as being structure-free, which means that there is no priori constraint to impose on it, and it will be specified during learning by other components in the BYY system. In contrast, a parametric $p(u|v) \in \mathcal{P}^S_{u|v}$ means that it comes from a family $\mathcal{P}^S_{u|v}$ with a pre-specified structure based on certain priori requirements or knowledge, and a particular density is specified by a set $\theta_{u|v}$ of parameters. That is, the function form of $p(u|v)$ is pre-specified, while the value of $\theta_{u|v}$ remains unknown. As suggested in [22.66, 22.62, 22.60], when u is a binary vector, a typical example is

$$p(u|v) = \prod_{j=1}^{m} \pi_j(v)^{u^{(j)}} (1 - \pi_j(v))^{1-u^{(j)}},$$
$$\pi(v) = [\pi_1(v), \cdots, \pi_m(v)]^T = S(Wv + c), \qquad (22.20)$$
$$S(y) = [s(y^{(1)}), \cdots, s(y^{(m)})]^T, \ 0 \leq s(r) \leq 1 \text{ is a sigmoid function.}$$

When u is a real vector, a typical example is

$$p(u|v) = \sum_{j=1}^{n} \beta_j(v)p_j(u|v), \ \sum_{j=1}^{n} \beta_j(v) = 1, \ \beta_j(v) \geq 0,$$
$$p_j(u|v) = G(u|f_j(v|\theta_{u|v,j}), \Sigma_{u|v,j}). \qquad (22.21)$$

Correspondingly, the function of a BYY system is featured by the representation types of y and the implementing architecture of a BYY system is featured by a combination of the specific structures of $p(\mathbf{y}|\mathbf{x})$ and $q(\mathbf{x}|\mathbf{y})$. As shown in Fig. 22.3(b), we have

- A B-architecture when focusing on only dependence structures of a generative type, as discussed in Sect. 22.2, with a structure-free $p(\mathbf{y}|\mathbf{x})$,
- An F-architecture when focusing on only the linear and nonlinear transform structures in Sect. 22.2, with a structure-free $q(\mathbf{x}|\mathbf{y})$,
- A BI-architecture when both types of dependence structures are explored together.

The architecture in which both $p(\mathbf{y}|\mathbf{x})$ and $q(\mathbf{x}|\mathbf{y})$ are structure-free is useless and, thus, ignored.

In a summary, our learning task includes two subtasks. One is parameter learning for determining the value of θ that consists of all the unknown parameters in $p(\mathbf{y}|\mathbf{x})$, $q(\mathbf{x}|\mathbf{y})$, and $q(\mathbf{y})$, as well as h (if applicable). The other is selecting the representation scales $\mathbf{k} = \{k, \{m_\ell\}\}$, called *model selection*, since a collection of specific BYY systems with different scales corresponds to a family of specific models that share the same system configuration but in different scales.

The *fundamental learning principle* is to make the Ying machine and Yang machine have the best harmony in a twofold sense:

- The difference between the two Bayesian representations in Eq. (22.19) should be minimized.
- The resulting BYY system should be of the least complexity.

22.6 BYY Harmony Learning

22.6.1 Kullback Divergence, Harmony Measure, and Z-Regularization

To implement the above harmony learning principle, we need to formalize it mathematically. One possible measure is the well known Kullback divergence

$$KL(p\|q) = \int p(u) \ln \frac{p(u)}{q(u)} du \geq 0, \ \ KL(p\|q) = 0, if f \ p(u) = q(u) \quad (22.22)$$

which is applicable to the cases where both p and q are discrete and continuous densities. The minimization of the Kullback divergence implements the first point well, and this is why it has been used in early stages of the BYY system [22.89, 22.82, 22.83, 22.85, 22.74, 22.75, 22.76, 22.77]. However, it is not able to implement the least complexity nature. In other words, the Kullback divergence can only be used for partial implementation. We need a measure that implements both the points.

We consider the following cross entropy

$$H(p\|q) = \sum_{t=1}^{N} p_t \ln q_t, \quad (22.23)$$

where both $p(u)$ and $q(u)$ are discrete densities of the form

$$q(u) = \sum_{t=1}^{N} q_t \delta(u - u_t), \ \ \sum_{t=1}^{N} q_t = 1. \quad (22.24)$$

The maximization of $H(p\|q)$ has two interesting natures:

- *Matching nature* with p fixed, $\max_q H(p\|q)$ pushes q toward

$$q_t = p_t, \ \text{for all } t. \quad (22.25)$$

- *Least complexity nature* with q fixed, $\max_p H(p\|q)$ pushes p toward its simplest form

$$p(u) = \delta(u - u_\tau), \text{or } p_t = \bar{\delta}_{t,\tau}, \text{ with } \tau = arg \max_t \ q_t, \quad (22.26)$$

where, and throughout this paper, $\bar{\delta}_{j,j^*} = 1$ when $j = j^*$, $\bar{\delta}_{j,j^*} = 0$ otherwise.

As discussed in [22.57, 22.58, 22.54, 22.55], Eq. (22.26) is a kind of the least complexity form from a statistical perspective. In other words, the maximization of the functional $H(p\|q)$ indeed implements the above harmony learning principle mathematically.

Moreover, as shown in [22.57, 22.58, 22.54, 22.55], either a discrete or a continuous density $q(u)$ can be represented in the form of Eq. (22.24) via the normalization

$$q_t = q(u_t)/z_q, \quad z_q = \sum_{t=1}^{N} q(u_t), \qquad (22.27)$$

based on a given set $\{u_t\}_{t=1}^{N}$ of samples.

Putting Eq. (22.27) into Eq. (22.23), it follows that we can further get a general form of the harmony measure [22.57]

$$H(p\|q) = \int p(u) \ln q(u) du - \ln z_q, \qquad (22.28)$$

which degenerates to Eq. (22.23) when $p(u)$ and $q(u)$ are discrete, as in Eq. (22.24).

To get a better insight, we further examine the maximization of $H(p\|q)$ from the aspects of its two natures discussed above.

Still, $\max_p H(p\|q)$, with q fixed, leads to the least complexity nature of the form in Eq. (22.26). It is not directly observable that this nature has any use, since $\delta(u - u_\tau)$ has a very limited representation ability. This may be the reason why the least complexity nature of Eq. (22.26) of the cross entropy in Eq. (22.23) has been rarely studied in the literature. However, as will be introduced in the next subsection, this nature makes a big difference on a BYY system because it enables model selection.

In contrast, the matching nature by $\max_q H(p\|q)$, with p fixed, behaves similarly in both the direct case of Eq. (22.28) and its use on a BYY system. The details are discussed as follows:

– When $p(u)$ is given by its empirical density in the form of Eq. (22.1), considering a crude approximation $z_q = 1$ will make $H(p\|q)$ become the likelihood

$$L(\theta) = \sum_{t=1}^{N} \ln q(u_t). \qquad (22.29)$$

That is, it becomes equivalent to the maximum likelihood (ML) learning.
– Considering Eq. (22.28) with the normalization term z_q in Eq. (22.27), it follows from $p(u)$, given by Eq. (22.1), and $q(u)$, given by Eq. (22.27), with u in the place of x, that

$$H(p\|q) = L(\theta) - \ln z_q, \quad z_q = \sum_{t=1}^{N} q(u_t). \qquad (22.30)$$

By comparing the gradients

$$\nabla_\theta L(\theta) = Gd(\gamma_t)|_{\gamma_t = \frac{1}{N}}, \quad \nabla_\theta H(p\|q) = Gd(\gamma_t)|_{\gamma_t = \frac{1}{N} - \tilde{q}(u_t|\theta)},$$
$$Gd(\gamma_t) = \sum_t \gamma_t \nabla_\theta \ln q(u_t|\theta), \quad \tilde{q}(u_t|\theta) = q(u_t|\theta)/\sum_\tau q(u_\tau|\theta), \qquad (22.31)$$

we see that the log-normalization term $\ln z_q$ causes a *conscience* de-learning on the ML learning, which is, thus, referred as *normalization learning*, in a sense that the degree of de-learning on learning u_t is proportional to the likelihood that $q(u|\theta)$ fits u_t. That is, the better it is fitted, the more conscience it makes during learning, which actually provides a regularization that prevents $q(u|\theta)$ from overfitting a data set of a finite size.

– Considering $p(u)$ given by the Parzen window estimate of Eq. (22.3), with u in the place of x, we can also approximate z_q under a weak constraint $\sum_{t=1}^{N} p(u_t) \approx \sum_{t=1}^{N} q(u_t)$, which leads to a regularized learning as shown by Eqs. (28) and (33) in [22.57]. That is, Eq. (22.28) becomes

$$H(p\|q) = \tilde{L}_S(\theta_h) + 0.5d \ln(2\pi h^2) + \ln N - J_H(h,k),$$

$$J_H(h,k) = \ln[\sum_{\tau=1}^{N} \sum_{t=1}^{N} e^{-0.5 \frac{\|u_t - u_\tau\|^2}{h^2}}], \qquad (22.32)$$

$$\tilde{L}_S(\theta_h) = \int p_h(u) \ln q(u) du \approx L(\theta) + 0.5 h^2 \pi_q,$$

$$\pi_q = \frac{1}{N} \sum_t Tr[\frac{\partial^2 \ln q(u|\theta)}{\partial u \partial u^T}]_{u=u_t}.$$

$\tilde{L}_S(\theta_h)$ regularizes the ML learning by smoothing each likelihood $\ln q(u_t)$ in the near-neighborhood of u_t, something referred to as *data smoothing*. It can also be observed that the role of h^2 is equivalent to that of the hyper-parameter in Tikhonov-type regularization [22.5]. What is new here is that the other terms in $H(p\|q)$ balance $\tilde{L}_S(\theta_h)$ such that an appropriate h can be learned together with θ by Eq. (34) [22.54].

The fact that $\max_\theta \int p_0(u) \ln q(u|\theta) du$ leads to the ML learning of Eq. (22.29) is well known in the literature. Moreover, $\max_\theta \int p^*(u) \ln q(u|\theta) du$, with $p^*(u)$ being the true distribution of samples, has also been studied in developing the AIC criterion [22.1]. What is new here is that the term z_q introduces a regularization to the ML learning, which is shortly called the z-regularization. This regularization is implemented by two techniques, i.e., either a conscience de-learning type or a Tikhonov-type term, with the hyper-parameter h learned in an easily implemented way.

A further insight can be obtained by returning to the Kullback divergence Eq. (22.22). When both $p(u)$ and $q(u)$ are discrete densities in the form of Eq. (22.24), from Eq. (22.22) we can directly get

$$KL(p\|q) = \sum_{t=1}^{N} p_t \ln \frac{p_t}{q_t} = -E_p - H(p\|q), E_p = -\sum_{t=1}^{N} p_t \ln p_t. (22.33)$$

Helped by the form of Eq. (22.27) for both p and q, similar to Eq. (22.28) we can get

$$KL(p\|q) = \sum_{t=1}^{N} \frac{p(u_t)}{z_p} \ln \frac{p(u_t)/z_p}{q(u_t)/z_q} \approx \int p(u) \ln \frac{p(u)/z_p}{q(u)/z_q} du,$$

$$\text{or } KL(p\|q) = -H(p\|q) - E_p, E_p = -\int p(u) \ln p(u) du + \ln z_p. \quad (22.34)$$

Obviously, $KL(p\|q)$ becomes as in Eq. (22.22) when we have $z_q = z_p$, e.g., both $p(u)$ and $q(u)$ are discrete densities or approximately have $z_q = z_p$, e.g., when both $p(u), q(u)$ are continuous densities with $p(u)$ given by the Parzen window estimate of Eq. (22.3) and $q(u)$ given by a continuous parametric model. That is, Eq. (22.22) is directly applicable to the cases where both

$p(u)$ and $q(u)$ are discrete densities and that both $p(u), q(u)$ are continuous densities. However, Eq. (22.22) is not directly applicable when $p(u)$ is discrete and $q(u)$ is continuous, with z_p being infinite and z_q remaining bounded. In this case, $p(u)$ should be replaced by its normalized version $\hat{p}(u) = p(u)/z_p$.

Also, it follows from Eq. (22.33) and Eq. (22.34) that $\min KL(p\|q)$ is different from $\max H(p\|q)$ in that it also maximizes the entropy E_p of $p(u)$, which prevents $p(u)$ from approaching the form of Eq. (22.26). This explains why Eq. (22.33) does not have the least complexity nature. Particularly, when $E_p = c$ is a constant that does not relate to learning, $\min KL(p\|q)$ become equivalent $\max_{s.t.\ E_p=c} H(p\|q)$. Moreover, when $p(u)$ is given by a Parzen window, we get a regularized learning via data smoothing in a way similar to Eq. (22.32) but with the team $J_H(h, k)$ replaced by a different term $J_{KL}(h, k)$ that takes a similar role [22.54]. Furthermore, when $p(u) = p_0(u)$ given by Eq. (22.1), we have $E_p = 0$ and, thus, $\min KL(p\|q)$ and $\max H(p\|q)$ become equivalent completely. A detailed relation of the two types of learning is referred to the next chapter.

22.6.2 BYY Harmony Learning

By putting Eq. (22.19) into Eq. (22.28), we have

$$H(p\|q) = \int p(\mathbf{y}|\mathbf{x})p(\mathbf{x}) \ln [q(\mathbf{x}|\mathbf{y})q(\mathbf{y})]d\mathbf{x}d\mathbf{y} - \ln z_q \qquad (22.35)$$

for the harmony learning on a BYY system. Here, we get a salient feature that is not shared by Eq. (22.28). Now, only $p(\mathbf{x})$ is fixed at a non-parametric estimate, and $p(\mathbf{y}|\mathbf{x})$ is either free in a B-architecture or a parametric form in a BI-architecture, and will be pushed into its least complexity form due to the nature Eq. (22.26). For example, in a B-architecture $p(\mathbf{y}|\mathbf{x})$ will be determined by $\max_{p(\mathbf{y}|\mathbf{x})} H(p\|q)$, resulting in the following least complexity form:

$$p(\mathbf{y}|\mathbf{x}) = \delta(\mathbf{y} - \mathbf{y}(\mathbf{x})), \quad \mathbf{y}(\mathbf{x}) = arg\max_{\mathbf{y}}[q(\mathbf{x}|\mathbf{y})q(\mathbf{y})]. \qquad (22.36)$$

On the other hand, the matching nature of harmony learning will further push $q(\mathbf{x}|\mathbf{y})$ and $q(\mathbf{y})$ toward their corresponding least complexity forms. In other words, the least complexity nature and the matching nature collaborate to make model selection possible such that $\mathbf{k}$ is appropriately determined.

Mathematically, the harmony learning is implemented by

$$\max_{\theta,\mathbf{k}} H(\theta, \mathbf{k}), \ H(\theta, \mathbf{k}) = H(p\|q), \qquad (22.37)$$

which is a combined task of continuous optimization for parameter learning and discrete optimization for model selection, both under the same cost function $H(\theta, \mathbf{k})$. This feature makes it possible to simultaneously implement parameter learning and model selection together. Actually, the least complexity nature of Eq. (22.26) makes it possible for us to implement parameter learning with automatic model selection.

To get a further insight, it follows from Eq. (22.16) that $\alpha_\ell = 0$ for some ℓ implies that k is reduced by one. Also, we observe that a form

$$q(y|\ell) = \delta(y^{(j)} - \mu_0)q(y^-|\ell), \quad y = [y^-, y^{(j)}]^T, \quad \mu_0 \text{ is a constant}, \quad (22.38)$$

implies that the dimension of $y^{(j)}$ is not actually in action, and, thus, the dimension m_ℓ is effectively reduced by one. Therefore, a value of θ with these two types of settings is equivalent to forcing k and $\{m_\ell\}$ to be effectively reduced to appropriate scales.

With Eq. (22.36) put into Eq. (22.35), we can observe that $\max H(p\|q)$ is equivalent to maximizing $\ln q(x_t|y_t, \ell)$, $\ln q(y_t|\ell)$, and $\ln \alpha_\ell$ for each sample. Specifically, in maximizing $\ln \alpha_\ell$,

$$\text{each extra } \alpha_\ell \text{ is pushed toward zero}, \quad (22.39)$$

and in maximizing $\ln q(y|\ell)$,

$$q(y^{(j)}|\ell) \text{ on each extra dimension is pushed toward } \delta(y^{(j)} - \mu_0). (22.40)$$

Therefore, fixing the scales of $\mathbf{k}$ large enough, we can implement the harmony learning by

$$\max_\theta \ H(\theta), \quad H(\theta) = H(\theta, \mathbf{k}), \quad (22.41)$$

which will let θ take a specific value such that $\mathbf{k} = \{k, m\}$ is effectively reduced to an appropriate scale. In other words, the least complexity nature of Eq. (22.26) automatically implies model selection during learning. As demonstrated in Fig. 22.4(b), the learning by Eq. (22.41) will push a set of extra parameters $\theta_2 = \theta_2^*$ such that a large k becomes effectively k^*, that is, a smallest value of k at which the maximum of $H(\theta)$ is reached.

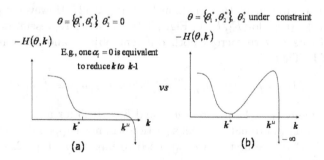

Fig. 22.4. (a) Model selection made after parameter learning on every k in a given interval $[k_d, \ k_u]$, (b) Automatic model selection with parameter learning on a value k of large enough.

The above feature is not shared by existing approaches in the literature. By the conventional approaches, parameter learning and model selection are made in a two-phase style. First, parameter learning is made usually under the maximum likelihood principle. Then, model selection is made by a different criterion, e.g., AIC, MDL, etc. These model selection criteria are usually not good for parameter learning, while the maximum likelihood criterion is not good for model selection, especially on a small size of training samples.

If one wants, the problem Eq. (22.37) can also be implemented in such a two-phase style. In the first phase, we enumerate $\mathbf{k} = \{k, \{m_\ell\}\}$ from small values incrementally. At each k and $\{m_\ell\}$, we perform parameter learning for seeking a best value of θ^*. In the second phase, we select a best $k^*, \{m_\ell^*\}$ by

$$\min_{1 \le k \le k^u, \ 1 \le m_\ell \le m_\ell^u, \ \forall \ell \in L} J(k, \{m_\ell\}), \quad J(k, \{m_\ell\}) = -H(\theta^*, \mathbf{k}), \quad (22.42)$$

where $k^u, \{m_\ell^u\}$ is a set of upper bounds. To get an insight, we consider a Gaussian mixture by Eq. (22.14) with $q(x|\ell) = G(x|\mu_\ell, \Sigma_\ell)$. In this case, we have [22.77]:

$$J(k) = \frac{0.5}{k} \sum_{\ell=1}^k \ln |\Sigma_\ell| + \ln k. \qquad (22.43)$$

As k increases the number of samples that are allocated to each Gaussian decreases. Thus, each $|\Sigma_\ell|$ decreases with k and the first term of $J(k)$ decreases with k. However, $\ln k$ increases with k. The two terms trade off such that $J(k)$ first decreases with k and reaches a minimum, and then increases due to $\ln k$. As k further increases, there will be few samples available to be allocated to a Gaussian such that Σ_ℓ becomes singular which brings $J(k)$ drops rapidly toward $-\infty$. That is, $J(k)$ generally has an inverse N shape as shown in Fig. 22.4(a). We use k^u to denote the smallest value of k that makes $J(k)$ rapidly tend to ∞. The best value k^* is selected by Eq. (22.42) within $1 \le k \le k^u$.

As above discussed, parameter learning by Eq. (22.41) usually (e.g., for F_1, F_2, F_3 in Fig. 22.2) leads to an automatic model selection and, thus, there is no need to implement the selection by Eq. (22.42). However, for certain learning tasks (e.g., for F_4, F_5 in Fig. 22.2), the inner representation is pre-specified to be uniform across both different objects and different dimensions [22.54]. That is,

$$\alpha_\ell = \frac{1}{k}, \int (y - \mu_{y,\ell})(y - \mu_{y,\ell})^T q(y|\ell) dy = b_0 I, \mu_{y,\ell} = \int y q(y|\ell) dy \ (22.44)$$

For example, $b_0 = 1$ for a real $y^{(j)}$ and $b_0 = 0.25$ for a binary $y^{(j)}$. Due to the constraint, automatic model selection will not happen during learning by Eq. (22.41) in the first phase, we need to implement Eq. (22.42) in the second phase.

Alternatively, we can also replace Eq. (22.41) by minimizing the Kullback divergence Eq. (22.22), i.e.,

$$\min_\theta KL(\theta) = \int p(\mathbf{y}|\mathbf{x})p(\mathbf{x}) \ln \frac{p(\mathbf{y}|\mathbf{x})p(\mathbf{x})}{q(\mathbf{x}|\mathbf{y})q(\mathbf{y})} d\mathbf{x}d\mathbf{y}, \qquad (22.45)$$

without the constraint by Eq. (22.44). Particularly, on a B-architecture, the minimization of the above $KL(\theta)$ with respect to a free $p(\mathbf{y}|\mathbf{x})$ will result in

$$p(\mathbf{y}|\mathbf{x}) = \frac{q(\mathbf{x}|\mathbf{y})q(\mathbf{y})}{q(\mathbf{x})}, \quad q(\mathbf{x}) = \int q(\mathbf{x}|\mathbf{y})q(\mathbf{y})d\mathbf{y},$$

$$KL(\theta) = \int p(\mathbf{x})\ln\frac{p(\mathbf{x})}{q(\mathbf{x})}d\mathbf{x}, \qquad (22.46)$$

which becomes equivalent to ML learning on $q(\mathbf{x})$ when $p(\mathbf{x}) = p_0(x)$ is given by Eq. (22.2) [22.88]. In this case, we actually implement the ML learning in the first phase and then model selection by Eq. (22.42) in the second phase.

No longer holding the least complexity nature by Eq. (22.26), the implementation of Eq. (22.45) will not lead to a case of Fig. 22.4(b), and, thus, there is no need to impose the assumption that $q(y)$ comes from a family with equal variances among components.

22.6.3 A Further Extension: From $\ln(r)$ to Convex Function

Extensions of the BYY harmony learning have been made either into temporal BYY system by taking temporal relation in consideration or into structural BYY system by considering the representation space $\mathbf{y}$ with certain structure [22.90, 22.74, 22.65, 22.61, 22.64, 22.59, 22.57, 22.53, 22.50]. Extension has also been made with the log function $\ln(r)$ replaced by the following convex function family [22.75, 22.77, 22.80, 22.81]

$$F_c = \{f(r): \frac{d^2 f(r)}{d^2} < 0, \ f(r) \text{ monotonically increases with } r\} \quad (22.47)$$

It is interesting to investigate what will happen when $\ln(r)$ is replaced with any $f(r) \in F_c$.

Let us return back to consider such an extension from Eq. (22.23). That is,

$$H_f(p\|q) = \sum_{t=1}^{N} p_t f(q_t), \qquad (22.48)$$

It can be observed that the maximization of this $H_f(p\|q)$ with respect to p_t still have the same least complexity nature as Eq. (22.26) while the match nature by Eq. (22.25) is modified into

$$q_t = f'(\frac{1}{p_t})/\sum_{t=1}^{N} f'(\frac{1}{p_t}), \ f'(r) = df(r)/dr, \qquad (22.49)$$

which returns back to Eq. (22.25) when $f'(r) = 1/r$. We can further classify other $f(r) \in F_c$ according to whether its $f'(r)$ decreases with r in a rate slower than $1/r$. If yes, it is said to be super-ln; otherwise it is said to be sub-ln.

A typical family of super-ln is the so-called α-function [22.69, 22.77, 22.80, 22.81] as follows:

$$f(r) = r^\alpha, 0 \le \alpha \le 1, \ and, \ thus, \ f'(r) = \alpha/r^{1-\alpha}. \qquad (22.50)$$

In this case, Eq. (22.49) can be rewritten as

$$q_t = p_t^{1-\alpha}/\sum_{t=1}^{N}p_t^{1-\alpha}, \tag{22.51}$$

which returns to Eq. (22.25) with $q_t = p_t$ when $\alpha = 0$ and becomes $q_t = 1/N$ when $\alpha = 1$. In other cases, p_t attempts to take q_t with lower probabilities increased but higher probabilities decreased in a certain degree that is controlled by α. That is, p_t becomes more spreading than q_t as α increases from 0 to 1. It becomes uniform when $\alpha = 1$. In other words, a super-ln function leads to a conscience de-learning regularization.

An example family of sub-ln is the following inverted and negated α-function as follows:

$$f(r) = -r^{\alpha}, \alpha < 0, \ and, \ thus, \ f'(r) = |\alpha|/r^{1+|\alpha|}. \tag{22.52}$$

which leads to

$$q_t = p_t^{1+|\alpha|}/\sum_{t=1}^{N}p_t^{1+|\alpha|}. \tag{22.53}$$

Now, p_t adopts q_t with higher probabilities increased but lower probabilities further decreased in a certain degree. That is, p_t becomes more concentrated than q_t as $|\alpha|$ increases. In other words, a sub-ln function leads to a competition effectively similar to the least complexity nature.

The above discussions also apply to the continuous case by Eq. (22.28), which becomes:

$$H(p\|q) = \int p(u)f(q(u)/z_q)du. \tag{22.54}$$

Now, not only the term z_q provides a regularization, which is referred to the next chapter in this book for a detailed discussion, but also z_q and a super-ln $f(r)$ jointly introduce the above discussed regularization.

Even without z_q (e.g., let $z_q = 1$), a super-ln $f(r)$ will also perform a regularization role. With $p(u)$ given by Eq. (22.1), Eq. (22.54) becomes a f-likelihood function $L_f = \frac{1}{N}\sum_{t=1}^{N}f(q(u_t))$ which differs from the log-likelihood $L = \frac{1}{N}\sum_{t=1}^{N}\ln q(u_t)$ in that $f(q(u_t))$ takes the position of $\ln q(u_t)$. Those unlikely samples with small $q(u_t)$ (e.g., outliers) locate within a drastic varying range of $\ln q(u_t)$ and, thus, contribute a big portion to affect L. That is, the ML learning is vulnerable to the disturbance by outliers. In contrast, for a super-ln $f(r)$ such as given by Eq. (22.50), the varying range of $f(q(u_t))$ with a small $q(u_t)$ is much smaller than that of $\ln q(u_t)$. Thus, L_f will be much less affected by those outliers with a small $q(u_t)$. In other words, we get a type of robust ML learning [22.77, 22.80].

The Kullback divergence can also be extended. There are two ways. One is simply replace $\ln r$ with $f(r)$, resulting in

$$KL_f(p\|q) = \int p(u)f(\frac{p(u)/z_p}{q(u)/z_q})du. \tag{22.55}$$

The other is considering its equivalent form of maximizing $H_{KL}(p\|q) = \int p(u)\ln\frac{q(u)/z_q}{p(u)/z_p}du$ and then replace $\ln r$ with $f(r)$, which results in the maximization of

$$H_{KL_f}(p\|q) = \int p(u)f(\frac{q(u)/z_q}{p(u)/z_p})du. \tag{22.56}$$

Being different from Eq. (22.55), it will return to exactly Eq. (22.54) when $p(u)$ is given by empirical density Eq. (22.1).

In the special case that $z_p = z_q$, we have

$$KL_f(p\|q) = \int p(u)f(\frac{p(u)}{q(u)})du, \quad H_{KL_f}(p\|q) = \int p(u)f(\frac{q(u)}{p(u)})du. \tag{22.57}$$

Detailed studies can be further made by considering certain specific features of $f(r)$. For example, for those $f(r) \in F_c$ that satisfy

$$(a) \ f(ab) = f(a)f(b), \quad (b) \ f(a^{-1}) = f^{-1}(a), \tag{22.58}$$

e.g., for the case of Eq. (22.50) we can rewrite Eq. (22.54) into

$$H(p\|q) = f^{-1}(z_q) \int p(u)f(q(u))du, \tag{22.59}$$

and Eq. (22.56) into

$$H_{KL_f}(p\|q) = \frac{f(z_p)}{f(z_q)} \int p(u)\frac{f(q(u))}{f(p(u))}du. \tag{22.60}$$

All the above results can be applied to a BYY system, what needs to do is simply put Eq. (22.19) into Eq. (22.54), Eq. (22.55), Eq. (22.56), Eq. (22.59), and Eq. (22.60). First, maximizing $H(p\|q)$ by Eq. (22.54) with respect to a free $p(\mathbf{y}|\mathbf{x})$ will still lead to Eq. (22.36). Also, both minimizing $KL_f(p\|q)$ by Eq. (22.55) and maximizing $H_{KL_f}(p\|q)$ by Eq. (22.56) with respect to a free $p(\mathbf{y}|\mathbf{x})$ will lead to Eq. (22.46) too. Second, we are also lead to what will be discussed in the next chapter on trading off the strength of regularization and the ability of model selection in help of designing a parametric model for $p(\mathbf{y}|\mathbf{x})$. Third, a new perspective we get here is that such a trading off may also be achieved via choosing $f(r)$ to be super-ln or sub-ln, in the matching process of a Ying machine $q(\mathbf{x}|\mathbf{y})q(\mathbf{y})$ to a Yang machine $p(\mathbf{x}|\mathbf{y})p(\mathbf{y})$ with $p(\mathbf{y}) = \int p(\mathbf{y}|\mathbf{x})p(\mathbf{x})d\mathbf{x}$ and $p(\mathbf{x}|\mathbf{y}) = p(\mathbf{y}|\mathbf{x})p(\mathbf{x})/p(\mathbf{y})$.

Learning can still be implemented by the Ying-Yang alternative procedure in Sect. 22.7. For the Ying step, we no longer have the separated integral format as in Eq. (22.63). However, we can still get the gradient format similar to those in Eq. (22.65). The only difference is that $\eta_t(y)$ is replaced by $\eta_t^f(y)\eta_t(y)$ with

$$\eta_t^f(y) = \begin{cases} f'(\frac{q(x_t|y)q(y)}{z_q})\frac{q(x_t|y)q(y)}{z_q}, & \text{Harmony learning by Eq. (22.54),} \\ f'(\frac{p(u)/z_p}{q(u)/z_q})\frac{p(u)/z_p}{q(u)/z_q}, & KL_f \text{ learning by Eq. (22.55),(22.61)} \\ f'(\frac{q(u)/z_q}{p(u)/z_p})\frac{q(u)/z_q}{p(u)/z_p}, & KL_f \text{ learning by Eq. (22.56),} \end{cases}$$

where $f'(r) = df(r)/dr$.

For the Yang step, the discussions on a B-architecture in Sect. 22.7 remains the same. While for a BI-architecture, $\eta_t(y)$ in Eq. (22.71) is also replaced by $\eta_t^f(y)\eta_t(y)$ with $\eta_t^f(y)$ given by Eq. (22.61).

22.7 Ying-Yang Alternative Procedure for Parameter Learning

As discussed in [22.88], learning on the Yang machine $p(\mathbf{x}, \mathbf{y})$ and the Ying machine $q(\mathbf{x}, \mathbf{y})$ can be implemented alternatively by

> Ying step: fixing $p(\mathbf{x}, \mathbf{y})$, update unknowns in $q(\mathbf{x}, \mathbf{y})$,
> Yang step: fixing $q(\mathbf{x}, \mathbf{y})$, update unknowns in $p(\mathbf{x}, \mathbf{y})$,   (22.62)

such that $-H(\theta)$ from Eq. (22.41), or $KL(\theta)$ from Eq. (22.45), gradually decreases until it converges.

This Ying-Yang alternative procedure also applies to the extension in the previous section, i.e., every appearance of $\ln r$ is replaced by a convex function $f(r) \in F_c$ by Eq. (22.47). For clarity, we focus on the function $\ln r$ in this section. However, all the discussions here directly applying to the cases with $\ln r$ replaced by a convex function $f(r)$.

(1) Ying step   With $p(\mathbf{y}|\mathbf{x})$ fixed and the regularization term z_q ignored, both Eq. (22.41) and Eq. (22.45) share the same updating format, as follows:

$$\max_{\theta_{x|y}} \ L(\theta_{x|y}), \ \ L(\theta_{x|y}) = \int p(\mathbf{y}|\mathbf{x})p(\mathbf{x}) \ln q(\mathbf{x}|\mathbf{y}) d\mathbf{x} d\mathbf{y},$$
$$\max_{\theta_y} \ L(\theta_y), \ \ L(\theta_y) = \int p(\mathbf{y}|\mathbf{x})p(\mathbf{x}) \ln q(\mathbf{y}) d\mathbf{x} d\mathbf{y}, \qquad (22.63)$$

where $\theta_{x|y}$ and θ_y consist of unknown parameters in $q(\mathbf{x}|\mathbf{y})$ and $q(\mathbf{y})$, respectively. We make the updates $\delta\theta_{x|y}$ and $\delta\theta_y$ such that $L(\theta_y + \delta\theta_y)$ and $L(\theta_{x|y} + \delta\theta_{x|y})$ either reach a local maximum or increase to certain extent. Typically, an updating $\delta\theta$ that increases an index $J(\theta)$ is a step size along a gradient-based direction $g_\theta = \nabla_\theta J(\theta)$, i.e.,

$$\delta\theta = \eta T g_\theta, \qquad (22.64)$$

where $\eta > 0$ is a small positive number that defines a step size. This $\delta\theta$ is along either the gradient direction when $T = I$ is a unit matrix or a direction that has a positive projection on the gradient direction when T is a positive definite matrix. Particularly, $\delta\theta$ is the well known natural gradient direction when T is the inverse of the metric tensor of $J(\theta)$.

When $p(\mathbf{x})$ is given by Eq. (22.1), the integral over $\mathbf{x}$ in both Eq. (22.35) and Eq. (22.45) will disappear. When $p(\mathbf{x})$ is given by Eq. (22.3), this integral over $\mathbf{x}$ can also be removed with the help of certain approximations [22.58, 22.54, 22.51, 22.52].

The integrals over y is either analytically solved when $q(x|y)$ and $q(y)$ are both Gaussian or becomes a computable summation when y takes one of discrete values $1, \cdots, k$ or is a binary vector $y = [y^{(1)}, \cdots, y^{(m)}]$. Otherwise, these integrals are difficult to compute. Still, even when it becomes a computable summation for a binary vector $y = [y^{(1)}, \cdots, y^{(m)}]$, the computing cost will increase exponentially with m.

The problem is tackled via letting the integral over the entire domain of y to be approximated by a summation on a set Y_t of a finite number of samples of y, which are obtained according to $p(y|x)$ in the Yang step. One typical case is that Y_t consists of only one sample $y_t = y(x_t)$. In this case, we can further make updating $\delta\theta_{x|y}, \delta\theta_y$ in the form of Eq. (22.64) with $\ln q(\mathbf{x}_t|\mathbf{y}_t)$ and $\ln q(\mathbf{y}_t)$ in the place of $J(\theta)$, respectively.

Without considering regularization, we have that both the harmony learning by Eq. (22.37) and the KL learning by Eq. (22.45) share a same format in the following gradients:

$$g_{\theta_{x|y}} = \frac{1}{N}\sum_{t=1}^{N}\int \eta_t(y)\nabla_{\theta_{x|y}}\ln q(x_t|y)dy, \tag{22.65}$$

$$g_{\theta_y} = \frac{1}{N}\sum_{t=1}^{N}\int \eta_t(y)\nabla_{\theta_y}\ln q(y)dy,$$

$$\eta_t(y) = \begin{cases} p(y|x_t) \text{ by Eq. (22.46),} & \text{the KL learning by Eq. (22.45),} \\ \delta(y - y(x_t)), & \text{the harmony learning by} \\ & \text{Eq. (22.37),} \end{cases}$$

which can be further put in Eq. (22.64) for updating.

(2) Yang step   This could be implemented in one of four typical ways, according not only to whether Eq. (22.41) or Eq. (22.45) is used for learning, but also whether a B-architecture or a BI-architecture is in consideration. The details are given as follows:

– In implementing the KL learning on a B-architecture with Eq. (22.45) for parameter learning or, equivalently, the ML learning on $q(\mathbf{x})$, the Yang step is to get Eq. (22.46). This is exactly the E-step by the well known EM algorithm [22.9, 22.38], with Eq. (22.63) being exactly the M-step. In other words, the EM algorithm is a specific case of the Ying-Yang alternative procedure given by Eq. (22.62). An integral over y has to be encountered for getting $q(x)$, which is either analytically solvable when $p(y|x)$, $p(x|y)$, and $p(y)$ are all Gaussian densities or becomes a computable summation when y takes one of discrete values $1, \cdots, k$ or is a binary vector $y = [y^{(1)}, \cdots, y^{(m)}]$. Moreover, even when it is computable for a binary vector $y = [y^{(1)}, \cdots, y^{(m)}]$, the computing cost will increase exponentially with m. In other cases, the integral is difficult to compute. It was as previously suggested in 1997 to be implemented approximately via a computationally expensive Monte Carlo simulation. That is, Y_t is obtained via randomly picking a set of samples of y according to $p(y|x_t)$, (see the choice (1) of Table 2(C) in [22.76] and the choice (a) in Eq. (24) and Sect. 3.1 in [22.66]) or even in a rough approximation according to $q(y)$ (see Step 1 in TableII(B) in [22.76]).
– In implementing the KL learning on a BI-architecture with Eq. (22.45) for parameter learning, the Yang step is to update $\theta_{y|x}$, which consists of all the parameters in $p(\mathbf{y}|\mathbf{x})$, by either Eq. (22.21) or Eq. (22.20). The update $\delta\theta_{y|x}$ is made such that $KL(\theta_{y|x} + \delta\theta_{y|x})$ either reaches a local minimum or reduces to a certain extent. Here, the integral over $\mathbf{y}$ for getting $q(\mathbf{x})$

by Eq. (22.46) is avoided. However, we encounter not only the integrals in Eq. (22.63) but also the following one

$$H(\theta_{y|x}) = \int p(\mathbf{y}|\mathbf{x})p(\mathbf{x}) \ln p(\mathbf{y}|\mathbf{x}) d\mathbf{x} d\mathbf{y}. \tag{22.66}$$

It was also firstly suggested in 1997 under the name of the mean field approximation (see the choice (3) of Table 2(C) in [22.76] and the choice (c) in Eq. (24) of [22.66]) that we get Y_t consisting of only one mean point

$$y_t = \int y p(y|x_t) dy \tag{22.67}$$

which is computable for a $p(y|x)$ given by either of Eq. (22.20) and Eq. (22.21), but not applicable to a B-architecture with $p(y|x)$ given by Eq. (22.46) since another integral has to be encountered to get $q(x)$.

- In implementing the harmony learning with Eq. (22.41) for parameter learning, the Yang step is simply getting Y_t that consists of only one peak point

$$y_t = arm \max_y p(y|x_t). \tag{22.68}$$

which was firstly suggested again in 1997 (see the choice (2) of Table 2(C) in [22.76] and the choice (b) in Eq. (24) of [22.66]) and then further encountered in Eq. (22.36) on a B-architecture. This nonlinear optimization is implemented in help of an iterative procedure:

$$\mathbf{y}^{new}(\mathbf{x}_t) = ITER(\mathbf{y}^{old}(\mathbf{x}_t)). \tag{22.69}$$

Specific algorithms of this type are proposed in [22.57, 22.54] to suit specific structures of $q(\mathbf{x}|\mathbf{y})$ and $q(\mathbf{y})$. The complexity of making a nonlinear optimization is considerably less than that of making the integrals over $\mathbf{y}$. Moreover, we usually need only a few iterations by Eq. (22.69) instead of waiting it to converge. This is another salient advantage that the least complexity nature of Eq. (22.36) provides us, in addition to making model selection possible.

- In implementing the harmony learning on a BI-architecture, the Yang step consists of two parts. One is simply implementing Eq. (22.68). For a $p(y|x)$ given by Eq. (22.21), we can get $y_t = y(x_t)$ via a simple comparison that reduces significantly the computational complexity for making a nonlinear optimization by Eq. (22.36). For a $p(y|x)$ given by Eq. (22.20), we can get either

$$y^{(j)} = \pi_j(x_t), \ or \ y^{(j)} = \begin{cases} 1, & \pi_j(x_t) \geq 0.5, \\ 0, & \text{otherwise.} \end{cases} \tag{22.70}$$

The second part attempts to increase $\ln [q(\mathbf{x}|\mathbf{y})q(\mathbf{y})]_{\mathbf{y}=f_{j^*(x)}(\mathbf{x}|\theta_{y|x,j^*(x)})}$ or $\ln [q(\mathbf{x}|\mathbf{y})q(\mathbf{y})]_{\mathbf{y}=\pi(x_t,\theta)}$. In help of the chain rule for gradient, it can be implemented via an update $\delta\theta_{y|x,j^*(x)}$ in the form of Eq. (22.64) as follows

$$\delta\theta_{y|x,j^*(x)} = \eta_t(y)g_{\theta_{y|x,j^*(x)}}^T[\psi(y) + \phi(y)],$$

$$\psi(y) = \frac{\partial \ln q(x|y)}{\partial y}, \ \ \phi(y) = \frac{\partial \ln p(y)}{\partial y}, \tag{22.71}$$

$$g_{\theta_{y|x,j^*(x)}} = \nabla_{\theta_{y|x,j^*(x)}} f_{j^*(x)}(x|\theta_{y|x,j^*(x)}), \ \ or \ \ g_\theta = \nabla_\theta \pi(x|\theta).$$

22.8 Learning Implementation: From Optimization Search to Accumulation Consensus

By Eq. (22.62), the process of implementing parameter learning can be regarded as an optimization process. Actually, many learning tasks are implemented in a two-stage way with its first stage to form an objective function $J(\theta)$ and with its second stage to maximize $J(\theta)$. At the second stage, any optimization techniques developed in the nonlinear programming literature are possible to be adopted for solving the problems. One typical class of techniques is the gradient-based search, that is,

$$\theta^{new} = \theta^{old} + \eta \nabla_\theta J(\theta), \tag{22.72}$$

where $\eta > 0$ is a small number as a pre-specified step size. The updating is iterated until it is converged. Such a process is featured by one candidate solution θ^{old} that is updated locally and iteratively, and also terminated according to certain local property (i.e., it is a local maximum). However, the local feature makes the process be easy to be trapped at a local maximum.

This problem may be remedied by implementing multiple search processes of the type by Eq. (22.72) either in parallel or subsequently. In parallel, it means that a set Θ_c of candidate solutions are considered with each being iterated via the same way as in Eq. (22.72). Subsequently, it means that once a convergence is reached, certain perturbation is made on the converged solution to get a new initial candidate of the next iterative process, and selecting the best among all the converged local maxima as the final solution. The chance of finding the global solution increases as the number of such search processes increases. However, the computational costs also increase significantly. In the worst case, an exhaustive search of the entire space of θ will guaranteed to find the global solution.

However, optimization should not be understood being same as learning. The key challenge of learning is to find a principle for designing an objective function such that the learning model can both fit a given set of training samples and generalize as good as possible on new samples that are different from the given training sample set but come from the same underlying distribution. Even assuming such an objective function already available, learning is also not simply an optimization.

Being different from an objective function $J(\theta)$ for optimization, an objective function $J(\theta|\mathbf{X})$ is not only a function of θ but also bases on a set $\mathbf{X}$ of samples. Of course, we can treat $J(\theta|\mathbf{X})$ as $J(\theta)$ with learning made as an optimization problem. However, we are not bounded to only this way. We can explore the specific structure of $J(\theta|\mathbf{X})$ for a different implementation. It can be observed, from not only Eq. (22.29) or Eq. (22.45) but also Eq. (22.30) and Eq. (22.35) with $z_q = 1$ as well as from many existing learning approaches, that an objective function formed under a learning principle usually has an additive decomposition form

$$J(\theta|\mathbf{X}) = \int p(\mathbf{x}|\mathbf{X})J(\theta|x)dx = \frac{1}{N}\sum_{t=1}^{N}J(\theta|x_t), \tag{22.73}$$

with $p(\mathbf{x})$ estimated by the empirical density Eq. (22.1).

First, this additive decomposition makes learning able to be implemented adaptively. Instead of forming the function $J(\theta|\mathbf{X})$ based on the entire set $\mathbf{X}$ and then making an optimization, a candidate solution is updated to adapt each current coming sample x_t such that the objective function $J(\theta|x_t)$ per each sample x_t increases by certain extent, e.g., via Eq. (22.72) but with $J(\theta)$ replaced by $J(\theta|x_t)$. This is usually called adaptive learning that is different from directly optimizing $J(\theta)$ in that the objective function $J(\theta|x_t)$ per each sample x_t actually varies. Since each sample x_t comes via random sampling, this learning process is also called stochastic approximation in the related literature [22.37, 22.24]. Extension can also be made by each time considering a set X_t that consists of several samples. Then, an adaptation is made by Eq. (22.72) but with $J(\theta)$ replaced by $\sum_{x \in X_t} J(\theta|x)$. With the learning step size η reduces as learning proceeds, such an iteration will be terminated. However, whether the resulted solution is same as a local or global maximum solution of $J(\theta|\mathbf{X})$ usually turned out a difficult problem, except some simple cases.

Second, the additive decomposition in Eq. (22.73) also provides a possibility to develop a completely different solving strategy called *accumulation consensus* that consists of the following ingredients:

(a) *Quantization*   Considering a bounded space of θ in which the solution is located, we quantize this bounded space into a set of discrete points with a pre-specified resolution $\delta > 0$. Each of these discrete points is considered as a candidate solution with an accumulator attached. All the accumulators forms an accumulation array.

(b) *Sampling*   Getting a sample x_t from the sample set $\mathbf{X}$.

(c) *Voting and accumulation*   At each time t, we let every accumulator in the accumulation array to be increased by a score of $J(\theta|x_t)$ with θ being the discrete point that the accumulator locates at.

(d) *Selection and Testing*   After a certain period, the accumulator with the largest score is selected among the accumulation array and the corresponding discrete point is tested by certain local statistical properties. If passed, it is taken as the solution.

The main problem of this accumulation procedure is that the size of the accumulation array increases rapidly with a fine resolution $\delta > 0$ and exponentially with the number of unknown parameters. Also, voting and accumulating should be made on all the accumulators.

However, to implement the BYY harmony learning by Eq. (22.35) on those finite mixture related learning tasks in Sect. 22.3.2, we are lead to a very different situation. With $q(\mathbf{y}) = q(y = \ell) = \alpha_\ell$ and $q(\mathbf{x}|\mathbf{y}) = q(x|\phi_\ell)$, the marginal density by Eq. (22.17) becomes equivalent to the finite mixture model by Eq. (22.14), which includes not only various Gaussian mixture based tasks for density estimation and clustering analysis [22.54, 22.58, 22.77] but also various multi-sets mixture based tasks for curve and object detection [22.52]. Moreover, we have that Eq. (22.36) becomes

$$p(\mathbf{y}|\mathbf{x} = x_t) = \delta(y - \ell_t), \quad \ell_t = arg\max_\ell[q(x_t|\phi_\ell)\alpha_\ell], \tag{22.74}$$

and we further have that Eq. (22.73) becomes

$$J(\theta|\mathbf{X}) = \tfrac{1}{N} \sum_{t=1}^{N} \ln\left[q(x_t|\phi_{\ell_t})\alpha_{\ell_t}\right], \quad \theta = \{\phi_\ell, \alpha_\ell\}_{\ell=1}^{k}. \tag{22.75}$$

As a result, the above voting and accumulating mechanism can be considerably simplified. Only one of k parameter sets $\{\phi_\ell, \alpha_\ell\}_{\ell=1}^{k}$ will be voted at each time t. Thus, it only needs to consider accumulating the votes on $\phi_{\ell_t}, \alpha_{\ell_t}$. Moreover, considering that the k sets of parameters share a same format ϕ, α and thus the dimension of accumulation array can be reduced by k times, we can simplify the above accumulation procedure as follows:

(a) *Quantization*   Considering a bounded space of ϕ, α that is quantized.

(b) *Sampling*   the same as above.

(c) *Voting and accumulation*   At each time t, we let every accumulator, that locates at $\theta = [\phi, \alpha]$, to be increased by a score of $\ln[q(x_t|\phi)\alpha]$.

(d) *Selection and Testing*   the same as above.

Specifically, the test can base on a known statistical property of $q(x_t|\phi)$. For simplicity, the test can also be ignored.

In the particular case that $\alpha_\ell = 1/k$ and each $q(x_t|\phi_\ell) = \delta(g(x_t, \phi_\ell))$ (i.e., $g(x_t, \phi_\ell) = 0$ is a deterministic equation), only the accumulators locate at those values of ϕ_ℓ that satisfing $g(x_t, \phi_\ell) = 0$ are voted and accumulated. This case leads us to the well known existing techniques developed in the literature of pattern recognition. Specifically, at each time t if every sample in the sample set $\mathbf{X}$ is enumerated, the above accumulation procedure will become equivalent to the well known Hough Transform (HT) for curve detection [22.20, 22.21]. For example, when x_t is of a 3-dimension and $g(x_t, \phi_\ell) = 0$ is a line, we are lead to a HT for detection lines on an image. Furthermore, if each x_t is sampled randomly and the period in (d) covers an enough number of samples, we are lead to a probabilistic variant of the HT [22.16].

The above accumulation procedure extends the HT and probabilistic HT. Not only α_ℓ is no longer constrained to be equally $1/k$ such that the number of pixels or samples on an object is considered together with their fitting errors. But also with a probabilistic $q(x_t|\phi_\ell)$ in place of the deterministic model $g(x_t, \phi_\ell) = 0$ such that random noises and variations are taken in consideration. For example, objects such lines, circles, ellipses, and even complicated shapes can be detected under strong noises. Moreover, we can further reduce the complexity for voting and accumulating because only those accumulators that locate on a smaller subset θ_t are increased by a score of $J(\theta|x_t)$, where θ_t is given as follows:

$$\theta_t = \{\phi, \alpha : q(x_t|\phi)\alpha \geq b, \quad b > 0 \text{ is a pre-specified threshold,}\} \quad (22.76)$$

which degenerates back to the entire space when $b = 0$.

One variant of the above accumulation procedure is obtained by modifying its sampling mechanism. At each t, a set X_t of several samples are sampled randomly from $\mathbf{X}$ instead of only one sample x_t is picked. In voting and accumulating, if the above hard-cutting is not considered, we simply let every accumulator in the accumulation array to be increased by a score $\sum_{x \in X_t} J(\theta|x)$ in place of $J(\theta|x_t)$. However, situation will become quite different by considering accumulators only on

$$\theta_t = \{\phi, \alpha : q(x|\phi)\alpha \geq b, \ \forall x \in X_t\}. \quad (22.77)$$

As $b > 0$ increases and the number of samples in X_t increase, the cardinality of θ_t will decrease, even possibly toward 1 or 0. Consequently, the voting mechanism changes from one-to-many diverging mapping to many-to-one or many-to-few converging mapping. As a result, the high space complexity of the above quantization based accumulation array can be replaced by a dynamic accumulator structure A. At each time t, we increase the accumulator at θ by a score $\sum_{x \in X_t} J(\theta|x)$. If such an accumulator has not been included in A yet, we add it in. Thus, the number of accumulators in A grows dynamically as the voting proceeds. Finally, solutions are selected among A either via those accumulators with largest votes or via the cluster centers of accumulators.

This modified accumulation procedure becomes equivalent to the randomized Hough Transform (RHT) [22.96, 22.95] in the particular case that each $q(x_t|\phi_\ell) = \delta(g(x_t, \phi_\ell))$ and $\alpha_\ell = 1/k$. The RHT was developed as an important advance along the research direction of the conventional HT. Moreover, the accumulation procedure here extends the RHT approach such that not only the number of pixels or samples on an object is considered together with their fitting errors but also random noises and variations are taken in consideration.

Further improvements are possible along the following directions:

(1) Instead of deciding solution only based on those accumulators with largest votes or the cluster centers of accumulators, a statistical test can be imposed by jointly considering the local fitting property via $q(x_t|\phi_\ell)$ and the global property via α_ℓ.

(2) Instead of implementing learning only based on either optimization search or accumulation consensus, possible combinations of two can be investigated. For example, the final results of the above accumulation procedure can also be used as initial points of a local adaptation based optimization search to further refine these solutions. Also, the accumulation consensus may

be implemented together with a parallel local adaptation based optimization search.

(3) Instead of pre-specifying a threshold $b > 0$, b may also be dynamically controlled during the above accumulation procedure.

22.9 Main Results and Bibliographic Remarks

22.9.1 Main Results on Typical Learning Problems

Bayesian Ying Yang (BYY) harmony learning was firstly proposed in 1995 [22.88] and then developed in past several years. New results can be summarized on two aspects. One aspect is on those advances of the BYY harmony learning as a general statistical learning framework with a new mechanism for model selection and regularization, the details will be introduced in next chapter of this book. The other aspect is on various specific cases of the BYY harmony learning, which lead to various specific learning algorithms as well as the detailed forms for implementing model selection and regularization, which covers three main statistical learning paradigms, namely, unsupervised learning, supervised learning, and temporal modeling. The major results are summarized into the following list.

(a) Gaussian mixture and multi-sets-mixture based structures

- Smoothed EM (see Eq. (18) in [22.75]), Robust EM (see Eq. (33) in [22.77]), Hard-cut EM (see [22.88, 22.82]), RPCL learning, Normalization RPCL-type EM [22.58, 22.54, 22.51].
- Elliptic clustering [22.58, 22.54] and multi-sets mixture based multiple object detection (see Sect. 3 in [22.82] and also see Sect. 3.3 in [22.52]).
- Criteria for the number k of clusters and of Gaussians [22.88, 22.82, 22.85, 22.77].
- Adaptive learning with automatic selection on k [22.88, 22.82, 22.77, 22.58, 22.54].
- Support vector based Parzen window estimate [22.58, 22.54].

(b) Independent structures

- Adaptive EM-like algorithms for independent binary or Bernoulli FA (BFA) with hidden factors selected in either of two ways, that is, selected either automatically during learning or alternatively via a criterion obtained from Eq. (22.42)(see Sect. 3.2(B) in [22.71], Sect. 4.2.2 & 4.2.4 and Fig. 2 in [22.66]).
- New insight on LMSER learning [22.94] for ICA and new adaptive algorithm with hidden factors selected in either of two ways (see Sect. 8 in [22.74], Sect. 4.3.4 in [22.66], and also the recent [22.54, 22.51]).
- Extensions of the LMSER learning (see Sect. 3 in [22.66] and Sect. 3 in [22.68]).

- Factor analysis and PCA with subspace dimension determined in either of two ways (see Sect. 3.2(B) in [22.71], also see Sect. 4.2.4 in [22.66]).
- Adaptive algorithms that are able to effectively implement the non-Gaussian factor analysis (NFA), with hidden factors selected in either of two ways [22.57, 22.54, 22.51, 22.53].
- A learned parametric model (LPM) ICA algorithm that it works on any combinations of super-Gaussian sources and sub-Gaussian sources [22.76, 22.78, 22.79, 22.72].
- The one-bit conjecture and its proof [22.26].

(c) Mixtures of independent structures

- Smoothed EM, Robust EM, Hard-cut EM, RPCL learning, Normalization RPCL-type EM, with the number of local models and the subspace dimension of each model determined in either of two ways [22.54, 22.51].
- Extensions of BFA, NFA, LMM-ICA, LMSER to their local versions, with both the number of local models and the number hidden factors selected in either of two ways (see Item 7.6 and Item 7.7 in [22.74], and the recent [22.54, 22.51, 22.52]).
- Extensions to self-organizing maps in collaboration with the BYY harmony learning [22.54].

(d) Mixture-of-experts (ME), RBF nets, and kernel regression

- Cooperative competitive learning (CCL) for ME learning, with the number of experts determined in either of two ways (see Sect. 4 in [22.82] and Table 7 in [22.76], also see Sect. 2.2 in [22.67]).
- An alternative ME model [22.92] with its ML learning exactly implemented by the EM algorithm. Criterion for the number of experts, as well as CCL learning algorithms and normalization RPCL-type EM, with the number of experts determined in either of two ways (see Sect. 4.3 in [22.83] and Sect. 3.2 in [22.67], also see [22.58, 22.54]).
- Variants on both the ME model and the alternative ME model via introducing regularization of either data smoothing or normalization.
- RBF nets are linked to the alternative ME model as special cases. Instead of the conventional two step learning, i.e., making clustering and then least square learning, learning is implemented not only by EM in exactly a maximum likelihood sense with the number of basis functions selected by a simple criterion from Eq. (22.42), but also by adaptive algorithms in a BYY harmony learning sense with the number of basis functions determined automatically during learning. Moreover, extensions are made by introducing regularization of either data smoothing or normalization [22.67, 22.58, 22.54].
- As the special cases of the normalized radial basis function (RBF), we can get a support vector based kernel regression [22.58, 22.54].

(e) Three layer nets and hidden units

- Regularized ML learning obtained via either normalization or data smoothing [22.51].
- Three layer nets with hidden units are linked to independent factor structures as extended cases [22.51, 22.53].
- Easy implementing criterion obtained from Eq. (22.42) for determining hidden units.
- An EM-like algorithms for a three layer net of stochastic hidden unit in multivariate Bernoulli, with the number of hidden units determined (see Table 4 and 5 in [22.76]).

(f) Temporal modeling

- Extensions of BFA to independent HMM and higher order independent HMM (see Item 4.1 in [22.75], Sect. 2.4 in [22.65], Eq. (26) in [22.61], and pages 839-849 of [22.57]).
- Extension of ICA to temporal ICA for blind separation of temporal sources (see Eq. (29a) and (29b) in [22.65], Sect. 3 in [22.65], Sect. IV(A) & (B) in [22.59], and Sect. III(B) in [22.57]).
- Extensions of FA to not only Kalman filter, but also temporal factor analysis (TFA) that has no rotation indeterminacy and, thus, can be used for both not only implementing a real BSS with noise and but also making model identification (see Sect. 2.3.1 in [22.65]), Sect. IV(C) in [22.59], and Sect. III(A) in [22.57]).
- Extensions of TFA to temporal NFA, with the driving noises of the autoregressive state space being non-Gaussian and modeled via Gaussian mixtures (see Sect. 4 in [22.61] and Sect. IV(B) in [22.59]).
- Easy implementing criterion obtained from Eq. (22.42) for determining hidden states.
- Adaptive algorithms with the number of states determined automatically during learning (see [22.57, 22.53]).
- Temporal extensions of LMSER, competitive ICA, local FA and local NFA to [22.57, 22.53].

These results have been obtained as the progress of studies on BYY system and harmony learning. Bibliographic remarks on the progress are provided in the following two subsections from both the aspect of BYY system with the KL learning and the aspect of computing techniques for implementing BYY learning. In the next chapter of the present book, bibliographic remarks will be also provided on the progress from the model selection and regularization aspects of BYY harmony learning.

22.9.2 Bibliographic Remarks on BYY System with KL Learning

As discussed in Sect. 22.5, acting as a unified framework for understanding several existing major learning models, the BYY learning was firstly published in 1995 with the following main issues [22.88]:

- The basic system as shown in Fig. 22.1 and three typical architectures as shown in Fig. 22.2 were firstly proposed and studied under the KL learning by Eq. (22.45).
- The equivalence of the KL learning on a B-architecture to the ML learning on $q(x)$ was found with Eq. (22.46), with the EM algorithm [22.9] revisited and an alternative but much simpler mathematical proof on its convergence (see Sect. 3.1 in [22.88], especially the part after Eq. (9) in [22.88]).
- The relationship of this KL learning on a BI-architecture to the Helmholtz machine learning and variational approximation [22.40, 22.18] was established (see Sect. 3.4 in [22.88]).
- The relationship of this KL learning on a F-architecture to the maximum information preservation were established (see Sect. 3.3 in [22.88]).

In the same year, the key ideas were also proposed for extending the BYY system under the KL learning by Eq. (22.45) to supervised learning (see Sect. 5 in [22.90]) and temporal modeling (see [22.89] and Sect. 4 in [22.90]). Specifically, advances of BYY system under the KL learning can be summarized along the following directions:

(1) The equivalence between the KL learning on a B-architecture and the ML learning on the marginal density $q(x)$ was further elaborated as Theorem 1 in [22.77]. The Ying-Yang procedure discussed in Sect. 22.7 provides the general form of the EM algorithm [22.9] from a new perspective, which not only leads us to revisit the detailed forms of the EM algorithm on Gaussian mixture [22.38], mixture of experts [22.13, 22.14, 22.92], factor analysis [22.39], etc., but also brings us new results on several typical learning models in two aspects:

- (a) On those models that have not taken the advantages of the EM algorithm yet, the detailed forms of the EM algorithm are developed for implementing the ML learning. The following are two typical examples.
  - The so-called multi-sets modeling is proposed in 1994 for modeling these objects in typical shapes such as lines, circles, and ellipses, as well as pre-specified templates [22.93] and [22.91]. At the beginning, its learning is formulized as an extension of the conventional mean square error (MSE) clustering analysis and then implemented via a generalized version of the well known KMEAN algorithm [22.10]. In help of a link, that was firstly built in [22.88], between the ML learning with the EM algorithm on Gaussian mixture and the MSE clustering with the KMEAN algorithm, this type of multi-sets is represented into

a finite mixture that is also recently called multi-sets-mixture [22.52], the detailed forms of the EM algorithm are proposed for implementing its ML learning (see Sect. 3.2 in [22.88] and Sect. 3 in [22.82], also see the recent elaboration in [22.52]).

- The alternative model [22.92] of the mixture-of-experts (ME) [22.12, 22.13] with the EM algorithm for its ML learning has been revisited. As special cases of this alternative ME model, the detailed forms of the EM algorithm is developed for implementing the ML learning on the normalized RBF nets, the extended normalized RBF nets, as well as their elliptic extensions (see Sect. 4 in [22.82] and Table 7 in [22.76], also see Sect. 2.2 in [22.67]), in place of the conventional sub-optimal two stage training algorithm (i.e., making clustering to locate basis functions and then making linear regression for the output layer).

(b) The above EM algorithms as well as those existing EM algorithms on several typical learning models are further extended to adaptive versions (i.e., adaptation is made per a sample). Typical examples include those adaptive algorithms on:

- Gaussian mixture (see Sect. 6.1 in [22.88]),
- factor analysis (see Sect. 3.2(B) in [22.71], also see Sect. 4.2.4 in [22.66]),
- binary factor analysis (see Sect. 4.2.2 and Fig. 2 in [22.66]),
- local PCA (see Item 7.6 and Item 7.7 in [22.74]),
- multi-sets mixture (see Sect. 3 in [22.82] and also see Sect. 3.3 in [22.52]),
- mixture-of-experts model and alternative mixture-of-experts model (see Sect. 4.3 in [22.83] and Sect. 3.2 in [22.67]).

(2) Following a link initially made in [22.88] on the relationship between the KL learning with a BI-architecture and the Helmholtz machine learning, the detailed relationship between BYY system and Helmholtz machine is further explored via a special BI-architecture for factorial encoding that leads both special cases of the Helmholtz machine with one hidden layer and a regularized version of LMSER [22.94] and auto-association [22.6]. This regularized LMSER actually performs jointly a noisy-ICA and a LMSER learning (see Sect. 8 in [22.74] and Sect. 4.3.4 in [22.66]). Its learning is implemented either via a Monte-Carlo technique (see Table 2 in [22.76]) or in help of a mean-field approximation based EM like algorithm (see Sect. 4.3.2 in [22.66]). With a binary inner representation replaced with a real non-Gaussian, two other advances have been achieved. One is a Monte-Carlo based EM algorithm for implementing independent factor analysis with non-Gaussian factors (see Sect. 3 in [22.66] and Sect. 3 in [22.68]) and the other is another extension of nonlinear LMSER learning with a real non-Gaussian inner representation (see Sect. 3.4 in [22.66]).

(3) Following a link initially made also in [22.88] on the relationship the relationship between the KL learning with a F-architecture and the maximum information preservation, subsequently it is further obtained that

- The KL learning on a joint special case of both BYY F-architecture and BYY BI-architecture leads the widely studied information theoretic based ICA [22.86].
- With $q(y)$ being a product of independent factors in a parametric model that is learned together with the F-architecture, we were motivated on getting an important improvement on the information theoretic based ICA, under the name Learned Parametric Mixture based ICA (shortly LMP-ICA) algorithm [22.87] that works well on any combination of super-Gaussian or sub-Gaussian sources. Each independent factor can be modeled via either a mixture of CDF [22.78, 22.79] or a mixture of Gaussian mixture (see Table 3 in [22.76] and [22.72]).
- This LMP-ICA has also been extended to the cases with observation noises via implementing jointly a noisy-ICA and either a non-Gaussian independent factor analysis or a LMSER learning (see Eq. (10.8) and Eq. (10.9) in [22.74]).

(4) The KL learning on a specific B-architecture also leads to the ML learning and particularly the mean square error learning on a three layer network (see Table 5(1) in [22.76]). With two types of stochastic hidden units, the ML learning on three layer networks can be made by an EM like algorithm (see Table 4 in [22.76]), which is implemented via either a Monte-Carlo technique (see Table 5(4) in [22.76]) or a mean-field based technique (see Table 5(3) in [22.76]).

(5) Studies on BYY systems for temporal modeling started from 1995 too (see [22.89] and Sect. 4 in [22.90]), in which a preliminary framework was proposed and the KL learning on temporal BYY system with a B-architecture was shown to be equivalent to the ML learning on a temporal process. To implement the learning, the following studies have been made:

- The KL measure on the entire temporal process has been turned into a summation of the KL measure at every time instance t weighted by a transfer probability from its past to the moment t.
- The integrals over each inner representation y_t has been approximated via a Monte-Carlo technique (see Sect. 5 in [22.75]).
- A fast implementation is further developed with the integrals over each y_t being approximated in help of a so-called CRP approximation that includes the mean-field approximation as one of three typical examples (see Sect. 2.1 in [22.65]). Then, this CRP approximation is mathematically justified by considering $\int p(u)T(u)du$ via the first order Taylor expansion of $T(u)$ around $\hat{u} = \int up(u)du$ (see Sect. 2.3 in [22.64]).
- A better approximation of the integrals over each y_t has also be considered via the above Taylor expansion of $T(u)$ up to the second order (see Sect. 2.3 in [22.64] and Sect. II(D) in [22.59]).

(6) Furthermore, two types of particular temporal models have been studied.

(a) One is that y is real and described by a state space model of an auto-regressive (AR) process (e.g., a first order AR process) even the first order auto-regressive process, with the following results:

- The well known Kalman filter on the state space is revisited as a special case with extensions obtained (see Item 4.2 in [22.75], Sect. 2.2 in [22.65] and [22.64], and Sect. III(B) in [22.57]).

- ICA is extended to temporal ICA (see Eq. (29a) and (29b) in [22.65]), and Sect. 3 in [22.65], also see Sect. IV(A) and (B) in [22.59] and Sect. III(B) in [22.57]).

- The temporal factor analysis (TFA) has been proposed not only as an extension of factor analysis with its rotation indeterminacy solved by temporal relation and but also as an extension of Kalman filter with the state space parameters determined via learning instead of requiring them known in advance (see Sect. 2.3.1 in [22.65]), Sect. IV(C) in [22.59], and Sect. III(A) in [22.57]).

- Temporal noisy-ICA that implements jointly a temporal ICA and a temporal extension of non-Gaussian independent factor analysis (see Sect. 4 in [22.61] and Sect. IV(B) in [22.59]).

(b) The other type is that y is discrete or binary and described by a Hidden Markov process, with the following results:

- The well known Hidden Markov model (HMM) is revisited when y takes a finite number of discrete labels, and the Ying-Yang procedure leads to the well known Baum-Welch algorithm (see Item 4.1 in [22.75] and Sect. 2.4 in [22.65]). Moreover, with the CRP approximation, a fast algorithm was suggested for approximately implementing the ML learning (see Sect. 2.4 in [22.65]).

- Temporal extension of binary FA is made into a type of independent HMM, with its learning implemented by an EM-like algorithm (see Eq. (26) in [22.61] and page 839–849 of [22.57]).

- When a B-architecture is replaced with a BI-architecture for directly mapping $x_t \rightarrow y_t$, temporal extension of the previously discussed regularized LMSER is made into another type of independent HMM, with its learning implemented by an EM-like algorithm (see Sect. 4 in [22.64], Sect. IV(D) in [22.59], and Sect. III(B) in [22.57]).

(7) Extensions of the KL learning by Eq. (22.45) has also been made with the KL-divergence replaced by f-divergence and the so-called weighted EM algorithm was firstly proposed (see Item 1.4 in [22.75] and Sect. 4 in [22.75]), supported by experiments on Gaussian mixture [22.77] and ICA problems [22.80] with advantage of being robust to outliers. Further studies have also been made in [22.69, 22.70, 22.62, 22.63], especially with a systematic summary in [22.60].

22.9.3 Bibliographic Remarks on Computing Techniques

During the studies on BYY learning, the following specific computing techniques have also been developed to support implementation:

- **Ying-Yang Alternative implementation**, as given by Eq. (22.62), provides a unified implementation procedure that facilitates the learning on such a paired system. It not only includes the EM algorithm [22.9] as a special case but also applies to various cases of learning on a BYY system under either KL-divergence or harmony measure as well as other learning costs.
- The technique of approximating $\int p(u)T(u)du$ via the first order Talyor expansion of $T(u)$ around $\hat{u} = \int up(u)du$ (see Sect. 2.3 in [22.64]) provides a very useful tool for tackling the integrals of $\int p(u)T(u)du$ type, which takes an important role in implementation of temporal BYY learning and data smoothing regularization.
- The peak finding problem in Eq. (22.36) takes an important part in implementation of BYY harmony learning, which is solved by techniques given in Table 1 of [22.57].
- An iterative updating on a covariance matrix Σ has to ensure its positive definite nature. It is usually not guaranteed by a updating rule with both learning and de-learning. Two techniques was firstly proposed in Table 1 of [22.58]. Namely, one bases on the decomposition $\Sigma = BB^T$, which have been adopted in [22.54, 22.56, 22.51, 22.52] and the other bases on the eigen-decomposition, which have been adopted in [22.27, 22.28, 22.29].
- The accumulation consensus technique discussed in Sect. 22.8 can be traced back to the RHT [22.96, 22.95] for detecting line and curves. The RHT opened one new direction on studies of the conventional HT, which has been widely studied and applied to various problems of curve detection and object detection.
- The multi-set mixture based learning was developed in [22.93, 22.91]. The link between RHT and multi-sets mixture based learning was built in [22.52] where the key ideas discussed in Sect. 22.8 were firstly initialized.

22.10 Conclusions

Various dependence structures among data are important to many tasks of statistical learning and data mining. Undertaking both a survey on major tasks of dependence structure mining and a summary on fundamentals and main results of BYY harmony learning, we observe that the BYY harmony learning provides a unified framework for various dependence structures, with new mechanisms for model selection and regularization on a finite size of samples.

References

22.1 H. Akaike: A new look at the statistical model identification, IEEE Tr. Automatic Control, 19, 714-723 (1974)

22.2 SI. Amari, A. Cichocki, HH. Yang: A new learning algorithm for blind separation of sources. In: DS Touretzky et al. (eds.) *Advances in Neural Information Processing 8*, MIT Press, 757-763 (1996)

22.3 TW. Anderson, H. Rubin: Statistical inference in factor analysis, *Proc. Berkeley Symp. Math. Statist. Prob. 3rd 5*, UC Berkeley, 111-150 (1956)

22.4 A. Bell, T. Sejnowski: An information maximization approach to blind separation and blind deconvolution, Neural Computation, 17, 1129-1159 (1995)

22.5 CM. Bishop: Training with noise is equivalent to Tikhonov regularization, Neural Computation 7, 108-116 (1995)

22.6 H. Bourlard, Y. Kamp: Auto-association by multilayer Perceptrons and singular value decomposition, Biol. Cyb. 59, 291-294 (1988)

22.7 P. Comon: Independent component analysis - a new concept? Signal Processing, 36, 287-314 (1994)

22.8 KY. Chan, WS. Chu, L. Xu: Experimental Comparison between two computational strategies for topological self-organization, *Proc. of IDEAL03*, Lecture Notes in Computer Science, LNCS 2690, Springer-Verlag, 410-414 (2003)

22.9 AP. Dempster, NM. Laird, DB. Rubin: Maximum-likelihood from incomplete data via the EM algorithm, J. Royal Statistical Society, B39, 1-38 (1977)

22.10 PA. Devijver, J. Kittler: Pattern Recognition: A Statistical Approach, Prentice-Hall (1982)

22.11 RO. Duda, PE. Hart: *Pattern classification and Scene analysis* (Wiley, 1973)

22.12 RA. Jacobs et al.: Adaptive mixtures of local experts, Neural Computation, 3, 79-87 (1991)

22.13 MI. Jordan, RA. Jacobs: Hierarchical mixtures of experts and the EM algorithm, Neural Computation, 6, 181-214 (1994)

22.14 MI. Jordan, L. Xu: Convergence results for the EM approach to mixtures of experts, Neural Networks, 8, 1409-1431 (1995)

22.15 C. Jutten, J. Herault: Independent Component Analysis versus Principal Component Analysis, *Proc. EUSIPCO88*, 643-646 (1988)

22.16 H. Kälviäinen, P. Hirvonen, L. Xu, E. Oja: Probabilistic and Nonprobabilistic Hough Transforms: Overview and Comparisons, *Image and Vision Computing*, Vol. 5, No. 4, pp. 239-252 (1995)

22.17 J. Han, M. Kamber: *Data Mining: Concepts and Techniques* (Morgan Kaufmann, 2001)

22.18 GE. Hinton, P. Dayan, BJ. Frey, RN. Neal: The wake-sleep algorithm for unsupervised learning neural networks, Science, 268, 1158-1160 (1995)

22.19 H. Hotelling: Simplified calculation of principal components, Psychometrika, 1, 27-35 (1936)

22.20 P.V.C. Hough: Method and means for recognizing complex patterns, *U.S. Patent 3069654* (Dec.18, 1962)

22.21 J. Illingworth, J. Kittler: A survey of the Hough Transform, Comput. Vision Graphics and Image Process, 43, 221-238 (1988)

22.22 FV. Jensen: *An introduction to Bayesian networks* (University of Collage London Press) (1996)

22.23 T. Kohonen: *Self-Organizing Maps* (Springer-Verlag, Berlin, 1995)

22.24 H. Kushner, D. Clark: *Stochastic approximation methods for constrained and unconstrained systems* (New York: Springer) (1998)

22.25 HY. Kwok, CM. Chen, L. Xu: Comparison between Mixture of ARMA and Mixture of AR Model with Application to Time Series Forecasting, *Proc. ICONIP'98*, Oct.21-23, 1998, Kitakyushu, Japan, Vol. 2, 1049-1052

22.26 ZY. Liu, KC. Chiu, L. Xu: The One-bit-Matching Conjecture for Independent Component Analysis, Neural Computation, Vol. 16, No. 2, pp. 383-399 (2003)

22.27 ZY. Liu, KC. Chiu, L. Xu: Strip Line Detection and Thinning by RPCL-Based Local PCA, Pattern Recognition Letters, 24, pp. 2335-2344 (2003)

22.28 ZY. Liu, KC. Chiu, L. Xu: Improved system for object detection and star/galaxy classification via local subspace analysis, Neural Networks, 16, 437-451 (2003)

22.29 ZY. Liu, L. Xu: Smoothed Local PCA by BYY data smoothing learning, *Proc ICCAS 2001*, Jeju, Korea, Oct.17-21, 2001, pp. 924-927

22.30 J. Ma, T. Wang, L. Xu: A gradient BYY harmony learning rule on Gaussian mixture with automated model selection, Neurocomputing, 56, 481-487 (2004)

22.31 Ch. von der Malsburg: Self-organization of orientation sensitive cells in the striate cortex, Kybernetik 14, 85-100 (1973)

22.32 R. McDonald: *Factor Analysis and Related Techniques* (Lawrence Erlbaum)

22.33 GJ. McLachlan, T. Krishnan: *The EM Algorithm and Extensions*, John Wiley & Son, INC (1997)

22.34 E. Oja: *Subspace Methods of Pattern Recognition* (Research Studies Press, UK 1983)

22.35 J. Pearl: *Probabilistic reasoning in intelligent systems: networks of plausible inference* (San Francisco, CA: Morgan Kaufman 1988)

22.36 L. Rabiner, BH. Juang: *Fundamentals of Speech Recognition*, Prentice Hall, Inc. (1993)

22.37 H. Robbins, S. Monro: A stochastic approximation method, Ann. Math. Statist., 22, 400-407 (1950)

22.38 RA. Redner, HF. Walker: Mixture densities, maximum likelihood, and the EM algorithm, SIAM Review, 26, 195-239 (1984)

22.39 D. Rubi, D. Thayer: EM algorithm for ML factor analysis, Psychometrika, 57, 69-76 (1976)

22.40 L. Saul, MI. Jordan: Exploiting tractable structures in intractable Networks, Advances in neural information processing systems, 8, MIT Press, 486-492 (1995)

22.41 C. Spearman: General intelligence domainively determined and measured, Am. J. Psychol. 15, 201-293 (2004)

22.42 A. Taleb, C. Jutten: Nonlinear source separation: The post-nonlinear Mixtures, *Proc. ESANN97*, 279-284 (1997)

22.43 H. Tang, KC. Chiu, L. Xu: Finite Mixture of ARMA-GARCH Model For Stock Price Prediction, to appear on *Proc. CIEF'2003*, NC, USA (Sept.26-30, 2003)

22.44 H. Tang, L. Xu: Mixture-Of-Expert ARMA-GARCH Models For Stock Price Prediction, *Proc. of ICCAS 2003*, Oct.22-25, 2003 Gyeongju, KOREA, pp. 402-407 (2003)

22.45 ME. Tipping, CM. Bishop: Mixtures of probabilistic principal component analysis, Neural Computation, 11, 443-482 (1999)

22.46 L. Tong, Y. Inouye, R. Liu: Waveform-preserving blind estimation of multiple independent sources, IEEE Tr on Signal Processing, 41, 2461-2470 (1993)

22.47 VN. Vapnik: *The Nature Of Statistical Learning Theory* (Springer-Verlag) (1995)

22.48 CS. Wong, WK. Li: On a mixture autoregressive model, Journal of the Royal Statistical Society Series B, Vol. 62, No. 1, pp. 95-115 (2000)

22.49 W. Wong, F. Yip, L. Xu: Financial Prediction by Finite Mixture GARCH Model, *Proc. ICONIP'98*, Oct.21-23, 1998, Kitakyushu, Japan, Vol. 3, pp. 1351-1354 (1998)

22.50 L. Xu: Temporal BYY Learning, Identifiable State Spaces, and Space Dimension Determination, *IEEE Trans on Neural Networks, Special Issue on Temporal Coding for Neural Information Processing*, in press (2004)

22.51 L. Xu: BYY Learning, Regularized Implementation, and Model Selection on Modular Networks with One Hidden Layer of Binary Units", *Neurocomputing*, Vol. 51, pp. 227-301 (2003)

22.52 L. Xu: Data smoothing regularization, multi-sets-learning, and problem solving strategies, Neural Networks, Vol. 15, No. 56, 817-825 (2003)

22.53 L. Xu: Independent Component Analysis and Extensions with Noise and Time: A Bayesian Ying Yang Learning Perspective, Neural Information Processing - Letters and Reviews, Vol. 1, No. 1, pp. 1-52 (2003)

22.54 L. Xu: BYY Harmony Learning, Structural RPCL, and Topological Self-Organizing on Mixture Models, Neural Networks, Vol. 15, nos. 8-9, 1125-1151 (2002)

22.55 L. Xu: Bayesian Ying Yang Harmony Learning, *The Handbook of Brain Theory and Neural Networks*, Second edition, (MA Arbib, Ed.), Cambridge, MA: The MIT Press, pp. 1231-1237 (2002)

22.56 L. Xu: Mining Dependence Structures from Statistical Learning Perspective. In: H, Yin et al. (eds.), *Proc. IDEAL2002: Lecture Notes in Computer Science, 2412*, Springer-Verlag, 285-306 (2002)

22.57 L. Xu: BYY Harmony Learning, Independent State Space and Generalized APT Financial Analyses, IEEE Tr on Neural Networks, 12(4), 822-849 (2001)

22.58 L. Xu: Best Harmony, Unified RPCL and Automated Model Selection for Unsupervised and Supervised Learning on Gaussian Mixtures, Three-Layer Nets and ME-RBF-SVM Models, *Intl J of Neural Systems* 11 (1), 43-69 (2001)

22.59 L. Xu: Temporal BYY Learning for State Space Approach, Hidden Markov Model and Blind Source Separation", IEEE Tr on Signal Processing 48, 2132-2144 (2000)

22.60 L. Xu: BYY Learning System and Theory for Parameter Estimation, Data Smoothing Based Regularization and Model Selection, Neural, Parallel and Scientific Computations, Vol. 8, pp. 55-82 (2000)

22.61 L. Xu: Temporal Bayesian Ying Yang Dependence Reduction, Blind Source Separation and Principal Independent Components, *Proc. IJCNN'99*, July 10-16, 1999, DC, USA, Vol. 2, pp. 1071-1076 (1999)

22.62 L. Xu: Bayesian Ying Yang Unsupervised and Supervised Learning: Theory and Applications, *Proc. of 1999 Chinese Conf. on Neural Networks and Signal Processing*, pp. 12-29, Shantou, China (Nov. 1999)

22.63 L. Xu: Bayesian Ying Yang Theory for Empirical Learning, Regularization and Model Selection: General Formulation, *Proc. IJCNN'99*, DC, USA, July 10-16, 1999, Vol. 1 of 6, pp. 552-557

22.64 L. Xu: Temporal BYY Learning and Its Applications to Extended Kalman Filtering, Hidden Markov Model, and Sensor-Motor Integration, *Proc. IJCNN'99*, DC, USA, July 10-16, 1999, vol.2 of 6, pp. 949-954

22.65 L. Xu: Bayesian Ying Yang System and Theory as a Unified Statistical Learning Approach :(V) Temporal Modeling for Temporal Perception and Control, *Proc. ICONIP'98*, Kitakyushu, Japan, Vol. 2, pp. 877-884 (1998)

22.66 L. Xu: Bayesian Kullback Ying-Yang Dependence Reduction Theory, Neurocomputing, 22 (1-3), 81-112 (1998)

22.67 L. Xu: RBF Nets, Mixture Experts, and Bayesian Ying Yang Learning, Neurocomputing, Vol. 19, No. 1-3, 223-257 (1998)

22.68 L. Xu: Bayesian Ying Yang Dependence Reduction Theory and Blind Source Separation on Instantaneous Mixture, *Proc. Intl ICSC Workshop I&ANN'98*, Feb.9-10, 1998, Tenerife, Spain, pp. 45-51 (1998)

22.69 L. Xu: Bayesian Ying Yang System and Theory as A Unified Statistical Learning Approach :(VI) Convex Divergence, Convex Entropy and Convex Likelihood? *Proc. IDEAL98*, Oct.14-16, 1998, Hong Kong, pp. 1-12 (1998)

22.70 L. Xu: Bayesian Ying Yang System and Theory as A Unified Statistical Learning Approach: (IV) Further Advances, *Proc. IJCNN98*, May 5-9, 1998, Anchorage, Alaska, Vol. 2, pp. 1275-1270 (1998)

22.71 L. Xu: BKYY Dimension Reduction and Determination, *Proc. IJCNN98*, May 5-9, 1998, Anchorage, Alaska, Vol. 3, pp. 1822-1827 (1998)

22.72 L. Xu, CC. Cheung, SI. Amari: Learned Parametric Mixture Based ICA Algorithm, Neurocomputing, 22 (1-3), 69-80 (1998)

22.73 L. Xu, CC. Cheung, SI. Amari: Further Results on Nonlinearity and Separation Capability of A Linear Mixture ICA Method and Learned Parametric Mixture Algorithm, *Proc. I&ANN'98*, Feb.9-10, 1998, Tenerife, Spain, pp. 39-44 (1998)

22.74 L. Xu: Bayesian Ying Yang System and Theory as A Unified Statistical Learning Approach: (I) Unsupervised and Semi-Unsupervised Learning. In: S. Amari, N. Kassabov (eds.), *Brain-like Computing and Intelligent Information Systems*, Springer-Verlag, pp. 241-274 (1997)

22.75 L. Xu: Bayesian Ying Yang System and Theory as A Unified Statistical Learning Approach (II): From Unsupervised Learning to Supervised Learning and Temporal Modeling. In: KM. Wong et al. (eds.), *Theoretical Aspects of Neural Computation: A Multidisciplinary Perspective*, Springer-Verlag, pp. 25-42 (1997)

22.76 L. Xu: Bayesian Ying Yang System and Theory as A Unified Statistical Learning Approach (III): Models and Algorithms for Dependence Reduction, Data Dimension Reduction, ICA and Supervised Learning. In: KM. Wong et al. (eds.), *Theoretical Aspects of Neural Computation: A Multidisciplinary Perspective*, Springer-Verlag, pp. 43-60 (1997)

22.77 L. Xu: Bayesian Ying Yang Machine, Clustering and Number of Clusters, Pattern Recognition Letters 18, No. 11-13, 1167-1178 (1997)

22.78 L. Xu, CC. Cheung, HH. Yang, SI. Amari: Independent Component Analysis by The Information-Theoretic Approach with Mixture of Density, *Proc. IJCNN97*, Vol. 3, 1821-1826 (1997)

22.79 L. Xu, CC. Cheung, J. Ruan, SI. Amari: Nonlinearity and Separation Capability: Further Justification for the ICA Algorithm with A Learned Mixture of Parametric Densities, *Proc. ESANN97*, Bruges, April 16-18, 1997, pp. 291-296 (1997)

22.80 L. Xu: Bayesian Ying Yang Learning Based ICA Models, *Proc. 1997 IEEE NNSP VII*, Sept.24-26, 1997, Florida, pp. 476-485 (1997)

22.81 L. Xu: New Advances on Bayesian Ying Yang Learning System with Kullback and Non-Kullback Separation Functionals, *Proc. IJCNN97*, June 9-12, 1997, Houston, TX, USA, Vol. 3, pp. 1942-1947 (1997)

22.82 L. Xu: Bayesian-Kullback YING-YANG Learning Scheme: Reviews and New Results, *Proc. ICONIP96*, Vol. 1, 59-67 (1996)

22.83 L. Xu: Bayesian-Kullback YING-YANG Machines for Supervised Learning, *Proc. WCNN96*, Sept.15-18, 1996, San Diego, CA, pp. 193-200 (1996)

22.84 L. Xu: A Maximum Balanced Mapping Certainty Principle for Pattern Recognition and Associative Mapping, *Proc. WCNN96*, Sept. 15-18, 1996, San Diego, CA, pp. 946-949 (1996)

22.85 L. Xu: How Many Clusters?: A YING-YANG Machine Based Theory for A Classical Open Problem in Pattern Recognition, *Proc. IEEE ICNN96*, June 2-6, 1996, DC, Vol. 3, pp. 1546-1551 (1996)

22.86 L. Xu, SI. Amari: A general independent component analysis framework based on Bayesian Kullback Ying Yang Learning, *Proc. ICONIP96*, 1253-1240 (1996)

22.87 L. Xu, HH. Yang, SI. Amari: Signal Source Separation by Mixtures Accumulative Distribution Functions or Mixture of Bell-Shape Density Distribution Functions, Research Proposal, presented at FRONTIER FORUM, organized by S. Amari, S. Tanaka, A. Cichocki, RIKEN, Japan (April 10, 1996)

22.88 L. Xu: Bayesian-Kullback Coupled YING-YANG Machines: Unified Learnings and New Results on Vector Quantization, *Proc. ICONIP95*, Oct 30-Nov.3, 1995, Beijing, China, pp. 977-988 (1995)

22.89 L. Xu: YING-YANG Machine for Temporal Signals, Keynote Talk, *Proc. 1995 IEEE Intl Conf. on Neural Networks and Signal Processing*, Dec. 10-13, 1995, Nanjing, Vol. 1, pp. 644-651 (1995)

22.90 L. Xu: New Advances on The YING-YANG Machine, *Proc. Intl. Symp. on Artificial Neural Networks*, Dec.18-20, 1995, Taiwan, pp. 07-12 (1995)

22.91 L. Xu: A unified learning framework: multisets modeling learning, *Proceedings of 1995 World Congress on Neural Networks*, Vol. 1, pp. 35-42 (1995)

22.92 L. Xu, MI. Jordan, GE. Hinton: An Alternative Model for Mixtures of Experts. In: JD. Cowan et al. (eds.), *Advances in Neural Information Processing Systems 7*, MIT Press, 633-640 (1995)

22.93 L. Xu: Multisets Modeling Learning: An Unified Theory for Supervised and Unsupervised Learning, Invited Talk, *Proc. IEEE ICNN94*, June 26-July 2, 1994, Orlando, Florida, Vol. 1, pp. 315-320 (1994)

22.94 L. Xu: Least mean square error reconstruction for self-organizing neuralnets, Neural Networks 6, 627-648, 1993. Its early version on *Proc. IJCNN91'Singapore*, 2363-2373 (1991)

22.95 L. Xu, E. Oja: Randomized Hough Transform (RHT): Basic Mechanisms, Algorithms and Complexities, *Computer Vision, Graphics, and Image Processing: Image Understanding*, Vol. 57, no.2, pp. 131-154 (1993)

22.96 L. Xu, E. Oja, P. Kultanen: A New Curve Detection Method: Randomized Hough Transform (RHT), Pattern Recognition Letters, 11, 331-338 (1990)

23. Bayesian Ying Yang Learning (II): A New Mechanism for Model Selection and Regularization

Lei Xu

Chinese University of Hong Kong, Hong Kong

Abstract

Efforts toward a key challenge of statistical learning, namely making learning on a finite size of samples with model selection ability, have been discussed in two typical streams. Bayesian Ying Yang (BYY) harmony learning provides a promising tool for solving this key challenge, with new mechanisms for model selection and regularization. Moreover, not only the BYY harmony learning is further justified from both an information theoretic perspective and a generalized projection geometry, but also comparative discussions are made on its relations and differences from the studies of minimum description length (MDL), the bit-back based MDL, Bayesian approach, maximum likelihood, information geometry, Helmholtz machines, and variational approximation. In addition, bibliographic remarks are made on the advances of BYY harmony learning studies.

23.1 Introduction: A Key Challenge and Existing Solutions

A key challenge to all the learning tasks is that learning is made on a finite size set $\mathcal{X}$ of samples from the world $\mathbf{X}$, while our ambition is to get the underlying distribution such that we can apply it to as many as possible new samples coming from $\mathbf{X}$.

Helped by certain pre-knowledge about $\mathbf{X}$ a learner, $\mathcal{M}$ is usually designed via a parametric family $p(x|\theta)$, with its density function form covering or being as close as possible to the function form of the true density $p_*(x|\cdot)$. Then, we obtain an estimator $\hat{\theta}(\mathcal{X})$ with a specific value for θ such that $p(x|\hat{\theta}(\mathcal{X}))$ is as close as possible to the true density $p_*(x|\theta_o)$, with the true value θ_o. This is usually obtained by determining a specific value of $\hat{\theta}(\mathcal{X})$ that minimizes a cost functional

$$\mathcal{F}(p(x|\theta), \mathcal{X}) \ or \ \mathcal{F}(p(x|\theta), q_{\mathcal{X}}(x)), \tag{23.1}$$

where $q_{\mathcal{X}}$ is an estimated density of x from $\mathcal{X}$, e.g., given by the empirical density:

$$p_0(x) = \tfrac{1}{N}\sum_{t=1}^{N}\delta(x - x_t), \quad \delta(x) = \begin{cases} \lim_{\delta \to 0}\frac{1}{\delta^d}, & x = 0, \\ 0, & x \neq 0, \end{cases} \tag{23.2}$$

where d is the dimension of x and $\delta > 0$ is a small number. With a given *smoothing parameter* $h > 0$, $q_{\mathcal{X}}$ can also be the following non-parametric Parzen window density estimate [23.21]:

$$p_h(x) = \tfrac{1}{N} \sum_{t=1}^{N} G(x|x_t, h^2 I), \tag{23.3}$$

When $p(x|\theta) = p_0(x)$, given by Eq. (23.2), a typical example of Eq. (23.1) is

$$\min_{\theta} -\mathcal{F}(p(x|\theta), \mathcal{X}) = -\int p_0(x) \ln p(x|\theta)\mu(dx), \tag{23.4}$$

where $\mu(.)$ is a given measure. It leads to the maximum likelihood (ML) estimator $\hat{\theta}(\mathcal{X})$. For a fixed N, we usually have $\hat{\theta}(\mathcal{X}) \neq \theta_o$ and $p(x|\hat{\theta}(\mathcal{X})) \neq p_*(x|\theta_o)$. Thus, though $p(x|\hat{\theta}(\mathcal{X}))$ best matches the sample set $\mathcal{X}$ in the sense of Eq. (23.1) or Eq. (23.4), $p(x|\hat{\theta}(\mathcal{X}))$ may not well apply to new samples from the same world $\mathbf{X}$.

However, if there is an oracle who tells us the function form of $p_*(x|\cdot)$, we can conveniently use it as the function form of $p(x|\cdot)$. In this case, it follows from the large number law in probability theory that the ML estimator $\hat{\theta}(\mathcal{X}) \to \theta_o$ and $p_*(x|\theta) \to p_*(x|\theta_o)$ as $N \to \infty$. Shortly, the estimator $\hat{\theta}(\mathcal{X})$ is said to be statistically consistent. Actually, this large number law can be regarded as the mathematical formalization of a fundamental philosophy or principle of modern science that a truth about the world exists independent of our perception, and that we will tend to and finally approach the truth as long as the evidences we collected about the truth become infinitely many. However, assuming knowing the true density form of $p_*(x|\cdot)$ implies actually a knowledge on a major structure of the world $\mathbf{X}$ and what to be precisely discovered are remaining details. In many realistic problems we have no such an oracle to tell us the knowledge on the true function form of $p_*(x|\cdot)$ and, thus, in these cases the large number law may fail even as $N \to \infty$.

To avoid the problem, we consider a family $\mathcal{F}$ of density function forms $p(x|\theta_j), j = 1, \cdots, k, \cdots$ with each sharing a same configuration but its structural scale increasing with k such that $\mathcal{P}_1 \subset \mathcal{P}_2 \subset \cdots \mathcal{P}_k \subset \cdots$, where $\mathcal{P}_j = \{p(x|\theta_j)|\forall \theta_j \in \Theta_j\}$. The task of learning is to decide both a best j^* and the corresponding best θ_{j*}^* for best describing the true $p_*(x|\theta_o)$. For a finite size set $\mathcal{X}$ of samples, $\mathcal{F}(p(x|\theta_j), \mathcal{X})$ by Eq. (23.1) will monotonically decrease and finally reach 0 as j increases. With a much larger scale, a $p(x|\theta_j)$ that reaches $\mathcal{F}(p(x|\theta_j), \mathcal{X}) = 0$ is usually far from $p_*(x|\theta_o)$. The smaller the sample size, the worse the situation is. Still, as $N \to \infty$ the resulted $p(x|\theta_j)$ will approach $p_*(x|\theta_o)$ if it is included in the family $\mathcal{F}$.

Unfortunately, this ML-type principle is challenged by the fact that the purpose of learning is to guide a learner $\mathcal{M}$ to interact with the world that is usually not only stochastic but also in changing dynamically. Thus, we are not able to collect enough samples either because not a plenty of resources or because not an enough speed to catch the dynamic changing of world.

Therefore, what $\mathcal{M}$ encounters is usually finite number N of samples and, thus, the large number law does not apply.

In past decades, many efforts have been made toward this critical challenge, forming roughly two main streams.

23.2 Existing Solutions

23.2.1 Efforts in the First Stream

By insisting in that there is a true underlying density $p_*(x|\theta_o)$ that apply to all the samples, we desire a best estimate by minimizing $\mathcal{F}(p(x|\theta), p_*(x|\theta_o))$. Unfortunately, this is not directly workable, since $p_*(x|\theta_o)$ is not known. Alternatively, a classic idea is to quantitatively estimate the discrepancy between $\mathcal{F}(p(x|\theta), p_*(x|\theta_o))$ and $\mathcal{F}(p(x|\theta), \mathcal{X})$ such that we have

$$\mathcal{F}(p(x|\theta), p_*(x|\theta_o)) = \mathcal{F}(p(x|\theta), \mathcal{X}) + \Delta(\theta, \theta_o, \mathcal{X}), \qquad (23.5)$$

where $\Delta(\theta, \theta_o, \mathcal{X})$ is an estimate of $\mathcal{F}(p(x|\theta), p_*(x|\theta_o)) - \mathcal{F}(p(x|\theta), \mathcal{X})$. This is usually difficult to accurately estimate without knowing $p_*(x|\theta_o)$. In the literature, Δ is usually an estimate on certain bounds of this discrepancy, which may be obtained from $\mathcal{X}$ and the structural features of $p(x|\theta)$, helped by some structural knowledge about $p_*(x|\theta_o)$. Using the bounds, we implement either one or both of the following two types of corrections on estimates from Eq. (23.1):

(a) *Model Selection*   We consider a number of candidate models $M_j, j = 1, \cdots, k$, each having its own density function $p(x|\theta_j)$. We estimate each bound Δ_j for the discrepancy between $\mathcal{F}(p(x|\theta_j, M_j), p_*(x|\theta_o))$ and $\mathcal{F}(p(x|\theta_j, M_j), \mathcal{X})$. Over all candidate models, we select the j^*-th model by

$$j^* = arg \min_j \left[\mathcal{F}(p(x|\theta_j, M_j), \mathcal{X}) + \Delta_j \right], \qquad (23.6)$$

which is referred as *Model Selection*. In the literature, model selection is usually made up of two stages. At the first stage, parameter learning takes place on determining θ_j^* by empirically minimizing $\mathcal{F}(p(x|\theta), \mathcal{X})$. At the second stage, selection of the best j^* takes place by Eq. (23.6). The estimated correcting term Δ_j relies on the complexity of the model M_j, while it does not contain any unknown variables of θ_j.

(b) *Regularization*   If we are able to estimate a tighter bound $\Delta(\theta)$ that varies with θ, we can directly get a corrected value θ^* by

$$\theta^* = arg \min_\theta \left[\Delta(\theta) + \mathcal{F}(p(x|\theta), \mathcal{X}) \right]. \qquad (23.7)$$

Such a type of effort is usually referred to as *regularization*, since it regularizes certain singularities caused by a finite number N of samples. Given a model with large enough scale, such a value θ^* makes the model act effectively as one with a reduced scale. This effective model may neither be identical to that resulted from the above model selection among a number of candidate models $M_j, j = 1, \cdots, k$, nor necessarily lead to $p_*(x|\theta_o)$. However, we have no particular reason to insist on which one is a true density $p_*(x|\theta_o)$ for a small size of samples that actually can be described by many models.

Several approaches have been developed in this stream. One typical example is the VC dimension-based learning theory [23.74], which considers $\mathcal{F}$ as the error or loss of performing a discrete nature task, such as classification or decision, on a set $\mathcal{X}$ of samples, with Δ_j estimated based on a complexity measure of the structure of M_j. The second type of example is AIC [23.1], as well as its extensions AICB, CAIC, etc. [23.2, 23.3, 23.67, 23.11, 23.12, 23.35, 23.36, 23.13], which usually consider a regression or modeling task, with Δ_j estimated as a bias of the likelihood $\int p_0(x) \ln p(x|\theta)\mu(dx)$ to the information measure $\int p_*(x|\theta_o) \ln p(x|\theta)\mu(dx)$. Another typical example is the so-called cross validation [23.64, 23.65, 23.66, 23.55]. Instead of estimating a bound of Δ_j, it targets at estimating $\mathcal{F}(p(x|\theta_j, M_j), p_*(x|\theta_o))$ via splitting $\mathcal{X}$ into a training subset $\mathcal{X}_t$ and a validation subset $\mathcal{X}_v$. First, one gets an estimate $\hat{\theta}_j$ by minimizing $\mathcal{F}(p(x|\theta_j, M_j), \mathcal{X}_t)$ and then estimates $\mathcal{F}(p(x|\theta_j, M_j), p_*(x|\theta_o))$ via jointly considering $\mathcal{F}(p(x|\hat{\theta}_j, M_j), \mathcal{X}_t)$ and $\mathcal{F}(p(x|\hat{\theta}_j, M_j), \mathcal{X}_v)$. Moreover, studies on cross validation relate closely to Jackknife and bootstrap techniques [23.24, 23.25].

Most of studies on these typical approaches are conducted on model selection only, since a rough bound Δ may already be able to give a correct selection among a series of individual models that are discretely different from each other, and, thus, have certain robustness on errors. In contrast, a rough bound $\Delta(\theta)$ usually makes $\min_\theta \Delta(\theta) + \mathcal{F}(p(x|\theta), \mathcal{X})$ lead to a poor performance. However, to get an appropriate bound, $\Delta(\theta)$ requires more knowledge on the true $p_*(x|\theta_o)$, which is usually difficult.

23.2.2 Efforts in the Second Stream

Instead of taking a true underlying density $p_*(x|\theta_o)$ as the target of considerations, the well known Ockham's principle of economy is used as the learning principle. If there are a number of choices for getting a model to fit a set $\mathcal{X}$ of samples, we use the one such that $p(x|\theta)$ not only matches $\mathcal{X}$ well but also has minimum complexity. This principle can be intuitively well understood. When $\mathcal{X}$ consists of a finite number N of samples, we can have infinite choices on $p(x|\theta)$ that describe or accommodate $\mathcal{X}$ well, or better as the complexity of $p(x|\theta)$ increases after satisfying a minimum requirement. That is, learning is a typical ill-posed problem, with intrinsic indeterminacy on its solution. The indeterminacy depends on how large the complexity of $p(x|\theta)$ is. The larger it is, the lower is the chance of getting the true underlying density $p_*(x|\theta_o)$, and, thus, the more likely that the learned choice generalizes poorly beyond the N samples in $\mathcal{X}$. Therefore, we choose the choice with the minimum complexity among all those that are able to describe $\mathcal{X}$ sufficiently well.

Based on this principle, approaches have been developed for both regularization and model selection.

(a) One example consists of various efforts either under the name 'regularization' or via certain equivalent techniques. One of most popular one is the well known Tikhonov regularization theory [23.30, 23.68], which minimizes $\mathcal{F}(p(x|\theta), \mathcal{X})$, with a so-called stabilizer that describes the irregularity or non-smoothness of $p(x|\theta)$. In the literature of both statistics and neural networks, there are many efforts that minimize $\mathcal{F}(p(x|\theta), \mathcal{X})$, with a penalty term in various forms. These heuristics take a role similar to that of the Tikhonov stabilizer [23.60, 23.22]. One critical weak point of these efforts is the lack of a systematic or quantitative way to guide how to choose the added term and to control the strength of the term in minimization. In the literature of statistics, the role of the added term is alternatively interpreted as controlling a tradeoff between bias and variance for an estimator [23.27, 23.73].

(b) The second type of efforts for implementing the Ockham's principle consists of those studies based on Bayesian approach. There are three major versions [23.42]. One is called maximum a posteriori probability (MAP), since it maximizes the posteriori probability

$$p(M_j, \theta_j | \mathcal{X}) = p(\mathcal{X}|\theta_j, M_j)p(\theta_j|M_j)p(M_j)/p(\mathcal{X}). \tag{23.8}$$

Specifically, its maximization with respect to θ_j is equivalent to maximizing $\ln[p(\mathcal{X}|\theta_j, M_j) p(\theta_j|M_j)] = \ln p(\mathcal{X}|\theta_j, M_j) + \ln p(\theta_j|M_j)$, with the first term being a specific example of $\mathcal{F}(p(x|\theta_j), \mathcal{X})$, and $\ln p(\theta_j|M_j)$ acting as a regularization term. That is, it provides a perspective that determines the Tikhonov stabilizer via a priori density $p(\theta_j|M_j)$. Moreover, model selection can be made by selecting j^* with the corresponding $p(\mathcal{X}|\hat{\theta}_j, M_j)p(\hat{\theta}_j|M_j)p(M_j)$ being the largest, where each a priori $p(M_j)$ is usually set uniformly and, thus, ignored, and $\hat{\theta}_j$ is given by either the above MAP regularization or an ML estimator, which is equivalent to using non-informative uniform prior as $p(\theta_j|M_j)$. However, an improperly selected $p(\theta_j|M_j)$ usually leads to a poor performance.

Instead of basing on a special value $\hat{\theta}_j$ of θ_j, the other version of the Bayesian approach makes model selection by selecting j^* as the largest of

$$p(M_j|\mathcal{X}) = p(\mathcal{X}|M_j)p(M_j)/p(\mathcal{X}),$$
$$p(\mathcal{X}|M_j) = \int p(\mathcal{X}|\theta_j, M_j)p(\theta_j|M_j)d\mu(\theta_j), \tag{23.9}$$

or, simply, the largest $p(\mathcal{X}|M_j)$, with $p(M_j)$ being regarded as uniform and, thus, ignored. The term $p(\mathcal{X}|M_j)$ is called the evidence (EV) or marginal likelihood, and, thus, it is also referred to as the EV approach. Typical studies include not only the so-called BIC and variants [23.59, 23.38, 23.48] that were proposed as a competitor of AIC and variants in the literature of statistics since the late 1970's, but also those renewed interests in the literature of neural networks in the last decade, exemplified by the study of [23.45, 23.46, 23.19].

Another version of the Bayesian approach is to use the Bayesian factor (BF)

$$BF_{ij} = p(\mathcal{X}|M_i)/p(\mathcal{X}|M_j) \ , \tag{23.10}$$

i.e., the ratio of evidences, for model comparison via hypothesis testing [23.26, 23.50, 23.40].

A common key problem in all three versions of Bayesian studies is how to get a priori density $p(\theta|M_j)$. Its choice reflects how much a priori knowledge is used. One widely used example is the Jeffery priori or a non-informative uniform priori [23.37, 23.9, 23.45, 23.46, 23.48, 23.42]. Moreover, the EV approach and the BF approach have the problem of how to compute the evidence accurately and efficiently, since it involves an integral. Stochastic simulation techniques such as the importance sampling approach and MCMC are usually used for implementations [23.48, 23.49, 23.14]. Certain comparisons are referred to [23.48, 23.23]. Recently, the Variational Bayes (VB) method has also been proposed in the literature of neural networks as an alternative way for efficient implementation [23.72, 23.28, 23.56].

The third type of efforts is made toward the implementation of Ockham's principle directly. One typical example is called the minimum message length (MML) theory [23.69, 23.70, 23.71], which was first proposed in the late 1960s' as an information measure for classification. The message length is defined via a two part message coding method. First, one needs a length for coding a hypothesis H (or equivalently called a model), described by $\log_2 P(H)$. Second, one needs a length for coding the residuals of using H to fit or interpret the observed set $\mathcal{X}$, described by $\log_2 P(\mathcal{X}|H)$. The two part message length

$$M_L = -\log_2 P(H) - \log_2 P(\mathcal{X}|H) \tag{23.11}$$

is minimized, which is conceptually equivalent to the posterior probability $P(H)P(\mathcal{X}|H)$, where H denotes either a specific parameter θ with a known probability function or a model M. The MML theory closely relates to the MAP approach Eq. (23.8) but actually has a difference. The MML theory considers the coding length of probability instead of considering density in the MAP approach [23.71].

The other typical example is the Minimum Description Length (MDL) theory [23.32, 23.52, 23.53, 23.54]. The basic idea is to represent a family of densities with an unknown parameter set θ, but a given density via a universal model that is able to imitate any particular density in the family. Such a universal model is described by a single probability distribution. Via the fundamental Kraft inequality, one constructs a code, e.g., a prefix code, for such a probability distribution, and, conversely, such a code defines a probability distribution. In this way, we can compare and select among different families by the code length of each family, which explains the name MDL.

A specific implementation of the MDL theory depends on how the code length is described. In the early stage, this length is actually specified via a two part coding method similar to the MML, and, thus, the corresponding implementation of the MDL is basically the same as the MML. Later, the mixture $p(\mathcal{X}|M_j)$ in Eq. (23.9) is used as the universal model for the family of M_j, and, thus, $\ln p(\mathcal{X}|M_j)$ is used as the code length. In this case, the corresponding implementation of the MDL is basically equivalent to the EV or BIC approach, as in Eq. (23.9). However, by selecting a non-informative uniform prior $p(\theta|M_j)$, and approximating the integral in getting the mixture $p(\mathcal{X}|M_j)$ via simplification, an MML code length and the average of all the MML code lengths for all distributions in a family become no different. Thus, this MDL implementation usually becomes identical to the MML. In the latest implementation of the MDL, a so-called normalized maximum likelihood (NML) model is used as the universal model, which leads to an improved code length and becomes different from both the MML and the EV/BIC approach [23.54]. Such a NML is also used to get a new estimate on the BF factor for model comparison.

Both the MML and the MDL can be regarded as specific implementations of more general algorithmic complexity, addressed by the celebrated Kolmogorov complexity. The connections discussed above between MML/MDL and MAP/EV actually reveal the deep relations between the fields of statistics, information theory, and computational complexity theory. Moreover, relations between the first two main types of efforts have also been explored in the past two decades, particularly on typical examples of the first type, such as AIC and cross validation, versus typical examples of the second type, such as MAP, EV/BIC, BF, etc. [23.62, 23.63, 23.64, 23.65, 23.66, 23.7, 23.15]. Furthermore, various applications of all the studies discussed above can be found in the literature, including linear regression, time series modeling, Markov chain [23.41], as well as complicated neural network modeling problems.

23.3 Bayesian Ying Yang Harmony Learning

The Bayesian Ying Yang (BYY) harmony learning was proposed in [23.109], and systematically developed in past years [23.84, 23.85, 23.82, 23.83, 23.80, 23.81, 23.76, 23.77]. This BYY harmony learning acts as a general statistical learning framework not only for understanding various dependence structures, such as generative structures, transform or mapping structures, finite mixture structures, temporal structures, and topological map structures, but also for tackling the key challenge, previously discussed, with a new learning mechanism that makes model selection either *automatically* implemented during parameter learning or *subsequently implemented after* parameter learning via a new class of model selection criteria obtained from this mechanism. Jointly with this BYY harmony learning, new types of regularization have also been proposed, namely a data smoothing technique that provides a new solution on the hyper-parameter in a Tikinov-like regularization [23.68], a normalization with a new conscience de-learning mechanism that has a nature similar to that of the rival penalized competitive learning (RPCL) [23.114], and a structural regularization imposing certain constraints by designing a specific forward structure in a BYY system, as well as a f-regularization by replacing $\ln(r)$ with a convex function $f(r)$. The details of the f-regularization has been introduced in the previous chapter in this book. The other three regularization approaches will be introduced in the following sections.

23.3.1 Bayesian Ying Yang Harmony Learning

As shown in Fig. 23.1, a BYY system considers coordinately learning two complement representations of the joint distribution $p(x, y)$:

$$p(u) = p(x, y) = p(y|x)p(x), \ q(u) = q(x, y) = q(x|y)q(y), \qquad (23.12)$$

basing on $p(x)$ that is estimated from a set of samples $\{x_t\}_{t=1}^N$, while $p(y|x)$, $q(x|y)$ and $q(y)$ are unknowns but subject to certain pre-specified structural constraints. In a compliment to the famous Chinese ancient Ying-Yang philosophy, the decomposition of $p(x, y)$ coincides the Yang concept with the visible domain by $p(x)$ regarded as a Yang space and the forward pathway by $p(y|x)$ as a Yang pathway. Thus, $p(x, y)$ is called Yang machine. Similarly, $q(x, y)$ is called Ying machine with the invisible domain by $q(y)$ regarded as a Ying space and the backward pathway by $q(x|y)$ as a Ying path.

On one hand, we can interpret that each x is generated from an invisible inner representation y via a backward path distribution $q(x|y)$ or called a generative model

$$q(x) = \int q(x|y)q(y)\mu(dy) \qquad (23.13)$$

that maps from an inner distribution $q(y)$. In this case, $p(y|x)$ is not explicitly specified or said being free to be specified, while two pre-specified parametric models $q(x|y)$ and $q(y)$ form a backward path to fix the observations of x. We say that the Ying-Yang system in this case has a backward architecture (shortly B-architecture).

On the other hand, we can interpret that each x is represented as being mapped into an invisible inner representation y via a forward path distribution $p(y|x)$ or called a representative model

$$p(y) = \int p(y|x)p(x)\mu(dx) \qquad (23.14)$$

that matches the inner density $q(y)$. In this case, $q(x|y)$ is not explicitly specified or said being free to be specified. Forming a forward path, $p(x)$ is estimated from a given set of samples and then is mapped via pre-specified parametric model $p(y|x)$ into $p(y)$ by Eq. (23.14) to match a pre-specified parametric model $q(y)$. We say that the Ying-Yang system in this case has a forward architecture (shortly F-architecture).

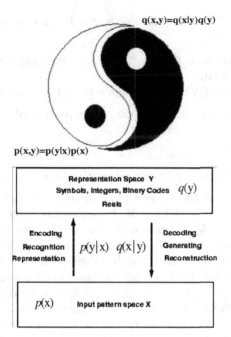

Fig. 23.1. Bayesian Ying Yang System

Moreover, the above two architectures can be combined with $p(y|x)$, $q(x|y)$ and $q(y)$ are all pre-specified parametric models. In this case, we say that the Ying-Yang system in this case has a Bi-directional architecture (shortly BI-architecture).

As discussed in [23.80] and in the previous chapter of this book, types of representation space of y specify types of learning functions that the BYY system can implement, while the above three architectures characterize the performances and computing costs of implementation.

The name of BYY system not just came for the above direct analogy between Eq. (23.12) and the Ying-Yang concept, but also is closely related to that the principle of making learning on Eq. (23.12) is motivated from the well known harmony principle of the Ying-Yang philosophy, which is different from making $p(x)$ by Eq. (23.13) fit a set of samples $\{x_t\}_{t=1}^{N}$ under the ML principle [23.57] or its approximation [23.58, 23.31, 23.17, 23.18] as well as simply the least mean square error criterion [23.115], and also different from making $q(y)$ by Eq. (23.15) satisfy certain pre-specified properties such as maximum entropy [23.8] or matching the following independent density [23.6]:

$$q(y) = \prod_{j=1}^{m} q(y^{(j)}). \tag{23.15}$$

Under this harmony principle, the Ying-Yang pair by Eq. (23.12) is learned coordinately such that the pair is matched in a compact way as the Ying-Yang sign shown in Fig. 23.1. In other words, the learning is made in a twofold sense that

- The difference between the two Bayesian representations in Eq. (23.12) should be minimized.
- The resulted entire BYY system should be of the least complexity.

Mathematically, this principle can be implemented by [23.109, 23.82, 23.80]

$$\max_{\theta,m} H(\theta, m), \tag{23.16}$$

$$H(\theta, m) = H(p\|q) = \int p(y|x)p(x) \ln [q(x|y)q(y)]\mu(dx)\mu(dy) - \ln z_q,$$

where θ consists of all the unknown parameters in $p(y|x)$, $q(x|y)$, and $q(y)$ as well as $p(x)$ (if any), while m is the scale parameter of the inner representation y. The task of determining θ is called *parameter learning*, and the task of selecting m is called *model selection* since a collection of specific BYY systems by Eq. (23.12) with different values of m corresponds to a family of specific models that share a same system configuration but in different scales. Furthermore, the term Z_q imposes regularization on learning [23.77, 23.80, 23.83], via two types of implementation. One is called data smoothing that provides a new solution to the hyper-parameter for a Tikinov-like regularization [23.68], and the other is called normalization that causes a new conscience de-learning mechanism similar to that of the rival penalized competitive learning (RPCL) [23.114, 23.83, 23.80].

As described in the previous chapter, considering the harmony measure

$$H(p\|q) = \int p(u) \ln q(u)\mu(du) - \ln z_q. \tag{23.17}$$

Least complexity nature means that $\max_p H(p\|q)$ with q fixed pushes p toward its simplest form

$$p(u) = \delta(u - u_\tau). \tag{23.18}$$

Now, only $p(x)$ is fixed at a non-parametric estimate but $p(y|x)$ is either free in a B-architecture or a parametric form in a BI-architecture and, thus, will be pushed into its least complexity form due to the least complexity nature by Eq. (23.18). For example, in a B-architecture $p(y|x)$ will be determined by $\max_{p(y|x)} H(p\|q)$, resulting in the following least complexity form:

$$p(y|x) = \delta(y - y(x)), \quad y(x) = arg \max_y [q(x|y)q(y)]. \tag{23.19}$$

On the other hand, the matching nature of harmony learning will further push $q(x|y)$ and $q(y)$ toward their corresponding least complexity forms. In other words, the least complexity nature and the matching nature collaborate to make model selection possible such that m is appropriately determined.

As described in [23.80], Eq. (23.16) introduces a new mechanism that makes model selection implemented

– either automatically during the following parameter learning with m initialized large enough:

$$\max_{\theta} \ H(\theta), \quad H(\theta) = H(\theta, m), \qquad (23.20)$$

which makes θ take a specific value such that m is effectively reduced to an appropriate one.

– or via the following type of model selection criteria obtained from this mechanism:

$$\min_{m} J(m), \quad J(m) = -H(\theta^*, m), \qquad (23.21)$$

which is implemented via parameter learning for θ^* at each value of m that is enumerated from a small value incrementally.

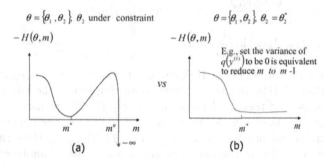

Fig. 23.2. (a) Model selection made after parameter learning on every m in a given interval $[m_d, m_u]$, (b) Automatic model selection with parameter learning on a value m of large enough.

The above feature is not shared by the existing approaches in literature. By the conventional approaches, parameter learning and model selection are made in a two-phase style. First, parameter learning is made usually under the maximum likelihood principle. Then, model selection is made by a different criterion, e.g., AIC, MDL, etc. These model selection criteria are usually not good for parameter learning, while the maximum likelihood criterion is not good for model selection, especially on a small size of training samples.

Specifically, the above parameter learning for getting θ^* can be implemented in help of either Eq. (23.20) or the following Kullback divergence based parameter learning:

$$\min_{\theta} \ KL(\theta) = \int p(y|x)p(x) \ln \frac{p(y|x)p(x)}{q(x|y)q(y)} \mu(dx)\mu(dy). \qquad (23.22)$$

Moreover, the implementation of both Eq. (23.20) and Eq. (23.22) can be made by alternatively performing the following two steps:

Ying step: fixing $p(x,y)$, update unknowns in $q(x,y)$,
Yang step: fixing $q(x,y)$, update unknowns in $p(x,y)$,   (23.23)

which is called the Ying-Yang alternative procedure. It is guaranteed that either of $-H(\theta)$ and $KL(\theta)$ gradually decreases until becomes converged. The details are referred to [23.80, 23.75].

As above discussed, parameter learning by Eq. (23.20) usually leads to an automatic model selection and, thus, there is no need to implement the selection by Eq. (23.21). However, for certain learning tasks, the inner representation is pre-specified to be uniform across both different objects and different dimensions [23.80]. In these cases, automatic model selection will not happen during learning by Eq. (23.20) in the first phase, we need to implement Eq. (23.21) in the second phase.

Particularly, on a B-architecture, the minimization of the above $KL(\theta)$ with respect to a free $p(y|x)$ will result in

$$p(y|x) = \frac{q(x|y)q(y)}{q(x)}, \ q(x) = \int q(x|y)q(y)\mu(dy),$$

$$KL(\theta) = \int p(x)\ln\frac{p(x)}{q(x)}\mu(dx), \tag{23.24}$$

which becomes equivalent to ML learning on $q(x)$ when $p(x) = p_0(x)$ is given by Eq. (23.3) [23.109]. In this case, we actually implement the ML learning in the first phase and then model selection by Eq. (23.21) in the second phase.

Without the least complexity nature by Eq. (23.18), the implementation of Eq. (23.22) will not lead to a case of Fig. 23.2(b), and, thus, there is no need to impose the assumption that $q(y)$ comes from a family with equal variances among components.

23.3.2 Structural Inner Representations

The inner representation by Eq. (23.15) is a typical example but not an only example. Actually it is a degenerated case of the multiple modular inner representation discussed by Eq. (1.16) in the previous chapter. That is,

$$q(\mathbf{y}) = q(y, \ell) = q(y|\ell)q(\ell), \ q(\ell) = \sum_{j=1}^{k} \alpha_j \bar{\delta}(\ell - j), \ \alpha_\ell \geq 0, \ \sum_{\ell=1}^{k} \alpha_\ell = 1,$$

$$q(y|\ell) = \prod_{j=1}^{m_\ell} q(y^{(j)}|\ell), \ y = [y^{(1)}, \cdots, y^{(m_\ell)}]^T,$$

$$\text{where } \bar{\delta}(u) = \begin{cases} 1, if \ u=0, \\ 0, otherwise; \end{cases} \tag{23.25}$$

from which we return to Eq. (23.15) when $k = 1$.

When $k \geq 2$, the above Eq. (23.25) also include the following two useful special cases:

$- \ q(\mathbf{y}) \ = \ \int q(y|\ell)q(\ell)dy \ = \ q(\ell)$ In this case, we have that $q(x)$ by Eq. (23.13) becomes the following finite mixture

$$q(x) = \sum_{\ell=1}^{k} q(\ell)q(x|\ell). \tag{23.26}$$

which is a weighted sum of k different component densities of $q(x|\ell)$. Moreover, $\max_{\theta,m} H(\theta,m)$ by Eq. (23.16) includes directly maximizing $\ln q(y) = \ln q(\ell)$ that not only contains the information of making k selected in a way similar to Eq. (23.21) but also can drive an extra α_ℓ toward zero such that an appropriate k can be automatically decided during learning.

$- q(y) = \sum_{\ell=1}^{k} q(y|\ell)q(\ell)$ In this case, we still have that $q(x)$ by Eq. (23.13) takes the format of Eq. (23.26) but now with

$$q(x|\ell) = \int q(x|y)q(y|\ell)dy. \tag{23.27}$$

In other words, the k different component densities share a common part $q(x|y)$ but with differences in $q(y|\ell)$. Now $\ln q(y)$ still contains the information that can select k in a way similar to Eq. (23.21), e.g., we have $\ln q(y) = -\ln k + \ln \sum_{\ell=1}^{k} q(y|\ell)$ when $q(\ell) = 1/k$. However, maximizing $\ln q(y)$ will not necessarily drive an extra individual α_ℓ toward zero. Thus, k will not be automatically decided during learning.

Generally, Eq. (23.25) covers a representation space with k modules and each module locally consists of m_ℓ independent components. As discussed in [23.80] and also the previous chapter in this book, $\max_{\theta,m} H(\theta,m)$ by Eq. (23.16) not only let $\{k, m_\ell\}$ to be selected in a way similar to Eq. (23.21) but also can drive both an extra α_ℓ and the variance of an extra component in $q(y|\ell)$ toward zero such that appropriate $\{k, m_\ell\}$ can be automatically decided during learning.

Specifically, when y is real and non-Gaussian, each component density can be modeled by a Gaussian mixture [23.103, 23.105, 23.82]

$$q(y^{(j)}|\ell) = \sum_{i=1}^{\kappa_{j,\ell}} \beta_{ji\ell} G(y^{(j)}|\mu_{ji\ell}, \lambda_{ji\ell}), \quad \sum_{i=1}^{\kappa_{j,\ell}} \beta_{ji\ell} = 1, 0 \le \beta_{ji\ell} \le 1. \tag{23.28}$$

In this case, the information about $\kappa_{j,\ell}$ is contained in $\beta_{ji\ell}$ and Eq. (23.21) can be used for selecting $\kappa_{j,\ell}$. For example, when $q(\ell) = 1/k$, and $\beta_{ji\ell} = 1/\kappa_{j,\ell} = \kappa_\ell$, we have $\ln q(\mathbf{y}) = \ln q(y, \ell)$ with

$$\ln q(y, \ell) = -\ln k - \sum_{j=1}^{m_\ell} \ln \kappa_\ell + \sum_{j=1}^{m_\ell} \ln \left[\sum_{i=1}^{\kappa} G(y^{(j)}|\mu_{ji\ell}, \lambda_{ji\ell}) \right], \tag{23.29}$$

which does include $-\ln k - \sum_{j=1}^{m_\ell} \ln \kappa_\ell$ that is in favor of smaller size of k, m_ℓ, κ_ℓ. However, maximizing $\ln q(y)$ is made via maximizing $\ln \left[\sum_{i=1}^{\kappa_{j,\ell}} \beta_{ji\ell} G(y^{(j)}|\mu_{ji\ell}, \lambda_{ji\ell}) \right]$ that will not necessarily drive an individual $\beta_{ji\ell}$ toward zero and, thus, $\kappa_{j,\ell}$ will not be automatically decided during learning.

Automatic selection on $\kappa_{j,\ell}$ can be made by introducing random variables $z_j = 1, \cdots, \kappa_{j,\ell}, j = 1, \cdots, m_\ell$ and re-organizing the structure of the inner representation space as follows

$$q(\mathbf{y}) = q(\ell) \prod_{j=1}^{m_\ell} q(y^{(j)}|\ell, z_j = i)q(z_j = i|\ell),$$

$$q(y^{(j)}|\ell, z_j = i) = G(y^{(j)}|\mu_{ji\ell}, \lambda_{ji\ell}), \quad q(z_j = i|\ell) = \beta_{ji\ell}. \tag{23.30}$$

In this case, maximizing $\ln q(\mathbf{y})$ consists of directly maximizing $\ln q(z_j = i|\ell)$ $= \ln \beta_{ji\ell}$ such that each $\kappa_{j,\ell}$ can be selected either via Eq. (23.21) or automatically driving an extra $\beta_{ji\ell}$ toward zero during learning.

It should be noted that different representation spaces also lead to differences on implementing Eq. (23.19). To get a further insight, we here focus on a special case that $k = 1$ and $q(x|y) = G(x|Ay, \Sigma)$, i.e., a linear factor model $x = Ay + e$ from real independent factors by Eq. (23.25) of nonGaussians. This is a typical model for the so-called noisy ICA [23.78]. It follows from Eq. (23.28) that $q(y^{(j)}) = \sum_{i=1}^{\kappa_j} \beta_{ji} G(y^{(j)}|\mu_{ji}, \lambda_{ji})$ and the problem of Eq. (23.19) is a continuous nonlinear optimization problem that has to be tackled by an iterative algorithm [23.82, 23.78]. While it follows from Eq. (23.30) and Eq. (23.19) that $arg\max_{\mathbf{y}}[q(x|\mathbf{y})q(\mathbf{y})] = arg\max_{\mathbf{y}}[\ln G(x|Ay, \Sigma) + \ln q(\mathbf{y})]$ and $\ln q(\mathbf{y}) = \sum_{j=1}^{m}[\ln G(y^{(j)}|\mu_{jz_j}, \lambda_{jz_j}) + \ln \beta_{jz_j}]$. Thus, Eq. (23.19) becomes

$$\max_z \{ \sum_{j=1}^{m} \ln \beta_{jz_j} + \max_y L_z(x, y) \}$$

$$\max_y L_z(x, y) = [\ln G(x|Ay, \Sigma) + \sum_{j=1}^{m} \ln G(y^{(j)}|\mu_{jz_j}, \lambda_{jz_j})], \tag{23.31}$$

where $z = [z_1, \cdots, z_m]^T$.

With x and $z_1, \cdots, z_m$ fixed, $\max_y L_z(x, y)$ is a quadratic optimization that can be analytically solved as follow:

$$y_z(x) = [\Lambda_z^{-1} + A^T \Sigma^{-1} A]^{-1}[A^T \Sigma^{-1}x + \Lambda_z^{-1}\mu_z],$$
$$\mu_z = [\mu_{1z_1}, \cdots, \mu_{mz_m}]^T, \quad \Lambda_z = diag[\lambda_{1z_1}, \cdots, \lambda_{mz_m}]. \tag{23.32}$$

Then, we can implement the following discrete optimization:

$$z^* = \max_z [L_z(x, y_z(x)) + \sum_{j=1}^{m} \ln \beta_{jz_j}]. \tag{23.33}$$

As a result, the solution of Eq. (23.19) is simply $y_{z^*}(x), z^*$.

Being different from an iterative algorithm [23.82, 23.78], the solution by Eq. (23.33) can be computed analytically. A direct implementation of Eq. (23.33) needs a number $\prod_{j=1}^{m} \kappa_j$ of comparisons. However, this cost can be further reduced by exploring the function structure of $L_z(x, y) + \sum_{j=1}^{m} \ln \beta_{jz_j}$.

23.4 Regularization Versus Model Selection

23.4.1 ML, HL, and Z-Regularization

As discussed in Sect. 23.1, regularization and model selection are two different strategies for tackling the problem of a finite size of samples. Model selection prefers a model that has least complexity for which a compact inner representation is aimed at such that extra representation space can be released. In contrast, regularization is imposed on a model that has a fixed scale of representation space with its complexity larger than needed such that inner representation can spread as uniformly as possible over all the representation space with a distribution that is as simple as possible, which, thus, becomes equivalent to a model with a reduced complexity.

The harmony learning by Eq. (23.20) attempts to compress the representation space via the least complexity that is demonstrated with a winner-take-all (WTA) competition by Eq. (23.19). This type of parameter learning aims at a compact inner representation with an automatic model selection by discarding extra representation space during parameter learning. However, there is no free lunch. The WTA operation by Eq. (23.19) locally per sample will make the learning become sensitive to the initialization of parameters and the way that samples are presented, resulting in that samples are over-aggregated in a small representation space. It usually leads to a local maximum solution for Eq. (23.20). Pre-specifying a uniform inner representation can regularizes the WTA operation. However, the feature of automatic model selection is also lost since the representation space scale is already fixed. Thus, model selection should be made by Eq. (23.21) in the second phase.

With a soft competition by Eq. (23.24) in place of the WTA competition by Eq. (23.19), the ML learning, or equivalently the Kullback divergence based learning (shortly KL learning) by Eq. (23.22) with a B-architecture and an empirical input density by Eq. (23.3), provides a more spread inner representation that improves the local maximum problem. Again, there is no free lunch since its model selection ability is weak, especially on a small size of samples. Thus, making model selection by Eq. (23.21) is needed in the second phase too.

Instead of the above two phase style, regularization to the WTA by Eq. (23.19) may also be imposed to the harmony learning (Shortly HL) by Eq. (23.20) such that automatic model selection still occurs via either some external help or certain internal mechanism.

Externally, we can combine the KL learning by Eq. (23.22) with the harmony learning by Eq. (23.20), in the following three ways:

- The simplest way is to make the KL learning by Eq. (23.22) with the resulted parameters as the initialization of the harmony learning by Eq. (23.20).
- The other way suggested in [23.83] is to let $H(\theta)$ in Eq. (23.20) replaced with $(1 - \lambda)H(\theta) - \lambda KL(\theta)$ in help of an appropriate λ.

– Moreover, with $\lambda > 0$ gradually reducing toward zero from a given value such that the regularization role by Eq. (23.22) takes effect at the beginning and then gradually decades as learning goes. That is, we combine KL and HL in a simulated annealing way such that KL is implemented in an early period of learning and is gradually switched to HL as learning goes [23.83, 23.77]. The disadvantage is that the computing cost is very high since parameter learning has to be repeatedly conducted.

Internally, regularization to the WTA by Eq. (23.19) is imposed during the HL learning by Eq. (23.20) via either a BI-architecture or the role of z_q.

Instead of letting $p(y|x)$ free to be decided by Eq. (23.19), we consider a BI-architecture with $p(y|x)$ designed in a structure such that it is not able to become the WTA by Eq. (23.19). Generally, for $p(y|x)$ in the form of

$$p(u|v) = \sum_{j=1}^{n} \beta_j(v)p_j(u|v), \ \sum_{j=1}^{n} \beta_j(v) = 1, \ \beta_j(v) \geq 0,$$
$$p_j(u|v) = G(u|f_j(v|\theta_{u|v,j}), \Sigma_{u|v,j}), \qquad (23.34)$$

the harmony learning by Eq. (23.20) will push it toward the following form of least complexity [23.84, 23.82, 23.80]

$$p(u|v) = \sum_{j=1}^{n} \beta_j(v)\delta(u - f_j(v, \theta_{u|v,j})), \ \sum_{j=1}^{n} \beta_j(v) = 1, \ \beta_j(v) = 0, \ or \ 1,$$

unless extra constraints are imposed to prevent $\Sigma_{u|v,j} \to \varepsilon^2 I$ and ε^2 tends to zero. Moreover, Eq. (23.19) is simplified into

$$p(y|x) = \delta(y - y(x)), \ y(x) = f_{j^*(x)}(x|\theta_{y|x,j^*}(x)),$$
$$j^*(x) = arg \max_j [q(x|y)q(y)]_{y=f_j(x|\theta_{y|x,j})}, \qquad (23.35)$$

where the maximum is searched by simply enumerating n possibilities. Thus, regularization can also be observed from the perspective that the number of local maxima considerably reduces in comparison with Eq. (23.19). However, there is no free lunch too. The problem is transferred to the difficulty of per-specifying the function form of each $y = f_j(x|\theta_{y|x,j})$. If each function is too simple, the representation ability of $p(y|x)$ is limited and is far from the optimal one. If it is too complicated with too much free parameters, it creates certain problems that need regularization to be imposed too.

Regularization to the WTA by Eq. (23.19) may also be imposed via the so-called z-regularization. This type of regularization can be implemented either by data smoothing or by normalization.

For data smoothing regularization, one simple way is only considering smoothing on x via $p(x) = p_{h_x}(x)$ by Eq. (23.3) with $z_q = \sum_{t=1}^{N} p_{h_x}(x_t)$. As discussed in [23.82, 23.83, 23.80], the regularization is made via $h_x^2 > 0$ while h is determined in help of $-\ln z_q$. Moreover, a smoothing can be imposed on y via modifying $p(y|x) = \delta(y - y(x))$ in Eq. (23.19). For example, in the case of only one object (i.e., $k = 1$), we let $p(y|x) = G(y|y_t, h_y^2 I)$ and $z_q = (2\pi h_y)^{-0.5m} \sum_{t=1}^{N} p_h(x_t)$.

For normalization regularization, we have also different implementations.

When y takes either a discrete value $1, \cdots, k$ or is a binary vector $y = [y^{(1)}, \cdots, y^{(m)}]$, and also when $q(x|y)$ and $q(y)$ are both Gaussian, we can consider the constraint $\sum_{t=1}^{N} \int \frac{q(x_t|y)q(y)}{z_q} \mu(dy) = 1$ because the integral over y is either a summation or analytically solvable. Thus, we have

$$z_q = \sum_{t=1}^{N} q(x_t), \quad q(x_t) = \int q(x_t|y)q(y)\mu(dy). \tag{23.36}$$

In other cases, this integral over y is difficult to compute. Even when it becomes a computable summation for a binary vector $y = [y^{(1)}, \cdots, y^{(m)}]$, the computing cost will increase exponentially with m.

One solution is to let the integral over the entire domain of y to be approximated by a summation on a set Y_t that consists of a few number of samples y_τ as follows:

$$q(x_t) = \gamma_t \sum_{y_\tau \in Y_t} q(x|y_\tau)q(y_\tau), \quad \gamma_t = 1/\sum_{y_\tau \in Y_t} q(y_\tau), \tag{23.37}$$

where γ_t makes $q(y_\tau)/\sum_{y_\tau \in Y_t} q(y_\tau)$ represent discrete probabilities that weight $q(x|y_\tau)$ such that $q(x_t)$ is closer to a marginal density.

One other solution is consider $\sum_{t=1}^{N} \sum_{y_\tau \in Y_t} \frac{q(x_t|y_\tau)q(y_\tau)}{z_q} \mu(dy) = 1$, which results in

$$z_q = \sum_{t=1}^{N} \sum_{y_\tau \in Y_t} q(x_t|y_\tau)q(y_\tau). \tag{23.38}$$

The set Y_t can be obtained according to $p(y|x)$. One way is randomly picking a set of samples of y according to $p(y|x)$. The other way is getting only one $y_t = y(x_t)$ for each x_t via either the peak point (e.g., by Eq. (23.19)) or the mean point (e.g., $y(x)$ by Eq. (23.35)) of $p(y|x)$.

In the cases that there is only one sample y_t in Y_t, it follows from Eq. (23.36) and Eq. (23.38) that

$$z_q = \begin{cases} \sum_{t=1}^{N} q(x_t|y_t)q(y_t), & \text{(a) by Eq. (23.38)}, \\ \sum_{t=1}^{N} q(x_t|y_t), & \text{(b) by Eq. (23.36)}. \end{cases} \tag{23.39}$$

Further with $p(x) = p_{h_x}(x)$ given by Eq. (23.3), $H(p\|q)$ by Eq. (23.17) either on a B-architecture with Eq. (23.19) or on a BI-architecture with Eq. (23.35) can be unified into the following representation

$$H(p\|q) = \tag{23.40}$$
$$\frac{1}{N}\sum_{t=1}^{N} \int \delta(y - y(x_t)) \ln [q(x_t|y)q(y)]\mu(dy) - \ln z_q + 0.5h_x^2 \pi_q,$$

$$\pi_q =$$
$$\frac{1}{N}\sum_{t=1}^{N} Tr[\frac{\partial^2 \ln q(x|y_t)}{\partial x \partial x^T}]_{x=x_t},$$

and we have the following gradient

$$\nabla_\theta H(p\|q) = 0.5h_x^2 \nabla_\theta \pi_q$$

$$+\tfrac{1}{N}\sum_{t=1}^{N}\int[\delta(y-y(x_t))-\eta_t(y)]\nabla_\theta\ln[q(x_t|y)q(y)]\mu(dy),$$

$$\eta_t(y)=\begin{cases} 0, & z_q=1, \\ \bar{\delta}(h_x)\sum_{y_\tau\in Y_t}\frac{q(x_t|y_\tau)q(y_\tau)}{z_q}\delta(y-y_\tau), & z_q \text{ by Eq. (23.36)}, \\ \bar{\delta}(h_x)\sum_{y_\tau\in Y_t}\gamma_t\frac{q(x_t|y_\tau)q(y_\tau)}{z_q}\delta(y-y_\tau), & z_q \text{ by Eq. (23.37)}, \end{cases}$$

$$\bar{\delta}(h_x)=\begin{cases} 1, & h_x=0, \\ 0, & h_x>0. \end{cases} \qquad\qquad (23.41)$$

For $h_x>0$, we have $\bar{\delta}(h_x)=0$ and $\eta_t(y)=\delta(y-y(x_t))$ for all the cases. In this case, the data smoothing regularization is imposed via $0.5h_x^2\nabla_\theta\pi_q$ and an appropriate regularization strength h_x^2 is determined via maximizing $-\ln z_q+0.5h_x^{2}\nabla_\theta\pi_q$ with $z_q=\sum_{t=1}^{N}p_{h_x}(x_t)$. The details are referred to [23.84, 23.85, 23.82, 23.83, 23.80, 23.81, 23.76, 23.77].

For $h_x=0$, we have $0.5h_x^{2}\nabla_\theta\pi_q=0$ and $\bar{\delta}(h_x)=1$. In this case, the normalization regularization is imposed via $-\ln z_q$, which can be observed via the difference of $\eta_t(y)$ in Eq. (23.41). It introduces a degree of conscience de-learning on each updating direction $\nabla_\theta\ln[q(x_t|y)q(y)]$ to avoid over-fitting on each sample pair x_t, y_τ. With and without $-\ln z_q$ in action, $\nabla_\theta H(p\|q)$ takes the same format, and also adaptive updating can be made in the form of $\eta_t(y)\nabla_\theta\ln[q(x_t|y)q(y)]$ per sample x_t.

23.4.2 KL-λ-HL Spectrum

The KL learning by Eq. (23.22) on a BYY system is not limited to just the ML learning. Even on a B-architecture with $p(y|x)$ determined by Eq. (23.24), letting $p(x)=p_h(x)$ by Eq. (23.3) will make the KL learning by Eq. (23.22) perform a regularized ML learning.

Moreover, the KL learning by Eq. (23.22) on a BI-architecture was suggested in [23.109] with $p(y|x)$ in a given parametric family $\mathcal{P}_{y|x}^S$. If the posteriori estimation by Eq. (23.24) is contained in this family $\mathcal{P}_{y|x}^S$, the situation will be equivalent to the KL learning by Eq. (23.22) with a B-architecture; if not, the posteriori estimation by Eq. (23.24) will be approximated by the closest one within the family $\mathcal{P}_{y|x}^S$. This architecture leads to an advantage that the computing difficulty on the integral in $p(y|x)$ by Eq. (23.24) is avoided by an easy implementing parametric model. In its sprit, this is equivalent to those approaches called variational approximation to the ML learning on $q(x)$ [23.58].

Beyond the approximation purpose, studies on the KL learning by Eq. (23.22) on a BI-architecture were also made along two directions since 1996 [23.110]. One is to design a parametric model that makes the inner representation more spreading than that of $p(y|x)$ by Eq. (23.24) such that ML learning is further regularized. The other is to design a parametric model that makes the inner representation more concentrated such that it tends to facilitate automatic model selection. One family of such designs is as follows:

$$p(y|x) = \frac{\psi(q(x|y), q(y))}{\int \psi(q(x|y), q(y))\mu(dy)}, \qquad (23.42)$$

which returns to $p(y|x)$ by Eq. (23.24) for the ML learning when $\psi(\xi, \eta) = \xi\eta$. It makes the inner representation either more spreading for a regularized ML learning, e.g., when $\psi(\xi, \eta) = \lambda_1 \ln \xi + \lambda_2 \ln \eta, \lambda_1 \geq 0, \lambda_2 \geq 0$, or more concentrated to facilitate model selection, e.g., when $\psi(\xi, \eta) = e^{\lambda_1 \xi} + e^{\lambda_2 \eta}, \lambda_1 \geq 0, \lambda_2 \geq 0$ that will make $p(y|x)$ tend to Eq. (23.19) as $\lambda_1 = \lambda_2 \to \infty$. A simply form can even be $\psi(\xi, \eta) = (\xi\eta)^\lambda$ which varies from spreading cases to concentrated cases as λ increases from 0 to ∞.

As discussed in the previous subsection, the HL learning with the WTA by Eq. (23.19) on a B-architecture will also be regularized by a BI-architecture with $p(y|x)$ in a more spreading representation. Except those extreme cases that become equivalent to the WTA by Eq. (23.19), e.g., when $\psi(\xi, \eta) = (\xi\eta)^\lambda$ with $\lambda \to \infty$, $p(y|x)$ by Eq. (23.42) generally leads to a regularized harmony learning. Even when $\psi(\xi, \eta) = \xi\eta$, we will not be lead to the ML learning but to a regularized HL learning in a B-architecture with a free $p(y|x)$ replaced by a posteriori estimation by Eq. (23.24).

From the above discussion, we observe that the KL learning by Eq. (23.22) and the HL learning by Eq. (23.20) become closely related via appropriately designing $p(y|x)$. The difference lays in the following term E_p:

$$E_p = -\int p(y|x)p(x) \ln [p(y|x)p(x)]\mu(dx)\mu(dy) + \ln z_p. \qquad (23.43)$$

When $p(y|x) = \delta(y - y(x))$ is deterministic and $p(x) = p_0(x)$ given by Eq. (23.2), we have $E_p = 0$. That is, the KL learning by Eq. (23.22) and the HL learning by Eq. (23.20) becomes equivalent on this special BI-directional architecture. When $p(y|x)$ is free to be determined via learning, the difference is that the HL learning by Eq. (23.20) automatically results in a deterministic type $p(y|x)$ by Eq. (23.19) while the KL learning by Eq. (23.22) will result in a non- deterministic type $p(y|x)$ by Eq. (23.24).

Moreover, the KL learning by Eq. (23.22) and the HL learning by Eq. (23.20) are also related for those $p(y|x)$ such that $E_p = c \neq 0$ becomes a constant irrelevant to any unknown parameters in θ, e.g., with $p(x) = p_0(x)$ given by Eq. (23.2) we have

$$E_p = 0.5m \ln (2\pi h_y^2) - \ln N, \ for \ p(y|x) = G(y|y(x), h_y^2 I). \qquad (23.44)$$

The KL learning by Eq. (23.22) and the HL learning by Eq. (23.20) are no longer equivalent when h_y, m are unknown to be determined. In the special case that h_y, m are prefixed in advance, the KL learning by Eq. (23.22) further becomes equivalent to

$$\max_{\theta, \ s.t. \ E_p = c \neq 0} H(\theta). \qquad (23.45)$$

The above discussions also apply to BYY systems with a F-architecture, with a free $p(x|y)$ decided by

$$p(x|y) = p(y|x)p(x)/p(y), \ p(y) = \int p(y|x)p(x)\mu(dx), \qquad (23.46)$$

such that we have

$$KL(\theta) = \int p(y) \ln \frac{p(y)}{q(y)} \mu(dy), \ H(\theta) = -KL(\theta) - E_p, \qquad (23.47)$$

with E_p given in Eq. (23.43). When $q(y)$ is a uniform distribution, minimizing this $KL(\theta)$ becomes equivalent to maximizing the entropy, which maximizes the information transfer from input data to its inner representation via the forward path. Generally, $-KL(\theta)$ describes the incremental of information contained in the representation of y after this information transfer. Thus, minimizing this $KL(\theta)$ is equivalent to making this incremental maximized via maximizing this information transfer against upon the information already in the inner representation. Particularly, when $p(y|x) = \delta(y - Wx)$, $q(y)$ by Eq. (23.15), and $p(x) = p_0(x)$ by Eq. (23.2), both the KL learning by Eq. (23.22) and the HL learning by Eq. (23.20) become equivalent to the minimum mutual information approach for ICA that was previously discussed after Eq. (23.15). All these cases are featured by maximum information transfer and, thus, shortly called as the Max-Inform approach.

Generally, the KL learning by Eq. (23.22) and the HL learning by Eq. (23.20) are different for those $p(y|x)$ that do not satisfy $E_p = c$. This difference can also be observed from the learning results in those cases that the KL learning by Eq. (23.22) results in only $p(x, y) = q(x, y)$, but does not the least complexity nature, while the HL learning by Eq. (23.20) results in not only $p(x, y) = q(x, y)$ but also a minimized entropy

$$H_q = -\int q(x, y) \ln q(x, y) \mu(dx) \mu(dy) \ or \ equivalently$$
$$H_p - \int p(x, y) \ln p(x, y) \mu(dx) \mu(dy), \qquad (23.48)$$

which makes model toward a least complexity.

In a summary, the family of KL learning by Eq. (23.22) and the family of HL learning by Eq. (23.20) do share an intersection that consists of interesting models. However, two families are different with each containing useful models outside this intersection. The union of the two families consists of a spectrum of learning models, ranging from regularized ML or Max-Inform versions to the original ML or Max-Inform versions, and then reaching regularized versions of HL learning and finally to the HL learning. In addition, as discussed previously in Sect. 23.4.1, regularized versions of ML or Max-Inform and the HL learning are also obtainable by the role of z_p and z_q via either data smoothing or normalization [23.77, 23.83, 23.84, 23.94].

This spectrum can be extended via a convex combination $\lambda KL(\theta) + (1 - \lambda)H(\theta)$, $0 \le \lambda \le 1$. Its minimization is equivalent to the KL learning when $\lambda = 1$ and then tends to the HL learning as λ decreases from 1 to 0. As λ varies from 0 to 1, the HL learning is regularized toward to the KL learning.

The combination may go beyond the above spectrum, which can be observed by considering a B-architecture with $p(y|x)$ free. It follows from $H(\theta) = -KL(\theta) - E_p$ that

$$\lambda KL(\theta) - (1 - \lambda)H(\theta) = \lambda[E_p + \frac{1}{\lambda}H(\theta)]. \tag{23.49}$$

Ignoring the regularization role of z_p and z_q by setting $z_p = 1, z_q = 1$, we can further get

$$E_p + \frac{1}{\lambda}H(\theta) = \int p(y|x)p(x)\ln \frac{p(y|x)p(x)}{[q(x|y)q(y)]^{\frac{1}{\lambda}}}\mu(dx)\mu(dy)$$
$$= \int p(y|x)p(x)\ln \frac{p(y|x)}{p_Q(y|x)}\mu(dx)\mu(dy) + \int p(x)\ln \frac{p(x)}{\hat{q}(x)}\mu(dx)\mu(dy),$$
$$p_Q(y|x) = [q(x|y)q(y)]^{\frac{1}{\lambda}}/\hat{q}(x), \ \hat{q}(x) = \int [q(x|y)q(y)]^{\frac{1}{\lambda}}\mu(dy). \tag{23.50}$$

which was firstly proposed in [23.101]. Its minimization with respect to a free $p(y|x)$ will lead to

$$p(y|x) = p_Q(y|x) = \frac{[q(x|y)q(y)]^{\frac{1}{\lambda}}}{\hat{q}(x)},$$
$$E_p + \frac{1}{\lambda}H(\theta) = \int p(x)\ln \frac{p(x)}{\hat{q}(x)}\mu(dx), \tag{23.51}$$

where $p(y|x)$ here is a special case of Eq. (23.42) with $\psi(\xi, \eta) = (\xi\eta)^{\frac{1}{\lambda}}$ that makes the inner representation more concentrated than $p(y|x)$ by Eq. (23.24). Minimizing $\int p(x)\ln \frac{p(x)}{\hat{q}(x)}\mu(dx)$ is different from both the KL learning by Eq. (23.22) with $p(y|x) = p_Q(y|x)$ and from the ML learning on $\hat{q}(x)$ since $\int \hat{q}(x)\mu(dx) \neq 1$.

The spectrum can be further extended by considering a linear combination $\lambda KL(\theta) - (1 - \lambda)H(\theta)$ with $\lambda > 1$, which is no longer a convex combination since $1 - \lambda < 0$. However, it is still meaningful by observing $E_p + \frac{1}{\lambda}H(\theta)$, and Eq. (23.51) still applies. The difference is that $1/\lambda < 1$ makes the inner representation more spreading than that of the ML learning, with the regularization strength increasing as λ increases. However, as λ becomes too large, a too strong regularization will make the system finally loose the ability of adapting input data.

23.5 An Information Transfer Perspective

In the past decade, extensive studies have been made on the minimum description length (MDL) [23.52, 23.54]. Sharing the common sprit of the minimum message length (MML) [23.69, 23.71], the BIC model selection criterion and variants [23.59, 23.48], and the celebrated Kolmogorov complexity [23.29], the key idea is to implement the well known Ockham's principle of economy to code a set of samples $\{x_t\}_{t=1}^N$ for being transferred from a sender to a receiver via a two part coding. One is the amount of bits for coding the residuals of using a parametric model $p(x|\theta)$ to fit a set of samples $\{x_t\}_{t=1}^N$. The second part is the amount of bits for coding the parameter set θ, provided that the function form of $p(x|\theta)$ has already known at the receiver and,

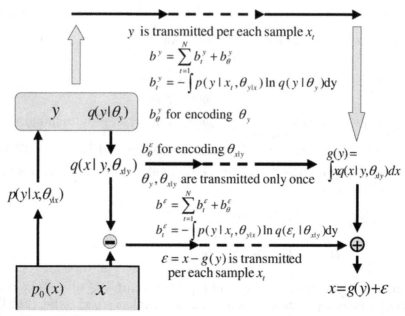

Fig. 23.3. BYY harmony learning from an Information-theoretic Perspective

thus, no need for being encoded. A best information transfer is reached when the bits for both the parts are minimized.

In the existing literature, given a density model $p(x|\theta)$ for a d dimensional real random vector x, the amount of bits per sample x_t to be transmitted is described by $b_t^\varepsilon = -\ln p(x_t|\theta) - d\ln\delta$, where $\delta > 0$ is a pre-specified constant resolution and usually ignored. The total amount of bits for the first part is $b^\varepsilon = \sum_{t=1}^N b_t^\varepsilon$. The amount b_θ^ε of bits for the second part is common to every sample of x_t, and, thus, only needs to be transmitted one time in advance. Thus, the average amount of bits to be transmitted is $\frac{1}{N}\sum_{t=1}^N b_t^\varepsilon + \frac{b_\theta^\varepsilon}{N}$. For a large size N of samples, the second term becomes very small and can be ignored. The minimization of the first term is actually equivalent to the ML learning. However, this term does not contain enough information to select an appropriate complexity (e.g., the number of parameters in θ) for $p(x|\theta)$.

In a contrary, for a finite size N of samples, we encounter a so-called over-fitting effect that the larger the complexity is, the smaller the residual of using $p(x|\theta)$ to fit the set $\{x_t\}_{t=1}^N$ is, and, thus, the smaller of the first term is. The second term takes its role that balances off the over-fitting effect since b_θ^ε increases as the complexity increases. However, b_θ^ε is described by $-\ln p(\theta)$. The priori distribution $p(\theta)$ is usually not available and can only be very roughly estimated, e.g., by a non-informative uniform prior or Jeffery priori [23.42, 23.37]. Instead of coding x_t directly for transmission, the MDL implementation with a bits-back strategy in [23.33, 23.32] maps x to y and then code y for transmission. However, as to be further discussed in the next subsection, this bits-back based MDL is actually equivalent to the ML learning and, thus, is still not good for model selection.

The BYY harmony learning by Eq. (23.16) can also be understood from an information transfer perspective, with a new insight on its ability for model selection and regularization. As shown in Fig. 23.3, we consider a system in which x is mapped to an inner representation y that is encoded and sent to the receiver, and the receiver then decodes y to reconstruct x. Learning is made to obtain the encoder $p(y|x)$ for getting y from x, the distribution $q(y)$ for the codes on y, and the decoder $q(x|y)$ for getting x from y.

Provided that the function form of $q(y|\theta_y)$ is already known at the receiver, the average amount of bits to be transmitted is $\frac{1}{N}\sum_{t=1}^{N} b_t^y + \frac{b_\theta^y}{N}$, with b_t^y being the amount of bits per sample for coding y and b_θ^y being the amount of bits for coding θ_y. To reconstruct x_t, one also needs the decoder $q(x|y)$ that should also be coded at the sender and then sent to the receiver. The decoder is also coded in two parts. One is coding the residual between the original x_t and its reconstruction by the decoder, and the amount of bits per sample is b_t^ε. The other part is the amount b_θ^ε of bits to code the parameter set $\theta_{x|y}$ of $q(x|y)$. Provided that the function form of $q(x|y)$ is already known at the receiver, the average amount of bits for $q(x|y)$ is $\frac{1}{N}\sum_{t=1}^{N} b_t^\varepsilon + \frac{b_\theta^\varepsilon}{N}$.

As a result, referred to [23.75], the entire amount of bits is $N[\frac{b_\theta^y + b_\theta^\varepsilon}{N} - H(\theta, m)]$, with $H(\theta, m)$ given in Eq. (23.16). That is, the BYY harmony learning by Eq. (23.16) attempts to maximizing the information transfer in a sense of minimizing the total coding bits after approximately ignoring $b_\theta^y + b_\theta^\varepsilon$.

Being different from the above discussed conventional MDL that degenerates back to he ML learning after discarding the bits b_θ/N, the harmony measure $-H(\theta, m)$ by Eq. (23.16) without b_θ^y/N and b_θ^ε/N will not disable the model selection ability. The role of b_θ has now been jointly shared by the bits b^y for encoding the inner representation y of x and the bits $b_\theta^y + b_\theta^\varepsilon$ as a counterpart of b_θ. Not only carrying the information about x, the bits b^y also encode the scales of representation that either indicates model complexity directly or includes the core part of model complexity. This provides an alternative insight on why the BYY harmony learning can make model selection.

The above difference also leads to an important difference in implementing model selection. To avoid an inappropriately chosen $q(\theta)$ to deteriorate learning considerably, only a non-informative uniform prior is used as $q(\theta)$ in MDL and thus has no effect on parameter learning for determining θ, which is still made by a ML learning as the first step. The MDL criterion comes in effect at the second step for model selection. This two step implementation costs heavily since parameters learning on getting θ has to be made on all the candidate models in consideration. By the BYY harmony learning, the job of model selection is also performed via a family of densities $q(y|\theta_y)$ with a given parametric structure but unknown parameters θ_y that is determined during learning process, which is a significant relaxation from solely relying on a priori density $q(\theta)$. As a result, not only parameter learning is performed more accurately but also model selection is made via the scale parameters of y that are determined automatically during learning parameters in θ_y.

The regularization role of Z_q in $-H(\theta, m)$ by Eq. (23.16) can also be understood from a more precise perspective of information transfer. Instead of considering a quantization by a pre-specified constant resolution $\delta > 0$ that is currently widely adopted in the MDL literature.

23.6 BYY Harmony Learning Versus Related Approaches

23.6.1 Relation and Difference to the Bits-Back Based MDL and Bayesian Approaches

The above information transfer perspective shares certain common part with the bits-back based MDL proposed in [23.33, 23.32]. However, there are two key differences.

First, the term Z_q replaces the role of a pre-fixed quantization resolution δ that is currently widely adopted in the MDL literature. Without considering what type of data distribution it is, manually setting a constant δ is simply because there is no a better solution available but it is clearly not a good solution. In the BYY harmony learning by Eq. (23.16), the term Z_q provides a better solution. In the data smoothing implementation, Z_q takes the input data distribution in consideration via the Parzen window estimator by Eq. (23.3) with a smoothing parameter h. This h takes a role similar to a quantization resolution δ, but now it is also learned to adapt the set of samples $\{x_t\}_{t=1}^{N}$. In the normalization implementation, Z_q takes the input data distribution in consideration indirectly via the learned parametric densities $q(x|y)$ and $q(y)$ as well as their values on the a set of samples $\{x_t\}_{t=1}^{N}$.

Second, an even fundamental difference is that BYY harmony learning does not adopt the bits-back strategy [23.33, 23.32]. Considering the dependence among the inner codes generated by $p(y|x)$, it has been argued in [23.33, 23.32] that the total amount of bits should be subtracted by the following amount of bits

$$H(\theta_{y|x}) = \int p(y|x)p(x) \ln p(y|x)\mu(dx)\mu(dy). \tag{23.52}$$

With this amount claimed back, the total amount of bits that has been considered by [23.33, 23.32] is actually equivalent to the Kullback divergence $KL(\theta)$ by Eq. (23.22), after discarding a term $H_x = \int p(x) \ln p(x) dx$ that is irrelevant to learning when $p(x) = p_0(x)$ by Eq. (23.2). In other words, the bits-back based MDL [23.33, 23.32] actually provides an interpretation to the Kullback learning by Eq. (23.22) from a information transfer perspective. In contrast, without including $H(\theta_{y|x})$ by Eq. (23.52), the discussion in Sect. 23.5 provides an interpretation to the BYY harmony learning by Eq. (23.16). As to be further discussed in the next subsection, the Kullback learning by Eq. (23.22) is equivalent to implementing parameter learning under the ML principle or its certain regularized variants in lack of model selection ability, while BYY harmony learning provides a new mechanism that makes model selection either after or during parameter learning.

An insight can also be obtained by further observing the role of the bits-back amount $-H(\theta_{y|x})$ by Eq. (23.52). With the dimension of y fixed, the Kullback learning by Eq. (23.22) implements a stochastic encoding by $p(y|x)$ that allows certain dependence among the resulted codes. This dependence generates a redundant amount $-H(\theta_{y|x})$ of bits that is suggested in [23.33, 23.32] to be subtracted from computing the total amount of bits. In a contrast, aiming at seeking an appropriate dimension for y, the BYY harmony learning by Eq. (23.16) actually minimizes [23.75]

$$-H(\theta, k) = KL(\theta) - H(\theta_{y|x}) + C_y . \tag{23.53}$$

Where $-H(\theta_{y|x}) + C_y \geq 0$. That is, $-H(\theta, k) \geq KL(\theta)$ is an upper bound of the total bits considered in [23.33, 23.32].

When $p(y|x)$ is free, $\max_{p(y|x)} H(p\|q)$ results in $p(y|x)$ as in Eq. (23.19). It happens similarly when $p(y|x)$ is parametric either directly in a form of $\delta(y - y(x))$ or tends to be pushed into this form via $\max_{p(y|x)} H(p\|q)$. In these cases, $-H(\theta_{y|x}) + C_y$ reaches its minimum value 0. Thus, the BYY harmony learning achieves the minimum total number of bits instead of one upper bound.

In other words, the BYY harmony learning reaches the optimal coding bits both by learning unknown parameters and by squeezing any stochastic redundancy that allows one x to share more than one inner codes of y. As a result, all the inner codes will occupy a representation space as compact as possible. That is, model selection occurs automatically during the process of approaching the optimal coding bits. On a contrary, the dimension for the inner codes of y is pre-specified for a bits-back based MDL case, and the task is learning unknown parameters under this fixed dimension (usually assumed to be large enough for what needed). Due to there is certain redundancy in the representation space, it is allowed that one x may be redundantly represented by more than one inner codes. Instead of squeezing out this dependence, the redundant bits of $-H(\theta_{y|x})$ by a stochastic $p(y|x)$ is not zero but discounted in counting the total amount of bits.

Though such a redundant coding makes information transfer more reliable, allowing redundancy in the representation space of y already means that this representation space is not in its minimum complexity.

Furthermore, the BYY harmony learning may also be related to Bayesian approaches by replacing y with a parameter set θ in that Eq. (23.19) becomes equivalent to the Bayesian learning by Eq. (23.8). Ignoring $-\ln z_q$, $H(\theta, m)$ by Eq. (23.16) is actually the MML description length by Eq. (23.11), while $-\ln z_q \neq 0$ provides a type of regularization similar to that discussed in

Sect. 23.4.1. Also, Eq. (23.13) becomes equivalent to the evidence given by Eq. (23.9) and it follows from Eq. (23.24) that $KL(\theta)$ becomes equivalent to the description length based on this evidence. The difference between the MML description length and the evidence based description length is actually the bits-back part by Eq. (23.52) with y replaced by θ. As discussed in Sect. 23.2.2, knowing a priori $q(\theta)$ is a difficult task and a rough estimate $q(\theta)$ may seriously affect the MAP solution by Eq. (23.8). Thus, the description length based on Eq. (23.9) is usually regarded as an improvement over that by Eq. (23.11) since the integral over θ can regularize in a certain extent the discrepancy caused by $q(\theta)$.

However, the BYY harmony learning is different from the above MML description length and the evidence based description length in that an inner representation y takes the place of θ to avoid the difficulty of getting $q(\theta)$, which brings us the following advantages:

- Instead of specifying a density $q(\theta)$, the BYY harmony learning only needs to specify a family of densities $q(y|\theta)$ with a given parametric structure but unknown parameters θ, while learning further specifies one among the family. Therefore, the difficulty of requiring a detailed priori knowledge has been relaxed significantly. Moreover, the above superiority of the evidence based description length due to the bits-back type regularization disappears. On a contrary, as discussed in the early part of this subsection, the bits-back type regularization actually weaken the model selection ability and lost the nature of automatic model selection.
- Instead of considering all the parameters in the description length, the BYY harmony learning focuses only at those useful scale parameters m, k, etc., via the structures of the inner representation space of y which avoids to handle the difficulty of and saves the computing costs on estimating those complexities that are unnecessary for determining m, k, etc.
- As discussed in above, the BYY harmony learning is able to make model selection automatically during learning parameters. In contrast, using the evidence based description length for model selection has to be made via a two stage implementation since the evidence based description length has to be estimated after parameter learning.

23.6.2 Relations to Information Geometry, Helmholtz Machine and Variational Approximation

The minimization of $KL(\theta)$ by Eq. (23.22) with respect to a free $p(y|x)$ will result in Eq. (23.24) and becomes equivalent to the ML learning on $q(x)$ when $p(x) = p_0(x)$ by Eq. (23.2) [23.109]. This case relates to the information geometry theory (IGT) [23.16, 23.4, 23.5] that is also equivalent to the ML learning on $q(x)$ by Eq. (23.13). Moreover, the well known EM algorithm [23.20, 23.51, 23.47] is reached by the em algorithm obtained in IGT.

Making parameter learning by Eq. (23.22) also relates to the Helmholtz machine learning (HML) when $p(x) = p_0(x)$ is given by Eq. (23.2) and both $p(y|x)$ and $q(x|y)$ are both given by the conditional independent densities by Eq. (23.54) as used in [23.31, 23.18]. That is, the densities are given with the following format

$$p(u|v) = \prod_{j=1}^{m} \pi_j(v)^{u^{(j)}} (1 - \pi_j(v))^{1-u^{(j)}}, \tag{23.54}$$

$$\pi(v) = [\pi_1(v), \cdots, \pi_m(v)]^T = S(Wv + c),$$
$$S(y) = [s(y^{(1)}), \cdots, s(y^{(m)})]^T, \ 0 \le s(r) \le 1 \text{ is a sigmoid function,}$$

where u is a binary vector. In this case, making parameter learning by Eq. (23.22) actually becomes equivalent to an one layer HML. Also, the well known wake-sleep algorithm for HML can be regarded as a simplified adaptive form of Eq. (23.23). With a general insight via Eq. (23.23), other specific algorithms for implementing the HML may also be developed.

It is also deserve to notice that making parameter learning by Eq. (23.22) with a parametric $p(y|x) \in \mathcal{P}_{y|x}(\theta_{y|x})$ is different from that a free $p(y|x) \in \mathcal{P}_{y|x}^0$ in that a parametric family $\mathcal{P}_{y|x}(\theta_{y|x})$ is a subset of the family $\mathcal{P}_{y|x}^0$ that consists of all the density functions in the form $p(y|x)$. Thus, we always have $\min_{p(y|x) \in \mathcal{P}_{y|x}(\theta_{y|x})} KL \ge \min_{p(y|x) \in \mathcal{P}_{y|x}^0} KL$. When $p(x) = p_0(x)$ is given by Eq. (23.2), it follows from Eq. (23.24) that the latter becomes equivalent to the ML learning on $q(x)$ by Eq. (23.13). In other words, making parameter learning by Eq. (23.22) with a parametric $p(y|x)$ actually implements a type of constrained ML learning on $q(x)$, which is also called a variational approximation to the ML learning on $q(x)$ [23.58, 23.56].

The BYY harmony learning is different from three existing approaches as follows. First, the BYY harmony learning minimizes the harmony measure $-H(p\|q)$ instead of the Kullback divergence $KL(p\|q)$ in Eq. (23.22), not only for parametric learning but also for model selection. Even using the Kullback learning by Eq. (23.22) for parameter learning, it is still followed by model selection via Eq. (23.21). In contrast, parameter learning via minimizing the Kullback divergence is the only target in IGT, HML, and variational approximation, while the issues of regularization and model selection are out of the scope of their studies.

Second, as discussed later in Eq. (23.66), the harmony learning may also be regarded as implementing a type of constrained ML learning, especially when $p(y|x) \in \mathcal{P}_{y|x}(\theta_{y|x})$ is parametric. However, it is different from the above discussed constrained ML learning via variational approximation [23.57, 23.56]. An additional constraint should be imposed on both types of learning to make them become equivalent.

Third, even focusing on the common part, i.e., parameter learning via minimizing Kullback divergence for implementing parameter learning, these studies are conducted from different perspectives with different purposes.

IGT studies the general properties possessed by Eq. (23.22) and alternative minimization for two general p and q from the perspectives of geometry structure [23.16] and differential geometry structure [23.4, 23.5]. HML and variational approximation consider developing efficient algorithms for implementing empirical parameter learning on a forward-backward net via an approximation of the ML learning on the marginal density $q(x)$ in Eq. (23.13). In contrast, the BYY learning studies two distributions in the two complementary Bayesian representations in Eq. (23.12) by systematically investigating not only three typical architectures for different learning tasks, but also regularization by either a conscience de-learning type via normalization or a Tikhonov-type via data smoothing with its smoothing parameter h estimated in sample way. While IGT, HML, and variational approximation have neither explicitly and systematically considered the two complementary representations in Eq. (23.12) nor the regularization of two such types.

23.6.3 A Projection Geometry Perspective

Projection Geometry in Vector Space. Through obtaining a quasi Pythagorean relation under the Kullback divergence

$$KL(p\|q) = \int p(u) \ln \frac{p(u)}{q(u)} du \geq 0, \ KL(p\|q) = 0, iff \ p(u) = q(u) \quad (23.55)$$

This divergence based learning has been further theoretically studied from the perspective of ordinary geometry and differential geometry under the name of information geometry [23.16, 23.4, 23.5]. Actually, neither the harmony measure by Eq. (23.17) nor the Kullback divergence by Eq. (23.55) satisfies all the properties of the conventional metric measure. Moreover, the harmony measure by Eq. (23.17) even does not satisfies a quasi Pythagorean relation that the Kullback divergence satisfies. In this section, we suggest to investigate both the harmony measure based learning and the Kullback divergence based learning by lowering down from a metric level to an even basic level, namely, a level of projection geometry.

We start at reviewing some basic properties in the conventional vector space R^d. We denote $U_c = \{u : u \in R^d \ and \ \|u\|^2 = c^2$, for a constant $c > 0\}$, which is a sphere shell with the radius c.

As shown in Fig. 23.4, for $u = ce^{\theta_u} \in U_c, v = c'e^{\theta_v} \in U_{c'}$, their inner product is

$$u^T v = cc' \cos (\theta_v - \theta_u), \quad (23.56)$$

which is symmetric to v and u and leads to a norm $\|u\|^2$ that further leads to the metric $\|u - v\|$.

Imposing the constraint $\|u\| = 1$, the inner product returns back to the projection of v on u as follows:

$$\Pi_u^v = c' \cos (\theta_v - \theta_u), \quad (23.57)$$

which has the following properties:

Inner product $u'v = cc'\cos(\theta_u - \theta_v)$

Projection of v on u: $\Pi_u^v = c'\cos(\theta_u - \theta_v) \le c'$ and $'='$ holds *if and only if* $\theta_u = \theta_v$

Projection of $q(u)$ on $p(u)$: $H = \int p(u)\ln q(u)\mu(du) + Z_q$

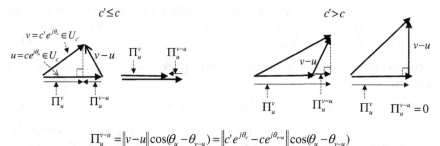

$$\Pi_u^{v-u} = \|v-u\|\cos(\theta_u - \theta_{v-u}) = \|c'e^{j\theta_v} - ce^{j\theta_{v-u}}\|\cos(\theta_u - \theta_{v-u})$$

When $c' \le c$, $|\Pi_u^{v-u}| \ge c - c'$
Equality holds *if and only if* $\theta_u = \theta_v$.

When $c' \ge c$, $|\Pi_u^{v-u}| \ge 0$
Equality holds *if and only if* $|\theta_u - \theta_{v-u}| = 90°$

Projection of $\dfrac{q(u)}{p(u)}$ on $p(u)$: $KL = \int p(u)\ln\dfrac{p(u)}{q(u)}\mu(du)$

Fig. 23.4. From an inner product back to a projection in the vector space

(a) The self-projection of u is simply the norm $\|u\|$.
(b) We have $-c' \le \Pi_u^v \le c'$ with the equality holding if and only if $\theta_v = \theta_u$. In other words, the projection Π_u^v is maximized when v is co-directional with u.
(c) The projection Π_u^v reaches its minimum 0 when $\theta_v - \theta_u = 0.5\pi$, which is said that v is orthogonal to u.
(d) When $c = c'$, $\theta_v = \theta_u$ implies $v = u$. That is, the maximal projection is equivalent to the equality $v = u$, when v, u are on the same shell U_c.
(e) $\theta_v = \theta_u$ can be achieved by rotating the directions of both v and u or the direction of either v or u. That is, the projection $v^T u$ has the symmetry property.

The error or residual $v - u = \|v - u\|e^{-\theta_{v-u}}$ also has a projection on u:

$$\Pi_u^{v-u} = \|v - u\| \cos(\theta_{v-u} - \theta_u), \tag{23.58}$$

with the following properties:

(f) As shown in Fig. 23.4, when $c' > c$, this residual projection $|\Pi_u^{v-u}|$ reaches its minimum 0 when $\theta_{v-u} - \theta_u = 0.5\pi$, with $\|v - u\| \ne 0$. In this case, the residual $v - u$ is said to be orthogonal to u, where the norm of u and the projection v on u becomes the same, i.e., $\Pi_u^v = c$ or $\cos(\theta_v - \theta_u) = c/c'$.

(g) When $\|v\| \leq \|u\|$, this residual projection $|\Pi_u^{v-u}|$ reaches its minimum $c-c'$ if and only if $\theta_v = \theta_u$. When $c = c'$, we have $u = v$ and the minimum value is 0.

In a summary, we have

> When v, u locate on a same shell U_c, the concepts of maximizing the projection v to u, minimizing the residual projection $(v - u)$ to u, of making residual $v - u$ being orthogonal to u, and the equality $v = u$ are all the same thing. (23.59)

Projection Geometry in a Functional Space. In an analogy, we consider a functional space

$$\mathcal{Q} = \{q(u) : q(u) \geq 0 \text{ and } \int q(u)\mu(du) < \infty\}, \qquad (23.60)$$

where $u \in S_u \subseteq R^d$ and μ is a given measure on the support S_u, and $\mu(du)$ only relates to du but to neither u nor $q(u)$. A useful subspace $\mathcal{P}_c \subset \mathcal{Q}$ is

$$\mathcal{P}_c = \{p(u) : p(u) \geq 0, \ \int p(u)\mu(du) = c, \text{ for a constant } c > 0\}. \quad (23.61)$$

Particularly, when $c = 1$, $\mathcal{P}_1$ is the probability density space.

Given $p(u) \in \mathcal{P}_c, q(u) \in \mathcal{P}_{c'}$, we define the projection of $q(u)$ on $p(u)$ by

$$\begin{aligned} H(p\|q) &= \int p(u)\mu(du) \ln (q(u)\mu(du)) = \int p(u) \ln q(u)\mu(du) - Z_q, \\ Z_q &= -\int p(u) \ln \mu(du)\mu(du) = -\ln \mu(du), \end{aligned} \qquad (23.62)$$

which can be regarded as the counterpart of Eq. (23.57) as shown in Fig. 23.4. It can be observed that Eq. (23.62) becomes the same as Eq. (23.16) and Z_q takes the same role as $-\ln z_q$, when

$$p(u) = p(x, y) = p(y|x)p(x), q(u) = q(x, y) = q(x|y)q(y). \qquad (23.63)$$

Considering $\int p(u)\mu(du) = 1$ and $p(u) = p_0(u)$ is the empirical density by Eq. (23.3) when $h_u = 0$, it follows that $\mu(du) = 1/\sum_{t=1}^N p(u_t)$ and $z_q = \sum_{t=1}^N q(u_t)$ from which and Eq. (23.63), we also get Eq. (23.36).

In correspondence to Eq. (23.57), we have the following properties:

(1) The self-projection of $p(u)$ is $H(p\|p) = \int p(u)\mu(du) \ln [p(u)\mu(du)]$, which can be regarded as a type of norm of p and it becomes the negative entropy of the probability distribution $p(u)\mu(du)$ when $p(u) \in \mathcal{P}_1$ is a density.
(2) $H(p\|q)$ is maximized if and only if $q(u) = \frac{c'}{c}p(u)$, i.e., $q(u)$ has the same shape as $p(u)$, because we have $\int \hat{p}(u) \ln \hat{q}(u)\mu(du) \leq \int \hat{p}(u) \ln \hat{p}(u)\mu(du)$ with $c\hat{p}(u) = p(u)$, $c'\hat{q}(u) = q(u)$ and $\hat{p}(u), \hat{q}(u) \in \mathcal{P}_1$.
(3) When $c = c'$, $H(p\|q)$ is maximized if and only if $q(u) = p(u)$.
(4) When $p(u)$ is free to be any choice in $\mathcal{P}_c$, the maximization of $H(p\|q)$ will also let $p(u)$ to become $c\delta(u - u^*)$, where $u^* = \arg\max_u q(u)$.

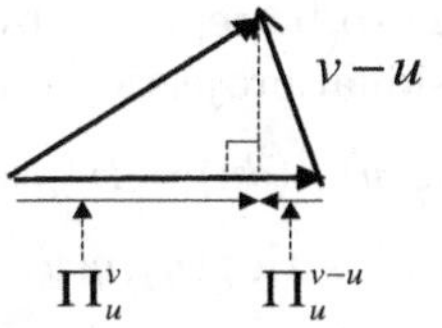

$$c = 1, c' = 1$$

$$\Pi_u^{v-u} = \|v - u\| \cos(\theta_u - \theta_{v-u})$$
$$= \left\| e^{j\theta_v} - e^{j\theta_{v-u}} \right\| \cos(\theta_u - \theta_{v-u})$$

$$\Pi_u^v = \cos(\theta_u - \theta_v)$$

$$\Pi_u^v = \max \; \text{if and only y if } \theta_u = \theta_v.$$

$\Longleftrightarrow$

$$\Pi_u^{v-u} = 0 \text{ if and only if } \theta_u = \theta_v.$$

$$\int p(u)\mu(du) = 1, \; \int q(u)\mu(du) = 1$$

$$H = \int p(u) \ln q(u)\mu(du) + Z_q$$

$$KL = \int p(u) \ln \frac{p(u)\mu(du)}{q(u)\mu(du)} \mu(du)$$
$$= \int p(u) \ln \frac{p(u)}{q(u)} \mu(du)$$

$\max_q$ H results in $q(u) = p(u)$

$\max_p$ H results in $p(u) = \delta(u - u^*)$

$\max_{p,q}$ H results in $q(u) = p(u) = \delta(u - u^*)$

$\min_q$ KL results in $q(u) = p(u)$

$\min_p$ KL results in $p(u) = q(u)$

$\min_{p,q}$ KL results in $q(u) = p(u)$

$\not\Leftrightarrow$

Fig. 23.5. Unit norm based projection: from the vector space to a functional space

In comparison with the situation of Eq. (23.57), there are three differences. One is that each density represents a point of infinite dimension. Second, each component is constrained to be nonnegative. Third, the constraint $\int p(u)\mu(du) = c$ is a first order linear constrained, instead of the quadratic constraint $\|u\|^2 = c^2$. These differences result in that the maximization of $H(p\|q)$ makes not only that $p(u)$ and $q(u)$ has a same shape in the sense $q(u) = \frac{c'}{c} p(u)$ but also that $p(u)$ prefers to have a simplest shape $c\delta(u - u^*)$. When $p(u)$ is free to be any choice in $\mathcal{P}_c$ and $q(u)$ is free to be any choice in $\mathcal{P}_{c'}$, the maximization of $H(p\|q)$ will finally let that both $p(u)$ and $q(u)$ become impulse functions. When $p(u) \in P, q(u) \in Q$ are constrained to be unable to become impulse functions, the maximization of $H(p\|q)$ will make that $p(u)$ and $q(u)$ become close in a shape of a least complexity but not able completely equal. Therefore, the maximization of $H(p\|q)$ on a BYY system Eq. (23.63) indeed implements the harmony principle given at the beginning of Sect. 23.3.1, while the maximization of the projection u to v only ensures u and v become co-directional but does not have such a least complexity.

In addition, $H(p\|q)$ does not share the symmetry by Π_u^v at $\|v\| = \|u\|$. If exchanging the positions of p, q, though $\max H(p\|q)$ still makes that $p(u)$ and $q(u)$ have a same shape, it is different in a sense that $q(u)$ but not $p(u)$ is now pushed to a shape of $c'\delta(u - u^*)$.

Moreover, if we use $p(u) \in \mathcal{P}_c$ to represent $q(u) \in \mathcal{P}_{c'}$ and define the discrepancy or residual [1] by $p(u) \ominus q(u) = p(u)\mu(du)/[q(u)\mu(du)] = p(u)/q(u)$,

---

[1] Under this definition, $p(u) \ominus q(u)$ is generally not guaranteed to still remain in $\mathcal{Q}$. For a subset $\mathcal{Q}_q \subset \mathcal{Q}$ with $\mathcal{Q}_q = \{q(u) : q(u) \in Q, \int_{D_u} q^2(u)\mu(du) <$

we let $q(u)\mu(du)$ in Eq. (23.62) to be replaced by the residual in this representation and get that this residual projection on $p(u)$ as follows

$$R(p\|q) = \int p(u) \ln [p(u)/q(u)]\mu(du) = H(p\|p) - H(p\|q). \qquad (23.64)$$

Since $p(u) = c\hat{p}(u)$, $q(u) = c'\hat{q}(u)$ with $\hat{p}(u), \hat{q}(u) \in \mathcal{P}_1$, it follows that

$$R(p\|q) = c[KL(\hat{p}\|\hat{q}) + \ln \frac{c}{c'}],$$
$$KL(\hat{p}\|\hat{q}) = \int \hat{p}(u) \ln [\hat{p}(u)/\hat{q}(u)]\mu(du). \qquad (23.65)$$

From which we can observe the following properties:

(5) Minimizing $R(p\|q)$ is equivalent to both minimizing the self-projection of $p(u)$ and maximizing the projection of $q(u)$ on $p(u)$. When the self-projection $H(p\|p)$ is fixed at a constant, minimizing the residual projection is equivalent to maximizing $H(p\|q)$.

(6) The residual $p(u) \ominus q(u)$ is said to be orthogonal to $p(u)$ when the residual projection $R(p\|q)$ becomes 0 that happens when the norm of p and the projection of q on p become the same, i.e., $H(p\|p) = H(p\|q)$.

(7) When $c = c'$, the minimum value of $R(p\|q)$ is 0 which is reached if and only if $p(u) = q(u)$. Moreover, when $c = c' = 1$, $p(u)$ and $q(u)$ are densities and $R(p\|q) = KL(p\|q)$.

From the above discussions, we see that the concepts of maximizing $H(p\|q)$ and of minimizing the residual projection $R(p\|q)$ are related, but not equivalent. Even when $c = c' = 1$, we do not have the equivalence between Π_u^v and Π_u^{v-u} as given in Eq. (23.59) for Eq. (23.57) and Eq. (23.58). This provides a geometry perspective on why and how the maximization of $H(p\|q)$ on a BYY system Eq. (23.63), which is a generalization of maximizing the projection for the co-directionality, is different from the minimization of Kullback divergence $KL(p\|q)$ on a BYY system Eq. (23.63) or equivalently the maximum likelihood learning, which is a generalization of minimizing the residual projection. Moreover, the latter does not have the least complexity nature that enables the former to make model selection.

However, imposing an additional constraint that $H(p\|p)$ is fixed at a constant H_0, we have

$$\max_{p \in P, q \in Q, \ s.t. \ H(p\|p)=H_0} H(p\|q) \text{ is equivalent to}$$
$$\min_{p \in P, q \in Q, \ s.t. \ H(p\|p)=H_0} KL(p\|q). \qquad (23.66)$$

$\infty, \int_{D_u} q^{-2}(u)\mu(du) < \infty, \int_{D_u} \mu(du) < \infty\}$, we can define the addition by $r(u) = p(u) \oplus q(u) = p(u)q(u)$ and have $r(u) \in \mathcal{Q}_q$. Also, we have the unit $1 = p(u)p^{-1}(u) \in \mathcal{Q}_q$ for $u \in S_u$ and the inverse $p^{-1}(u) = 1/p(u) \in \mathcal{Q}_q$.

In this case, it follows that the induced minus operation $p(u) \ominus q(u) = p(u)/q(u)$ is still in $\mathcal{Q}_q$. That is, we get $\mathcal{Q}_q$ as an Abel group. Moreover, on an appropriate subset $\mathcal{Q}_l$ we can further define the dot product $\alpha \circ p(u) = p(u)^\alpha \in \mathcal{Q}_l$ for $\alpha \in R$ and, thus, get $\mathcal{Q}_l$ as a linear functional space. Furthermore, we can introduce the geometrical concepts of the projection Eq. (23.62), the residual projection Eq. (23.64) and the corresponding orthogonality to $\mathcal{Q}_q, \mathcal{Q}_l$.

With $p(x)$ given by Eq. (23.3), the constraint $H(p\|p) = H_0$ means certain constraint imposed on $p(y|x)$. In other words, Eq. (23.66) happens on a class of BI-directional architectures, and can also be regarded as implementing a type of constrained ML learning, which is different from those of variational approximation [23.57, 23.56] that implements $\min_{p\in P, q\in Q} KL(p\|q)$ with $p(y|x)$ in a constrained structure but without requiring the constraint $H(p\|p) = H_0$.

In addition, the above discussions on the geometry properties of $p(u) \in \mathcal{P}_c$ and $q(u) \in \mathcal{P}_{c'}$ with $c \neq 1, c' \neq 1$ may also be extended beyond probability densities. Also, with $R(p\|q) = 0$ we can get the concept of the orthogonality of the residual $p(u) \ominus q(u)$ to $p(u)$.

23.7 Bibliographic Remarks

In the previous chapter of the present book, main results of using BYY system and harmony learning on typical learning problems have been summarized. Also, bibliographic remarks have been made on the progress of these studies from both the aspect of BYY system with the KL learning and the aspect of computing techniques for implementing BYY learning. In this section, further bibliographic remarks will be made on the progress from the model selection and regularization aspects of BYY harmony learning.

23.7.1 On BYY Harmony Learning (I): Model Selection Criteria vs. Automatic Model Selection

As discussed in Sect. 23.3.1, maximizing the harmony measure by Eq. (23.16) that makes model selection either automatically during parameter learning by Eq. (23.20) or via a selection criterion Eq. (23.21) after parameter learning.

In help of the so-called 'hard-cut' treatment of posteriori probabilities, this harmony measure with $z_q = 1$ was firstly obtained in 1995 both at its special case of both Gaussian mixture (see Eq. (20) and (22) in [23.109], Eq. (13) and (14) in [23.113], and Eq. (13) in [23.112]), and at a special case of finite mixture (see Eq. (7) in [23.112]). Companying with this measure, two types of detailed studies were conducted as follows:

- One is called two-phase style learning in Sect. 23.3.1. That is, model selection is made via $\min_k J(k)$ after parameter learning. This $J(k)$ is a simplified version of the harmony measure after discarding irrelevant terms. Typical examples include $J(k)$ by Eq. (24) in [23.109] and $J(k)$ by Eq. (13) in [23.111].
- The other type of studies is on parameter learning with automatic model selection by Eq. (23.16) in Sect. 23.3.1. It was suggested (see Sect. 5.2 and the footnote on page 986 in [23.109], also see Sect. 3 in [23.111] and the second part of Sect. 5 in [23.112]) that an appropriate value of k can be

automatically determined during parameter learning by the so-called hard-cut EM algorithm (see the algorithm after Eq. (20) in [23.109], also see the algorithm after Eq. (15) in [23.112]), via discarding a Gaussian component $\alpha_j G(x|m_j, \Sigma_j)$ if either or both of Σ_j and $\alpha_j = P(y = j)$ become zero.

It should also be noticed that studies of BYY system with the harmony measure based learning by Eq. (23.16) and the Kullback divergence based learning by Eq. (23.22) were conducted jointly, with the following relations found:

- The additive relationship $KL = -H - E_p + D$ by

$$KL(p\|q) = -H(p\|q) - E_p, E_p = -\int p(u)\ln p(u)\mu(du) + \ln z_p.(23.67)$$

  or equivalently $H = -KL - E_p + D$ by Eq. (23.47) that was firstly presented in [23.109], where D is a term that is only related to the smoothing parameter h for a $p(x)$ given by Eq. (23.3).
- The term D becomes irrelevant to learning when $h = 0$ or equivalently $p(x)$ is given by Eq. (23.2). In these cases, D can be discarded and we can simply consider $KL = -H - E_p$ or equivalently $H = -KL - E_p$.
- The inequality relation $KL \leq -H$ was also firstly observed in [23.112]. The equality $KL = -H$ holds when $E_p = 0$, where the KL learning and the harmony learning become equivalent as discussed in Sect. 23.4.2.
- As discussed in Sect. 23.6.1, the harmony learning is different from the KL learning in that the minimization of $-H = KL + E_p$ attempts to push $E_p \geq 0$ toward its minimum $E_p = 0$ such that a minimum coding length is reached via minimizing the model complexity.

23.7.2 On BYY Harmony Learning (II): Model Selection Criteria

After the initial results obtained in 1995 [23.109], various specific $J(k)$ forms of Eq. (23.21) have been subsequently obtained from Eq. (23.16) for model selection in typical learning models, with main progresses summarized as follows:

(1) Not only $J(k)$ by Eq. (24) in [23.109] and by Eq. (13) in [23.111] was further studied experimentally in 1996 [23.108] and both theoretically and experimentally in 1997 [23.104], but also Eq. (7) in [23.112] (i.e, the special form of Eq. (23.16) with $z_q = 1$ on a finite mixture) is reiterated via Eq. (10) in [23.107] and Eq. (18) in [23.104], and then applied to multi-sets mixture learning by Eq. (15) in [23.107]).

(2) Started from 1997, the harmony measure by Eq. (23.16) with $z_q = 1$ is further suggested under the notation $J_2(k)$ as a general criterion for model selection (see Eqs. (3.8) and (3.9) in [23.101], Eqs. (13) and (15) in [23.102], and Eq. (12) in [23.104]). Recently, the superiority of this criterion has been further supported by experiments made in comparison with classic criteria including AIC, CAIC and MDL [23.34].

(3) In 1998, extending the relation $-H = KL + E_p$, not only the weighted sum by Eq. (23.49) was firstly suggested (see Eq. (48) in [23.92]), but also its variant $KL+\lambda E_p$ was also suggested (see Eq. (8) in [23.90], Eq. (22) in [23.91], Eq. (7c) in [23.100], Eqs. (17) and (18) in [23.93], Eq. (6f) in [23.94], as well as Eq. (8b) in [23.96]). The form $KL + \lambda E_p$ returns to $-H = KL + E_p$ when $\lambda = 1$. This form makes it possible to be further extended with the function ln(.) replaced by a general convex function $f(.)$ (see Eq. (15) and (10) in [23.97] and Sect. 4 in [23.87], also Sect. II(B) in [23.84] and Sect. 2.5 in [23.85]).

(4) Also started from 1997, typical forms of the harmony measure by Eq. (23.16) with $z_q = 1$ and $p(x)$ given by Eq. (23.2) have also been developed as model selection criteria for the following learning models:

– PCA and FA (see Eq. (9.13) in [23.101], Eq. (56) in [23.92], Eq. (33) and (37) in [23.93], as well as Eq. (13) and (18b) in [23.100]).
– Principal ICA that extends ICA to noise situation (see Eq. (10.9) in [23.101], Eq. (11) in [23.89], Eq. (56) in [23.86] and Eq. (55) in [23.85]).
– Binary LMSER (see Eq. (8.10) and (8.13) in [23.101]).
– Logistic LMSER (see Eq. (38) in [23.76]).
– Regularized LMSER via minimizing the variances of hidden units (see Eq. (8.10) and (8.13) in [23.101], Eq. (40) and (41) in [23.93], Eq. (20b) and (20c) in [23.100]).
– Mixture of experts with an approximated criterion proposed firstly (see Table 6(5) in [23.103], Sect. 4.3(1)&(2) in [23.92]), and then a much improved version (see Eq. (84) in [23.83]).
– Alternative ME (see Table 6(5) in [23.103], Sect. 4.3(1)&(2) in [23.92]);
– RBF nets (see Table 7(4) in [23.103], Sect. 4.3(3) in [23.92]).
– Three layer networks, with not only some approximate criteria proposed for both binary stochastic hidden units [23.103, 23.94, 23.95] (e.g., see Eq. (56) in [23.92]) and deterministic real hidden units (see Eqn(88) & (89) in [23.86]), but also improved versions for binary stochastic hidden units (see Type (b) of Eq. (67) in [23.83], also see Eq. (47) in [23.76]), stochastic Gaussian hidden units (see the real y case of Eq. (9a) in [23.99]), and deterministic real hidden units (see Eqn(139) in [23.78]).
– Temporal factor analysis (see Eq. (23b) in [23.90], Eq. (49) in [23.82], the case (a) of Eq. (82) & Eq. (83) in [23.80]).
– Hidden Markov model (HMM) (see Eq. (34) in [23.90]).
– Independent HMM (see Eq. (51) in [23.82], Eq. (85) and (86) in [23.80]).
– Temporal extensions of binary LMSER (see Eq. (46) in [23.82], Eq. (93) in [23.80]).

(5) Started from 2001, the above criteria have been further extended via z-regularization with z_q as discussed in Sect. 23.4.1 and $p(x)$ by Eq. (23.3).

– In [23.83], we got criteria for model selection by Eq. (41) for Gaussian mixture and various special cases, by Eq. (67) for three layer networks, by

Eq. (84) for mixture-of-experts and alternative ME, as well as by Eq. (85) for RBF nets.

- In [23.82], we got criteria for model selection given by the cases (b) & (c) of Eq. (82) & (83) for temporal factor analysis.
- In [23.80], Table 3 provides a systematic summary of criteria, ranging from empirical learning to z-regularization for model selection on various Gaussian mixtures and Gaussian mixture of experts, and Table 4 provides a systematic summary of criteria, also ranging from empirical learning to z-regularization for model selection on various and non-Gaussian mixtures.
- In [23.76], criteria for model selection are also given by Eq. (33) on modular binary factor analyses, and by Eq. (48) on modular binary LMSER.

(6) The last but not least, the relation between the status of observation noise and model selection has been elaborated. For a backward model $x = Ay$ with no noise, the dimension m of y can be determined via the rank of the covariance matrix x. For a forward model $y = Wx$, the dimension m of y is actually pre-given instead of being decided by model selection. In other words, model selection are necessary only for a B-architecture and a BI-architecture, where an observation noise is considered via its backward or generative path (see pp841-843 of [23.82] and pp 1148-1149 of [23.80]).

23.7.3 On BYY Harmony Learning (III): Automatic Model Selection

As discussed in Sect. 23.7.1, also started from 1995 on Gaussian mixture, an appropriate value of k is automatically determined via discarding a Gaussian component $\alpha_j G(x|m_j, \Sigma_j)$ if either or both of Σ_j and $\alpha_j = P(y = j)$ become zero, during implementing a hard-cut EM algorithm for the maximization of harmony measure by Eq. (23.16) [23.109, 23.111, 23.112]. Main progresses along this direction are summarized as follows:

(1) This hard-cut EM algorithm based automatic model selection was not only experimentally demonstrated [23.108] but also further extended to learning on alternative mixture of experts (see Sect. 3.3 in [23.107]).

(2) An adaptive version of this hard-cut EM algorithm was linked to the winner-take-all (WTA) competitive learning. An adaptive version of the EM algorithm was also heuristically proposed and shown to demonstrate a type of rival penalized competitive learning (RPCL) [23.114] mechanism (see Sect. 6 in [23.109]).

(3) Not only adaptive version of the hard-cut EM algorithm is used in implementing learning, but also the rival penalized competitive learning (RPCL) [23.114] is used in place of the hard-cut EM algorithm such that the advantage of RPCL on learning with automatic model selection is adopted. Moreover, the original RPCL learning has been further extended into two types of general forms (i.e., TYPE A and Type B) for learning on Gaussian mixture, multisets modeling (local PCA, local MCA, local subspace, etc.),

mixture of experts, and RBF net (see Sect. 5 in [23.106], Sect. 4.3 in [23.107], and [23.98]).

(4) Started from 1999, the general form of using the harmony measure for both parameter learning and model selection, as shown in Eq. (23.16), has been studied (see Eq. (11) and (12) in [23.86] and [23.85]). Moreover, making parameter learning with automatic model selection by Eq. (23.20) was further made systematically in 2001 [23.83]. Not only the role of the least complexity nature Eq. (23.18) in model selection has been understood, but also the side-effect of the WTA competition by Eq. (23.19), i.e., making the maximization of Eq. (23.20) easy to be trapped at local maximums, is tackled by introducing certain regularization (see Sect. 2.5 in [23.83]). Moreover, four types of regularization have been proposed, as summarized into the following two groups:

(a) Harmony measure + regularization term, that is, a regularization is introduced additively. Specifically, the regularization term can be one of the following three choices:

   − The normalization term as discussed in Sect. 23.4.1 was proposed as regularization term (see the second part on page 52 in [23.83]). It was first time revealed that the harmony learning by Eq. (23.20) with normalization regularization acts as a general RPCL learning framework that implements a floating RPCL learning mechanism (see Sect. 3.2 in [23.83]), which not only justifies the heuristically proposed RPCL learning from the BYY harmony learning perspective but also provides a guide for automatically controlling the ratio of learning and de-learning that was a difficult task in the original RPCL learning [23.114].

   − The normalization term is λE_p, that is, we have $H + \lambda E_p$ that returns to the harmony measure H alone when $\lambda = 0$ and becomes $-KL = H + E_p$ when $\lambda = 1$ (see Eq. (42) and (43) in [23.83]). We can simply choose one appropriate value for λ or let λ to decrease from 1 to 0 gradually in a simulated annealing way.

(b) A regularization can also be introduced in a non-additive way. As will be further discussed in the next subsection, we have two typical techniques as follows (see Sect. 3.2 in [23.83]):

   − regularization is structural and introduced via a BI-architecture,
   − regularization is introduced via data smoothing.

(5) The above four types of combining the roles of the harmony measure and regularization can also be understood from the perspective of competitive learning [23.79]. The nature by Eq. (23.18) encourages a WTA competition by Eq. (23.19), while each of them acts in different manners. Data smoothing penalizes the winner, while both the λE_p and the structural regularization penalize the winner but compensates other participants. However, all these competition-penalty mechanisms makes the WTA effect weaken but encourage gain diversification among participants in competition.

(6) Similarly, the detailed forms of the two groups were also proposed for implementing the harmony learning Eq. (23.20) on mixture of experts, alternative ME, RBF nets, three layer net, as well as SVM type kernel regression (see Sect. 4 in [23.83]).

In the past two years, BYY harmony learning on various Gaussian/ non-Gaussian mixture and mixture-of-experts as well as modular networks with one hidden layer have been systematically studied in [23.80] and [23.76], respectively, with the following main results:

- A systematic summary and further elaboration of BYY harmony learning and RPCL learning on various details of Gaussian mixture and Gaussian mixture of experts, including MSE clustering, elliptic clustering, subspace clustering, alternative ME, RBF nets with automatic model selection (see Sect. 3 in [23.80]).
- BYY harmony learning algorithms for learning with automatic hidden factor determination on modular binary FA, local LMSER, competitive ICA (see Sect. 4 in [23.80] and Sect. 4 in [23.76]), as well as on three layer networks (see Sect. 5 in [23.76]).
- Extension of the harmony learning by Eq. (23.20) to the so-called f-harmony learning (see Sect. 2.3.2 in [23.80]).

23.7.4 On Regularization Methods

Several regularization methods have also been developed during the studies on BYY learning. Not only each of them can improve the learning performances on a BYY system in the case of a small size of samples, but also some of the methods remain useful even being independent of BYY system. Main results are summarized as follows:

Data smoothing regularization, which came from replacing the empirical density $p_0(x)$ by Eq. (23.2), that is equivalent to directly use a set of training samples, via a Parzen window density $p_h(x)$ by Eq. (23.3) with a smoothing parameter $h > 0$. The idea started from suggesting the use of $p_h(x)$ by Eq. (23.3) in a BYY system (see Eq. (5) in [23.109] and Eq. (1) in [23.111], also see Sect. 1 in [23.106] and Eq. (1) in [23.107]). In 1997, it was further proposed under the name of data smoothing (see Eq. (16) in [23.107] and Eq. (3.10) in [23.101]) that an appropriate h is also learned via implementing the KL learning by Eq. (23.22), which becomes equivalent to

$$\min_{\theta,h} KL(\theta,h), \quad KL(\theta,h) = \int p_h(x)\ln\frac{p_h(x)}{q(x|\theta)}\mu(dx), \tag{23.68}$$

which was firstly presented by Eq. (7) in [23.103]. In a BYY system, $q(x|\theta) = \int q(x|y)q(y)dy$ is the marginal density represented by the Ying machine. Generally, being independent of BYY system, $q(x|\theta)$ can be any parametric model for density estimation. Also in [23.102], the data smoothing regularization is suggested on $q(z|x, \theta_{z|x})$ for supervised learning of three layer forward net and mixture of experts.

Data smoothing introduces a Tikhonov-type regularization [23.10] into parameter learning, with the role h^2 being equivalent to the hyper-parameter in a Tikhonov regularization. What is new here is that an appropriate h can be learned via an easy implementation. Several advances have been made on implementing data smoothing since 1997, including

- A smoothed EM algorithm from learning on Gaussian mixture (see Eq. (18) in [23.102]).
- Three techniques for computing the integral $\int G(x|x_t, h^2 I)F(x)\mu(dx)$, namely stochastic approximation and mean-field [23.94] as well as the following second order approximation (see Sect. 2.4 in [23.88] and Sect. 2.3 in [23.86]):

$$\int G(x|x_t, h^2 I)F(x)\mu(dx) \approx F(x_t) + 0.5h^2 Tr[H_F],$$
$$\text{with the Hessian matrix } H_F = \frac{\partial^2 F(x)}{\partial x \partial x^T}]_{x=x_t}. \tag{23.69}$$

- Four approaches for solving h, i.e., quantization based enumeration, stochastic approximation, iterative updating, and solving a second order algebraic equation [23.94, 23.86, 23.85, 23.82, 23.80].
- In independent factor model of non-Gaussian real factors, mixture of experts, alternative ME, RBF nets, and three layer networks, different smoothing parameters are provided for input data, output data, and inner representation, respectively [23.94, 23.86, 23.85, 23.82, 23.80, 23.76].
- Two types of data smoothing mechanisms are provided, with one for the KL learning and the other for the harmony learning (see Sect. II(A) & (B) in [23.82] and Sect. 2 in [23.80]).

Details are further referred to [23.85, 23.82, 23.80] as well as a recent summary given in [23.77].

Normalization regularization, which came from the normalization term z_q. Firstly proposed in [23.82, 23.83], this normalization term causes a conscience de-learning that not only introduces a regularization to the ML learning, but also makes BYY harmony learning behave similar to the RPCL learning [23.114]. The details of the normalization role and its implementation on Gaussian mixture can be found in Sect. 3.2 of [23.83] and Sect. II(E) of [23.82]. Further results on Gaussian mixture, Gaussian mixture of experts, non-Gaussian mixture of experts, as well as modular networks with one hidden layer of binary units can be found in [23.76] and [23.78].

Structural regularization, which happens in a BYY system where certain regularization to a B-architecture or a F-architecture is imposed via its free part being replaced with an appropriately chosen parametric model. This was firstly suggested in 1997 (see Item 3.4 in [23.101], also see Item 2.5 and Item 2.6 in [23.102]). Typical examples are $p(y|x)$ given by Eq. (23.24) for a B-architecture and $p(x|y)$ given by Eq. (23.46). For example,

- It was suggested (see Sect. 2.5 in [23.83]) that the local maximum side-effect of the WTA competition by Eq. (23.19) with a B-architecture can be regularized with an appropriate parametric $p(y|x)$ by Eq. (23.24). Recently it have been experimentally shown in [23.43] that such a regularization makes BYY harmony learning on Gaussian mixture also demonstrate a RPCL mechanism with automatic selection on k.

- The previously discussed principal ICA that extends ICA to noise situation (see Eq. (10.9) in [23.101], Eq. (11) in [23.89]) can also be regarded as that an ICA $y = Wx$ is regularized by $G(x|Wy, \sigma^2 I)$.

- In comparison with the above first two types of regularization, one major advantage of structural regularization is easy to be implemented via an adaptive algorithm. However, we can not avoid computational difficulty of the integral in $p(y|x)$ by Eq. (23.24) when y is real and non-Gaussian. Moreover, choosing a parametric model instead of $p(y|x)$ by Eq. (23.24) is not easy if there is not enough a priori knowledge.

Annealing Procedure   As discussed in Sect. 23.4.2, the KL learning can be regarded as a regularized version of the HL learning. The advantage of two can be combined by Eq. (23.49) with the regularization strength gradually decreasing as λ decreases in a simulated annealing procedure. As discussed in the previous subsection, the local maximum side-effect of the WTA competition by Eq. (23.19) can be solved via such a simulated annealing (see Eq. (42) and (43) in [23.83]), which has been further supported by experiments on Gaussian mixture [23.44].

f-function   regularization can also be imposed with $\ln(r)$ replaced by a convex function f, which has also been supported by experimental demonstrations on Gaussian mixture [23.104] and ICA problems [23.105]. Readers are referred to a detailed introduction provided in the previous chapter in this same book.

23.8 Conclusions

Efforts of making learning on a finite size of samples have been discussed in three typical streams. BYY harmony learning provides a new mechanisms for model selection and regularization, which has been further justified from both an information theoretic perspective and a generalized projection geometry. Further insights have also been obtained via discussions on its relations and differences from major existing approaches.

References

23.1 H. Akaike: A new look at the statistical model identification, IEEE Tr. Automatic Control, 19, 714-723 (1974)
23.2 H. Akaike: Likelihood of a model and information criteria, Journal of Econometrics, 16, 3-14 (1981)
23.3 H. Akaike: Factor analysis and AIC, Psychometrika, 52, 317-332 (1987)

23.4 S. Amari: Differential geometry methods in statistics, Lecture Notes in Statistics 28, Springer (1985)

23.5 S. Amari: Information geometry of the EM and em algorithms for neural networks, Neural Networks, 8, No. 9, 1379-1408 (1995)

23.6 SI. Amari, A. Cichocki, HH. Yang: A new learning algorithm for blind separation of sources. In: DS. Touretzky et al. (eds.), *Advances in Neural Information Processing 8*, MIT Press, 757-763 (1996)

23.7 AC. Atkinson: Likelihood ratios, posterior odds and information criteria, Journal of Econometrics, 16, 15-20 (1981)

23.8 A. Bell, T. Sejnowski: An information maximization approach to blind separation and blind deconvolution, Neural Computation, 17, 1129-1159 (1995)

23.9 J. Berger: *Statistical Decision Theory and Bayesian Analyses* (Springer-Verlag, New York) (1985)

23.10 CM. Bishop: Training with noise is equivalent to Tikhonov regularization, Neural Computation, 7, 108-116 (1995)

23.11 H. Bozdogan: Model Selection and Akaike's Information Criterion: The general theory and its analytical extension, Psychometrika, 52, 345-370 (1987)

23.12 H. Bozdogan, DE. Ramirez: FACAIC: Model selection algorithm for the orthogonal factor model using AIC and FACAIC, Psychometrika, 53 (3), 407-415 (1988)

23.13 JE. Cavanaugh: Unifying the derivations for the Akaike and corrected Akaike information criteria, Statistics and Probability Letters, 33, 201-208 (1997)

23.14 S. Chib: Marginal likelihood from the Gibbs output, Journal of the American Statistical Association, 90 (432), 1313-1321 (1995)

23.15 GC. Chow: A comparison of the information and posterior probability criteria for model selection, Journal of Econometrics, 16, 21-33 (1981)

23.16 I. Csiszar, G. Tusnady: Information geometry and alternating minimization procedures, Statistics and Decisions, Supplementary Issue, No. 1, 205-237 (1984)

23.17 P. Dayan, GE. Hinton: The Helmholtz machine, Neural Computation 7, No. 5, 889-904 (1995)

23.18 P. Dayan, GE. Hinton: Varieties of Helmholtz machine, Neural Networks, 9, No. 8, 1385-1403 (1996)

23.19 G. Cooper, E. Herskovitz: A Bayesian method for the induction of probabilistic networks from data, Machine Learning, 9, 309-347 (1992)

23.20 AP. Dempster, NM. Laird, DB. Rubin: Maximum-likelihood from incomplete data via the EM algorithm, J. Royal Statistical Society, 39, 1-38 (1977)

23.21 PA. Devijver, J. Kittler: Pattern Recognition: A Statistical Approach (Prentice-Hall) (1982)

23.22 L. Devroye et al.: *A Probability Theory of Pattern Recognition* (Springer) (1996)

23.23 TJ. DiCiccio et al.: Computing Bayes factors by combining simulations and asymptotic Approximations, J. American Statistical Association, 92 (439), 903-915 (1997)

23.24 B. Efron: Estimating the error rate of a prediction rule: Improvement on cross-validation, J. American Statistical Association, 78, 316-331 (1983)

23.25 B. Efron, R. Tibshirani: *An Introduction to the Bootstrap* (Chaoman and Hall, New York) (1993)

23.26 AE. Gelfand, DK. Dey: Bayesian model choice: Asymptotic and exact calculations, Journal of the Royal Statistical Society B, 56 (3), 501-514 (1994)

23.27 S. Geman, E. Bienenstock, R. Doursat: Neural Networks and the bias-variance dilemma, Neural Computation, 4, 1-58 (1992)

23.28 Z. Ghahramani, MJ. Beal: Variational inference for Bayesian mixture of factor analysis. In: SA. Solla, TK. Leen, KR, Muller, (eds.), *Advances in Neural Information Processing Systems 12*, Cambridge, MA: MIT Press, 449-455 (2000)

23.29 A. Gammerman, V. Vovk: Kolmogorov complexity, Computer Journal, 42 (4) (1999)

23.30 F. Girosi et al.: Regularization theory and neural architectures, Neural Computation, 7, 219-269 (1995)

23.31 GE. Hinton, P. Dayan, BJ. Frey, RN. Neal: The wake-sleep algorithm for unsupervised learning neural networks, Science, 268, 1158-1160 (1995)

23.32 GE. Hinton, RS. Zemel: Autoencoders, minimum description length and Helmholtz free energy, Advances in NIPS, 6, 3-10 (1994)

23.33 GE. Hinton, D. van Camp: Keeping neural networks simple by minimizing the description length of the weights, *Sixth ACM Conference on Computational Learning Theory*, Santa Cruz, July, 1993 (1993)

23.34 XL. Hu, L. Xu: A Comparative Study of Several Cluster Number Selection Criteria, *Proc. of IDEAL03*, Lecture Notes in Computer Science 2690, Springer-Verlag, pp. 195-202 (2003)

23.35 CM. Hurvich, CL. Tsai: Regression and time series model in samll samples, Biometrika, 76, 297-307 (1989)

23.36 CM. Hurvich, CL. Tsai: A corrected Akaike information criterion for vector autoregressive model selection, J. of Time Series Analysis, 14, 271-279 (1993)

23.37 H. Jeffreys: *Theory of Probability* (Clarendon Press, Oxford 1939)

23.38 RL. Kashyap: Optimal choice of AR and MA parts in autoregressive and moving-average models, IEEE Trans. PAMI, 4, 99-104 (1982)

23.39 RE. Kass, AE. Raftery: Bayes factors, Journal of the American Statistical Association, 90 (430), 773-795 (1995)

23.40 RE. Kass, L. Wasserman: The selection of prior distributions by formal rules, J. American Statistical Association, 91 (435), 1343-1370 (1996)

23.41 RW. Katz: On some criteria for estimating the order of a Markov chain, Technometrics, 23 (3), 243-249 (1981)

23.42 P. Kontkanen et al.: Bayesian and Information-Theoretic priors for Bayeisan network parameters, Machine Learning: ECML-98, Lecture Notes in Artificial Intelligence, Vol. 1398, 89-94, Springer-Verlag (1998)

23.43 J. Ma, T. Wang, L. Xu: A gradient BYY harmony learning rule on Gaussian mixture with automated model selection, in press, Neurocomputing (2003)

23.44 J. Ma, T. Wang, L. Xu: The Annealing EM Algorithm for Gaussian Mixture with Automated Model Selection (submitted) (2003)

23.45 D. Mackey: A practical Bayesian framework for back-propagation, Neural Computation, 4, 448-472 (1992)

23.46 D. Mackey: Bayesian Interpolation, Neural Computation, 4, 405-447 (1992)

23.47 GJ. McLachlan, T. Krishnan: *The EM Algorithm and Extensions* (John Wiley and Son, INC. 1997)

23.48 AA. Neath, JE. Cavanaugh: Regression and Time Series model selection using variants of the Schwarz information criterion, Communications in Statistics A, 26, 559-580 (1997)

23.49 MA. Newton, AE. Raftery: Approximate Bayesian inference with the weighted likelihood Bootstrap, J. Royal Statistical Society B, 56 (1), 3-48 (1994)

23.50 A. O'Hagan: Fractional Bayes factors for model comparison, J. Royal Statistical Society B, 57 (1), 99-138 (1995)

23.51 RA. Redner, HF. Walker: Mixture densities, maximum likelihood, and the EM algorithm, SIAM Review, 26, 195-239 (1984)

23.52 J. Rissanen: Stochastic complexity and modeling, Annals of Statistics, 14 (3), 1080-1100 (1986)

23.53 J. Rissanen: *Stochastic Complexity in Statistical Inquiry* (World Scientific, Singapore 1989)

23.54 J. Rissanen: Hypothesis selection and testing by the MDL principle, Computer Journal, 42 (4), 260-269 (1999)

23.55 I. Rivals, L. Personnaz: On Cross Validation for Model Selection, Neural Computation, 11, 863-870 (1999)

23.56 M. Sato: Online model selection based on the vairational Bayes, Neural Computation, 13, 1649-1681 (2001)

23.57 E. Saund: A multiple cause mixture model for unsupervised learning, Neural Computation, Vol. 7, pp. 51-71 (1995)

23.58 L. Saul, MI. Jordan: Exploiting tractable structures in intractable Networks, *Advances in NIPS 8*, MIT Press, 486-492 (1995)

23.59 G. Schwarz: Estimating the dimension of a model, Annals of Statistics, 6, 461-464 (1978)

23.60 SL. Sclove: Application of model-selection criteria to some problems in multivariate analysis, Psychometrika, 52 (3), 333-343 (1987)

23.61 C. Spearman: General intelligence domainively determined and measured, Am. J. Psychol. 15, 201-293 (2004)

23.62 M. Stone: Cross-validatory choice and assessment of statistical prediction, *J. Royal Statistical Society B*, 36, 111-147 (1974)

23.63 M. Stone: Asymptotics for and against cross-validation, *Biometrika*, 64 (1), 29-35 (1977)

23.64 M. Stone: An asymptotic equivalence of choice of model by cross-validation and Akaike's criterion, *J. Royal Statistical Society B*, 39 (1), 44-47 (1977)

23.65 M. Stone: Cross-validation: A review, *Math. Operat. Statist.* , 9, 127-140 (1978)

23.66 M. Stone: Comments on model selection criteria of Akaike and Schwartz. *J. Royal Statistical Society B*, 41 (2), 276-278 (1979)

23.67 N. Sugiura: Further analysis of data by Akaike's infprmation criterion and the finite corrections, *Communications in Statistics A*, 7, 12-26 (1978)

23.68 AN. Tikhonov, VY. Arsenin: *Solutions of Ill-posed Problems*, Winston and Sons (1977)

23.69 CS. Wallace, DM. Boulton: An information measure for classification, *Computer Journal*, 11, 185-194 (1968)

23.70 CS. Wallace, PR. Freeman: Estimation and inference by compact coding, *J. of the Royal Statistical Society*, 49 (3), 240-265 (1987)

23.71 CS. Wallace, DR. Dowe: Minimum message length and Kolmogorov complexity, *Computer Journal*, 42 (4), 270-280 (1999)

23.72 S. Waterhouse et al.: Bayesian method for mixture of experts. In: DS. Touretzky et al. (eds.), Advances in NIPS 8, 351-357 (1996)

23.73 DH. Wolpert: On Bias Plus Variance, Neural Computation, 9 (1997)

23.74 VN. Vapnik: *The Nature Of Statistical Learning Theory* (Springer-Verlag) (1995)

23.75 L. Xu: Advances on BYY Harmony Learning: Information Theoretic Perspective, Generalized Projection Geometry, and Independent Factor Auto-determination, in press, IEEE Trans on Neural Networks (2004)

23.76 L. Xu: BYY Learning, Regularized Implementation, and Model Selection on Modular Networks with One Hidden Layer of Binary Units, *Neurocomputing*, Vol. 51, pp. 227-301 (2003)

23.77 L. Xu: Data smoothing regularization, multi-sets-learning, and problem solving strategies, *Neural Networks, Vol. 15, No. 5-6*, 817-825 (2003)

23.78 L. Xu: Independent Component Analysis and Extensions with Noise and Time: A Bayesian Ying Yang Learning Perspective, *Neural Information Processing - Letters and Reviews*, Vol. 1, No. 1, pp1-52 (2003)

23.79 L. Xu: Data-Smoothing Regularization, Normalization Regularization, and Competition-Penalty Mechanism for Statistical Learning and Multi-Agents, *Proc. IJCNN '03*, July 20-24, 2003, Portland, Oregon, pp. 2649-2654 (2003)

23.80 L. Xu: BYY Harmony Learning, Structural RPCL, and Topological Self-Organizing on Mixture Models, *Neural Networks, Vol. 15, No. 8-9*, 1125-1151 (2002)

23.81 L. Xu: Bayesian Ying Yang Harmony Learning, *The Handbook of Brain Theory and Neural Networks*, Second edition, (MA Arbib, Ed.), Cambridge, MA: The MIT Press, pp. 1231-1237 (2002)

23.82 L. Xu: BYY Harmony Learning, Independent State Space and Generalized APT Financial Analyses, *IEEE Tr on Neural Networks*, 12 (4), 822-849 (2001)

23.83 L. Xu: Best Harmony, Unified RPCL and Automated Model Selection for Unsupervised and Supervised Learning on Gaussian Mixtures, Three-Layer Nets and ME-RBF-SVM Models, *Intl J of Neural Systems*, 11 (1), 43-69 (2001)

23.84 L. Xu: Temporal BYY Learning for State Space Approach, Hidden Markov Model and Blind Source Separation, *IEEE Tr on Signal Processing 48*, 2132-2144 (2000)

23.85 L. Xu: BYY Learning System and Theory for Parameter Estimation, Data Smoothing Based Regularization and Model Selection, *Neural, Parallel and Scientific Computations*, Vol. 8, pp. 55-82 (2000)

23.86 L. Xu: Bayesian Ying Yang Unsupervised and Supervised Learning: Theory and Applications, *Proc. of 1999 Chinese Conference on Neural Networks and Signal Processing*, pp. 12-29, Shantou, China, Nov. 1999 (1999)

23.87 L. Xu: Bayesian Ying Yang Theory for Empirical Learning, Regularization and Model Selection: General Formulation, *Proc. IJCNN'99*, DC, USA, July 10-16, 1999, Vol. 1 of 6, pp. 552-557 (1999)

23.88 L. Xu: BYY Data Smoothing Based Learning on A Small Size of Samples, *Proc. IJCNN'99*, DC, USA, July 10-16, 1999, Vol. 1 of 6, pp. 546-551 (1999)

23.89 L. Xu: Data Mining, Unsupervised Learning and Bayesian Ying Yang Theory, *Proc. IJCNN'99*, DC, USA, July 10-16, 1999, Vol. 4 of 6, pp. 2250-2525 (1999)

23.90 L. Xu: Bayesian Ying Yang System and Theory as a Unified Statistical Learning Approach:(V) Temporal Modeling for Temporal Perception and Control, *Proc. ICONIP'98*, Oct.21-23, 1998, Kitakyushu, Japan, Vol. 2, pp. 877-884 (1998)

23.91 L. Xu: Bayesian Kullback Ying-Yang Dependence Reduction Theory, Neurocomputing, 22 (1-3), 81-112 (1998)

23.92 L. Xu: RBF Nets, Mixture Experts, and Bayesian Ying Yang Learning, Neurocomputing, Vol. 19, No. 1-3, 223-257 (1998)

23.93 L. Xu: Bayesian Ying Yang Learning Theory For Data Dimension Reduction and Determination, *Journal of Computational Intelligence in Finance*, Finance & Technology Publishing, Vol. 6, No. 5, pp, 6-18 (1998)

23.94 L. Xu: Bayesian Ying Yang System and Theory as A Unified Statistical Learning Approach (VII): Data Smoothing, *Proc. ICONIP'98*, Oct. 21-23, 1998, Kitakyushu, Japan, Vol. 1, pp. 243-248 (1998)

23.95 L. Xu: BKYY Three Layer Net Learning, EM-Like Algorithm, and Selection Criterion for Hidden Unit Number, *Proc. ICONIP'98*, Oct. 21-23, 1998, Kitakyushu, Japan, Vol. 2, pp. 631-634 (1998)

23.96 L. Xu: Bayesian Ying Yang Dependence Reduction Theory and Blind Source Separation on Instantaneous Mixture, *Proc. Intl ICSC Workshop I&ANN'98*, Feb.9-10, 1998, Tenerife, Spain, pp. 45-51 (1998)

23.97 L. Xu: Bayesian Ying Yang System and Theory as A Unified Statistical Learning Approach: (VI) Convex Divergence, Convex Entropy and Convex Likelihood, *Proc. IDEAL98*, Hong Kong, pp. 1-12 (1998)

23.98 L. Xu: Rival Penalized Competitive Learning, Finite Mixture, and Multisets Clustering, *Proc. of IJCNN98*, Anchorage, Alaska, Vol. 2, pp. 2525-2530 (1998)

23.99 L. Xu: Bayesian Ying Yang System and Theory as A Unified Statistical Learning Approach: (IV) Further Advances, *Proc. IJCNN98*, Anchorage, Alaska, Vol. 2, pp. 1275-1270 (1998)

23.100 L. Xu: BKYY Dimension Reduction and Determination, *Proc. IJCNN98*, Anchorage, Alaska, Vol. 3, pp. 1822-1827 (1998)

23.101 L. Xu: Bayesian Ying Yang System and Theory as A Unified Statistical Learning Approach: (I) Unsupervised and Semi-Unsupervised Learning. In: S. Amari, N. Kassabov (eds.), *Brain-like Computing and Intelligent Information Systems*, Springer-Verlag, pp. 241-274 (1997)

23.102 L. Xu: Bayesian Ying Yang System and Theory as A Unified Statistical Learning Approach (II): From Unsupervised Learning to Supervised Learning and Temporal Modeling. In: KM. Wong et al. (eds.), *Theoretical Aspects of Neural Computation: A Multidisciplinary Perspective*, Springer-Verlag, pp. 25-42 (1997)

23.103 L. Xu: Bayesian Ying Yang System and Theory as A Unified Statistical Learning Approach (III): Models and Algorithms for Dependence Reduction, Data Dimension Reduction, ICA and Supervised Learning. In: KM. Wong et al. (eds.), *Theoretical Aspects of Neural Computation: A Multidisciplinary Perspective*, Springer-Verlag, pp. 43-60 (1997)

23.104 L. Xu: Bayesian Ying Yang Machine, Clustering and Number of Clusters, *Pattern Recognition Letters 18*, No. 11-13, 1167-1178 (1997)

23.105 L. Xu: Bayesian Ying Yang Learning Based ICA Models, *Proc. IEEE NNSP97*, Sept. 24-26, 1997, Florida, pp. 476-485 (1997)

23.106 L. Xu: Bayesian-Kullback YING-YANG Learning Scheme: Reviews and New Results, *Proc. ICONIP96*, Sept.24-27, 1996, Hong Kong, Vol. 1, 59-67 (1996)

23.107 L. Xu: Bayesian-Kullback YING-YANG Machines for Supervised Learning, *Proc. WCNN96*, Sept.15-18, 1996, San Diego, CA, pp. 193-200 (1996)

23.108 L. Xu: How Many Clusters?: A YING-YANG Machine Based Theory for A Classical Open Problem in Pattern Recognition, *Proc IEEE ICNN96*, June 2-6, 1996, DC, Vol. 3, pp. 1546-1551 (1996)

23.109 L. Xu: Bayesian-Kullback Coupled YING-YANG Machines: Unified Learnings and New Results on Vector Quantization, *Proc. ICONIP95*, Oct 30-Nov.3, 1995, Beijing, China, pp. 977-988 (1995)

23.110 L. Xu: YING-YANG Machine for Temporal Signals, Keynote Talk, *Proc. IEEE Intl Conf. on NNSP95*, Dec.10-13, 1995, Nanjing, Vol. 1, pp. 644-651 (1995)

23.111 L. Xu: New Advances on The YING-YANG Machine, *Proc. Intl. Symp. on Artificial Neural Networks*, Dec.18-20, 1995, Taiwan, ppIS07-12 (1995)

23.112 L. Xu: Cluster Number Selection, Adaptive EM Algorithms and Competitive Learnings, Invited Talk, *Proc. IEEE Intl Conf.. on NNSP95*, Dec.10-13, 1995, Nanjing, Vol. 2, pp. 1499-1502 (1995)

23.113 L. Xu: New Advances on The YING-YANG Machine, Invited Talk, *Proc. of 1995 Intl Symp. on Artificial Neural Networks*, Dec.18-20, 1995, Taiwan, ppIS07-12 (1995)

23.114 L. Xu, A. Krzyzak, E. Oja: Rival Penalized Competitive Learning for Clustering Analysis, RBF net and Curve Detection, IEEE Tr. on Neural Networks, 4, 636-649 (1993)

23.115 L. Xu: Least mean square error reconstruction for self-organizing neural-nets, Neural Networks, 6, 627-648, 1993. Its early version on *Proc. IJCNN91'Singapore*, 2363-2373 (1991)

Author Index

Amardeilh, Florence, 243

Beautement, Patrick, 361
Bradshaw, Jeffrey M., 361
Breedy, Maggie R., 361
Buccafurri, Francesco, 311
Bunch, Larry, 361

Cannataro, Mario, 19
Cole, Richard, 243
Congiusta, Antonio, 19

Doan, AnHai, 135
Drakunov, Sergey V., 361
Dubois, Vincent, 265

Eklund, Peter, 243

Feltovich, Paul J., 361

Gao, Jianfeng, 169
Greco, Salvatore, 513, 553

Han, Jiawei, 135
Hoffman, Robert R., 361
Huang, Jiajin, 109

Jeffers, Renia, 361
Jeon, Jongwoo, 67
Jin, Xiaolong, 291
Johnson, Matthew, 361

Kim, Jinseog, 67
Kim, Yongdai, 67
Kulkarni, Shriniwas, 361
Kumar, Vipin, 193

Ling, Charles X., 169
Liu, Chunnian, 109
Liu, Jiming, 1, 291
Lott, James, 361

Mastroianni, Carlo, 19
Matarazzo, Benedetto, 513, 553
Missikoff, Michele, 223
Mitra, Pabitra, 475
Mitra, Sushmita, 475
Morik, Katharina, 47

Navigli, Roberto, 223

Ou, Chuangxin, 109

Pal, Sankar K., 475
Palopoli, Luigi, 311
Pugliese, Andrea, 19

Qian, Minglun, 409
Quafafou, Mohamed, 265

Raj, Anil K., 361
Rosaci, Domenico , 311

Sarnè, Giuseppe M.L., 311
Scholz, Martin, 47
Skowron, Andrzej, 433
Slowinski, Roman, 513, 553
Suri, Niranjan, 361
Suzuki, Einoshin, 89

Talia,Domenico, 19
Tan, Pang-Ning, 193
Trunfio, Paolo, 19

Uszok, Andrzej, 361

Velardi, Paola, 223

Wah, Benjamin W., 409

Xu, Lei, 615, 661

Yang, Qiang, 169

Yao, Yiyu, 109
Yu, Hwanjo, 135

Zhang, Chengqi, 333
Zhang, Zili, 333
Zhong, Ning, 1, 109

Subject Index

actionable knowledge, 169
actionable Web log mining, 189
Adaptive Learning, 316
adjustable autonomy, 367
agent cooperation, 312
agent networks, 293
Agent-Oriented Methodologies, 335
agents for cooperation, 312
analysis, 478
artificial neural networks, 411
association rule, 496
association rules, 114
attribute weights, 118
augmented cognition, 397
authorizations, 370
autonomy, 366
autonomy oriented satisfiability solving,
   293
axiomatic foundations, 596

Bagging, 69
Bayesian Ying Yang (BYY), 661
Boosting, 70

chaotic time series, 410
characteristic path length, 291
CHEM, 67
classification, 196, 478
classifier, 452
clause-based representation, 300
closeness relations, 442
clustering, 479
clustering coefficient, 291
cognitive and robotic prostheses, 395
Common Knowledge Representation
   Language (CKRL), 48
complexity, 293
computational complexity, 293
computing with words and perceptions
   (CWP), 433

concept relations in text, 265
conceptual data model, 54
conjoint measurement, 597
constraint satisfaction problems, 292
cooperation among agents, 330
correlation analysis, 215
cross validation, 412, 431
customer relationship management
   (CRM), 109

Data Mining Grid, 8, 128
decile analysis, 123
decision class, 441
decision rules, 514
decision table, 441
decision under risk, 575
decomposition problem, 466
dependence structure mining, 654
descriptor, 440
DIG, 100
direct marketing, 109
direction setting rule, 100
discovery of unexpected rules , 92
distributed and parallel data mining, 3
distributed data mining, 21
distributed problem solving, 291, 292
domain ontology, 223
dominance cone, 519
dominance principle, 515
dominance relation, 517
dominance-based decision rules, 565
dominance-based rough set, 521
Dominance-based Rough Set Approach
   (DRSA), 516

E-business intelligence, 127
e-business intelligence, 5
Email Harvester, 195
Ensemble, 75
Exception rule/group discovery, 90

feature vector, 210
foreign language, 469
Formal Concept Analysis, 248
fuzzy DRSA, 554
Fuzzy Logic, 475
fuzzy matching-based interestingness, 92
fuzzy rough approximation, 554
fuzzy rough set, 575

general impressions, 93
Generalization, 494
Genetic Algorithm, 503
gradual decision rules, 565
gradual rules, 554
granular computing, 434, 516
Granular Information Retrieval, 504
granule decomposition, 437
Graph Structures, 265
Grid computing, 20
Grid Services, 22

harmony learning, 661
hierarchical decision making, 553
high frequency random noise, 413
HTTP protocol, 199
hybrid intelligent systems, 334
hybrid systems, 334
hybridization, 3

inclusion relation, 436
Information Extraction, 243, 494
information extraction and the Web, 243
information fusion, 434
information granulation, 433
information granule, 433
information processing systems, 136
information retrieval, 477
information system, 433
Intelligent decision support, 589
intelligent enterprise portals, 7
intelligent information technology (iIT), 1
inter-agent computations, 300
interval order, 584
intra-agent computational complexity, 300

KAoS, 370
knowledge discovery, 513
Knowledge Discovery in Databases (KDD), 19, 48, 333

knowledge discovery in text collections, 265
Knowledge discovery process, 50
Knowledge Grid, 20

link checker, 200
lower approximation, 440

MASSAT, 292
MEPRO, 95
meta pattern, 98
minimum description length (MDL), 661
mining market value functions, 118
model selection, 630
Modular Rough-fuzzy, 483
Multi-agent oriented constraint satisfaction, 309
multi-agent systems, 292, 311, 338
multi-agent-based representations, 293
multi-aspect analysis, 127
multi-layer perceptron, 509
multi-phase process, 3
multicriteria choice, 516
multicriteria classification, 525
multicriteria decision making, 547

navigate and explore the Web, 136
Neural Networks, 114, 475
Nomads, 377
norms, 367

object matching, 153
obligations, 367
offline browser, 197
OntoLearn system, 227
ontology, 223
outranking relation, 554
OWL, 380

page rank, 501
parameterized approximation space, 439
pattern, 438
Pattern recognition, 475
policy, 366
predictive models, 219
preference model, 529
preference order, 514
preference relation, 517
prior knowledge, 514
Problem Solver Markup Language (PSML), 7

random agent networks, 293
recurrent neural networks, 415
referrer, 195
regression, 486
relative interestingness measure, 102
Response models, 112
Robot Exclusion Standard, 197
robust decomposition problem, 468
rough mereology, 435
rough neuron, 453
rough set, 482
Rough set theory, 114
rough sets, 436
rough-fuzzy granule, 451
rule generation, 67, 485

satisfiability problems, 292
search engine, 200
security, 376
Segmentation models, 112
semantic social networks, 7
Semantic Web, 223
semantic web services, 377
semi-structured texts, 245
session identification, 196
session labeling, 207
signature, 441
simultaneous reliability evaluation, 98
small world topology, 291
Specific Domain Ontology (SDO), 224
stationary time series, 410
statistical learning, 615, 667
stochastic dominance, 575
subgroup discovery, 94
subjective measure of interestingness,
    92
syntax and semantics, 441
synthesis problem, 463
Syskill & Webert, 499

targeted marketing, 109
targeted relationship marketing, 109
teamwork, 374
technical aspects of agent acceptability,
    379
the genetic model, 112
The logit and probit model, 113
the market value functions model, 112
the peculiarity-oriented model, 115
threshold scheduling, 104
time series, 409
time series predictions, 409
trust, 361

unstructured Web texts, 243
upper approximation, 438
User Agent, 195
User Profiles, 171
utility function, 596
utility functions, 118

variable-based representation, 300
violation guided back-propagation, 420
visualization algorithm, 284

Web content mining, 265
Web farming, 7
Web Intelligence (WI), 1
Web mining, 5, 135, 477
Web page classification, 141
Web Prefetching, 170
Web Query Log Mining, 178
Web robot, 193
Web-based targeted marketing, 127
WEBSOM, 499
Wisdom Web-based computing, 9